LONDON MATHEMATICAL SOCIETY LECTURE NOTE SERIES

Managing Editor: Professor Endre Süli, Mathematical Institute, Univer
Woodstock Road, Oxford OX2 6GG, United Kingdom

The titles below are available from booksellers, or from Cambridge Ur
www.cambridge.org/mathematics

T0292917

London Mathematical Society Lecture Note Series: 459

Integrable Systems and Algebraic Geometry

A Celebration of Emma Previato's 65th Birthday

Volume 2

Edited by

RON DONAGI
University of Pennsylvania

TONY SHASKA
Oakland University, Michigan

CAMBRIDGE
UNIVERSITY PRESS

University Printing House, Cambridge CB2 8BS, United Kingdom

One Liberty Plaza, 20th Floor, New York, NY 10006, USA

477 Williamstown Road, Port Melbourne, VIC 3207, Australia

314–321, 3rd Floor, Plot 3, Splendor Forum, Jasola District Centre, New Delhi – 110025, India

79 Anson Road, #06–04/06, Singapore 079906

Cambridge University Press is part of the University of Cambridge.

It furthers the University's mission by disseminating knowledge in the pursuit of education, learning, and research at the highest international levels of excellence.

www.cambridge.org
Information on this title: www.cambridge.org/9781108715775
DOI: 10.1017/9781108773355

First published 2020

Printed and bound in Great Britain by Clays Ltd, Elcograf S.p.A.

A catalogue record for this publication is available from the British Library.

ISBN – 2 Volume Set 978-1-108-78549-5 Paperback
ISBN – Volume 1 978-1-108-71574-4 Paperback
ISBN – Volume 2 978-1-108-71577-5 Paperback

Contents of Volume 2

Contents of Volume 1

Algebraic Geometry

A Celebration of Emma Previato's 65th Birthday

Ron Donagi and Tony Shaska

1 A Word from the Editors

These two volumes celebrate and honor Emma Previato, on the occasion of her 65th birthday. The present volume consists of 16 articles in algebraic geometry and its surrounding fields, emphasizing the connections to integrable systems which are so central to Emma's work. The companion volume focuses on Emma's interests within integrable systems. The articles were contributed by Emma's coauthors, colleagues, students, and other researchers who have been influenced by Emma's work over the years. They present a very attractive mix of expository articles, historical surveys and cutting edge research.

Emma Previato is a mathematical pioneer, working in her two chosen areas, algebraic geometry and integrable systems. She has been among the first women to do research, in both areas. And her work in both areas has been deep and influential. Emma received a Bachelor's degree from the University of Padua in Italy, and a PhD from Harvard University under the direction of David Mumford in 1983. Her thesis was on hyperelliptic curves and solitons. The work on hyperelliptic curves has evolved and expanded into Emma's life-long interest in algebraic geometry, which is the subject of the present volume. The work on solitons has led to her ongoing research on integrable systems, which is the subject of the first volume.

Emma Previato has been a faculty member at the Department of Mathematics at Boston University since 1983. She has published nearly a hundred research articles, edited six books, and directed seven PhD dissertations. Her broader impact extends through her renowned teaching and her extensive mentoring activities. She runs AFRAMATH, an annual outreach symposium, and works tirelessly on several on- and off-campus mentoring programs. She has also founded and has been leading the activities of the Boston University chapters of MAA and AWM. She serves on numerous advisory boards.

Algebraic geometry is an ancient subject that has been at the center of mathematical research for at least the past two centuries, attracting many of the greatest mathematical minds. Previato's work has made important contributions in many directions within algebraic geometry, and especially so

1

in areas that connect with the theory of integrable systems. We review some of her accomplishments in section 2 below.

We tried to collect in this volume a broad range of articles covering most of these areas. We summarize these contributions in section 3.

We would like to quote David Mumford, Emma's PhD advisor, who wrote in the context of this volume:

"One of the greatest pleasures in a Professor's life is when one of your PhD students goes on to become a world leader in the area of their thesis. In fact, all my seven female PhD students have done extremely well. Emma was the last in algebraic geometry and it makes me sad that when my own research shifted, I have not followed hers. But reading the list of her papers and their topics completely blows me away. The scope of her work is astonishing: it is the whole area where algebraic geometry meets differential equations and the special functions and spectral curves (and surfaces) that arise. Congratulations Emma on your 65th birthday, you have come such a long way since we shared gelato in Harvard Square and I am deeply happy to see what you have done."

The editors and many of the authors have enjoyed years of fruitful interactions with Emma Previato. We all join in wishing her many more years of health, productivity, and great mathematics.

2 Emma Previato's Contributions

Emma Previato works in different areas, using methods from algebra, algebraic geometry, mechanics, differential geometry, analysis, and differential equations. The bulk of her research belongs to integrable equations. She is noted for often finding unexpected connections between integrability and many other areas, often including various branches of algebraic geometry.

Early Activity

As an undergraduate at the University of Padua, Italy, Emma wrote a dissertation on group lattices, followed by six journal publications [1, 2, 3, 4, 5, 6]. With methods from algebra, initiated by Dedekind in the 19th century, this area's goal is to relate the group structure to the lattice of subgroups, and provide classifications for certain properties: an excellent overview is the article by Freese [7], a review of the definitive treatise by R. Schmidt, where results from all of Emma's papers are used to give one example, a lattice criterion for a finitely generated group to be solvable.

PhD Thesis and Main Area

Emma's thesis [8], submitted at Harvard in 1983 under the supervision of D.B. Mumford [9], is still her most cited paper. Her thesis advisor was among the pioneers of this beautiful area, integrable equations, which grew and unified disparate parts of mathematics over the next twenty years, and is still very active. Emma's original tool for producing exact solutions to large classes of nonlinear PDEs, the Riemann theta function, remained one of her main interests.

Theta Functions

She later pursued more theoretical aspects of special functions, such as Prym theta functions [10, 11, 12, 13, 14] also surprisingly related to numerical results in conformal field theory, the Schottky problem [15], and Thetanulls [16].

Algebraically Completely Integrable Systems

The area of integrable PDEs is surprisingly related to algebraically completely integrable Hamiltonian systems, or ACIS, in the sense that algebro-geometric (aka finite-gap) solutions of integrable hierarchies linearize on Abelian varieties, which can be organized into angle variables for an ACIS over a suitable base, typically a subset of the moduli space of curves whose Jacobian is the fiber [12, 17]. Thanks to this discovery, the area integrates with classical geometric invariant theory, surface theory, and other traditional studies of algebraic geometry. With the appearance of the moduli spaces of vector bundles and Higgs bundles over a curve, at the hands of N. Hitchin in the 1980s, large families of ACIS were added to the examples, as well as theoretical algebro-geometric techniques. In [13, 18, 19, 20], Emma took up the challenge of generalizing the connection between ACIS and integrable hierarchies to curves beyond hyperelliptic. In [21], the families of curves are organized as divisors in surfaces.

Higher Rank and Higher-Dimensional Spectra

On the PDE side, the challenges were of two types. When the ring of functions on the (affine) spectral curve can be interpreted as differential operators with a higher-dimensional space of common eigenfunctions, the fiber of the integrable system is no longer a Jacobian: it degenerates to a moduli space of higher-rank vector bundles, possibly with some auxiliary structures [22]. Neither the PDEs nor the integrable systems have been made explicit in higher rank in general. Some cases, however, are worked out in [23, 24, 25, 26, 27].

The other challenge is to increase the dimension of the spectral variety, for example from curve to surface. Despite much work, this problem too has arguably no explicit solution in general. An attempt to set up a general theory over a multi-dimensional version of the formal Universal Grassmann Manifold of Sato which hosts all linear flows of solutions of integrable hierarchies, is given in [28], and more concrete special settings are mentioned below, under the heading of "Differential Algebra".

Special Solutions: Coverings of Curves

An important aspect of theta functions is their reducibility, a property whose investigation goes back to Weierstrass and his student S. Kowalevski. Given their special role in integrability, reducible theta functions are invaluable for applied mathematicians to approximate solutions, or even derive exact expressions and periods in terms of elliptic functions. To the algebro-geometric theory of Elliptic Solitons, initiated by I.M. Krichever and developed by A. Treibich and his thesis supervisor J.-L. Verdier, Emma contributed [29, 30, 31, 32, 33, 34, 11, 35], while [36, 37] generalize the reduction to hyperelliptic curves or Abelian subvarieties. More general aspects of elliptic (sub)covers are taken up in [38].

Another type of special solution is the one obtained by self-similarity [39]; the challenge here is to find an explicit relationship between the PDE flows and the deformation in moduli that obeys Painlevé-type equations: this is one reason why Emma's work has turned to a special function which is associated to Riemann's theta function but only exists on Jacobians: the sigma function (cf. the eponymous section below).

Poncelet and Billiards: Generalizing ACIS

Classical theorems of projective geometry can be generalized to ACIS [40, 41], while the challenge of matching them with integrable hierarchies is still ongoing [42].

Generalizing ACIS: Hitchin Systems

Explicit Hamiltonians for the Hitchin system are only available in theory: they are given explicit algebraic expression in [43] (cf. also [44], which led to work on the geometry of the moduli space of bundles [45]). An explicit integration in terms of special functions leads to the problem of non-commutative theta functions [46].

Differential Algebra

Differential Algebra is younger than Algebraic Geometry, but it has many features in common. Mumford gives credit to J.L. Burchnall and T.W. Chaundy for the first spectral curve, the Spectrum of a commutative ring of differential operators [47]. This is arguably the reason behind algebro-geometric solutions to integrable hierarchies. On the differential-algebra setting, Emma published [48, 49], connecting geometric properties of the curve with differential resultants, a major topic of elimination theory which is currently being worked out [50, 51] and naturally leads to the higher-rank solutions: their Grassmannian aspects are taken up in [52, 53, 54, 55, 56] the higher-dimensional spectral varieties are addressed in [57]. Other aspects of differential algebra are connected to integrability in [58] (the action of an Abelian vector field on the meromorphic functions of an Abelian variety) and [59] (a p-adic analog); in [60], the deformations act on modular forms.

The Sigma Function

Klein extended the definition of the (genus-one) Weierstrass sigma function to hyperelliptic curves and curves of genus three. H.F. Baker developed an in-depth theory of PDEs satisfied by the hyperelliptic sigma function, which plays a key role in recent work on integrable hierarchies (KdV-type, e.g.). Beginning in the 1990s, this theory of Kleinian sigma functions was revisited, originally by V.M. Buchstaber, V.Z. Enolskii, and D.V. Leykin, much extended in scope, eventually to be developed for "telescopic" curves (a condition on the Weierstrass semigroup at a point). Previato goes beyond the telescopic case in [61, 62], while she investigates the higher-genus analog of classical theorems in [63, 64, 65, 66, 67, 68, 69, 70] and their connections with integrability in [71] and [72], which gives the first algebro-geometric solutions to a dispersionless integrable hierarchy. It is not a coincidence that its integrable flow on the Universal Grassmann Manifold 'cut across' the Jacobian flows of traditional hierarchies, and this is where the two variables of the sigma function (the Jacobian, and the modular ones) should unite to explain the mystery of the Painlevé equations.

Algebraic Coding Theory

Emma's primary contribution to this area is through mentoring undergraduate and graduate thesis or funded-research projects. In fact, this research strand began at the prompting of students in computer science who asked her to give a course on curves over fields of prime characteristic, which she ran for years as a vertically-integrated seminar. Together with her PhD student Drue Coles, she published research papers pursuing Trygve Johnsen's innovative idea of error-correction for Goppa codes implemented via vector bundles [73, 74, 75], then she pursued overviews and extensions of Goppa codes to surfaces [76].

Other

Emma edited or co-edited four books [77, 78, 79, 80]. In addition to book and journal publication, Emma published reviews (BAMS, SIAM), entries in mathematical dictionaries or encyclopaedias, teaching manuals and online research or teaching materials; she also published on the topic of mentoring in the STEAM disciplines.

3 Articles in this Volume

Several of the articles in this volume explore arithmetic aspects of integrable systems:

Buium's article is an exposition of a massive body of work establishing arithmetic analogues of ODEs that preserve the underlying geometry, and exploring arithmetic versions of integrable systems.

The article by Dubrovin explores **Q**-structures on spectral curves and their Jacobians. Under some natural assumptions he proves that certain combinations of logarithmic derivatives of the Riemann theta-function of an arbitrary order ≥ 3 all take rational values at the point of the Jacobi variety specified by the line bundle of common eigenvectors.

Kovacs extends to arbitrary characteristic two theorems of Sommese regarding Abelian varieties: that an Abelian variety cannot be an ample divisor in a smooth projective variety, and that a cone over an Abelian variety of dimension at least two is not smoothable.

Horozov proves that every multiple Dedekind zeta value over any number field K is a period of a mixed Tate motive. Moreover, if K is a totally real number field, then we can choose a cone C so that every multiple Dedekind zeta associated to the pair (K;C) is unramified over the ring of algebraic integers in K. (Multiple Dedekind zeta values associated to a number field K and a cone C were defined previously by the author as number theoretic analogues of multiple zeta values.)

A second group of articles explores various aspects of curves or Riemann surfaces: The points of order 2 in the Jacobian of a curve form a group, and the halves of any divisor of degree 0, e.g. the difference $P - Q$ of two points, form a torsor under this group. Zarhin's work describes this torsor explicitly when the curve is hyperelliptic and the point P is a Weierstrass point.

Komeda and Matsutani investigate curves given by a natural generalization of Weierstrass' equation and show the Jacobi inversion formulae of the strata of its Jacobian.

Spera presents a survey of some recent topological applications of Riemann surface theory, and especially of the associated theta functions, to geometric quantization in mathematical physics. These include classical and quantum monodromy of 2d-integrable systems, the construction of unitary Riemann surface braid group representations, and the noncommutative version of theta functions due to A. Schwarz.

An emphasis on special functions attached to a Riemann surface is the common theme of the next group:

Buchstaber, Enolski, and Leykin consider the dependence of multi-variable sigma function of genus g hyperelliptic curves on both Jacobian variables and parameters of the curve. They develop new representations of periods in terms of theta-constants.

Korotkin contributed a review of the role played by Bergman tau-functions in various areas: theory of isomonodromic deformations, solutions of Einstein's equations, theory of Dubrovin-Frobenius manifolds, geometry of moduli spaces, and spectral theory of Riemann surfaces. These tau-functions are natural generalizations of Dedekinds eta-function to higher genus.

Cogdell, Jorgenson, and Smajlov construct a generalization of the Eisenstein elliptic series attached to finite volume Riemann surfaces. Their input is a pair (X, D), where X is a smooth, compact, projective Kahler variety and D a divisor of a holomorphic form F, assumed smooth up to codimension two. The original case is recovered when X is the quotient of a symmetric space.

Other algebraic aspects of integrable systems appear in the next four papers. In two of them the emphasis is on the Kummer surfaces that appear in genus-2 systems.

Bates and Churchill focus on Lax pairs. These are useful, but often hard to find. The authors describe an elementary systematic construction which has proven successful in several contexts of practical interest. Specific examples are presented.

Retakh, Rubtsov, and Sharygin present their theory of noncommutative cross-ratios, Schwarz derivatives and their connections and relations to the operator cross-ratio. They apply the theory to noncommutative elementary geometry and relate it to noncommutative integrable systems.

The work of Clingher and Malmendier determines normal forms for the Kummer surfaces associated with Abelian surfaces of several polarization of types. It also produces explicit formulas for coordinates and moduli parameters in terms of Theta functions of genus two.

Francoise and Tarama wrote an expository article about the completely integrable Hamiltonian system of the Clebsch top under a special condition introduced by Weber. This led them to Kummer surfaces and to linearization on genus 2 Jacobians.

Finally, two works connect with other aspects of Emma's research:

Carvalho and Neumann present both a survey and some new results in coding theory, extending the well-known work of Delsarte, Goethals, and Mac Williams to a class of Reed-Muller type codes defined on a product of (possibly distinct) finite field the same characteristic.

Schmidt's work aims at a lattice-theoretic characterization of various groups. First he does this for some classes of infinite soluble groups. Then for a finite group G he tries to determine in its subgroup lattice L(G), the Fitting length of G, and properties defined by arithmetical conditions.

References

[1] Emma Previato, *Some families of simple groups whose lattices are complemented*, Boll. Un. Mat. Ital. B (6) **1** (1982), no. 3, 1003–1014. MR683488

[2] _____, *A lattice-theoretic characterization of finitely generated solvable groups*, Istit. Veneto Sci. Lett. Arti Atti Cl. Sci. Mat. Natur. **136** (1977/78), 7–11. MR548255

[3] _____, *Groups in whose dual lattice the Dedekind relation is transitive*, Rend. Sem. Mat. Univ. Padova **58** (1977), 287–308 (1978). MR543147

[4] _____, *Sui sottogruppi di Dedekind nei gruppi infiniti*, Atti Accad. Naz. Lincei Rend. Cl. Sci. Fis. Mat. Natur. (8) **60** (1976), no. 4, 388–394. MR0491981

[5] _____, *Una caratterizzazione dei sottogruppi di Dedekind di un gruppo finito*, Atti Accad. Naz. Lincei Rend. Cl. Sci. Fis. Mat. Natur. (8) **59** (1975), no. 6, 643–650 (1976). MR0480738

[6] _____, *Gruppi in cui la relazione di Dedekind è transitiva*, Rend. Sem. Mat. Univ. Padova **54** (1975), 215–229 (1976). MR0466319

[7] R. Freese, *Subgroup lattices of groups by roland schmidt*, Review, 1994.

[8] Emma Previato, *Hyperelliptic quasiperiodic and soliton solutions of the nonlinear Schrödinger equation*, Duke Math. J. **52** (1985), no. 2, 329–377. MR792178

[9] _____, *HYPERELLIPTIC CURVES AND SOLITONS*, ProQuest LLC, Ann Arbor, MI, 1983, Thesis (PhD)–Harvard University. MR2632885

[10] Bert van Geemen and Emma Previato, *Heisenberg action and Verlinde formulas*, Integrable systems (Luminy, 1991), 1993, pp. 61–80. MR1279817

[11] Emma Previato and Jean-Louis Verdier, *Boussinesq elliptic solitons: the cyclic case*, Proceedings of the Indo-French Conference on Geometry (Bombay, 1989), 1993, pp. 173–185. MR1274502

[12] Emma Previato, *Geometry of the modified KdV equation*, Geometric and quantum aspects of integrable systems (Scheveningen, 1992), 1993, pp. 43–65. MR1253760

[13] Letterio Gatto and Emma Previato, *A remark on Griffiths' cohomological interpretation of Lax equations: higher-genus case*, Atti Accad. Sci. Torino Cl. Sci. Fis. Mat. Natur. **126** (1992), no. 3-4, 63–70. MR1231817

[14] Bert van Geemen and Emma Previato, *Prym varieties and the Verlinde formula*, Math. Ann. **294** (1992), no. 4, 741–754. MR1190454

[15] Christian Pauly and Emma Previato, *Singularities of 2Θ-divisors in the Jacobian*, Bull. Soc. Math. France **129** (2001), no. 3, 449–485. MR1881203

[16] E. Previato, T. Shaska, and G. S. Wijesiri, *Thetanulls of cyclic curves of small genus*, Albanian J. Math. **1** (2007), no. 4, 253–270. MR2367218

[17] Emma Previato, *A particle-system model of the sine-Gordon hierarchy*, Phys. D **18** (1986), no. 1-3, 312–314, Solitons and coherent structures (Santa Barbara, Calif., 1985). MR838338

[18] M. R. Adams, J. Harnad, and E. Previato, *Isospectral Hamiltonian flows in finite and infinite dimensions. I. Generalized Moser systems and moment maps into loop algebras*, Comm. Math. Phys. **117** (1988), no. 3, 451–500. MR953833

[19] Emma Previato, *Generalized Weierstrass ℘-functions and KP flows in affine space*, Comment. Math. Helv. **62** (1987), no. 2, 292–310. MR896099

[20] ———, *Flows on r-gonal Jacobians*, The legacy of Sonya Kovalevskaya (Cambridge, Mass., and Amherst, Mass., 1985), 1987, pp. 153–180. MR881461

[21] Silvio Greco and Emma Previato, *Spectral curves and ruled surfaces: projective models*, The Curves Seminar at Queen's, Vol. VIII (Kingston, ON, 1990/1991), 1991, pp. Exp. F, 33. MR1143110

[22] Emma Previato and George Wilson, *Vector bundles over curves and solutions of the KP equations*, Theta functions—Bowdoin 1987, Part 1 (Brunswick, ME, 1987), 1989, pp. 553–569. MR1013152

[23] E. Previato, *Burchnall-Chaundy bundles*, Algebraic geometry (Catania, 1993/Barcelona, 1994), 1998, pp. 377–383. MR1651105

[24] Geoff A. Latham and Emma Previato, *KP solutions generated from KdV by "rank 2" transference*, Phys. D **94** (1996), no. 3, 95–102. MR1392449

[25] ———, *Darboux transformations for higher-rank Kadomtsev-Petviashvili and Krichever-Novikov equations*, Acta Appl. Math. **39** (1995), no. 1-3, 405–433, KdV '95 (Amsterdam, 1995). MR1329574

[26] Geoff Latham and Emma Previato, *Higher rank Darboux transformations*, Singular limits of dispersive waves (Lyon, 1991), 1994, pp. 117–134. MR1321199

[27] Emma Previato and George Wilson, *Differential operators and rank 2 bundles over elliptic curves*, Compositio Math. **81** (1992), no. 1, 107–119. MR1145609

[28] Min Ho Lee and Emma Previato, *Grassmannians of higher local fields and multivariable tau functions*, The ubiquitous heat kernel, 2006, pp. 311–319. MR2218024

[29] J. Chris Eilbeck, Victor Z. Enolski, and Emma Previato, *Spectral curves of operators with elliptic coefficients*, SIGMA Symmetry Integrability Geom. Methods Appl. **3** (2007), Paper 045, 17. MR2299846

[30] E. Previato, *Jacobi varieties with several polarizations and PDE's*, Regul. Chaotic Dyn. **10** (2005), no. 4, 531–543. MR2191376

[31] J. C. Eilbeck, V. Z. Enolskii, and E. Previato, *Reduction theory, elliptic solitons and integrable systems*, The Kowalevski property (Leeds, 2000), 2002, pp. 247–270. MR1916786

[32] ———, *Varieties of elliptic solitons*, J. Phys. A **34** (2001), no. 11, 2215–2227, Kowalevski Workshop on Mathematical Methods of Regular Dynamics (Leeds, 2000). MR1831289

[33] Emma Previato, *Monodromy of Boussinesq elliptic operators*, Acta Appl. Math. **36** (1994), no. 1-2, 49–55. MR1303855

[34] E. Colombo, G. P. Pirola, and E. Previato, *Density of elliptic solitons*, J. Reine Angew. Math. **451** (1994), 161–169. MR1277298

[35] Emma Previato, *The Calogero-Moser-Krichever system and elliptic Boussinesq solitons*, Hamiltonian systems, transformation groups and spectral transform methods (Montreal, PQ, 1989), 1990, pp. 57–67. MR1110372

[36] È. Previato and V. Z. Ènol′ skiĭ, *Ultra-elliptic solitons*, Uspekhi Mat. Nauk **62** (2007), no. 4(376), 173–174. MR2358755

[37] Ron Y. Donagi and Emma Previato, *Abelian solitons*, Math. Comput. Simulation **55** (2001), no. 4-6, 407–418, Nonlinear waves: computation and theory (Athens, GA, 1999). MR1821670

[38] Robert D. M. Accola and Emma Previato, *Covers of tori: genus two*, Lett. Math. Phys. **76** (2006), no. 2-3, 135–161. MR2235401

[39] G. N. Benes and E. Previato, *Differential algebra of the Painlevé property*, J. Phys. A **43** (2010), no. 43, 434006, 14. MR2727780

[40] Emma Previato, *Some integrable billiards*, SPT 2002: Symmetry and perturbation theory (Cala Gonone), 2002, pp. 181–195. MR1976669

[41] _____, *Poncelet's theorem in space*, Proc. Amer. Math. Soc. **127** (1999), no. 9, 2547–2556. MR1662198

[42] Yuji Kodama, Shigeki Matsutani, and Emma Previato, *Quasi-periodic and periodic solutions of the Toda lattice via the hyperelliptic sigma function*, Ann. Inst. Fourier (Grenoble) **63** (2013), no. 2, 655–688. MR3112844

[43] Bert van Geemen and Emma Previato, *On the Hitchin system*, Duke Math. J. **85** (1996), no. 3, 659–683. MR1422361

[44] E. Previato, *Dualities on $\mathcal{T}^* \mathcal{SU}_X(2, \mathcal{O}_X)$*, Moduli spaces and vector bundles, 2009, pp. 367–387. MR2537074

[45] W. M. Oxbury, C. Pauly, and E. Previato, *Subvarieties of $\mathcal{SU}_C(2)$ and 2θ-divisors in the Jacobian*, Trans. Amer. Math. Soc. **350** (1998), no. 9, 3587–3614. MR1467474

[46] Emma Previato, *Theta functions, old and new*, The ubiquitous heat kernel, 2006, pp. 347–367. MR2218026

[47] _____, *Seventy years of spectral curves: 1923–1993*, Integrable systems and quantum groups (Montecatini Terme, 1993), 1996, pp. 419–481. MR1397276

[48] Jean-Luc Brylinski and Emma Previato, *Koszul complexes, differential operators, and the Weil-Tate reciprocity law*, J. Algebra **230** (2000), no. 1, 89–100. MR1774759

[49] Emma Previato, *Another algebraic proof of Weil's reciprocity*, Atti Accad. Naz. Lincei Cl. Sci. Fis. Mat. Natur. Rend. Lincei (9) Mat. Appl. **2** (1991), no. 2, 167–171. MR1120136

[50] Alex Kasman and Emma Previato, *Factorization and resultants of partial differential operators*, Math. Comput. Sci. **4** (2010), no. 2-3, 169–184. MR2775986

[51] _____, *Commutative partial differential operators*, Phys. D **152/153** (2001), 66–77, Advances in nonlinear mathematics and science. MR1837898

[52] _____, *Differential algebras with Banach-algebra coefficients I: from C^*-algebras to the K-theory of the spectral curve*, Complex Anal. Oper. Theory **7** (2013), no. 4, 739–763. MR3079828

[53] _____, *Curvature of universal bundles of Banach algebras*, Topics in operator theory. Volume 1. Operators, matrices and analytic functions, 2010, pp. 195–222. MR2723277

[54] _____ , *Differential algebras with Banach-algebra coefficients II: The operator cross-ratio tau-function and the Schwarzian derivative*, Complex Anal. Oper. Theory **7** (2013), no. 6, 1713–1734. MR3129889

[55] Emma Previato and Mauro Spera, *Isometric embeddings of infinite-dimensional Grassmannians*, Regul. Chaotic Dyn. **16** (2011), no. 3-4, 356–373. MR2810984

[56] Maurice J. Dupré, James F. Glazebrook, and Emma Previato, *A Banach algebra version of the Sato Grassmannian and commutative rings of differential operators*, Acta Appl. Math. **92** (2006), no. 3, 241–267. MR2266488

[57] Emma Previato, *Multivariable Burchnall-Chaundy theory*, Philos. Trans. R. Soc. Lond. Ser. A Math. Phys. Eng. Sci. **366** (2008), no. 1867, 1155–1177. MR2377688

[58] _____ , *Lines on abelian varieties*, Probability, geometry and integrable systems, 2008, pp. 321–344. MR2407603

[59] Alexandru Buium and Emma Previato, *Arithmetic Euler top*, J. Number Theory **173** (2017), 37–63. MR3581908

[60] Eleanor Farrington and Emma Previato, *Symbolic computation for Rankin-Cohen differential algebras: a case study*, Math. Comput. Sci. **11** (2017), no. 3-4, 401–415. MR3690055

[61] Jiryo Komeda, Shigeki Matsutani, and Emma Previato, *The Riemann constant for a non-symmetric Weierstrass semigroup*, Arch. Math. (Basel) **107** (2016), no. 5, 499–509. MR3562378

[62] _____ , *The sigma function for Weierstrass semigoups $\langle 3, 7, 8 \rangle$ and $\langle 6, 13, 14, 15, 16 \rangle$*, Internat. J. Math. **24** (2013), no. 11, 1350085, 58. MR3143604

[63] Shigeki Matsutani and Emma Previato, *The al function of a cyclic trigonal curve of genus three*, Collect. Math. **66** (2015), no. 3, 311–349. MR3384012

[64] _____ , *Jacobi inversion on strata of the Jacobian of the C_{rs} curve $y^r = f(x)$, II*, J. Math. Soc. Japan **66** (2014), no. 2, 647–692. MR3201830

[65] E. Previato, *Sigma function and dispersionless hierarchies*, XXIX Workshop on Geometric Methods in Physics, 2010, pp. 140–156. MR2767999

[66] Shigeki Matsutani and Emma Previato, *A generalized Kiepert formula for C_{ab} curves*, Israel J. Math. **171** (2009), 305–323. MR2520112

[67] _____ , *Jacobi inversion on strata of the Jacobian of the C_{rs} curve $y^r = f(x)$*, J. Math. Soc. Japan **60** (2008), no. 4, 1009–1044. MR2467868

[68] J. C. Eilbeck, V. Z. Enolski, S. Matsutani, Y. Ônishi, and E. Previato, *Abelian functions for trigonal curves of genus three*, Int. Math. Res. Not. IMRN (2008), no. 1, Art. ID rnm 140, 38. MR2417791

[69] _____ , *Addition formulae over the Jacobian pre-image of hyperelliptic Wirtinger varieties*, J. Reine Angew. Math. **619** (2008), 37–48. MR2414946

[70] J. C. Eilbeck, V. Z. Enolskii, and E. Previato, *On a generalized Frobenius-Stickelberger addition formula*, Lett. Math. Phys. **63** (2003), no. 1, 5–17. MR1967532

[71] Shigeki Matsutani and Emma Previato, *From Euler's elastica to the mKdV hierarchy, through the Faber polynomials*, J. Math. Phys. **57** (2016), no. 8, 081519, 12. MR3541543

[72] _____ , *A class of solutions of the dispersionless KP equation*, Phys. Lett. A **373** (2009), no. 34, 3001–3004. MR2559804

[73] Drue Coles and Emma Previato, *Decoding by rank-2 bundles over plane quartics*, J. Symbolic Comput. **45** (2010), no. 7, 757–772. MR2645976

[74] Emma Previato, *Vector bundles in error-correcting for geometric Goppa codes*, Algebraic aspects of digital communications, 2009, pp. 42–80. MR2605297

[75] Drue Coles and Emma Previato, *Goppa codes and Tschirnhausen modules*, Advances in coding theory and cryptography, 2007, pp. 81–100. MR2440171

[76] Brenda Leticia De La Rosa Navarro, Mustapha Lahyane, and Emma Previato, *Vector bundles with a view toward coding theory*, Algebra for secure and reliable communication modeling, 2015, pp. 159–171. MR3380380

[77] David A. Ellwood and Emma Previato (eds.), *Grassmannians, moduli spaces and vector bundles*, Clay Mathematics Proceedings, vol. 14, American Mathematical Society, Providence, RI; Clay Mathematics Institute, Cambridge, MA, 2011, Papers from the Clay Mathematics Institute (CMI) Workshop on Moduli Spaces of Vector Bundles, with a View Towards Coherent Sheaves held in Cambridge, MA, October 6–11, 2006. MR2809924

[78] Emma Previato (ed.), *Dictionary of applied math for engineers and scientists*, Comprehensive Dictionary of Mathematics, CRC Press, Boca Raton, FL, 2003. MR1966695

[79] _____ (ed.), *Advances in algebraic geometry motivated by physics*, Contemporary Mathematics, vol. 276, American Mathematical Society, Providence, RI, 2001. MR1837106

[80] R. Donagi, B. Dubrovin, E. Frenkel, and E. Previato, *Integrable systems and quantum groups*, Lecture Notes in Mathematics, vol. 1620, Springer-Verlag, Berlin; Centro Internazionale Matematico Estivo (C.I.M.E.), Florence, 1996, Lectures given at the First 1993 C.I.M.E. Session held in Montecatini Terme, June 14–22, 1993, Edited by M. Francaviglia and S. Greco, Fondazione CIME/CIME Foundation Subseries. MR1397272

[81] Emma Previato, *Soliton equations and their algebro-geometric solutions. Vol. I [book review of mr1992536]*, Bull. Amer. Math. Soc. (N.S.) **45** (2008), no. 3, 459–467. MR3077138

[82] E. Previato, *Poncelet's porism and projective fibrations*, Higher genus curves in mathematical physics and arithmetic geometry, 2018, pp. 157–169. MR3782465

[83] _____, *Complex algebraic geometry applied to integrable dynamics: concrete examples and open problems*, Geometric methods in physics XXXV, 2018, pp. 269–280. MR3803645

[84] Alexandru Buium and Emma Previato, *The Euler top and canonical lifts*, J. Number Theory **190** (2018), 156–168. MR3805451

[85] E. Previato, *Curves in isomonodromy and isospectral deformations: Painlevé VI as a case study*, Algebraic curves and their applications, 2019, pp. 247–265. MR3916744

[86] Jiryo Komeda, Shigeki Matsutani, and Emma Previato, *The sigma function for trigonal cyclic curves*, Lett. Math. Phys. **109** (2019), no. 2, 423–447. MR3917350

Ron Donagi
Mathematics Department, UPenn, Philadelphia, PA, 19104-6395

Tony Shaska
Department of Mathematics and Statistics, Oakland University, Rochester, MI. 48309

1

Arithmetic Analogues of Hamiltonian Systems

Alexandru Buium

Dedicated to Emma Previato's 65th birthday

Abstract. The paper reviews various arithmetic analogues of Hamiltonian systems introduced in [10, 11, 13, 15, 16], and presents some new facts suggesting ways to relate/unify these examples.

1 Introduction

1.1 Aim of the Paper

In a series of papers starting with [5] an arithmetic analogue of the concept of (not necessarily linear) ordinary differential equation was developed; cf. the monographs [9] and [10] for an exposition of (and references for) this theory and [6, 8, 14, 17] for some purely arithmetic applications. (There is a version for partial differential equations [19, 20, 21, 10] which we will not discuss here.) The rough idea behind this theory is to replace differentiation operators

$$y \mapsto \delta_x y := \frac{dy}{dx},$$

acting on smooth functions $y = y(x)$ in one variable x, by *Fermat quotient operators* with respect to an odd rational prime p,

$$z \mapsto \delta_p z := \frac{\phi_p(z) - z^p}{p},$$

acting on a complete discrete valuation ring A with maximal ideal generated by p and perfect residue field; here we denoted by ϕ_p the unique Frobenius lift on A, i.e. the unique ring endomorphism of A whose reduction mod p is the p-power Frobenius. Then classical differential equations,

$$F(y, \delta_x y, \ldots, \delta_x^n y) = 0,$$

13

where F is a smooth function, are replaced by *arithmetic differential equations*,

$$F(z, \delta_p z, \ldots, \delta_p^n z) = 0,$$

where F is a p-adic limit of polynomials with coefficients in A. More generally one can consider systems of such equations. One would then like to introduce arithmetic analogues of some remarkable classical ordinary differential equations such as:

Linear differential equations, Riccati equations, Weierstrass equation satisfied by elliptic functions, Painlevé equations, Schwarzian equations satisfied by modular forms, Ramanujan differential equations satisfied by Einsenstein series, Euler equations for the rigid body, Lax equations, etc.

A first temptation, in developing the theory, would be to consider the function(s) F in each of these classical equations and formally replace, in the corresponding equation, the quantities $\delta_x^i y$ by the quantities $\delta_p^i z$. This strategy of preserving F, i.e., the *shape of the equations*, turns out to destroy the underlying *geometry of the equations* and, therefore, seems to lead to a dead end. Instead, what turns out to lead to an interesting theory (and to interesting applications) is to discover how to change the shape of the equations (i.e. change F) so that the geometry of the equations is (in some sense) preserved. The first step in then to "geometrize" the situation by introducing *arithmetic jet spaces* [5] which are arithmetic analogues of the classical jet spaces of Lie and Cartan and to seek a conceptual approach towards arithmetic analogues of the classical equations we just mentioned; this has been done in the series of papers and in the monographs mentioned in the beginning of this paper, for the list of classical equations we just mentioned. In achieving this one encounters various obstacles and surprises. For instance, the question, *"What is the arithmetic analogue of linear differential equations?"* is already rather subtle; indeed, linearity of arithmetic differential equations turns out to be not an absolute but, rather, a relative concept; more precisely there is no concept of linearity for *one* arithmetic differential equation but there is a concept of an arithmetic differential equation being *linear* with respect to another arithmetic differential equation. We refer to [10, 11, 12] for the problem of linearity. The question we would like to address in this paper is a different one (although a related one), namely:

"What is the arithmetic analogue of a Hamiltonian system?"

A number of remarkable classical differential equations admit Hamiltonian structures; this is the case, for instance, with the following 3 examples: Painlevé VI equations, Euler equations, and (certain special) Lax equations. Arithmetic analogues of these 3 types of equations have been developed, in 3 separate frameworks, in a series of papers as follows: the Painlevé case in [13];

the Euler case in [15, 16]; and the Lax case in [11], respectively. Cf. also [10]. One is tempted to believe that these 3 examples are pieces of a larger puzzle. The aim of the present paper is to review these 3 examples and attempt to give hints as to a possible unification of these 3 pictures by proving some new facts (and providing some new comments) that connect some of the dots. The task of setting up a general framework (and a more general array of examples) for an arithmetic Hamiltonian formalism is still elusive; we hope that the present paper contains clues as to what this general formalism could be.

1.2 Structure of the Paper

Each of the following sections contains 2 subsections. In each section the first of the subsections offers a treatment of the classical differential setting while the second subsection offers a treatment of the arithmetic differential setting. Section 2 of the paper is devoted to an exposition of the main general concepts. Sections 3, 4, 5 are devoted to the main examples under consideration: the Painlevé equations, the Euler equations, and the Lax equations respectively.

The main new results of the paper are Theorems 4.6 and 4.7 which hint towards a link between the formalisms in the Painlevé and Euler examples.

1.3 Acknowledgement

The present work was partially supported by the Institut des Hautes Études Scientifiques in Bures sur Yvette, and by the Simons Foundation (award 311773). The author would also like to express his debt to Emma Previato for an inspiring collaboration and a continuing interaction.

2 General Concepts

2.1 The Classical Case

We begin by reviewing some of the main concepts in the theory of classical (ordinary) differential equations. We are interested in the purely algebraic aspects of the theory so we will place ourselves in the context of differential algebra [24]; our exposition follows [4, 10, 13].

Let us start with a ring A equipped with a derivation $\delta^A : A \to A$ (i.e., an additive map satisfying the Leibniz rule). For simplicity we assume A is Noetherian. Let X be a scheme of finite type over A. One defines the *jet spaces* $J^n(X)$ of X (over A) as follows. If X is affine,

$$X = Spec\ A[x]/(f), \tag{1}$$

with x an N-tuple of indeterminates and f a tuple of polynomials, then one sets

$$J^n(X) := Spec\ A[x, x', \ldots, x^{(n)}]/(f, \delta f, \ldots, \delta^n f) \qquad (2)$$

where $x', \ldots, x^{(n)}$ are new N-tuples of indeterminates and $\delta = \delta^{univ}$ is the unique ("universal") derivation on the ring of polynomials in infinitely many indeterminates,

$$A[x, x', \ldots, x^{(n)}, \ldots],$$

extending δ^A, and satisfying

$$\delta x = x', \ldots, \delta x^{(n)} = x^{(n+1)}, \ldots \qquad (3)$$

One gets induced derivations

$$\delta = \delta^{univ} : \mathcal{O}(J^n(X)) \to \mathcal{O}(J^{n+1}(X)).$$

If X is not necessarily affine then one defines $J^n(X)$ by gluing $J^n(X_i)$ where $X = \cup X_i$ is a Zariski affine open cover. The family $(J^n(X))_{n \geq 0}$ has the structure of a projective system of schemes depending functorially on X, with $J^0(X) = X$. If X is smooth and descends to (a scheme over) *the ring of δ-constants* of A,

$$A^\delta = \{a \in A;\ \delta a = 0\},$$

then $J^1(X)$ identifies with the (total space of the) tangent bundle $T(X)$ of X; if we drop the condition that X descend to A^δ then $J^1(X)$ is only a torsor under $T(X)$.

If G is a group scheme over A then $(J^n(G))_{n \geq 0}$ forms a projective system of group schemes over A; if A is a field of characteristic zero, say, the kernels of the homomorphisms $J^n(G) \to J^{n-1}(G)$ are isomorphic as algebraic groups to powers of the additive group scheme $\mathbb{G}_a$.

By a *differential equation* on X we understand a closed subscheme of some $J^n(X)$. By a δ^A-*flow* on X we will understand a derivation δ^X on the structure sheaf of X, extending δ^A; giving a δ^A-flow is equivalent to giving a section of the canonical projection $J^1(X) \to X$, and hence, to giving a differential equation $Z \subset J^1(X)$ for which the projection $Z \to X$ is an isomorphism. A *prime integral* for a δ^A-flow δ^X is a function $H \in \mathcal{O}(X)$ such that $\delta^X H = 0$. For any A-point

$$P \in X(A), \quad P : Spec\ A \to X,$$

one defines the jets

$$J^n(P) \in J^n(X)(A), \quad J^n(P) : Spec\ A \to J^n(X),$$

as the unique morphisms lifting P that are compatible with the actions of δ^{univ} and δ^A. A *solution* (in A) for a differential equation $Z \subset J^n(X)$ is an A-point $P \in X(A)$ such that $J^n(P)$ factors through Z. If P is a solution to the differential equation defined by a δ^A-flow and if H is a prime integral for that δ^A-flow then $\delta^A(H(P)) = 0$; intuitively H is "constant" along any solution. If X is affine and $Z \subset J^1(X)$ is the differential equation corresponding to a δ^A-flow δ^X on X then a point $P \in X(A)$ is a solution to Z if and only if the ring homomorphism $P^* : \mathcal{O}(X) \to A$ defined by P satisfies $\delta^A \circ P^* = P^* \circ \delta^X$.

For any smooth affine scheme $Spec\ B$ over A we may consider the algebraic deRham complex of abelian groups,

$$B \xrightarrow{d} \Omega^1_{B/A} \xrightarrow{d} \Omega^2_{B/A} \xrightarrow{d} \dots, \tag{4}$$

where $\Omega^i_{B/A} := \wedge^i \Omega_{B/A}$, $\Omega_{B/A}$ the (locally free) B-module of Kähler differentials.

Recall that for any A-algebra B a *Poisson structure/bracket* on B (or on $Spec\ B$) is an A-bilinear map

$$\{\,,\,\} : B \times B \to B$$

which is a derivation in each of the arguments and defines a structure of Lie A-algebra on B.

In what follows we would like to review the classical concept of *Hamiltonian* system/equation from a purely algebro-geometric viewpoint; later we will introduce its arithmetic analogues. We will restrict ourselves to the case of affine surfaces since our main examples fit into this setting.

Let $S = Spec\ B$ be a smooth affine surface (i.e. smooth scheme of relative dimension 2) over a Noetherian ring A.

By a *symplectic form* we will understand a basis η of the B-module $\Omega^2_{B/A}$. (The usual condition that this form be closed is automatically satisfied because S is a surface.) By a *contact form* we will understand an element $v \in \Omega^1_{B/A}$ such that dv is a symplectic form. Given a symplectic form η on $S = Spec\ B$ one can define a Poisson structure on S by the formula:

$$\{\,,\,\}_\eta : B \times B \to B, \quad \{f, g\}_\eta := \frac{df \wedge dg}{\eta}.$$

Assume now that we are given a derivation $\delta^A : A \to A$. Recall that by a δ^A-*flow* on S we mean a derivation $\delta := \delta^B : B \to B$ extending our derivation $\delta^A : A \to A$. Recall that δ^B induces then unique additive maps, referred to as *Lie derivatives*,

$$\delta = \delta^{\Omega^i} : \Omega^i_{B/A} \to \Omega^i_{B/A}, \quad i = 0, 1, 2,$$

such that $\delta^{\Omega^0} = \delta^B$, δ commute with d, and δ induce a derivation on the exterior algebra $\wedge \Omega_{B/A}$.

Recall that a function $H \in B$ is called a *prime integral* for δ^B (or a δ^B-*constant*) if $\delta^B H = 0$.

Let us say that a δ^A-flow $\delta = \delta^B$ on S is *Hamiltonian* (or, more accurately, *symplectic-Hamiltonian*) with respect to a symplectic form η on S if

$$\delta\eta = 0. \tag{5}$$

Let us say that δ^A-flow $\delta = \delta^B$ on S is *Hamiltonian* (or, more accurately, *Poisson-Hamiltonian*) with respect to a Poisson structure $\{ , \}$ on $B = \mathcal{O}(S)$ if S descends to a smooth scheme $S_0 = Spec\ B_0$ over $A_0 := A^\delta$, with $\{B_0, B_0\} \subset B_0$, and there exists a function (called *Hamiltonian*) $H \in B_0$ such that

$$\delta f = \{f, H\}, \quad \text{for all } f \in B_0. \tag{6}$$

A direct computation with étale coordinates shows that if a δ^A-flow $\delta = \delta^B$ on S is *Hamiltonian* with respect to the Poisson structure $\{ , \}_\eta$ on B attached to a symplectic form η on S that comes from S_0 then δ is Hamiltonian with respect to η; moreover if H is a Hamiltonian then, trivially, H is a prime integral for δ. As we shall see the examples of Painlevé and Euler equations are symplectic-Hamiltonian but *not* Poisson-Hamiltonian simply because the surfaces S on which these equations "live" do not descend to surfaces S_0 over the constants. A large class of examples coming from Lax equations are Poisson-Hamiltonian. Both symplectic-Hamiltonian and Poisson-Hamiltonian equations have arithmetic analogues.

The discussion above has, of course, a higher dimensional analogue in which S is a smooth affine scheme of arbitrary dimension; the Poisson-Hamiltonian picture is valid word for word; the symplectic-Hamiltonian picture has to be modified by asking that S have relative dimension $2d$, η be a closed 2-form, and η^d be invertible.

Going back to the case when S is a surface, let ν be a contact form, let $\delta = \delta^S$ be a δ^A-flow on S, consider the symplectic form $\eta := d\nu$, and define the *Euler-Lagrange* form

$$\epsilon := \delta\nu \in \Omega^1_{B/A}.$$

Since d and δ commute we have that δ^S is Hamiltonian with respect to η if and only if ϵ is closed, i.e. $d\epsilon = 0$. If in addition ϵ is exact, i.e. $\epsilon = d\mathcal{L}$ for some $\mathcal{L} \in \mathcal{O}(S)$, we call $\mathcal{L}$ a *Lagrangian* for (ν, δ^S).

A special case that plays a role in the theory is that in which our surface S is the first jet space of a smooth curve Y over A,

$$S = J^1(Y).$$

In this case a 1-form ν on S is called *canonical* if $\nu = f\beta$ where $f \in B = \mathcal{O}(S)$ and β is a pull-back of a 1-form on Y. Assume ν is a canonical contact form, assume $\delta = \delta^S$ is Hamiltonian with respect to $\eta := dv$, assume ϵ is exact with Lagrangian $\mathcal{L}$, and assume $x \in \mathcal{O}(Y)$ is an étale coordinate on Y. Assume in addition that $x, \delta x$ are étale coordinates on S (which is "generically the case" and is automatic, for instance, for the *canonical* δ^A-flows to be introduced below). Then there are unique A-derivations $\frac{\partial}{\partial x}, \frac{\partial}{\partial \delta x}$ on $\mathcal{O}(S)$ sending $x, \delta x$ into $1, 0$ and $0, 1$ respectively. It is then trivial to check that $\delta \left(\frac{\partial \mathcal{L}}{\partial \delta x} \right) = \frac{\partial \mathcal{L}}{\partial x}$ in $\mathcal{O}(S)$. In particular if $Z \subset J^1(S)$ is the differential equation corresponding to the δ^A-flow δ^S on S then any solution in $S(A)$ to Z is a solution to the *Euler-Lagrange equation* $EL(Z) \subset J^1(S)$ defined by

$$\delta^{\text{univ}} \left(\frac{\partial \mathcal{L}}{\partial \delta x} \right) - \frac{\partial \mathcal{L}}{\partial x} \in \mathcal{O}(J^1(S)).$$

Contact forms v that are canonical should be viewed as generalizing the *canonical contact forms* on cotangent bundles in differential geometry (and classical mechanics); also our Lagrangians and Euler-Lagrange equation correspond, formally, to the Lagrangians and Euler-Lagrange equation in classical mechanics. Note however the following discrepancy with the usual definition in differential geometry: our $J^1(Y)$ is related to (is a torsor under) the *tangent* bundle while in classical differential geometry *canonical forms* live on the *cotangent* bundle. This discrepancy is resolved, in usual differential geometry, by identifying the tangent and the cotangent bundle via dv; in our setting (when $J^1(Y)$ is not a trivial torsor) no such identification is available. By the way, as we shall explain, it is the definition of *canonical contact form* that we just gave above (and not the usual definition in differential geometry) that will have an arithmetic analogue.

Finally, we make the following definition: a *canonical* δ^A-*flow* on $S = J^1(Y)$ is a δ^A-flow $\delta^{J^1(Y)}$ on $J^1(Y)$ with the property that the composition of

$$\delta^{J^1(Y)} : \mathcal{O}(J^1(Y)) \to \mathcal{O}(J^1(Y))$$

with the pull back map

$$\mathcal{O}(Y) \to \mathcal{O}(J^1(Y))$$

equals the universal derivation

$$\delta^{\text{univ}} : \mathcal{O}(Y) \to \mathcal{O}(J^1(Y)).$$

By the way, notice that one has a natural closed embedding

$$\iota : J^2(Y) \to J^1(J^1(Y)).$$

Then one checks that a δ^A-flow $\delta^{J^1(Y)}$ on $J^1(Y)$ is canonical if and only if the section $J^1(Y) \to J^1(J^1(Y))$ defined by $\delta^{J^1(Y)}$ factors through ι. Also notice that if $x \in \mathcal{O}(Y)$ is an étale coordinate and $\delta = \delta^{J^1(Y)}$ is a canonical flow then x, δx are étale coordinates on S. The concept of canonical δ^A-flow is an algebraic version of a classical concept related to second order ODEs (for instance Painlevé equations) and has an arithmetic analogue.

2.2 The Arithmetic Case

Let p be a rational odd prime. If B is a ring a *Frobenius lift* on B is a ring endomorphism $\phi = \phi^B : B \to B$ whose reduction mod p is the p-power Frobenius on B/pB. Similarly if X is a scheme or a p-adic formal scheme a *Frobenius lift* on X is an endomorphism $\phi = \phi^X : X \to X$ whose reduction mod p is the p-power Frobenius on the reduction of X mod p. Let A be a complete discrete valuation ring with maximal ideal generated by p and perfect residue field $k = A/pA$; we fix this A once and for all in the discussion below. Such an A is uniquely determined up to isomorphism by k and possesses a unique Frobenius lift $\phi = \phi^A : A \to A$. For any A-algebra B and any scheme or p-adic formal scheme X over A Frobenius lifts on B or X will be tacitly assumed to be compatible with the Frobenius lift on A. For any Noetherian A-algebra B and Noetherian scheme X over A we denote by $\widehat{B}$ and $\widehat{X}$ the p-adic completions of B and X respectively. We also define the Kähler differentials on the formal scheme $\widehat{X}$ by

$$\Omega_{\widehat{X}} = \varprojlim \Omega_{X_n/A_n} \tag{7}$$

where $A_n = A/p^n A$, $X_n = X \otimes A_n$. If X is smooth over A and ϕ is a Frobenius lift on $\widehat{X}$ then ϕ naturally induces additive maps

$$\frac{\phi^*}{p^i} : \Omega^i_{\widehat{X}} \to \Omega^i_{\widehat{X}}, \tag{8}$$

where $\Omega^i_{\widehat{X}} := \wedge^i \Omega_{\widehat{X}}$.

Given a ring B which is p-torsion free (i.e., p is a non-zero divisor in B) a map of sets

$$\delta = \delta^B : B \to B$$

will be called a *p-derivation* if the map

$$\phi = \phi^B : B \to B, \quad \phi(b) := b^p + p\delta b$$

is a ring homomorphism (equivalently a Frobenius lift); we say that δ and ϕ are *attached* to each other. We view p-derivations as arithmetic analogues of derivations; cf. [5, 23]. Then we view 8 as analogues of Lie derivatives with respect to p-derivations. Similarly, if X is a p-adic formal scheme over A, a *p-derivation* on X (or an *arithmetic δ^A-flow* on X) is a map of sheaves of sets $\delta = \delta^X : \mathcal{O}_X \to \mathcal{O}_X$ such that the map of sheaves of sets $\phi = \phi^X : \mathcal{O}_X \to \mathcal{O}_X$, $\phi(b) = b^p + p\delta b$, is a map of sheaves of rings (and hence induces a Frobenius lift $\phi = \phi^X : X \to X$). We again say that δ and ϕ are *attached* to each other. As we will see later the above concept of *arithmetic δ^A-flow* is not flexible enough to accommodate some of the interesting examples of the theory; in the case of the Painlevé equations we will need a generalization of the concept of *arithmetic δ^A-flow* which will be referred to as *generalized arithmetic δ^A-flow*.

Let δ be a p-derivation on some p-adically complete p-torsion free ring B. An element $c \in B$ is called a *δ-constant* if $\delta c = 0$. The set $B^\delta \subset B$ of δ-constants is a multiplicative submonoid (but not a subring) of B. Let $\mathbb{Z}[B^\delta]$ be the subring of B generated by B^δ and let Σ be the multiplicative system $\Sigma := B^\times \cap \mathbb{Z}[B^\delta]$. An element of B is called a *pseudo-δ-constant* if it is a p-adic limit in B of elements in the ring of fractions $\Sigma^{-1}\mathbb{Z}[B^\delta]$. So the set of pseudo-$\delta$-constants in B is a subring of B. One can easily check that if B/pB is perfect (i.e., the p-power Frobenius on B/pB is surjective) then any element in B is congruent mod p to an element of B^δ and, consequently, any element of B is pseudo-δ-constant; in particular any element of A is a pseudo-δ-constant. Conversely, if an element $b \in B$ is congruent mod p to an element in B^δ then δb is congruent mod p to a p-th power in B.

One can introduce arithmetic analogues of jet spaces as follows; cf. [5]. Let X be a scheme of finite type over A or the p-adic completion of such a scheme. Say first that X is affine,

$$X = Spec\ A[x]/(f) \quad \text{or} \quad X = Spf\ \widehat{A[x]}/(f),$$

with x and f tuples. Then define the *p-jet spaces* of X to be the p-adic formal schemes

$$J^n(X) := Spf\ \widehat{A[x, x', \dots, x^{(n)}]}/(f, \delta f, \dots, \delta^n f) \qquad (9)$$

where $x', \dots, x^{(n)}$ are new tuples of indeterminates and $\delta = \delta^{univ}$ is the unique p-derivation on $A[x, x', \dots, x^{(n)}, \dots]$ extending δ^A and satisfying $\delta x = x', \dots, \delta x^{(n)} = x^{(n+1)}, \dots$ We denote, as usual, by $\phi = \phi^{univ}$ the Frobenius lift attached to δ^{univ}; it induces ring homomorphisms

$$\phi = \phi^{univ} : \mathcal{O}(J^n(X)) \to \mathcal{O}(J^{n+1}(X)).$$

If X is not necessarily affine then, again, one defines $J^n(X)$ by gluing $J^n(X_i)$ where $X = \cup X_i$ is a Zariski affine open cover. (In the gluing process one uses the fact that we are dealing with formal schemes rather than schemes. There is a more global approach, avoiding gluing, that leads to functorially constructed algebraizations of our p-jet spaces; cf. [1, 2]. We will not need these algebraized p-jet spaces in what follows.) Then $(J^n(X))_{n \geq 0}$ has, again, a structure of projective system of p-adic formal schemes depending functorially on X, with $J^0(X) = \widehat{X}$.

If G is a group in the category of schemes or p-adic formal schemes over A then $(J^n(G))_{n \geq 0}$ has, again, a structure of projective system of groups in the category of p-adic formal schemes; however, even if G/A is smooth, the kernels of the homomorphisms $J^n(G) \to J^{n-1}(G)$ are *generally not* isomorphic as groups to powers of $\widehat{\mathbb{G}_a}$, although they are always isomorphic as formal schemes to some completed affine space $\widehat{\mathbb{A}^d}$. (By the way, these kernels are commutative if and only if G itself is commutative!)

By an *arithmetic differential equation* on X we will understand a closed formal subscheme of some $J^n(X)$. An *arithmetic δ^A-flow* on X will mean an arithmetic δ^A-flow on $\widehat{X}$. To give an arithmetic δ^A-flow on $\widehat{X}$ is equivalent to giving a section of the canonical projection $J^1(X) \to \widehat{X}$, i.e. to giving a differential equation $Z \subset J^1(X)$ for which the projection $Z \to \widehat{X}$ is an isomorphism. A *prime integral* for an arithmetic δ^A-flow δ^X is a δ^X-constant in $\mathcal{O}(\widehat{X})$, i.e., a function $H \in \mathcal{O}(\widehat{X})$ such that $\delta^X H = 0$. For any A-point $P \in X(A)$, one defines the jets $J^n(P) \in J^n(X)(A)$ as the unique morphisms lifting P that are compatible with the actions of δ^{univ} and δ^A. A *solution* (in A) for a differential equation $Z \subset J^n(X)$ is, again, an A-point $P \in X(A)$ such that $J^n(P)$ factors through Z. If P is a solution to the differential equation defined by an arithmetic δ^A-flow and if H is a prime integral for that arithmetic δ^A-flow then, again, $\delta^A(H(P)) = 0$; so, again, intuitively H is "constant along any solution". If X is affine and $Z \subset J^1(X)$ is the arithmetic differential equation corresponding to an arithmetic δ^A-flow δ^X on X then a point $P \in X(A)$ is a solution to Z if and only if the ring homomorphism $P^* : \mathcal{O}(\widehat{X}) \to A$ defined by P satisfies $\delta^A \circ P^* = P^* \circ \delta^X$.

Let, in what follows, $S = Spec\ B$ be a smooth affine surface. Then, by the discussion in the previous subsection, we have a notion of *symplectic* form $\eta \in \Omega^2_{B/A}$ and associated *Poisson structure*, $\{ \ , \ \}_\eta$. For an arithmetic δ^A-flow $\delta = \delta^B : \widehat{B} \to \widehat{B}$ on S the analogue of Lie derivatives will be the maps

$$\frac{\phi^*}{p^i} : \Omega^i_{\widehat{S}} \to \Omega^i_{\widehat{S}}, \quad i = 0, 1, 2. \tag{10}$$

At this point we would like to define what it means for an arithmetic δ^A-flow on S to be *Hamiltonian with respect to a symplectic form η on S*. One is tempted to make the following definition: an arithmetic δ^A-flow $\delta = \delta^S$ on S is *Hamiltonian* with respect to the symplectic form η on S if

$$\frac{\phi^*}{p^2}\eta = \lambda \cdot \eta, \tag{11}$$

where $\lambda \in \mathcal{O}(\widehat{S})$ is a pseudo-δ-constant. The concept we just defined is, however, not flexible enough to accommodate our examples. In particular, for the Painlevé equations one will need to replace arithmetic δ^A-flows with what we will call *generalized arithmetic δ^A-flows*; while for the Euler equations one will need to replace equality in 11 by a congruence mod p. In view of the above we will *not* adopt, in what follows, the above attempted definition of the *Hamiltonian* property but rather postpone the discussion of the Hamiltonian-related concepts to the next sections where the main examples of the theory are discussed; there we will encounter arithmetic analogues of the Hamiltonian property, canonical contact forms, canonical (generalized) arithmetic δ^A-flows, Euler-Lagrange forms, etc., each of which will be adapted to their specific context.

3 Painlevé Equations

3.1 The Classical Case

As a step in his proof of the Mordell conjecture over function fields of characteristic zero [25] Manin introduced a differential algebraic map now referred to as the *Manin map*. A different, but equivalent, construction of this map was given in [3]. On the other hand Manin showed in [26] (cf. also [27], p. 71) how the Painlevé VI equation can be understood as a "deformation of the Manin map"; he attributes this viewpoint to Fuchs. In [26] Manin also explained how this viewpoint leads to a Hamiltonian structure for the Painlevé VI equation. We quickly review here the "deformation of the Manin map" interpretation of Painlevé VI in [26] and refer to [26] for the Hamiltonian picture.

Let A be the algebraic closure of a function field of one variable, $A = \overline{\mathbb{C}(t)}$, and $\delta^A = d/dt$. Let $\mathcal{E}$ be an elliptic curve (i.e., a smooth projective curve of genus one) over A which does not descend to $\mathbb{C}$. Then the second jet space $J^2(\mathcal{E})$ is easily seen to possess a non-zero group homomorphism of algebraic groups, unique up to multiplication by an element of A,

$$\psi : J^2(\mathcal{E}) \to \mathbb{G}_a, \tag{12}$$

into the additive group $\mathbb{G}_a$ over A. The map 12 is an incarnation of the Manin map, as explained, in a more general setting, in [3]. Let us view ψ as an element of $\mathcal{O}(J^2(\mathcal{E}))$ and, for any open set $Y \subset \mathcal{E}$, let us view $\mathcal{O}(Y)$ and $\mathcal{O}(J^2(\mathcal{E}))$ as subrings of $\mathcal{O}(J^2(Y))$ via pull-backs. Also let us recall that the classical Painlevé VI equation is a family, depending on 4 parameters in $\mathbb{C}$, of differential equations. Then Manin's analysis in [26] shows that each of the differential equations in the Painlevé VI family can be interpreted (in our language introduced above) as the closed subscheme Z of $J^2(Y)$ defined by

$$f := \psi - r \in \mathcal{O}(J^2(Y)) \tag{13}$$

where Y is the complement in $\mathcal{E}$ of the set $\mathcal{E}[2]$ of 2-torsion points in $\mathcal{E}(A)$ and $r \in \mathcal{O}(Y)$ is an appropriate function. More precisely r is a suitable $\mathbb{C}$-linear combination of the 4 translates, by the 4 points in $\mathcal{E}[2]$, of the y-function on $\mathcal{E}\backslash\mathcal{E}[2]$ in a representation

$$\mathcal{E}\backslash\mathcal{E}[2] = Spec \; A[x, y, y^{-1}]/(y^2 - (x^3 + ax + b));$$

the complex coefficients of this linear combination are related to the 4 complex parameters in the corresponding classical Painlevé equation. Moreover one can easily show that

Theorem 3.1 *For any function $r \in \mathcal{O}(Y)$ the differential equation $Z \subset J^2(Y)$ given by 13 defines a canonical δ^A-flow on $S := J^1(Y)$.*

In particular the equations in the Painlevé VI family are defined by canonical δ^A-flows on $J^1(Y)$. By the way notice that $J^1(\mathcal{E})$, on which Painlevé equations "live", is an $\mathbb{A}^1$-fibration over the elliptic curve $\mathcal{E}$; and actually, over Y, this fibration is trivial, so we have an isomorphism

$$J^1(Y) \simeq Y \times \mathbb{A}^1.$$

For details on the Hamiltonian picture we refer to [26].

3.2 The Arithmetic Case

The construction of the Manin map in [3] was shown in [5] to have an arithmetic analogue. Then, in [13], an arithmetic analogue of the Painlevé VI equation was introduced and a *Hamiltonian structure* was shown to exist for it. We explain this in what follows.

Let A be a complete discrete valuation ring with maximal ideal generated by p and perfect residue field. Recall that elliptic curves over A do not generally admit Frobenius lifts; an elliptic curve that admits a Frobenius lift is automatically with complex multiplication.

Theorem 3.2 [5] *Let $\mathcal{E}$ be an elliptic curve over A that admits no Frobenius lift. There exists a non-zero group homomorphism, in the category of p-adic formal schemes,*

$$\psi : J^2(\mathcal{E}) \to \widehat{\mathbb{G}_a}, \tag{14}$$

which is unique up to multiplication by a constant in $A^{\times}$.

We view 14 as an arithmetic analogue of the Manin map 12. Given an invertible 1-form ω on $\mathcal{E}$ one can normalize ψ with respect to ω; we will need, and review, this normalization later. The normalized ψ can be referred to as the *canonical δ-character* on $\mathcal{E}$; cf. [9], Definition 7.24. One can view ψ as an element of $\mathcal{O}(J^2(\mathcal{E}))$. By the way one has:

Theorem 3.3 [18] *Let $\mathcal{E}$ be an elliptic curve over A that admits no Frobenius lift. Then the following hold:*

1) $\mathcal{O}(J^1(\mathcal{E})) = R$.
2) $J^1(\mathcal{E})$ admits no Frobenius lift.

Assertion 1 in Theorem 3.3 shows that "order 2 in Theorem 3.2 is optimal." Assertion 2 in Theorem 3.3 is equivalent to saying that the projection

$$J^1(J^1(\mathcal{E})) \rightarrow J^1(\mathcal{E})$$

does not admit a section in the category of p-adic formal schemes; equivalently, there is no arithmetic δ^A-flow on $J^1(\mathcal{E})$! This justifies our generalization of the notion of arithmetic δ^A-flow below. To introduce this let us assume, for a moment that Y is any smooth affine curve over A and assume we are given an arithmetic differential equation

$$Z \subset \mathcal{O}(J^r(Y)). \tag{15}$$

Then one can consider the module

$$\Omega_Z = \varprojlim \Omega_{Z_n/A_n}, \tag{16}$$

$A_n := A/p^n A$, $Z_n := Z \otimes A_n$, and the module

$$\Omega_J = \varprojlim \Omega_{J_n/A_n}, \tag{17}$$

where $J := J^r(Y)$. On the other hand put

$$\Omega'_Z := \frac{\Omega_J}{\langle I_Z \Omega_J, dI_Z \rangle} \tag{18}$$

where I_Z is the ideal of Z in $J^r(Y)$. Moreover define $\Omega_Z'^i$ to be the i-th wedge power $\wedge^i \Omega'_Z$. Under quite general hypotheses the modules 7 and 18 coincide; we will not discuss this here but, rather, refer to [10], Lemma 3.165.

Going back to Z as in 15, for each $s \leq r$, there is a natural map

$$\pi_{r,s} : Z \rightarrow J^s(Y).$$

We also have natural maps

$$\frac{\phi^{\mathrm{univ}*}}{p^i} : \Omega^i_{J^{r-1}(Y)} \rightarrow \Omega^i_{J^r(Y)},$$

inducing maps which we will denote by

$$\frac{\phi^*_Z}{p^i} : \Omega^i_{J^1(Y)} \rightarrow \Omega_Z'^i.$$

We say that $Z \subset J^2(Y)$ defines a *generalized δ-flow* on $J^1(Y)$, if the induced map

$$\pi_{2,1}^* \Omega_{J^1(Y)} \to \Omega_Z'$$

is injective, and its cokernel is annihilated by a power of p. Under quite general conditions, if Z defines an arithmetic δ^A-flow on $J^1(Y)$ then Z defines a generalized arithmetic δ^A-flow on $J^1(Y)$; again we will not need this so we will not discuss these conditions here; but see, again, [10], Lemma 3.165.

Now let S be a smooth surface over A or the p-adic completion of such a surface. Recall that a *symplectic form* on S is an invertible 2-form on X over A; a *contact form* on S is a 1-form on X over S such that dv is symplectic; and for $S = J^1(Y)$ with Y a smooth curve over A, a 1-form v on S is called *canonical* if $v = f\beta$, where $f \in \mathcal{O}(S)$ and β is an 1-form lifted from Y.

Let Y be a smooth affine curve over A and let $f \in \mathcal{O}(J^2(Y))$ be a function whose zero locus defines a generalized arithmetic δ^A-flow on $S := J^1(Y)$. The respective generalized arithmetic δ^A-flow is called *Hamiltonian* with respect to the symplectic form η on S, if

$$\frac{\phi_Z^*}{p}\eta = \lambda \cdot \eta$$

in $\Omega_Z'^2$ for some $\lambda \in A$; note that any element in A, hence in particular λ, is a pseudo-δ-constant; so the definition we just gave is a generalized version of the definition we proposed in 11. Assume, moreover, that $\eta = dv$ for some canonical 1-form v on S. Then we call

$$\epsilon := \frac{\phi_Z^*}{p}v - \lambda v \in \Omega_Z' \tag{19}$$

the *Euler-Lagrange form* attached to v.

Now let $\mathcal{E}$ be an elliptic curve over A that does not admit a Frobenius lift and let $\psi \in \mathcal{O}(J^2(\mathcal{E}))$ be the canonical δ-character with respect to an invertible 1-form ω on $\mathcal{E}$. Consider the symplectic form

$$\eta = \omega \wedge \frac{\phi^{univ*}}{p}\omega$$

on $J^1(\mathcal{E})$. Let $Y \subset \mathcal{E}$ be an affine open set possessing an étale coordinate; this latter condition is satisfied, for instance, if $Y = \mathcal{E} \setminus \mathcal{E}[2]$. By the way, notice that $J^1(\mathcal{E})$ is an $\widehat{\mathbb{A}^1}$-fibration over the elliptic curve $\mathcal{E}$; and actually this fibration is trivial over Y, hence we have an isomorphism of formal schemes,

$$J^1(Y) \simeq \widehat{Y} \times \widehat{\mathbb{A}^1}.$$

Theorem 3.4 [13]

1) *There exists a canonical contact form v on $S := J^1(Y)$ such that $dv = \eta$.*
2) *For any function $r \in \mathcal{O}(Y)$ the differential equation $Z \subset J^2(Y)$ given by the zero locus of the function*

$$f = \psi - \phi^{univ}(r) \in \mathcal{O}(J^2(Y))$$

defines a generalized arithmetic δ^A-flow on S which is Hamiltonian with respect to η.

In particular the symplectic form η is exact and the Euler-Lagrange form ϵ is closed. The function f in assertion 2 of the Theorem is our analogue of the Painlevé VI equation. By the Theorem it defines a generalized arithmetic δ-flow on S; however, it does not define an arithmetic δ-flow on S which is our motivation for generalizing the definition of arithmetic δ-flow. Note the discrepancy with the classical case coming from replacing r by $\phi^{univ}(r)$ in the expression of f in Theorem 3.4. Another discrepancy comes from the absence, in the arithmetic setting, of an analogue of the 4 constant parameters in the classical Painlevé equations.

4 Euler Equations

In Manin's picture [26] we have just reviewed the Painlevé VI equation "lives" on an $\mathbb{A}^1$-fibration over an elliptic curve. On the other hand, the *Euler equation* describing the motion of a rigid body with a fixed point, which we are discussing next, "lives" on an elliptic fibration over $\mathbb{A}^1$. This already suggests an analogy between the geometries underlying these differential equations and their arithmetic analogues; we will make such analogies/links more precise below.

4.1 The Classical Case

We begin by reviewing the classical Euler equations from a purely algebraic point of view. Let A be either a field or a discrete valuation ring and assume 2 is invertible in A. Let x_1, x_2, x_3 and z_1, z_2 be variables and let $a_1, a_2, a_3 \in A$ be such that

$$(a_1 - a_2)(a_2 - a_3)(a_3 - a_1) \in A^\times.$$

We consider the quadratic forms,

$$H_1 := \sum_{i=1}^{3} a_i x_i^2 \in A[x_1, x_2, x_3], \quad H_2 := \sum_{i=1}^{3} x_i^2 \in A[x_1, x_2, x_3].$$

Also we consider the affine spaces $\mathbb{A}^2 = \text{Spec } A[z_1, z_2]$, $\mathbb{A}^3 = \text{Spec } A[x_1, x_2, x_3]$ and the morphism $\mathcal{H} : \mathbb{A}^3 \to \mathbb{A}^2$ defined by $z_1 \mapsto H_1$, $z_2 \mapsto H_2$. For $i = 1, 2, 3$ denote by $Z_i \subset \mathbb{A}^3$ the x_i-*coordinate plane* and let

$$L_1 = Z_2 \cap Z_3, \quad L_2 = Z_3 \cap Z_1, \quad L_3 = Z_1 \cap Z_2$$

be the x_i-*coordinate axes*. Then $\mathcal{H}$ is smooth on the complement of $L_1 \cup L_2 \cup L_3$. For any A-point $c = (c_1, c_2) \in \mathbb{A}^2 = \mathbb{A}^2(A)$ we set

$$E_c := \mathcal{H}^{-1}(c) = \text{Spec } A[x_1, x_2, x_3]/(H_1 - c_1, H_2 - c_2),$$

and we let $i_c : E_c \to \mathbb{A}^3$ be the inclusion. Consider the polynomial

$$N(z_1, z_2) = \prod_{i=1}^{3}(z_1 - a_i z_2) \in A[z_1, z_2].$$

Then, for $N(c_1, c_2) \in A^\times$, E_c is disjoint from $L_1 \cup L_2 \cup L_3$ and, in particular, E_c is smooth over A: it is an affine elliptic curve. Moreover E_c comes equipped with a global 1-form given by

$$\omega_c = i_c^* \frac{dx_1}{(a_2 - a_3)x_2 x_3} = i_c^* \frac{dx_2}{(a_3 - a_1)x_3 x_1} = i_c^* \frac{dx_3}{(a_1 - a_2)x_1 x_2}. \qquad (20)$$

If one considers the smooth projective model $\mathcal{E}_c$ of E_c then ω_c extends to an invertible 1-form on the whole of $\mathcal{E}_c$. In the discussion below a certain plane quartic will play a role; let us review this next. Consider two more indeterminates x, y, and consider the polynomial

$$F := ((a_2 - a_3)x^2 + z_1 - a_2 z_2)((a_3 - a_1)x^2 - z_1 + a_1 z_2) \in A[z_1, z_2][x]. \qquad (21)$$

For any $c = (c_1, c_2) \in A^2$ set

$$E_c' := \text{Spec } A[x, y]/(y^2 - F(c_1, c_2, x)).$$

Then we have a morphism $\pi : E_c \to E_c'$ given by $x \mapsto x_3$, $y \mapsto (a_1 - a_2)x_1 x_2$. If $N(c_1, c_2) \in A^\times$, E_c' is smooth over A and

$$\pi^* \left(\frac{dx}{y} \right) = i_c^* \frac{dx_3}{(a_1 - a_2)x_1 x_2} = \omega_c.$$

For A a perfect field and c_1, c_2 satisfying $N(c_1, c_2) \neq 0$ we have that E_c' is a smooth plane curve. If $\mathcal{E}_c'$ is its smooth projective model then we have an induced isogeny of elliptic curves, $\mathcal{E}_c \to \mathcal{E}_c'$.

Assume now, until further notice, that A is a field of characteristic zero (classically $A = \mathbb{C}$, the complex field), viewed as equipped with the trivial derivation $\delta^A = 0$, and consider the A-derivation $\delta = \delta^B$ on the polynomial ring $B = A[x_1, x_2, x_3]$ given by

$$\delta x_1 = (a_2 - a_3)x_2x_3, \quad \delta x_2 = (a_3 - a_1)x_3x_1, \quad \delta x_3 = (a_1 - a_2)x_1x_2. \quad (22)$$

We refer to the derivation δ as the *classical Euler flow* on $\mathbb{A}^3$.

For any $c = (c_1, c_2) \in A^2$ with $N(c_1, c_2) \neq 0$ denote by δ_c the derivation on $\mathcal{O}(E_c)$ induced by the derivation δ on B. We have the following trivially checked classical fact:

Theorem 4.1

1) H_1 and H_2 are prime integrals for the classical Euler flow, i.e.,

$$\delta H_1 = \delta H_2 = 0.$$

2) For any $c = (c_1, c_2) \in A^2$ with $N(c_1, c_2) \neq 0$ the Lie derivative δ_c on $\Omega^1_{\mathcal{O}(E_c)/A}$ annihilates the 1-form ω_c on E_c:

$$\delta_c \omega_c = 0.$$

Condition 2 can be viewed as a *linearization* condition for the δ^A-flow δ_c on E_c. It is equivalent to δ_c having an extension to a vector field on the compactification $\mathcal{E}_c$. It is the condition in 2 and *not* the "extension to the compactification" property that will have an arithmetic analogue.

The classical Euler flow fits into the Hamiltonian paradigm. We explain this in what follows. Since we will later need a discussion of these concepts in the arithmetic case as well we revert in what follows to the case when A is either a field or a discrete valuation ring. Let $c_2 \in A^\times$ and set

$$S_{c_2} := Spec \, A[x_1, x_2, x_3]/(H_2 - c_2) \subset \mathbb{A}^3,$$

the *sphere of radius* $c_2^{1/2}$. Then S_{c_2} is the scheme theoretic pullback via $\mathcal{H}$ of the line in $\mathbb{A}^2$ defined by $z_2 - c_2$ and hence the map

$$H : S_{c_2} \to \mathbb{A}^1 = Spec \, A[z_1]$$

induced by $z_1 \to H_1$ is smooth above the complement of the closed subscheme defined by

$$N(z_1, c_2) = (z_1 - c_2a_1)(z_1 - c_2a_2)(z_1 - c_2a_3).$$

Now consider the 2-forms

$$\eta_1 = \frac{dx_2 \wedge dx_3}{x_1}, \quad \eta_2 = \frac{dx_3 \wedge dx_1}{x_2}, \quad \eta_3 = \frac{dx_1 \wedge dx_2}{x_3}$$

defined on the complements in S_{c_2} of Z_1, Z_2, Z_3, respectively. These forms glue together defining a symplectic form η_{c_2} on S_{c_2}. If one considers the Poisson structure $\{\,,\,\}$ on $\mathcal{O}(\mathbb{A}^3) = A[x_1, x_2, x_3]$ defined by

$$\{x_1, x_2\} = x_3, \quad \{x_2, x_3\} = x_1, \quad \{x_3, x_1\} = x_2,$$

then H_2 is a *Casimir* i.e., $\{H_2, -\} = 0$, so this Poisson structure induces a Poisson structure $\{\,,\,\}_{c_2}$ on each $\mathcal{O}(S_{c_2})$. On $\mathcal{O}(S_{c_2})$ we have

$$\{x_1, x_2\}_{c_2} = \frac{dx_1 \wedge dx_2}{\eta_{c_2}}, \quad \{x_2, x_3\}_{c_2} = \frac{dx_2 \wedge dx_3}{\eta_{c_2}}, \quad \{x_3, x_1\}_{c_2} = \frac{dx_3 \wedge dx_1}{\eta_{c_2}},$$

hence the Poisson structure $\{\,,\,\}_{c_2}$ on $\mathcal{O}(S_{c_2})$ coincides with the Poisson structure $\{\,,\,\}_{\eta_{c_2}}$ on $\mathcal{O}(S_{c_2})$ defined by the symplectic form η_{c_2} (because the two Poisson structures coincide on the generators x_1, x_2, x_3 of $\mathcal{O}(S_{c_2})$). In other words S_{c_2} are *symplectic leaves* for our Poisson structure on $\mathcal{O}(\mathbb{A}^3)$, with corresponding symplectic forms η_{c_2}. Furthermore, if δ is the *classical Euler flow* 22 then δ induces a derivation δ_{c_2} on each $\mathcal{O}(S_{c_2})$ and the Lie derivative on 2-forms,

$$\delta_{c_2} : \Omega^2_{\mathcal{O}(S_{c_2})/A} \to \Omega^2_{\mathcal{O}(S_{c_2})/A}$$

is trivially seen to satisfy

$$\delta_{c_2} \eta_{c_2} = 0. \tag{23}$$

In other words we have:

Theorem 4.2 *The δ^A-flow δ_{c_2} on S_{c_2} is symplectic with respect to η_{c_2}.*

The link between the 2-forms η_{c_2} and the 1-forms ω_c is as follows. Consider the 1-forms

$$\omega_1 = \frac{dx_1}{(a_2 - a_3)x_2x_3}, \quad \omega_2 = \frac{dx_2}{(a_3 - a_1)x_3x_1}, \quad \omega_3 = \frac{dx_3}{(a_1 - a_2)x_1x_2}$$

defined on

$$S_{c_2}\backslash(Z_2 \cup Z_3), \quad S_{c_2}\backslash(Z_3 \cup Z_1), \quad S_{c_2}\backslash(Z_1 \cup Z_2),$$

respectively. Recall that for any $c = (c_1, c_2)$ with $N(c_1, c_2) \in A^\times$ the restrictions of $\omega_1, \omega_2, \omega_3$ to E_c glue to give the form ω_c on E_c. A trivial computation then gives the following equalities of 2-forms on $S_{c_2}\backslash(Z_1 \cup Z_2 \cup Z_3)$ which will play a role later:

$$\eta_{c_2} = -dH_1 \wedge \omega_1 = -dH_1 \wedge \omega_2 = -dH_1 \wedge \omega_3. \tag{24}$$

By the way, the equalities 24 imply that for all $c = (c_1, c_2)$ with $N(c_1, c_2) \in A^\times$ the form ω_c on E_c satisfies

$$\omega_c = -P.R.\left(\frac{\eta_{c_2}}{H_1 - c_1}\right), \tag{25}$$

where

$$P.R. : \Omega^2_{S_{c_2}/A}(E_c) \to \Omega^1_{E_c/A}$$

is the Poincaré residue map [22], p. 147; we will not need this interpretation in what follows.

4.2 The Arithmetic Case

In what follows A is a complete discrete valuation ring with maximal ideal generated by an odd prime p and perfect residue field $k = A/pA$. Let $F \in A[z_1, z_2][x]$ be the polynomial in 21. Define the *Hasse invariant* to be the coefficient $A_{p-1} \in A[z_1, z_2]$ of x^{p-1} in the polynomial $F^{\frac{p-1}{2}}$. In addition to the quantities defined in the previous subsection we also consider the following polynomial

$$Q := x_1 x_2 \cdot N(H_1, H_2) \cdot A_{p-1}(H_1, H_2) \in A[x_1, x_2, x_3],$$

and the open subscheme of $\mathbb{A}^3$ defined by

$$X = Spec \, A[x_1, x_2, x_3][1/Q].$$

Assume in addition that $c = (c_1, c_2) \in A^2$ satisfies

$$\delta c_1 = \delta c_2 = 0 \quad \text{and} \quad N(c_1, c_2) \cdot A_{p-1}(c_1, c_2) \in A^\times$$

and let δ^X be any arithmetic δ^A-flow on $\widehat{X}$ satisfying $\delta^X H_1 = \delta^X H_2 = 0$. Then the Frobenius lift ϕ^X on $\mathcal{O}(\widehat{X})$ induces a Frobenius lift $\phi_c := \phi^{E_c^0}$ on $\widehat{E_c^0}$ where E_c^0 is the open set of E_c given by $E_c^0 := E_c \cap X$. We refer to ϕ_c as the *Frobenius lift* on $\widehat{E_c^0}$ attached to δ^X. On the other hand, the global 1-form ω_c in 20 restricted to E_c^0 will be referred to as the *canonical* 1-form on E_c^0 and will still be denoted by ω_c.

The following provides an arithmetic analogue of the classical Euler flow: assertions 1 and 2 below are arithmetic analogues of assertions 1 and 2 in Theorem 4.1 respectively.

Theorem 4.3 [15] *There exists an arithmetic δ^A-flow δ^X on $\widehat{X}$ such that:*

1) H_1 *and* H_2 *are prime integrals for* δ^X, *i.e., the following holds in* $\mathcal{O}(\widehat{X})$:

$$\delta^X H_1 = \delta^X H_2 = 0;$$

2) *For any point* $c = (c_1, c_2) \in A^2$ *with*

$$\delta c_1 = \delta c_2 = 0, \quad \text{and} \quad N(c_1, c_2) \cdot A_{p-1}(c_1, c_2) \in A^\times$$

the Frobenius lift ϕ_c on $\widehat{E_c^0}$ attached to δ^X and the canonical 1-form ω_c on E_c^0 satisfy the following congruence in $\Omega^1_{\widehat{E_c^0}}$:

$$\frac{\phi_c^*}{p}\omega_c \equiv A_{p-1}(c_1, c_2)^{-1} \cdot \omega_c \quad mod \quad p.$$

By the way, one can ask if the open set X in Theorem 4.3 can be taken to be the whole of $\mathbb{A}^3$. In contrast with the classical case (Theorem 4.1), the answer to this is negative; indeed we have the following "singularity theorem":

Theorem 4.4 [15] *If $X \subset \mathbb{A}^3$ is an open set such that $\widehat{X}$ possesses an arithmetic δ^A-flow δ^X with $\delta^X H_1 = \delta^X H_2 = 0$ and if $\delta a_i \in A^\times$ for some $i \in \{1, 2, 3, \}$ then $\widehat{X}$ cannot meet the coordinate axis $\widehat{L_i}$.*

Another question one can ask is whether it is possible to extend the Frobenius lifts ϕ_c in Theorem 4.3 to the compactifications $\mathcal{E}_c$ of E_c. In contrast with the classical case (Theorem 4.1), the answer to this is, again, negative, cf. Theorem 4.5 below. For this theorem we fix, for every rational prime p, a complete discrete valuation ring R_p with maximal ideal generated by p and algebraically closed residue field. We also fix a number field F, with ring of integers $\mathcal{O}_F$, a rational integer M, and, for each $p >> 0$ we fix an embedding of $\mathcal{O}_F[1/M]$ into R_p.

Theorem 4.5 [16] *Let $a_1, a_2, a_3 \in \mathcal{O}_F[1/M]$. Then, if $p >> 0$, there is no triple (K, X, ϕ^X) with*

- $K \in \mathcal{O}(\widehat{\mathbb{A}^2}) = R_p[z_1, z_2]\widehat{\,}$, $K \not\equiv 0 \bmod p$,
- $X \subset \mathbb{A}^3$ *an open set over R_p,*
- ϕ^X *a Frobenius lift on $\widehat{X}$,*

satisfying the following two conditions:

1) H_1 *and H_2 are prime integrals for the arithmetic δ^A-flow δ^X attached to ϕ^X, i.e., the following holds in $\mathcal{O}(\widehat{X})$:*

$$\delta^X H_1 = \delta^X H_2 = 0;$$

2) *for all $c \in R_p^2$ with $\delta c = 0$, $N(c)K(c) \in R_p^\times$, one has $\widehat{E_c} \cap \widehat{X} \neq \emptyset$ and*

   ϕ_c *extends to an endomorphism of the compactification $\mathcal{E}_c$ of E_c,*

   where ϕ_c is the Frobenius lift on $\widehat{E_c} \cap \widehat{X}$ induced by ϕ^X.

Interestingly, the proof of Theorem 4.5 is based on a variant of a Diophantine result in [14] which, in its turn, is proved using, again, arguments involving arithmetic differential equations.

In what follows we use Theorem 4.3 to derive an arithmetic analogue of the Hamiltonian picture; cf. Theorem 4.2.

Let $c_2 \in A^\times$ be such that $\delta c_2 = 0$. Then the Frobenius lift ϕ^X attached to the arithmetic δ^A-flow δ^X in Theorem 4.3 induces a Frobenius lift ϕ_{c_2} on $\widehat{S^0_{c_2}}$ where $S^0_{c_2} := S_{c_2} \cap X$. Recall that it follows from equations 6.1 and 6.2 in [15] that the function $A_{p-1}(H_1, c_2)$ is invertible on $\widehat{S^0_{c_2}}$. Set

$$\lambda := \frac{H_1^{p-1}}{A_{p-1}(H_1, c_2)} \in \mathcal{O}(\widehat{S^0_{c_2}})$$

and note that λ is a pseudo-δ-constant in $\mathcal{O}(\widehat{S^0_{c_2}})$ because H_1 is a δ-constant and all elements of A are pseudo-δ-constants. We will prove the following result which can be interpreted as a relaxation of the condition defining the *Hamiltonian* property in 11:

Theorem 4.6 *The following holds in* $\Omega^2_{\widehat{S^0_{c_2}}}$:

$$\frac{\phi^*_{c_2}}{p^2} \eta_{c_2} \equiv \lambda \eta_{c_2} \quad mod \quad p.$$

Proof. Consider the form θ on $\widehat{S^0_{c_2}}$ defined by

$$\theta := \frac{\phi^*_{c_2}}{p^2} \eta_{c_2} - \frac{H_1^{p-1}}{A_{p-1}(H_1, c_2)} \eta_{c_2}.$$

By 24 and $\phi_{c_2}(H_1) = H_1^p$ we have

$$\theta = -H_1^{p-1} dH_1 \wedge \frac{\phi^*_{c_2}}{p} \omega_1 + \frac{H_1^{p-1}}{A_{p-1}(H_1, c_2)} dH_1 \wedge \omega_1 = -H_1^{p-1} dH_1 \wedge \beta$$

where

$$\beta := \frac{\phi^*_{c_2}}{p} \omega_1 - \frac{1}{A_{p-1}(H_1, c_2)} \omega_1$$

Let $i_c : E^0_c = E_c \cap X \to S^0_{c_2}$ be the inclusion. Then, if $\delta c_1 = 0$, $N(c_1, c_2) \in A^\times$, $A_{p-1}(c_1, c_2) \in A^\times$, by Theorem 4.3,

$$i^*_c \beta = \frac{\phi^*_c}{p} \omega_c - \frac{1}{A_{p-1}(c_1, c_2)} \omega_c \equiv 0 \quad mod \quad p.$$

Let us denote by an upper bar the operation of reduction mod p. Since any element in k can be lifted to an element c_1 of A killed by δ (this lift is the Teichmüller lift) it follows that

$$\overline{i^*_c \beta} = 0$$

for all except finitely many $\bar{c}_1 \in k$, where $\bar{i}_{\bar{c}} : \overline{E_c^0} \to \overline{S_{c_2}^0}$ is the inclusion. Recall from [15] that H_1, H_2, x_3 are étale coordinates on X; so H_1, x_3 are étale coordinates on $S_{c_2}^0$. Write

$$\beta = b_1 dH_1 + b_2 dx_3,$$

on $\widehat{S_{c_2}^0}$, with $b_1, b_2 \in \mathcal{O}(\widehat{S_{c_2}^0})$. Since

$$\bar{i}_{\bar{c}}^* \bar{\beta} = \bar{i}_{\bar{c}}^* \bar{b}_1 \cdot d\overline{c_1} + \bar{i}_{\bar{c}}^* \bar{b}_2 \cdot d\overline{x_3} = \bar{i}_{\bar{c}}^* \bar{b}_2 \cdot d\overline{x_3}$$

it follows that

$$\bar{i}_{\bar{c}}^* \bar{b}_2 = 0.$$

Since this is true for all except finitely many $\bar{c}_1$ it follows that $\bar{b}_2 = 0$. But then

$$\bar{\theta} = -\overline{H_1}^{p-1} d\overline{H_1} \wedge \bar{\beta} = -\overline{H_1}^{p-1} d\overline{H_1} \wedge \bar{b}_2 d\overline{x_3} = 0.$$

We conclude that the congruence in the statement of the Theorem holds on an open set of $\widehat{S_{c_2}^0}$ and hence on the whole of $\widehat{S_{c_2}^0}$. □

We next deduce a result that establishes a *link* between the Painlevé paradigm and the Euler paradigm. Assume we are under the hypotheses and notation of Theorem 4.3, with $a_1, a_2, a_3 \in \mathbb{Z}_p$. (So morally we are in the "Euler paradigm".) Assume moreover that c_1, c_2 in assertion 2 of that Theorem belong to $\mathbb{Z}_p$ and are such that $\mathcal{E}_c$ does not have a Frobenius lift. Examples of this situation are abundant; cf. the last Remark in [16]. Let, furthermore, $\phi_c : \widehat{E_c^0} \to \widehat{E_c^0}$ be as in Theorem 4.3 and let $\sigma_c^n : \widehat{E_c^0} \to J^n(E_c^0)$ be the sections of the projections $J^n(E_c^0) \to \widehat{E_c^0}$ induced by ϕ_c. On the other hand let $\psi_c \in J^2(\mathcal{E}_c)$ be the canonical δ-character. (The latter belongs, as we saw, to the "Painlevé paradigm".) Assume, for simplicity, that the field $k := A/pA$ is algebraically closed and let K_c be function field of $\mathcal{E}_c \otimes k$. We will prove:

Theorem 4.7 *The image of $\sigma_c^{2*} \psi_c$ in K_c is a p-th power in K_c.*

Proof. Recall by [9], Corollary 7.28, that

$$d\psi_c = \lambda_2 \left(\frac{\phi_c^{univ*}}{p}\right)^2 \omega_c + \lambda_1 \frac{\phi_c^{univ*}}{p} \omega_c + \lambda_0 \omega_c \qquad (26)$$

in $\Omega_{J^2(\mathcal{E}_c)}$ where $\phi_c^{univ} : J^n(\mathcal{E}_c) \to J^{n-1}(\mathcal{E}_c)$ are the universal Frobenius lifts and $\lambda_i \in A$, $\lambda_2 = p$. By the way, the above holds without the assumption that $a_i, c_j \in \mathbb{Z}_p$; also the equality $\lambda_2 = p$ is precisely the definition of ψ being *normalized* with respect to ω_c. With the additional assumption that $a_i, c_j \in \mathbb{Z}_p$ we have that $\mathcal{E}_c$ descends to an elliptic curve $\mathcal{E}_{c/\mathbb{Z}_p}$ over $\mathbb{Z}_p$; then, by [7], Theorem 1.10, and [9], Theorem 7.22, we actually also have

$$\lambda_1 = -a_p, \quad \lambda_0 = 1,$$

where a_p in the trace of Frobenius acting on the reduction mod p of $\mathcal{E}_{c/\mathbb{Z}_p}$. Similarly $\mathcal{E}'_c$ descends to an elliptic curve $\mathcal{E}'_{c/\mathbb{Z}_p}$ over $\mathbb{Z}_p$. Since there is a separable isogeny between $\mathcal{E}'_{c/\mathbb{Z}_p}$ and $\mathcal{E}_{c/\mathbb{Z}_p}$ it follows that a_p is also the trace of Frobenius acting on the reduction mod p of $\mathcal{E}'_{c/\mathbb{Z}_p}$. On the other hand, by [28], p. 141–142,

$$a_p \equiv 1 - |\mathcal{E}'_{c/\mathbb{Z}_p}(\mathbb{F}_p)| \mod p.$$

Now by [15], Lemma 5.2,

$$|\mathcal{E}'_{c/\mathbb{Z}_p}(\mathbb{F}_p)| \equiv 1 - A_{p-1}(c_1, c_2) \mod p.$$

It follows that

$$\lambda_1 \equiv -A_{p-1}(c_1, c_2) \mod p.$$

Let us view the maps $\mathcal{O}(J^n(\mathcal{E}_c)) \to \mathcal{O}(J^{n+1}(\mathcal{E}_c))$ induced by the natural projections as inclusions. Then ϕ_c equals the composition $\phi_c^{\text{univ}} \circ \sigma_c^1$; hence

$$\sigma_c^{2*} \phi^{\text{univ}*} = \sigma_c^{1*} \phi^{\text{univ}*} = \phi_c^*.$$

(Note, by the way, that ϕ_c^2 is not equal to $(\phi_c^{\text{univ}})^2 \circ \sigma_c^2$!) Taking σ_c^{2*} in 26 we get

$$d(\sigma_c^{2*} \psi_c) = \sigma_c^{2*} d\psi_c \equiv -A_{p-1}(c_1, c_2)\frac{\phi_c^*}{p} \omega_c + \omega_c \equiv 0 \mod p.$$

If K_c is the function field of $\mathcal{E}_c \otimes k$ we have $K_c = k(x, \gamma)$ with x a variable and γ quadratic over $k(x)$. Since $k(x, \gamma) = k(x, \gamma^p)$ we may write

$$\overline{\sigma_c^{2*} \psi_c} = u + v\gamma^p \in K_c, \quad u, v \in k(x)$$

hence

$$0 = d(\overline{\sigma_c^{2*} \psi_c}) = \left(\frac{du}{dx} + \frac{dv}{dx}\gamma^p\right) dx \in \Omega_{K_c/k} = K_c dx,$$

hence $\frac{du}{dx} = \frac{dv}{dx} = 0$ which implies that $u, v \in k(x^p)$ as one can see by considering the simple fraction decomposition of u and v. Consequently $\overline{\sigma_c^{2*} \psi_c} \in K_c^p$. □

5 Lax equations

5.1 The Classical Case

Let A be a Noetherian ring, let $B := A[x_1, \ldots, x_N]$ be a polynomial ring, and consider the affine space $\mathbb{A}^n = \text{Spec } B$. Let L be an A-Lie algebra, free as an A-module, with basis $e_1, \ldots, e_N$, and write

$$[e_i, e_j] = \sum_k c_{ijk} e_k, \quad c_{ijk} \in A.$$

Then there is a unique Poisson structure $\{\ ,\ \}$ on B, (or on $\mathbb{A}^N = Spec\ B$) called the *Lie-Poisson* structure attached to $(L, (e_i))$, such that

$$[x_i, x_j] = \sum_k c_{ijk} x_k.$$

In particular we may consider the variables $x_1, \ldots, x_N$, with $N = n^2$, to be the entries of a matrix of indeterminates $x = (x_{ij})$, we may consider the affine space $\mathfrak{g} := \mathbb{A}^{n^2} = Spec\ B$, $B := A[x]$, and we may consider the Lie algebra $L := \mathfrak{g}(A)$ of $n \times n$ matrices with coefficients in A, with respect to the commutator, with basis e_{ij} the matrices that have 1 on position (i, j) and 0 everywhere else. One may consider then the Lie-Poisson structure $\{\ ,\ \}$ on B (equivalently on $\mathfrak{g}$) attached to $(L, (e_{ij}))$.

On the other hand let $\delta = \delta^A$ be a derivation on A, and let $\delta = \delta^{\mathfrak{g}}$ be a δ^A-flow on $\mathfrak{g}$, i.e. $\delta^{\mathfrak{g}}$ is a derivation on $B = A[x]$, extending δ^A. Say that $\delta = \delta^{\mathfrak{g}}$ is a *Lax δ^A-flow* if we have an equality of matrices with B-coefficients

$$\delta x = [M, x]$$

for some matrix $M = (m_{ij})$ with B-coefficients, i.e.,

$$\delta x_{ij} = \sum_k (m_{ik} x_{kj} - x_{ik} m_{kj}).$$

It is trivial to check that any δ^A-flow on $A[x]$ that is Hamiltonian with respect to the Lie-Poisson structure on $A[x]$ is a Lax δ^A-flow on $\mathfrak{g}$: if the Hamiltonian is H then M can be taken to be the matrix $\frac{\partial H}{\partial x} := \left(\frac{\partial H}{\partial x_{ij}}\right)$. Let us say that a Lax δ^A-flow on $\mathfrak{g}$ is *Hamiltonian* (or more accurately *Poisson-Hamiltonian*) if it is Hamiltonian with respect to the Lie-Poisson structure on $A[x]$, equivalently if

$$\delta x = \left[\frac{\partial H}{\partial x}, x\right]$$

for some $H \in B$.

On the other hand, assuming for simplicity that A is an algebraically closed field of characteristic zero, any Lax δ^A-flow δ is *isospectral*, by which we understand that:

Theorem 5.1 *The following diagram is commutative:*

$$
\begin{array}{ccc}
B & \xrightarrow{\delta} & B \\
\mathcal{P} \uparrow & & \uparrow \mathcal{P} \\
A[z] & \xrightarrow{\delta_0} & A[z]
\end{array}
$$

where $A[z] = A[z_1, \ldots, z_n]$, *a polynomial ring in n variables*, $\delta_0 : A[z] \to A[z]$ *is the unique derivations extending* δ^A *with* $\delta_0 z_j = 0$, *and* $\mathcal{P} : A[z] \to B$ *is the A-algebra homomorphism with* $\mathcal{P}(z_j) = \mathcal{P}_j(x)$, *where*

$$\det(s \cdot 1_n - x) = \sum_{j=0}^{n} (-1)^j \mathcal{P}_j(x) s^{n-j}.$$

In the above 1_n is the identity matrix, s is a variable, $\mathcal{P}_0 = 1$, and, for $j = 1, \ldots, n$, $\mathcal{P}_j(x)$ are, of course, the coefficients of the characteristic polynomial of x:

$$\mathcal{P}_1(x) = \mathrm{tr}(x), \ldots, \mathcal{P}_n(x) = \det(x).$$

The commutativity of the above diagram implies $\delta^A(\mathcal{P}_j(x)) = 0$, i.e., $\mathcal{P}_j(x)$ are prime integrals for any Lax δ^A-flow. This implies that the characteristic polynomial of any solution to a Lax δ^A-flow has δ-constant coefficients; equivalently, the spectrum of any solution consists of δ-constants. (This equivalence will fail in the arithmetic case.)

5.2 The Arithmetic Case

As usual we consider a complete discrete valuation ring A with maximal ideal generated by an odd rational prime p and perfect residue field and we view A as equipped with its unique p-derivation $\delta = \delta^A$. There are two arithmetic analogues of Lax δ^A-flows: one for which the characteristic polynomial of any solution has δ-constant coefficients; and another one for which the solutions have δ-constant spectrum. These two conditions are not equivalent because, for a monic polynomial

$$\sum_{j=0}^{n} a_j s^j = \prod_{j=1}^{n} (x - r_j) \in A[s],$$

with all its roots r_j in A, the condition that $\delta a_j = 0$ for all j is *not* equivalent to the condition $\delta r_j = 0$ for all j; these two conditions are equivalent for δ a derivation on a field of characteristic zero but *not* for δ our p-derivation on A. In what follows we explain these two analogues of Lax equations following [10].

First let $T \subset G := GL_n$ be the diagonal maximal torus,

$$T = Spec \, A[t_1, t_1{}^{-1}, \ldots, t_n, t_n^{-1}], \quad G = Spec \, A[x, \det(x)^{-1}],$$

with embedding given by $x_{jj} \mapsto t_j$ and $x_{ij} \mapsto 0$ for $j \neq i$, and consider the map

$$\mathcal{C} : T \times G \to G, \quad \mathcal{C}(h, g) = g^{-1} h g.$$

We have the following:

Theorem 5.2 [10] *There exists an open set G^* of $G = GL_n$ and a unique Frobenius lift ϕ^{G^*} on $\widehat{G^*}$ such that the following diagram is commutative:*

$$
\begin{array}{ccc}
\widehat{T^*} \times \widehat{G} & \xrightarrow{\phi_0^{T^*} \times \phi_0^G} & \widehat{T^*} \times \widehat{G} \\
\mathcal{C} \downarrow & & \downarrow \mathcal{C} \\
\widehat{G^*} & \xrightarrow{\phi^{G^*}} & \widehat{G^*}
\end{array}
$$

where $T^* := T \cap G^*$, $\mathcal{C}(T^* \times G) \subset G^*$, $\phi_0^{T^*}$ is induced by the unique Frobenius lift on T that sends $t_j \mapsto t_j^p$, and ϕ_0^G is the Frobenius lift on $\widehat{G}$ that sends $x_{ij} \mapsto x_{ij}^p$.

Cf. [10], Theorem 4.50. By the way, in contrast with the classical case (Theorem 5.1), and in analogy with the arithmetic Euler paradigm (Theorem 4.4) we have the following "singularity theorem":

Theorem 5.3 [10] *For $n \geq 3$, G^* in Theorem 5.2 cannot be taken to be the whole of G.*

Cf. [10], Theorem 4.54. Also, it was shown in [10], Theorem 4.60, that any solution to the arithmetic δ^A-flow δ^{G^*} attached to ϕ^{G^*}, with spectrum contained in A, has the property that its spectrum consists of δ^A-constants. More generally, the same property holds if one replaces the Frobenius lift ϕ^{G^*} on $\widehat{G^*}$ by any Frobenius lift $\phi^{G^*(\alpha)}$ on $\widehat{G^*}$ that is *conjugate* to ϕ^{G^*} in the sense that

$$
\phi^{G^*(\alpha)}(x) := \epsilon(x)^{-1} \cdot \phi^{G^*}(x) \cdot \epsilon(x),
$$

where $\epsilon(x) = 1 + p\alpha(x)$, $\alpha(x)$ any $n \times n$ matrix with coefficients in $\mathcal{O}(\widehat{G^*})$. By the way, for α with coefficients in A (rather than in $\mathcal{O}(\widehat{G^*})$) the arithmetic δ^A-flows corresponding to ϕ^{G^*} and $\phi^{G^*(\alpha)}$ are *linear* with respect to each other in the sense of [10]; we will not review this concept of linearity here. We refer to loc. cit. for details.

On the other hand we have:

Theorem 5.4 [10] *There exists an open set G^{**} of $G = GL_n$ and a Frobenius lift $\phi^{G^{**}}$ on $\widehat{G^{**}}$ such that the following diagram is commutative:*

$$
\begin{array}{ccc}
\widehat{G^{**}} & \xrightarrow{\phi^{G^{**}}} & \widehat{G^{**}} \\
\mathcal{P} \downarrow & & \downarrow \mathcal{P} \\
\widehat{\mathbb{A}^n} & \xrightarrow{\phi_0^{\mathbb{A}^n}} & \widehat{\mathbb{A}^n},
\end{array}
$$

where $\phi_0^{\mathbb{A}^n}$ is induced by the unique Frobenius lift on $\mathbb{A}^n = \text{Spec } A[z_1, \ldots, z_n]$ which sends $z_j \mapsto z_j^p$.

Cf. [10], Theorem 4.56. The polynomials $\mathcal{P}_j(x)$ are then prime integrals for the arithmetic δ^A-flow $\delta^{G^{**}}$ attached to $\phi^{G^{**}}$. In particular the characteristic polynomial of any solution to the arithmetic δ^A-flow $\delta^{G^{**}}$ are δ^A-constant. More generally, the same property holds if one replaces the Frobenius lift $\phi^{G^{**}}$ on $\widehat{G^{**}}$ by any Frobenius lift $\phi^{G^{**}(\alpha)}$ on $\widehat{G^{**}}$ that is *conjugate* to $\phi^{G^{**}}$ in the same sense as before, namely that

$$\phi^{G^{**}(\alpha)}(x) := \epsilon(x)^{-1} \cdot \phi^{G^{**}}(x) \cdot \epsilon(x),$$

where $\epsilon(x) = 1 + p\alpha(x)$, $\alpha(x)$ any $n \times n$ matrix with coefficients in $\mathcal{O}(\widehat{G^{**}})$. Again, for α with coefficients in A (rather than in $\mathcal{O}(\widehat{G^{**}})$) the arithmetic δ^A-flows corresponding to $\phi^{G^{**}}$ and $\phi^{G^{**}(\alpha)}$ are *linear* with respect to each other in the sense of [10]. We refer to loc. cit. for details. The $\phi^{G^{**}}$ in Theorem 5.4 is not unique; one can further subject it to appropriate constraints that make it unique; we will not go into this here.

In view of the above mentioned "isospectrality-type" properties for the arithmetic δ^A-flows δ^{G^*} and $\delta^{G^{**}}$ in Theorems 5.2 and 5.4 respectively one may see these arithmetic flows as analogues of the classical Lax δ^A-flows. One is then tempted to ask for an arithmetic analogue of the condition that a Lax δ^A-flow be Poisson-Hamiltonian, i.e., an arithmetic analogue of the condition that the matrix M in the classical equation $\delta x = [M, x]$ is of the form $M = \frac{\partial H}{\partial x}$ for some $H \in B$. The matrix M itself does not have an obvious arithmetic analogue so the problem needs to be approached on a more conceptual level.

References

[1] J. Borger, *The basic geometry of Witt vectors, I*: the affine case, Algebra and Number Theory 5 (2011), no. 2, pp 231–285.

[2] J. Borger, *The basic geometry of Witt vectors, II: Spaces*, Mathematische Annalen 351 (2011), no. 4, pp 877–933.

[3] A. Buium, *Intersections in jet spaces and a conjecture of S.Lang*, Annals of Math. 136 (1992) 557–567.

[4] A. Buium, *Geometry of differential polynomial functions I: algebraic groups*, Amer J. Math. 115, 6 (1993), 1385–1444.

[5] A. Buium, *Differential characters of abelian varieties over p-adic fields*, Invent. Math. 122, 2 (1995), 309–340.

[6] A. Buium, *Geometry of p-jets*, Duke Math. J., 82, 2, (1996), 349–367.

[7] A. Buium, *Differential characters and characteristic polynomial of Frobenius*, Crelle J., 485 (1997), 209–219.

[8] A.Buium, *Differential Modular Forms*, Crelle J., 520 (2000), 95–167.

[9] A. Buium, *Arithmetic Differential Equations*, Mathematical Surveys and Monographs 118, AMS, 2005.

[10] A. Buium, *Foundations of Arithmetic Differential Geometry*, Mathematical Surveys and Monographs 222, AMS, 2017.

[11] A.Buium, T. Dupuy, *Arithmetic differential equations on GL_n, II: arithmetic Lie-Cartan theory*, Selecta Math. 22, 2, (2016), 447–528.

[12] A. Buium, T. Dupuy, *Arithmetic differential equations on GL_n, III: Galois groups*, Selecta Math. 22, 2, (2016), 529–552.

[13] A. Buium, Yu. I. Manin, *Arithmetic differential equations of Painlevé VI type*, in: Arithmetic and Geometry, London Mathematical Society Lecture Note Series: 420, L. Dieulefait, G. Faltings, D. R. Heath-Brown, Yu. V. Manin, B. Z. Moroz and J.-P. Wintenberger (eds), Cambridge University Press, 2015, pp. 114–138.

[14] A. Buium, B. Poonen, *Independence of points on elliptic curves arising from special points on modular and Shimura curves, II: local results*, Compositio Math., 145 (2009), 566–602.

[15] A. Buium, E. Previato, *Arithmetic Euler top*, J. Number Theory, 173 (2017), 37–63.

[16] A. Buium, E. Previato, *The Euler top and canonical lifts*, J. Number Theory, 190 (2018), 156–168.

[17] A. Buium, A. Saha, *The ring of differential Fourier expansions*, J. of Number Theory 132 (2012), 896–937.

[18] A. Buium, A. Saha, *The first p-jet space of an elliptic curve: global functions and lifts of Frobenius*, Math. Res. Letters, Vol. 21, Number 04 (2014), 677–689.

[19] A. Buium, S. R. Simanca, *Arithmetic Laplacians*, Advances in Math. 220 (2009), 246–277.

[20] A. Buium, S. R. Simanca, *Arithmetic partial differential equations I*, Advances in Math. 225 (2010), 689–793.

[21] A. Buium, S. R. Simanca, *Arithmetic partial differential equations II*, Advances in Math., 225 (2010), 1308–1340.

[22] P. Griffiths, J. Harris, *Principles of Algebraic Geometry*, A. Wiley, New York, 1978.

[23] A. Joyal, *δ-anneaux et vecteurs de Witt*, C.R. Acad. Sci. Canada, Vol. VII, No. 3, (1985), 177–182.

[24] E. R. Kolchin, *Differential Algebra and Algebraic Groups*, Academic Press, New York, 1973.

[25] Yu. I. Manin, *Rational points on algebraic curves over function fields*, Izv. Acad. Nauk USSR, 27 (1963), 1395–1440.

[26] Yu. I. Manin, *Sixth Painlevé equation, universal elliptic curve, and mirror of $\mathbb{P}^2$*, arXiv:9605010.

[27] Yu.I. Manin, *Frobenius Manifolds, Quantum Cohomology, and Moduli Spaces*, Colloquium Publications 47, AMS 1999.

[28] J. H. Silverman, *The Arithmetic of Elliptic curves*, GTM 106, Springer, 1986.

Alexandru Buium
Department of Mathematics and Statistics
University of New Mexico
Albuquerque, NM 87131, USA
E-mail address: buium@math.unm.edu

2

Algebraic Spectral Curves over $\mathbb{Q}$ and their Tau-Functions

Boris Dubrovin

To Emma Previato with admiration

Abstract. Let $W(z)$ be an $n \times n$ matrix polynomial with rational coefficients. Denote C the *spectral curve* $\det(w \cdot \mathbf{1} - W(z)) = 0$. Under some natural assumptions about the structure of $W(z)$ we prove that certain combinations of logarithmic derivatives of the Riemann theta-function of C of an arbitrary order starting from the third one all take rational values at the point of the Jacobi variety $J(C)$ specified by the line bundle of eigenvectors of $W(z)$.

1 Introduction

Let

$$W(z) = B^0 z^m + B^1 z^{m-1} + \cdots + B^m, \quad B^i \in Mat_n(\mathbb{C}),$$
$$B^0 = \mathrm{diag}(b_1^0, \ldots, b_n^0) \tag{1}$$

be an $n \times n$ matrix polynomial. Consider an algebraic curve C defined by the characteristic equation

$$R(z, w) := \det(w \cdot \mathbf{1} - W(z)) = 0. \tag{2}$$

It will be called an *algebraic spectral curve* associated with the matrix-valued polynomial $W(z)$. Consider the generic situation when the leading coefficient B^0 has pairwise distinct eigenvalues. In that case the Riemann surface $C \xrightarrow{z} \mathbf{P}^1$ has n distinct infinite points $P_1 \cup \cdots \cup P_n = z^{-1}(\infty)$ labelled by the eigenvalues. It will also be assumed that the affine part of the curve (2) is smooth irreducible. In what follows only such spectral curves will be considered. The genus g of such a spectral curve is uniquely determined by the numbers m and n, see eq. (4) below.

41

Assuming $g > 0$ choose a canonical basis of cycles $a_1, \ldots, a_g, b_1, \ldots,$ $b_g \in H_1(C, \mathbb{Z})$,

$$a_i \circ a_j = b_i \circ b_j = 0, \quad a_i \circ b_j = \delta_{ij}, \quad i, j = 1, \ldots, g. \tag{3}$$

Let $\omega_1, \ldots, \omega_g$ be the basis of holomorphic differentials on C normalized by the conditions

$$\oint_{a_j} \omega_k = 2\pi \sqrt{-1}\, \delta_{jk}, \quad j, k = 1, \ldots, g. \tag{4}$$

Denote

$$B_{jk} = \oint_{b_j} \omega_k, \quad j, k = 1, \ldots, g \tag{5}$$

and let

$$\theta(\mathbf{u}) = \sum_{\mathbf{n} \in \mathbb{Z}^g} \exp \left\{ \frac{1}{2} \langle \mathbf{n}, B\mathbf{n} \rangle + \langle \mathbf{n}, \mathbf{u} \rangle \right\} \tag{6}$$

be the Riemann theta-function of the curve C associated with the chosen basis of cycles. Here $\mathbf{u} = (u_1, \ldots, u_g)$ is the vector of independent complex variables $u_1, \ldots, u_g$, $\mathbf{n} = (n_1, \ldots, n_g) \in \mathbb{Z}^g$,

$$\langle \mathbf{n}, B\mathbf{n} \rangle = \sum_{j,k=1}^{g} B_{jk} n_j n_k, \quad \langle \mathbf{n}, \mathbf{u} \rangle = \sum_{k=1}^{g} n_k u_k.$$

Also recall that the Jacobi variety, or simply Jacobian of the curve C is defined as the quotient

$$J(C) = \mathbb{C}^g / \{ 2\pi \sqrt{-1}\, \mathbf{m} + B\mathbf{n} \,|\, \mathbf{m}, \mathbf{n} \in \mathbb{Z}^g \}.$$

There is a natural line bundle of degree $g + n - 1$ on the spectral curve (2) given by the eigenvectors of the matrix $W(z)$. Let D_0 be the divisor of poles of a section of this line bundle and denote

$$\mathbf{u}_0 = D_0 - D_\infty - \Delta \in J(C) \tag{7}$$

the point of the Jacobian corresponding to the line bundle. Here D_∞ is the divisor of poles of the function $z : C \to \mathbf{P}^1$, Δ is the Riemann divisor (see [17] for the definition). Here and below we identify divisors of degree 0 with their images in the Jacobian by means of the Abel–Jacobi map $A : C \to J(C)$,

$$A(P - Q) = \left(\int_Q^P \omega_1, \ldots, \int_Q^P \omega_g \right).$$

Our interest is in the N-differentials

$$\sum_{k_1, \ldots, k_N = 1}^{g} \frac{\partial^N \log \theta(\mathbf{u}_0)}{\partial u_{k_1} \ldots \partial u_{k_N}} \omega_{k_1}(Q_1) \ldots \omega_{k_N}(Q_N), \quad Q_1, \ldots, Q_N \in C \tag{8}$$

for any $N \geq 3$. For $N = 2$ instead of (8) we will be looking at the following bi-differential

$$\frac{\theta \, (P - Q - \mathbf{u}_0) \, \theta \, (P - Q + \mathbf{u}_0)}{\theta^2(\mathbf{u}_0) E(P, Q)^2}, \quad P, \, Q \in C \tag{9}$$

that differs from the second logarithmic derivative of the form (8) by the fundamental *normalized* bi-differential (see Corollary 2.12 in J. Fay's book [17]). Here $E(P, Q)$ is the prime-form,

$$E(P, Q) = \frac{\theta[\nu](P - Q)}{\sqrt{\sum \omega_i(P) \partial_{u_i} \theta[\nu](0)} \sqrt{\sum \omega_j(Q) \partial_{u_j} \theta[\nu](0)}}$$

where ν is a non-degenerate odd half-period. The goal is to express these multi-differentials on a spectral curve in terms of the associated matrix polynomial $W(z)$. The expression will involve the following matrix-valued function on the spectral curve C

$$\Phi(P) \equiv \Phi(z, w) = \frac{R\,(z, W(z)) - R(z, w)}{W(z) - w}, \quad P = (z, w) \in C. \tag{10}$$

In the right hand side it is understood that first one has to cancel the factor $(W - w)$ common for the numerator and denominator in the ratio $\frac{R(z,W) - R(z,w)}{W - w}$ and then to replace W with $W(z)$.

Theorem 1.1 (Main Theorem) *For the spectral curve (2) and for arbitrary points $Q_1 = (z_1, w_1), \ldots, Q_N = (z_N, w_N) \in C$ the following expressions hold true*

$$\frac{\theta \, (Q_1 - Q_2 - \mathbf{u}_0) \, \theta \, (Q_1 - Q_2 + \mathbf{u}_0)}{\theta^2(\mathbf{u}_0) E(Q_1, Q_2)^2}$$
$$= \operatorname{tr} \frac{\Phi(Q_1) \Phi(Q_2)}{(z_1 - z_2)^2} \frac{dz_1}{R_w(z_1, w_1)} \frac{dz_2}{R_w(z_2, w_2)} \tag{11}$$

and

$$\sum_{k_1, \ldots, k_N = 1}^{g} \frac{\partial^N \log \theta(\mathbf{u}_0)}{\partial u_{k_1} \ldots \partial u_{k_N}} \, \omega_{k_1}(Q_1) \ldots \omega_{k_N}(Q_N) = \frac{(-1)^{N+1}}{N}$$

$$\times \sum_{s \in S_N} \frac{\operatorname{tr} \left[\Phi \left(Q_{s_1} \right) \ldots \Phi \left(Q_{s_N} \right) \right]}{\left(z_{s_1} - z_{s_2} \right) \ldots (z_{s_{N-1}} - z_{s_N})(z_{s_N} - z_{s_1})} \frac{dz_1}{R_w(z_1, w_1)} \cdots \frac{dz_N}{R_w(z_N, w_N)} \tag{12}$$

for any $N \geq 3$.

In these equations $R_w(z, w)$ is the partial derivative of the characteristic polynomial $R(z, w)$ with respect to w; the summation in (12) is taken over all permutations of $\{1, 2, \ldots, N\}$.

For a given m, n denote $\mathcal{W}_{m,n}$ the space of $n \times n$ matrix polynomials of the form (1). Functions on $\mathcal{W}_{m,n}$ will be denoted by $f\,([W])$. By $\mathbb{Z}\left[\mathcal{W}_{m,n}\right]$ we denote the ring

$$\mathbb{Z}\left[\mathcal{W}_{m,n}\right] := \mathbb{Z}\left[b_1^0, \ldots, b_n^0, B_{ij}^1, \ldots, B_{ij}^m, \prod_{i<j}(b_i^0 - b_j^0)^{-1}\right].$$

Let $C_{m,n}$ be the space of algebraic curves of the form (2), (23). There is a natural fibration

$$\begin{array}{c} \mathcal{W}_{m,n} \\ \downarrow \\ C_{m,n} \end{array} \qquad (13)$$

assigning to a matrix polynomial $W(z)$ its spectral curve C. The fiber over the point C is isomorphic to the affine part of the generalized Jacobian (see below) of the singularized curve C_{sing} obtained from C by identifying its infinite points, cf. [15], [22], [24], [30], [28], [18]. The quotient of the fiber over the diagonal conjugations

$$W(z) \mapsto D^{-1}W(z)\,D, \qquad D = \text{diag}(d_1, \ldots, d_n).$$

is naturally isomorphic to $J(C) \setminus (\theta)$.

Define a function

$$F\,([W]; \mathbf{t}) \in \mathbb{Z}\left[\mathcal{W}_{m,n}\right] \otimes \mathbb{Z}[[\mathbf{t}]], \qquad \mathbf{t} = (t_k^a), \qquad a = 1, \ldots, n, \qquad k \geq 0 \quad (14)$$

by the infinite sum

$$F\,([W]; \mathbf{t}) = \sum_{N=2}^{\infty} \frac{1}{N!} \sum_{a_1, \ldots, a_N = 1}^{n} \sum_{k_1, \ldots, k_N = 0}^{\infty} F_{k_1 \ldots k_N}^{a_1 \ldots a_N}[W]\, t_{k_1}^{a_1} \ldots t_{k_N}^{a_N} \quad (15)$$

where the coefficients $F_{k_1 \ldots k_N}^{a_1 \ldots a_N}[W] \in \mathbb{Q}\left[\mathcal{W}_{m,n}\right]$ are defined by the following generated series

$$\sum_{k_1, k_2 = 0}^{\infty} \frac{F_{k_1 k_2}^{a_1 a_2}[W]}{z_1^{k_1} z_2^{k_2}} = \frac{\text{tr}\,[\Pi_{a_1}(z_1)\Pi_{a_2}(z_2)]}{(z_1 - z_2)^2} - \frac{\delta_{a_1, a_2}}{(z_1 - z_2)^2} \quad (16)$$

for any $a_1, a_2 = 1, \ldots, n$ and

$$\sum_{k_1, \ldots, k_N = 0}^{\infty} \frac{F_{k_1 \ldots k_N}^{a_1 \ldots a_N}[W]}{z_1^{k_1+2} \ldots z_N^{k_N+2}} = -\frac{1}{N} \sum_{s \in S_N} \frac{\text{tr}\left[\Pi_{s_1}(z_{s_1}) \ldots \Pi_{s_N}(z_{s_N})\right]}{(z_{s_1} - z_{s_2}) \ldots (z_{s_{N-1}} - z_{s_N})(z_{s_N} - z_{s_1})} \quad (17)$$

for any $N \geq 3$ and any $a_1, \ldots, a_N = 1, \ldots, n$. Here the matrix-valued series

$$\Pi_a(z) = \Pi_a([W]; z) \in Mat_n\left(\mathbb{Z}[\mathcal{W}_{m,n}]\right) \otimes \mathbb{Z}[[z^{-1}]]$$

are defined by the expansions

$$\frac{\Phi(P)}{R_w(z, w)} = \Pi_a(z), \quad P = (z, w) \to P_a, \quad a = 1, \ldots, n. \quad (18)$$

Observe that the function $F([W]; \mathbf{t})$ is invariant with respect to diagonal conjugations.

Main Lemma. *Let $W(z) \in \mathcal{W}_{m,n}$ be any matrix-valued polynomial such that its spectral curve (2) has n distinct points at infinity and it is nonsingular. Denote $\theta(\mathbf{u})$ the theta-function of the curve with respect to some basis of cycles and $\mathbf{u}_0$ the point (7) of the Jacobian specified by the line bundle of eigenvectors of $W(z)$. Then the following equality of formal series in $\mathbf{t}$ takes place*

$$e^{F([W];\mathbf{t})} = e^{\alpha + \sum \beta_{a,i} t_i^a + \frac{1}{2} \sum \gamma_{a,i;b,j} t_i^a t_j^b} \theta \left(\sum t_k^a \mathbf{V}^{(a,k)} - \mathbf{u}_0 \right) \quad (19)$$

for some coefficients α, $\beta_{a,i}$, $\gamma_{a,i;b,j} \in \mathbb{C}$. Here the vectors $\mathbf{V}^{(a,k)} = \left(V_1^{(a,k)}, \ldots, V_g^{(a,k)} \right)$ come from the coefficients of the expansion

$$\omega_i(P) = \sum_{k=0}^{\infty} \frac{V_i^{(a,k)}}{z^{k+2}} dz, \quad z = z(P), \quad P \to P_a \quad (20)$$

of the holomorphic differentials $\omega_i(P)$ near infinity, $i = 1, \ldots, g$, $a = 1, \ldots, n$.

Corollary 1.2 *For any matrix-valued polynomial $W(z) \in \mathcal{W}_{m,n}$ satisfying the conditions of Main Lemma the series (15) has a nonzero radius of convergence when restricted onto a finite number of the indeterminates $\mathbf{t}$.*

Definition. *We say that $\det(w \cdot \mathbf{1} - W(z)) = 0$ is an algebraic spectral curve over $\mathbb{Q}$ if $W(z) \in Mat_n(\mathbb{Q}) \otimes \mathbb{Q}[z]$.*

Corollary 1.3 *For an algebraic spectral curve over $\mathbb{Q}$ all coefficients of expansions of the multi-differentials (8), (9) near infinity are rational numbers.*

One can also derive from the Main Theorem some identities between sums of products of Riemann theta-function and its logarithmic derivatives.

Corollary 1.4 *Let C be an arbitrary compact Riemann surface of genus $g > 0$. For any $N \geq 3$ the following identity holds true for its theta-function along with the normalized holomorphic differentials on C*

$$\theta^N(\mathbf{u}) \sum_{i_1,\ldots,i_N=1}^{g} \frac{\partial^N \log \theta(\mathbf{u})}{\partial u_{i_1} \ldots \partial u_{i_N}} \omega_{i_1}(Q_1) \ldots \omega_{i_N}(Q_N) = -\frac{1}{N} \sum_{s \in S_N}$$

$$\times \frac{\theta(Q_{s_1} - Q_{s_2} + \mathbf{u})\theta(Q_{s_2} - Q_{s_3} + \mathbf{u}) \ldots \theta(Q_{s_{N-1}} - Q_{s_N} + \mathbf{u})\theta(Q_{s_N} - Q_{s_1} + \mathbf{u})}{E(Q_{s_1}, Q_{s_2})E(Q_{s_2}, Q_{s_3}) \ldots E(Q_{s_{N-1}}, Q_{s_N})E(Q_{s_N}, Q_{s_1})}$$

$$(21)$$

for arbitrary points $Q_1, Q_2, \ldots, Q_N \in C$ and arbitrary $\mathbf{u} \in J(C) \setminus (\theta)$.

See below eqs. (106), (107) for the explicit spelling of the above identity for the cases $N = 3$ and $N = 4$. For $N = 4$ and $\mathbf{u} = 0$ the identity (21) appeared in [17] (see Proposition 2.14 there). We did not find in the literature other identities of the form (21).

The constructions of the present paper can be extended to algebraic spectral curves with an arbitrary ramification profile at infinity. This will be done[1] in a subsequent publication.

Before we proceed to the precise constructions and to the proofs let us say a few words about the main ideas behind them. First, it is the connection between spectral curves and their theta-functions with particular classes of solutions to integrable systems, see e.g. [21, 12]. Second, it is the remarkable idea that goes back to M. Sato *et al.* that suggests considering tau-functions of integrable systems as partition functions of quantum field theories, see e.g. [6, 29]. The time variables of the integrable hierarchy play the role of coupling constants. So, the logarithmic derivatives of tau-functions can be considered as the connected correlators of the underlined quantum field theory. For the algebro-geometric solutions the tau-function essentially coincides with the theta-function of the spectral curve, up to multiplication by exponential of a quadratic form. There were quite a few interesting results in the theory of theta-functions inspired by this connection, see e.g. [20, 23]. The novelty of the present work is that we are looking more on the correlators than on the tau-function. The main tool used to prove the statements formulated above is the algorithm of [2, 3, 4] developed for efficient computation of correlators in cohomological field theories. This algorithm applied to tau-functions of algebraic spectral curves readily produces the explicit expressions for the correlators given above.

2 Main Constructions and Proofs

2.1 Matrix Polynomials ↔ Spectral Curves + Divisors

For a given $n \geq 2$, $m \geq 1$ for $n > 2$ or $m \geq 2$ for $n = 2$ consider the space $\mathcal{W}$ of matrix polynomials of the form

---

[1] For hyperelliptic curves with a branch point at infinity this has already been done in [14].

$$W = \{W(z) = z^m B + \text{lower degree terms}\} \tag{22}$$

where $B = \text{diag}(b_1, \ldots, b_n)$ is an arbitrary diagonal matrix. For any $W(z) \in W$ the corresponding spectral curve C is of the form

$$\det(w \cdot \mathbf{1} - W(z)) = w^n + a_1(z)w^{n-1} + \cdots + a_n(z) = 0,$$
$$\deg a_i(z) = m\,i, \quad i = 1, \ldots, n. \tag{23}$$

Assume the entries $b_1, \ldots, b_n$ of the matrix B to be pairwise distinct. Then the Riemann surface (23) has n distinct points $P_1, \ldots, P_n$ at infinity,

$$P_a = \left\{ z \to \infty, \ w \to \infty, \ \frac{w}{z^m} \to b_a \right\}, \quad a = 1, \ldots, n.$$

For the algebraic curve (23) this condition translates as follows. Let $a_i(z) = \alpha_i z^{m\,i} + \ldots, i = 1, \ldots, n$. Then the roots of the equation

$$b^n + \alpha_1 b^{n-1} + \cdots + \alpha_n = 0 \tag{24}$$

must be pairwise distinct.

Assuming smoothness of the finite part of the curve we compute its genus

$$g = \frac{(n-1)(m\,n - 2)}{2}. \tag{25}$$

We have a natural line bundle $\mathcal{L}$ over C of the eigenvectors

$$W(z)\boldsymbol{\psi}(P) = w\,\boldsymbol{\psi}(P), \quad P = (z, w) \in C, \quad \boldsymbol{\psi}(P) = (\psi^1(P), \ldots, \psi^n(P))^T \tag{26}$$

(the symbol $(.)^T$ stands for the transposition) of the matrix $W(z)$. We will associate with this line bundle a point in the *generalized Jacobian* $J(C; P_1, \ldots, P_n)$ that can be considered as an analogue of the Jacobi variety for the singular curve obtained by gluing together all infinite points $P_1, \ldots, P_n$, see, e.g., [18]. It can be represented by classes of *relative linear equivalence* of divisors of degree zero on the curve. By definition two divisors D_1 and D_2 of the same degree belong to the same relative linear equivalence class if there exists a rational function f on the curve C with $(f) = D_1 - D_2$ satisfying $f(P_1) = f(P_2) = \cdots = f(P_n)$. There is a natural fibration

$$J(C; P_1, \ldots, P_n) \to J(P) \tag{27}$$

associating with any divisor its class of linear equivalence.

With the line bundle $\mathcal{L}$ we associate a divisor D_0 on the spectral curve defined by

$$D_0 = \{P \in C \mid \psi^1(P) + \cdots + \psi^n(P) = 0\} \tag{28}$$

for a nonzero eigenvector. It can be considered as the divisor of poles of the section normalized by the condition

$$\psi^1 + \cdots + \psi^n = 1. \tag{29}$$

The components of the eigenvector can be represented as

$$\psi^j(P) = \frac{\Delta_{ij}(z, w)}{\sum_{s=1}^{n} \Delta_{is}(z, w)}, \quad P = (z, w) \in C, \quad j = 1, \ldots, n \qquad (30)$$

for any i. Here $\Delta_{ij}(z, w)$ is the (i, j)-cofactor of the matrix $w \cdot \mathbf{1} - W(z)$. So at infinity the normalized (29) eigenvector behaves as

$$\psi^i(P) = \delta_{ia} + \mathcal{O}\left(\frac{1}{z(P)}\right), \quad P \to P_a. \qquad (31)$$

Lemma 2.1 *The divisor D_0 of poles of the eigenvector normalized by eq. (29) is a nonspecial divisor on $C \setminus (P_1 \cup \cdots \cup P_n)$ of degree $g + n - 1$. Conversely, for any nonsingular curve C of the form*

$$w^n + a_1(z)w^{n-1} + \cdots + a_n(z) = 0, \quad \deg a_i(z) = m\, i, \quad i = 1, \ldots, n \qquad (32)$$

with n distinct ordered points $P_1, \ldots, P_n$ at infinity and an arbitrary nonspecial divisor $D_0 \subset C \setminus (P_1 \cup \cdots \cup P_n)$ of degree $g + n - 1$ there exists a unique matrix polynomial $W(z)$ of the form (22) with the spectral curve coinciding with C and the divisor of poles of the normalized eigenvector coinciding with D_0.

Proof. To prove the first part of the Lemma we will use an algorithm developed in [9] for computing the poles of eigenvectors of matrix polynomials adjusting it to the present situation. Denote $e^* = (1, 1, \ldots, 1) \in \mathbb{C}^{n*}$ and define a polynomial

$$D(z) = e^* \wedge e^* W(z) \wedge \cdots \wedge e^* W^{n-1}(z) \in \wedge^n \mathbb{C}^{n*} \otimes \mathbb{C}[z]. \qquad (33)$$

We also define polynomials $q_{ij}(z)$, $i, j = 1, \ldots, n$ as coefficients of the expansion of sums of cofactors $\Delta_{ij}(z, w)$ of the matrix $w \cdot \mathbf{1} - W(z)$

$$\sum_{s=1}^{n} \Delta_{is}(z, w) = w^{n-1} + q_{i\,2}(z)w^{n-2} + \cdots + q_{i,n-1}(z)w + q_{i,n}(z),$$

$$i = 1, \ldots, n. \qquad (34)$$

($q_{i1}(z) = 1$ for any $i = 1, \ldots, n$).

Proposition 2.2 *1) For any matrix polynomial of the form*

$$W(z) = z^m B + \text{lower degree terms},$$

$$B = \mathrm{diag}(b_1, \ldots, b_n), \quad b_i \neq b_j \quad \text{for} \quad i \neq j$$

the polynomial (33) has degree $\frac{mn(n-1)}{2} = g + n - 1$.

2) Assume that the spectral curve of $W(z)$ is nonsingular and the roots $z_1, \ldots, z_{g+n-1}$ of the polynomial $D(z)$ are pairwise distinct. Then the rank of the rectangular matrix

$$C(z) = \left(q_{ij}(z) \right)_{1 \le i \le n, \; 1 \le j \le n-1}$$

evaluated at $z = z_k$, $k = 1, \ldots, g + n - 1$, is equal to $n - 1$.

3) For a given $k \in \{1, 2, \ldots, g + n - 1\}$ let C_k be a nonzero $(n-1)$-minor of the matrix $C(z_k)$,

$$C_k = \left(q_{i_s, j}(z_k) \right)_{1 \le s, \; j \le n-1}.$$

Denote $\hat{C}_k$ the matrix obtained from C_k by changing the last column

$$q_{i_s, n-1}(z_k) \mapsto q_{i_s, n}(z_k), \quad s = 1, \ldots, n - 1$$

and put

$$w_k = -\frac{\det \hat{C}_k}{\det C_k}, \quad k = 1, \ldots, g + n - 1. \tag{35}$$

Then the poles of the eigenvector of the matrix $W(z)$ normalized by (29) are at the points

$$Q_1 = (z_1, w_1), \ldots, Q_{g+n-1} = (z_{g+n-1}, w_{g+n-1}) \in C \setminus \{P_1 \cup \ldots P_n\}. \tag{36}$$

Let us prove that the divisor D_0 is nonspecial, i.e., that $l(D_0) = n$, where $l(D) = \dim H^0 (C, \mathcal{O}(D))$. Indeed, if $l(D_0) > n$ then there exists a non-constant rational function f on the curve C with poles[2] at D_0 satisfying $f(P_1) = \cdots = f(P_n) = 1$. Consider the vector-function

$$\tilde{\psi}(P) = \frac{1}{f(P)} \psi(P).$$

Clearly it is again an eigenvector of the matrix $W(z)$ satisfying the normalization (29). Due to uniqueness it must coincide with ψ, so f must be identically equal to 1. Such a contradiction completes the proof of the first part of Lemma 2.1.

Let us now explain the reconstruction procedure of the polynomial matrix $W(z)$ starting from a pair (C, D_0) consisting of a curve C of the form (23) smooth for $|z| < \infty$ and a nonspecial positive divisor D_0 on $C \setminus (P_1 \cup \cdots \cup P_n)$ of degree $g + n - 1$. As by assumption $l(D_0) = n$, there exist n rational functions $\psi^1(P), \ldots, \psi^n(P)$ on C with poles at D_0 satisfying

$$\psi^i(P_j) = \delta_{ij}, \quad i, j = 1, \ldots, n. \tag{37}$$

---

[2] Here and below we will say that a rational function f on the curve C has poles at the points of a divisor D if $(f) + D \ge 0$.

Let R be a sufficiently large number such that neither of the ramification points of the divisor D_0 occur for $|z| > R$. For any such z denote $(z, 1), \ldots, (z, n)$ the preimages of z on the spectral curve with respect to the natural projection $z : C \mapsto \mathbf{P}^1$ ordered in an arbitrary way. Define an $n \times n$ matrix $\Psi(z)$ whose i-th row is $\left(\psi^i ((z, 1)), \ldots, \psi^i ((z, n)) \right)$. The matrix $\Psi(z)$ is invertible for any $R < |z| \leq \infty$. Denote

$$w_a(z) = w ((z, a)) = b_a z^m + \ldots, a = 1, \ldots, n$$

the branches of the algebraic function $\mu(P)$, $P \in C$ and put

$$\hat{w}(z) = \mathrm{diag}(w_1(z), \ldots, w_n(z)), \quad |z| > R.$$

The matrix-valued function $\Psi(z)\hat{w}(z)\Psi^{-1}(z)$ is analytic for $|z| > R$ having an m-th order pole at infinity. Observe that it does not depend on the ordering of the preimages of z, so it can be extended to a rational function on the complex plane. Consider its Laurent expansion at infinity

$$\Psi(z)\hat{w}(z)\Psi^{-1}(z) = B z^m + B_1 z^{m-1} + \cdots + B_m + \mathcal{O}\left(\frac{1}{z}\right)$$

where $B = \mathrm{diag}(b_1, \ldots, b_n)$ and $B_1, \ldots, B_m$ are some $n \times n$ matrices. Put

$$W(z) = B z^m + B_1 z^{m-1} + \cdots + B_m$$

and consider the vector-valued function $\tilde{\psi}(P)$ on the curve defined by

$$\tilde{\psi}(P) = W(z(P)) \psi(P) - w(P)\psi(P).$$

It has poles only at the points of the divisor D_0. From the definition of the matrix $W(z)$ it readily follows that $\tilde{\psi}(P)$ vanishes at $P_1, \ldots, P_n$. Hence it equals zero due to the nonspeciality of the divisor D_0. This implies that C coincides with the spectral curve of the matrix polynomial $W(z)$. To complete the reconstruction procedure it remains to observe that the function

$$\psi^1(P) + \cdots + \psi^n(P) - 1$$

having poles at D_0 vanishes at $P_1, \ldots, P_n$. Hence it is identically equal to 0, that is, the eigenvector $\psi(P)$ of the matrix $W(z)$ satisfies the normalization (29). $\square$

Remark 2.3   Changing the divisor $D_0 \to D_0' \sim D_0$ in the class of linear equivalence yields a conjugation of the matrix $W(z)$ by a diagonal matrix

$$W(z) \to F W(z) F^{-1}, \quad F = \mathrm{diag}\left(f(P_1), \ldots, f(P_n) \right)$$

where f is a rational function on C with the divisor $D_0 - D_0'$.

One can repeat the above constructions dealing with the *dual line bundle* $\mathcal{L}^\dagger$ over C coming from the left eigenvectors of the matrix $W(z)$

$$\boldsymbol{\psi}^\dagger(P) W(z) = w \, \boldsymbol{\psi}^\dagger(P), \quad P = (z, w) \in C,$$

$$\boldsymbol{\psi}^\dagger(P) = (\psi_1^\dagger(P), \dots, \psi_n^\dagger(P)). \tag{38}$$

Let us use the same normalization

$$\psi_1^\dagger + \cdots + \psi_n^\dagger = 1 \tag{39}$$

so

$$\psi_i^\dagger(P) = \frac{\Delta_{ij}(z, w)}{\sum_{s=1}^{n} \Delta_{sj}(z, w)}, \quad P = (z, w) \in C, \quad i = 1, \dots n \tag{40}$$

for any choice of j (cf. eq. (30)). Denote $D_0^\dagger$ the divisor of poles of the dual eigenvector (38) normalized by the condition (39). The following statement is an analogue of Lemma 2.1 for the dual eigenvectors.

Lemma 2.4 *The divisor $D_0^\dagger$ of poles of the eigenvector (38) normalized by eq. (39) is a nonspecial divisor on $C \setminus (P_1 \cup \cdots \cup P_n)$ of degree $g + n - 1$. Conversely, for any nonsingular curve C of the form*

$$w^n + a_1(z)w^{n-1} + \cdots + a_n(z) = 0, \quad \deg a_i(z) = m\, i, \quad i = 1, \dots, n \tag{41}$$

and an arbitrary nonspecial divisor $D_0^\dagger \subset C \setminus (P_1 \cup \cdots \cup P_n)$ of degree $g + n - 1$ there exists a unique matrix polynomial $W(z)$ of the form (22) with the spectral curve coinciding with C and the divisor of the normalized eigenvector (38) coinciding with D_0.

The proof is similar to that of Lemma 2.1, so it will be omitted.

Remark 2.5 Of course the divisors D_0 and $D_0^\dagger$ do depend on each other. The nature of this dependence will be clarified below.

We will now look at the *spectral projectors* of the matrix polynomial $W(z)$. Consider the matrix-valued function

$$\Pi(P) = \frac{\Phi(P)}{R_w(z, w)}, \quad P = (z, w) \in C \tag{42}$$

where $R(z, w) = \det(w \cdot \mathbf{1} - W(z))$ is the characteristic polynomial of the matrix $W(z)$ and $\Phi(P)$ is defined by (10). So

$$\Pi(z, w) = \frac{1}{R_w(z, w)} \frac{R(z, W) - R(z, w)}{W - w} = \frac{\sum_{i=0}^{n-1} b_i(z) w^{n-i-1}}{R_w(z, w)}$$

$$b_i(z) = \sum_{j=0}^{i} a_j(z) W^{i-j}(z). \tag{43}$$

Let $z \in \mathbb{C}$ be a point such that C is not ramified over it. Denote $w_1(z), \ldots,$ $w_n(z)$ the points above z on the spectral curve C and put

$$\Pi_i(z) = \Pi(z, w_i(z)), \quad i = 1, \ldots, n. \tag{44}$$

We will prove that these matrices are the spectral projectors

$$\Pi_i(z) : \mathbb{C}^n \to \text{Ker}\, (W(z) - w_i(z) \cdot \mathbf{1}), \quad i = 1, \ldots, n$$

of $W(z)$.

Lemma 2.6 *The matrices* $\Pi_1(z), \ldots, \Pi_n(z)$ *are basic idempotents of the matrix* $W(z)$, *i.e.*

$$\Pi_i^2 = \Pi_i, \quad \Pi_i \cdot \Pi_j = 0 \quad \text{for} \quad i \neq j, \quad i, j = 1, \ldots, n \tag{45}$$

$$\sum_{i=1}^{n} \Pi_i(z) = \mathbf{1}, \quad \sum_{i=1}^{n} w_i(z)\Pi_i(z) = W(z). \tag{46}$$

Proof. Let us first prove that

$$\sum_{i=1}^{n} w_i^r(z)\Pi_i(z) = W^r(z) \quad \text{for any} \quad 0 \leq r. \tag{47}$$

We have

$$\sum_{i=1}^{n} w_i^r(z) \frac{\sum_{k=0}^{n-1} b_k(z)w_i^{n-k-1}(z)}{R_w(z, w_i(z))} = \sum_{i=1}^{n} \operatorname*{res}_{w=w_i(z)} w^r \frac{\sum_{k=0}^{n-1} b_k(z)w^{n-k-1}}{R(z, w)}$$

$$= - \operatorname*{res}_{w=\infty} w^r \frac{\sum_{k=0}^{n-1} b_k(z)w^{n-k-1}}{R(z, w)}.$$

Using the explicit expression (43) along with the obvious identity

$$\left(1 - \frac{W}{w}\right)\left(1 + \frac{b_1}{w} + \cdots + \frac{b_{n-1}}{w^{n-1}}\right) = 1 + \frac{a_1}{w} + \cdots + \frac{a_n}{w^n}$$

we arrive at

$$\frac{\sum_{k=0}^{n-1} b_k(z)w^{n-k-1}}{R(z, w)} = \frac{1}{w - W}, \quad |w| \to \infty. \tag{48}$$

Thus

$$\operatorname*{res}_{w=\infty} w^r \frac{\sum_{k=0}^{n-1} b_k(z)w^{n-k-1}}{R(z, w)} = -W^r.$$

This proves (47) and, hence (46).

To prove (45) we solve the system (47) for $r = 0, 1, \ldots, n - 1$ with respect to $\Pi_1(z), \ldots \Pi_n(z)$ to obtain

$$\Pi_i(z) = \frac{\prod_{j \neq i}(W(z) - w_j(z))}{\prod_{j \neq i}(w_i(z) - w_j(z))}, \quad i = 1, \ldots, n.$$

Rewriting these matrices in the basis of eigenvectors of $W(z)$ we readily get (45). □

It will be convenient to also consider a matrix-valued differential with the matrix entries

$$\Omega^i_j(P) = \Pi^i_j(P)dz, \quad i, j = 1, \ldots, n \tag{49}$$

where the matrix $\Pi(P) = \left(\Pi^i_j(P)\right)_{1 \leq i, j \leq n}$ is given by (42), (43).

Proposition 2.7 *For every* $i \neq j$ *the differential* $\Omega^i_j(P)$ *is holomorphic on* $C \setminus \left(P_i \cup P_j\right)$ *with simple poles at* $P = P_i$ *and* $P = P_j$. *The differential* $\Omega^i_i(P)$ *is holomorphic on* $C \setminus P_i$ *having a double pole at* $P = P_i$ *such that*

$$\Omega^i_i(P) = dz + \text{regular terms}, \quad P \to P_i \tag{50}$$

for every $i = 1, \ldots, n$.

Proof. As it follows from the explicit expression (10) the entries of the matrix $\Phi(P)$ are holomorphic on the affine part $P = (z, w) \in C$, $|z| < \infty$. The differential $dz/R_w(z, w)$ is holomorphic on C. Hence the differentials $\Pi_{ij}(P)$ are holomorphic on the affine part of C. Let us look at their behaviour at infinity. To this end we will use the standard realization of the spectral projectors of an $n \times n$ complex matrix with pairwise distinct eigenvalues that in our case can be formulated in the following way. Let $z \in \mathbb{C}$ be not a ramification point with respect to the projection $z : C \to \mathbb{C}$. Order the points $(z, w_1(z)), \ldots, (z, w_n(z))$ in the preimage. Like in the proof of Lemma 2.1 produce a matrix $\Psi(z)$ whose k-th column is given by the eigenvector $\left(\psi^1(z, w_k(z)), \ldots, \psi^n(z, w_k(z))\right)^T$ of $W(z)$ normalized by the condition (29). Then

$$\Pi(P)|_{P=(z,w_k(z))} = \Psi(z)E_k\Psi^{-1}(z) \tag{51}$$

where the $n \times n$ matrix E_k has only one nonzero entry

$$(E_k)^i_j = \delta^i_k \delta^k_j.$$

Consider now sufficiently large R; choose the order of eigenvalues over the disk $|z| > R$ in such a way that $w_k(z) \sim b_k z^m$ for $|z| \to \infty$. Then

$$\Psi(z) = 1 + \mathcal{O}\left(\frac{1}{z}\right), \quad \Psi^{-1}(z) = 1 + \mathcal{O}\left(\frac{1}{z}\right), \quad z \to \infty \tag{52}$$

due to eq. (31). The behaviour of the differentials $\Omega^i_j(P)$ at infinity easily follows from (51), (52). □

Consider now the differentials

$$\Omega_j(P) = \sum_{i=1}^{n} \Omega^i_j(P) \tag{53}$$

$$\Omega^i(P) = \sum_{j=1}^{n} \Omega^i_j(P). \tag{54}$$

Lemma 2.8 *The differential* $\Omega_j(P)$ *has zeros at the points of the divisor* D *and at some divisor* $D^\dagger_j$ *of degree* g. *It has poles at the points of the divisor* $P_j + \sum_{s=1}^{n} P_s$. *In a similar way the differential* $\Omega^i(P)$ *has zeros at the points of the divisor* $D^\dagger$ *and at some divisor* D_i *of degree* g. *It has poles at the points of the divisor* $P_i + \sum_{s=1}^{n} P_s$.

Proof. Using the representation (51) along with the normalization (29) we immediately conclude that the sum (53) vanishes at the points of the divisor D. The configuration of poles of this differential can be easily recovered from Proposition 2.7. The degree counting

$$\deg D + \deg D^\dagger_j - \deg\left(P_j + \sum_{s=1}^{n} P_s\right) = 2g - 2$$

yields $\deg D^\dagger_j = g$. To derive similar statements about the differentials $\Omega^i(P)$ we use an alternative representation of the projector matrix

$$\Pi(P)|_{P=(z,w_k(z))} = \Psi^\dagger(z)^{-1} E_k \Psi^\dagger(z) \tag{55}$$

where the k-th row of the matrix $\Psi^\dagger(z)$ is given by the left eigenvector

$$\left(\psi^\dagger_1(z, w_k(z)), \ldots, \psi^\dagger_n(z, w_k(z))\right),$$

normalized by the condition (39). □

Corollary 2.9 *The divisor of zeros of the differential* $\Omega^i_j(P)$ *on* $C \setminus (P_1 \cup \cdots P_n)$ *coincides with* $D_i + D^\dagger_j$, $i, j = 1, \ldots, n$.

Proof. The matrix $\Omega^i_j(P)$ has rank one. Its columns are eigenvectors of the matrix $W(z)$ with the same eigenvalue. Normalizing any column we obtain the same vector function $\psi(P)$

$$\psi^i(P) = \frac{\Omega^i_j(P)}{\sum_{k=1}^{n} \Omega^k_j(P)}, \quad i = 1, \ldots, n \tag{56}$$

for any j. According to Lemma the denominator vanishes at the points of the divisor $D^\dagger_j$. Hence also the numerator must vanish at the points of this divisor.

In a similar way, the rows of $\Omega^i_j(P)$ are left eigenvectors of the same matrix $W(z)$. Normalizing them one obtains

$$\psi^\dagger_j(P) = \frac{\Omega^i_j(P)}{\sum_{k=1}^n \Omega^i_k(P)}, \quad j = 1, \ldots, n \tag{57}$$

for any i. Hence $\Omega^i_j(P)$ vanishes also at the points of the divisor D_i. Since degree of the divisor of poles of this differential equals two and $\deg(D_i + D^\dagger_j) = 2g$, there are no other zeros. $\qquad\square$

Corollary 2.10 *The zeros of the differential*

$$\Omega(P) = \sum_{i, j=1}^n \Omega^i_j(P) \tag{58}$$

are at the points of the divisor $D + D^\dagger$. It has double poles at the infinite points $P_1, \ldots, P_n$ and

$$\Omega(P) = \left(1 + \mathcal{O}\left(\frac{1}{z}\right)\right) dz, \quad P \to P_k, \quad k = 1, \ldots, n. \tag{59}$$

The corollary suggests the following way of determining the dual divisor $D^\dagger$ starting from D. For a nonspecial divisor D of degree $g+n-1$ there exists a unique differential $\Omega(P)$ vanishing at the points of D and having double poles of the form (59) at infinity. The remaining zeros of the differential give the points of the divisor $D^\dagger$.

Corollary 2.11 *The differentials $\Omega^i_j(P)$ admit the following representation*

$$\Omega^i_j(P) = \psi^i(P)\psi^\dagger_j(P)\Omega(P), \quad i, j = 1, \ldots, n. \tag{60}$$

Proof. Due to the previous Corollary the product (60) is holomorphic on $C \setminus (P_1 \cup \cdots \cup P_n)$. From Corollary 2.9 it follows that the divisor of zeros of this product coincides with $D_i + D^\dagger_j$. Finally, using (59) along with the asymptotics of $\psi^i(P)$ and $\psi^\dagger_j(P)$ at infinity we complete the proof. $\qquad\square$

2.2 Generalized Jacobian and Theta-Functions. Proof of eq. (11)

The generalized Jacobian $J(C; P_1, \ldots, P_n)$ can be realized as a fiber bundle over $J(C)$ with $(n-1)$-dimensional fiber. For $n = 2$ the construction was already given in [17]; it is quite similar also for arbitrary n.

Define $J(C; P_1, \ldots, P_n)$ as the set of all pairs

$$(\mathbf{u}, \lambda), \quad \mathbf{u} = (u_1, \ldots, u_g) \in \mathbb{C}^g, \quad \lambda = (\lambda_1, \ldots, \lambda_n) \in \left(\mathbb{C}^*\right)^n$$

modulo the following equivalence relation $(\mathbf{u}, \lambda) \sim (\mathbf{u}', \lambda')$ if

$$\mathbf{u}' = \mathbf{u} + 2\pi i M + BN, \quad \lambda_k' = c \lambda_k e^{\langle N, A(P_k)\rangle},$$

$$k = 1, \dots, n, \quad c \in \mathbb{C}^*, \quad M, N \in \mathbb{Z}^g. \tag{61}$$

The fibration (27) is realized by the map $(\mathbf{u}, \lambda) \mapsto \mathbf{u}$.

To define an analogue of the Abel map

$$C \to J(C; P_1, \dots, P_n) \tag{62}$$

fix a pair of distinct points $P_0, Q_0 \in C \setminus (P_1 \cup \cdots \cup P_n)$ and put

$$P \mapsto (\mathbf{u}(P), \lambda(P)), \quad u_i(P) = A_i(P) = \int_{P_0}^{P} \omega_i,$$

$$\lambda_k(P) = e^{\alpha_k(P)}, \quad \alpha_k(P) = \int_{P_0}^{P} \Omega_{P_k Q_0}. \tag{63}$$

Here and below Ω_{PQ} is the third kind differential on C having simple poles at the points P and Q with residues $+1$ and -1 respectively and vanishing a-periods. The map is extended linearly/multiplicatively on the group of divisors of a given degree. The following statement is an analogue of the Abel–Jacobi theorem.

Proposition 2.12 *Two divisors* D, D' *of the same degree are relatively equivalent iff*

$$(\mathbf{u}(D), \lambda(D)) \sim (\mathbf{u}(D'), \lambda(D'))$$

modulo equivalence (61).

An analogue of the Riemann theorem about zeros of theta-function is given by

Proposition 2.13 *For a given* $\mathbf{u} \in \mathbb{C}^g$ *such that* $\theta(\mathbf{u}) \neq 0$ *and* $\lambda \in (\mathbb{C}^*)^n$ *consider the function*

$$F(P) = \sum_{s=1}^{n} \lambda_s \, \theta(P - P_s - \mathbf{u}) \frac{E(P, P_0)(d\zeta_0)^{1/2}}{E(P, P_s)(d\zeta_s)^{1/2}} \tag{64}$$

on the $4g$-*gon* $\tilde{C}$ *obtained by cutting the curve* C *along the chosen basis of cycles* $a_1, \dots, a_g, b_1, \dots, b_g$. *Here* $\zeta_0(P) = z(P) - z(P_0)$, $\zeta_s(P) = 1/z(P)$ *are local parameters[3] near* P_0, P_s *respectively. This function has simple poles at* $P_1, \dots, P_n$, *a simple zero at* P_0 *and also zeros at some points* $Q_1, \dots, Q_{g+n-1}$ *of a divisor* $D \subset C \setminus (P_1 \cup \cdots \cup P_n)$ *satisfying*

---

[3] For simplicity we assume that $P_0 \in C$ is not a branch point.

$$\sum_{i=1}^{g+n-1} A_j(Q_i) - \sum_{s=1}^{n} A_j(P_s) = u_j - \mathcal{K}_j, \quad j = 1, \ldots, g \qquad (65)$$

$$\exp \sum_{i=1}^{g+n-1} \alpha_s(Q_i) = \lambda_s e^{-\kappa_s}, \quad s = 1, \ldots, n \qquad (66)$$

where $\mathcal{K}_1, \ldots, \mathcal{K}_g$ are Riemann constants

$$\mathcal{K}_j = \frac{2\pi i + B_{jj}}{2} - \frac{1}{2\pi i} \sum_{k \neq j} \int_{a_k} A_j(P)\omega_k(P),$$

$\kappa_1, \ldots, \kappa_n$ are given by analogous formulae,

$$\kappa_s = -\log E(P_s, P_0) - \frac{1}{2\pi i} \sum_{k=1}^{g} \int_{a_k} \alpha_s(P)\omega_k(P). \qquad (67)$$

Recall [17] that the combination

$$\Delta := -\mathcal{K} + (g-1)P_0 \qquad (68)$$

does not depend on the choice of the base point P_0. This gives rise to definition of the Riemann divisor Δ that already appeared above. In particular eq. (65) can be rewritten in the form

$$D - \sum_{s=1}^{n} P_s - \Delta = \mathbf{u}. \qquad (69)$$

Observe that, for a given $(\mathbf{u}, \lambda)$ the divisor of zeros of the function (64) determined by eqs. (65), (66) does not depend on the choice of the basepoint P_0.

Proof. The function $F(P)$ does not change its value when the point P crosses the cut along the cycle b_k while crossing the cut a_k it is multiplied by

$$\exp\left(-\frac{1}{2}B_{kk} - \int_{P_0}^{P} \omega_k + u_k\right).$$

So the total logarithmic residue

$$\frac{1}{2\pi i} \oint_{\partial \tilde{C}} d \log F(P)$$

is equal to g. Hence $F(P)$ has $g + n - 1$ zeros on $C \setminus (P_1 \cup \cdots \cup P_n)$, counted with multiplicity. To prove the first equality (65) one has to compute the contour integral

$$\frac{1}{2\pi i} \oint_{\partial \tilde{C}} A_j(P) d \log F(P) = A_j(Q_1) + \cdots + A_j(Q_{g+n-1})$$
$$- A_j(P_1) - \cdots - A_j(P_n).$$

To prove eq. (66) we need to add more cuts: a cut from P_0 to Q_0 and also cuts from Q_0 to the infinite points $P_1, \ldots, P_n$. Denote $\tilde{C}'$ the resulting polygon. Then

$$\sum_{i=1}^{g+n-1} \alpha_s(P_i) = \frac{1}{2\pi i} \oint_{\partial \tilde{C}'} \alpha_s(P) d \log F(P), \quad s = 1, \ldots, n$$

up to an s-independent shift. The integral in the right hand side must be regularized at $P \to \infty$ or $P \to Q_0$. After such regularization we arrive at eqs. (66) up to equivalence (61). $\square$

We will now express the differentials (49) in terms of the coordinates $(\mathbf{u}_0, \lambda_0)$.

Proposition 2.14 *Let* $Q_1 + \cdots + Q_{g+n-1} = D \subset C \setminus (P_1 \cup \cdots \cup P_n)$ *be the divisor of poles of the eigenvector of the matrix $W(z)$ normalized by the condition (29). Denote $(\mathbf{u}_0, \lambda_0)$ the corresponding point on the generalized Jacobian (61)*

$$\mathbf{u}_0 = D_0 - \sum_{a=1}^{n} P_a - \Delta$$
$$\lambda_j^0 = \exp\left\{ \sum_{s=1}^{g+n-1} \int_{P_0}^{Q_s} \Omega_{P_j Q_0} + \kappa_j \right\}, \quad j = 1, \ldots, n \quad (70)$$

(cf. eqs. (65), (66)). Then the differentials $\Omega_j^i(P)$ of the form (49) are given by the following equation

$$\Omega_j^i(P) = \frac{\lambda_i^0}{\lambda_j^0} \frac{\theta(P - P_i - \mathbf{u}_0)\theta(P - P_j + \mathbf{u}_0)}{\theta^2(\mathbf{u}_0)E(P_i, P)E(P, P_j)\sqrt{d\zeta_i}\sqrt{d\zeta_j}}. \quad (71)$$

Proof. Any differential $\Omega_j^i(P)$ having, for $i \neq j$ simple poles at $P = P_i$ and $P = P_j$ and, for $i = j$ a double pole of the form (50) at $P = P_i$ can be written [17] as follows

$$\Omega_j^i(P) = \alpha_{ij} \frac{\theta(P - P_i - \mathbf{u}_{ij})\theta(P - P_j + \mathbf{u}_{ij})}{\theta^2(\mathbf{u}_{ij})E(P_i, P)E(P, P_j)\sqrt{d\zeta_i}\sqrt{d\zeta_j}}$$

for some $\mathbf{u}_{ij} \in J(C) \setminus (\theta)$ and some nonzero constants α_{ij} satisfying $\alpha_{ii} = 1$. According to Riemann theorem the zeros of the function $\theta(P - P_j + \mathbf{u}_{ij})$ are at the points of a divisor $\mathcal{D}$ of degree g satisfying

$$\mathcal{D} - P_j - \Delta = -\mathbf{u}_{ij}.$$

According to the Corollary 2.9 it must coincide with the divisor $D_j^\dagger$. From Lemma 2.8 we derive the following linear equivalence

$$D_0 + D_j^\dagger - P_j - \sum_{a-1}^{n} P_a = K_C.$$

Substituting $K_C = 2\Delta$ we can rewrite it as follows

$$D_j^\dagger - P_j - \Delta = -D_0 + \sum_{a=1}^{n} P_a + \Delta = -\mathbf{u}_0.$$

Hence the condition $\mathcal{D} = D_j^\dagger$ implies $\mathbf{u}_{ij} = \mathbf{u}_0$ on $Jac(C)$.

It remains to fix the constants α_{ij}. Since the rank of the matrix $\Omega_j^i(P)$ must be equal to one, we conclude that $\alpha_{ij} = \alpha_i \beta_j$ for some nonzero constants α_i, β_j. As $\alpha_{ii} = 1$ then $\beta_j = \alpha_j^{-1}$. The last condition to be used is that the sum $\sum_{i=1}^{n} \Omega_j^i(P)$ must vanish at the points $Q_1, \ldots, Q_{g+n-1}$ of the divisor D. This implies that $\alpha_i = \lambda_i^0$, up to a common factor. $\square$

Corollary 2.15 *The eigenvector $\psi(P)$ of the matrix $W(z)$ normalized by the condition (29) is*

$$\psi^i(P) = \frac{\lambda_i^0 \frac{\theta(P-P_i-\mathbf{u}_0)}{E(P,P_i)(d\zeta_i)^{1/2}}}{\sum_{b=1}^{n} \lambda_b^0 \frac{\theta(P-P_b-\mathbf{u}_0)}{E(P,P_b)(d\zeta_b)^{1/2}}}, \quad i = 1, \ldots, n. \tag{72}$$

The dual eigenvector $\psi^\dagger(P)$ is given by a similar formula

$$\psi_i^\dagger(P) = \frac{\frac{1}{\lambda_i^0} \frac{\theta(P-P_i+\mathbf{u}_0)}{E(P_i,P)(d\zeta_i)^{1/2}}}{\sum_{b=1}^{n} \frac{1}{\lambda_b^0} \frac{\theta(P-P_b+\mathbf{u}_0)}{E(P_b,P)(d\zeta_b)^{1/2}}}, \quad i = 1, \ldots, n. \tag{73}$$

Proof. Use (71) along with (56), (57). $\square$

Remark 2.16 Observe that the change

$$D_0 \mapsto D_0^\dagger$$

corresponds to the involution

$$(\mathbf{u}_0, \lambda_1^0, \ldots, \lambda_n^0) \mapsto \left(-\mathbf{u}_0, 1/\lambda_1^0, \ldots, 1/\lambda_n^0\right) \tag{74}$$

on the generalized Jacobian.

We are now in a position to prove the first equation (11) of the Main Theorem.

Proposition 2.17 *Let C be a compact Riemann surface of positive genus and $z : C \to \mathbf{P}^1$ a rational function with n simple poles at the points $P_1, \dots, P_n$. Introduce the following matrix of Abelian differentials on C*

$$\boldsymbol{\Omega}(P) = \left(\Omega^i_j(P) \right)_{1 \le i, j \le n},$$
$$\Omega^i_j(P) = \frac{\lambda_i}{\lambda_j} \frac{\theta(P - P_i - \mathbf{u})\theta(P - P_j + \mathbf{u})}{\theta^2(\mathbf{u})E(P_i, P)E(P, P_j)\sqrt{d\zeta_i}\sqrt{d\zeta_j}} \tag{75}$$

where $\mathbf{u} \in J(C) \backslash (\theta)$ is an arbitrary point and $\lambda_1, \dots, \lambda_n$ are arbitrary nonzero numbers. Then for an arbitrary pair of distinct points $P, Q \in C$ the following equation holds true

$$\mathrm{tr} \frac{\boldsymbol{\Omega}(P)\boldsymbol{\Omega}(Q)}{(z(P) - z(Q))^2} = \frac{\theta(P - Q - \mathbf{u})\theta(P - Q + \mathbf{u})}{\theta^2(\mathbf{u})E^2(P, Q)}. \tag{76}$$

Proof. The trace of the product of the matrices $\boldsymbol{\Omega}(P)$ and $\boldsymbol{\Omega}(Q)$ factorizes as follows

$$\mathrm{tr}\,\boldsymbol{\Omega}(P)\boldsymbol{\Omega}(Q) = \sum_{i=1}^{n} \frac{\theta(P - P_i - \mathbf{u})\theta(Q - P_i + \mathbf{u})}{\theta^2(\mathbf{u})E(P_i, P)E(Q, P_i)d\zeta_i}$$
$$\cdot \sum_{j=1}^{n} \frac{\theta(Q - P_j - \mathbf{u})\theta(P - P_j + \mathbf{u})}{\theta^2(\mathbf{u})E(P_j, Q)E(P, P_j)d\zeta_j}. \tag{77}$$

For a fixed pair of distinct points $P, Q \in C$ consider the differential

$$H_{PQ}(Z) = \frac{\theta(P - Z - \mathbf{u})\theta(Q - Z + \mathbf{u})}{\theta^2(\mathbf{u})E(Z, P)E(Q, Z)\sqrt{dz(P)}\sqrt{dz(Q)}}, \quad Z \in C. \tag{78}$$

It has simple poles at $Z = P$ and $Z = Q$ with residues

$$\operatorname*{res}_{Z=P} H_{PQ}(Z) = -\operatorname*{res}_{Z=Q} H_{PQ}(Z) = \frac{\theta(P - Q - \mathbf{u})}{\theta(\mathbf{u})E(Q, P)\sqrt{dz(P)}\sqrt{dz(Q)}}.$$

We now consider the product $z(Z)H_{PQ}(Z)$. Vanishing of the sum of residues of this differential yields

$$\sum_{i=1}^{n} \frac{\theta(P - P_i - \mathbf{u})\theta(Q - P_i + \mathbf{u})}{\theta^2(\mathbf{u})E(P_i, P)E(Q, P_i)\sqrt{dz(P)}\sqrt{dz(Q)}} d\zeta_i = \sum_{i=1}^{n} \operatorname*{res}_{Z=P_i} z(Z) H_{PQ}(Z)$$

$$= [z(P) - z(Q)] \frac{\theta(P - Q - \mathbf{u})}{\theta(\mathbf{u})E(Q, P)\sqrt{dz(P)}\sqrt{dz(Q)}}$$

$$\sum_{j=1}^{n} \frac{\theta(Q - P_j - \mathbf{u})\theta(P - P_j + \mathbf{u})}{\theta^2(\mathbf{u})E(P_j, Q)E(P, P_j)\sqrt{dz(P)}\sqrt{dz(Q)}} d\zeta_j = \sum_{j=1}^{n} \operatorname*{res}_{Z=P_j} z(Z) H_{QP}(Z)$$

$$= [z(P) - z(Q)] \frac{\theta(P - Q + \mathbf{u})}{\theta(\mathbf{u})E(Q, P)\sqrt{dz(P)}.\sqrt{dz(Q)}} \tag{79}$$

Therefore the bi-differential in the right hand side of eq. (77) becomes

$$\sum_{i=1}^{n} \frac{\theta(P - P_i - \mathbf{u})\theta(Q - P_i + \mathbf{u})}{\theta^2(\mathbf{u})E(P_i, P)E(Q, P_i)d\zeta_i} \cdot \sum_{j=1}^{n} \frac{\theta(Q - P_j - \mathbf{u})\theta(P - P_j + \mathbf{u})}{\theta^2(\mathbf{u})E(P_j, Q)E(P, P_j)d\zeta_j}$$

$$= \sum_{i=1}^{n} \operatorname*{res}_{Z=P_i} z(Z) H_{PQ}(Z) \cdot \sum_{j=1}^{n} \operatorname*{res}_{Z=P_j} z(Z) H_{QP}(Z) \, dz(P)dz(Q)$$

$$= [z(P) - z(Q)]^2 \frac{\theta(P - Q - \mathbf{u})\theta(P - Q + \mathbf{u})}{\theta^2(\mathbf{u})E^2(P, Q)}.$$

$\square$

Equation (11) immediately follows from Proposition 2.14 and eq. (76).

2.3 Algebro-Geometric Solutions to the *n*-wave Hierarchy and their Tau-Functions. Proof of eq. (12)

According to the original idea of S.P. Novikov [25] algebro-geometric (aka *finite gap*) solutions to integrable systems of PDEs are obtained by considering stationary points of a linear combination of the commuting flows. Here we will be dealing with the *n*-wave system of nonlinear evolution PDEs represented in the form

$$[L_{a,k}, L_{b,l}] = 0$$

$$L_{a,k} = \frac{\partial}{\partial t_k^a} - U_{a,k}(\mathbf{t}; z), \quad a = 1, \dots, n, \quad k \geq -1$$

where

$$U_{a,k}(\mathbf{t}; z) = z^{k+1} E_a + \text{lower degree terms}$$

is an $n \times n$ matrix-valued polynomial in z of degree $k + 1$ depending on the infinite number of independent variables $\mathbf{t} = (t_k^a)$. The independent variable z

is often called *spectral parameter*. The above equations hold true identically in z. Here the diagonal $n \times n$ matrix E_a has only one nonzero entry

$$(E_a)_{ij} = \delta_{ia}\delta_{aj}.$$

For example, for $k = -1$

$$U_{a,-1} = E_a$$

and, for $k = 0$

$$U_{a,0} = z\,E_a - [E_a, Y], \quad a = 1, \ldots, n$$

where the diagonal entries of the $n \times n$ matrix $Y = \vec{Y}(\mathbf{t})$ vanish. It turns out that the coefficients of the matrix polynomials $U_{a,k}(\mathbf{t}; z)$ can be represented as polynomials in the entries of the matrix $Y(\mathbf{t})$ and its derivatives in the variables $t_0^1, \ldots, t_0^n$. Thus the n-wave system can be considered as an infinite family of partial differential equations for the matrix $Y(\mathbf{t})$. See Appendix A below for the details about the structure of the n-wave hierarchy.

The following statement is crucial for computing tau-functions of solutions to the n-wave hierarchy.

Proposition 2.18 *1) For any solution to the n-wave hierarchy there exists a unique n-tuple of matrix-valued series*

$$M_b(\mathbf{t}, z) = E_b + \sum_{k \geq 1} \frac{B_{b,k}(\mathbf{t})}{z^k}, \quad b = 1, \ldots, n$$

satisfying

$$\left[L_{a,k}, M_b\right] = 0 \quad \Leftrightarrow \quad \frac{\partial M_b(\mathbf{t}, z)}{\partial t_k^a} = \left[U_{a,k}(\mathbf{t}, z), M_b(\mathbf{t}, z)\right]$$

$$\forall a = 1, \ldots, n, \quad k \geq -1$$

and also

$$M_a(\mathbf{t}, z)M_b(\mathbf{t}, z) = \delta_{ab}M_a(\mathbf{t}, z), \quad M_1(\mathbf{t}, z) + \cdots + M_n(\mathbf{t}, z) = \mathbf{1}.$$

2) The (principal) tau-function $\tau(\mathbf{t})$ of this solution is determined from the following generating series in independent variables z_1, z_2 for its second logarithmic derivatives

$$\sum_{p,q=0}^{\infty} \frac{1}{z_1^{p+2}} \frac{1}{z_2^{q+2}} \frac{\partial^2 \log \tau(\mathbf{t})}{\partial t_q^b \partial t_p^a} = \mathrm{tr}\frac{M_a(\mathbf{t}, z_1)M_b(\mathbf{t}, z_2)}{(z_1 - z_2)^2} - \frac{\delta_{ab}}{(z_1 - z_2)^2} \quad (80)$$

for any a, $b = 1, \ldots, n$.

3) The logarithmic derivatives of higher orders $N \geq 3$ of the same tau-function can be determined by the following generating series

$$\sum_{k_1,\ldots,k_N \geq 0} \frac{\partial^N \log \tau(\mathbf{t})}{\partial t_{k_1}^{a_1} \cdots \partial t_{k_N}^{a_N}} \frac{1}{z_1^{k_1+2} \cdots z_N^{k_N+2}}$$

$$= -\frac{1}{N} \sum_{s \in S_N} \frac{\operatorname{tr}\left[M_{a_{s_1}}(\mathbf{t}, z_{s_1}) \cdots M_{a_{s_N}}(\mathbf{t}, z_{s_N}) \right]}{(z_{s_1} - z_{s_2}) \cdots (z_{s_{N-1}} - z_{s_N})(z_{s_N} - z_{s_1})}. \qquad (81)$$

Clearly the tau-function is determined by (80) uniquely up to

$$\tau(\mathbf{t}) \mapsto e^{\alpha + \sum \beta_{a,k} t_k^a} \tau(\mathbf{t})$$

for some constants α and $\beta_{a,k}$.

The above proposition about construction of tau-functions of solutions to the n-wave integrable hierarchy is an extension to this case of the approach of [2]–[4] based on the theory of the so-called matrix resolvents. For the proofs see Appendix A below.

Remark 2.19 The construction of tau-function given in the Proposition differs from the original definition of [19], [7], [31] (we recall this definition in Appendix B below). One can prove equivalence of the two definitions following the scheme of [2]. We will not do it here as it is not needed for the proofs of the results of this paper.

We will now apply the Proposition to the finite-gap solutions of the n-wave hierarchy. According to the Novikov's recipe mentioned above for arbitrary choice of constants $c_{a,k}$, $a = 1, \ldots, n$, $-1 \leq k \leq N$ for any integer $N \geq 0$, we obtain a family of algebro-geometric solutions $Y(\mathbf{t})$ satisfying

$$\sum_{a=1}^{n} \sum_{k=-1}^{N} c_{a,k} \frac{\partial Y(\mathbf{t})}{\partial t_k^a} = 0.$$

For any such solution define a matrix polynomial

$$U(\mathbf{t}, z) = \sum_{a=1}^{n} \sum_{k=-1}^{N} c_{a,k} U_{a,k}(\mathbf{t}, z).$$

Recall that the coefficients of the matrix polynomial belong to the space $\mathcal{Y}$. From commutativity (123) of the flows it readily follows that the matrix $U = U(\mathbf{t}, z)$ satisfies

$$\frac{\partial U}{\partial t_l^b} = [U_{b,l}, U] \quad \text{for any} \quad b = 1, \ldots, n, \quad l \geq -1. \qquad (82)$$

Therefore the characteristic polynomial

$$R(z, w) = \det(w \cdot \mathbf{1} - U(\mathbf{t}, z))$$

does not depend on $\mathbf{t}$. Its coefficients can be considered as first integrals of the differential equations (82). For a given N, assuming the coefficients $c_{1,N}, \ldots, c_{n,N}$ to be pairwise distinct the matrix polynomial $U(\mathbf{t}, z)$ for any $\mathbf{t}$ belongs to the family $\mathcal{W}$ of polynomials of the form (22) with $m = N + 1$. Therefore the spectral curves $C = \{R(z, w) = 0\}$ are of the form (23).

Our nearest goal is to construct an algebro-geometric solution to the n-wave system such that the matrix polynomial $U(\mathbf{t}, z)$ satisfies the initial condition

$$W(z) = U(0, z) \tag{83}$$

so C is the spectral curve of the matrix polynomial $W(z)$. The construction is rather standard for the theory of integrable systems. Namely, for the matrix polynomial $W(z)$ we have constructed the spectral curve C with a nonspecial divisor D_0. Starting from these data one can construct an algebro-geometric solution of the n-wave system following I.M. Krichever's scheme [21]. We will use a *vector-valued Baker–Akhiezer function* $\boldsymbol{\psi}(\mathbf{t}, P) = \left(\psi^1(\mathbf{t}, P), \ldots, \psi^n(\mathbf{t}, P)\right)^T$ meromorphic on the spectral curve $C \setminus (P_1 \cup \cdots \cup P_n)$ with poles at the points of the divisor D_0 and having essential singularities at $P = P_1, \ldots, P = P_n$ of the form

$$\psi^i(\mathbf{t}, P) = \left(\delta_{ij} + \mathcal{O}\left(\frac{1}{z}\right)\right) e^{\phi_j(\mathbf{t}, z)}, \quad P \to P_j, \quad i, j = 1, \ldots, n.$$

It is a standard fact of the theory of Baker–Akhiezer functions that $\boldsymbol{\psi}(\mathbf{t}, P)$ exists for sufficiently small $|\mathbf{t}|$ and is unique. It also exists the unique dual Baker–Akhiezer function $\boldsymbol{\psi}^{\dagger}(\mathbf{t}, P) = \left(\psi_1^{\dagger}(\mathbf{t}, P), \ldots, \psi_n^{\dagger}(\mathbf{t}, P)\right)$ with poles at the divisor $D_0^{\dagger}$ and essential singularities of the form

$$\psi_i^{\dagger}(\mathbf{t}, P) = \left(\delta_{ij} + \mathcal{O}\left(\frac{1}{z}\right)\right) e^{-\phi_j(\mathbf{t}, z)}, \quad P \to P_j, \quad i, j = 1, \ldots, n.$$

The needed algebro-geometric solution to the n-wave system is uniquely specified by the condition that its wave function is expressed in terms of $\boldsymbol{\psi}(\mathbf{t}, P)$. We will now obtain an explicit expression of this solution in terms of theta-functions of the spectral curve.

Proposition 2.20 *Let $W(z)$ be a matrix polynomial of the form (22) with a nonsingular spectral curve C. Let D_0 be the divisor of poles of the eigenvector of $W(z)$ normalized by the condition (29) and denote $(\mathbf{u}_0, \boldsymbol{\lambda}^0) \in J(C, P_1, \ldots, P_n)$ the corresponding point (70) of the generalized Jacobian. Introduce Abelian differentials*

$$\Omega^i_j(\mathbf{t}, P) = \frac{\lambda_i(\mathbf{t})}{\lambda_j(\mathbf{t})} \frac{\theta(P - P_i - \mathbf{u}(\mathbf{t}))\theta(P - P_j + \mathbf{u}(\mathbf{t}))}{\theta^2(\mathbf{u}(\mathbf{t}))E(P_i, P)E(P, P_j)\sqrt{d\zeta_i}\sqrt{d\zeta_j}} \tag{84}$$

$$\mathbf{u}(\mathbf{t}) = \mathbf{u}_0 - \sum t^a_k \mathbf{V}^{(a,k)},$$

$$\lambda_i(\mathbf{t}) = \exp\left\{\sum t^a_k \int_{P_i}^{P_0} \Omega^{(k)}_a\right\} \lambda^0_i, \quad i, j = 1, \dots, n.$$

Here and below

$$\zeta_a = \zeta_a(Q) = \frac{1}{z(Q)}, \quad Q \in C, \quad Q \to P_a, \quad a = 1, \dots, n$$

is a natural local parameter near P_a. The principal values of the integrals are defined by the following limits

$$\fint_{P_i}^{P_0} \Omega^{(k)}_a = \lim_{Q \to P_i} \left(\int_Q^{P_0} \Omega^{(k)}_a + z^{k+1}(Q)\right). \tag{85}$$

Define matrix-valued power series in z^{-1} by expanding the differentials at infinity

$$M_a(\mathbf{t}, z) = \left(\frac{\Omega^i_j(\mathbf{t}, P)}{dz}\right)_{1 \le i, j \le n}, \quad P = (z, w_a(z)) \to P_a, \quad a = 1, \dots, n \tag{86}$$

and put

$$U_{a,k}(\mathbf{t}, z) = \left(z^{k+1} M_a(\mathbf{t}, z)\right)_+. \tag{87}$$

This collection of matrix polynomials is an algebro-geometric solution to the n-wave system with the corresponding matrix $U(\mathbf{t}, z)$ satisfying (82) given by

$$U(\mathbf{t}, z) = w_1(z)M_1(\mathbf{t}, z) + \dots + w_n(z)M_n(\mathbf{t}, z). \tag{88}$$

In this formula $w_a(z)$ is the Laurent expansion of the algebraic function $w(z)$ near $P = P_a$, $a = 1, \dots, n$. This matrix polynomial $U(\mathbf{t}, z)$ satisfies the initial condition (83).

Proof. Let $\Omega(P)$ be the differential (58) on C constructed above. We first prove that the differentials (84) coincide with

$$\Omega^i_j(\mathbf{t}, P) = \psi^i(\mathbf{t}, P)\Omega(P)\psi^\dagger_j(\mathbf{t}, P), \quad i, j = 1, \dots, n. \tag{89}$$

To this end we use the following expressions of the Baker–Akhiezer functions $\psi(\mathbf{t}, P)$ and $\psi^\dagger(\mathbf{t}, P)$

$$\psi^i(\mathbf{t}, P) = \exp\left(\sum t_k^a \int_{P_i}^{P} \Omega_a^{(k)}\right) \frac{\lambda_i^0 \dfrac{\theta(P - P_i + \sum t_k^a V^{(a,k)} - \mathbf{u}_0)}{\theta(\sum t_k^a V^{(a,k)} - \mathbf{u}_0)E(P, P_i)(d\zeta_i)^{1/2}}}{\sum_{b=1}^{n} \lambda_b^0 \dfrac{\theta(P - P_b - \mathbf{u}_0)}{\theta(\mathbf{u}_0)E(P, P_b)(d\zeta_b)^{1/2}}},$$

$$i = 1, \ldots, n \tag{90}$$

and

$$\psi_i^\dagger(\mathbf{t}, P) = \exp\left(-\sum t_k^a \int_{P_i}^{P} \Omega_a^{(k)}\right) \frac{\dfrac{1}{\lambda_i^0} \dfrac{\theta(P - P_i - \sum t_k^a V^{(a,k)} + \mathbf{u}_0)}{\theta(\sum t_k^a V^{(a,k)} - \mathbf{u}_0)E(P_i, P)(d\zeta_i)^{1/2}}}{\sum_{b=1}^{n} \dfrac{1}{\lambda_b^0} \dfrac{\theta(P - P_b + \mathbf{u}_0)}{\theta(\mathbf{u}_0)E(P_b, P)(d\zeta_b)^{1/2}}},$$

$$i = 1, \ldots, n. \tag{91}$$

Here $\Omega_a^{(k)}$ is the normalized second kind differential on C with a unique pole at P_a of order $k + 2$

$$\Omega_a^{(k)}(P) = dz^{k+1} + \text{regular terms}, \quad P \to P_a$$

$$\oint_{a_i} \Omega_a^{(k)} = 0, \quad V_i^{(a,k)} = \oint_{b_i} \Omega_a^{(k)}, \quad i = 1, \ldots, g \tag{92}$$

for a chosen canonical basis a_i, b_j in $H_1(C, \mathbb{Z})$. Recall [17] that the b-periods $V_i^{(a,k)}$ of the differentials $\Omega_a^{(k)}$ coincide with the coefficients of expansions (20) of holomorphic differentials $\omega_i(P)$ at $P \to P_a$. Observe that for $\mathbf{t} = 0$ the functions $\psi(0, P)$ and $\psi^\dagger(0, P)$ coincide with the right and left eigenvectors of $W(z)$.

The derivation of the representations (90), (91) is standard for the theory of Baker–Akhiezer functions: we check that (90), (91) are well-defined meromorphic functions on $C \setminus (P_1 \cup \cdots \cup P_n)$ with essential singularities at infinity of the needed form having poles at the points of the divisors D_0 and $D_0^\dagger$ respectively (for the claim about the location of poles use Proposition 2.13 and Remark 2.16). With the help of these expressions it is easy to verify validity of eq. (89).

Define now a matrix-valued function $\Psi(\mathbf{t}, z)$ such that its i-th row is given by the expansions of $\psi^i(\mathbf{t}, P)$ at the infinite points $P_1, \ldots, P_n$. We will prove that $\Psi(\mathbf{t}, z)$ is the wave function of the solution (87) to the n-wave system. To this end we will first verify that the definition (86) of the matrices $M_a(z)$ can be rewritten in the form (150).

Introduce another matrix-valued function $\Psi^\dagger(\mathbf{t}, z)$ in a similar way: its i-th column is given by the expansions of $\psi_i^\dagger(\mathbf{t}, P)$ at the infinite points $P_1, \ldots, P_n$. Let us prove that this matrix is inverse to $\Psi(\mathbf{t}, z)$ up to multiplication on the left by a nondegenerate diagonal matrix.

Lemma 2.21 *Define*

$$\hat{\rho}(z) = \text{diag}\,(\rho_1(z), \ldots, \rho_n(z)), \quad \rho_a(z) = \left(\frac{\Omega(P)}{dz}\right)_{P=(z,w_a(z))}$$

$$= 1 + \mathcal{O}\left(\frac{1}{z}\right), \quad a = 1, \ldots, n. \tag{93}$$

Then

$$\Psi(\mathbf{t}, z)\hat{\rho}(z)\Psi^{\dagger}(\mathbf{t}, z) = \mathbf{1}. \tag{94}$$

Proof. The differentials (84) are holomorphic on $C \setminus (P_1 \cup \cdots \cup P_n)$. At infinity they behave in the same way as the differentials $\Omega^i_j(P)$ (see Proposition 2.7 above). For an arbitrary complex number z away from the ramification points of C consider the sum

$$\Omega^i_j(\mathbf{t}, (z, w_1(z))) + \cdots + \Omega^i_j(\mathbf{t}, (z, w_n(z)))$$

$$= \Psi^i_1(\mathbf{t}, z)\rho_1(z)\Psi^{\dagger 1}_j(\mathbf{t}, z)dz + \cdots + \Psi^i_n(\mathbf{t}, z)\rho_n(z)\Psi^{\dagger n}_j(\mathbf{t}, z)dz.$$

This is a well-defined differential on $\mathbf{P}^1$. It can have poles only at $z = \infty$, namely, a simple pole for $i \neq j$ and a double pole $\sim dz$ for $i = j$. Therefore the above sum is equal to $\delta^i_j dz$. $\qquad\square$

Corollary 2.22 *The matrix series $M_a(\mathbf{t}, z)$ coincide with*

$$M_a(\mathbf{t}, z) = \Psi(\mathbf{t}, z)E_a\Psi^{-1}(\mathbf{t}, z), \quad a = 1, \ldots n. \tag{95}$$

Lemma 2.23 *1) The matrix $\Psi(\mathbf{t}, z)$ satisfies*

$$\frac{\partial}{\partial t^a_k}\Psi(\mathbf{t}, z) = U_{a,k}(\mathbf{t}, z)\Psi(\mathbf{t}, z) \quad \forall\, a, b = 1, \ldots, n, \quad k \geq -1 \tag{96}$$

where $U_{a,k}(\mathbf{t}, z)$ are given by (87).
2) The matrix polynomials $U_{a,k}(\mathbf{t}, z)$ satisfy eqs. (123) of the n-wave hierarchy.
3) The matrix series $M_b(\mathbf{t}, z)$ satisfy

$$\frac{\partial M_b(z)}{\partial t^a_k} = \left[U_{a,k}(z), M_b(z)\right]. \tag{97}$$

Proof. Let $\Psi(\mathbf{t}, z) = A(\mathbf{t}, z)e^{\phi(\mathbf{t}, z)}$ with $A(\mathbf{t}, z) = 1 + \mathcal{O}\left(\frac{1}{z}\right)$ and denote

$$\tilde{U}_{a,k}(\mathbf{t}, z) = \left(z^{k+1}M_a(\mathbf{t}, z)\right)_-.$$

It is a power series in z^{-1}. We have

$$\frac{\partial}{\partial t^a_k}\Psi(\mathbf{t}, z) - U_{a,k}(\mathbf{t}, z)\Psi(\mathbf{t}, z) = \left(\frac{\partial A(\mathbf{t}, z)}{\partial t^a_k} \cdot A^{-1}(\mathbf{t}, z) + \tilde{U}_{a,k}(\mathbf{t}, z)\right)\Psi(\mathbf{t}, z).$$

As the expression in the parenthesis contains only negative powers of z, the right hand side is a Baker–Akhiezer function on the curve C with the same divisor of poles and with expansion at infinity of the form

$$\left(\frac{\partial A(\mathbf{t}, z)}{\partial t_k^a} \cdot A^{-1}(\mathbf{t}, z) + \tilde{U}_{a,k}(\mathbf{t}, z) \right) \Psi(\mathbf{t}, z) = \mathcal{O}\left(\frac{1}{z} \right) e^{\phi(\mathbf{t}, z)}.$$

Hence this Baker–Akhiezer function identically vanishes. This proves the first part of Lemma.

The equations (123) readily follow from the compatibility

$$\frac{\partial}{\partial t_k^a} \frac{\partial}{\partial t_l^b} \Psi(\mathbf{t}, z) = \frac{\partial}{\partial t_l^b} \frac{\partial}{\partial t_k^a} \Psi(\mathbf{t}, z).$$

Finally the eq. (97) follows from (95) and (96). $\qquad\qquad\qquad\qquad\square$

In a similar way one can verify validity of eq. (82) for the matrix $U(\mathbf{t}, z)$ defined by (88). It remains to prove that this matrix is polynomial in z satisfying the initial condition (83). To this end we consider the differentials

$$U_j^i(\mathbf{t}, z)dz = w_1(z)\Omega_{\cdot j}^i(\mathbf{t}, (z, w_1(z))) + \cdots + w_n(z)\Omega_{\cdot j}^i(\mathbf{t}, (z, w_n(z))).$$

Like in the proof of Lemma 2.21 this is a differential on $\mathbf{P}^1$ with poles only at infinity. Hence it must be a polynomial. Since $\Omega_{\cdot j}^i(0, P) = \Omega_{\cdot j}^i(P)$ we have $U(0, z) = W(z)$. The Proposition is proved. $\qquad\qquad\qquad\qquad\square$

We are now ready to compute the tau-function of the algebro-geometric solution. Define numbers $q_{a,k;b,l}$ as coefficients of expansions of the second kind differentials $\Omega_a^{(k)}(P)$ at $P \to P_l$

$$\Omega_a^{(k)}(P) = \delta_{ab}d\left(z^{k+1} \right) + \sum_{l \geq 0} \frac{q_{a,k;b,l}}{z^{l+2}} dz, \quad P \to P_b. \qquad (98)$$

Alternatively these coefficients can be recovered from the expansions of the normalized second kind bi-differential [17]

$$\omega(P, Q) = d_P d_Q \log E(P, Q) = \left[\frac{\delta_{ab}}{(z_1 - z_2)^2} + \sum_{k, l \geq 0} \frac{q_{a,k;b,l}}{z_1^{k+2} z_2^{l+2}} \right] dz_1 dz_2 \qquad (99)$$

$$z_1 = z(P), \ z_2 = z(Q), \quad P \to P_a, \ Q \to P_b.$$

Proposition 2.24 *Tau-function of the algebro-geometric solution constructed in Proposition 2.20 is equal to*

$$\tau(\mathbf{t}) = e^{\frac{1}{2} \sum q_{a,k;b,l} t_k^a t_l^b} \theta\left(\sum t_k^a \mathbf{V}^{(a,k)} - \mathbf{u}_0 \right) \qquad (100)$$

up to multiplication by exponential of a linear function. Here

$$\mathbf{V}^{(a,k)} = \left(V_1^{(a,k)}, \ldots, V_g^{(a,k)} \right). \qquad (101)$$

Proof. We have to compute the generating function (155) of the second logarithmic derivatives of the tau-function

$$\sum \frac{\partial^2 \log \tau(\mathbf{t})}{\partial t_k^a \partial t_l^b} \frac{dz_1}{z_1^{k+2}} \frac{dz_2}{z_2^{l+2}}$$
$$= \frac{\text{tr}[M_a(\mathbf{t}, z_1) M_b(\mathbf{t}, z_2)]}{(z_1 - z_2)^2} dz_1 dz_2 - \frac{\delta_{ab}}{(z_1 - z_2)^2} dz_1 dz_2 \qquad (102)$$

where $M_1(\mathbf{t}, z)$, $M_b(\mathbf{t}, z)$ are solutions to the equations (97) in the class of matrix-valued power series in z^{-1} uniquely specified by the conditions (95). Using the representation (86) we can rewrite the previous equation in the form

$$\sum \frac{\partial^2 \log \tau(\mathbf{t})}{\partial t_k^a \partial t_l^b} \frac{dz_1}{z_1^{k+2}} \frac{dz_2}{z_2^{l+2}} = \frac{\text{tr}[\mathbf{\Omega}(\mathbf{t}, P) \mathbf{\Omega}(\mathbf{t}, Q)]}{(z(P) - z(Q))^2} - \frac{\delta_{ab}}{(z_1 - z_2)^2} dz_1 dz_2$$

$$z_1 = z(P), \quad z_2 = z(Q), \quad P \to P_a, \quad Q \to P_b$$

where the matrix entries of $\mathbf{\Omega}(\mathbf{t}, P)$ are equal to (84). According to Proposition 2.17 the right hand side can be rewritten in the form

$$\frac{\text{tr}[\mathbf{\Omega}(\mathbf{t}, P) \mathbf{\Omega}(\mathbf{t}, Q)]}{(z(P) - z(Q))^2} - \frac{\delta_{ab}}{(z_1 - z_2)^2} dz_1 dz_2$$
$$= \frac{\theta(P - Q - \mathbf{u}(\mathbf{t})) \theta(P - Q + \mathbf{u}(\mathbf{t}))}{\theta^2(\mathbf{u}(\mathbf{t})) E^2(P, Q)} - \frac{\delta_{ab}}{(z_1 - z_2)^2} dz_1 dz_2. \qquad (103)$$

We will now use the following important identity [17]

$$\frac{\theta(P - Q - \mathbf{u}(\mathbf{t})) \theta(P - Q + \mathbf{u})}{\theta^2(\mathbf{u}) E^2(P, Q)} = \sum_{i, j = 1}^n \frac{\partial^2 \log \theta(\mathbf{u})}{\partial u_i \partial u_j} \omega_i(P) \omega_j(Q) + \omega(P, Q)$$

where $\omega(P, Q)$ is the normalized bi-differential (99). Using this identity we can expand the right hand side of eq. (103) at $P \to P_a$, $Q \to P_b$

$$\frac{\theta(P - Q - \mathbf{u}(\mathbf{t})) \theta(P - Q + \mathbf{u}(\mathbf{t}))}{\theta^2(\mathbf{u}(\mathbf{t})) E^2(P, Q)} - \frac{\delta_{ab}}{(z_1 - z_2)^2} dz_1 dz_2$$

$$= \sum_{i, j = 1}^n \frac{\partial^2 \log \theta(\mathbf{u}(\mathbf{t}))}{\partial u_i \partial u_j} \sum_{k \geq 0} \frac{V_i^{(a,k)}}{z_1^{k+2}} \sum_{l \geq 0} \frac{V_j^{(b,l)}}{z_2^{l+2}} dz_1 dz_2 + \sum_{k, l \geq 0} \frac{q_{a,k;b,l}}{z_1^{k+2} z_2^{k+2}} dz_1 dz_2$$

$$= \sum_{k, l \geq 0} \frac{\partial^2 \log \theta(\mathbf{u}(\mathbf{t}))}{\partial t_k^a \partial t_l^b} \frac{dz_1}{z_1^{k+2}} \frac{dz_2}{z_2^{l+2}} + \sum_{k, l \geq 0} \frac{q_{a,k;b,l}}{z_1^{k+2} z_2^{k+2}} dz_1 dz_2.$$

Comparing this expansion with (102) we arrive at the proof of the Proposition. □

Remark 2.25 An expression similar to (100) is well known in the theory of KP equation and its reductions [29], [16], [23]. We emphasize that here our main task was to prove that eq. (100) is in agreement with the construction of the tau-function given in terms of Proposition 2.18.

Let us now proceed to the proof of eq. (12). The expression (162) for the N-th order logarithmic derivatives of the tau-function will be applied to the tau-function (100) of an algebro-geometric solution. Due to the previous Proposition the tau-function in the left hand side of eq. (162) for $N \geq 3$ can be replaced with the theta-function

$$
\sum_{k_1,\ldots,k_N \geq 0} \frac{\partial^N \log \tau(\mathbf{t})}{\partial t_{k_1}^{a_1} \cdots \partial t_{k_N}^{a_N}} \frac{dz_1 \cdots dz_N}{z_1^{k_1+2} \cdots z_N^{k_N+2}}
$$

$$
= \sum_{k_1,\ldots,k_N \geq 0} \frac{\partial^N \log \theta(\mathbf{u}(\mathbf{t}))}{\partial t_{k_1}^{a_1} \cdots \partial t_{k_N}^{a_N}} \frac{dz_1 \cdots dz_N}{z_1^{k_1+2} \cdots z_N^{k_N+2}}
$$

$$
= (-1)^N \sum_{k_1,\ldots,k_N \geq 0} \sum_{i_1,\ldots,i_N=1}^{g} V_{i_1}^{(a_1,k_1)} \cdots V_{i_N}^{(a_N,k_N)} \frac{\partial^N \log \theta(\mathbf{u}(\mathbf{t}))}{\partial u_{i_1} \cdots \partial u_{i_N}}
$$

$$
\times \frac{dz_1 \cdots dz_N}{z_1^{k_1+2} \cdots z_N^{k_N+2}}
$$

$$
= (-1)^N \sum_{i_1,\ldots,i_N=1}^{g} \frac{\partial^N \log \theta(\mathbf{u}(\mathbf{t}))}{\partial u_{i_1} \cdots \partial u_{i_N}} \omega_{i_1}(Q_1) \cdots \omega_{i_N}(Q_N),
$$

$$
Q_1 \to P_{a_1}, \ldots, Q_N \to P_{a_N}
$$

where the last multi-differential is considered as its expansion in negative powers of

$$
z_1 = z(Q_1), \ldots, z_N = z(Q_N).
$$

Let us now consider the right hand side of eq. (162) multiplying it, like above, by $dz_1 \cdots dz_N$

$$
-\frac{1}{N} \sum_{s \in S_N} \frac{\text{tr}\left[M_{a_{s_1}}(\mathbf{t}, z_{s_1}) \cdots M_{a_{s_N}}(\mathbf{t}, z_{s_N}) \right]}{(z_{s_1} - z_{s_2}) \cdots (z_{s_{N-1}} - z_{s_N})(z_{s_N} - z_{s_1})} dz_1 \cdots dz_N =
$$

$$
= -\frac{1}{N} \sum_{s \in S_N} \frac{\text{tr}\left[\Omega(\mathbf{t}, Q_{s_1}) \cdots \Omega(\mathbf{t}, Q_{s_N}) \right]}{(z(Q_{s_1}) - z(Q_{s_2})) \cdots (z(Q_{s_{N-1}}) - z(Q_{s_N}))(z(Q_{s_N}) - z(Q_{s_1}))}
$$

where $Q_1, \ldots, Q_N$ are arbitrary points of C such that $Q_1 \to P_{a_1}, \ldots Q_N \to P_{a_N}$. So eq. (162) implies that the two N-differentials coincide when the points

$Q_1, \ldots, Q_N$ go to infinity in all possible ways. Therefore these N-differentials coincide

$$\sum_{i_1,\ldots,i_N=1}^{g} \frac{\partial^N \log \theta(\mathbf{u(t)})}{\partial u_{i_1} \cdots \partial u_{i_N}} \omega_{i_1}(Q_1) \cdots \omega_{i_N}(Q_N) = \frac{(-1)^{N-1}}{N}$$

$$\times \sum_{s \in S_N} \frac{\text{tr}\left[\boldsymbol{\Omega}(\mathbf{t}, Q_{s_1}) \cdots \boldsymbol{\Omega}(\mathbf{t}, Q_{s_N})\right]}{(z(Q_{s_1}) - z(Q_{s_2})) \cdots (z(Q_{s_{N-1}}) - z(Q_{s_N}))(z(Q_{s_N}) - z(Q_{s_1}))}.$$

$$(104)$$

To complete the derivation of eq. (12) we set $\mathbf{t} = 0$ where

$$\mathbf{u}(0) = \mathbf{u}_0, \quad \boldsymbol{\Omega}(0, P) = \Phi(P) \frac{dz}{R_w(z, w)}.$$

The Main Lemma and Main Theorem are proved. $\qquad \square$

2.4 Proof of Corollary 1.4

Let C be a compact Riemann surface of genus $g > 0$ and n, m a pair of positive integers.

Proposition 2.26 *For sufficienly large n, m and an arbitrary collection of n pairwise distinct points $P_1, \ldots, P_n$ there exist two rational functions z, w on C such that*

(i) the function z has simple poles at $P_1, \ldots, P_n$
(ii) the function w has poles of order m at the same points and $w \neq P(z)$ for any polynomial P.

This is an easy consequence of the Riemann–Roch theorem.

Corollary 2.27 *An arbitrary compact Riemann surface of genus $g > 0$ can be represented as the spectral curve of a matrix $W(z)$ of the form (1) for sufficiently large n and m.*

According to the Corollary we can rewrite eq. (12) in the form

$$\sum_{i_1,\ldots,i_N=1}^{g} \frac{\partial^N \log \theta(\mathbf{u})}{\partial u_{i_1} \cdots \partial u_{i_N}} \omega_{i_1}(Q_1) \cdots \omega_{i_N}(Q_N) = \frac{(-1)^{N-1}}{N}$$

$$\times \sum_{s \in S_N} \frac{\text{tr}\left[\boldsymbol{\Omega}(Q_{s_1}) \cdots \boldsymbol{\Omega}(Q_{s_N})\right]}{(z(Q_{s_1}) - z(Q_{s_2})) \cdots (z(Q_{s_{N-1}}) - z(Q_{s_N}))(z(Q_{s_N}) - z(Q_{s_1}))}$$

$$(105)$$

(cf. (104)). Here $\mathbf{u} \in J(C) \setminus (\theta)$ is the point of the Jacobian corresponding to the matrix $W(z)$, the matrix-valued differential $\mathbf{\Omega}(Q)$ equals

$$\mathbf{\Omega}(Q) = \Phi(Q)\frac{dz}{R_w(z,w)}.$$

Using the representation (75) of the matrix entries of this differential we can rewrite the numerator, for an arbitrary permutation $s \in S_N$ as follows

$$\operatorname{tr}\left[\mathbf{\Omega}\left(Q_{s_1}\right)\cdots\mathbf{\Omega}\left(Q_{s_N}\right)\right]$$

$$= \sum_{i_1,\dots,i_N=1}^{n} \frac{\theta(Q_{s_1}-P_{i_1}-\mathbf{u})\theta(Q_{s_1}-P_{i_2}+\mathbf{u})}{\theta^2(\mathbf{u})E(P_{i_1},Q_{s_1})E(Q_{s_1},P_{i_2})\sqrt{d\zeta_{i_1}}\sqrt{d\zeta_{i_2}}}$$

$$\times \frac{\theta(Q_{s_2}-P_{i_2}-\mathbf{u})\theta(Q_{s_2}-P_{i_3}+\mathbf{u})}{\theta^2(\mathbf{u})E(P_{i_2},Q_{s_2})E(Q_{s_2},P_{i_3})\sqrt{d\zeta_{i_2}}\sqrt{d\zeta_{i_3}}}$$

$$\cdots \frac{\theta(Q_{s_{N-1}}-P_{i_{N-1}}-\mathbf{u})\theta(Q_{s_{N-1}}-P_{i_N}+\mathbf{u})}{\theta^2(\mathbf{u})E(P_{i_{N-1}},Q_{s_{N-1}})E(Q_{s_{N-1}},P_{i_N})\sqrt{d\zeta_{i_{N-1}}}\sqrt{d\zeta_{i_N}}}$$

$$\times \frac{\theta(Q_{s_N}-P_{i_N}-\mathbf{u})\theta(Q_{s_N}-P_{i_1}+\mathbf{u})}{\theta^2(\mathbf{u})E(P_{i_N},Q_{s_N})E(Q_{s_N},P_{i_N})\sqrt{d\zeta_{i_N}}\sqrt{d\zeta_{i_1}}}$$

$$= \sum_{i_2=1}^{n} \frac{\theta(Q_{s_2}-P_{i_2}-\mathbf{u})\theta(Q_{s_1}-P_{i_2}+\mathbf{u})}{\theta^2(\mathbf{u})E(P_{i_2},Q_{s_2})E(Q_{s_1},P_{i_2})\,d\zeta_{i_2}}$$

$$\times \sum_{i_3=1}^{n} \frac{\theta(Q_{s_3}-P_{i_3}-\mathbf{u})\theta(Q_{s_2}-P_{i_3}+\mathbf{u})}{\theta^2(\mathbf{u})E(P_{i_3},Q_{s_3})E(Q_{s_2},P_{i_3})\,d\zeta_{i_3}}$$

$$\cdots \sum_{i_N=1}^{n} \frac{\theta(Q_{s_N}-P_{i_N}-\mathbf{u})\theta(Q_{s_{N-1}}-P_{i_N}+\mathbf{u})}{\theta^2(\mathbf{u})E(P_{i_N},Q_{s_N})E(Q_{s_{N-1}},P_{i_N})\,d\zeta_{i_N}}$$

$$\times \sum_{i_1=1}^{n} \frac{\theta(Q_{s_1}-P_{i_1}-\mathbf{u})\theta(Q_{s_N}-P_{i_1}+\mathbf{u})}{\theta^2(\mathbf{u})E(P_{i_1},Q_{s_1})E(Q_{s_N},P_{i_1})\,d\zeta_{i_1}}$$

$$= \sum_{i_2=1}^{n} \operatorname*{res}_{Z=P_{i_2}} z(Z)H_{Q_{s_2}Q_{s_1}}(Z)\cdots \sum_{i_N=1}^{n} \operatorname*{res}_{Z=P_{i_N}} z(Z)H_{Q_{s_N}Q_{s_{N-1}}}(Z)$$

$$\times \sum_{i_1=1}^{n} \operatorname*{res}_{Z=P_{i_1}} z(Z)H_{Q_{s_1}Q_{s_N}}(Z)$$

$$= (z_{s_2}-z_{s_1})\cdots(z_{s_N}-z_{s_{N-1}})(z_{s_1}-z_{s_N})$$

$$\times \frac{\theta(Q_{s_2}-Q_{s_1}-\mathbf{u})\cdots\theta(Q_{s_N}-Q_{s_{N-1}}-\mathbf{u})\theta(Q_{s_1}-Q_{s_N}-\mathbf{u})}{\theta^N(\mathbf{u})E(Q_{s_1},Q_{s_2})\cdots E(Q_{s_{N-1}},Q_{s_N})E(Q_{s_N},Q_{s_1})}.$$

In the above computation we have used the differential H_{PQ} defined by (78). The computation of residues is analogous to the one in (79). In the last line we use the short notation $z_s := z(Q_s)$. Corollary 1.4 is proved. $\square$

For $N = 3$ the identity (21) takes the following explicit form

$$\theta^3(\mathbf{u}) \sum_{i,j,k=1}^{g} \frac{\partial^3 \log \theta(\mathbf{u})}{\partial u_i \partial u_j \partial u_k} \omega_i(Q_1) \omega_j(Q_2) \omega_k(Q_3) =$$

$$\frac{\theta(Q_1 - Q_2 - \mathbf{u})\theta(Q_2 - Q_3 - \mathbf{u})\theta(Q_3 - Q_1 - \mathbf{u}) - \theta(Q_1 - Q_2 + \mathbf{u})\theta(Q_2 - Q_3 + \mathbf{u})\theta(Q_3 - Q_1 + \mathbf{u})}{E(Q_1, Q_2)E(Q_2, Q_3)E(Q_3, Q_1)}$$

$$(106)$$

and for $N = 4$

$$\theta^4(\mathbf{u}) \sum_{i,j,k,l=1}^{g} \frac{\partial^4 \log \theta(\mathbf{u})}{\partial u_i \partial u_j \partial u_k \partial u_l} \omega_i(Q_1) \omega_j(Q_2) \omega_k(Q_3) \omega_l(Q_4)$$

$$= V_{\mathbf{u}}(Q_1, Q_2, Q_3, Q_4) + V_{\mathbf{u}}(Q_1, Q_3, Q_2, Q_4) + V_{\mathbf{u}}(Q_1, Q_3, Q_4, Q_2)$$
$$(107)$$

where

$$V_{\mathbf{u}}(Q_1, Q_2, Q_3, Q_4)$$
$$= -\frac{\theta(Q_1 - Q_2 + \mathbf{u}) \cdots \theta(Q_4 - Q_1 + \mathbf{u}) + \theta(Q_1 - Q_2 - \mathbf{u}) \cdots \theta(Q_4 - Q_1 - \mathbf{u})}{E(Q_1, Q_2)E(Q_2, Q_3)E(Q_3, Q_4)E(Q_4, Q_1)}.$$

3 Examples

3.1 Hyperelliptic Case

Consider a 2×2 matrix polynomial of the form

$$W(z) = \begin{pmatrix} a(z) & b(z) \\ c(z) & -a(z) \end{pmatrix}, \quad \deg a(z) = g + 1, \quad \deg b(z) = \deg c(z) = g,$$
$$(108)$$

the polynomial $a(z)$ is monic. The spectral curve

$$w^2 = Q(z), \quad Q(z) = -\det W(z) = z^{2g+2} + q_1 z^{2g+1} + \cdots + q_{2g+2} \quad (109)$$

is hyperelliptic. It has two distinct points $P_\pm$ at infinity,

$$w = \pm z^{g+1} + \ldots, \quad (z, w) \to P_\pm.$$

We have

$$\Pi(z, w) = \frac{1}{2} \frac{w + W(z)}{w} \quad (110)$$

so the basic idempotents of the matrix $W(z)$ take the form

$$M_\pm(z) = \frac{1}{2} \pm \frac{1}{2} \frac{W(z)}{w(z)}, \quad w(z) = z^{g+1} \sqrt{1 + \frac{q_1}{z} + \cdots + \frac{q_{2g+2}}{z^{2g+2}}}.$$

Thus eq. (11) for the bi-differential takes the following form

$$\frac{\theta\,(Q_1 - Q_2 - \mathbf{u}_0)\,\theta\,(Q_1 - Q_2 + \mathbf{u}_0)}{\theta^2(\mathbf{u}_0)\,E(Q_1, Q_2)^2}$$
$$= \frac{b(z_1)c(z_2) + 2a(z_1)a(z_2) + b(z_2)c(z_1) + 2w_1w_2}{4(z_1 - z_2)^2 w_1 w_2}\,dz_1\,dz_2,$$

$Q_1 = (z_1, w_1)$, $Q_2 = (z_2, w_2)$.
   Since

$$M_+(z) - M_-(z) = \frac{W(z)}{w(z)}$$

and the time-derivatives satisfy

$$\frac{\partial}{\partial t_k^+} + \frac{\partial}{\partial t_k^-} = 0$$

we introduce

$$\frac{\partial}{\partial t_k} = \frac{\partial}{\partial t_k^+} - \frac{\partial}{\partial t_k^-}.$$

So eq. (162) for the tau-function of the spectral curve (109) reduces to

$$\sum_{k_1,\dots,k_N} \frac{\frac{\partial^N \log \tau\,(0)}{\partial t_{k_1} \cdots \partial t_{k_N}}}{z_1^{k_1+2} \cdots z_N^{k_N+2}} = -\frac{1}{N} \frac{1}{w(z_1) \cdots w(z_N)}$$

$$\times \sum_{s \in S_N} \frac{\operatorname{tr}\left[W(z_{s_1}) \cdots W(z_{s_N})\right]}{(z_{s_1} - z_{s_2}) \cdots (z_{s_{N-1}} - z_{s_N})(z_{s_N} - z_{s_1})} - \frac{2\delta_{N,2}}{(z_1 - z_2)^2} \quad (111)$$

for any $N \geq 2$. First few logarithmic derivatives $F_{i_1 \dots i_N} := \partial^N \log \tau\,(0)/\partial t_{k_1} \cdots \partial t_{k_N}$ read

$$F_{00} = -b_1 c_1, \quad F_{01} = 2a_1 b_1 c_1 - b_2 c_1 - b_1 c_2,$$
$$F_{11} = \frac{1}{2}(-8a_1^2 b_1 c_1 + 4a_2 b_1 c_1 + 6a_1 b_2 c_1 - 2b_3 c_1 + b_1^2 c_1^2$$
$$+ 6a_1 b_1 c_2 - 4b_2 c_2 - 2b_1 c_3)$$
$$F_{000} = 2(b_1 c_2 - b_2 c_1), \quad F_{001} = 2(a_1 b_2 c_1 - b_3 c_1 - a_1 b_1 c_2 + b_1 c_3)$$
$$F_{0000} = 4(2a_2 b_1 c_1 - a_1 b_2 c_1 - b_3 c_1 - a_1 b_1 c_2 + 2b_2 c_2 - b_1 c_3).$$

We do not specify the genus: the above expressions are valid for any $g \geq 2$; for $g = 1$ one has to set $a_3 = b_3 = c_3 = 0$.

The theta-function of the spectral curve is related to the tau-function by the equation[4]

$$\tau(\mathbf{t}) = e^{\frac{1}{2}\sum_{i,j} q_{ij} t_i t_j} \theta\left(\sum_k t_k \mathbf{V}^{(k)} - \mathbf{u}_0\right), \quad \mathbf{t} = (t_0, t_1, \ldots) \qquad (112)$$

with suitable coefficients q_{ij} (cf. eq. (99) above). The vectors $\mathbf{V}^{(k)} = \left(V_1^{(k)}, \ldots, V_g^{(k)}\right)$ have the form

$$V_i^{(k)} = \alpha_{i1} r_k + \alpha_{i2} r_{k-1} + \cdots + \alpha_{ig} r_{k-g+1}, \quad i = 1, \ldots, g, \quad k \geq 0 \quad (113)$$

where the $g \times g$ matrix (α_{ij}) is the inverse, up to a factor $2\pi\sqrt{-1}$ to the matrix of a-periods of the following holomorphic differentials

$$(\alpha_{ij}) = 2\pi\sqrt{-1}\left(\oint_{a_j} z^{g-i} \frac{dz}{2w}\right)^{-1} \qquad (114)$$

and the rational numbers r_k come from the expansion

$$\left(1 + \frac{q_1}{z} + \cdots + \frac{q_{2g+2}}{z^{2g+2}}\right)^{-1/2} = \sum_{k \geq 0} \frac{r_k}{z^k}.$$

The point $\mathbf{u}_0$ is given by

$$\mathbf{u}_0 = \sum_{j=1}^{g+1}\left(\int_{P_+}^{Q_j} \omega_1, \ldots, \int_{P_+}^{Q_j} \omega_g\right) - \boldsymbol{\varpi} \qquad (115)$$

$$\omega_i = (\alpha_{i1} z^{g-1} + \cdots + \alpha_{ig})\frac{dz}{2w}, \quad i = 1, \ldots, g \qquad (116)$$

and the half-period $\boldsymbol{\varpi}$ for a suitable choice of the basis of cycles (see details in [17]) has the form

$$\boldsymbol{\varpi} = \pi\sqrt{-1}(1, 0, 1, 0, \ldots) + \frac{1}{2}\sum_{i=1}^{g}(B_{1i}, B_{2i}, \ldots, B_{gi}). \qquad (117)$$

Points of the divisor $D_0 = Q_1 + \cdots + Q_{g+1}$ of poles of the normalized eigenvector of the matrix $W(z)$ have the form $Q_i = (z_i, w_i)$ where $z_1, \ldots, z_{g+1}$ are roots of the equation

$$a(z) = \frac{1}{2}(b(z) + c(z)) \qquad (118)$$

and

$$w_i = \frac{1}{2}(c(z_i) - b(z_i)), \quad i = 1, \ldots, g+1. \qquad (119)$$

Corollary 1.4 in this particular case takes the following form.

---

[4] Like above the eq. (112) holds true up to multiplication by exponential of a linear function of t_i.

Corollary 3.1 *Assume rationality of coefficients of the polynomials $a(z)$, $b(z)$, $c(z)$. Then for any $N \geq 3$ and an arbitrary choice of indices $k_1, \ldots, k_N \geq 0$ one has*

$$\sum_{i_1,\ldots,i_N=1}^{g} V_{i_1}^{(k_1)} \cdots V_{i_N}^{(k_N)} \frac{\partial^N \log \theta(\mathbf{u}_0)}{\partial u_{i_1} \cdots \partial u_{i_N}} \in \mathbb{Q}.$$

Here $\theta = \theta(\mathbf{u}|B)$ is the theta-function (6) of the hyperelliptic curve (109) and the point $\mathbf{u}_0$ is given by (115), (118), (119).

3.2 Three-Sheet Riemann Surfaces

Let

$$W(z) = z^m B^0 + z^{m-1} B^1 + \cdots + B^m, \quad B^k = \left(B_{ij}^k\right)_{1 \leq i, j \leq 3},$$
$$B^0 = \mathrm{diag}(b_1^0, b_2^0, b_3^0) \tag{120}$$

be a 3×3 matrix polynomial satisfying $b_i^0 \neq b_j^0$ for $i \neq j$ and $\mathrm{tr}\, W(z) = 0$. Let

$$w^3 + p(z)w + q(z) = \det(w \cdot \mathbf{1} - W(z))$$

be the characteristic polynomial. The genus of the spectral curve C is equal to $g = 3m - 2$. The spectral projectors of $W(z)$ are given by branches of the algebraic function

$$\Pi(z, w) = \frac{W^2 + w\, W + w^2 + p(z)}{3w^2 + p(z)}, \quad (z, w) \in C. \tag{121}$$

The expression for the bi-differential (11) takes the following form

$$\frac{\theta\,(Q_1 - Q_2 - \mathbf{u}_0)\,\theta\,(Q_1 - Q_2 + \mathbf{u}_0)}{\theta^2(\mathbf{u}_0)E(Q_1, Q_2)^2} =$$

$$\frac{\mathrm{tr}\left[W_1^2 W_2^2 + (w_1 W_2 + w_2 W_1)W_1 W_2 + w_1 w_2 W_1 W_2\right] - 2(p_1 p_2 + w_1^2 p_2 + w_2^2 p_1 + 3w_1^2 w_2^2)}{(3w_1^2 + p_1)(3w_2^2 + p_2)(z_1 - z_2)^2} dz_1 dz_2$$

$$\tag{122}$$

where $Q_i = (z_i, w_i)$ and we use short notations

$$W_i = W(z_i), \quad p_i = p(z_i), \quad i = 1, 2.$$

The first few logarithmic derivatives of the tau-function (19) read as follows

$$F_{00}^{ij}[W] = \frac{b_{ij}^1 b_{ji}^1}{(b_i^0 - b_j^0)^2}, \quad i \neq j$$

$$F_{10}^{ij}[W] = \frac{b_{ij}^2 b_{ji}^1 + b_{ji}^2 b_{ij}^1}{(b_i^0 - b_j^0)^2} - 2(b_{ii}^1 - b_{jj}^1)\frac{b_{ij}^1 b_{ji}^1}{(b_i^0 - b_j^0)^3}, \quad i \neq j$$

$$F_{000}^{123}[W] = \frac{b_{12}^1 b_{23}^1 b_{31}^1 - b_{13}^1 b_{32}^1 b_{21}^1}{(b_1^0 - b_2^0)(b_2^0 - b_3^0)(b_3^0 - b_1^0)}$$

$$F_{000}^{iij}[W] = \frac{b_{ij}^1 b_{jk}^1 b_{ki}^1 - b_{ik}^1 b_{kj}^1 b_{ji}^1}{(b_i^0 - b_j^0)^2(b_k^0 - b_i^0)} + \frac{b_{ij}^2 b_{ji}^1 - b_{ji}^2 b_{ij}^1}{(b_i^0 - b_j^0)^2}, \quad i \neq j, \quad k \neq i, j$$

etc. One can also compute the derivatives of the above type for $i = j$ using the identities

$$\sum_{a=1}^{n} \frac{\partial}{\partial t_k^a} = 0, \quad \forall k \geq 0.$$

Appendix A. Tau-Function of the *n*-wave Integrable System

The (complexified) *n*-wave system [26] is an infinite family of pairwise commuting systems of nonlinear PDEs for $n(n-1)$ functions y_{ij}, $i \neq j$ of infinite number of independent variables t_k^a, $a = 1, \ldots, n$, $k \geq 0$ called *times*. We will often use an alternative notation for the variables $t_0^a =: x^a$, $a = 1, \ldots, n$ that will be called *spatial variables*.

The equations of the *n*-wave hierarchy (also called AKNS-D hierarchy, see [8]) are written as conditions of commutativity of linear differential operators

$$L_{a,k} = \frac{\partial}{\partial t_k^a} - U_{a,k}(\mathbf{y}; z), \quad a = 1, \ldots, n, \quad k \geq 0$$

$$[L_{a,k}, L_{b,l}] = 0 \tag{123}$$

where $U_{a,k}(\mathbf{y}; z)$ is an $n \times n$ matrix-valued polynomial in z of degree $k+1$ depending polynomially on the functions y_{ij} and their derivatives in $x^1, \ldots, x^n$. For $k = 0$ one has

$$U_{a,0} = z E_a - [E_a, Y] \tag{124}$$

where the diagonal $n \times n$ matrix E_a has only one nonzero entry

$$(E_a)_{ij} = \delta_{ia}\delta_{aj},$$

the $n \times n$ matrix Y has the form

$$Y = (y_{ij}), \quad y_{ii} = 0.$$

The commutativity

$$[L_{a,0}, L_{b,0}] = 0 \tag{125}$$

implies the system of constraints

$$\sum_{k=1}^{n} \frac{\partial y_{ij}}{\partial x^k} = 0$$

$$\frac{\partial y_{ij}}{\partial x^k} = y_{ik} y_{kj}, \quad \text{the indices } i, j, k \text{ are pairwise distinct.} \tag{126}$$

For $n = 2$ the second part of the constraints is empty.

In order to construct the matrix polynomials $U_{a,k}$ for $k > 0$ we will use the following procedure [11] that can be considered as a generalization of the well-known AKNS construction developed for $n = 2$ in the seminal paper [1].

Consider an arbitrary function[5] $Y(\mathbf{x})$ satisfying the system (126). We are looking for solutions to the following system of linear differential equations

$$\frac{\partial M}{\partial x^a} = [U_{a,0}, M] \quad \Leftrightarrow \quad [L_{a,0}, M] = 0, \quad a = 1, \ldots, n \tag{127}$$

for a matrix-valued function $M = M(\mathbf{x}, z)$ of the form

$$M(\mathbf{x}, z) = \sum_{k \geq 0} \frac{M_k(\mathbf{x})}{z^k}. \tag{128}$$

Compatibility of this overdetermined system of differential equations follows from (125). Observe that the coefficients of the characteristic polynomial $\det(M(\mathbf{x}, z) - w \cdot \mathbf{1})$ of the matrix $M(\mathbf{x}, z)$ are first integrals of the system (127).

Proposition A.1 *For an arbitrary solution $Y(\mathbf{x})$ to the system (125), (126) there exist unique matrix series of the form*

$$M_a(\mathbf{x}, z) = E_a + \sum_{k \geq 1} \frac{B_{a,k}(\mathbf{x})}{z^k}, \quad a = 1, \ldots, n \tag{129}$$

satisfying (127) as well as the following equations

$$M_a(\mathbf{x}, z) M_b(\mathbf{x}, z) = \delta_{ab} M_a(\mathbf{x}, z), \quad M_1(\mathbf{x}, z) + \cdots + M_n(\mathbf{x}, z) = \mathbf{1}. \tag{130}$$

Proof. We begin with the recursion procedure for computing the coefficients of the expansion (128). Clearly M_0 must be a constant diagonal matrix. Other coefficients can be determined by the following procedure.

---

[5] Here and below saying "functions" we have in mind just formal power series in the independent variables.

Lemma A.2 *For an arbitrary solution $Y = Y(\mathbf{x})$ to eqs. (126) and an arbitrary diagonal matrix $B = \mathrm{diag}(b_1, \ldots, b_n)$ with pairwise distinct diagonal entries there exists a unique solution*

$$M = M_B(\mathbf{x}, z) = B + \sum_{k \geq 1} \frac{M_{B,k}(\mathbf{x})}{z^k} \tag{131}$$

to the system (127) normalized by the condition

$$\det(M_B(\mathbf{x}, z) - w \cdot \mathbf{1}) = \det(B - w \cdot \mathbf{1}). \tag{132}$$

Proof. Split every coefficient into its diagonal and off-diagonal part

$$M_{B,k} = D_k + C_k, \quad k \geq 1.$$

Vanishing of the constant term in (127) implies

$$[E_a, M_{B,1}] = [B, [E_a, Y]] = [E_a, [B, Y]], \quad a = 1, \ldots, n.$$

This system uniquely determines the off-diagonal part of the matrix $M_{B,1}$

$$C_1 = [B, Y].$$

To determine the diagonal part D_1 we use the coefficient of $1/z$ of eq. (132). The off-diagonal part C_1 does not contribute to this coefficient, so we obtain

$$\sum_{m=1}^{n} D_{1mm} \prod_{s \neq m} (b_s - w) = 0 \quad \Rightarrow \quad D_1 = 0.$$

We proceed by induction. Assume that the matrices $D_1, \ldots, D_{k-1}, C_1, \ldots, C_{k-1}$ are already computed so that equations (127), (132) hold true modulo $\mathcal{O}(1/z^{k-1})$ and $\mathcal{O}(1/z^k)$ respectively. From the coefficient of $1/z^{k-1}$ in (127) we have

$$[E_a, C_k] = \frac{\partial C_{k-1}}{\partial x^a} + \big[[E_a, Y], C_{k-1}\big]_{\text{off-diag}} - \big[E_a, [D_{k-1}, Y]\big].$$

From this equation we can compute for any $i \neq a$ the (a, i)- and (i, a)-entries of the matrix C_k. Since a is an arbitrary number between 1 and n we obtain the full off-diagonal matrix C_k. Equating to zero the coefficient of $1/z^k$ in (132) we obtain the diagonal matrix D_k

$$D_{kmm} = - \prod_{s \neq m} (b_s - b_m)^{-1}$$

$$\times \text{ coefficient of} \frac{1}{z^k} \text{ in } \det \left(B + \sum_{i \leq k-1} \frac{D_i + C_i}{z^i} - b_m \cdot \mathbf{1} \right).$$

$\square$

We will now prove that there exists a matrix-valued series

$$A(\mathbf{x}, z) = 1 + \sum_{k \geq 1} \frac{A_k(\mathbf{x})}{z^k} \tag{133}$$

such that

$$A^{-1}(\mathbf{x}, z) M_B(\mathbf{x}, z) A(\mathbf{x}, z) = B \tag{134}$$

for any diagonal matrix B.

Lemma A.3 *For any $Y(\mathbf{x})$ satisfying eqs. (125) there exists a solution*

$$\Psi(\mathbf{x}, z) = A(\mathbf{x}, z) e^{z \, \mathrm{diag}(x^1, \dots, x^n)} \tag{135}$$

to the following system of linear differential equations

$$\frac{\partial \Psi}{\partial x^a} = U_{a,0}(\mathbf{x}, z) \Psi(\mathbf{x}, z), \quad a = 1, \dots, n \tag{136}$$

where the matrix series $A(\mathbf{x}, z)$ has the form (133).

Proof. For the coefficients $A_k = A_k(\mathbf{x})$ we obtain

$$[E_a, A_k] = \frac{\partial A_{k-1}}{\partial x^a} + [E_a, Y] A_{k-1}, \quad a = 1, \dots, n.$$

From this system we uniquely determine the off-diagonal part of the matrix A_k. Using the next equation $k \mapsto k + 1$ we arrive at

$$\frac{\partial}{\partial x^a} (A_k)_{\mathrm{diag}} = -([E_a, Y] A_k)_{\mathrm{diag}}.$$

The off-diagonal part of A_k does not contribute to the right hand side. So the matrix A_k is determined uniquely up to adding a constant diagonal matrix. $\square$

Remark A.4 From the proof it follows that the matrix $\Psi(\mathbf{x}, z)$ is determined by eqs. (136) uniquely up to a multiplication on the right by a diagonal matrix series in $1/z$

$$\Psi(\mathbf{x}, z) \mapsto \Psi(\mathbf{x}, z) \Delta(z), \quad \Delta(z) = 1 + \sum_{k=0}^{\infty} \frac{\Delta^k}{z^{k+1}}, \quad \Delta^k = \mathrm{diag}\left(\Delta_1^k, \dots, \Delta_n^k\right). \tag{137}$$

Lemma A.5 *For any diagonal matrix B the solution $M_B(\mathbf{x}, z)$ to the equations (127), (132) can be represented in the form*

$$M_B(\mathbf{x}, z) = A(\mathbf{x}, z) B \, A^{-1}(\mathbf{x}, z) \tag{138}$$

where the matrix $A(\mathbf{x}, z)$ is defined in the previous Lemma.

Proof. Since

$$A(\mathbf{x}, z)B A^{-1}(\mathbf{x}, z) = \Psi(\mathbf{x}, z)B \Psi^{-1}(\mathbf{x}, z)$$

the matrix (138) satisfies eqs. (127). Obviously it also satisfies (132). Due to uniqueness of such a solution to (127), (132) the Lemma is proved. □

We are now in a position to complete the proof of Proposition A.1. Due to uniqueness the matrix $M_B(\mathbf{x}, z)$ depends linearly on $B = \text{diag}(b_1, \ldots, b_n)$. So the construction can be extended to an arbitrary diagonal matrix B (see also eq. (138)). Put

$$M_a(\mathbf{x}, z) = M_{E_a}(\mathbf{x}, z), \quad a = 1, \ldots, n. \tag{139}$$

These matrices clearly satisfy eqs. (127) and (130). It remains to prove uniqueness.

Since the matrices $M_1(\mathbf{x}, z), \ldots, M_n(\mathbf{x}, z)$ commute pairwise due to (130) and $M_a \to E_a$ for $z \to \infty$, we can look for their common eigenvectors in $\mathbb{C}^n \otimes \mathbb{C}\left[[z^{-1}]\right]$. Every matrix $M_a = M_a(\mathbf{x}, z)$ has only one nonzero eigenvalue; the corresponding eigenvector

$$M_a \mathbf{f}_a = \mathbf{f}_a$$

can be normalized in such a way that $(\mathbf{f}_a)_b = \delta_{ab} + \mathcal{O}(1/z)$. It is determined uniquely up to multiplication

$$\mathbf{f}_a \mapsto c_a(z)\mathbf{f}_a, \quad c_a(z) = 1 + \mathcal{O}\left(\frac{1}{z}\right) \in \mathbb{C}\left[z^{-1}\right].$$

Denote $A(\mathbf{x}, z)$ the matrix whose columns are the eigenvectors $\mathbf{f}_1, \ldots, \mathbf{f}_n$. According to the previous arguments this matrix is uniquely defined up to a multiplication on the right by $\text{diag}(c_1(z), \ldots, c_n(z))$ and satisfies

$$M_a(\mathbf{x}, z) = A(\mathbf{x}, z)E_a A^{-1}(\mathbf{x}, z), \quad a = 1, \ldots, n.$$

The Proposition is proved. □

We will now slightly modify the setting of Proposition A.1 in order to apply it to the construction of the n-wave hierarchy. Denote $\mathcal{Y}$ the ring of polynomials in variables y_{ij}, $\partial y_{ij}/\partial x^k$, $\partial^2 y_{ij}/\partial x^k \partial x^l$ etc. satisfying the constraints (126) along with their differential consequences. Elements of this ring will be denoted like $P(\mathbf{y})$ where P is a polynomial. The commuting derivations $\partial/\partial x^1, \ldots, \partial/\partial x^n$ naturally act on this ring.

So, consider equations of the form (127)

$$\frac{\partial M}{\partial x^a} = [zE_a - [E_a, Y], M], \quad a = 1, \ldots, n \tag{140}$$

as equations for matrices

$$M = M(\mathbf{y}, z) = M_0 + \sum_{k \geq 1} \frac{M_k(\mathbf{y})}{z^k}.$$

Proposition A.6 *There exists a unique collection of matrix series*

$$M_a(\mathbf{y}, z) = E_a + \sum_{k \geq 1} \frac{B_{a,k}(\mathbf{y})}{z^k} \in Mat_n(\mathcal{Y}) \otimes \mathbb{C}\left[z^{-1}\right], \quad a = 1, \ldots, n$$

$$(141)$$

satisfying (140) and also

$$M_a(\mathbf{y}, z) M_b(\mathbf{y}, z) = \delta_{ab} M_a(\mathbf{y}, z), \quad M_1(\mathbf{y}, z) + \cdots + M_n(\mathbf{y}, z) = \mathbf{1}. \quad (142)$$

The proof essentially repeats the above arguments so it will be omitted.

Remark A.7   In practical computations of the coefficients $B_{a,k}(\mathbf{y})$ instead of the normalization (132) one can alternatively use the following one:

$$B_{a,k}(0) = 0, \quad a = 1, \ldots, n, \quad k \geq 1.$$

Explicitly

$$M_a = E_a + \frac{B_{a,1}}{z} + \frac{B_{a,2}}{z^2} + \frac{B_{a,3}}{z^3} + \mathcal{O}\left(\frac{1}{z^4}\right)$$

$$B_{a,1} = -[E_a, Y]$$

$$(B_{a,2})_{ij} = \begin{cases} -\frac{\partial y_{ij}}{\partial x^a}, & i \neq j \\ -y_{ia} y_{ai}, & j = i \neq a \\ \sum_s y_{as} y_{sa}, & i = j = a \end{cases}$$

$$(B_{a,3})_{ij} = \begin{cases} \frac{\partial y_{ia}}{\partial x^a} y_{ai} - y_{ia} \frac{\partial y_{ai}}{\partial x^a}, & i \neq a, \ j \neq a \\ -\frac{\partial^2 y_{aj}}{\partial x^{a2}} - 2 y_{aj} \sum_s y_{as} y_{sa}, & i = a, \ j \neq a \\ \frac{\partial^2 y_{ia}}{\partial x^{a2}} + 2 y_{ia} \sum_s y_{as} y_{sa}, & i \neq a, \ j = a \\ \sum_s y_{sa} \frac{\partial y_{as}}{\partial x^a} - \frac{\partial y_{sa}}{\partial x^a} y_{as}, & i = j = a \end{cases}.$$

Define matrix-valued polynomials

$$U_{a,k}(\mathbf{y}, z) = \left(z^{k+1} M_a(\mathbf{y}, z)\right)_+ \in Mat_n(\mathcal{Y}) \otimes \mathbb{C}[z],$$

$$a = 1, \ldots, n, \quad k \geq -1. \quad (143)$$

Here and below the notation $(\ )_+$ will be used for the polynomial part of a Laurent series in $1/z$. The matrix-valued polynomials $U_{a,k}$ are exactly those that appear in the formulation (123) of equations of the n-wave hierarchy that can be rewritten in the following form

$$\frac{\partial Y}{\partial t_k^a} = \left(B_{a,k+2}(\mathbf{y})\right)_{\text{off-diagonal}} . \tag{144}$$

For $n = 2$ it coincides with the complexified nonlinear Schrödinger hierarchy, also known as the AKNS hierarchy. Observe that the t_{-1}^a-flows generate just conjugations by diagonal matrices

$$y_{ij} \mapsto \frac{\lambda_i}{\lambda_j} y_{ij}, \quad i, j = 1, \dots, n. \tag{145}$$

Such transformations are symmetries of the n-wave hierarchy (144).

Remark A.8 The dependence of the functions $y_{ij}(\mathbf{x})$ is uniquely determined by their restriction onto any line

$$x^i = a_i x, \quad i = 1, \dots, n$$

for arbitrary pairwise distinct constants $a_1, \dots, a_n$. Indeed, we reconstruct all partial derivatives in $x^1, \dots, x^n$

$$\frac{\partial y_{ij}}{\partial x^k} = \begin{cases} y_{ik} y_{kj}, & k \neq i, j \\ \frac{y'_{ij}}{a_i - a_j} + \sum_s \frac{a_j - a_s}{a_i - a_j} y_{is} y_{sj}, & k = i \\ \frac{y'_{ij}}{a_j - a_i} + \sum_s \frac{a_i - a_s}{a_j - a_i} y_{is} y_{sj}, & k = j \end{cases}$$

starting from the derivatives $y'_{ij} = dy_{ij}/dx$ in x. So for every pair of indices (a, k) the equation (144) can be considered as a system of $n(n-1)$ partial differential equations with one space variable x and one time variable t_k^a.

Let $Y(\mathbf{t})$ be a solution to the n-wave hierarchy. Then the matrices $M_a(\mathbf{y}, z)$ become well-defined functions $M_a(\mathbf{t}, z)$ of $\mathbf{t}$. So do the matrix-valued polynomials $U_{a,p} = U_{a,p}(\mathbf{t}, z)$.

Proposition A.9 *The matrix-valued series* $M_b = M_b(\mathbf{t}, z)$ *satisfy*

$$\left[L_{a,k}, M_b\right] = 0 \iff \frac{\partial M_b(\mathbf{t}, z)}{\partial t_k^a} = \left[U_{a,k}(\mathbf{t}, z), M_b(\mathbf{t}, z)\right]$$

$$\forall\, a, b = 1, \dots, n, \quad k \geq -1. \tag{146}$$

Proof. It suffices to verify validity of eq. (146) for the series $M_b = M_b(\mathbf{y}, z)$ and polynomials $U_{a,k}(\mathbf{y}, z)$. Let

$$\tilde{M}_b = \frac{\partial M_b}{\partial t_k^a} - [U_{a,k}, M_b].$$

It is easy to check that this matrix-valued Laurent series satisfies

$$\frac{\partial \tilde{M}_b}{\partial x^c} = \left[U_{c,0}, \tilde{M}_b \right]$$

for any $c = 1, \ldots, n$. Let us now check that the expansion of $\tilde{M}_b$ contains only strictly negative powers of z. To this end define the matrix series

$$V_{a,k}(\mathbf{y}, z) = \left(z^{k+1} M_a(\mathbf{y}, z) \right)_-$$

so that

$$z^{k+1} M_a(\mathbf{y}, z) = U_{a,k}(\mathbf{y}, z) + V_{a,k}(\mathbf{y}, z).$$

Therefore

$$\tilde{M}_b = \frac{\partial M_b}{\partial t_k^a} + [V_{a,k}, M_b] \in Mat_n(\mathcal{Y}) \otimes z^{-1} \mathbb{C} \left[z^{-1} \right].$$

So, the series $\tilde{M}_b = \tilde{M}_b(\mathbf{y}, z)$ satisfies eqs. (140) and contains only strictly negative powers of z. Due to uniqueness it is equal to zero. $\qquad\square$

Lemma A.10 *The matrices* $M_a(\mathbf{t}, z)$ *satisfy the identities (142).*

Proof. Using eq. (146) we prove that

$$\frac{\partial}{\partial t_k^c} (M_a M_b - \delta_{ab} M_a) = 0.$$

$\qquad\square$

Definition A.11 *A wave function* $\Psi = \Psi(\mathbf{t}, z)$ *of the solution* $Y(\mathbf{t})$ *is a solution to the infinite family of systems of linear differential equations*

$$\frac{\partial}{\partial t_p^a} \Psi = U_{a,p}(\mathbf{t}, z)\Psi, \quad a = 1, \ldots, n, \quad p \geq -1 \qquad (147)$$

of the form

$$\Psi(\mathbf{t}, z) = A(\mathbf{t}, z)e^{\phi(\mathbf{t}, z)}$$

$$A(\mathbf{t}, z) = 1 + \frac{A^0(\mathbf{t})}{z} + \frac{A^1(\mathbf{t})}{z^2} + \ldots,$$

$$\phi(\mathbf{t}, z) = \sum_{k=0}^{\infty} \operatorname{diag}\left(t_k^1, \ldots, t_k^n \right) z^{k+1} \qquad (148)$$

where $A^0(\mathbf{t}), A^1(\mathbf{t})$ *etc. are* $n \times n$ *matrix-valued functions of* $\mathbf{t}$.

The wave function is determined by a solution $Y(\mathbf{t})$ uniquely up to multiplication on the right by a constant diagonal matrix-valued series

$$\Psi(\mathbf{t}, z) \mapsto \Psi(\mathbf{t}, z)\Delta(z), \quad \Delta(z) = 1 + \sum_{k=0}^{\infty} \frac{\Delta^k}{z^{k+1}}, \quad \Delta^k = \mathrm{diag}\left(\Delta_1^k, \ldots, \Delta_n^k\right)$$
(149)

Lemma A.12 *Let* $(Y(\mathbf{t}), \Psi(\mathbf{t}, z))$ *be a solution to the equations of the hierarchy (144) and its wave function. Then the matrix-valued series* $M_1(\mathbf{t}, z)$, *..., $M_n(\mathbf{t}, z)$ can be represented in the form*

$$M_a(\mathbf{t}, z) = \Psi(\mathbf{t}, z)E_a\Psi^{-1}(\mathbf{t}, z), \quad a = 1, \ldots, n.$$
(150)

Proof. As

$$\Psi(\mathbf{t}, z)E_a\Psi^{-1}(\mathbf{t}, z) = A(\mathbf{t}, z)E_a A^{-1}(\mathbf{t}, z) = E_a + \mathcal{O}\left(\frac{1}{z}\right),$$

the right hand side of (150) is a series in inverse powers of z. It satisfies the differential equations (146). Due to uniqueness it coincides with $M_a(\mathbf{t}, z)$. □

Introduce the following generating series for the time derivatives

$$\nabla_a(z) = \sum_{k \geq -1} \frac{1}{z^{k+2}} \frac{\partial}{\partial t_k^a}, \quad a = 1, \ldots, n.$$
(151)

Lemma A.13 *The following formula holds true*

$$\nabla_a(w)\Psi(\mathbf{t}, z) = \frac{M_a(\mathbf{t}, w)\Psi(\mathbf{t}, z)}{w - z}.$$
(152)

The *proof* is straightforward by using (143) and (147). □

We now proceed to the definition of tau-function. It is based on the following statement (cf. [2]).

Proposition A.14 *For any solution* $Y(t)$ *to the system (147) and its wave function (148) there exists a function* $\log \tau(\mathbf{t})$ *such that*

$$\frac{\partial \log \tau(\mathbf{t})}{\partial t_p^a} = -\operatorname*{res}_{z=\infty} \operatorname{tr}\left(A_z(\mathbf{t}, z)E_a A^{-1}(\mathbf{t}, z)\right) z^{p+1} dz.$$
(153)

Proof. We need to prove symmetry of the second derivatives

$$\frac{\partial^2 \log \tau(\mathbf{t})}{\partial t_p^a \partial t_q^b} = \frac{\partial^2 \log \tau(\mathbf{t})}{\partial t_q^b \partial t_p^a}$$

or, equivalently

$$\nabla_a(z_1)\nabla_b(z_2) \log \tau(\mathbf{t}) = \nabla_b(z_2)\nabla_a(z_1) \log \tau(\mathbf{t}).$$

Using eq. (152) one can represent the generating series for the logarithmic derivatives of $\tau(\mathbf{t})$ in the following form

$$\nabla_a(z) \log \tau(\mathbf{t}) = \mathrm{tr}\left(A_z(\mathbf{t}, z) E_a A^{-1}(\mathbf{t}, z)\right)$$
$$= \mathrm{tr}\left(\Psi_z(\mathbf{t}, z) E_a \Psi^{-1}(\mathbf{t}, z)\right) - \phi_z^a(\mathbf{t}, z) \qquad (154)$$

where $\phi(\mathbf{t}, z) = \mathrm{diag}\left(\phi^1(\mathbf{t}, z), \ldots, \phi^n(\mathbf{t}, z)\right)$ (see eq. (148) above).

Lemma A.15 *The second order logarithmic derivatives of the tau-function (153), (154) can be computed from the following generating series*

$$\nabla_a(z_1)\nabla_b(z_2) \log \tau(\mathbf{t}) = \frac{\mathrm{tr}\, M_a(\mathbf{t}, z_1) M_b(\mathbf{t}, z_2) - \delta_{ab}}{(z_1 - z_2)^2}. \qquad (155)$$

Before we proceed to the proof let us observe that, using $M_a(z)M_b(z) = \delta_{ab} M_a(z)$ (see eq. (142) above) it readily follows that the numerator in (155) vanishes at $z_1 = z_2$. Hence, due to its symmetry in z_1, z_2 it is divisible by $(z_1 - z_2)^2$. Thus the right hand side is a series in inverse powers of z_1, z_2.

Proof. Using the second part of eq. (154) we obtain

$$\sum_{p,q=0}^{\infty} \frac{1}{w^{q+2}} \frac{1}{z^{p+2}} \frac{\partial^2 \log \tau(\mathbf{t})}{\partial t_q^b \partial t_p^a}$$
$$= \nabla_b(w) \,\mathrm{tr}\left(\Psi_z(\mathbf{t}, z) E_a \Psi^{-1}(\mathbf{t}, z)\right) - \nabla_b(w)\phi_z^a(\mathbf{t}, z).$$

Obviously

$$\nabla_b(w)\phi^a(\mathbf{t}, z) = \frac{\delta_{ab}}{w - z} \quad \Rightarrow \quad \nabla_b(w)\phi_z^a(\mathbf{t}, z) = \frac{\delta_{ab}}{(w - z)^2}.$$

From (152) obtain

$$\nabla_b(w)\Psi_z(\mathbf{t}, z) = \frac{M_b(\mathbf{t}, w)\Psi_z(\mathbf{t}, z)}{w - z} + \frac{M_b(\mathbf{t}, w)\Psi(\mathbf{t}, z)}{(w - z)^2}$$

and

$$\nabla_b(w)\Psi^{-1}(\mathbf{t}, z) = -\frac{\Psi^{-1}(\mathbf{t}, z) M_b(\mathbf{t}, w)}{w - z}.$$

Therefore

$$\nabla_b(w) \,\mathrm{tr}\left(\Psi_z(\mathbf{t}, z) E_a \Psi^{-1}(\mathbf{t}, z)\right) = \mathrm{tr}\frac{M_b(\mathbf{t}, w)\Psi(\mathbf{t}, z) E_a \Psi^{-1}(\mathbf{t}, z)}{(w - z)^2}$$
$$= \mathrm{tr}\frac{M_b(\mathbf{t}, w) M_a(\mathbf{t}, z)}{(z - w)^2}.$$

Summarizing we arrive at

$$\sum_{p,q=0}^{\infty} \frac{1}{w^{q+2}} \frac{1}{z^{p+2}} \frac{\partial^2 \log \tau(\mathbf{t})}{\partial t_q^b \partial t_p^a} = \mathrm{tr}\frac{M_b(\mathbf{t}, w) M_a(\mathbf{t}, z)}{(z-w)^2} - \frac{\delta_{ab}}{(w-z)^2}$$

that completes the proof of Lemma and, therefore of the Proposition. $\quad\square$

Remark A.16 The definition (153), (154) does depend on the normalization of the wave function. A change of the normalization

$$\Psi(\mathbf{t}, z) \mapsto \Psi(\mathbf{t}, z)\Delta(z), \quad \Delta(z) = \mathrm{diag}\left(\Delta_1(z), \ldots, \Delta_n(z)\right),$$

$$\Delta_a(z) \in \mathbb{C}[[z^{-1}]], \quad a = 1, \ldots, n$$

yields

$$\mathrm{tr}\left(\Psi_z(\mathbf{t}, z) E_a \Psi^{-1}(\mathbf{t}, z)\right) \mapsto \mathrm{tr}\left(\Psi_z(\mathbf{t}, z) E_a \Psi^{-1}(\mathbf{t}, z)\right) + \frac{d}{dz}\log \Delta_a(z).$$

The tau-function will change as follows

$$\tau(\mathbf{t}) \mapsto e^{\sum c_{a,p} t_p^a} \tau(\mathbf{t}), \quad \frac{d}{dz}\log \Delta_a(z) = \sum \frac{c_{a,p}}{z^{p+2}}. \tag{156}$$

We see that the logarithmic derivatives of the tau-function of order two (and, therefore, of any higher order) belong to the ring $\mathcal{Y}$. In particular they do not depend on the choice of a wave function of a solution $Y(\mathbf{t})$ of the hierarchy (144). Explicitly,

$$\frac{\partial^2 \log \tau}{\partial t_0^a \partial t_0^b} = \begin{cases} -y_{ab}y_{ba}, & b \neq a \\ \sum_s y_{as}y_{sa}, & b = a \end{cases} \tag{157}$$

$$\frac{\partial^2 \log \tau}{\partial t_0^a \partial t_1^b} = \begin{cases} \frac{\partial y_{ab}}{\partial x^b}y_{ba} - y_{ab}\frac{\partial y_{ba}}{\partial x^b}, & b \neq a \\ \sum_s y_{as}\frac{\partial y_{sa}}{\partial x^s} - \frac{\partial y_{as}}{\partial x^s}y_{sa}, & b = a \end{cases} \tag{158}$$

$$\frac{\partial^2 \log \tau}{\partial t_0^a \partial t_2^b} = \begin{cases} -y_{ab}\frac{\partial^2 y_{ba}}{\partial x^{b2}} - y_{ba}\frac{\partial^2 y_{ab}}{\partial x^{b2}} + \frac{\partial y_{ab}}{\partial x^b}\frac{\partial y_{ba}}{\partial x^b} - 3y_{ab}y_{ba}\sum_s y_{bs}y_{sb}, & b \neq a \\ -\sum_{s \neq a}\frac{\partial^2 \log \tau}{\partial t_0^s \partial t_2^a}, & b = a \end{cases}$$

$$\tag{159}$$

etc.

For computation of the derivatives of order three and higher we will need the following.

Lemma A.17 *The following equations hold true for all $a, b = 1, \ldots, n$*

$$\nabla_a(z_1) M_b(z_2) = \frac{[M_a(z_1), M_b(z_2)]}{z_1 - z_2}. \tag{160}$$

Here and below we omit the explicit dependence on **t** of the matrix-valued functions $M_a(\mathbf{t}, z)$.

Proof. It easily follows from (146). ◻

Proposition A.18 *The following equation holds true*

$$\nabla_a(z_1)\nabla_b(z_2)\nabla_c(z_3) \log \tau(\mathbf{t}) = -\operatorname{tr} \frac{[M_a(z_1), M_b(z_2)]M_c(z_3)}{(z_1 - z_2)(z_2 - z_3)(z_3 - z_1)}. \quad (161)$$

Proof. It can be easily obtained by applying the operator $\nabla_c(z_3)$ at both sides of eq. (155) with the help of (160) and then using invariance of the trace of product of matrices with respect to cyclic permutations. ◻

Higher order logarithmic derivatives of the tau-function can be computed using the following.

Proposition A.19 *For the logarithmic derivatives of order $N \geq 3$ of the tau-function of any solution $Y(\mathbf{t})$ to the n-wave hierarchy (144) the following expression holds true*

$$\sum_{k_1,\ldots,k_N \geq 0} \frac{\partial^N \log \tau(\mathbf{t})}{\partial t_{k_1}^{a_1} \cdots \partial t_{k_N}^{a_N}} \frac{1}{z_1^{k_1+2} \cdots z_N^{k_N+2}}$$

$$= -\frac{1}{N} \sum_{s \in S_N} \frac{\operatorname{tr}\left[M_{a_{s_1}}(z_{s_1}) \cdots M_{a_{s_N}}(z_{s_N})\right]}{(z_{s_1} - z_{s_2}) \cdots (z_{s_{N-1}} - z_{s_N})(z_{s_N} - z_{s_1})}. \quad (162)$$

Proof. For $N = 3$ eq. (162) coincides with (161). For higher N the proof is obtained by induction using (160). It does not differ from the proof of a similar equation given in [4], so we omit the details. ◻

Appendix B. Another Definition of the Principal Tau-Function

Proposition B.1 *[19] For a given pair $(Y(\mathbf{t}), \Psi(\mathbf{t}, z))$ consisting of a solution to the hierarchy (144) and its wave function there exists a function $\tau(\mathbf{t})$ such that*

$$\nabla_a(z) \log \tau(\mathbf{t}) = \left(\frac{\partial}{\partial z} - \nabla_a(z)\right) \log\left[A(\mathbf{t}, z)_{aa}\right]. \quad (163)$$

It is easy to see that the diagonal entries of the matrix $A(\mathbf{t}, z)$ do not depend on the variables t_{-1}^a. So, according to this definition $\frac{\partial \log \tau(\mathbf{t})}{\partial t_{-1}^a} \equiv 0$.

For example,

$$\frac{\partial \log \tau}{\partial t_0^a} = -A_{aa}^0$$

$$\frac{\partial \log \tau}{\partial t_1^a} = -2A_{aa}^1 + \left(A_{aa}^1\right)^2 - \frac{\partial A_{aa}^0}{\partial t_0^a}.$$

Definition B.2 *The function* $\tau(\mathbf{t})$ *will be called* the principal tau-function *of the pair* $(Y(\mathbf{t}), \Psi(\mathbf{t}, z))$.

Clearly the principal tau-function of a given pair (Y, Ψ) is determined uniquely up to a nonzero constant factor.

Remark B.3 There are other tau-functions in the theory of the n-wave hierarchy described in [19]. The principal one is selected by the following property: it is invariant with respect to diagonal conjugations

$$Y(\mathbf{t}) \mapsto \Lambda\, Y(\mathbf{t})\, \Lambda^{-1}, \quad \Psi(\mathbf{t}, z) \mapsto \Lambda\, \Psi(\mathbf{t}, z)\, \Lambda^{-1}, \quad \Lambda = \mathrm{diag}(\lambda_1, \dots, \lambda_n),$$
$$\tau(\mathbf{t}) \mapsto \tau(\mathbf{t}).$$

In other words, it does not depend on the time variables t_{-1}^a. Thus its logarithmic derivatives, starting from the second one, are combinations of the functions $y_{ij}(\mathbf{t})$ and their derivatives invariant with respect to the diagonal conjugations (145).

References

[1] M. Ablowitz, D. Kaup, A. Newell, H. Segur, The inverse scattering transform–Fourier analysis for nonlinear problems. *Studies in Appl. Math.* **53** (1974), no. 4, 249–315.

[2] M. Bertola, B. Dubrovin, D. Yang, Correlation functions of the KdV hierarchy and applications to intersection numbers over $\overline{\mathcal{M}}_{g,n}$. *Physica D: Nonlinear Phenomena* **327** (2016) 30–57.

[3] M. Bertola, B. Dubrovin, D. Yang, Simple Lie algebras and topological ODEs. *IMRN* **2018** no. 5, 1368–1410.

[4] M. Bertola, B. Dubrovin, D. Yang (2016). Simple Lie algebras, Drinfeld–Sokolov hierarchies, and multipoint correlation functions, arXiv:1610.07534.

[5] V.M. Buchtaber, D.V. Leykin, V.Z. Ènolski, Uniformization of Jacobi varieties of trigonal curves and nonlinear differential equations. *Funktsional. Anal. i Prilozhen.* **34** (2000), Issue 3, 1–16.

[6] E. Date, M. Jimbo, M. Kashiwara, T. Miwa, Transformation groups for soliton equations. *Nonlinear Integrable Systems–Classical Theory and Quantum Theory*, M. Jimbo and T. Miwa (eds.), World Scientific, Singapore, 1983, 39–119.

[7] L.A. Dickey, On Segal–Wilson's definition of the τ-function and hierarchies of AKNS-D and mcKP. *Integrable systems* (Luminy, 1991), 147–161, Progr. Math., **115**, Birkhäuser Boston, Boston, MA, 1993.

[8] L.A. Dickey, *Soliton equations and Hamiltonian systems.* Second edition. Advanced Series in Mathematical Physics, **26**. World Scientific Publishing Co., Inc., River Edge, NJ, 2003.

[9] P. Diener, B. Dubrovin, Algebraic-geometrical Darboux coordinates in R-matrix formalism. Preprint SISSA 88/94 FM, 1994.

[10] V.G. Drinfeld, V.V. Sokolov, Lie algebras and equations of Korteweg–de Vries type. (Russian) *Current problems in mathematics* **24** (1984) 81–180. *Journal of Soviet Mathematics* **30:2** (1985) 197–2036.

[11] B. Dubrovin, Completely integrable Hamiltonian systems associated with matrix operators, and Abelian varieties. (Russian) *Funkcional. Anal. i Prilozen.* **11** (1977), no. 4, 28–41, 96.

[12] B. Dubrovin, Theta-functions and nonlinear equations. *Russ. Math. Surveys* **36:2** (1981) 11–92.

[13] B. Dubrovin, Matrix finite-gap operators. In: *Sovremennye Problemy Matematiki*, VINITI **23** (1983) 77–123. English transl.: *J. Soviet Math.* **28** (1985) 20–50.

[14] B. Dubrovin, Approximating tau-functions by theta-functions, arXiv:1807.03377.

[15] B. Dubrovin, S.P. Novikov, Periodic Korteweg–de Vries and Sturm–Liouville problems. Their connection with algebraic geometry. *Sov. Math. Dokl.* **219:3** (1974).

[16] V. Enolski, J. Harnad, Schur function expansions of KP tau functions associated with algebraic curves. (Russian) *Uspekhi Mat. Nauk* **66** (2011), no. 4(400), 137–178; translation in *Russian Math. Surveys* **66** (2011), no. 4, 767–807.

[17] J. Fay. *Theta-Functions on Riemann Surfaces.* Springer Lecture Notes in Mathematics **352**, 1973.

[18] L. Gavrilov, Jacobians of singularized spectral curves and completely integrable systems. *The Kowalevski property* (Leeds, 2000), 59–68, CRM Proc. Lecture Notes, **32**, Amer. Math. Soc., Providence, RI, 2002.

[19] V.G. Kac, J.W. van de Leur, The n-component KP hierarchy and representation theory. *J. Math. Phys.* **44** (2003) 3245.

[20] N. Kawamoto, Y. Namikawa, A. Tsuchiya, Y. Yamada, Geometric realization of conformal field theory on Riemann surfaces. *Comm. Math. Phys.* **116** (1988) 247–308.

[21] I.M. Krichever, Methods of algebraic geometry in the theory of nonlinear equations. *Russ. Math. Surveys* **32** (1977) 183–208.

[22] D. Mumford, *Tata Lectures on Theta*, vols, I, II. Birkhäuser, Boston (1983).

[23] A. Nakayashiki, Tau function approach to theta functions, IMRN **2016**, Issue 17 (2016) 5202–5248.

[24] A. Nakayashiki, F. Smirnov, Cohomologies of affine Jacobi varieties and integrable systems. *Comm. Math. Phys.* **217** (2001) 623–652.

[25] S.P. Novikov, The periodic problem for the Korteweg–de Vries equation. *Funct. Anal. Appl.* **8:3** (1974) 236–246.

[26] S. Novikov, S. Manakov, L. Pitaevskij, V. Zakharov, *Theory of solitons. The inverse scattering method.* Translated from the Russian. Contemporary Soviet Mathematics. Consultants Bureau [Plenum], New York, 1984.

[27] S.P. Novikov, S.P. Veselov, Poisson brackets compatible with the algebraic geometry and the dynamics of the Korteweg–de Vries equation on the set of finite-gap potentials. (Russian) *Dokl. Akad. Nauk SSSR* **266** (1982), no. 3, 533–537.

[28] M.J. Dupré, J.F. Glazebrook, E. Previato, Differential algebras with Banach-algebra coefficients I: From C^*-algebras to the K-theory of the spectral curve. *Complex Anal. and Operator Theory* **7** (2013) 739–763.

[29] G. Segal, G. Wilson, Loop groups and equations of KdV type. *Publ. Math. IHES* **61** (1985) 5–65.

[30] F.A. Smirnov, V. Zeitlin, Affine Jacobians of spectral curves and integrable models. arXiv:math-ph/0203037.

[31] G. Wilson, The τ-function of the gAKNS equations. *Integrable systems* (Luminy, 1991), 131–145, Progr. Math., **115**, Birkhauser Boston, Boston, MA, 1993.

3

Frobenius Split Anticanonical Divisors

Sándor J. Kovács

Dedicated to Emma Previato on the occasion of her 65th birthday

Abstract. In this note I extend two theorems of Sommese regarding abelian varieties to arbitrary characteristic; that an abelian variety cannot be an ample divisor in a smooth projective variety and that a cone over an abelian variety of dimension at least two is not smoothable.

1 Introduction

The main goal of this note is to extend to arbitrary characteristic two theorems of Sommese regarding abelian varieties; that an abelian variety cannot be an ample divisor in a smooth projective variety [Som76] and that a cone over an abelian variety of dimension at least two is not smoothable [Som79]. Note that the latter statement has already been extended to arbitrary characteristic in [KK18], but the proof given here is different and arguably more direct. I give a new proof and a slightly stronger version of both results of Sommese already in the characteristic zero case. The main technical ingredient in positive characteristic is a sort of lifting theorem (for the definition of F-split see Definition 2.2):

Theorem 1.1 (cf. Corollary 3.5) *Let X be a smooth projective variety over an algebraically closed field k of* char $k > 0$ *and $D \subseteq X$ an effective anti-canonical divisor, i.e., such that $\omega_X \simeq \mathcal{O}_X(-D)$. If D is F-split then so is X.*

This in turn is used to prove the vanishing of several cohomology groups:

Theorem 1.2 (cf. Theorem 4.1) *Let X be a smooth projective variety over an algebraically closed field k of* char $k > 0$ *and $D \subseteq X$ an effective ample divisor such that $\omega_D \simeq \mathcal{O}_D$. If D is F-split and $\dim X \geq 3$, then*

(i) $H^i(X, \mathcal{O}_X) = 0$ *for* $i > 0$, *and*

(ii) $H^j(D, \mathcal{O}_D) = 0$ *for* $0 < j < \dim D$.

Remark 1.3 See Corollary 3.5 and Theorem 4.1 for stronger versions of these statements.

Finally, these vanishing results and Kawamata-Viehweg vanishing [Kaw82, Vie82] in characteristic zero are used to prove characteristic independent versions of Sommese's theorems in Corollary 4.3 and Corollary 5.4.

Acknowledgment *I would like to thank Max Lieblich for useful conversations and the referee for helpful comments.*

This work was supported in part by NSF Grant DMS-1565352 and the Craig McKibben and Sarah Merner Endowed Professorship in Mathematics at the University of Washington.

2 Frobenius Splitting and Vanishing

The following notation will be used throughout the article.

Notation 2.1 Let k be an algebraically closed field and X a scheme over k. If char $k = p > 0$ then let $F : X \to X$ denote the (absolute) Frobenius morphism. Recall that F is the identity on the underlying space X and its comorphism on the sheaf of regular functions is the pth power map: $\mathcal{O}_X \to F_*\mathcal{O}_X$ given by $f \mapsto f^p$.

Definition 2.2 *A scheme X over k of* char $k > 0$ *is called* Frobenius split *or* F-split *if the natural morphism* $\eta : \mathcal{O}_X \to F_*\mathcal{O}_X$ *has a left inverse, i.e.,* $\exists \eta' : F_*\mathcal{O}_X \to \mathcal{O}_X$ *such that* $\eta' \circ \eta = \mathrm{id}_{\mathcal{O}_X}$.

It was proved in [MR85, Prop. 2] that Kodaira vanishing [Kod53] holds on smooth projective F-split varieties. In fact, Mehta-Ramanathan's proof works in a slightly more general setting:

Theorem 2.3 (Mehta-Ramanathan) *Let X be an equidimensional projective Cohen-Macaulay scheme over k of* char $k > 0$ *and let $\mathscr{L}$ be an ample line bundle on X. If X is F-split , then*

$$H^j(X, \mathscr{L}^{-1}) = 0$$

for $j < \dim X$.

Proof. This follows directly from Serre duality [Har77, 7.6(b)] and [MR85, Prop. 1]. $\square$

The proof of the following simple lemma uses the usual trick of obtaining a more precise vanishing statement from Serre vanishing and surjective maps.

Lemma 2.4 *Let X be an equidimensional projective Cohen-Macaulay scheme over k (of arbitrary characteristic) and $D \subseteq X$ an effective, ample Cartier divisor. Fix an $m_0 \in \mathbb{Z}$ and a $j \in \mathbb{N}$ such that $j < \dim X$ and let $\mathscr{E}$ be a locally free sheaf on X. Assume that $H^j(D, (\mathscr{E}(-mD))|_D) = 0$ for each $m \geq m_0$. Then $H^j(X, \mathscr{E}(-mD)) = 0$ for each $m \geq m_0$.*

Proof. Consider the following short exact sequence:

$$0 \longrightarrow \mathscr{E}(-(m+1)D) \longrightarrow \mathscr{E}(-mD) \longrightarrow (\mathscr{E}(-mD))|_D \longrightarrow 0.$$

It follows from the assumption that the induced morphism

$$H^j(X, \mathscr{E}(-(m+1)D)) \longrightarrow H^j(X, \mathscr{E}(-mD))$$

is surjective for each $m \geq m_0$. By iterating this step we obtain that the induced morphism

$$H^j(X, \mathscr{E}(-(m+l)D)) \longrightarrow H^j(X, \mathscr{E}(-mD))$$

is surjective for any $l \in \mathbb{N}$. However, $H^j(X, \mathscr{E}(-(m+l)D)) = 0$ for $l \gg 0$ by Serre duality [Har77, 7.6(b)] which implies the desired statement. □

Corollary 2.5 *Let X be an equidimensional projective Gorenstein scheme over k and $D \subseteq X$ an effective, ample Cartier divisor such that $\omega_D \simeq \mathscr{O}_D$. Assume that if $\operatorname{char} k > 0$ then D is F-split and if $\operatorname{char} k = 0$ then X has rational singularities. Then*

$$H^j(X, \omega_X(-mD)) = 0$$

for $j < \dim X - 1$ and each $m \in \mathbb{N}$.

Proof. First note that by the adjunction formula the assumption implies that $\omega_X(D)|_D \simeq \mathscr{O}_D$. Fix a $j < \dim X - 1 = \dim D$. Then

$$H^j(D, \omega_X(-mD)|_D) \simeq H^j(D, (\omega_X(D) \otimes \mathscr{O}_X(-(m+1)D))|_D) \simeq$$

$$\simeq H^j(D, \mathscr{O}_D(-(m+1)D|_D)) = 0$$

for each $m \in \mathbb{N}$ by Kawamata-Viehweg vanishing [Kaw82, Vie82] in case $\operatorname{char} k = 0$ and by Theorem 2.3 if $\operatorname{char} k > 0$. Hence the statement follows from Lemma 2.4 by taking $\mathscr{E} = \omega_X$ and $m_0 = 0$. □

Theorem 2.6 *Let X be an equidimensional projective Gorenstein scheme over k and $D \subseteq X$ an effective, ample Cartier divisor such that $\omega_D \simeq \mathscr{O}_D$. Assume that $\dim X \geq 3$ and if $\operatorname{char} k > 0$ then assume further that D is F-split. Then $\omega_X \simeq \mathscr{O}_X(-D)$.*

Proof. Consider the following short exact sequence:

$$0 \longrightarrow \omega_X \longrightarrow \omega_X(D) \longrightarrow \omega_D \simeq \mathcal{O}_D \longrightarrow 0.$$

Observe that $H^j(X, \omega_X) = 0$ for $j < \dim X - 1$ by Corollary 2.5. In particular this holds for $j = 1$, i.e., $H^1(X, \omega_X) = 0$, and so the induced morphism

$$H^0(X, \omega_X(D)) \longrightarrow H^0(D, \omega_D) \simeq H^0(D, \mathcal{O}_D)$$

is surjective. It follows that there exists a section $0 \neq s \in H^0(X, \omega_X(D))$ such that $(s = 0) \cap D = \emptyset$. Because D is ample, this implies that $(s = 0) = \emptyset$ and hence that $\omega_X(D) \simeq \mathcal{O}_X$. This proves the statement. $\square$

3 Lifting Frobenius Splittings

The following is a simple criterion for Frobenius splitting, probably well-known to experts. A proof is included for the convenience of the reader.

Proposition 3.1 *Let X be a projective Cohen-Macaulay scheme of equidimension n over k of char $k > 0$. Then X is F-split if and only if there exists a morphism $\sigma : \omega_X \to F_*\omega_X$ such that the induced morphism*

$$H^n(\sigma) : H^n(X, \omega_X) \xrightarrow{\neq 0} H^n(X, F_*\omega_X)$$

is non-zero. Any morphism σ satisfying the above criterion will be a called a dual splitting morphism of X.

Remark 3.2 Note that an important feature of this criterion is that there is no assumption of functoriality or any other constraints on σ, only that $H^n(\sigma) \neq 0$.

Proof. If X is F-split, then letting σ be the Grothendieck dual of the splitting morphism $\eta' : F_*\mathcal{O}_X \to \mathcal{O}_X$ given by the definition shows the "only if" part of the claim.

To show the other direction, first notice that since both $H^n(X, \omega_X)$ and $H^n(X, F_*\omega_X)$ are 1-dimensional, the assumption is equivalent to saying that $H^n(\sigma)$ is an isomorphism.

Next observe that by Serre duality the morphism

$$H^n(\tau) : H^n(X, F_*\omega_X) \to H^n(X, \omega_X)$$

induced by the Grothendieck trace map $\tau : F_*\omega_X \to \omega_X$ is also non-zero, and again, since both $H^n(X, \omega_X)$ and $H^n(X, F_*\omega_X)$ are 1-dimensional, it is an isomorphism.

It follows that the composition $\tau \circ \sigma : \omega_X \to \omega_X$, which factors through $F_*\omega_X$, induces an isomorphism on $H^n(X, \omega_X)$. In particular, $\tau \circ \sigma \neq 0$. Now let $\eta : \mathcal{O}_X \to F_*\mathcal{O}_X$ be the comorphism of the Frobenius, which is of course the Grothendieck dual of τ and let $\eta' : F_*\mathcal{O}_X \to \mathcal{O}_X$ be the Grothendieck dual of σ. Then we see that $\eta' \circ \eta : \mathcal{O}_X \to \mathcal{O}_X$ cannot be zero, since otherwise so would be its Grothendieck dual, $\tau \circ \sigma$. However, if $\eta' \circ \eta \neq 0$, then it must be an isomorphism. Replacing η' with itself composed with the inverse of this isomorphism we obtain that X is F-split . $\qquad\square$

Definition 3.3 *Let X be an equidimensional projective Cohen-Macaulay scheme over k of* char $k > 0$ *and $D \subseteq X$ a non-empty effective Cartier divisor. Assume that D is F-split and let $\alpha : \omega_X(D) \to \omega_D$ denote the adjunction morphism. Then we have the following diagram:*

$$
\begin{array}{ccc}
\omega_X(D) & \xrightarrow{\ \alpha\ } & \omega_D \\
{\scriptstyle ?}\Big\downarrow{\scriptstyle \lambda} & & \Big\downarrow{\scriptstyle \sigma} \\
F_*(\omega_X(D)) & \xrightarrow{\ F_*\alpha\ } & F_*\omega_D.
\end{array}
\tag{1}
$$

Here $\sigma : \omega_D \to F_*\omega_D$ is a dual splitting morphism of D provided by Proposition 3.1. Note that the morphism λ does not always exist. If a morphism λ making the diagram commutative does exist then we will say that the dual splitting morphism σ can be lifted to X.

Theorem 3.4 *Let X be an equidimensional projective Cohen-Macaulay scheme over k of* char $k > 0$ *and $D \subseteq X$ an effective (non-empty) Cartier divisor. Assume that D is F-split and that $\sigma : \omega_D \to F_*\omega_D$, a dual splitting morphism of D, can be lifted to X (cf. Definition 3.3). Then X is also F-split.*

Proof. I will use the notation of Definition 3.3. Let $\lambda : \omega_X(D) \to F_*(\omega_X(D))$ be a lifting of σ, i.e., such that λ completes (1) to a commutative diagram. Note that $\ker \alpha \simeq \omega_X$ and hence setting $\sigma_X := \lambda|_{\omega_X}$ we have the following commutative diagram of short exact sequences:

$$
\begin{array}{ccccccccc}
0 & \longrightarrow & \omega_X & \longrightarrow & \omega_X(D) & \longrightarrow & \omega_D & \longrightarrow & 0 \\
& & {\scriptstyle \sigma_X}\Big\downarrow & & \Big\downarrow{\scriptstyle \lambda} & & \Big\downarrow{\scriptstyle \sigma} & & \\
0 & \longrightarrow & F_*\omega_X & \longrightarrow & F_*(\omega_X(D)) & \longrightarrow & F_*\omega_D & \longrightarrow & 0.
\end{array}
$$

Let $n := \dim X$ and consider the induced diagram of long exact sequences of cohomology:

$$
\begin{array}{ccccc}
\cdots \longrightarrow H^{n-1}(D, \omega_D) & \longrightarrow & H^n(X, \omega_X) & \longrightarrow & H^n(X, \omega_X(D)) \\
{\scriptstyle H^{n-1}(\sigma)}\Big\downarrow & & \Big\downarrow{\scriptstyle H^n(\sigma_X)} & & \\
\cdots \longrightarrow H^{n-1}(D, F_*\omega_D) & \longrightarrow & H^n(X, F_*\omega_X) & \longrightarrow & H^n(X, F_*(\omega_X(D)))
\end{array}
$$

Observe that $H^n(X, F_*(\omega_X(D))) \simeq H^n(X, \omega_X(D)) = 0$ by Serre duality and hence we have a commutative square where the horizontal maps are isomorphisms:

$$
\begin{array}{ccc}
H^{n-1}(D, \omega_D) & \xrightarrow{\simeq} & H^n(X, \omega_X) \\
{\scriptstyle H^{n-1}(\sigma)}\downarrow & & \downarrow{\scriptstyle H^n(\sigma_X)} \\
H^{n-1}(D, F_*\omega_D) & \xrightarrow{\simeq} & H^n(X, F_*\omega_X).
\end{array}
$$

This implies that $H^n(\sigma_X) \neq 0$, hence X is F-split by Proposition 3.1. $\qquad\square$

Corollary 3.5 *Let X be an equidimensional projective Gorenstein scheme over k of char $k > 0$ and $D \subseteq X$ an effective (non-empty) anti-canonical divisor, i.e., such that $\omega_X \simeq \mathcal{O}_X(-D)$. If D is F-split then so is X.*

Proof. Since D is an anti-canonical divisor it follows that $\omega_X(D) \simeq \mathcal{O}_X$ and $\omega_D \simeq \mathcal{O}_D$, so the adjunction short exact sequence

$$
0 \longrightarrow \omega_X \longrightarrow \omega_X(D) \longrightarrow \omega_D \longrightarrow 0
$$

becomes

$$
0 \longrightarrow \mathcal{O}_X(-D) \longrightarrow \mathcal{O}_X \longrightarrow \mathcal{O}_D \longrightarrow 0
$$

If D is F-split , then the natural morphism $\eta_D : \mathcal{O}_D \to F_*\mathcal{O}_D$ splits and hence $H^n(\eta_D) \neq 0$. In other words, η_D is a dual splitting morphism of D. The natural morphism $\eta_X : \mathcal{O}_X \to F_*\mathcal{O}_X$ is a lifting of η_D to X and hence X is F-split by Theorem 3.4. $\qquad\square$

Corollary 3.6 *Let X be an equidimensional projective Gorenstein scheme over k of char $k > 0$ and $D \subseteq X$ an effective (non-empty) anti-canonical divisor, i.e., such that $\omega_X \simeq \mathcal{O}_X(-D)$. If D is F-split then Kodaira vanishing holds on X, i.e., for any ample line bundle $\mathscr{L}$ on X,*

$$
H^j(X, \mathscr{L}^{-1}) = 0
$$

for $j < \dim X$.

Proof. This is a direct consequence of Corollary 3.5 and Theorem 2.3. $\qquad\square$

4 Frobenius Split Anti-canonical Divisors on Fano Varieties

We are now ready to prove the main result.

Theorem 4.1 *Let X be an equidimensional projective Gorenstein scheme over k and $D \subseteq X$ an effective, ample Cartier divisor such that $\omega_D \simeq \mathcal{O}_D$. Assume* *that* dim $X \geq 3$ *and also that if* char $k > 0$ *then* D *is F-split and if* char $k = 0$ *then* X *has rational singularities. Then*

(i) *if* char $k > 0$ *then* X *is F-split,*
(ii) *Kodaira vanishing holds on* X,
(iii) $\omega_X \simeq \mathscr{O}_X(-D)$,
(iv) $H^i(X, \mathscr{O}_X) = 0$ *for* $i > 0$, *and*
(v) $H^j(D, \mathscr{O}_D) = 0$ *for* $0 < j < \dim D$.

Proof. Theorem 2.6 implies (iii), which combined with Corollary 3.5 implies (i) and if char $k > 0$ then that combined with Theorem 2.3 implies (ii). Of course, if char $k = 0$ then (ii) is well-known by Kawamata-Viehweg vanishing [Kaw82, Vie82]. Since $\mathscr{O}_X \simeq \omega_X \otimes \mathscr{O}_X(D)$ and D is ample, Serre duality and (ii) implies (iv). Finally, consider the short exact sequence,

$$0 \longrightarrow \mathscr{O}_X(-D) \longrightarrow \mathscr{O}_X \longrightarrow \mathscr{O}_D \longrightarrow 0.$$

As $H^j(X, \mathscr{O}_X(-D)) = 0$ for $j < \dim X$ by (ii), (v) follows from (iv). □

As a consequence of Theorem 4.1 we will obtain the generalization of the main result of [Som76] to ordinary abelian varieties in positive characteristic promised in the introduction. Note that by [MS87, 1.1] an abelian variety is ordinary if and only if it is F-split. In particular, the methods of this paper do not say anything about what happens for non-ordinary abelian varieties.

For the definition of ordinary varieties in general the reader is referred to [BK86] although the definition of ordinariness will not be used directly. We will only use the following properties, proved respectively by Illusie [Ill90], and Joshi and Rajan [JR03].

Proposition 4.2 *Let* Z *be an ordinary smooth projective variety over a field* k *of positive characteristic. Then*

(i) [Ill90, Prop. 1.2] *any small deformation of* Z *is also ordinary, and*
(ii) [JR03, Prop. 3.1] *if in addition* $\omega_Z \simeq \mathscr{O}_Z$, *then* Z *is F-split.*

Corollary 4.3 *Let* A *be an abelian variety of dimension at least* 2 *over* k. *If* char $k > 0$ *assume that* A *is ordinary. Suppose* A *is an ample divisor on* X. *Then* X *cannot be an equidimensional projective Gorenstein scheme if* char $k > 0$ *and an equidimensional projective Gorenstein scheme with only rational singularities if* char $k = 0$.

Proof. If char $k > 0$ then A is F-split by [MS87, 1.1] or Proposition 4.2(ii). By the dimension assumption $H^1(A, \mathscr{O}_A) \neq 0$ and hence the statement follows from Theorem 4.1(v). □

Remark 4.4 If char $k = 0$, essentially the same statement as Corollary 4.3 was proved in [Som76]. However, the proof given here is quite different even in the char $k = 0$ case. In particular, [Som76] relied on topological arguments and it only stated that A could not be an ample divisor on a smooth projective variety.

5 Non-Smoothable Singularities

In this section I prove Sommese's second theorem which is an application of Theorem 4.1 showing that certain singularities are not smoothable.

Next recall that the Betti numbers of smooth projective varieties are defined as the dimension of ℓ-adic cohomology groups, i.e., let Z be a smooth projective variety over k and let ℓ be a prime different from char k. Then $b_i(Z): = \dim H^i_{\acute{e}t}(Z, \mathbb{Q}_\ell)$. Note that this definition is valid in all characteristics. In characteristic zero these numbers are the same as the dimension of singular cohomology groups of the underlying topological space of Z, which is the more common definition in this case.

The following lemma is folklore. For the reader's convenience a proof is provided. It also follows from various stronger statements which we will not need here.

Lemma 5.1 *Let Z be a smooth projective variety over an algebraically closed field k. If $b_1(Z) \neq 0$ then $H^1(Z, \mathscr{O}_Z) \neq 0$.*

Proof. If $b_1(Z) \neq 0$ then by the expression of étale cohomology as a limit implies that there are arbitrarily high order torsion elements in Pic Z (cf. [Tam94, 4.4.4]). It follows that then Pic$^\circ Z$ cannot be finite by the Theorem of the Base [Sev34, Nér52]. Hence Pic$^\circ Z$ is positive dimensional and then so is $H^1(Z, \mathscr{O}_Z) = T_{[\mathscr{O}_Z]}$ Pic$^\circ Z$. $\qquad\square$

I will use the notion of smoothability used in [Som79] and [Har74]. This is slightly more restrictive than the one used in [KK18].

Definition 5.2 (cf. [Har74, Som79])) *For a morphism $f : \mathscr{X} \to T$ and $t \in T$, the fibre of f over t will be denoted by $\mathscr{X}_t = f^{-1}(t)$. Let X be a closed subscheme of a scheme P over an algebraically closed field k. A deformation of X in P is a morphism $(f : \mathscr{X} \to T)$ where*

(i) T is a connected positive dimensional scheme of finite type over k,
(ii) $\mathscr{X} \subseteq P \times T$ is a closed subscheme which is flat over T,
(iii) there exists a closed point $0 \in T$, such that $\mathscr{X}_0 \simeq X$, and
(iv) f is the restriction of the projection morphism $P \times T \to T$ to $\mathscr{X}$.

A deformation $(f : \mathscr{X} \to T)$ *of X in P will be called a* Gorenstein deformation *if for all* $t \in T$, $t \neq 0$, *the fibre* $\mathscr{X}_t$ *is Gorenstein. It will be called a* smooth deformation *if for all* $t \in T$, $t \neq 0$, *the fibre* $\mathscr{X}_t$ *is smooth over* $k(t)$. *In this latter case we also say that X is* smoothable *in P.*

A somewhat weaker statement than the following was proved in [Som79, 2.1.1] in characteristic zero and a somewhat more general statement, as a consequence of much deeper results, was established in [KK18, Cor. 8.7] in all characteristics.

Theorem 5.3 *Let* $X \subseteq \mathbb{P}^n$ *be a projective variety over an algebraically closed field k and let* $H \subseteq \mathbb{P}^n$ *be a hypersurface such that* $Z = X \cap H$ *is smooth,* $\dim Z > 1$, $\omega_Z \simeq \mathscr{O}_Z$, *and* $b_1(Z) \neq 0$ *(e.g., Z is an abelian variety of dimension at least 2). Further assume that if* char $k > 0$ *then Z is ordinary. Let* $(f : \mathscr{X} \to T)$ *be a deformation of X in* $\mathbb{P}^n$ *and assume that if* char $k = 0$ *then* $\mathscr{X}_t$ *has rational singularities for* $t \neq 0$. *Then there exists a non-empty open neighbourhood* $0 \in U \subseteq T$ *such that* $\mathscr{X}_t$ *has isolated non-Gorenstein singularities for every* $t \in U$. *Consequently X does not admit a Gorenstein deformation in* $\mathbb{P}^n$ *and in particular, it is not smoothable in* $\mathbb{P}^n$.

Proof. Observe that there exists a non-empty open set $0 \in U \subseteq T$ such that $\mathscr{X}_t$ has the same properties as $X = \mathscr{X}_0$, that is, there exists a hypersurface $H_t \subseteq \mathbb{P}^n$ such that $Z_t = \mathscr{X}_t \cap H_t$ is smooth, $\dim Z_t > 1$, $\omega_{Z_t} \simeq \mathscr{O}_{Z_t}$, and $b_1(Z_t) \neq 0$. We may also assume that if char $k > 0$ then Z_t is ordinary by Proposition 4.2(i) and note that if char $k = 0$ then $\mathscr{X}_t$ has rational singularities for $t \neq 0$ by assumption.

It follows that Z_t is F-split by Proposition 4.2(ii) and $H^1(Z_t, \mathscr{O}_{Z_t}) \neq 0$ by Lemma 5.1 and hence $\mathscr{X}_t$ is not Gorenstein by Theorem 4.1(v).

On the other hand, as a hypersurface section of $\mathscr{X}_t$ is smooth, its singular set must be zero-dimensional and hence $\mathscr{X}_t$ must have isolated non-Gorenstein singularities. $\square$

Finally, this implies the following:

Corollary 5.4 *Let* $A \subseteq \mathbb{P}^{n-1}$ *be an abelian variety of dimension at least 2 over k and let X denote the cone over A in* $\mathbb{P}^n$. *If* char $k > 0$ *further assume that A is ordinary. Then X is not smoothable in* $\mathbb{P}^n$.

References

[BK86] S. BLOCH AND K. KATO: *p-adic étale cohomology*, Inst. Hautes Études Sci. Publ. Math. (1986), no. 63, 107–152. MR 849653

[Har74] R. HARTSHORNE: *Topological conditions for smoothing algebraic singularities*, Topology **13** (1974), 241–253. MR 0349677 (50 #2170)

[Har77] R. HARTSHORNE: *Algebraic geometry*, Springer-Verlag, New York, 1977, Graduate Texts in Mathematics, No. 52. MR 0463157

[Ill90] L. ILLUSIE: *Ordinarité des intersections complètes générales*, The Grothendieck Festschrift, Vol. II, Progr. Math., vol. 87, Birkhäuser Boston, Boston, MA, 1990, pp. 376–405. MR 1106904

[JR03] K. JOSHI AND C. S. RAJAN: *Frobenius splitting and ordinarity*, Int. Math. Res. Not. (2003), no. 2, 109–121. MR 1936581

[Kaw82] Y. KAWAMATA: *A generalization of Kodaira-Ramanujam's vanishing theorem*, Math. Ann. **261** (1982), no. 1, 43–46. MR 675204 (84i:14022)

[Kod53] K. KODAIRA: *On a differential-geometric method in the theory of analytic stacks*, Proc. Nat. Acad. Sci. U. S. A. **39** (1953), 1268–1273. MR 0066693 (16,618b)

[KK18] J. KOLLÁR AND S. J. KOVÁCS: *Deformation of log canonical and F-pure singularities.* arXiv:1807.07417

[MR85] V. B. MEHTA AND A. RAMANATHAN: *Frobenius splitting and cohomology vanishing for Schubert varieties*, Ann. of Math. (2) **122** (1985), no. 1, 27–40. MR 799251

[MS87] V. B. MEHTA AND V. SRINIVAS: *Varieties in positive characteristic with trivial tangent bundle*, Compositio Math. **64** (1987), no. 2, 191–212, With an appendix by Srinivas and M. V. Nori. MR 916481

[Nér52] A. NÉRON: *La théorie de la base pour les diviseurs sur les variétés algébriques*, Deuxième Colloque de Géométrie Algébrique, Liège, 1952, Georges Thone, Liège; Masson & Cie, Paris, 1952, pp. 119–126. MR 0052154

[Sev34] F. SEVERI: *La base per le varietà algebriche di dimensione qualunque contenute in una data e la teoria generale delle corrispondénze fra i punti di due superficie algebriche*, Mem. Accad. Ital. (1934), no. 5, 239–283.

[Som79] A. J. SOMMESE: *Nonsmoothable varieties*, Comment. Math. Helv. **54** (1979), no. 1, 140–146. MR 522037

[Som76] A. J. SOMMESE: *On manifolds that cannot be ample divisors*, Math. Ann. **221** (1976), no. 1, 55–72. MR 0404703

[Tam94] G. TAMME: *Introduction to étale cohomology*, Universitext, Springer-Verlag, Berlin, 1994, Translated from the German by Manfred Kolster. MR 1317816

[Vie82] E. VIEHWEG: *Vanishing theorems*, J. Reine Angew. Math. **335** (1982), 1–8. MR 667459 (83m:14011)

Sándor J. Kovács
University of Washington
Department of Mathematics
Seattle, WA 98195, USA
skovacs@uw.edu

4

Halves of Points of an Odd Degree Hyperelliptic Curve in its Jacobian

Yuri G. Zarhin

Abstract. Let $f(x)$ be a degree $(2g + 1)$ monic polynomial with coefficients in an algebraically closed field K with $\mathrm{char}(K) \neq 2$ and without repeated roots. Let $\mathfrak{R} \subset K$ be the $(2g + 1)$-element set of roots of $f(x)$. Let $\mathcal{C} : y^2 = f(x)$ be an odd degree genus g hyperelliptic curve over K. Let J be the jacobian of $\mathcal{C}$ and $J[2] \subset J(K)$ the (sub)group of points of order dividing 2. We identify $\mathcal{C}$ with the image of its canonical embedding into J (the infinite point of $\mathcal{C}$ goes to the identity element of J). Let $P = (a, b) \in \mathcal{C}(K) \subset J(K)$ and

$$M_{1/2, P} = \{\mathfrak{a} \in J(K) \mid 2\mathfrak{a} = P\} \subset J(K),$$

which is $J[2]$-torsor. In a previous work we established an explicit bijection between the sets $M_{1/2, P}$ and

$$\mathfrak{R}_{1/2, P} := \{\mathfrak{r} : \mathfrak{R} \to K \mid \mathfrak{r}(\alpha)^2 = a - \alpha \; \forall \alpha \in \mathfrak{R}; \; \prod_{\alpha \in \mathfrak{R}} \mathfrak{r}(\alpha) = -b\}.$$

The aim of this paper is to describe the induced action of $J[2]$ on $\mathfrak{R}_{1/2, P}$ (i.e., how signs of square roots $r(\alpha) = \sqrt{a - \alpha}$ should change).

1 Introduction

Let K be an algebraically closed field of characteristic different from 2, g a positive integer, $\mathfrak{R} \subset K$ a $(2g + 1)$-element set,

$$f(x) = f_{\mathfrak{R}}(x) := \prod_{\alpha \in \mathfrak{R}} (x - \alpha)$$

2010 *Mathematics Subject Classification.* 14H40, 14G27, 11G10

Key words and phrases. Hyperelliptic curves, jacobians, Mumford representations

Partially supported by Simons Foundation Collaboration grant # 585711.

This paper was started during my stay in May-July 2018 at the Max-Planck-Institut für Mathematik (Bonn, Germany), whose hospitality and support are gratefully acknowledged.

a degree $(2g+1)$ polynomial with coefficients in K and without repeated roots, $C : y^2 = f(x)$ the corresponding genus g hyperelliptic curve over K, and J the jacobian of C. We identify C with the image of its canonical embedding

$$C \hookrightarrow J, \ P \mapsto \mathrm{cl}((P) - (\infty))$$

into J (the infinite point ∞ of C goes to the identity element of J). Let $J[2] \subset J(K)$ be the kernel of multiplication by 2 in $J(K)$, which is a $2g$-dimensional $\mathbb{F}_2$-vector space. All the $(2g + 1)$ points

$$\mathfrak{W}_\alpha := (\alpha, 0) \in \mathcal{C}(K) \subset J(K) \ (\alpha \in \mathfrak{R})$$

lie in $J[2]$ and generate it as the $2g$-dimensional $\mathbb{F}_2$-vector space; they satisfy the only relation

$$\sum_{\alpha \in \mathfrak{R}} \mathfrak{W}_\alpha = 0 \in J[2] \subset J(K).$$

This leads to a well known canonical isomorphism [4] between $\mathbb{F}_2$-vector spaces $J[2]$ and

$$(\mathbb{F}_2{}^{\mathfrak{R}})^0 = \left\{ \phi : \mathfrak{R} \to \mathbb{F}_2 \mid \sum_{\alpha \in \mathfrak{R}} \phi(\alpha) = 0 \right\}.$$

Namely, each function $\phi \in (\mathbb{F}_2{}^{\mathfrak{R}})^0$ corresponds to

$$\sum_{\alpha \in \mathfrak{R}} \phi(\alpha) \mathfrak{W}_\alpha \in J[2].$$

For example, for each $\beta \in \mathfrak{R}$ the point $\mathfrak{W}_\beta = \sum_{\alpha \neq \beta} \mathfrak{W}_\alpha$ corresponds to the function $\psi_\beta : \mathfrak{R} \to \mathbb{F}_2$ that sends β to 0 and all other elements of $\mathfrak{R}$ to 1.

If $\mathfrak{b} \in J(K)$ then the finite set

$$M_{1/2, \mathfrak{b}} := \{\mathfrak{a} \in J(K) \mid 2\mathfrak{a} = \mathfrak{b}\} \subset J(K)$$

consists of 2^{2g} elements and carries the natural structure of a $J[2]$-torsor.

Let

$$P = (a, b) \in \mathcal{C}(K) \subset J(K).$$

Let us consider, the set

$$\mathfrak{R}_{1/2, P} := \left\{ \mathfrak{r} : \mathfrak{R} \to K \mid \mathfrak{r}(\alpha)^2 = a - \alpha \ \forall \alpha \in \mathfrak{R}; \ \prod_{\alpha \in \mathfrak{R}} \mathfrak{r}(\alpha) = -b \right\}.$$

Changes of signs in the (even number of) square roots provide $\mathfrak{R}_{1/2, P}$ with the natural structure of a $(\mathbb{F}_2{}^{\mathfrak{R}})^0$-torsor. Namely, let

$$\chi : \mathbb{F}_2 \to K^*$$

be the **additive character** such that

$$\chi(0) = 1, \chi(1) = -1.$$

Then the result of the action of a function $\phi : \mathfrak{R} \to \mathbb{F}_2$ from $(\mathbb{F}_2{}^{\mathfrak{R}})^0$ on $\mathfrak{r} : \mathfrak{R} \to K$ from $\mathfrak{R}_{1/2,P}$ is just the product

$$\chi(\phi)\mathfrak{r} : \mathfrak{R} \to K, \ \alpha \mapsto \chi(\phi(\alpha))\mathfrak{r}(\alpha).$$

On the other hand, I constructed in [9] an explicit *bijection* of finite sets

$$\mathfrak{R}_{1/2,P} \cong M_{1/2,P}, \ \mathfrak{r} \mapsto \mathfrak{a}_{\mathfrak{r}} \in M_{1/2,P} \subset J(K).$$

Identifying (as above) $J[2]$ and $(\mathbb{F}_2{}^{\mathfrak{R}})^0$, we obtain a second structure of a $(\mathbb{F}_2{}^{\mathfrak{R}})^0$-torsor on $\mathfrak{R}_{1/2,P}$. Our main result asserts that these two structures actually coincide. In down-to-earth terms this means the following.

Theorem 1.1 *Let* $\mathfrak{r} \in \mathfrak{R}_{1/2,P}$ *and* $\beta \in \mathfrak{R}$. *Let us define* $\mathfrak{r}^\beta \in \mathfrak{R}_{1/2,P}$ *as follows.*

$$\mathfrak{r}^\beta(\beta) = \mathfrak{r}(\beta), \ \mathfrak{r}^\beta(\alpha) = -\mathfrak{r}(\alpha) \ \forall \alpha \in \mathfrak{R} \setminus \{\beta\}.$$

Then

$$\mathfrak{a}_{\mathfrak{r}^\beta} = \mathfrak{a}_{\mathfrak{r}} + \mathfrak{W}_\beta = \mathfrak{a}_{\mathfrak{r}} + \left(\sum_{\alpha \neq \beta} \mathfrak{W}_\alpha \right).$$

Remark 1.2   In the case of elliptic curves (i.e., when $g = 1$) the assertion of Theorem 1.1 was proven in [2, Th. 2.3(iv)].

Example 1.3 *If* $P = \mathfrak{W}_\beta = (\beta, 0)$ *then*

$$\mathfrak{a}_{\mathfrak{r}} + \mathfrak{W}_\beta = \mathfrak{a}_{\mathfrak{r}} - \mathfrak{W}_\beta = \mathfrak{a}_{\mathfrak{r}} - 2\mathfrak{a}_{\mathfrak{r}} = -\mathfrak{a}_{\mathfrak{r}}$$

while

$$-\mathfrak{a}_{\mathfrak{r}} = \mathfrak{a}_{-\mathfrak{r}}$$

(see [9, Remark 3.5]). On the other hand, $\mathfrak{r}(\beta) = \sqrt{\beta - \beta} = 0$ *for all* $\mathfrak{r}$ *and*

$$\mathfrak{r}^\beta = -\mathfrak{r} : \alpha \mapsto -\mathfrak{r}(\alpha) \ \forall \alpha \in \mathfrak{R}.$$

This implies that

$$\mathfrak{a}_{\mathfrak{r}^\beta} = \mathfrak{a}_{-\mathfrak{r}} = \mathfrak{a}_{\mathfrak{r}} + \mathfrak{W}_\beta.$$

This proves Theorem 1.1 in the special case $P = \mathfrak{W}_\beta$.

The paper is organized as follows. In Section 2 we recall basic facts about Mumford representations of points of $J(K)$ and review results of [9], including an explicit description of the bijection between $\mathfrak{R}_{1/2,P}$ and $M_{1/2,P}$. In Section 3 we give explicit formulas for the Mumford representation of $\mathfrak{a} + \mathfrak{W}_\beta$

when $\mathfrak{a}$ lies neither on the theta divisor of J nor on its translation by $\mathfrak{W}_\beta$, assuming that we know the Mumford representation of $\mathfrak{a}$. In Section 4 we prove Theorem 1.1, using auxiliary results from commutative algebra that are proven in Section 5.

2 Halves and Square Roots

Let C be the smooth projective model of the smooth affine plane K-curve

$$y^2 = f(x) = \prod_{\alpha \in \mathfrak{R}} (x - \alpha)$$

where $\mathfrak{R}$ is a $(2g + 1)$-element subset of K. In particular, $f(x)$ is a monic degree $(2g + 1)$ polynomial without repeated roots. It is well known that C is a genus g hyperelliptic curve over K with precisely one *infinite* point, which we denote by ∞. In other words,

$$C(K) = \left\{ (a, b) \in K^2 \mid b^2 = \prod_{\alpha \in \mathfrak{R}} (a - \alpha_i) \right\} \bigsqcup \{\infty\}.$$

Clearly, x and y are nonconstant rational functions on C, whose only pole is ∞. More precisely, the polar divisor of x is $2(\infty)$ and the polar divisor of y is $(2g + 1)(\infty)$. The zero divisor of y is $\sum_{\alpha \in \mathfrak{R}} (\mathfrak{W}_\alpha)$. In particular, y is a *local parameter* at (every) $\mathfrak{W}_\alpha$.

We write ι for the hyperelliptic involution

$$\iota : C \to C, (x, y) \mapsto (x, -y), \infty \mapsto \infty.$$

The set of fixed points of ι consists of ∞ and all $\mathfrak{W}_\alpha$ ($\alpha \in \mathfrak{R}$). It is well known that for each $P \in C(K)$ the divisor $(P) + \iota(P) - 2(\infty)$ is principal. More precisely, if $P = (a, b) \in C(K)$ then $(P) + \iota(P) - 2(\infty)$ is the divisor of the rational function $x - a$ on C. In particular, if $P = \mathfrak{W}_\alpha = (\alpha, 0)$ then

$$2(\mathfrak{W}_\alpha) - 2(\infty) = \text{div}(x - \alpha).$$

In particular, $x - \alpha$ has a double zero at $\mathfrak{W}_\alpha$ (and no other zeros). If D is a divisor on C then we write $\text{supp}(D)$ for its *support*, which is a finite subset of $C(K)$.

Recall that the jacobian J of C is a g-dimensional abelian variety over K. If D is a degree zero divisor on C then we write $\text{cl}(D)$ for its linear equivalence class, which is viewed as an element of $J(K)$. Elements of $J(K)$ may be described in terms of so called **Mumford representations** (see [4, Sect. 3.12], [8, Sect. 13.2] and Subsection 2.2 below).

We will identify C with its image in J with respect to the canonical regular map $C \hookrightarrow J$ under which ∞ goes to the identity element of J. In other words, a point $P \in C(K)$ is identified with $\mathrm{cl}((P) - (\infty)) \in J(K)$. Then the action of the hyperelliptic involution ι on $C(K) \subset J(K)$ coincides with multiplication by -1 on $J(K)$. In particular, the list of points of order 2 on C consists of all $\mathfrak{W}_\alpha$ ($\alpha \in \mathfrak{R}$).

2.1 Since K is algebraically closed, the commutative group $J(K)$ is divisible. It is well known that for each $\mathfrak{b} \in J(K)$ there are exactly 2^{2g} elements $\mathfrak{a} \in J(K)$ such that $2\mathfrak{a} = \mathfrak{b}$. In [9] we established explicitly the following bijection $\mathfrak{r} \mapsto \mathfrak{a}_\mathfrak{r}$ between the 2^{2g}-element sets $\mathfrak{R}_{1/2,P}$ and $M_{1/2,P}$.

If $\mathfrak{r} \in \mathfrak{R}_{1/2,P}$ then for each positive integer $i \leq 2g + 1$ let us consider $s_i(\mathfrak{r}) \in K$ defined as the value of ith basic symmetric function at $(2g + 1)$ elements $\{\mathfrak{r}(\alpha) \mid \alpha \in \mathfrak{R}\}$ (notice that all $\mathfrak{r}(\alpha)$ are distinct, since their squares $\mathfrak{r}(\alpha)^2 = a - \alpha$ are distinct). Let us consider the degree g monic polynomial

$$U_\mathfrak{r}(x) = (-1)^g \left[(a - x)^g + \sum_{j=1}^{g} s_{2j}(\mathfrak{r})(a - x)^{g-j} \right],$$

and the polynomial

$$V_\mathfrak{r}(x) = \sum_{j=1}^{g} \left(s_{2j+1}(\mathfrak{r}) - s_1(\mathfrak{r})s_{2j}(\mathfrak{r}) \right) (a - x)^{g-j}$$

whose degree is *strictly less* than g. Let $\{c_1, \ldots, c_g\} \subset K$ be the collection of all g roots of $U_\mathfrak{r}(x)$, i.e.,

$$U_\mathfrak{r}(x) = \prod_{j=1}^{g} (x - c_j) \in K[x].$$

Let us put

$$d_j = V_\mathfrak{r}(c_j) \ \forall j = 1, \ldots, g.$$

It is proven in [9, Th. 3.2] that $Q_j = (c_j, d_j)$ lies in $C(K)$ for all j and

$$\mathfrak{a}_\mathfrak{r} := \mathrm{cl} \left(\left(\sum_{j=1}^{g} (Q_j) \right) - g(\infty) \right) \in J(K)$$

satisfies $2\mathfrak{a}_\mathfrak{r} = P$, i.e., $\mathfrak{a}_\mathfrak{r} \in M_{1/2,P}$. In addition, **none of Q_j coincides with any $\mathfrak{W}_\alpha$**, i.e.,

$$U_\mathfrak{r}(\alpha) \neq 0, \ c_j \neq \alpha, \ d_j \neq 0.$$

The main result of [9] asserts that the map

$$\mathfrak{R}_{1/2,P} \to M_{1/2,P}, \ \mathfrak{r} \mapsto \mathfrak{a}_\mathfrak{r}$$

is a **bijection**.

Remark 2.1 Notice that one may express explicitly $\mathfrak{r}$ in terms of $U_{\mathfrak{r}}(x)$ and $V_{\mathfrak{r}}(x)$. Namely [9, Th. 3.2], *none* of $\alpha \in \mathfrak{R}$ is a root of $U_{\mathfrak{r}}(x)$ and

$$\mathfrak{r}(\alpha) = s_1(\mathfrak{r}) + (-1)^g \frac{V_{\mathfrak{r}}(\alpha)}{U_{\mathfrak{r}}(\alpha)} \quad \text{for all } \alpha \in \mathfrak{R}. \tag{1}$$

In order to determine $s_1(\mathfrak{r})$, let us fix two *distinct* roots $\beta, \gamma \in \mathfrak{R}$. Then [9, Cor. 3.4]

$$\frac{V_{\mathfrak{r}}(\gamma)}{U_{\mathfrak{r}}(\gamma)} \neq \frac{V_{\mathfrak{r}}(\beta)}{U_{\mathfrak{r}}(\beta)}$$

and

$$s_1(\mathfrak{r}) = \frac{(-1)^g}{2} \times \frac{\left(\beta + \left(\frac{V_{\mathfrak{r}}(\beta)}{U_{\mathfrak{r}}(\beta)}\right)^2\right) - \left(\gamma + \left(\frac{V_{\mathfrak{r}}(\gamma)}{U_{\mathfrak{r}}(\gamma)}\right)^2\right)}{\frac{V_{\mathfrak{r}}(\gamma)}{U_{\mathfrak{r}}(\gamma)} - \frac{V_{\mathfrak{r}}(\beta)}{U_{\mathfrak{r}}(\beta)}}. \tag{2}$$

2.2 Mumford Representations (see [4, Sect. 3.12], [8, Sect. 13.2, pp. 411–415, especially, Prop. 13.4, Th. 13.5 and Th. 13.7]). Recall [8, Sect. 13.2, p. 411] that if D is an effective divisor on $\mathcal{C}$ of (nonnegative) degree m, whose support does *not* contain ∞, then the degree zero divisor $D - m(\infty)$ is called *semi-reduced* if it enjoys the following properties.

- If $\mathfrak{W}_\alpha$ lies in supp(D) then it appears in D with multiplicity 1.
- If a point Q of $\mathcal{C}(K)$ lies in supp(D) and does not coincide with any of $\mathfrak{W}_\alpha$ then $\iota(Q)$ does *not* lie in supp(D).

If, in addition, $m \le g$ then $D - m(\infty)$ is called *reduced*.

It is known ([4, Ch. 3a], [8, Sect. 13.2, Prop. 3.6 on p. 413]) that for each $\mathfrak{a} \in J(K)$ there exist *exactly one* nonnegative m and (effective) degree m divisor D such that the degree zero divisor $D - m(\infty)$ is *reduced* and $\mathrm{cl}(D - m(\infty)) = \mathfrak{a}$. If

$$m \ge 1, \quad D = \sum_{j=1}^{m} (Q_j) \text{ where } Q_j = (a_j, b_j) \in \mathcal{C}(K) \text{ for all } j = 1, \ldots, m$$

(here Q_j do *not* have to be distinct) then the corresponding

$$\mathfrak{a} = \mathrm{cl}(D - m(\infty)) = \sum_{j=1}^{m} Q_j \in J(K).$$

The *Mumford representation* of $\mathfrak{a} \in J(K)$ is the pair $(U(x), V(x))$ of polynomials $U(x), V(x) \in K[x]$ such that

$$U(x) = \prod_{j=1}^{m} (x - a_j)$$

is a degree m monic polynomial while $V(x)$ has degree $< m = \deg(U)$, the polynomial $V(x)^2 - f(x)$ is divisible by $U(x)$, and

$$b_j = V(a_j), \quad Q_j = (a_j, V(a_j)) \in C(K) \quad \text{for all} \quad j = 1, \dots m.$$

(Here (a_j, b_j) are as above.) Such a pair always exists, is unique, and (as we have just seen) uniquely determines not only $\mathfrak{a}$ but also divisors D and $D - m(\infty)$.

Conversely, if $U(x)$ is a monic polynomial of degree $m \le g$ and $V(x)$ a polynomial such that $\deg(V) < \deg(U)$ and $V(x)^2 - f(x)$ is divisible by $U(x)$ then there exists exactly one $\mathfrak{a} = \mathrm{cl}(D - m(\infty))$ where $D - m(\infty)$ is a reduced divisor and $(U(x), V(x))$ is the Mumford representation of $\mathfrak{a} = \mathrm{cl}(D - m(\infty))$.

2.3 In the notation of Subsect. 2.1, let us consider the effective degree g divisor

$$D_{\mathfrak{r}} := \sum_{j=1} (Q_j)$$

on C. Then $\mathrm{supp}(D_{\mathfrak{r}})$ (obviously) does contain *neither* ∞ nor any of $\mathfrak{W}_\alpha$'s. It is proven in [9, Th. 3.2] that the divisor $D_{\mathfrak{r}} - g(\infty)$ is *reduced* and the pair $(U_{\mathfrak{r}}(x), V_{\mathfrak{r}}(x))$ is the Mumford representation of

$$\mathfrak{a}_{\mathfrak{r}} := \mathrm{cl}(D_{\mathfrak{r}} - g(\infty)).$$

In particular, if $Q \in C(K)$ lies in $\mathrm{supp}(D)$ (i.e., is one of Q_j's) then $\iota(Q)$ does *not*.

Lemma 2.2 *Let D be an effective divisor on C of degree $m > 0$ such that $m \le 2g + 1$ and $\mathrm{supp}(D)$ does not contain ∞. Assume that the divisor $D - m(\infty)$ is principal.*

(1) *Suppose that m is odd. Then:*

    (i) *$m = 2g + 1$ and there exists exactly one polynomial $v(x) \in K[x]$ such that the divisor of $y - v(x)$ coincides with $D - (2g + 1)(\infty)$. In addition, $\deg(v) \le g$.*

    (ii) *If $\mathfrak{W}_\alpha$ lies in $\mathrm{supp}(D)$ then it appears in D with multiplicity 1.*

    (iii) *If b is a nonzero element of K and $P = (a, b) \in C(K)$ lies in $\mathrm{supp}(D)$ then $\iota(P) = (a, -b)$ does not lie in $\mathrm{supp}(D)$.*

    (iv) *$D - (2g + 1)(\infty)$ is semi-reduced (but not reduced).*

(2) *Suppose that $m = 2d$ is even. Then:*

    (i) *there exists exactly one monic degree d polynomial $u(x) \in K[x]$ such that the divisor of $u(x)$ coincides with $D - m(\infty)$;*

    (ii) *every point $Q \in C(K)$ appears in $D - m(\infty)$ with the same multiplicity as $\iota(Q)$;*

    (iii) *every $\mathfrak{W}_\alpha$ appears in $D - m(\infty)$ with even multiplicity.*

Proof. All the assertions except (2)(iii) are already proven in [9, Lemma 2.2]. In order to prove the remaining one, let us split the polynomial $v(x)$ into a product $v(x) = (x - \alpha)^d v_1(x)$ where d is a nonnegative integer and $v_1(x) \in K[x]$ satisfies $v_1(\alpha) \neq 0$. Then $\mathfrak{W}_\alpha$ appears in $D - m(\infty)$ with multiplicy $2d$, because $(x - \alpha)$ has a double zero at $\mathfrak{W}_\alpha$. (See also [5].) $\qquad\square$

Let $d \leq g$ be a positive integer and $\Theta_d \subset J$ be the image of the regular map

$$C^d \to J, \ (Q_1, \ldots, Q_d) \mapsto \sum_{i=1}^{d} Q_i \subset J.$$

It is well known that Θ_d is an irreducible closed d-dimensional subvariety of J that coincides with C for $d = 1$ and with J if $d = g$; in addition, $\Theta_d \subset \Theta_{d+1}$ for all $d < g$. Clearly, each Θ_d is stable under multiplication by -1 in J. We write Θ for the $(g - 1)$-dimensional *theta divisor* Θ_{g-1}.

Theorem 2.3 (See Th. 2.5 of [9]) *Suppose that $g > 1$ and let*

$$C_{1/2} := 2^{-1}C \subset J$$

be the preimage of C with respect to multiplication by 2 in J. Then the intersection of $C_{1/2}(K)$ and Θ consists of points of order dividing 2 on J. In particular, the intersection of C and $C_{1/2}$ consists of ∞ and all $\mathfrak{W}_\alpha$'s.

3 Adding Weierstrass Points

In this section we discuss how to compute a sum $\mathfrak{a} + \mathfrak{W}_\beta$ in $J(K)$ when $\mathfrak{a} \in J(K)$ lies neither on Θ nor on its translation $\Theta + \mathfrak{W}_\beta$. Let $D - g(\infty)$ be the reduced divisor on C, whose class represents $\mathfrak{a}$. Here

$$D = \sum_{j=1}^{g}(Q_j) \text{ where } Q_j = (a_j, b_j) \in C(K) \setminus \{\infty\}$$

is a degree g effective divisor. Let $(U(x), V(x))$ be the Mumford representation of $\mathrm{cl}(D - g(\infty))$. We have

$$\deg(U) = g > \deg(V),$$

$$U(x) = \prod_{j=1}^{g}(x - a_j), \ b_j = V(a_j) \ \forall j$$

and $f(x) - V(x)^2$ is divisible by $U(x)$.

Example 3.1 *Assume additionally that none of Q_j coincides with $\mathfrak{W}_\beta = (\beta, 0)$, i.e.,*

$$U(\beta) \neq 0.$$

Let us find explicitly the Mumford representation $(U^{[\beta]}(x), V^{[\beta]}(x))$ of the sum

$$\mathfrak{a} + \mathfrak{W}_\beta = \mathrm{cl}(D - m(\infty)) + \mathrm{cl}((\mathfrak{W}_\beta) - (\infty))$$
$$= \mathrm{cl}((D + (\mathfrak{W}_\beta)) - (g+1)(\infty)) = \mathrm{cl}(D_1 - (g+1)(\infty)).$$

where

$$D_1 := D + (\mathfrak{W}_\beta) = \left(\sum_{j=1}^{g}(Q_j)\right) + (\mathfrak{W}_\beta)$$

is a degree $(g+1)$ effective divisor on C. (We will see that $\deg(\tilde{U}^{[\beta]}) = g$.) Clearly, $D_1 - (g+1)(\infty)$ is semi-reduced but not reduced.

Let us consider the polynomials

$$U_1(x) = (x - \beta)U(x), \ \ V_1(x) = V(x) - \frac{V(\beta)}{U(\beta)}U(x) \in K[x].$$

Then U_1 is a degree $(g+1)$ monic polynomial, $\deg(V_1) \leq g$,

$$V_1(\beta) = 0, \ \ V_1(a_j) = V(a_j) = b_j \ \forall j$$

and $f(x) - V_1(x)^2$ is divisible by $U_1(x)$. (The last assertion follows from the divisibility of both $f(x)$ and $V_1(x)$ by $x - \beta$ combined with the divisibility of $f(x) - V(x)^2$ by $U(x)$.) If we put

$$a_{g+1} = \beta, \ b_{g+1} = 0, \ Q_{g+1} = \mathfrak{W}_\beta = (\beta, 0)$$

then

$$U_1(x) = \prod_{j=1}^{g+1}(x - a_j), \ D_1 = \sum_{j=1}^{g+1}(Q_j) \text{ where } Q_j = (a_j, b_j) \in C(K), \ b_j = V_1(a_j) \forall j$$

and $f(x) - V_1(x)^2$ is divisible by $U_1(x)$. In particular, $(U_1(x), V_1(x))$ is the pair of polynomials that corresponds to semi-reduced $D_1 - (g+1)(\infty)$ as described in [8, Prop. 13.4 and Th. 3.5]. In order to find the Mumford representation of $\mathrm{cl}(D_1 - (g+1)(\infty))$, we use an algorithm described in [8, Th. 13.9]. Namely, let us put

$$\tilde{U}(x) = \frac{f(x) - V_1(x)^2}{U_1(x)} \in K[x].$$

Since $\deg(V_1(x)) \leq g$ and $\deg(f) = 2g + 1$, we have

$$\deg\left(V_1(x)^2\right) \leq 2g, \ \deg\left(f(x) - V_1(x)^2\right) = 2g + 1, \ \deg\left(\tilde{U}(x)\right) = g.$$

Since $f(x)$ is monic, $f(x) - V_1(x)^2$ is also monic and therefore $\tilde{U}(x)$ is also monic, because $U_1(x)$ is monic. By [8, Th. 13.9], $U^{[\beta]}(x) = \tilde{U}(x)$ (since the latter is monic and has degree $g \leq g$) and $V^{[\beta]}(x)$ is the remainder of $-V_1(x)$ with respect to division by $\tilde{U}(x)$. Let us find this remainder. We have

$$-V_1(x) = -\left(V(x) - \frac{V(\beta)}{U(\beta)}U(x)\right) = -V(x) + \frac{V(\beta)}{U(\beta)}U(x).$$

Recall that

$$\deg(V) < g = \deg(U) = \deg(\tilde{U}).$$

This implies that the coefficient of $-V_1$ at x^g equals $V(\beta)/U(\beta)$ and therefore

$$V^{[\beta]}(x) = \left(-V(x) + \frac{V(\beta)}{U(\beta)}U(x)\right) - \frac{V(\beta)}{U(\beta)}\tilde{U}(x)$$

$$= -V(x) + \frac{V(\beta)}{U(\beta)}\left(U(x) - \tilde{U}(x)\right).$$

Using formulas above for U_1, V_1, $\tilde{U}$, we obtain that

$$U^{[\beta]}(x) = \frac{f(x) - \left(V(x) - \frac{V(\beta)}{U(\beta)}U(x)\right)^2}{(x - \beta)U(x)}, \tag{3}$$

$$V^{[\beta]}(x) = -V(x) + \frac{V(\beta)}{U(\beta)}\left(U(x) - \frac{f(x) - \left(V(x) - \frac{V(\beta)}{U(\beta)}U(x)\right)^2}{(x - \beta)U(x)}\right). \tag{4}$$

Remark 3.2 There is an algorithm of David Cantor [8, Sect. 13.3] that explains how to compute the Mumford representation of a sum of arbitrary divisor classes (elements of $J(K)$) when their Mumford representations are given.

Remark 3.3 Suppose that $\mathfrak{a} \in J(K)$ and $P = 2\mathfrak{a}$ lies in $\mathcal{C}(K)$ but is not the zero of the group law. Then $\mathfrak{a}$ does not lie on the theta divisor (Theorem 2.3) and satisfies the conditions of Example 3.1 for all $\beta \in \mathfrak{R}$ (see Subsect. 2.1).

4 Proof of Main Theorem

Let us choose an order on $\mathfrak{R}$. This allows us to identify $\mathfrak{R}$ with $\{1, \ldots, 2g, 2g + 1\}$ and list elements of $\mathfrak{R}$ as $\{\alpha_1, \ldots, \alpha_{2g}, \alpha_{2g+1}\}$. Then

$$f(x) = \prod_{i=1}^{2g+1} (x - \alpha_i)$$

and the affine equation for $C \setminus \{\infty\}$ is

$$y^2 = \prod_{i=1}^{2g+1} (x - \alpha_i).$$

Slightly abusing notation, we denote $\mathfrak{W}_{\alpha_i}$ by $\mathfrak{W}_i$.

Let us consider the closed affine K-subset $\tilde{C}$ in the affine K-space $\mathbb{A}^{2g+1}$ with coordinate functions $z_1, \ldots, z_{2g}, z_{2g+1}$ that is cut out by the system of quadratic equations

$$z_1^2 + \alpha_1 = z_2^2 + \alpha_2 = \cdots = z_{2g+1}^2 + \alpha_{2g+1}.$$

We write x for the regular function $z_i^2 + \alpha_i$ on $\tilde{C}$, which does *not* depend on a choice of i. By Hilbert's Nullstellensatz, the K-algebra $K[\tilde{C}]$ of regular functions on $\tilde{C}$ is canonically isomorphic to the following K-algebra. First, we need to consider the quotient A of the polynomial $K[x]$-algebra $K[x][T_1, \ldots, T_{2g+1}]$ by the ideal generated by all quadratic polynomials $T_i^2 - (x - \alpha_i)$. Next, $K[\tilde{C}]$ is canonically isomorphic to the quotient $A/\mathcal{N}(A)$ where $\mathcal{N}(A)$ is the nilradical of A. In the next section (Example 5.4) we will prove that A has no zero divisors (in particular, $\mathcal{N}(A) = \{0\}$) and therefore $\tilde{C}$ is *irreducible*. (See also [3].) We write y for the regular function

$$y = -\prod_{i=1}^{2g} z_i \in K[\tilde{C}].$$

Clearly, $y^2 = \prod_{i=1}^{2g}(x - \alpha_i)$ in $K[\tilde{C}]$. The pair (x, y) gives rise to the finite regular map of affine K-varieties (actually, curves)

$$\mathfrak{h} : \tilde{C} \to C \setminus \{\infty\}, \ (r_1, \ldots, r_{2g}, r_{2g+1}) \mapsto (a, b) = \left(r_1^2 + \alpha_1, -\prod_{i=1}^{2g+1} r_i \right)$$

$$(5)$$

of degree 2^{2g}. For each

$$P = (a, b) \in K^2 = \mathbb{A}^2(K) \text{ with } b^2 = \prod_{i=1}^{2g+1} (a - \alpha_i)$$

the fiber $\mathfrak{h}^{-1}(P) = \mathfrak{R}_{1/2, P}$ consists of (familiar) collections of square roots

$$\mathfrak{r} = \{ r_i = \sqrt{a - \alpha_i} \mid 1 \le i \le 2g + 1 \}$$

with $\prod_{i=1}^{2g+1} r_i = -b$. Each such $\mathfrak{r}$ gives rise to $\mathfrak{a}_{\mathfrak{r}} \in J(K)$ such that

$$2\mathfrak{a}_{\mathfrak{r}} = P \in C(K) \subset J(K)$$

(see [9, Th. 3.2]). On the other hand, for each $\mathfrak{W}_l = (\alpha_l, 0)$ (with $1 \leq l \leq 2g+1$) the sum $\mathfrak{a}_\mathfrak{r} + \mathfrak{W}_l$ is also a half of P and therefore corresponds to a certain collection of square roots. Which one? The answer is given by Theorem 1.1. We repeat its statement, using the new notation.

Theorem 4.1 *Let $P = (a, b)$ be a K-point on C and $\mathfrak{r} = (r_1, \ldots, r_{2g}, r_{2g+1})$ be a collection of square roots $r_i = \sqrt{a - \alpha_i} \in K$ such that $\prod_{i=1}^{2g+1} r_i = -b$. Let l be an integer that satisfies $1 \leq l \leq 2g + 1$ and let*

$$\mathfrak{r}^{[l]} = \left(r_1^{[l]}, \ldots, r_{2g}^{[l]}, r_{2g+1}^{[l]} \right) \in \mathfrak{h}^{-1}(P) \subset \tilde{C}(K) \tag{6}$$

be the collection of square roots $r_i^{[l]} = \sqrt{a - \alpha_i}$ such that

$$r_l^{[l]} = r_l, \; r_i^{[l]} = -r_i \; \forall \, i \neq l. \tag{7}$$

Then

$$\mathfrak{a}_\mathfrak{r} + \mathfrak{W}_l = \mathfrak{a}_{\mathfrak{r}^{[l]}}.$$

Example 4.2 *Let us take as P the point $\mathfrak{W}_l = (\alpha_l, 0)$. Then*

$$r_l = \sqrt{\alpha_l - \alpha_l} = 0 \; \forall \, \mathfrak{r} = (r_1, \ldots, r_{2g}, r_{2g+1}) \in \mathfrak{h}^{-1}(\mathfrak{W}_l)$$

and therefore

$$\mathfrak{r}^{[l]} = (-r_1, \ldots, -r_{2g}, -r_{2g+1}) = -\mathfrak{r}.$$

It follows from Example 1.3 (if we take $\beta = \alpha_l$) that

$$\mathfrak{a}_\mathfrak{r} + \mathfrak{W}_l = \mathfrak{a}_\mathfrak{r} - \mathfrak{W}_l = \mathfrak{a}_\mathfrak{r} - 2\mathfrak{a}_\mathfrak{r} = -\mathfrak{a}_\mathfrak{r} = \mathfrak{a}_{\mathfrak{r}^{[l]}}.$$

This proves Theorem 4.1 in the case of $P = \mathfrak{W}_l$. We are going to deduce the general case from this special one.

4.1 Before starting the proof of Theorem 4.1, let us define for each collections of signs

$$\varepsilon = \left\{ \epsilon_i = \pm 1 \mid 1 \leq i \leq 2g + 1, \; \prod_{i=1}^{2g+1} \epsilon_i = 1 \right\}$$

the biregular automorphism

$$T_\varepsilon : \tilde{C} \to \tilde{C}, \; z_i \mapsto \epsilon_i z_i \; \forall i.$$

Clearly, all T_ε constitute a finite automorphism group of $\tilde{C}$ that leaves invariant every K-fiber of $\mathfrak{h} : \tilde{C} \to C \setminus \{\infty\}$, acting on it **transitively**. Notice that if T_ε leaves invariant all the points of a certain fiber $\mathfrak{h}^{-1}(P)$ with $P \in C(K)$ then all the $\epsilon_i = 1$, i.e., T_ε is the identity map.

Proof of Theorem 4.1. Let us put

$$\beta := \alpha_l.$$

Then we have

$$\mathfrak{W}_l = (\alpha_l, 0) = (\beta, 0).$$

Let us consider the automorphism (involution)

$$\mathfrak{s}^{[l]} : \tilde{C} \to \tilde{C}, \ \mathfrak{r} \mapsto \mathfrak{r}^{[l]}$$

of $\tilde{C}$ defined by (6) and (7). We need to define another (actually, it will turn out to be the same) involution (and therefore an automorphism)

$$\mathfrak{t}^{[l]} : \tilde{C} \to \tilde{C}$$

that is defined by

$$\mathfrak{a}_{\mathfrak{t}^{[l]}(\mathfrak{r})} = \mathfrak{a}_{\mathfrak{r}} + \mathfrak{W}_l$$

as a composition of the following **regular** maps. First, $\mathfrak{r} \in \tilde{C}(K)$ goes to the pair of polynomials $(U_{\mathfrak{r}}(x), V_{\mathfrak{r}}(x))$ as in Remark 2.1, which is the Mumford representation of $\mathfrak{a}_{\mathfrak{r}}$ (see Subsect. 2.3). Second, $(U_{\mathfrak{r}}(x), V_{\mathfrak{r}}(x))$ goes to the pair of polynomials $(U^{[\beta]}(x), V^{[\beta]}(x))$ defined by formulas (3) and (3) in Section 3, which is the Mumford representation of $\mathfrak{a}_{\mathfrak{r}} + \mathfrak{W}_l$. Third, applying formulas (1) and (2) in Remark 2.1 to $(U^{[\beta]}(x), V^{[\beta]}(x))$ (instead of $(U(x), V(x))$), we get at last $\mathfrak{t}^{[l]}(\mathfrak{r}) \in \tilde{C}(K)$ such that

$$\mathfrak{a}_{\mathfrak{t}^{[l]}(\mathfrak{r})} = \mathfrak{a}_{\mathfrak{r}} + \mathfrak{W}_l.$$

Clearly, $\mathfrak{t}^{[l]}$ is a regular selfmap of $\tilde{C}$ that is an involution, which implies that $\mathfrak{t}^{[l]}$ is a biregular automorphism of $\tilde{C}$. It is also clear that both $\mathfrak{s}^{[l]}$ and $\mathfrak{t}^{[l]}$ leave invariant every fiber of $\mathfrak{h} : \tilde{C} \to C \setminus \{\infty\}$ and coincide on $\mathfrak{h}^{-1}(\mathfrak{W}_l)$, thanks to Example 4.2. This implies that $\mathfrak{u} := \left(\mathfrak{s}^{[l]}\right)^{-1} \mathfrak{t}^{[l]}$ is a biregular automorphism of $\tilde{C}$ that leaves invariant every fiber of $\mathfrak{h} : \tilde{C} \to C \setminus \{\infty\}$ and acts as the identity map on $\mathfrak{h}^{-1}(\mathfrak{W}_l)$. The invariance of each fiber of $\mathfrak{h}$ implies that $\tilde{C}(K)$ coincides with the finite union of its closed subsets $\tilde{C}_\varepsilon$ defined by the condition

$$\tilde{C}_\varepsilon := \{Q \in \tilde{C}(K) \mid \mathfrak{u}(Q) = T_\varepsilon(Q)\}.$$

Since $\tilde{C}$ is irreducible, the whole $\tilde{C}(K)$ coincides with one of $\tilde{C}_\varepsilon$. In particular, the fiber

$$\mathfrak{h}^{-1}(\mathfrak{W}_l) \subset \tilde{C}_\varepsilon$$

and therefore T_ε acts identically on all points of $\mathfrak{h}^{-1}(\mathfrak{W}_l)$. In light of arguments of Subsect. 4.1, T_ε is the *identity map* and therefore $\mathfrak{u}$ acts identically on the whole $\tilde{C}(K)$. This means that $\mathfrak{s}^{[l]} = \mathfrak{t}^{[l]}$, i.e.,

$$\mathfrak{a}_\mathfrak{r} + \mathfrak{W}_l = \mathfrak{a}_{\mathfrak{r}^{[l]}}. \qquad \square$$

4.2 Let $\phi : \mathfrak{R} \to \mathbb{F}_2$ be a function that satisfies $\sum_{\alpha \in \mathfrak{R}} \phi(\alpha) = 0$, i.e. $\phi \in (\mathbb{F}_2^{\mathfrak{R}})^0$. Then the finite subset

$$\text{supp}(\phi) = \{\alpha \in \mathfrak{R} \mid \phi(\alpha) \neq 0\} \subset \mathfrak{R}$$

has *even* cardinality and the corresponding point of $J[2]$ is

$$\mathfrak{T}_\phi = \sum_{\alpha \in \mathfrak{R}} \phi(\alpha)\mathfrak{W}_\alpha = \sum_{\alpha \in \text{supp}(\phi)} \mathfrak{W}_\alpha = \sum_{\gamma \notin \text{supp}(\phi)} \mathfrak{W}_\gamma.$$

Theorem 4.3 *Let* $\mathfrak{r} \in \mathfrak{R}_{1/2,P}$. *Let us define* $\mathfrak{r}^{(\phi)} \in \mathfrak{R}_{1/2,P}$ *as follows.*

$$\mathfrak{r}^{(\phi)}(\alpha) = -\mathfrak{r}(\alpha) \; \forall \alpha \in \text{supp}(\phi); \quad \mathfrak{r}^{(\phi)}(\gamma) = \mathfrak{r}(\gamma) \; \forall \gamma \notin \text{supp}(\phi).$$

Then

$$\mathfrak{a}_\mathfrak{r} + \mathfrak{T}_\phi = \mathfrak{a}_{\mathfrak{r}^{(\phi)}}.$$

Remark 4.4 If ϕ is identically zero then

$$\mathfrak{T}_\phi = 0 \in J[2], \quad \mathfrak{r}^{(\phi)} = \mathfrak{r}$$

and the assertion of Theorem 4.3 is obviously true. If $\alpha_l \in \mathfrak{R}$ and $\phi = \psi_{\alpha_l}$, i.e. $\text{supp}(\phi) = \mathfrak{R} \setminus \{\alpha_l\}$ then

$$\mathfrak{T}_\phi = \mathfrak{W}_l \in J[2], \quad \mathfrak{r}^{(\phi)} = \mathfrak{r}^{[l]}$$

and the assertion of Theorem 4.3 follows from Theorem 4.1.

Proof of Theorem 4.3. We may assume that ϕ is *not* identically zero. We need to apply Theorem 4.1 d times where d is the (even) cardinality of $\text{supp}(\phi)$ in order to get $\mathfrak{r}' \in \mathfrak{R}_{1/2,P}$ such that

$$\mathfrak{a}_\mathfrak{r} + \sum_{\alpha \in \text{supp}(\phi)} \mathfrak{W}_\alpha = \mathfrak{a}_{\mathfrak{r}'}.$$

Let us check how many times do we need to change the sign of each $\mathfrak{r}(\beta)$. First, if $\beta \notin \text{supp}(\phi)$ then we need to change to sign of $\mathfrak{r}(\beta)$ at every step, i.e., we do it exactly d times. Since d is even, the sign of $\mathfrak{r}(\beta)$ remains the same, i.e.,

$$\mathfrak{r}'(\beta) = \mathfrak{r}(\beta) \; \forall \beta \notin \text{supp}(\phi).$$

Now if $\beta \in \text{supp}(\phi)$ then we need to change the sign of $\mathfrak{r}(\beta)$ every time when we add W_α with $\alpha \neq \beta$ and it occurs exactly $(d-1)$ times. On the other hand, when we add $\mathfrak{W}_\beta$, we don't change the sign of $\mathfrak{r}(\beta)$. So, we change the sign of $\mathfrak{r}(\beta)$ exactly $(d-1)$ times, which implies that

$$\mathfrak{r}'(\beta) = -\mathfrak{r}(\beta) \ \forall \beta \in \mathrm{supp}(\phi).$$

Combining the last two displayed formula, we obtained that

$$\mathfrak{r}' = \mathfrak{r}^{(\phi)}. \qquad \square$$

5 Useful Lemma

As usual, we define the Kronecker delta δ_{ik} as 1 if $i = k$ and 0 if $i \neq k$.

The following result is probably well known but I did not find a suitable reference. (However, see [3, Lemma 5.10] and [1, pp. 425–427].)

Lemma 5.1 *Let n be a positive integer, E a field provided with n distinct discrete valuation maps*

$$v_i : E^* \twoheadrightarrow \mathbb{Z}, \ (i = 1, \ldots, n).$$

For each i let $O_{v_i} \subset E$ the discrete valuation ring attached to v_i and $\pi_i \in O_{v_i}$ its uniformizer, i.e., a generator of the maximal ideal in O_{v_i}. Suppose that for each i we are given a prime number p_i such that the characteristic of the residue field O_{v_i}/π_i is different from p_k for all $k \neq i$. Let us assume also that

$$v_i(\pi_k) = \delta_{ik} \ \forall i, k = 1, \ldots n,$$

i.e, each π_i is a v_k-adic unit if $i \neq k$.

Then the quotient $B = E[T_1, \ldots, T_n]/(T_1^{p_1} - \pi_1, \ldots, T_n^{p_n} - \pi_n)$ of the polynomial E-algebra $E[T_1, \ldots T_n]$ by the ideal generated by all $T_i^{p_i} - \pi_i$ is a field that is an algebraic extension of E of degree $\prod_{i=1}^n p_i$. In addition, the set of monomials

$$S = \left\{ \prod_{i=1}^n T_i^{e_i} \mid 0 \leq e_i \leq p_i - 1 \right\} \subset E[T_1, \ldots T_n]$$

maps injectively into B and its image is a basis of the E-vector space B.

Remark 5.2   By definition of a uniformizer, $v_i(\pi_i) = 1$ for all i.

Proof of Lemma 5.1. First, the cardinality of S is $\prod_{i=1}^n p_i$ and the image of S generates B as the E-vector space. This implies that if the E-dimension of B is $\prod_{i=1}^n p_i$ then the image of S is a basis of the E-vector space B. Second, notice that for each i the polynomial $T^{p_i} - \pi_i$ is irreducible over E, thanks to the Eisenstein criterion applied to v_i and therefore $E[T_i]/(T^{p_i} - \pi_i)$ is a field that is an algebraic degree p_i extension of E. In particular, the E-dimension of $E[T_i]/(T^{p_i} - \pi_i)$ is p_i. This proves Lemma for $n = 1$.

Induction by n. Suppose that $n > 1$ and consider the finite degree p_i field extension $E_n = E[T_n]/(T^{p_n} - \pi_n)$ of E.

Clearly, the E-algebra B is isomorphic to the quotient $E_n[T_1, \ldots T_{n-1}]/$ $(T_1^{p_1} - \pi_1, \ldots, T_{n-1}^{p_{n-1}} - \pi_{n-1})$ of the polynomial ring $E_n[T_1, \ldots T_{n-1}]$ by the ideal generated by all polynomials $T_i^{p_i} - \pi_i$ with $i < n$. Our goal is to apply the induction assumption to E_n instead of E. In order to do that, let us consider for each $i < n$ the integral closure $\tilde{O}_i$ of O_{v_i} in E_n. It is well known that $\tilde{O}_i$ is a Dedekind ring. Our conditions imply that E_n/E is *unramified* at all v_i for all $i < n$. This means that if $\mathcal{P}_i$ is a maximal ideal of $\tilde{O}_i$ that contains $\pi_i \tilde{O}_i$ (such an ideal always exists) and

$$\mathrm{ord}_{\mathcal{P}_i} : E_n^* \twoheadrightarrow \mathbb{Z}$$

is the discrete valuation map attached to $\mathcal{P}_i$ then the restriction of $\mathrm{ord}_{\mathcal{P}_i}$ to E^* coincides with v_i. This implies that for all positive integers $i, k \leq n - 1$

$$\mathrm{ord}_{\mathcal{P}_i}(\pi_k) = v_i(\pi_k) = \delta_{ik}.$$

In particular,

$$\mathrm{ord}_{\mathcal{P}_i}(\pi_i) = v_i(\pi_i) = 1,$$

i.e., π_i is a uniformizer in the corresponding discrete valuation (sub)ring $O_{\mathrm{ord}_{\mathcal{P}_i}}$ of E_n attached to $\mathrm{ord}_{\mathcal{P}_i}$. Now the induction assumption applied to E_n and its $(n - 1)$ discrete valuation maps $\mathrm{ord}_{\mathcal{P}_i}$ $(1 \leq i \leq n - 1)$ implies that B/E_n is a field extension of degree $\prod_{i=1}^{n-1} p_i$. This implies that the degree

$$[B : E] = [B : E_n][E_n : E] = \left(\prod_{i=1}^{n-1} p_i\right) p_n = \prod_{i=1}^{n} p_i.$$

This means that the E-dimension of B is $\prod_{i=1}^{n} p_i$ and therefore the image of S is a basis of the E-vector space B. $\qquad \square$

Corollary 5.3 *We keep the notation and assumptions of Lemma 5.1. Let R be a subring of E that contains 1 and all π_i $(1 \leq i \leq n)$. Then the quotient $B_R = R[T_1, \ldots, T_n]/(T_1^{p_1} - \pi_1, \ldots, T_n^{p_n} - \pi_n)$ of the polynomial R-algebra $R[T_1, \ldots, T_n]$ by the ideal generated by all $T_i^{p_i} - \pi_i$ has no zero divisors.*

Proof. There are the natural homomorphisms of R-algebras

$$R[T_1, \ldots T_n] \twoheadrightarrow B_R \to B$$

such that the first homomorphism is surjective and the *injective* image of

$$S \subset R[T_1, \ldots T_n] \subset E[T_1, \ldots T_n]$$

in B is a basis of the E-vector space B. On the other hand, the image of S generates B_R as R-module. It suffices to prove that $B_R \to B$ is injective, since B is a field by Lemma 5.1.

Suppose that $u \in B_R$ goes to 0 in B. Clearly, u is a linear combination of (the images of) elements of S with coefficients in R. Since the image of u in B is 0, all these coefficients are zeros, i.e., $u = 0$ in B_R. $\qquad\square$

Example 5.4 *We use the notation of Section 4. Let us put $n = 2g + 1$, $R = K[x]$, $E = K(x)$, $\pi_i = x - \alpha_i$, $p_i = 2$ and let*

$$v_i : E^* = K(x)^* \twoheadrightarrow \mathbb{Z}$$

be the discrete valuation map of the field of rational functions $K(x)$ attached to α_i. Then $K[\tilde{C}] = B_R/\mathcal{N}(B_R)$ where $\mathcal{N}(B_R)$ is the nilradical of B_R. It follows from Corollary 5.3 that $\mathcal{N}(B_R) = \{0\}$ and $K[\tilde{C}]$ has no zero divisors, i.e., $\tilde{C}$ is irreducible.

References

[1] T. Bandman, S. Garion, F. Grunewald, *On the surjectivity of Engel words on* PSL(2, q). Groups Geom. Dyn. **6** (2012), 409–439.

[2] B.M. Bekker, Yu.G. Zarhin, *The divisibility by 2 of rational points on elliptic curves*. Algebra i Analiz **29:4** (2017), 196–239; St. Petersburg Math. J. **29** (2018), 683–713.

[3] N. Bruin and E.V. Flynn, *Towers of 2-covers of hyperelliptic curves*. Trans. Amer. Math. Soc. **357** (2005), no. 11, 4329–4347.

[4] D. Mumford, Tata Lectures on Theta. II. Progress in Math. **43**, Birkhäuser, Boston Basel Stutgart, 1984.

[5] M. Stoll, *Arithmetic of Hyperelliptic Curves*. Available at Summer Semester 2014, University of Bayreuth. http://www.mathe2.uni-bayreuth.de/stoll/teaching/ ArithHypKurven-SS2014/Skript-ArithHypCurves-pub-screen.pdf.

[6] E. Schaefer, *2-descent on the Jacobians of hyperelliptic curves*. J. Number Theory **51** (1995), no. 2, 219–232.

[7] J.-P. Serre, Algebraic groups and class fields. Graduate Texts in Math. **117**, Springer-Verlag, New York, 1988.

[8] L.C. Washington, Elliptic Curves: Number Theory and Cryptography. Second edition. Chapman & Hall/CRC Press, Boca Raton London New York, 2008.

[9] Yu. G. Zarhin, *Division by 2 on odd-degree hyperelliptic curves and their jacobians*. Izvestiya RAN: Ser. Mat. **83:3** (2019), 93–112; Izvestiya: Mathematics **83:3** (2019), 501–520.

Yuri G. Zarhin
Pennsylvania State University,
Department of Mathematics,
University Park, PA 16802, USA
E-mail address: zarhin@math.psu.edu

5

Normal Forms for Kummer Surfaces

Adrian Clingher and Andreas Malmendier

Abstract. We determine normal forms for the Kummer surfaces associated with abelian surfaces of polarization of type $(1, 1)$, $(1, 2)$, $(2, 2)$, $(2, 4)$, and $(1, 4)$. Explicit formulas for coordinates and moduli parameters in terms of Theta functions of genus two are also given. The normal forms in question are closely connected to the generalized Riemann identities for Theta functions of Mumford's.

1 Introduction

Let $\mathbf{A}$ denote an abelian surface defined over the field of complex numbers and let $-\mathbb{I}$ be its minus identity involution. The quotient $\mathbf{A}/\langle -\mathbb{I} \rangle$ has sixteen ordinary double points and its minimum resolution, denoted $\mathrm{Kum}(\mathbf{A})$, is known as the *Kummer surface* of $\mathbf{A}$. The rich geometry of these surfaces, as well as their strong connection with Theta functions have been the subject of multiple studies [19, 34, 10, 26, 18, 25, 8] over the last century and a half.

Polarizations on an abelian surface $\mathbf{A} \cong \mathbb{C}^2/\Lambda$ are known to correspond to positive definite hermitian forms H on $\mathbb{C}^2$, satisfying $E = \mathrm{Im}\, H(\Lambda, \Lambda) \subset \mathbb{Z}$. In turn, such a hermitian form determines a line bundle $\mathcal{L}$ in the Néron-Severi group $\mathrm{NS}(\mathbf{A})$. One may always then choose a basis of Λ such that E is given by a matrix $\left(\begin{smallmatrix} 0 & D \\ -D & 0 \end{smallmatrix} \right)$ with $D = \left(\begin{smallmatrix} d_1 & 0 \\ 0 & d_2 \end{smallmatrix} \right)$ where $d_1, d_2 \in \mathbb{N}$, $d_1, d_2 \geq 0$, and d_1 divides d_2. The pair (d_1, d_2) gives the *type* of the polarization.

If $\mathbf{A} = \mathrm{Jac}(\mathcal{C})$ is the Jacobian of a smooth curve $\mathcal{C}$ of genus two, the hermitian form associated to the divisor class $[\mathcal{C}]$ is a polarization of type $(1, 1)$ - a *principal polarization*. Conversely, a principally polarized abelian surface is either the Jacobian of a smooth curve of genus two or the product of two complex elliptic curves, with the product polarization.

2010 *Mathematics Subject Classification.* 14J28

The present work focuses on Kummer surfaces Kum(**A**) associated with abelian surfaces **A** of principal polarization, as well as of polarizations of type $(1, 2)$, $(2, 2)$, $(2, 4)$, and $(1, 4)$. We present a detailed description of moduli parameters for these surfaces, as well as several *normal forms*, i.e., explicit projective equations describing Kummer surfaces Kum(**A**). The crucial ingredient in our considerations is given by the theory of classical Theta functions of genus two, as well as their associated generalized Riemann identities, as derived by Mumford in [29].

2 Theta Functions and Mumford Identities

An effective way to understand the geometry of Kummer surfaces is to use the Siegel modular forms and Theta functions of genus two. Let us enumerate here the main such forms that will be relevant to the present paper. For detailed references, we refer the reader to the classical papers of Igusa [20, 21]; for some further applications see also [9, 7].

2.1 Theta Functions and Abelian Surfaces

The Siegel three-fold is a quasi-projective variety of dimension 3 obtained from the Siegel upper half-plane of degree two which by definition is the set of two-by-two symmetric matrices over $\mathbb{C}$ whose imaginary part is positive definite, i.e.,

$$\mathbb{H}_2 = \left\{ \tau = \begin{pmatrix} \tau_{11} & \tau_{12} \\ \tau_{12} & \tau_{22} \end{pmatrix} \middle| \tau_{11}, \tau_{22}, \tau_{12} \in \mathbb{C}, \operatorname{Im}(\tau_{11}) \operatorname{Im}(\tau_{22}) \right.$$

$$\left. > \operatorname{Im}(\tau_{21})^2, \operatorname{Im}(\tau_{22}) > 0 \right\}, \tag{1}$$

divided by the action of the modular transformations $\Gamma_2 := \operatorname{Sp}_4(\mathbb{Z})$, i.e.,

$$\mathcal{A}_2 = \mathbb{H}_2 / \Gamma_2 . \tag{2}$$

Each $\tau \in \mathbb{H}_2$ determines the principally polarized complex abelian surface $\mathbf{A}_\tau = \mathbb{C}^2 / \langle \mathbb{Z}^2 \oplus \tau \mathbb{Z}^2 \rangle$ with period matrix $(\mathbb{I}_2, \tau) \in \operatorname{Mat}(2, 4; \mathbb{C})$. The canonical principal polarization $\mathcal{L}$ of $\mathbf{A}_\tau$ is determined by the Riemann form $E(x_1 + x_2\tau, y_1 + y_2\tau) = x_1^t \cdot y_2 - y_1^t \cdot x_2$ on $\mathbb{Z}^2 \oplus \tau \mathbb{Z}^2$. Two abelian surfaces $\mathbf{A}_\tau$ and $\mathbf{A}_{\tau'}$ are isomorphic if and only if there is a symplectic matrix

$$M = \begin{pmatrix} A & B \\ C & D \end{pmatrix} \in \Gamma_2 \tag{3}$$

such that $\tau' = M(\tau) := (A\tau + B)(C\tau + D)^{-1}$. Since M preserves the Riemann form, it follows that the Siegel three-fold $\mathcal{A}_2$ is the set of isomorphism classes of principally polarized abelian surfaces. Similarly, we define the subgroup $\Gamma_2(2n) = \{M \in \Gamma_2 | M \equiv \mathbb{I} \mod 2n\}$ and Igusa's congruence subgroups $\Gamma_2(2n, 4n) = \{M \in \Gamma_2(2n) | \operatorname{diag}(B) = \operatorname{diag}(C) \equiv \mathbb{I} \mod 4n\}$ with corresponding Siegel modular threefolds $\mathcal{A}_2(2)$, $\mathcal{A}_2(2, 4)$, and $\mathcal{A}_2(4, 8)$ such that

$$\Gamma_2 / \Gamma_2(2) \cong S_6, \quad \Gamma_2(2) / \Gamma_2(2, 4) \cong (\mathbb{Z}/2\mathbb{Z})^4,$$
$$\Gamma_2(2, 4) / \Gamma_2(4, 8) \cong (\mathbb{Z}/2\mathbb{Z})^9, \tag{4}$$

where S_6 is the permutation group of six elements. $\mathcal{A}_2(2)$ is the three-dimensional moduli space of principally polarized abelian surfaces with level-two structure. The meaning of $\mathcal{A}_2(2, 4)$ and $\mathcal{A}_2(4, 8)$ will be discussed below.

For an elliptic variable $z \in \mathbb{C}^2$ and modular variable $\tau \in \mathbb{H}_2$, Riemann's Theta function is defined by setting

$$\vartheta(z, \tau) = \sum_{u \in \mathbb{Z}^2} e^{\pi i (u^t \cdot \tau \cdot u + 2u^t \cdot z)}.$$

The Theta function is holomorphic on $\mathbb{C}^2 \times \mathbb{H}_2$. For rational-valued vector $\vec{a}, \vec{b} \in \mathbb{Q}^2$, a Theta function is defined by setting

$$\vartheta \begin{bmatrix} \vec{a} \\ \vec{b} \end{bmatrix} (z, \tau) = \sum_{u \in \mathbb{Z}^2} e^{\pi i \left((u+\vec{a})^t \cdot \tau \cdot (u+\vec{a}) + 2(u+\vec{a})^t \cdot (z+\vec{b}) \right)}.$$

For a rational matrix $\begin{pmatrix} a_1 & a_2 \\ b_1 & b_2 \end{pmatrix}$, which we call a *theta characteristic*, we set $\vec{a} = \langle a_1, a_2 \rangle / 2$ and $\vec{b} = \langle b_1, b_2 \rangle / 2$ and define

$$\theta \begin{bmatrix} a_1 & a_2 \\ b_1 & b_2 \end{bmatrix} (z, \tau) = \vartheta \begin{bmatrix} \vec{a} \\ \vec{b} \end{bmatrix} (z, \tau).$$

In this way, all standard Theta functions have characteristics with coefficients in $\mathbb{F}_2$. This will make it easier to relate them to the description of the 16_6 configuration in finite geometry in Section 3.1.1. For $\begin{pmatrix} a_1 & a_2 \\ b_1 & b_2 \end{pmatrix} \in \mathbb{F}_2^4$ – where $\mathbb{F}_2$ denotes the finite field with two elements – there are sixteen corresponding Theta functions; 10 are even and 6 are odd functions according to

$$\theta \begin{bmatrix} a_1 & a_2 \\ b_1 & b_2 \end{bmatrix} (-z, \tau) = (-1)^{a^t \cdot b} \theta \begin{bmatrix} a_1 & a_2 \\ b_1 & b_2 \end{bmatrix} (z, \tau). \tag{5}$$

We denote the even Theta functions by

$$\theta_1 = \theta \begin{bmatrix} 0 & 0 \\ 0 & 0 \end{bmatrix}, \; \theta_2 = \theta \begin{bmatrix} 0 & 0 \\ 1 & 1 \end{bmatrix}, \; \theta_3 = \theta \begin{bmatrix} 0 & 0 \\ 1 & 0 \end{bmatrix}, \; \theta_4 = \theta \begin{bmatrix} 0 & 0 \\ 0 & 1 \end{bmatrix}, \; \theta_5 = \theta \begin{bmatrix} 1 & 0 \\ 0 & 0 \end{bmatrix},$$
$$\theta_6 = \theta \begin{bmatrix} 1 & 0 \\ 0 & 1 \end{bmatrix}, \; \theta_7 = \theta \begin{bmatrix} 0 & 1 \\ 0 & 0 \end{bmatrix}, \; \theta_8 = \theta \begin{bmatrix} 1 & 1 \\ 0 & 0 \end{bmatrix}, \; \theta_9 = \theta \begin{bmatrix} 0 & 1 \\ 1 & 0 \end{bmatrix}, \; \theta_{10} = \theta \begin{bmatrix} 1 & 1 \\ 1 & 1 \end{bmatrix}.$$

We denote the odd Theta functions by

$$\theta_{11} = \theta\begin{bmatrix}0 & 1 \\ 0 & 1\end{bmatrix}, \ \theta_{12} = \theta\begin{bmatrix}0 & 1 \\ 1 & 1\end{bmatrix}, \ \theta_{13} = \theta\begin{bmatrix}1 & 1 \\ 0 & 1\end{bmatrix},$$

$$\theta_{14} = \theta\begin{bmatrix}1 & 0 \\ 1 & 0\end{bmatrix}, \ \theta_{15} = \theta\begin{bmatrix}1 & 0 \\ 1 & 1\end{bmatrix}, \ \theta_{16} = \theta\begin{bmatrix}1 & 1 \\ 1 & 0\end{bmatrix}.$$

A scalar obtained by evaluating a Theta function at $z = 0$ is called a Theta constant. The six odd Theta functions give trivial Theta constants. We write

$$\theta_i(z) \quad \text{instead of} \quad \theta\begin{bmatrix}a_1^{(i)} & a_2^{(i)} \\ b_1^{(i)} & b_2^{(i)}\end{bmatrix}(z, \tau) \quad \text{where } i = 1, \dots, 16, \qquad (6)$$

and $\theta_i = \theta_i(0)$ such that $\theta_i = 0$ for $i = 11, \dots, 16$.

According to [32, Sec. 3] we have the following *Frobenius identities* relating Theta constants:

$$
\begin{array}{rclcrcl}
\theta_5^2\theta_6^2 & = & \theta_1^2\theta_4^2 - \theta_2^2\theta_3^2, & \quad & \theta_5^4 + \theta_6^4 & = & \theta_1^4 - \theta_2^4 - \theta_3^4 + \theta_4^4, \\
\theta_7^2\theta_9^2 & = & \theta_1^2\theta_3^2 - \theta_2^2\theta_4^2, & \quad & \theta_7^4 + \theta_9^4 & = & \theta_1^4 - \theta_2^4 + \theta_3^4 - \theta_4^4, \quad (7) \\
\theta_8^2\theta_{10}^2 & = & \theta_1^2\theta_2^2 - \theta_3^2\theta_4^2, & \quad & \theta_8^4 + \theta_{10}^4 & = & \theta_1^4 + \theta_2^4 - \theta_3^4 - \theta_4^4.
\end{array}
$$

2.1.1 Doubling Formulas for Theta Constants

We will also introduce the Theta functions that are evaluated at 2τ. Under duplication of the modular variable, the Theta functions $\theta_1, \theta_5, \theta_7, \theta_8$ play a role dual to $\theta_1, \theta_2, \theta_3, \theta_4$. We renumber the former and use the symbol Θ to mark the fact that they are evaluated at the isogenous abelian variety. That is, we will denote Theta functions with doubled modular variable by

$$\Theta_i(z) \quad \text{instead of} \quad \theta\begin{bmatrix}b_1^{(i)} & b_2^{(i)} \\ a_1^{(i)} & a_2^{(i)}\end{bmatrix}(z, 2\tau) \quad \text{where } i = 1, \dots, 16, \qquad (8)$$

and $\Theta_i = \Theta_i(0)$. The following identities are called the *second principal transformations of degree two* [20, 21] for Theta constants:

$$
\begin{array}{rclcrcl}
\theta_1^2 & = & \Theta_1^2 + \Theta_2^2 + \Theta_3^2 + \Theta_4^2, & \quad & \theta_2^2 & = & \Theta_1^2 + \Theta_2^2 - \Theta_3^2 - \Theta_4^2, \\
\theta_3^2 & = & \Theta_1^2 - \Theta_2^2 - \Theta_3^2 + \Theta_4^2, & \quad & \theta_4^2 & = & \Theta_1^2 - \Theta_2^2 + \Theta_3^2 - \Theta_4^2.
\end{array}
$$
$$(9)$$

We also have the following identities:

$$
\begin{array}{rclcrcl}
\theta_5^2 & = & 2\left(\Theta_1\Theta_3 + \Theta_2\Theta_4\right), & \quad & \theta_6^2 & = & 2\left(\Theta_1\Theta_3 - \Theta_2\Theta_4\right), \\
\theta_7^2 & = & 2\left(\Theta_1\Theta_4 + \Theta_2\Theta_3\right), & \quad & \theta_8^2 & = & 2\left(\Theta_1\Theta_2 + \Theta_3\Theta_4\right), \quad (10) \\
\theta_9^2 & = & 2\left(\Theta_1\Theta_4 - \Theta_2\Theta_3\right), & \quad & \theta_{10}^2 & = & 2\left(\Theta_1\Theta_2 - \Theta_3\Theta_4\right).
\end{array}
$$

Similarly, we have identities for Theta functions with non-vanishing elliptic argument:

$$\theta_1\,\theta_1(z) = \Theta_1(z)^2 + \Theta_2(z)^2 + \Theta_3(z)^2 + \Theta_4(z)^2\,,$$

$$\theta_2\,\theta_2(z) = \Theta_1(z)^2 + \Theta_2(z)^2 - \Theta_3(z)^2 - \Theta_4(z)^2\,,$$

$$\theta_3\,\theta_3(z) = \Theta_1(z)^2 - \Theta_2(z)^2 - \Theta_3(z)^2 + \Theta_4(z)^2\,,$$

$$\theta_4\,\theta_4(z) = \Theta_1(z)^2 - \Theta_2(z)^2 + \Theta_3(z)^2 - \Theta_4(z)^2\,, \tag{11}$$

and

$$4\,\Theta_1\,\Theta_1(2z) = \theta_1(z)^2 + \theta_2(z)^2 + \theta_3(z)^2 + \theta_4(z)^2\,,$$

$$4\,\Theta_2\,\Theta_2(2z) = \theta_1(z)^2 + \theta_2(z)^2 - \theta_3(z)^2 - \theta_4(z)^2\,,$$

$$4\,\Theta_3\,\Theta_3(2z) = \theta_1(z)^2 - \theta_2(z)^2 - \theta_3(z)^2 + \theta_4(z)^2\,,$$

$$4\,\Theta_4\,\Theta_4(2z) = \theta_1(z)^2 - \theta_2(z)^2 + \theta_3(z)^2 - \theta_4(z)^2\,. \tag{12}$$

The following is a well-known fact:

Remark 2.1   For the principally polarized abelian surface $(A_\tau, \mathcal{L})$ defined above, a basis of sections for $\mathcal{L}^2$, called Theta functions of level 2, is given by $\Theta_i(2z)$ or, alternatively, $\theta_i^2(z)$ for $1 \leq i \leq 4$, and the point $[\Theta_1 : \Theta_2 : \Theta_3 : \Theta_4] \in \mathbb{P}^3$ is called the Theta null point of level 2 of A_τ. Similarly, a basis of sections for $\mathcal{L}^4$, called the Theta functions of level $(2,2)$, is given by $\theta_i(z)$ for $11 \leq i \leq 16$, and $[\theta_1 : \cdots : \theta_{10}] \in \mathbb{P}^9$ is called the Theta null point of level $(2,2)$ of A_τ.

2.2 Mumford Identities for Theta Functions

To obtain all quadratic relations between Theta functions, we apply the generalized Riemann identity for Theta functions derived by Mumford in [29, p.214]. His master equation (R_{CH}) generating all Theta relations – which we adjusted to reflect our convention for Theta functions – is given by

$$4 \prod_{\epsilon,\epsilon' \in \{\pm 1\}} \theta \begin{bmatrix} \frac{a+\epsilon b+\epsilon' c+\epsilon\epsilon' d}{2} \\ \frac{e+\epsilon f+\epsilon' g+\epsilon\epsilon' h}{2} \end{bmatrix} \left(\frac{x + \epsilon y + \epsilon' u + \epsilon\epsilon' v}{2} \right)$$

$$= \sum_{\alpha,\beta \in \mathbb{Z}/2\mathbb{Z}} e^{-\frac{\pi i}{2} \beta^t (a+b+c+d)} \theta \begin{bmatrix} a+\alpha \\ e+\beta \end{bmatrix} (x)\, \theta \begin{bmatrix} b+\alpha \\ f+\beta \end{bmatrix} (y)\, \theta \begin{bmatrix} c+\alpha \\ g+\beta \end{bmatrix} (u)$$

$$\times \theta \begin{bmatrix} d+\alpha \\ h+\beta \end{bmatrix} (v). \tag{13}$$

We first consider all four-term relations between squares of Theta functions, we set $\xi_i = \theta_i^2(z)$ for $1 \leq i \leq 16$. We have the following:

Proposition 2.2 *The ideal of linear equations relating squares of Theta functions are generated by 12 equations: three equations relating odd Theta functions*

$$\theta_6^2 \xi_{11} - \theta_4^2 \xi_{13} - \theta_9^2 \xi_{14} + \theta_3^2 \xi_{16} = 0,$$

$$\theta_6^2 \xi_{12} - \theta_2^2 \xi_{13} - \theta_7^2 \xi_{14} + \theta_1^2 \xi_{16} = 0,$$

$$\theta_{10}^2 \xi_{13} + \theta_5^2 \xi_{14} - \theta_6^2 \xi_{15} - \theta_8^2 \xi_{16} = 0, \qquad (14)$$

and nine equations relating even and odd Theta functions

$$\theta_6^2 \xi_1 - \theta_1^2 \xi_6 + \theta_7^2 \xi_{13} - \theta_2^2 \xi_{14} = 0,$$

$$\theta_6^2 \xi_2 - \theta_2^2 \xi_6 - \theta_1^2 \xi_{14} + \theta_7^2 \xi_{16} = 0,$$

$$\theta_6^2 \xi_3 - \theta_3^2 \xi_6 + \theta_9^2 \xi_{13} - \theta_4^2 \xi_{14} = 0,$$

$$\theta_6^2 \xi_4 - \theta_4^2 \xi_6 - \theta_3^2 \xi_{14} + \theta_9^2 \xi_{16} = 0,$$

$$\theta_6^2 \xi_5 - \theta_5^2 \xi_6 + \theta_8^2 \xi_{13} - \theta_{10}^2 \xi_{16} = 0,$$

$$\theta_7^2 \xi_6 - \theta_6^2 \xi_7 - \theta_1^2 \xi_{13} + \theta_2^2 \xi_{16} = 0,$$

$$\theta_8^2 \xi_6 - \theta_6^2 \xi_8 - \theta_5^2 \xi_{13} + \theta_{10}^2 \xi_{14} = 0,$$

$$\theta_9^2 \xi_6 - \theta_6^2 \xi_9 - \theta_3^2 \xi_{13} + \theta_4^2 \xi_{16} = 0,$$

$$\theta_{10}^2 \xi_6 - \theta_6^2 \xi_{10} + \theta_8^2 \xi_{14} - \theta_5^2 \xi_{16} = 0. \qquad (15)$$

Proof. We follow the strategy outlined in [22, Sec. 3.2] where fifteen quadrics of rank four involving only odd Theta functions were derived. Using [29, p.214, Eq. (R_{CH})], one generates all rank-four quadrics relating squares of Theta function. We obtain 142 equations. One then uses Equations (9) and Equations (10) to write all coefficients in terms of $\{\Theta_1, \Theta_2, \Theta_3, \Theta_4\}$ and determines a generating set of quadrics. □

Next, we consider all three-term relations between bi-monomial combinations of Theta functions. We set $\xi_{i,j} = \theta_i(z)\theta_j(z)$ for $1 \le i < j \le 16$. We have the following:

Proposition 2.3 *The ideal of linear equations relating $\xi_{i,j}$ for $1 \le i < j \le 16$ is generated by the following 60 equations:*

$$\theta_1\theta_2\xi_{1,2} - \theta_3\theta_4\xi_{3,4} - \theta_8\theta_{10}\xi_{8,10} = 0, \ \theta_3\theta_4\xi_{1,2} - \theta_1\theta_2\xi_{3,4} - \theta_8\theta_{10}\xi_{13,16} = 0, \ \dots$$

(A complete generating set of 60 equations is given in Equation (108).) (16)

Proof. Using [29, p.214, Eq. (R_{CH})], one generates all three-term bi-monomial combinations of Theta function. □

Remark 2.4 The map $z \mapsto [\theta_1(z) : \cdots : \theta_{16}(z)]$ given by a all Theta functions provides a high-dimensional embedding of the abelian variety $\mathbf{A}_\tau$ into $\mathbb{P}^{15}$ [30, Sec. 3] whose defining equations were determined explicitly in [14]. The image in $\mathbb{P}^{15}$ is given as the intersection of the 72 conics given by Equations (14), (15), and (108).

2.3 Theta Functions and Genus-Two Curves

Let C be an irreducible, smooth, projective curve of genus two, defined over the complex field $\mathbb{C}$. Let $\mathcal{M}_2$, be the coarse moduli space of smooth curves of genus two. We denote by $[C]$ the isomorphism class of C, i.e., the corresponding point in $\mathcal{M}_2$. For a genus-two curve C given as sextic $Y^2 = f_6(X, Z)$ in weighted projective space $\mathbb{P}(1, 3, 1)$, we send three roots $\lambda_4, \lambda_5, \lambda_6$ to $0, 1, \infty$ to get an isomorphic curve in Rosenhain normal form, i.e.,

$$C: \quad Y^2 = X Z \left(X - Z\right) \left(X - \lambda_1 Z\right) \left(X - \lambda_2 Z\right) \left(X - \lambda_3 Z\right). \tag{17}$$

The ordered tuple $(\lambda_1, \lambda_2, \lambda_3)$ where the λ_i are all distinct and different from $0, 1, \infty$ determines a point in $\mathcal{M}_2(2)$, the moduli space of genus-two curves with level-two structure.

Torelli's theorem states that the map sending a curve C to its Jacobian $\mathrm{Jac}(C)$ is injective and defines a birational map $\mathcal{M}_2 \dashrightarrow \mathcal{A}_2$. In fact, if the point τ is not equivalent with respect to Γ_2 to a point with $\tau_{12} = 0$, the Θ-divisor is a non-singular curve C of genus-two and $\mathbf{A}_\tau = \mathrm{Jac}(C)$ is its Jacobian.

Thomae's formula is a formula introduced by Thomae relating Theta constants to the branch points. The three λ parameters appearing in the Rosenhain normal form of a genus-two curve C in Equation (17) are ratios of even Theta constants. There are 720 choices for such expressions since the forgetful map $\mathcal{M}_2(2) \rightarrow \mathcal{M}_2$ is a Galois covering of degree $720 = |S_6|$ where S_6 acts on the roots of C by permutations. Any of the 720 choices may be used, we choose the one also used in [32, 17]:

Lemma 2.5 *For any period point $\tau \in \mathcal{A}_2(2)$, there is a genus-two curve $C \in \mathcal{M}_2(2)$ with level-two structure and Rosenhain roots $\lambda_1, \lambda_2, \lambda_3$ such that*

$$\lambda_1 = \frac{\theta_1^2 \theta_3^2}{\theta_2^2 \theta_4^2}, \quad \lambda_2 = \frac{\theta_3^2 \theta_8^2}{\theta_4^2 \theta_{10}^2}, \quad \lambda_3 = \frac{\theta_1^2 \theta_8^2}{\theta_2^2 \theta_{10}^2}. \tag{18}$$

Similarly, the following expressions are perfect squares of Theta constants:

$$\lambda_1 - 1 = \frac{\theta_7^2 \theta_9^2}{\theta_2^2 \theta_4^2}, \qquad \lambda_2 - 1 = \frac{\theta_5^2 \theta_9^2}{\theta_4^2 \theta_{10}^2}, \qquad \lambda_3 - 1 = \frac{\theta_5^2 \theta_7^2}{\theta_2^2 \theta_{10}^2},$$

$$\lambda_2 - \lambda_1 = \frac{\theta_3^2 \theta_6^2 \theta_9^2}{\theta_2^2 \theta_4^2 \theta_{10}^2}, \qquad \lambda_3 - \lambda_1 = \frac{\theta_1^2 \theta_6^2 \theta_7^2}{\theta_2^2 \theta_4^2 \theta_{10}^2}, \qquad \lambda_3 - \lambda_2 = \frac{\theta_5^2 \theta_6^2 \theta_8^2}{\theta_2^2 \theta_4^2 \theta_{10}^2}.$$

$$(19)$$

Conversely, given a smooth genus-two curve $\mathcal{C} \in \mathcal{M}_2(2)$ with three distinct complex numbers $(\lambda_1, \lambda_2, \lambda_3)$ different from $0, 1, \infty$, there is complex abelian surface $\mathbf{A}_\tau$ with period matrix $(\mathbb{I}_2, \tau)$ and $\tau \in \mathcal{A}_2(2)$ such that $\mathbf{A}_\tau = \operatorname{Jac} \mathcal{C}$ and the fourth powers of the even Theta constants are given by

$$\theta_1^4 = R\,\lambda_3\lambda_1(\lambda_2 - 1)(\lambda_3 - \lambda_1), \qquad \theta_2^4 = R\,\lambda_2(\lambda_2 - 1)(\lambda_3 - \lambda_1),$$

$$\theta_3^4 = R\,\lambda_2\lambda_1(\lambda_2 - \lambda_1)(\lambda_3 - \lambda_1), \qquad \theta_4^4 = R\,\lambda_3(\lambda_3 - 1)(\lambda_2 - \lambda_1),$$

$$\theta_5^4 = R\,\lambda_1(\lambda_2 - 1)(\lambda_3 - 1)(\lambda_3 - \lambda_2), \qquad \theta_6^4 = R\,(\lambda_3 - \lambda_2)(\lambda_3 - \lambda_1)(\lambda_2 - \lambda_1),$$

$$\theta_7^4 = R\,\lambda_2(\lambda_3 - 1)(\lambda_1 - 1)(\lambda_3 - \lambda_1), \qquad \theta_8^4 = R\,\lambda_2\lambda_3(\lambda_3 - \lambda_2)(\lambda_1 - 1),$$

$$\theta_9^4 = R\,\lambda_3(\lambda_2 - 1)(\lambda_1 - 1)(\lambda_2 - \lambda_1), \qquad \theta_{10}^4 = R\,\lambda_1(\lambda_1 - 1)(\lambda_3 - \lambda_2),$$

$$(20)$$

where $R \in \mathbb{C}^*$ is a non-zero constant.

The characterization of the Siegel modular threefolds $\mathcal{A}_2(2,4)$, and $\mathcal{A}_2(4,8)$ as projective varieties and their Satake compactifications was given in [22, Prop. 2.2]:

Lemma 2.6

(1) *The holomorphic map $\Xi_{2,4} : \mathbb{H}_2 \to \mathbb{P}^3$ given by*
    $\tau \mapsto [\Theta_1 : \Theta_2 : \Theta_3 : \Theta_4]$ induces an isomorphism between the Satake
    compactification $\overline{\mathcal{A}_2(2,4)}$ and $\mathbb{P}^3$.

(2) *The holomorphic map $\Xi_{4,8} : \mathbb{H}_2 \to \mathbb{P}^9$ given by $\tau \mapsto [\theta_1 : \cdots : \theta_{10}]$*
    induces an isomorphism between the Satake compactification $\overline{\mathcal{A}_2(4,8)}$
    and the closure of $\Xi_{4,8}$ in $\mathbb{P}^9$.

(3) *We have the following commutative diagram:*

$$
\begin{array}{ccc}
\overline{\mathcal{A}_2(4,8)} & \xrightarrow{\;\;\Xi_{4,8}\;\;} & \mathbb{P}^9 \\
\Big\downarrow{\scriptstyle \pi} & & \Big\downarrow{\scriptstyle \mathrm{Sq}} \\
\overline{\mathcal{A}_2(2,4)} & \xrightarrow{\Xi_{2,4}} \mathbb{P}^3 \xrightarrow{\mathrm{Ver}} & \mathbb{P}^9 \\
\Big\downarrow{\scriptstyle \mathrm{Ros}} & & \\
\overline{\mathcal{A}_2(2)} & &
\end{array}
$$

Here, π is the covering map with deck transformations $\Gamma_2(2, 4)/\Gamma_2(4, 8)$
$\cong (\mathbb{Z}/2\mathbb{Z})^9$, the map Sq is the square map $[\theta_1 : \cdots : \theta_{10}] \mapsto$
$[\theta_1^2 : \cdots : \theta_{10}^2]$, the map Ver is the Veronese type map defined by the
quadratic relations (9) and (10), and the map Ros is the covering map
with the deck transformations $\Gamma_2(2)/\Gamma_2(2, 4) \cong (\mathbb{Z}/2\mathbb{Z})^3$ given by
plugging the quadratic relations (9) and (10) into Equations (18).

Remark 2.7 Lemma 2.5 shows that $\Gamma_2(2)$ is the group of isomorphisms
which fix the 4th power of the Theta constants θ_i for $1 \leq i \leq 10$, $\Gamma_2(2, 4)$ fixes
their 2nd power, and $\Gamma_2(4, 8)$ the Theta constants of level $(2, 2)$ themselves.

3 Jacobians and Two-Isogenies

3.1 16_6 Configuration on the Jacobian

On the Jacobian $\mathbf{A} = \text{Jac}(\mathcal{C})$ the sixteen Theta divisors together with the
sixteen order-two points form a 16_6 configuration in the following way; see
[11]. For the genus-two curve $\mathcal{C}$ in Equation (17), we denote the Weierstrass
points by $\mathfrak{p}_i : [X : Y : Z] = [\lambda_i : 0 : 1]$ for $1 \leq i \leq 3$, $\mathfrak{p}_4 : [X : Y : Z] = [0 :$
$0 : 1]$, $\mathfrak{p}_5 : [X : Y : Z] = [1 : 0 : 1]$, and $\mathfrak{p}_6 : [X : Y : Z] = [1 : 0 : 0]$; we
also denote the space of two torsion points on an abelian variety $\mathbf{A}$ by $\mathbf{A}[2]$.
In the case of the Jacobian of a genus-two curve, every nontrivial two-torsion
point can be expressed using differences of Weierstrass points of $\mathcal{C}$. Concretely,
the sixteen order-two points of $\mathbf{A}[2]$ are obtained using the embedding of the
curve into the connected component of the identity in the Picard group, i.e.,
$\mathcal{C} \hookrightarrow \mathbf{A} \cong \text{Pic}^0(\mathcal{C})$ with $\mathfrak{p} \mapsto [\mathfrak{p} - \mathfrak{p}_6]$. In particular, we obtain all elements of
$\mathbf{A}[2]$ as

$$\mathfrak{p}_{i6} = [\mathfrak{p}_i - \mathfrak{p}_6] \text{ for } 1 \leq i < 6, \qquad \mathfrak{p}_{ij} = [\mathfrak{p}_i + \mathfrak{p}_j - 2\mathfrak{p}_6] \text{ for } 1 \leq i < j \leq 5,$$
$$(21)$$

where we set $\mathfrak{p}_0 = \mathfrak{p}_{66} = [0]$. For $\{i, j, k, l, m, n\} = \{1, \ldots 6\}$, the group law
on $\mathbf{A}[2]$ is given by the relations

$$\mathfrak{p}_0 + \mathfrak{p}_{ij} = \mathfrak{p}_{ij}, \quad \mathfrak{p}_{ij} + \mathfrak{p}_{ij} = \mathfrak{p}_0, \quad \mathfrak{p}_{ij} + \mathfrak{p}_{kl} = \mathfrak{p}_{mn}, \quad \mathfrak{p}_{ij} + \mathfrak{p}_{jk} = \mathfrak{p}_{ik}.$$
$$(22)$$

The space $\mathbf{A}[2]$ of two torsion points on an abelian variety $\mathbf{A}$ admits a
symplectic bilinear form, called the *Weil pairing*. The Weil pairing is induced
by the pairing

$$\langle [\mathfrak{p}_i - \mathfrak{p}_j], [\mathfrak{p}_k - \mathfrak{p}_l] \rangle = \#\{\mathfrak{p}_i, \mathfrak{p}_j\} \cap \{\mathfrak{p}_k, \mathfrak{p}_l\} \mod 2,$$

such that the following table gives all possible pairings:

$\langle \bullet, \bullet \rangle$	$\mathfrak{p}_0$	$\mathfrak{p}_{i6}$	$\mathfrak{p}_{j6}$	$\mathfrak{p}_{ij}$	$\mathfrak{p}_{il}$	$\mathfrak{p}_{kl}$
$\mathfrak{p}_0$	0	0	0	0	0	0
$\mathfrak{p}_{i6}$	0	0	1	1	1	0
$\mathfrak{p}_{j6}$	0	1	0	1	0	0
$\mathfrak{p}_{ij}$	0	1	1	0	1	0
$\mathfrak{p}_{il}$	0	1	0	1	0	1
$\mathfrak{p}_{kl}$	0	0	0	0	1	0

$$(23)$$

We call a two-dimensional, maximal isotropic subspace of $\mathbf{A}[2]$ with respect to the Weil pairing, i.e., a subspace such that the symplectic form vanishes on it, a *Göpel group* in $\mathbf{A}[2]$. Such a maximal subgroup is isomorphic to $(\mathbb{Z}/2\mathbb{Z})^2$. We have the following:

Lemma 3.1 *There are 15 distinct Göpel groups in* $\mathbf{A}[2]$ *given by*

$$\{\mathfrak{p}_0, \mathfrak{p}_{12}, \mathfrak{p}_{34}, \mathfrak{p}_{56}\}, \ \{\mathfrak{p}_0, \mathfrak{p}_{12}, \mathfrak{p}_{35}, \mathfrak{p}_{46}\}, \ \{\mathfrak{p}_0, \mathfrak{p}_{12}, \mathfrak{p}_{36}, \mathfrak{p}_{45}\}, \ \{\mathfrak{p}_0, \mathfrak{p}_{13}, \mathfrak{p}_{24}, \mathfrak{p}_{56}\},$$
$$\{\mathfrak{p}_0, \mathfrak{p}_{13}, \mathfrak{p}_{25}, \mathfrak{p}_{46}\}, \ \{\mathfrak{p}_0, \mathfrak{p}_{13}, \mathfrak{p}_{26}, \mathfrak{p}_{45}\}, \ \{\mathfrak{p}_0, \mathfrak{p}_{14}, \mathfrak{p}_{23}, \mathfrak{p}_{56}\}, \ \{\mathfrak{p}_0, \mathfrak{p}_{14}, \mathfrak{p}_{25}, \mathfrak{p}_{36}\},$$
$$\{\mathfrak{p}_0, \mathfrak{p}_{14}, \mathfrak{p}_{26}, \mathfrak{p}_{35}\}, \ \{\mathfrak{p}_0, \mathfrak{p}_{15}, \mathfrak{p}_{23}, \mathfrak{p}_{46}\}, \ \{\mathfrak{p}_0, \mathfrak{p}_{15}, \mathfrak{p}_{24}, \mathfrak{p}_{36}\}, \ \{\mathfrak{p}_0, \mathfrak{p}_{15}, \mathfrak{p}_{26}, \mathfrak{p}_{34}\},$$
$$\{\mathfrak{p}_0, \mathfrak{p}_{16}, \mathfrak{p}_{23}, \mathfrak{p}_{45}\}, \ \{\mathfrak{p}_0, \mathfrak{p}_{16}, \mathfrak{p}_{24}, \mathfrak{p}_{35}\}, \ \{\mathfrak{p}_0, \mathfrak{p}_{16}, \mathfrak{p}_{25}, \mathfrak{p}_{34}\}.$$

Moreover, there are 60 distinct affine spaces in $\mathbf{A}[2]$ *obtained from the four translates of each Göpel group.*

$\square$

We define a *Rosenhain group* to be a subgroup in $\mathbf{A}[2]$ isomorphic to $(\mathbb{Z}/2\mathbb{Z})^2$ of the from $\{\mathfrak{p}_0, \mathfrak{p}_{ij}, \mathfrak{p}_{ik}, \mathfrak{p}_{jk}\}$ with $1 \leq i < j, k \leq 6$. Note that a Rosenhain group is not an isotropic subspace of $\mathbf{A}[2]$. We have the following:

Lemma 3.2 *There are 20 different Rosenhain groups in* $\mathbf{A}[2]$ *given by*

$$\{\mathfrak{p}_0, \mathfrak{p}_{12}, \mathfrak{p}_{13}, \mathfrak{p}_{23}\}, \ \{\mathfrak{p}_0, \mathfrak{p}_{12}, \mathfrak{p}_{14}, \mathfrak{p}_{24}\}, \ \{\mathfrak{p}_0, \mathfrak{p}_{12}, \mathfrak{p}_{15}, \mathfrak{p}_{25}\}, \ \{\mathfrak{p}_0, \mathfrak{p}_{12}, \mathfrak{p}_{16}, \mathfrak{p}_{26}\},$$
$$\{\mathfrak{p}_0, \mathfrak{p}_{13}, \mathfrak{p}_{14}, \mathfrak{p}_{34}\}, \ \{\mathfrak{p}_0, \mathfrak{p}_{13}, \mathfrak{p}_{15}, \mathfrak{p}_{35}\}, \ \{\mathfrak{p}_0, \mathfrak{p}_{13}, \mathfrak{p}_{16}, \mathfrak{p}_{36}\}, \ \{\mathfrak{p}_0, \mathfrak{p}_{14}, \mathfrak{p}_{15}, \mathfrak{p}_{45}\},$$
$$\{\mathfrak{p}_0, \mathfrak{p}_{14}, \mathfrak{p}_{16}, \mathfrak{p}_{46}\}, \ \{\mathfrak{p}_0, \mathfrak{p}_{15}, \mathfrak{p}_{16}, \mathfrak{p}_{56}\}, \ \{\mathfrak{p}_0, \mathfrak{p}_{23}, \mathfrak{p}_{24}, \mathfrak{p}_{34}\}, \ \{\mathfrak{p}_0, \mathfrak{p}_{23}, \mathfrak{p}_{25}, \mathfrak{p}_{35}\},$$
$$\{\mathfrak{p}_0, \mathfrak{p}_{23}, \mathfrak{p}_{26}, \mathfrak{p}_{36}\}, \ \{\mathfrak{p}_0, \mathfrak{p}_{24}, \mathfrak{p}_{25}, \mathfrak{p}_{45}\}, \ \{\mathfrak{p}_0, \mathfrak{p}_{24}, \mathfrak{p}_{26}, \mathfrak{p}_{46}\}, \ \{\mathfrak{p}_0, \mathfrak{p}_{25}, \mathfrak{p}_{26}, \mathfrak{p}_{56}\},$$
$$\{\mathfrak{p}_0, \mathfrak{p}_{34}, \mathfrak{p}_{35}, \mathfrak{p}_{45}\}, \ \{\mathfrak{p}_0, \mathfrak{p}_{34}, \mathfrak{p}_{36}, \mathfrak{p}_{46}\}, \ \{\mathfrak{p}_0, \mathfrak{p}_{35}, \mathfrak{p}_{36}, \mathfrak{p}_{56}\}, \ \{\mathfrak{p}_0, \mathfrak{p}_{45}, \mathfrak{p}_{46}, \mathfrak{p}_{56}\}.$$

Moreover, there are 80 distinct affine spaces in $\mathbf{A}[2]$ *comprised of four translates of each Rosenhain group.*

$\square$

In general, for a principally polarized abelian variety $\mathbf{A}$ the line bundle $\mathcal{L}$ defining its principal polarization is ample and satisfies $h^0(\mathcal{L}) = 1$. Then, there exists an effective divisor Θ such that $\mathcal{L} = \mathcal{O}_{\mathbf{A}}(\Theta)$, uniquely defined only up to a translation. The divisor Θ is called a *Theta divisor* associated with the polarization. It is known that the abelian surface $\mathbf{A}$ is not the product of two elliptic curves if and only if Θ is an irreducible divisor. In this case, Θ is a smooth curve of genus two and $\mathbf{A} = \mathrm{Jac}(\mathcal{C})$. The standard Theta divisor $\Theta = \Theta_6 = \{[\mathfrak{p}-\mathfrak{p}_0] \mid \mathfrak{p} \in \mathcal{C}\}$ contains the six order-two points $\mathfrak{p}_0$, $\mathfrak{p}_{i6}$ for $1 \leq i \leq 5$. Likewise for $1 \leq i \leq 5$, the five translates $\Theta_i = \mathfrak{p}_{i6} + \Theta$ contain $\mathfrak{p}_0$, $\mathfrak{p}_{i6}$, $\mathfrak{p}_{ij}$ with $j \neq i, 6$, and the ten translates $\Theta_{ij6} = \mathfrak{p}_{ij} + \Theta$ for $1 \leq i < j \leq 5$ contain $\mathfrak{p}_{ij}$, $\mathfrak{p}_{i6}$, $\mathfrak{p}_{j6}$, $\mathfrak{p}_{kl}$ with $k, l \neq i, j, 6$ and $k < l$. Conversely, each order-two point lies on exactly six of the divisors, namely

$$\mathfrak{p}_0 \in \Theta_i \qquad \text{for } i = 1, \ldots, 6, \tag{24}$$

$$\mathfrak{p}_{i6} \in \Theta_i, \Theta_6, \Theta_{ij6} \quad \text{for } i = 1, \ldots, 5 \text{ with } j \neq i, 6, \tag{25}$$

$$\mathfrak{p}_{ij} \in \Theta_i, \Theta_j, \Theta_{kl6} \quad \text{for } 1 \leq i < j \leq 5 \text{ with } k, l \neq i, j, 6 \text{ and } k < l. \tag{26}$$

We call the divisors $\{\Theta_i\}$ and $\{\Theta_{jk6}\}$ with $1 \leq i \leq 6$ and $1 \leq j < k < 6$, the six *odd* and the ten *even* Theta divisors, respectively. The odd Theta divisors can be identified with the six translates of the curve $\mathcal{C}$ by $\mathfrak{p}_{i6}$ with $1 \leq i \leq 6$, and thus with the six Weierstrass points $\mathfrak{p}_i$ on the curve $\mathcal{C}$. Furthermore, there is a one-to-one correspondence between pairs of odd symmetric Theta divisors and two-torsion points on $\mathbf{A}[2]$ since $\Theta_i \cap \Theta_j = \{\mathfrak{p}_0, \mathfrak{p}_{ij}\}$, and, in turn, unordered pairs $\{\mathfrak{p}_i, \mathfrak{p}_j\}$ of Weierstrass points since $\mathfrak{p}_{ij} = \mathfrak{p}_{i6} + \mathfrak{p}_{j6}$.

3.1.1 Relation to Finite Geometry

The 16_6 configuration on $\mathbf{A}[2]$ is effectively described in the context of finite geometry; see [12]. We denote elements of the vector space $\mathbb{F}_2^4$ as matrices $A = \begin{pmatrix} a_1 & a_2 \\ a_3 & a_4 \end{pmatrix}$. It is easy to show that these matrices form a group with $16 = 2^{2g}$ elements for $g = 2$, and a group structure defined by addition modulo two. A symplectic form on $\mathbb{F}_2^4$ is defined by $(A, A') \mapsto \mathrm{Tr}\left(A^t \cdot \begin{pmatrix} 0 & 1 \\ -1 & 0 \end{pmatrix} \cdot A'\right)$. We say that two matrices A, A' are *syzygetic* if $\mathrm{Tr}\left(A^t \cdot \begin{pmatrix} 0 & 1 \\ -1 & 0 \end{pmatrix} \cdot A'\right) \equiv 0 \mod 2$. A *Göpel group* in $\mathbb{F}_2^4$ is a subgroup of four elements such that every two elements are syzygetic. It is well known [32] that (i) each Göpel group in $\mathbb{F}_2^4$ is maximal and isomorphic to $(\mathbb{Z}/2\mathbb{Z})^2$, (ii) the number of different Göpel groups in $\mathbb{F}_2^4$ is 15, (iii) each Göpel group in $\mathbb{F}_2^4$ has $2^{2g-2} = 4$ cosets which are called Göpel systems. Moreover, singular planes in finite geometry are indexed by points in $\mathbb{F}_2^4$ as follows: a plane indexed by $\begin{pmatrix} b_1 & b_2 \\ b_3 & b_4 \end{pmatrix} \in \mathbb{F}_2^4$ contains the points $\begin{pmatrix} a_1 & a_2 \\ a_3 & a_4 \end{pmatrix} \in \mathbb{F}_2^4$ that satisfy either $\begin{pmatrix} a_1 \\ a_3 \end{pmatrix} = \begin{pmatrix} b_1 \\ b_3 \end{pmatrix}$ and $\begin{pmatrix} a_2 \\ a_4 \end{pmatrix} \neq \begin{pmatrix} b_2 \\ b_4 \end{pmatrix}$ or $\begin{pmatrix} a_1 \\ a_3 \end{pmatrix} \neq \begin{pmatrix} b_1 \\ b_3 \end{pmatrix}$ and $\begin{pmatrix} a_2 \\ a_4 \end{pmatrix} = \begin{pmatrix} b_2 \\ b_4 \end{pmatrix}$. The following was proved in [3]:

Lemma 3.3 *The points* $\begin{pmatrix} a_1 & a_2 \\ a_3 & a_4 \end{pmatrix} \in \mathbb{F}_2^4$ *and the singular planes indexed by* $\begin{pmatrix} b_1 & b_2 \\ b_3 & b_4 \end{pmatrix} \in \mathbb{F}_2^4$ *form a* 16_6 *configuration on* $\mathbb{F}_2^4$:

(1) *Any point is contained in exactly six singular planes.*

(2) *Any singular plane indexed contains exactly six points.*

The automorphism group of the 16_6 configuration is $\mathbb{F}_2^4 \rtimes \mathrm{Sp}(4, \mathbb{F}_2)$ and is given by translations by order-two points and rotations preserving the symplectic form.

$\square$

We construct an isomorphism between $\mathbf{A}[2]$ and $\mathbb{F}_2^4$ such that the point $\mathfrak{p}_0$ is mapped to $\left(\begin{smallmatrix} 0 & 0 \\ 0 & 0 \end{smallmatrix}\right)$. Following Lemma 3.3, each Theta divisor Θ_i or Θ_{ijk}, respectively, can also be identified with a singular plane given by points $\mathbb{F}_2^4$. We have the following:

Proposition 3.4 *Table 5.1 provides an isomorphism between points and planes of the 16_6 configurations on $\mathbf{A}[2]$ and $\mathbb{F}_2^4$ such that Göpel groups and their translates in $\mathbf{A}[2]$ – given in Lemma 3.1 – are mapped bijectively to Göpel systems in $\mathbb{F}_2^4$.*

Proof. Since $\mathfrak{p}_0$ is mapped to the matrix $\left(\begin{smallmatrix} 0 & 0 \\ 0 & 0 \end{smallmatrix}\right)$, the divisors Θ_i for $i = 1, \ldots, 6$ must be mapped to the six matrices $\left(\begin{smallmatrix} 0 & b_2 \\ 0 & b_4 \end{smallmatrix}\right)$ or $\left(\begin{smallmatrix} b_1 & 0 \\ b_3 & 0 \end{smallmatrix}\right)$. Making a choice (cf. Remark 3.5) for these, we obtain the images of all points $\mathfrak{p}_{ij}$ with $1 \leq i < j \leq 6$, since we have $\Theta_i \cap \Theta_j = \{\mathfrak{p}_0, \mathfrak{p}_{ij}\}$. Using the properties of the 16_6 configuration, we obtain the matrices indexing the meaning divisors Θ_{ij6}. Finally, one checks by an explicit computation that the Göpel groups and their translates in $\mathbf{A}[2]$ given in Lemma 3.1 coincide with the Göpel systems in $\mathbb{F}_2^4$. $\square$

Remark 3.5 Following the proof of Proposition 3.4 we can say even more. Table 5.1 is the unique isomorphism such that the odd Theta divisors are mapped to translates – namely translates by the fixed element $\left(\begin{smallmatrix} 1 & 1 \\ 1 & 1 \end{smallmatrix}\right)$ – of the characteristics of the odd Theta functions introduced in Section 2.1. We will prove in Lemma 4.11 that this is precisely the property required to make the identification compatible with the Mumford identities for Theta functions while at the same time also mapping odd Theta divisor to odd Theta functions. All Theta divisors and Theta characteristics are then paired up according to Table 5.1.

$\Theta_\bullet$	$\left(\begin{smallmatrix} a_1 & a_2 \\ a_3 & a_4 \end{smallmatrix}\right) + \left(\begin{smallmatrix} 1 & 1 \\ 1 & 1 \end{smallmatrix}\right)$	$\theta\left[\begin{smallmatrix} a_1 & a_2 \\ a_3 & a_4 \end{smallmatrix}\right]$	$\Theta_\bullet$	$\left(\begin{smallmatrix} a_1 & a_2 \\ a_3 & a_4 \end{smallmatrix}\right) + \left(\begin{smallmatrix} 1 & 1 \\ 1 & 1 \end{smallmatrix}\right)$	$\theta\left[\begin{smallmatrix} a_1 & a_2 \\ a_3 & a_4 \end{smallmatrix}\right]$
Θ_1	$\left(\begin{smallmatrix} 1 & 0 \\ 1 & 1 \end{smallmatrix}\right) + \left(\begin{smallmatrix} 1 & 1 \\ 1 & 1 \end{smallmatrix}\right)$	$\theta_{15}(z)$	Θ_2	$\left(\begin{smallmatrix} 0 & 1 \\ 1 & 1 \end{smallmatrix}\right) + \left(\begin{smallmatrix} 1 & 1 \\ 1 & 1 \end{smallmatrix}\right)$	$\theta_{12}(z)$
Θ_3	$\left(\begin{smallmatrix} 0 & 1 \\ 0 & 1 \end{smallmatrix}\right) + \left(\begin{smallmatrix} 1 & 1 \\ 1 & 1 \end{smallmatrix}\right)$	$\theta_{11}(z)$	Θ_4	$\left(\begin{smallmatrix} 1 & 1 \\ 1 & 0 \end{smallmatrix}\right) + \left(\begin{smallmatrix} 1 & 1 \\ 1 & 1 \end{smallmatrix}\right)$	$\theta_{16}(z)$
Θ_5	$\left(\begin{smallmatrix} 1 & 0 \\ 1 & 0 \end{smallmatrix}\right) + \left(\begin{smallmatrix} 1 & 1 \\ 1 & 1 \end{smallmatrix}\right)$	$\theta_{14}(z)$	Θ_6	$\left(\begin{smallmatrix} 0 & 1 \\ 0 & 1 \end{smallmatrix}\right) + \left(\begin{smallmatrix} 1 & 1 \\ 1 & 1 \end{smallmatrix}\right)$	$\theta_{13}(z)$
Θ_{126}	$\left(\begin{smallmatrix} 0 & 0 \\ 0 & 1 \end{smallmatrix}\right) + \left(\begin{smallmatrix} 1 & 1 \\ 1 & 1 \end{smallmatrix}\right)$	$\theta_4(z)$	Θ_{136}	$\left(\begin{smallmatrix} 0 & 0 \\ 1 & 1 \end{smallmatrix}\right) + \left(\begin{smallmatrix} 1 & 1 \\ 1 & 1 \end{smallmatrix}\right)$	$\theta_2(z)$
Θ_{146}	$\left(\begin{smallmatrix} 1 & 0 \\ 0 & 0 \end{smallmatrix}\right) + \left(\begin{smallmatrix} 1 & 1 \\ 1 & 1 \end{smallmatrix}\right)$	$\theta_5(z)$	Θ_{156}	$\left(\begin{smallmatrix} 1 & 1 \\ 0 & 0 \end{smallmatrix}\right) + \left(\begin{smallmatrix} 1 & 1 \\ 1 & 1 \end{smallmatrix}\right)$	$\theta_8(z)$
Θ_{236}	$\left(\begin{smallmatrix} 1 & 1 \\ 1 & 1 \end{smallmatrix}\right) + \left(\begin{smallmatrix} 1 & 1 \\ 1 & 1 \end{smallmatrix}\right)$	$\theta_{10}(z)$	Θ_{246}	$\left(\begin{smallmatrix} 0 & 0 \\ 0 & 0 \end{smallmatrix}\right) + \left(\begin{smallmatrix} 1 & 1 \\ 1 & 1 \end{smallmatrix}\right)$	$\theta_7(z)$
Θ_{256}	$\left(\begin{smallmatrix} 0 & 0 \\ 0 & 0 \end{smallmatrix}\right) + \left(\begin{smallmatrix} 1 & 1 \\ 1 & 1 \end{smallmatrix}\right)$	$\theta_1(z)$	Θ_{346}	$\left(\begin{smallmatrix} 0 & 1 \\ 1 & 0 \end{smallmatrix}\right) + \left(\begin{smallmatrix} 1 & 1 \\ 1 & 1 \end{smallmatrix}\right)$	$\theta_9(z)$
Θ_{356}	$\left(\begin{smallmatrix} 0 & 0 \\ 1 & 0 \end{smallmatrix}\right) + \left(\begin{smallmatrix} 1 & 1 \\ 1 & 1 \end{smallmatrix}\right)$	$\theta_3(z)$	Θ_{456}	$\left(\begin{smallmatrix} 1 & 0 \\ 0 & 1 \end{smallmatrix}\right) + \left(\begin{smallmatrix} 1 & 1 \\ 1 & 1 \end{smallmatrix}\right)$	$\theta_6(z)$

Table 5.1. *Isomorphism between 16_6 configurations*

$\mathfrak{p}_\bullet \in A[2]$	$A \in \mathbb{F}_2^4$	$\Theta_\bullet \in NS(A)$	$\mathfrak{p}_\bullet \in \Theta_\bullet$
$\mathfrak{p}_0$	$\begin{bmatrix} 0 & 0 \\ 0 & 0 \end{bmatrix}$	$\Theta_{236} = \mathfrak{p}_{23} + \Theta$	$\mathfrak{p}_{14}, \mathfrak{p}_{15}, \mathfrak{p}_{23}, \mathfrak{p}_{26}, \mathfrak{p}_{36}, \mathfrak{p}_{45}$
$\mathfrak{p}_{45}$	$\begin{bmatrix} 0 & 1 \\ 0 & 0 \end{bmatrix}$	$\Theta_1 = \mathfrak{p}_{16} + \Theta$	$\mathfrak{p}_0, \mathfrak{p}_{12}, \mathfrak{p}_{13}, \mathfrak{p}_{14}, \mathfrak{p}_{15}, \mathfrak{p}_{16}$
$\mathfrak{p}_{36}$	$\begin{bmatrix} 1 & 0 \\ 0 & 0 \end{bmatrix}$	$\Theta_2 = \mathfrak{p}_{26} + \Theta$	$\mathfrak{p}_0, \mathfrak{p}_{12}, \mathfrak{p}_{23}, \mathfrak{p}_{24}, \mathfrak{p}_{25}, \mathfrak{p}_{26}$
$\mathfrak{p}_{26}$	$\begin{bmatrix} 1 & 0 \\ 1 & 0 \end{bmatrix}$	$\Theta_3 = \mathfrak{p}_{36} + \Theta$	$\mathfrak{p}_0, \mathfrak{p}_{13}, \mathfrak{p}_{23}, \mathfrak{p}_{34}, \mathfrak{p}_{35}, \mathfrak{p}_{36}$
$\mathfrak{p}_{15}$	$\begin{bmatrix} 0 & 0 \\ 0 & 1 \end{bmatrix}$	$\Theta_4 = \mathfrak{p}_{46} + \Theta$	$\mathfrak{p}_0, \mathfrak{p}_{14}, \mathfrak{p}_{24}, \mathfrak{p}_{34}, \mathfrak{p}_{45}, \mathfrak{p}_{46}$
$\mathfrak{p}_{14}$	$\begin{bmatrix} 0 & 1 \\ 0 & 1 \end{bmatrix}$	$\Theta_5 = \mathfrak{p}_{56} + \Theta$	$\mathfrak{p}_0, \mathfrak{p}_{15}, \mathfrak{p}_{25}, \mathfrak{p}_{35}, \mathfrak{p}_{45}, \mathfrak{p}_{56}$
$\mathfrak{p}_{23}$	$\begin{bmatrix} 0 & 0 \\ 1 & 0 \end{bmatrix}$	$\Theta_6 = \mathfrak{p}_0 + \Theta$	$\mathfrak{p}_0, \mathfrak{p}_{16}, \mathfrak{p}_{26}, \mathfrak{p}_{36}, \mathfrak{p}_{46}, \mathfrak{p}_{56}$
$\mathfrak{p}_{16}$	$\begin{bmatrix} 0 & 1 \\ 1 & 0 \end{bmatrix}$	$\Theta_{456} = \mathfrak{p}_{45} + \Theta$	$\mathfrak{p}_{12}, \mathfrak{p}_{13}, \mathfrak{p}_{23}, \mathfrak{p}_{45}, \mathfrak{p}_{46}, \mathfrak{p}_{56}$
$\mathfrak{p}_{13}$	$\begin{bmatrix} 1 & 1 \\ 1 & 0 \end{bmatrix}$	$\Theta_{126} = \mathfrak{p}_{12} + \Theta$	$\mathfrak{p}_{12}, \mathfrak{p}_{16}, \mathfrak{p}_{26}, \mathfrak{p}_{34}, \mathfrak{p}_{35}, \mathfrak{p}_{45}$
$\mathfrak{p}_{12}$	$\begin{bmatrix} 1 & 1 \\ 0 & 0 \end{bmatrix}$	$\Theta_{136} = \mathfrak{p}_{13} + \Theta$	$\mathfrak{p}_{13}, \mathfrak{p}_{16}, \mathfrak{p}_{24}, \mathfrak{p}_{25}, \mathfrak{p}_{36}, \mathfrak{p}_{45}$
$\mathfrak{p}_{24}$	$\begin{bmatrix} 1 & 0 \\ 0 & 1 \end{bmatrix}$	$\Theta_{346} = \mathfrak{p}_{34} + \Theta$	$\mathfrak{p}_{12}, \mathfrak{p}_{15}, \mathfrak{p}_{25}, \mathfrak{p}_{34}, \mathfrak{p}_{36}, \mathfrak{p}_{46}$
$\mathfrak{p}_{34}$	$\begin{bmatrix} 1 & 0 \\ 1 & 1 \end{bmatrix}$	$\Theta_{246} = \mathfrak{p}_{24} + \Theta$	$\mathfrak{p}_{13}, \mathfrak{p}_{15}, \mathfrak{p}_{24}, \mathfrak{p}_{26}, \mathfrak{p}_{35}, \mathfrak{p}_{46}$
$\mathfrak{p}_{56}$	$\begin{bmatrix} 0 & 1 \\ 1 & 1 \end{bmatrix}$	$\Theta_{146} = \mathfrak{p}_{14} + \Theta$	$\mathfrak{p}_{14}, \mathfrak{p}_{16}, \mathfrak{p}_{23}, \mathfrak{p}_{25}, \mathfrak{p}_{35}, \mathfrak{p}_{46}$
$\mathfrak{p}_{25}$	$\begin{bmatrix} 1 & 1 \\ 0 & 1 \end{bmatrix}$	$\Theta_{356} = \mathfrak{p}_{35} + \Theta$	$\mathfrak{p}_{12}, \mathfrak{p}_{14}, \mathfrak{p}_{24}, \mathfrak{p}_{35}, \mathfrak{p}_{36}, \mathfrak{p}_{56}$
$\mathfrak{p}_{35}$	$\begin{bmatrix} 1 & 1 \\ 1 & 1 \end{bmatrix}$	$\Theta_{256} = \mathfrak{p}_{25} + \Theta$	$\mathfrak{p}_{13}, \mathfrak{p}_{14}, \mathfrak{p}_{25}, \mathfrak{p}_{26}, \mathfrak{p}_{34}, \mathfrak{p}_{56}$
$\mathfrak{p}_{46}$	$\begin{bmatrix} 0 & 0 \\ 1 & 1 \end{bmatrix}$	$\Theta_{156} = \mathfrak{p}_{15} + \Theta$	$\mathfrak{p}_{15}, \mathfrak{p}_{16}, \mathfrak{p}_{23}, \mathfrak{p}_{24}, \mathfrak{p}_{34}, \mathfrak{p}_{56}$

3.2 (2, 2)-Isogenies of Abelian Surfaces

The translation of the Jacobian $A = \mathrm{Jac}(\mathcal{C})$ by a two-torsion point is an isomorphism of the Jacobian and maps the set of two-torsion points to itself.

For any isotropic two-dimensional subspace K of $\mathbf{A}[2]$, i.e., a Göpel group in $\mathbf{A}[2]$, it is well-known that $\hat{\mathbf{A}} = \mathbf{A}/\mathsf{K}$ is again a principally polarized abelian surface [30, Sec. 23]. Therefore, the isogeny $\psi : \mathbf{A} \to \hat{\mathbf{A}}$ between principally polarized abelian surfaces has as its kernel the two-dimensional isotropic subspace K of $\mathbf{A}[2]$. We call such an isogeny ψ a $(2, 2)$-*isogeny*. Concretely, given any choice of K the $(2, 2)$-isogeny is analytically given by the map

$$\psi : \mathbf{A} = \mathbb{C}^2/\langle \mathbb{Z}^2 \oplus \tau\mathbb{Z}^2 \rangle \to \hat{\mathbf{A}} = \mathbb{C}^2/\langle \mathbb{Z}^2 \oplus 2\tau\mathbb{Z}^2 \rangle$$

$$(z, \tau) \mapsto (z, 2\tau). \tag{27}$$

We have the following lemma:

Lemma 3.6 *Let* K *and* K' *be two maximal isotropic subgroup of* $\mathbf{A}[2]$ *such that* $\mathsf{K} + \mathsf{K}' = \mathbf{A}[2]$, $\mathsf{K} \cap \mathsf{K}' = \{\mathfrak{p}_0\}$. *Set* $\hat{\mathbf{A}} = \mathbf{A}/\mathsf{K}$, *and denote the image of* K' *in* $\hat{\mathbf{A}}$ *by* $\hat{\mathsf{K}}$. *Then it follows that* $\hat{\mathbf{A}}/\hat{\mathsf{K}} \cong \mathbf{A}$, *and the composition of* $(2, 2)$-*isogenies* $\hat{\psi} \circ \psi$ *is multiplication by two on* $\mathbf{A}$, *i.e.,* $(z, \tau) \mapsto (2z, \tau)$.

Proof. By construction K is a finite subgroup of $\mathbf{A}$, $\hat{\mathbf{A}} = \mathbf{A}/\mathsf{K}$ a complex torus, and the natural projection $\psi : \mathbf{A} \to \hat{\mathbf{A}} \cong \mathbf{A}/\mathsf{K}$ and isogeny. The order of the kernel is two, hence it is a degree-four isogeny. The same applies to the map $\hat{\psi} : \hat{\mathbf{A}} \to \hat{\mathbf{A}}/\hat{\mathsf{K}}$. Therefore, the composition $\hat{\psi} \circ \psi$ is an isogeny with kernel $\mathsf{K} + \mathsf{K}' = \mathbf{A}[2]$. Thus, $\hat{\mathbf{A}}/\hat{\mathsf{K}} \cong \mathbf{A}$ and the map $\hat{\psi} \circ \psi$ is the group homomorphism $z \mapsto 2z$ whose kernel are the two-torsion points. $\square$

In the case $\mathbf{A} = \mathrm{Jac}(\mathcal{C})$ of the Jacobian of a smooth genus-two curve, one may ask whether $\hat{\mathbf{A}} = \mathrm{Jac}(\hat{\mathcal{C}})$ for some other curve $\hat{\mathcal{C}}$ of genus two, and what the precise relationship between the moduli of $\mathcal{C}$ and $\hat{\mathcal{C}}$ is. The geometric moduli relationship between the two curves of genus two was found by Richelot [33]; see [5].

Because of the isomorphism $S_6 \cong \mathrm{Sp}(4, \mathbb{F}_2)$, the $(2, 2)$-isogenies induce an action of the permutation group of the set of six Theta divisors containing a fixed two-torsion point. There is a classical way, called *Richelot isogeny*, to describe the 15 inequivalent $(2, 2)$-isogenies on the Jacobian $\mathrm{Jac}(\mathcal{C})$ of a generic curve $\mathcal{C}$ of genus-two. If we choose for $\mathcal{C}$ a sextic equation $Y^2 = f_6(X, Z)$, then any factorization $f_6 = A \cdot B \cdot C$ into three degree-two polynomials A, B, C defines a genus-two curve $\hat{\mathcal{C}}$ given by

$$\Delta_{ABC} \cdot Y^2 = [A, B][A, C][B, C] \tag{28}$$

where we have set $[A, B] = A'B - AB'$ with A' the derivative of A with respect to X and Δ_{ABC} is the determinant of A, B, C with respect to the basis X^2, XZ, Z^2. It was proved in [5] that $\mathrm{Jac}(\mathcal{C})$ and $\mathrm{Jac}(\hat{\mathcal{C}})$ are $(2, 2)$-isogenous, and that there are exactly 15 different curves $\hat{\mathcal{C}}$ that are obtained this way.

It follows that this construction yields all principally polarized abelian surfaces $(2, 2)$-isogenous to $\mathbf{A} = \mathrm{Jac}(\mathcal{C})$.

3.2.1 An Explicit Model Using Theta Functions

We provide an explicit model for a pair of $(2, 2)$-isogenies in terms of Theta functions: For the Göpel groups $\mathsf{K} = \{\mathfrak{p}_0, \mathfrak{p}_{15}, \mathfrak{p}_{23}, \mathfrak{p}_{46}\}$ and $\mathsf{K}' = \{\mathfrak{p}_0, \mathfrak{p}_{12}, \mathfrak{p}_{34}, \mathfrak{p}_{56}\}$, we have $\mathsf{K}+\mathsf{K}' = \mathbf{A}[2]$, $\mathsf{K}\cap\mathsf{K}' = \{\mathfrak{p}_0\}$. We set $\hat{\mathbf{A}} = \mathbf{A}/\mathsf{K}$. We will use Theta functions to determine explicit formulas relating the Rosenhain roots of $\mathcal{C}$ in Equation (17) – given by $\lambda_1, \lambda_2, \lambda_3$ and $\lambda_4 = 0$, $\lambda_5 = 1$, and $\lambda_6 = \infty$ – to the roots of a curve $\hat{\mathcal{C}}$ given by the sextic curve

$$\hat{\mathcal{C}}: \quad y^2 = x z \left(x - z\right) \left(x - \Lambda_1 z\right) \left(x - \Lambda_2 z\right) \left(x - \Lambda_3 z\right). \tag{29}$$

Since the Theta functions Θ_i play a role dual to θ_i for the isogenous abelian variety, the Rosenhain roots of $\hat{\mathcal{C}}$ are given by

$$\Lambda_1 = \frac{\Theta_1^2 \Theta_3^2}{\Theta_2^2 \Theta_4^2}, \quad \Lambda_2 = \frac{\Theta_3^2 \Theta_8^2}{\Theta_4^2 \Theta_{10}^2}, \quad \Lambda_3 = \frac{\Theta_1^2 \Theta_8^2}{\Theta_2^2 \Theta_{10}^2}. \tag{30}$$

Rosenhain roots can be expressed in terms of just four Theta constants whose characteristics form a Göpel group in $\mathbb{F}_2^4$. We will write the roots $\lambda_1, \lambda_2, \lambda_3$ and $\Lambda_1, \Lambda_2, \Lambda_3$ in terms of the Theta constants $\{\Theta_1^2, \Theta_2^2, \Theta_3^2, \Theta_4^2\}$ and $\{\theta_1^2, \theta_2^2, \theta_3^2, \theta_4^2\}$, respectively. Their Theta characteristics are form Göpel groups, namely

$$\begin{pmatrix} 0 & 0 \\ 0 & 0 \end{pmatrix}, \begin{pmatrix} 1 & 0 \\ 0 & 0 \end{pmatrix}, \begin{pmatrix} 1 & 1 \\ 0 & 0 \end{pmatrix}, \begin{pmatrix} 0 & 1 \\ 0 & 0 \end{pmatrix}, \text{ and } \begin{pmatrix} 0 & 0 \\ 0 & 0 \end{pmatrix}, \begin{pmatrix} 0 & 0 \\ 1 & 1 \end{pmatrix}, \begin{pmatrix} 0 & 0 \\ 1 & 0 \end{pmatrix}, \begin{pmatrix} 0 & 0 \\ 0 & 1 \end{pmatrix}.$$

We have the following:

Lemma 3.7 *The Rosenhain roots $\lambda_1, \lambda_2, \lambda_3$ and $\Lambda_1, \Lambda_2, \Lambda_3$ are rational functions of the Theta functions $\{\Theta_1^2, \Theta_2^2, \Theta_3^2, \Theta_4^2\}$ and $\{\theta_1^2, \theta_2^2, \theta_3^2, \theta_4^2\}$, respectively. Over $\mathcal{A}_2(2, 4)$ the rational function*

$$l = \frac{(\Theta_1\Theta_2 - \Theta_3\Theta_4)(\Theta_1^2 + \Theta_2^2 + \Theta_3^2 + \Theta_4^2)(\Theta_1^2 - \Theta_2^2 - \Theta_3^2 + \Theta_4^2)}{(\Theta_1\Theta_2 + \Theta_2\Theta_4)(\Theta_1^2 - \Theta_2^2 + \Theta_3^2 - \Theta_4^2)(\Theta_1^2 + \Theta_2^2 - \Theta_3^2 - \Theta_4^2)}, \tag{31}$$

satisfies $l^2 = \lambda_1\lambda_2\lambda_3$. A similar statement applies to L such that $L^2 = \Lambda_1\Lambda_2\Lambda_3$.

We claim that there is a Richelot isogeny realizing the $(2, 2)$-isogeny ψ : $\mathbf{A} = \mathrm{Jac}(\mathcal{C}) \to \hat{\mathbf{A}} = \mathrm{Jac}(\hat{\mathcal{C}})$, for the maximal isotropic subgroup K. We have the following:

Lemma 3.8 *Taking the quotient by the Göpel group $\mathsf{K} = \{\mathfrak{p}_0, \mathfrak{p}_{15}, \mathfrak{p}_{23}, \mathfrak{p}_{46}\}$ corresponds to the Richelot isogeny acting on the genus-two curve $\mathcal{C}$ in*

Equation (17) by pairing the linear factors according $A = (x - \lambda_1)(x - \lambda_5)$, $B = (x - \lambda_2)(x - \lambda_3)$, $C = (x - \lambda_4)(x - \lambda_6)$ in Equation (28).

Proof. We compute the Richelot-isogeny in Equation (28) obtained from pairing the roots according to $(\lambda_1, \lambda_5 = 1)$, (λ_2, λ_3), $(\lambda_4 = 0, \lambda_6 = \infty)$. For this new curve we compute its Igusa-invariants which are in fact rational functions of the Theta functions $[\theta_1 : \theta_2 : \theta_3 : \theta_4]$. We then compute the Igusa invariants for the quadratic twist $\hat{\mathcal{C}}^{(\mu)}$ of the curve in Equation (29) with

$$\mu = \frac{(\theta_1\theta_2 - \theta_3\theta_4)^2(\theta_1^2 + \theta_2^2 - \theta_3^2 - \theta_4^2)(\theta_1^2 - \theta_2^2 + \theta_3^2 - \theta_4^2)}{4\,\theta_1\theta_2\theta_3\theta_4(\theta_1^2 + \theta_2^2 + \theta_3^2 + \theta_4^2)(\theta_1^2 - \theta_2^2 - \theta_3^2 + \theta_4^2)}.$$

They again are rational functions of the Theta functions $[\theta_1 : \theta_2 : \theta_3 : \theta_4]$ since the Rosenhain roots of $\hat{\mathcal{C}}$ are determined by the equations

$$\Lambda_1 = \frac{(\theta_1^2 + \theta_2^2 + \theta_3^2 + \theta_4^2)(\theta_1^2 - \theta_2^2 - \theta_3^2 + \theta_4^2)}{(\theta_1^2 + \theta_2^2 - \theta_3^2 - \theta_4^2)(\theta_1^2 - \theta_2^2 + \theta_3^2 - \theta_4^2)},$$

$$\Lambda_2 = \frac{(\theta_1^2 - \theta_2^2 - \theta_3^2 + \theta_4^2)(\theta_1^2\theta_2^2 + \theta_3^2\theta_4^2 + 2\theta_1\theta_2\theta_3\theta_4)}{(\theta_1^2 - \theta_2^2 + \theta_3^2 - \theta_4^2)(\theta_1^2\theta_2^2 - \theta_3^2\theta_4^2)},$$

$$\Lambda_3 = \frac{(\theta_1^2 + \theta_2^2 + \theta_3^2 + \theta_4^2)(\theta_1^2\theta_2^2 + \theta_3^2\theta_4^2 + 2\theta_1\theta_2\theta_3\theta_4)}{(\theta_1^2 + \theta_2^2 - \theta_3^2 - \theta_4^2)(\theta_1^2\theta_2^2 - \theta_3^2\theta_4^2)}. \tag{32}$$

The two sets of Igusa invariants are identical. $\qquad\square$

     Lemma 3.8 proves that the genus-two curve $\hat{\mathcal{C}}$ is isomorphic to the curve obtained by Richelot isogeny using Equation (28) with

$$[B, C] = x^2 - \lambda_1, [A, C] = x^2 - \lambda_2\lambda_3,$$
$$[A, B] = (1 + \lambda_1 - \lambda_2 - \lambda_3)x^2 - 2(\lambda_1 - \lambda_2\lambda_3)x + \lambda_1\lambda_2$$
$$+ \lambda_1\lambda_3 - \lambda_2\lambda_3 - \lambda_1\lambda_2\lambda_3.$$

Recall that for the Göpel group $\mathsf{K}' = \{\mathfrak{p}_0, \mathfrak{p}_{12}, \mathfrak{p}_{34}, \mathfrak{p}_{56}\}$ we have $\mathsf{K} + \mathsf{K}' = \mathbf{A}[2]$, $\mathsf{K} \cap \mathsf{K}' = \{\mathfrak{p}_0\}$. We denote the image of K' in $\hat{\mathbf{A}}$ by $\hat{\mathsf{K}}$. We have the following:

Lemma 3.9 *Taking the quotient by the Göpel group $\hat{\mathsf{K}} = \{\hat{\mathfrak{p}}_0, \hat{\mathfrak{p}}_{12}, \hat{\mathfrak{p}}_{34}, \hat{\mathfrak{p}}_{56}\}$ corresponds to the Richelot isogeny acting on the curve $\Delta_{ABC} Y^2 = \hat{A} \cdot \hat{B} \cdot \hat{C}$ isomorphic to $\hat{\mathcal{C}}$ by pairing linear factors according to $\hat{A} = [B, C]$, $\hat{B} = [A, C]$, $\hat{C} = [A, B]$.*

Proof. The proof is analogous to the proof of Lemma 3.8. $\qquad\square$

     To see the symmetric relation between the moduli of the isogenous curves $\mathcal{C}$ and $\hat{\mathcal{C}}$ directly, we introduce new moduli $\lambda_1' = (\lambda_1 + \lambda_2\lambda_3)/l$, $\lambda_2' = (\lambda_2 + \lambda_1\lambda_3)/l$, and $\lambda_3' = (\lambda_3 + \lambda_1\lambda_2)/l$, and similarly $\Lambda_1', \Lambda_2', \Lambda_3'$ with $l^2 = \lambda_1\lambda_2\lambda_3$ and $L^2 = \Lambda_1\Lambda_2\Lambda_3$. One checks by explicit computation the following:

Lemma 3.10 *The parameters* $\{\lambda_i'\}$ *and* $\{\Lambda_i'\}$ *are rational functions in the squares of Theta constants* $\{\theta_1^2, \theta_2^2, \theta_3^2, \theta_4^2\}$ *and* $\{\Theta_1^2, \Theta_2^2, \Theta_3^2, \Theta_4^2\}$*, respectively.*

Proof. The proof follows by direct computation. □

Moreover, we have the following relations:

Proposition 3.11 *The moduli of the genus-two curve* $\mathcal{C}$ *in Equation (17) and the* $(2, 2)$*-isogenous genus-two curve* $\hat{\mathcal{C}}$ *in Equation (29) are related by*

$$
\begin{aligned}
\Lambda_1' &= 2\frac{2\lambda_1' - \lambda_2' - \lambda_3'}{\lambda_2' - \lambda_3'}, & \lambda_1' &= 2\frac{2\Lambda_1' - \Lambda_2' - \Lambda_3'}{\Lambda_2' - \Lambda_3'}, \\
\Lambda_2' - \Lambda_1' &= -\frac{4(\lambda_1' - \lambda_2')(\lambda_1' - \lambda_3')}{(\lambda_1' + 2)(\lambda_2' - \lambda_3')}, & \lambda_2' - \lambda_1' &= -\frac{4(\Lambda_1' - \Lambda_2')(\Lambda_1' - \Lambda_3')}{(\Lambda_1' + 2)(\Lambda_2' - \Lambda_3')}, \\
\Lambda_3' - \Lambda_1' &= -\frac{4(\lambda_1' - \lambda_2')(\lambda_1' - \lambda_3')}{(\lambda_1' - 2)(\lambda_2' - \lambda_3')}, & \lambda_3' - \lambda_1' &= -\frac{4(\Lambda_1' - \Lambda_2')(\Lambda_1' - \Lambda_3')}{(\Lambda_1' - 2)(\Lambda_2' - \Lambda_3')}.
\end{aligned}
$$

$$(33)$$

Proof. The proof follows by direct computation. □

In summary, we proved that there is a Richelot isogeny realizing the $(2, 2)$-isogeny $\psi : \mathbf{A} = \text{Jac}(\mathcal{C}) \to \hat{\mathbf{A}} = \text{Jac}(\hat{\mathcal{C}}) = \mathbf{A}/\mathsf{K}$, for the maximal isotropic subgroup K. For the isogenous abelian $\hat{\mathbf{A}}$ variety with period matrix $(\mathbb{I}_2, 2\tau)$, the moduli are rational functions in the roots of the $(2, 2)$-isogenous curve $\hat{\mathcal{C}}$ in Equation (29). We consider these Rosenhain roots $\Lambda_1, \Lambda_2, \Lambda_3$ the coordinates on the moduli space of $(2, 2)$-Isogenous abelian varieties which we denote by $\hat{\mathcal{A}}_2(2)$. Following Lemma 2.6, the holomorphic map $\xi_{2,4} : \mathbb{H}_2 \to \mathbb{P}^3$ given by $\tau \mapsto [\theta_1 : \theta_2 : \theta_3 : \theta_4]$ induces an isomorphism between the Satake compactification of $\hat{\mathcal{A}}_2(2, 4)$ and $\mathbb{P}^3$.

4 Principally Polarized Kummer Surfaces

In this section we discuss various normal forms for Kummer surfaces with principal polarization and their rich geometry. One can always choose such a Theta divisor to satisfy $(-\mathbb{I})^*\Theta = \Theta$, that is, to be a *symmetric Theta divisor*. The abelian surface $\mathbf{A}$ then maps to the complete linear system $|2\Theta|$, and the rational map $\varphi_{\mathcal{L}^2} : \mathbf{A} \to \mathbb{P}^3$ associated with the line bundle $\mathcal{L}^2$ factors via an embedding through the projection $\mathbf{A} \to \mathbf{A}/\langle -\mathbb{I} \rangle$; see [3]. In this way, we can identify $\mathbf{A}/\langle -\mathbb{I} \rangle$ with its image in $\mathbb{P}^3$, a singular quartic surface with sixteen ordinary double points, called a *singular Kummer variety*.

4.1 The Shioda Normal Form

We start with two copies of a smooth genus-two curve $\mathcal{C}$ in Rosenhain normal form in Equation (17). The ordered tuple $(\lambda_1, \lambda_2, \lambda_3)$ – where the λ_i are pairwise distinct and different from $(\lambda_4, \lambda_5, \lambda_6) = (0, 1, \infty)$ – determines a point in $\mathcal{M}_2(2)$, the moduli space of genus-two curves with level-two structure. The symmetric product of the curve $\mathcal{C}$ is given by $\mathcal{C}^{(2)} = (\mathcal{C} \times \mathcal{C})/\langle \sigma_{\mathcal{C}^{(2)}} \rangle$ where $\sigma_{\mathcal{C}^{(2)}}$ interchanges the two copies of $\mathcal{C}$. We denote the hyperelliptic involution on $\mathcal{C}$ by $\iota_{\mathcal{C}}$. The variety $\mathcal{C}^{(2)}/\langle \iota_{\mathcal{C}} \times \iota_{\mathcal{C}} \rangle$ is given in terms of the variables $z_1 = Z^{(1)} Z^{(2)}$, $z_2 = X^{(1)} Z^{(2)} + X^{(2)} Z^{(1)}$, $z_3 = X^{(1)} X^{(2)}$, and $\tilde{z}_4 = Y^{(1)} Y^{(2)}$ with $[z_1 : z_2 : z_3 : \tilde{z}_4] \in \mathbb{P}(1, 1, 1, 3)$ by the equation

$$\tilde{z}_4^2 = z_1 z_3 (z_1 - z_2 + z_3) \prod_{i=1}^{3} \left(\lambda_i^2 z_1 - \lambda_i z_2 + z_3 \right). \tag{34}$$

Definition 4.1 *The hypersurface in $\mathbb{P}(1, 1, 1, 3)$ given by Equation (34) is called Shioda sextic and was described in [35].*

Lemma 4.2 *The Shioda sextic in Equation (34) is birational to the Kummer surface* $\mathrm{Kum}(\mathrm{Jac}\, \mathcal{C})$ *associated with the Jacobian* $\mathrm{Jac}(\mathcal{C})$ *of a genus-two curve $\mathcal{C}$ in Rosenhain normal form (17).*

Proof. $\mathrm{Kum}(\mathrm{Jac}\, \mathcal{C})$ is birational to the quotient of the Jacobian by the involution $-\mathbb{I}$. For a smooth genus-two curve $\mathcal{C}$, we identify the Jacobian $\mathrm{Jac}(\mathcal{C})$ with $\mathrm{Pic}^2(\mathcal{C})$ under the map $x \mapsto x + K_{\mathcal{C}}$. Since $\mathcal{C}$ is a hyperelliptic curve with involution $\iota_{\mathcal{C}}$ a map from the symmetric product $\mathcal{C}^{(2)}$ to $\mathrm{Pic}^2(\mathcal{C})$ given by $(p, q) \mapsto p + q$ is the blow down of the graph of the hyperelliptic involution to the canonical divisor class. Thus, the Jacobian $\mathrm{Jac}(\mathcal{C})$ is birational to the symmetric square $\mathcal{C}^{(2)}$. The involution $-\mathbb{I}$ restricts to the hyperelliptic involution on each factor of $\mathcal{C}$ in $\mathcal{C}^{(2)}$.   □

Remark 4.3 Equation (34) defines a double cover of $\mathbb{P}^2 \ni [z_1 : z_2 : z_3]$ branched along six lines given by

$$z_1 = 0, \ z_3 = 0, \ z_1 - z_2 + z_3 = 0, \ \lambda_i^2 z_1 - \lambda_i z_2 + z_3 = 0 \text{ with } 1 \le i \le 3. \tag{35}$$

The six lines are tangent to the common conic $z_2^2 - 4 z_1 z_3 = 0$. Conversely, any six lines tangent to a common conic can always be brought into the form of Equations 35. A picture is provided in Figure 5.1.

4.2 The Cassels-Flynn Normal Form

We call a surface in complex projective space a *nodal surface* if its only singularities are nodes. For a quartic surface in $\mathbb{P}^3$, it is known that the maximal number of simple nodes is sixteen.

In $\mathbb{P}^3$ we use the projective coordinates $[z_1 : z_2 : z_3 : z_4] \in \mathbb{P}^3$. We consider the morphism $\pi : \mathbb{P}^3 \to \mathbb{P}(1, 1, 1, 3)$ defined by $z_4 \mapsto \tilde{z}_4 = (2K_2 z_4 + K_1)/4$

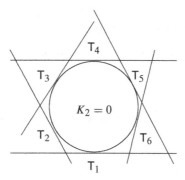

Figure 5.1 Double cover branched along reducible sextic

with coefficients $K_j = K_j(z_1, z_2, z_3)$ homogeneous of degree $4 - j$ with $j = 0, 1, 2$ and given by

$$K_2 = z_2^2 - 4 z_1 z_3 ,$$

$$K_1 = ((4\lambda_1 + 4\lambda_2 + 4\lambda_3 + 4) z_1 - 2 z_2) z_3^2$$
$$+ \left((4\lambda_1\lambda_2\lambda_3 + 4\lambda_1\lambda_2 + 4\lambda_1\lambda_3 + 4\lambda_2\lambda_3) z_1^2 \right.$$
$$\left. + (-2\lambda_1\lambda_2 - 2\lambda_1\lambda_3 - 2\lambda_2\lambda_3 - 2\lambda_1 - 2\lambda_2 - 2\lambda_3) z_2 z_1 \right) z_3$$
$$- 2\lambda_1\lambda_2\lambda_3 z_1^2 z_2 , \tag{36}$$

$$K_0 = z_3^4 - 2 (\lambda_1\lambda_2 + \lambda_1\lambda_3 + \lambda_2\lambda_3 + \lambda_1 + \lambda_2 + \lambda_3) z_1 z_3^3$$
$$+ \left((\lambda_1^2\lambda_2^2 - 2\lambda_1^2\lambda_2\lambda_3 + \lambda_1^2\lambda_3^2 - 2\lambda_1\lambda_2^2\lambda_3 - 2\lambda_1\lambda_2\lambda_3^2 \right.$$
$$+ \lambda_2^2\lambda_3^2 - 2\lambda_1^2\lambda_2 - 2\lambda_1^2\lambda_3 - 2\lambda_1\lambda_2^2 - 8\lambda_1\lambda_2\lambda_3$$
$$- 2\lambda_1\lambda_3^2 - 2\lambda_2^2\lambda_3 - 2\lambda_2\lambda_3^2 + \lambda_1^2 - 2\lambda_1\lambda_2$$
$$\left. - 2\lambda_1\lambda_3 + \lambda_2^2 - 2\lambda_2\lambda_3 + \lambda_3^2) z_1^2 \right.$$
$$\left. + \left(4\lambda_1\lambda_2\lambda_3 + 4\lambda_1\lambda_2 + 4\lambda_1\lambda_3 + 4\lambda_2\lambda_3 \right) z_1 z_2 \right) z_3^2$$
$$+ \left(-4\lambda_1\lambda_2\lambda_3 z_1 z_2^2 + 4\lambda_1\lambda_2\lambda_3 (\lambda_1 + \lambda_2 + \lambda_3 + 1) z_1^2 z_2 \right.$$
$$\left. - 2\lambda_1\lambda_2\lambda_3 (\lambda_1\lambda_2 + \lambda_1\lambda_3 + \lambda_2\lambda_3 + \lambda_1 + \lambda_2 + \lambda_3) z_1^3 \right) z_3 + \lambda_1^2\lambda_2^2\lambda_3^2 z_1^4 .$$

We have the following result:

Lemma 4.4 *The map* $\pi : \mathbb{P}^3 \to \mathbb{P}(1, 1, 1, 3)$ *blows down the double cover of the special conic* $K_2 = 0$, *and is an isomorphism elsewhere. In particular, the* *proper transform of $\pi^{-1}\mathcal{C}^{(2)}/\langle \iota_\mathcal{C} \times \iota_\mathcal{C}\rangle$ is a nodal quartic surface in $\mathbb{P}^3 \ni [z_1 : z_2 : z_3 : z_4]$ with sixteen nodes given by*

$$K_2(z_1, z_2, z_3)\, z_4^2 + K_1(z_1, z_2, z_3)\, z_4 + K_0(z_1, z_2, z_3) = 0 . \qquad (37)$$

Proof. We observe that

$$\frac{1}{16}\left(2\,K_2\,z_4 + K_1\right)^2 = \frac{1}{16}\left(K_1^2 - 4\,K_0 K_2\right) = z_1 z_3 \prod_{i=1}^{4}\left(\lambda_i^2\, z_1 - \lambda_i\, z_2 + z_3\right) ,$$

and Equation (34) is equivalent to

$$0 = K_2\left(K_2 z_4^2 + K_1 z_4 + K_0\right) .$$

The preimage contains the additional point $[z_1 : z_2 : z_3 : z_4] = [0 : 0 : 0 : 1]$ for which $K_2 = K_1 = K_0 = 0$. We then obtain sixteen nodes on the quartic surface given by Equation (37). All singular points are listed in Table 5.2. $\qquad \square$

Definition 4.5 *The quartic surface in $\mathbb{P}^3$ given by Equation (37) appeared in Cassels and Flynn [6, Sec. 3] and is called the Cassels-Flynn quartic.*

We have the immediate:

Corollary 4.6 *The Cassels-Flynn quartic in Equation (37) is isomorphic to the singular Kummer variety $\mathrm{Jac}(\mathcal{C})/\langle -\mathbb{I}\rangle$ associated with the Jacobian $\mathrm{Jac}(\mathcal{C})$ of the genus-two curve $\mathcal{C}$ in Rosenhain form (17).* $\qquad \square$

The map from the Jacobian $\mathrm{Jac}(\mathcal{C})$ onto its image in $\mathbb{P}^3$ is two-to-one, except on sixteen points of order two where it is injective. This can be seen as follows: We label the sixteen singular points p_0, p_{ij} with $1 \le i < j \le 6$ where p_0 is located at $[0 : 0 : 0 : 1]$. Fifteen nodes for Equation (34), namely p_{ij} with $1 \le i < j \le 6$, are given by

$$[z_1 : z_2 : z_3 : \tilde{z}_4] = [1 : \lambda_i + \lambda_j : \lambda_i \lambda_j : 0] ,$$

where we have used $\lambda_4 = 0$, $\lambda_5 = 1$. These points are obtained by combining the Weierstrass points $\mathfrak{p}_i$ and $\mathfrak{p}_j$ on $\mathcal{C}$ given by

$$[X^{(1)} : Y^{(1)} : Z^{(1)}] = [\lambda_i : 0 : 1] \text{ and } [X^{(2)} : Y^{(2)} : Z^{(2)}] = [\lambda_j : 0 : 1]. \qquad (38)$$

Similarly, five nodes p_{i6} for $1 \le i \le 5$ are given by

$$[z_1 : z_2 : z_3 : \tilde{z}_4] = [0 : 1 : \lambda_i : 0]$$

and obtained by combining Weierstrass points $\mathfrak{p}_i$ and $\mathfrak{p}_6$ on $\mathcal{C}$ given by

$$[X^{(1)} : Y^{(1)} : Z^{(1)}] = [\lambda_i : 0 : 1] \text{ and } [X^{(2)} : Y^{(2)} : Z^{(2)}] = [1 : 0 : 0]. \quad (39)$$

Table 5.2. *Nodes on a generic Jacobian Kummer surface*

p	n	e	$[z_1 : z_2 : z_3 : z_4]$
p_0	n_0	e_0	$[0 : 0 : 0 : 1]$
p_{16}	n_3	e_{14}	$[0 : 1 : \lambda_1 : \lambda_1^2]$
p_{26}	n_4	e_{15}	$[0 : 1 : \lambda_2 : \lambda_2^2]$
p_{36}	n_5	e_{16}	$[0 : 1 : \lambda_3 : \lambda_3^2]$
p_{46}	n_1	e_{12}	$[0 : 1 : 0 : 0]$
p_{56}	n_2	e_{13}	$[0 : 1 : 1 : 1]$
p_{14}	n_{13}	e_{24}	$[1 : \lambda_1 : 0 : \lambda_2\lambda_3]$
p_{24}	n_{14}	e_{25}	$[1 : \lambda_2 : 0 : \lambda_1\lambda_3]$
p_{34}	n_{15}	e_{26}	$[1 : \lambda_3 : 0 : \lambda_1\lambda_2]$
p_{45}	n_{12}	e_{23}	$[1 : 1 : 0 : \lambda_1\lambda_2\lambda_3]$
p_{15}	n_{23}	e_{34}	$[1 : \lambda_1 + 1 : \lambda_1 : \lambda_1(\lambda_2 + \lambda_3)]$
p_{25}	n_{24}	e_{35}	$[1 : \lambda_2 + 1 : \lambda_2 : \lambda_2(\lambda_1 + \lambda_3)]$
p_{35}	n_{25}	e_{36}	$[1 : \lambda_3 + 1 : \lambda_3 : \lambda_3(\lambda_1 + \lambda_2)]$
p_{13}	n_{35}	e_{46}	$[1 : \lambda_1 + \lambda_3 : \lambda_1\lambda_3 : (\lambda_2 + 1)\lambda_1\lambda_3]$
p_{23}	n_{45}	e_{56}	$[1 : \lambda_2 + \lambda_3 : \lambda_2\lambda_3 : (\lambda_1 + 1)\lambda_1\lambda_2]$
p_{12}	n_{34}	e_{45}	$[1 : \lambda_1 + \lambda_2 : \lambda_1\lambda_2 : (\lambda_3 + 1)\lambda_1\lambda_2]$

Therefore, the singular points p_{ij} are the images of the two-torsion points $\mathfrak{p}_{ij} \in$ $A[2]$ in Equation (21). Table 5.2 lists all points p_{ij} corresponding to points $\mathfrak{p}_{ij} \in A[2]$ and relate our notation to the notation $n_\bullet$ used by Kumar in [23] and $e_\bullet$ used by Mehran in [28].

Conversely, we can start with a singular Kummer surface in $\mathbb{P}^3$ and reconstruct the configuration of six lines; see [13]. For a fixed singular point p_0, we identify the lines in $\mathbb{P}^3$ through the point p_0 with $\mathbb{P}^2$ and map any line in the tangent cone of p_0 to itself. Any projective line through p_0 meets the quartic surface generically in two other points and with multiplicity two at the other nodes. In this way we obtain a double cover of $\mathbb{P}^2$ branched along a plane curve of degree six where all nodes of the quartic surface different from p_0 map to nodes of the sextic. By the genus-degree formula, the maximal number of nodes on a sextic curve is attained when the curve is a union of six lines, in which case we obtain the fifteen remaining nodes apart from p_0. Since p_0 is a node, the tangent cone to this point is mapped to a conic, and this conic is tangent to the six lines. In summary, the branch locus of the double cover to $\mathbb{P}^2$ is a reducible plane sextic curve, namely the union of six lines tangent to a special conic. By Remark 4.3 this proves the following:

Corollary 4.7 *Every nodal quartic surface with sixteen nodes is isomorphic to the singular Kummer variety of the Jacobian of a genus-two curve.*   □

4.3 Intersection Model for a Kummer Surface

We consider the principally polarized abelian surface $\mathbf{A} = \mathrm{Jac}(\mathcal{C})$ with the standard Theta divisor $\Theta \cong [\mathcal{C}]$. The image of each two-torsion point is a singular point on the Kummer surface, called a *node*. The nodes $\mathfrak{p}_{ij}$ are the images of the two-torsion points $\mathfrak{p}_{ij} \in \mathbf{A}[2]$ in Equation (21). Any Theta divisor is mapped to the intersection of the Kummer quartic with a plane in $\mathbb{P}^3$. We call such a singular plane a *trope*. Hence, we have a configuration of sixteen nodes and sixteen tropes in $\mathbb{P}^3$, where each contains six nodes, and such that the intersection of each two is along two nodes. This configuration is called the 16_6 configuration on the Kummer surface.

In the complete linear system $|2\Theta|$ on $\mathrm{Jac}(\mathcal{C})$, the odd symmetric Theta divisors Θ_i on $\mathrm{Jac}(\mathcal{C})$ introduced in Section 3.1 are mapped to six tropes T_i. Equations for the six tropes are now easily found by inspection: for a given integer i with $1 \leq i \leq 5$, the nodes $\mathfrak{p}_0$ and $\mathfrak{p}_{ij}$ or $\mathfrak{p}_{ji}$ all lie on the plane

$$\mathsf{T}_i: \quad \lambda_i^2 z_1 - \lambda_i z_2 + z_3 = 0\,, \tag{40}$$

where we have set $\mathsf{T}_6: z_1 = 0$. Thus, we obtain

$$\tilde{z}_4^2 = \mathsf{T}_1 \mathsf{T}_2 \mathsf{T}_3 \mathsf{T}_4 \mathsf{T}_5 \mathsf{T}_6\,,$$

and the Kummer surface $\mathrm{Kum}(\mathrm{Jac}\,\mathcal{C})$ is the minimal resolution of the double cover of $\mathbb{P}^2$ branched along a reducible plane sextic curve – the union of six lines all tangent to a conic. In fact, the trope T_i is tangent to the conic $K_2 = 0$ at $[1 : 2\lambda_i : \lambda_i^2 : 0]$ for $i = 1, \ldots, 5$, and T_6 is tangent to $K_2 = 0$ at $[0 : 0 : 1 : 0]$.

The remaining 10 tropes T_{ijk} with $1 \leq i < j < k \leq 6$ correspond to partitions of $\{1, \ldots, 6\}$ into two unordered sets of three, say $\{i, j, k\}\,\{l, m, n\}$. We use the formulas for T_{ijk} from [6, Sec. 3.7] paying careful attention to the fact that we have moved the root λ_6 to infinity. For example, we have

$$\mathsf{T}_{246}: \quad -\lambda_1\lambda_3\, z_1 - \lambda_2\, z_3 + z_4 = 0\,,$$

$$\mathsf{T}_{346}: \quad -\lambda_1\lambda_2\, z_1 - \lambda_3\, z_3 + z_4 = 0\,,$$

$$\mathsf{T}_{236}: \quad -(\lambda_1 + 1)\lambda_2\lambda_3\, z_1 + \lambda_2\lambda_3\, z_2 - (\lambda_2 + \lambda_3)\, z_3 + z_4 = 0\,. \tag{41}$$

We also have the following:

Lemma 4.8 *The ideal of linear relations between the 16 tropes is generated by 12 equations given by*

$$T_1 = (1 - \lambda_1)T_4 + \lambda_1 T_5 + \lambda_1(\lambda_1 - 1)T_6,$$

$$T_2 = (1 - \lambda_2)T_4 + \lambda_2 T_5 + \lambda_2(\lambda_2 - 1)T_6,$$

$$T_3 = (1 - \lambda_3)T_4 + \lambda_3 T_4 + \lambda_3(\lambda_3 - 1)T_6, \tag{42}$$

and

$$
\begin{aligned}
T_{126} &= (\lambda_1 - 1)(\lambda_2 - 1)T_4 & -\lambda_1\lambda_2 T_5 & & +T_{456}, \\
T_{136} &= (\lambda_1 - 1)(\lambda_3 - 1)T_4 & -\lambda_1\lambda_3 T_5 & & +T_{456}, \\
T_{146} &= (1 - \lambda_1)T_4 & & +\lambda_2\lambda_3(\lambda_1 - 1)T_6 & +T_{456}, \\
T_{156} &= & -\lambda_1 T_5 & +\lambda_1(\lambda_2 - 1)(\lambda_- 1)T_6 & +T_{456}, \\
T_{236} &= (1 - \lambda_2)(1 - \lambda_3)T_4 & -\lambda_2\lambda_3 T_5 & & +T_{456}, \\
T_{246} &= (1 - \lambda_2)T_4 & & +\lambda_1\lambda_3(\lambda_2 - 1)T_6 & +T_{456}, \\
T_{256} &= & -\lambda_2 T_5 & +\lambda_2(\lambda_1 - 1)(\lambda_3 - 1)T_6 & +T_{456}, \\
T_{346} &= (1 - \lambda_3)T_4 & & +\lambda_1\lambda_2(\lambda_3 - 1)T_6 & +T_{456}, \\
T_{356} &= & -\lambda_3 T_5 & +\lambda_3(\lambda_1 - 1)(\lambda_2 - 1)T_6 & +T_{456}.
\end{aligned}
\tag{43}
$$

Proof. Using the explicit form of the tropes given in Table 5.3, one checks that there are 240 tuples of four tropes such that the four tropes satisfy a linear relation. Solving these equations in terms of $\{T_4, T_5, T_6, T_{456}\}$ yields the result. $\qquad\square$

There are fifteen linear four-term relations that involve only the tropes $T_1, \ldots, T_6$, i.e., the images of the odd symmetric Theta divisors Θ_i on $\mathrm{Jac}(\mathcal{C})$:

Corollary 4.9 *There are fifteen linear four-terms relations that involve only the tropes* $T_1, \ldots, T_6$. *They are given by*

$$
\begin{aligned}
-(\lambda_j - \lambda_k)(\lambda_j - \lambda_l)(\lambda_k - \lambda_l)T_i + (\lambda_i - \lambda_k)(\lambda_i - \lambda_l)(\lambda_k - \lambda_l)T_j \\
-(\lambda_i - \lambda_j)(\lambda_i - \lambda_l)(\lambda_j - \lambda_l)T_k + (\lambda_i - \lambda_j)(\lambda_i - \lambda_k)(\lambda_j - \lambda_k)T_l = 0,
\end{aligned}
\tag{44}
$$

with $1 \leq i < j < k < l \leq 5$ *and*

$$
\begin{aligned}
-(\lambda_j - \lambda_k)T_i + (\lambda_i - \lambda_k)T_j - (\lambda_i - \lambda_j)T_k \\
+ (\lambda_i - \lambda_j)(\lambda_i - \lambda_k)(\lambda_j - \lambda_k)T_6 = 0.
\end{aligned}
\tag{45}
$$

Table 5.3. *Tropes on a generic Jacobian Kummer surface*

$\mathbf{T}$	T	$[z_1 : z_2 : z_3 : z_4]$	Contained nodes
T_1	T_3	$[\lambda_1^2 : -\lambda_1 : 1 : 0]$	$p_0 , p_{12}, p_{13}, p_{14}, p_{15}, p_{16}$
T_2	T_4	$[\lambda_2^2 : -\lambda_2 : 1 : 0]$	$p_0 , p_{12}, p_{23}, p_{24}, p_{25}, p_{26}$
T_3	T_5	$[\lambda_3^2 : -\lambda_3 : 1 : 0]$	$p_0 , p_{13}, p_{23}, p_{34}, p_{35}, p_{36}$
T_4	T_1	$[0 : 0 : 1 : 0]$	$p_0 , p_{14}, p_{24}, p_{34}, p_{45}, p_{46}$
T_5	T_2	$[1 : -1 : 1 : 0]$	$p_0 , p_{15}, p_{25}, p_{35}, p_{45}, p_{56}$
T_6	T_0	$[1 : 0 : 0 : 0]$	$p_0 , p_{16}, p_{26}, p_{36}, p_{46}, p_{56}$
T_{146}	T_{13}	$[-\lambda_2\lambda_3 : 0 : -\lambda_1 : 1]$	$p_{14}, p_{16}, p_{23}, p_{25}, p_{35}, p_{46}$
T_{246}	T_{14}	$[-\lambda_1\lambda_3 : 0 : -\lambda_2 : 1]$	$p_{13}, p_{15}, p_{24}, p_{26}, p_{35}, p_{46}$
T_{346}	T_{15}	$[-\lambda_1\lambda_2 : 0 : -\lambda_3 : 1]$	$p_{12}, p_{15}, p_{25}, p_{34}, p_{36}, p_{46}$
T_{456}	T_{12}	$[-\lambda_1\lambda_2\lambda_3 : 0 : -1 : 1]$	$p_{12}, p_{13}, p_{23}, p_{46}, p_{45}, p_{56}$
T_{156}	T_{23}	$[-\lambda_1(\lambda_2+\lambda_3) : \lambda_1 : -(\lambda_1+1) : 1]$	$p_{15}, p_{16}, p_{23}, p_{24}, p_{34}, p_{56}$
T_{256}	T_{24}	$[-\lambda_2(\lambda_1+\lambda_3) : \lambda_2 : -(\lambda_2+1) : 1]$	$p_{13}, p_{14}, p_{25}, p_{26}, p_{34}, p_{56}$
T_{356}	T_{25}	$[-\lambda_3(\lambda_1+\lambda_2) : \lambda_3 : -(\lambda_3+1) : 1]$	$p_{12}, p_{14}, p_{24}, p_{35}, p_{36}, p_{56}$
T_{136}	T_{35}	$[-(\lambda_2+1)\lambda_1\lambda_3 : \lambda_1\lambda_3 : -(\lambda_1+\lambda_3) : 1]$	$p_{13}, p_{16}, p_{24}, p_{25}, p_{36}, p_{45}$
T_{236}	T_{45}	$[-(\lambda_1+1)\lambda_2\lambda_3 : \lambda_2\lambda_3 : -(\lambda_2+\lambda_3) : 1]$	$p_{14}, p_{15}, p_{23}, p_{26}, p_{36}, p_{45}$
T_{126}	T_{34}	$[-(\lambda_3+1)\lambda_1\lambda_2 : \lambda_1\lambda_2 : -(\lambda_1+\lambda_2) : 1]$	$p_{12}, p_{16}, p_{26}, p_{34}, p_{35}, p_{45}$

Moreover, these fifteen equations have rank three, and a generating set is given by

$$T_1 = (1-\lambda_1)T_4 + \lambda_1 T_5 - \lambda_1(1-\lambda_1)T_6,$$
$$T_2 = (1-\lambda_2)T_4 + \lambda_2 T_5 - \lambda_2(1-\lambda_2)T_6,$$
$$T_3 = (1-\lambda_3)T_4 + \lambda_3 T_5 - \lambda_3(1-\lambda_3)T_6. \tag{46}$$

We consider the blow up $p : \tilde{\mathbf{A}} \to \mathbf{A}$ of the sixteen two-torsion points with the exceptional curves $E_1, \ldots, E_{16}$. The linear system $|4p^*\Theta - \sum E_i|$ determines a morphism of degree two from $\tilde{\mathbf{A}}$ to a complete intersection of three quadrics in $\mathbb{P}^5$. In particular, the image is the Kummer surface $\mathrm{Kum}(\mathbf{A}) = \tilde{\mathbf{A}}/\langle -\mathbb{I}\rangle$. The net spanned by these quadrics is isomorphic to $\mathbb{P}^2$ with a discriminant locus corresponding to the union of six lines. We have the following:

Proposition 4.10 *The Kummer surface* $\mathrm{Kum}(\mathrm{Jac}\,\mathcal{C})$ *associated with the Jacobian* $\mathrm{Jac}(\mathcal{C})$ *of a genus-two curve* $\mathcal{C}$ *in Rosenhain normal form (17) is the complete intersection of three quadrics in* $\mathbb{P}^5$. *For* $[t_1 : t_2 : t_3 : t_4 : t_5 : t_6] \in \mathbb{P}^5$ *with* $t_i^2 = T_i$ *and* $1 \le i \le 6$, *the Kummer surface* $\mathrm{Kum}(\mathrm{Jac}\,\mathcal{C})$ *is given by the intersection of the three quadrics*

$$t_1^2 = (1-\lambda_1)t_4^2 + \lambda_1 t_5^2 - \lambda_1(1-\lambda_1)t_6^2,$$
$$t_2^2 = (1-\lambda_2)t_4^2 + \lambda_2 t_5^2 - \lambda_2(1-\lambda_2)t_6^2,$$
$$t_3^2 = (1-\lambda_3)t_4^2 + \lambda_3 t_5^2 - \lambda_3(1-\lambda_3)t_6^2. \tag{47}$$

Proof. We proved that the Shioda sextic in Lemma 4.2 is given by the double cover of $\mathbb{P}^2$ branched along the reducible sextic that is the union of the six lines given by the tropes $T_1, \ldots, T_6$, i.e.,

$$\tilde{z}_4^2 = T_1 T_2 T_3 T_4 T_5 T_6 .$$

The tropes $T_1, \ldots, T_6$ satisfy the fifteen linear relations in Lemma 4.8 of rank three equivalent to Equations (42). Introducing $T_i = t_i^2$ such that $\tilde{z}_4 = t_1 t_2 t_3 t_4 t_5 t_6$, the Kummer surface $\text{Kum}(\mathbf{A}) = \tilde{\mathbf{A}}/\langle -\mathbb{I} \rangle$ is the complete intersection of three quadrics in $\mathbb{P}^5 \ni [t_1 : t_2 : t_3 : t_4 : t_5 : t_6]$ given by Equations (47). □

The coordinates $[t_1 : t_2 : t_3 : t_4 : t_5 : t_6] \in \mathbb{P}^5$ are related to the odd Theta functions. We have the following:

Lemma 4.11 *Given Thomae's formula (18), there is a unique bijection between the six tropes $T_1, \ldots, T_6$, i.e., the images of the odd symmetric Theta divisors Θ_i on $\text{Jac}(C)$, and squares of the odd Theta functions $\theta_{11}^2(z), \ldots, \theta_{16}^2(z)$ such that Equations (46) coincide with the Mumford relations in Equations (14). It is given by setting*

$$
\begin{aligned}
T_1 &= R^2 \theta_1^2 \theta_3^2 \theta_6^2 \theta_7^2 \theta_9^2 \theta_{10}^2 \, \theta_{15}^2(z), & T_2 &= R^2 \theta_2^2 \theta_3^2 \theta_5^2 \theta_6^2 \theta_8^2 \theta_9^2 \, \theta_{12}^2(z), \\
T_3 &= R^2 \theta_1^2 \theta_4^2 \theta_5^2 \theta_6^2 \theta_7^2 \theta_8^2 \, \theta_{11}^2(z), & T_4 &= R^2 \theta_1^2 \theta_2^2 \theta_3^2 \theta_4^2 \theta_8^2 \theta_{10}^2 \, \theta_{16}^2(z), \\
T_5 &= R^2 \theta_2^2 \theta_4^2 \theta_5^2 \theta_7^2 \theta_9^2 \theta_{10}^2 \, \theta_{14}^2(z), & T_6 &= R^2 \theta_2^4 \theta_4^4 \theta_{10}^4 \, \theta_{13}^2(z),
\end{aligned}
$$

$$(48)$$

where $R \in \mathbb{C}^*$ is a non-zero constant.

Proof. Using a computer algebra system, we showed that Equations (48) is the only solution such that Equations (42) coincide with Equations (14). □

We introduce expressions t_i, that is sections of appropriate lines bundles over $\mathbf{A}_\tau$ (see Remarks 2.1 and 2.7), such that $t_i^2 = T_i$ for $1 \le i \le 6$. We set

$$t_1 = R \, \theta_1 \theta_3 \theta_6 \theta_7 \theta_9 \theta_{10} \, \theta_{15}(z), \quad t_2 = R \, \theta_2 \theta_3 \theta_5 \theta_6 \theta_8 \theta_9 \, \theta_{12}(z), \quad \text{etc.} \quad (49)$$

We have the following:

Theorem 4.12 *For $[\tau] \in \mathcal{A}_2(2, 4)$ the Kummer surface $\text{Kum}(\mathbf{A}_\tau)$ associated with the principally polarized abelian surfaces $\mathbf{A}_\tau$ with period matrix $(\mathbb{I}_2, \tau)$ is isomorphic to the image of the odd Theta functions $[\theta_{11}(z) : \cdots : \theta_{16}(z)]$ in $\mathbb{P}^5$ which satisfy the Mumford relations (14).*

Proof. Using Lemma 4.11 we obtain the linear map of $\mathbb{P}^5$ given by $[T_1 : \cdots : T_6] \mapsto [\theta_{11}(z) : \cdots : \theta_{16}(z)]$ which is well defined over $\mathcal{A}_2(2, 4)$. Equations (42) then coincide with Equations (14). □

In Lemma 4.11 we determined the unique bijection between the tropes $T_1, \ldots, T_6$, i.e., the images of the odd symmetric Theta divisors Θ_i on $\mathrm{Jac}(\mathcal{C})$, and squares of the odd Theta functions $\theta_{11}^2(z), \ldots, \theta_{16}^2(z)$ such that Equations (46) coincide with the Mumford identities in Equations (14). We now extend the bijection to all tropes. We have the following:

Proposition 4.13 *Given Thomae's formula (18), there is a unique bijection between the tropes $T_1, \ldots, T_6$ and $T_{126}, \ldots, T_{456}$ and the squares of the Theta functions $\theta_1^2(z), \ldots, \theta_{16}^2(z)$ such that Equations (46) coincide with the Mumford identities in Equations (14) and Equations (15). It is given by Equations (48) and by*

$$T_{126} = k\,\theta_1^2\theta_2^2\theta_3^2\theta_5^2\theta_6^2\theta_7^2\theta_8^2\theta_9^2\theta_{10}^2\,\theta_4^2(z), \quad T_{136} = k\,\theta_1^2\theta_3^2\theta_4^2\theta_5^2\theta_6^2\theta_7^2\theta_8^2\theta_9^2\theta_{10}^2\,\theta_2^2(z),$$

$$T_{146} = k\,\theta_1^2\theta_2^2\theta_3^2\theta_4^2\theta_6^2\theta_7^2\theta_8^2\theta_9^2\theta_{10}^2\,\theta_5^2(z), \quad T_{156} = k\,\theta_1^2\theta_2^2\theta_3^2\theta_4^2\theta_5^2\theta_6^2\theta_7^2\theta_9^2\theta_{10}^2\,\theta_8^2(z),$$

$$T_{236} = k\,\theta_1^2\theta_2^2\theta_3^2\theta_4^2\theta_5^2\theta_6^2\theta_7^2\theta_8^2\theta_9^2\,\theta_{10}^2(z), \quad T_{246} = k\,\theta_1^2\theta_2^2\theta_3^2\theta_4^2\theta_5^2\theta_6^2\theta_8^2\theta_9^2\theta_{10}^2\,\theta_7^2(z),$$

$$T_{256} = k\,\theta_2^2\theta_3^2\theta_4^2\theta_5^2\theta_6^2\theta_7^2\theta_8^2\theta_9^2\theta_{10}^2\,\theta_1^2(z), \quad T_{346} = k\,\theta_1^2\theta_2^2\theta_3^2\theta_4^2\theta_5^2\theta_6^2\theta_7^2\theta_8^2\theta_{10}^2\,\theta_9^2(z),$$

$$T_{356} = k\,\theta_1^2\theta_2^2\theta_4^2\theta_5^2\theta_6^2\theta_7^2\theta_8^2\theta_9^2\theta_{10}^2\,\theta_3^2(z), \quad T_{456} = k\,\theta_1^2\theta_2^2\theta_3^2\theta_4^2\theta_5^2\theta_7^2\theta_8^2\theta_9^2\theta_{10}^2\,\theta_6^2(z),$$

$$(50)$$

where $k \in \mathbb{C}^*$ is a non-zero constant, and $R^2 = -\theta_2^2\theta_4^2\theta_{10}^2\,k$ in Equations (50).

Proof. Using a computer algebra system, we showed that Equations (50) constitute the only solution such that Equations (42) and Equations (43) coincide with Equations (14) and Equations (15), respectively. □

Remark 4.14 The isomorphism between $\mathbf{A}[2]$ and $\mathbb{F}_2^4$ in Table 5.1 in Lemma 4.11 was determined. It was the unique isomorphism that mapped the Theta divisors to the translates of the characteristics of the Theta functions such that the symmetric Theta divisors are the tropes on the Kummer surface given by the squares of Theta functions.

We also introduce sections $t_{a,b}$ which are bi-monomial expressions in terms of Theta functions, such that $t_{a,b}^2 = T_a T_b$ for $a, b \in \{k, ij6\}$ with $1 \le i < j < 6$ and $1 \le k < 6$ and the following consistent choice for the sign of the 120 square roots:

$$t_{1,2} = k\,\theta_1\theta_2^3\theta_3^2\theta_4^2\theta_5\theta_6^2\theta_7\theta_8\theta_9^2\theta_{10}^3\,\theta_{12}(z)\theta_{15}(z), \quad \ldots, \quad t_{5,6} = k \ldots,$$

$$t_{1,126} = ik\,\theta_1^2\theta_2^2\theta_3^2\theta_4\theta_5\theta_6^2\theta_7^2\theta_8\theta_9^2\theta_{10}^3\,\theta_4(z)\theta_{15}(z), \quad \ldots, \quad t_{5,456} = ik \ldots,$$

$$t_{6,126} = -ik\,\theta_1\theta_2^4\theta_3\theta_4^3\theta_5\theta_6\theta_7\theta_8\theta_9\theta_{10}^4\,\theta_4(z)\theta_{13}(z), \quad \ldots, \quad t_{6,456} = -ik \ldots,$$

$$t_{126,136} = k\,\theta_1^2\theta_2\theta_3^2\theta_4\theta_5\theta_6^2\theta_7^2\theta_8^2\theta_9^2\theta_{10}^2\,\theta_2(z)\theta_4(z), \quad \ldots, \quad t_{356,456} = k \ldots.$$

$$(51)$$

We have the following:

Proposition 4.15 *In terms of the sections* $t_{a,b}$ *with* $t_{a,b}^2 = T_a T_b$ *introduced above, the 72 Equations (14), (15), and (108) are given by Equations (42), (43), and*

$$(\lambda_2 - \lambda_3)(\lambda_1 - 1)t_{4,6} + t_{136,256} - t_{126,356} = 0, \ldots$$
$$(\lambda_1 - \lambda_2)(\lambda_3 - 1)t_{126,356} - (\lambda_1 - \lambda_3)(\lambda_2 - 1)t_{136,256}$$
$$+(\lambda_2 - \lambda_3)(\lambda_1 - 1)t_{156,236} = 0,$$

(A complete generating set of 60 equations is given in Equation (109).) (52)

In particular, all coefficients of the 72 conics are polynomials in $\mathbb{Z}[\lambda_1, \lambda_2, \lambda_3]$.

We have the following:

Corollary 4.16 *The image of the embedding of* $\mathrm{Jac}(\mathcal{C}) \hookrightarrow \mathbb{P}^{15}$ *of the Jacobian* $\mathrm{Jac}(\mathcal{C})$ *of a genus-two curve* $\mathcal{C}$ *in Equation (17) given by* $z \mapsto [t_1 : \cdots : t_{16}]$ *is the intersection of the 72 conics given by Equations (42), (43), and (109). Similarly, the image of the image of the embedding of* $\mathrm{Kum}(\mathrm{Jac}\,\mathcal{C}) \hookrightarrow \mathbb{P}^5$ *given by* $z \mapsto [t_1 : \cdots : t_6]$ *is the intersection of the 3 conics given by Equations (42). In particular, the images are defined over* $\mathcal{A}_2(2)$, *i.e., in terms of the Rosenhain parameters of* $\mathcal{C}$ *in Equation (17).* □

4.4 Tetrahedra and Even Eights

We now use the Cassels-Flynn quartic in Equation (37) as model for a singular Kummer variety to derive special relations between the tropes. The 16_6 configuration on the Jacobian descends to the Kummer surface given by Equation (37) as follows: any trope contains exactly six nodes. Any node is contained in exactly six tropes. Any two different tropes have exactly two nodes in common. This is easily verified using Table 5.3. Tetrahedra in $\mathbb{P}^3$ whose faces are tropes are given by tuples $\{T_a, T_b, T_c, T_d\}$. They are called *Rosenhain tetrahedra* if all vertices are nodes. They are called *Göpel tetrahedra* if none of the vertices are nodes. We also remind the reader that we call the tropes $\{T_i\}$ and $\{T_{jk6}\}$, with $1 \leq i \leq 6$ and $1 \leq j < k < 6$, the six *odd* and the ten *even* tropes, respectively. We have the following:

Lemma 4.17 *There are 60 Göpel tetrahedra on the quartic surface given by Equation (37) which are obtained for* $\{i, j, k, l, m\} = \{1, \ldots, 5\}$ *as follows:*

(1) *30 tetrahedra are of the form* $\{T_i, T_j, T_{ik6}, T_{jk6}\}$;
(2) *15 tetrahedra are of the form* $\{T_m, T_6, T_{ij6}, T_{kl6}\}$;
(3) *15 tetrahedra are of the form* $\{T_{ij6}, T_{kl6}, T_{ik6}, T_{jl6}\}$.

In particular, 15 Göpel tetrahedra contain only even tropes, and 45 contain two even and two odd tropes.

Proof. One checks that the given list contains all sets of four tropes $\{T_a, T_b, T_c, T_d\}$ whose vertices are not nodes. □

Lemma 4.18 *There are 80 Rosenhain tetrahedra $\{T_a, T_b, T_c, T_d\}$ on the surface given by Equation (37). The Rosenhain tetrahedra fall into five subsets $R^{(1)}, \ldots, R^{(5)}$ such that all Rosenhain tetrahedra are obtained for $\{i, j, k, l, m\} = \{1, \ldots, 5\}$ as follows:*

notation	#	T_a	T_b	T_c	T_d
$R^{(1)}$	10	T_i	T_j	T_k	T_{lm6}
$R^{(2)}$	10	T_{jk6}	T_{ik6}	T_{ij6}	T_6
$R^{(3)}$	10	T_i	T_j	T_{ij6}	T_6
$R^{(4)}$	30	T_i	T_{ij6}	T_{ik6}	T_{lm6}
$R^{(5)}$	20	T_i	T_{jk6}	T_{jl6}	T_{jm6}

$$(53)$$

In particular, there are 60 Rosenhain tetrahedra with one odd and three even tropes, and 20 that contain one even and three odd tropes.

Proof. One checks that the given list contains all sets of four tropes $\{T_a, T_b, T_c, T_d\}$ whose vertices are nodes. □

One easily checks the following:

Lemma 4.19 *Under the isomorphism in Table 5.1, the Göpel and Rosenhain tetrahedra are bijectively mapped to the Göpel groups and their translates in Lemma 3.1 and the Rosenhain groups and their translates in Lemma 3.2, respectively.* □

By explicit computation one also checks:

Lemma 4.20 *There are 30 quadratic relations involving eight tropes. The relations can be written in the form*

$$\mu\nu\rho T_a T_{a'} + \gamma\delta\mu T_b T_{b'} + \beta\delta\nu T_c T_{c'} + \beta\gamma\rho T_d T_{d'} = 0, \qquad (54)$$

where $\beta, \gamma, \delta, \mu, \nu, \gamma \in \mathbb{C}[\lambda_1, \lambda_2, \lambda_3]$ with $\mu + \nu + \rho = 0$ and $\beta\mu + \gamma\nu + \delta\rho = 0$, and $\{T_a, T_b, T_c, T_d\}$ and $\{T_{a'}, T_{b'}, T_{c'}, T_{d'}\}$ are two disjoint Rosenhain tetrahedra. In particular, all quadratic relations are obtained for $\{i, j, k, l, m\} = \{1, \ldots, 5\}$ as follows:

#	R's	T_a	T_b	T_c	T_d	$T_{a'}$	$T_{b'}$	$T_{c'}$	$T_{d'}$
10	$R^{(3)}, R^{(4)}$	T_6	T_i	T_j	T_{ij6}	T_{lm6}	T_{jk6}	T_{ik6}	T_k
		\multicolumn{8}{l}{$\mu = \lambda_j - \lambda_k,\ \nu = \lambda_k - \lambda_i,\ \rho = \lambda_i - \lambda_j$}							
		\multicolumn{8}{l}{$\beta = \gamma = \delta = 1$}							
10	$R^{(4)}, R^{(4)}$	T_i	T_{jm6}	T_{ik6}	T_{il6}	T_j	T_{im6}	T_{jk6}	T_{jl6}
		\multicolumn{8}{l}{$\mu = \lambda_l - \lambda_k,\ \nu = \lambda_m - \lambda_l,\ \rho = \lambda_k - \lambda_m$}							
		\multicolumn{8}{l}{$\beta = \gamma = \delta = 1$}							
5	$R^{(1)}, R^{(5)}$	T_{lm6}	T_i	T_j	T_k	T_l	T_{im6}	T_{jm6}	T_{km6}
		\multicolumn{8}{l}{$\mu = \lambda_j - \lambda_k,\ \nu = \lambda_k - \lambda_i,\ \rho = \lambda_i - \lambda_j$}							
		\multicolumn{8}{l}{$\beta = \lambda_i - \lambda_l,\ \gamma = \lambda_j - \lambda_l,\ \delta = \lambda_k - \lambda_l$}							
5	$R^{(2)}, R^{(5)}$	T_6	T_{jk6}	T_{ik6}	T_{ij6}	T_m	T_{il6}	T_{jl6}	T_{kl6}
		\multicolumn{8}{l}{$\mu = (\lambda_j - \lambda_k)(\lambda_i - \lambda_l),\ \nu = (\lambda_k - \lambda_i)(\lambda_j - \lambda_l),\ \rho = (\lambda_i - \lambda_j)(\lambda_k - \lambda_l)$}							
		\multicolumn{8}{l}{$\beta = \gamma = \delta = 1$}							

$$(55)$$

Proof. The proof follows from direct computation using the explicit equations for the tropes in Table 5.3. $\qquad\square$

Lemma 4.21

(1) *Every set of eight tropes given in Lemma 4.20 can be decomposed into three different pairs of disjoint Göpel tetrahedra.*

(2) *Every Göpel tetrahedron is contained in three sets of eight tropes.*

(3) *Every set of eight tropes given in Lemma 4.20 can be decomposed into four different pairs of disjoint Rosenhain tetrahedra. Eight nodes arise as the vertices of each such pair of Rosenhain tetrahedra. Moreover, the set of eight nodes is independent of the chosen decomposition.*

(4) *Every Rosenhain tetrahedron is contained in three sets of eight tropes.*

Proof. The proof follows from direct computation. $\qquad\square$

On a singular Kummer variety, sets of eight distinct nodes are called *even eights*. After minimal resolution, an even eight is a set of eight disjoint smooth rational curves on the smooth Kummer surface whose divisors add up to an even element in the Néron-Severi group. Nikulin proved [31] that on a Kummer surface there are 30 even eights. If we fix a node, say p_0, the 15 even eights not containing p_0 are enumerated by nodes p_{ij} as follows

$$\Delta_{ij} = \{p_{1i}, \ldots, \widehat{p_{ij}}, \ldots, p_{i6}, p_{1j}, \ldots, \widehat{p_{ij}}, \ldots, p_{j6}\},$$

where $p_{11} = 0$ and a hat indicates a node that is not part of the even eight; see Mehran [28]. On the other hand, sets of eight distinct nodes arose in Lemma 4.20.(3) as the vertices of the two Rosenhain tetrahedra forming each set of eight tropes in Lemma 4.20. We have the following:

Proposition 4.22 *The 30 sets of eight tropes satisfying a quadratic relation in Lemma 4.20 are in one-to-one correspondence with the 30 even eights where we denote an even eight not containing p_0 by Δ_{ij} and its complement by Δ_{ij}^C:*

Δ_{12}	$\{T_1, T_{136}, T_{146}, T_{156}, T_2, T_{236}, T_{246}, T_{256}\}$	Δ_{12}^C	$\{T_{126}, T_3, T_{346}, T_{356}, T_4, T_{456}, T_5, T_6\}$
Δ_{13}	$\{T_1, T_{126}, T_{146}, T_{156}, T_{236}, T_3, T_{346}, T_{356}\}$	Δ_{13}^C	$\{T_{136}, T_2, T_{246}, T_{256}, T_4, T_{456}, T_5, T_6\}$
Δ_{14}	$\{T_1, T_{126}, T_{136}, T_{156}, T_{246}, T_{346}, T_4, T_{456}\}$	Δ_{14}^C	$\{T_{146}, T_2, T_{236}, T_{256}, T_3, T_{356}, T_5, T_6\}$
Δ_{15}	$\{T_1, T_{126}, T_{136}, T_{146}, T_{256}, T_{356}, T_{456}, T_5\}$	Δ_{15}^C	$\{T_{156}, T_2, T_{236}, T_{246}, T_3, T_{346}, T_4, T_6\}$
Δ_{16}	$\{T_1, T_{236}, T_{246}, T_{256}, T_{346}, T_{356}, T_{456}, T_6\}$	Δ_{16}^C	$\{T_{126}, T_{136}, T_{146}, T_{156}, T_2, T_3, T_4, T_5\}$
Δ_{23}	$\{T_{126}, T_{136}, T_2, T_{246}, T_{256}, T_3, T_{346}, T_{356}\}$	Δ_{23}^C	$\{T_1, T_{146}, T_{156}, T_{236}, T_4, T_{456}, T_5, T_6\}$
Δ_{24}	$\{T_{126}, T_{146}, T_2, T_{236}, T_{256}, T_{346}, T_4, T_{456}\}$	Δ_{24}^C	$\{T_1, T_{136}, T_{156}, T_{246}, T_3, T_{356}, T_5, T_6\}$
Δ_{25}	$\{T_{126}, T_{156}, T_2, T_{236}, T_{246}, T_{356}, T_{456}, T_5\}$	Δ_{25}^C	$\{T_1, T_{136}, T_{146}, T_{256}, T_3, T_{346}, T_4, T_6\}$
Δ_{26}	$\{T_{136}, T_{146}, T_{156}, T_2, T_{346}, T_{356}, T_{456}, T_6\}$	Δ_{26}^C	$\{T_1, T_{126}, T_{236}, T_{246}, T_{256}, T_3, T_4, T_5\}$
Δ_{34}	$\{T_{136}, T_{146}, T_{236}, T_{246}, T_3, T_{356}, T_4, T_{456}\}$	Δ_{34}^C	$\{T_1, T_{126}, T_{156}, T_2, T_{256}, T_{346}, T_5, T_6\}$
Δ_{35}	$\{T_{136}, T_{156}, T_{236}, T_{256}, T_3, T_{346}, T_{456}, T_5\}$	Δ_{35}^C	$\{T_1, T_{126}, T_{146}, T_2, T_{246}, T_{356}, T_4, T_6\}$
Δ_{36}	$\{T_{126}, T_{146}, T_{156}, T_{246}, T_{256}, T_3, T_{456}, T_6\}$	Δ_{36}^C	$\{T_1, T_{136}, T_2, T_{236}, T_{346}, T_{356}, T_4, T_5\}$
Δ_{45}	$\{T_{146}, T_{156}, T_{246}, T_{256}, T_{346}, T_{356}, T_4, T_5\}$	Δ_{55}^C	$\{T_1, T_{126}, T_{136}, T_2, T_{236}, T_3, T_{456}, T_6\}$
Δ_{46}	$\{T_{126}, T_{136}, T_{156}, T_{236}, T_{256}, T_{356}, T_4, T_6\}$	Δ_{46}^C	$\{T_1, T_{146}, T_2, T_{246}, T_3, T_{346}, T_{456}, T_5\}$
Δ_{56}	$\{T_{126}, T_{136}, T_{146}, T_{236}, T_{246}, T_{346}, T_5, T_6\}$	Δ_{56}^C	$\{T_1, T_{156}, T_2, T_{256}, T_3, T_{356}, T_4, T_{456}\}$

In particular, the sets of eight tropes corresponding to even eights Δ_{ij} contain 6 even tropes T_{kl6} and 2 odd tropes T_m (second and fourth case in Equation (55)).

Proof. Sets of eight distinct nodes arise in Lemma 4.20.(3) as the vertices of the two Rosenhain tetrahedra forming each set of eight tropes. Thus, we have two ways of constructing even eights. One checks that this defines a one-to-one correspondence between even eights and sets of eight tropes satisfying a quadratic relation. This one-to-one correspondence maps the even eights Δ_{14}, Δ_{15}, Δ_{23}, Δ_{26}, Δ_{36}, Δ_{45} precisely to the sets of eight tropes obtained using the duality map between Theta divisors and Theta characteristics in Remark 3.5. For the remaining even eights one has to take the complement. The proof then follows from a direct computation and Lemmas 4.20 and 4.21. $\qquad\Box$

One easily checks the following:

Lemma 4.23
(1) *Every Göpel tetrahedron with all even tropes is contained in three sets of eight tropes corresponding to an even eight Δ_{ij}.*
(2) *Every Rosenhain tetrahedron with one odd and three even tropes is contained in two sets of eight tropes corresponding to an even eight Δ_{ij}.*

Proof. The proof follows from direct computation. $\qquad\Box$

Lemma 4.24 *The 15 sets of eight tropes corresponding to even eights* Δ_{ij}

Corollary 4.25 *For every Rosenhain or Göpel tetrahedron, the remaining twelve tropes are linear functions of the tropes contained in the tetrahedron with coefficients in* $\mathbb{C}(\lambda_1, \lambda_2, \lambda_3)$.

Proof. For every Rosenhain or Göpel tetrahedron, the projective coordinates $[z_1 : z_2 : z_3 : z_4] \in \mathbb{P}^3$ are linear functions of the tropes contained in the tetrahedron with coefficients that are rational functions in $\mathbb{C}[\lambda_1, \lambda_2, \lambda_3]$. But all tropes are linear functions of the coordinates $[z_1 : z_2 : z_3 : z_4] \in \mathbb{P}^3$ and were given in Table 5.3. □

4.5 The Irrational Kummer Normal Form

The following lemma shows that every Göpel tetrahedron can be completed into one of the sets of eight tropes given by Lemma 4.20:

Corollary 4.26 *Every Göpel tetrahedron arises as union of two two-tuples chosen from*

$$\left\{ \{\mathsf{T}_a, \mathsf{T}_{a'}\}, \{\mathsf{T}_b, \mathsf{T}_{b'}\}, \{\mathsf{T}_c, \mathsf{T}_{c'}\}, \{\mathsf{T}_d, \mathsf{T}_{d'}\} \right\},$$

where $\{\mathsf{T}_a, \mathsf{T}_b, \mathsf{T}_c, \mathsf{T}_d\}$ *and* $\{\mathsf{T}_{a'}, \mathsf{T}_{b'}, \mathsf{T}_{c'}, \mathsf{T}_{d'}\}$ *are two disjoint Rosenhain tetrahedra given in Lemma 4.20. Moreover, every Göpel tetrahedron is contained in exactly three different sets of eight tropes (out of 30) in Lemma 4.20. Conversely, every set of eight tropes given in Lemma 4.20 can be decomposed into pairs of disjoint Göpel tetrahedra in three different ways.* □

Having already established quadratic relations between these sets of eight tropes, we can re-write the equation of a singular Kummer surface. We have the following:

Lemma 4.27 *For every set of eight tropes given in Lemma 4.20, the equation*

$$\left(\mu^2 \mathsf{T}_b \mathsf{T}_{b'} + \nu^2 \mathsf{T}_c \mathsf{T}_{c'} - \rho^2 \mathsf{T}_d \mathsf{T}_{d'} \right)^2 = 4\mu^2 \nu^2 \mathsf{T}_b \mathsf{T}_{b'} \mathsf{T}_c \mathsf{T}_{c'} \tag{56}$$

is equivalent to the Cassels-Flynn quartic in Equation (37).

Proof. The proof follows from direct computation using the explicit equations for the tropes in Table 5.3. □

Remark 4.28   Because of Equation (54), $\mu + \nu + \rho = 0$, and $\beta = \gamma = \delta = 1$ or $\beta = \delta - \nu, \gamma = \delta + \mu$, Equation (56) is identical to any of the following equations:

$$\left(\mu^2 T_b T_{b'} + \nu^2 T_c T_{c'} - \rho^2 T_d T_{d'} \right)^2 = 4\mu^2 \nu^2 T_b T_{b'} T_c T_{c'}$$

$$\left(\mu^2 T_b T_{b'} - \nu^2 T_c T_{c'} + \rho^2 T_d T_{d'} \right)^2 = 4\mu^2 \rho^2 T_b T_{b'} T_d T_{d'},$$

$$\left(\mu^2 T_b T_{b'} - \nu^2 T_c T_{c'} - \rho^2 T_d T_{d'} \right)^2 = 4\nu^2 \rho^2 T_c T_{c'} T_d T_{d'},$$

$$\left(\rho^2 T_a T_{a'} + \gamma^2 T_b T_{b'} - \beta^2 T_c T_{c'} \right)^2 = 4\gamma^2 \rho^2 T_a T_{a'} T_b T_{b'},$$

$$\left(\nu^2 T_a T_{a'} - \delta^2 T_b T_{b'} + \beta^2 T_d T_{d'} \right)^2 = 4\beta^2 \nu^2 T_a T_{a'} T_d T_{d'},$$

$$\left(\mu^2 T_a T_{a'} + \delta^2 T_c T_{c'} - \gamma^2 T_d T_{d'} \right)^2 = 4\delta^2 \mu^2 T_a T_{a'} T_c T_{c'}. \qquad (57)$$

We have the following:

Proposition 4.29   *For every Göpel tetrahedron $\{T_a, T_{a'}, T_b, T_{b'}\}$, the Cassels-Flynn quartic in Equation (37) is equivalent to an equation of the form*

$$\left(\rho^2 T_a T_{a'} + \gamma^2 T_b T_{b'} - \beta^2 T_c T_{c'} \right)^2 = 4\gamma^2 \rho^2 T_a T_{a'} T_b T_{b'},$$

where β, γ, ρ are polynomials in $\mathbb{C}[\lambda_1, \lambda_2, \lambda_3]$, and $T_c, T_{c'}$ are two tropes that are linear functions of $T_a, T_{a'}, T_b, T_{b'}$ with coefficients in $\mathbb{C}(\lambda_1, \lambda_2, \lambda_3)$.

Proof.   By Corollary 4.26, each Göpel tetrahedron on a Kummer quartic is obtained from two pairs of Rosenhain tetrahedra in Lemma 4.20 such that Equations (57) are equivalent to the Cassels-Flynn quartic in Equation (37). By Corollary 4.25, the remaining twelve tropes are linear functions of the tropes contained in the Göpel tetrahedron with coefficients that are rational functions in $\mathbb{C}[\lambda_1, \lambda_2, \lambda_3]$. $\square$

We give the following example:

Example 4.30   *If we choose the two Rosenhain tetrahedra*

$$\{T_{lm6}, T_i, T_j, T_k\} \quad and \quad \{T_6, T_{jk6}, T_{ik6}, T_{ij6}\}$$

in Lemma 4.8, then Equation (56) is equivalent to

$$\varphi^2 = 4 \left(\lambda_i - \lambda_k \right)^2 (\lambda_j - \lambda_k)^2 T_i T_j T_{ik6} T_{jk6}, \qquad (58)$$

with

$$\varphi = (\lambda_j - \lambda_k)^2 T_i T_{jk6} + (\lambda_i - \lambda_k)^2 T_j T_{ik6} - (\lambda_i - \lambda_j)^2 T_k T_{ij6}.$$

T_k and T_{ij6} *are linear functions of the Göpel tetrahedron* $\{\mathsf{T}_i, \mathsf{T}_j, \mathsf{T}_{ik6}, \mathsf{T}_{jk6}\}$
given by

$$
\begin{aligned}
\mathsf{T}_k &= \frac{(\lambda_k-\lambda_m)(\lambda_k-\lambda_l)(\lambda_j-\lambda_k)\mathsf{T}_i}{(\lambda_i-\lambda_j)(\lambda_i\lambda_j-\lambda_i\lambda_k-\lambda_j\lambda_k+\lambda_k\lambda_l)} + \frac{(\lambda_k-\lambda_m)(\lambda_k-\lambda_l)(\lambda_i-\lambda_k)\mathsf{T}_j}{(\lambda_j-\lambda_i)(\lambda_i\lambda_j-\lambda_i\lambda_k-\lambda_j\lambda_k+\lambda_k\lambda_l)} \\
&+ \frac{(\lambda_i-\lambda_k)(\lambda_j-\lambda_k)\mathsf{T}_{ik6}}{(\lambda_i-\lambda_j)(\lambda_i\lambda_j-\lambda_i\lambda_k-\lambda_j\lambda_k+\lambda_k\lambda_l)} + \frac{(\lambda_i-\lambda_k)(\lambda_j-\lambda_k)\mathsf{T}_{jk6}}{(\lambda_j-\lambda_i)(\lambda_i\lambda_j-\lambda_i\lambda_k-\lambda_j\lambda_k+\lambda_k\lambda_l)}, \\
\mathsf{T}_{ij6} &= \frac{(\lambda_j-\lambda_m)(\lambda_j-\lambda_l)(\lambda_j-\lambda_k)(\lambda_i-\lambda_k)\mathsf{T}_i}{(\lambda_j-\lambda_i)(\lambda_i\lambda_j-\lambda_i\lambda_k-\lambda_j\lambda_k+\lambda_k\lambda_l)} + \frac{(\lambda_i-\lambda_m)(\lambda_i-\lambda_l)(\lambda_i-\lambda_k)(\lambda_j-\lambda_k)\mathsf{T}_j}{(\lambda_i-\lambda_j)(\lambda_i\lambda_j-\lambda_i\lambda_k-\lambda_j\lambda_k+\lambda_k\lambda_l)} \\
&+ \frac{(\lambda_j-\lambda_m)(\lambda_j-\lambda_l)(\lambda_i-\lambda_k)\mathsf{T}_{ik6}}{(\lambda_j-\lambda_i)(\lambda_i\lambda_j-\lambda_i\lambda_k-\lambda_j\lambda_k+\lambda_k\lambda_l)} + \frac{(\lambda_i-\lambda_m)(\lambda_i-\lambda_l)(\lambda_j-\lambda_k)\mathsf{T}_{jk6}}{(\lambda_i-\lambda_j)(\lambda_i\lambda_j-\lambda_i\lambda_k-\lambda_j\lambda_k+\lambda_k\lambda_l)}.
\end{aligned}
$$

$$(59)$$

Kummer [24] introduced the notion of formal square roots of Equations (57) which he called *irrational Kummer normal form*. Irrational Kummer normal forms of a singular Kummer surface also appeared in a slightly different form in [19]. We make the following:

Remark 4.31   For every Göpel tetrahedron, four formal square roots giving rise to Equations (57) are as follows:

$$
\begin{aligned}
\mu\sqrt{\mathsf{T}_b\mathsf{T}_{b'}} + \nu\sqrt{\mathsf{T}_c\mathsf{T}_{c'}} + \rho\sqrt{\mathsf{T}_d\mathsf{T}_{d'}} &= 0, \\
\rho\sqrt{\mathsf{T}_a\mathsf{T}_{a'}} + \gamma\sqrt{\mathsf{T}_b\mathsf{T}_{b'}} - \beta\sqrt{\mathsf{T}_c\mathsf{T}_{c'}} &= 0, \\
\nu\sqrt{\mathsf{T}_a\mathsf{T}_{a'}} - \delta\sqrt{\mathsf{T}_b\mathsf{T}_{b'}} + \beta\sqrt{\mathsf{T}_d\mathsf{T}_{d'}} &= 0, \\
\mu\sqrt{\mathsf{T}_a\mathsf{T}_{a'}} + \delta\sqrt{\mathsf{T}_c\mathsf{T}_{c'}} - \gamma\sqrt{\mathsf{T}_d\mathsf{T}_{d'}} &= 0.
\end{aligned}
$$

$$(60)$$

A particular compatible choice for the signs of the coefficients in Equations (60) has to be made such that two equations are linear combinations of the other two equations using $\mu + \nu + \rho = 0$ and $\beta\mu + \gamma\nu + \delta\rho = 0$.

If we replace the formal expressions $\sqrt{\mathsf{T}_b\mathsf{T}_{b'}}$ in Equations (57) by the well-defined sections $\mathsf{t}_{b,b'}$ introduced in Equations (51), we obtain the following:

Proposition 4.32   *The irrational Kummer normal forms coincide with the Mumford Theta relations in Proposition 2.3. In particular, Equations (109) or, equivalently, Equations (108), generate the same ideal as the following 60 equations given by*

$$
\mu\,\mathsf{t}_{b,b'} + \nu\,\mathsf{t}_{c,c'} + \rho\,\mathsf{t}_{d,d'} = 0, \qquad \rho\,\mathsf{t}_{a,a'} + \gamma\,\mathsf{t}_{b,b'} - \beta\,\mathsf{t}_{c,c'} = 0, \qquad (61)
$$

where $\{a, b, c, d, a', b', c', d'\}$ *and* $\beta, \gamma, \mu, \nu, \gamma$ *run over all cases in Lemma 4.20.*

Proof. The proof follows from direct computation using the explicit equations for the tropes in terms of Theta functions in Equations (49) and

Equations (51). The ideal generated by the irrational Kummer normal forms then coincides with the one generated by the Mumford Theta relations in Proposition 2.3. ☐

4.6 The Göpel and Göpel-Hudson Normal Forms

The following definition is due to Kummer [24] and Borchardt [4]:

Definition 4.33 *A Göpel quartic is the surface in* $\mathbb{P}^3(P, Q, R, S)$ *given by*

$$\Phi^2 - 4\delta^2 SPQR = 0 \,, \tag{62}$$

with

$$\Phi = P^2 + Q^2 + R^2 + S^2 - \alpha(PS + QR) - \beta(PQ + RS) - \gamma(PR + QS), \tag{63}$$

where $\alpha, \beta, \gamma, \delta \in \mathbb{C}$ *such that*

$$\delta^2 = \alpha^2 + \beta^2 + \gamma^2 + \alpha\beta\gamma - 4 \,. \tag{64}$$

We have the following:

Proposition 4.34 *Every Göpel tetrahedron determines an isomorphism between the Cassels-Flynn quartic in Equation (37) and a Göpel quartic in Equation (62). In particular, for generic parameters* $(\alpha, \beta, \gamma, \delta)$ *satisfying Equation (64), the Göpel quartic in Equation (62) is isomorphic to the singular Kummer variety associated with a principally polarized abelian variety.*

Proof. First, one realizes every Göpel tetrahedron $\{T_a, T_{a'}, T_b, T_{b'}\}$ using two disjoint Rosenhain tetrahedra given in Lemma 4.20. Using Proposition 4.29, the Cassels-Flynn quartic in Equation (37) is identical to the equation

$$\left(\rho^2 T_a T_{a'} + \gamma^2 T_b T_{b'} - \beta^2 T_c T_{c'}\right)^2 - 4\gamma^2\rho^2 T_a T_{a'} T_b T_{b'} = 0, \tag{65}$$

with $\beta, \gamma, \rho \in \mathbb{C}[\lambda_1, \lambda_2, \lambda_3]$. Using Corollary 4.25, the two additional tropes $T_c, T_{c'}$ are linear functions of $\{T_a, T_{a'}, T_b, T_{b'}\}$ with coefficients in $\mathbb{C}(\lambda_1, \lambda_2, \lambda_3)$. One substitutes

$$[T_a, T_{a'}, T_b, T_{b'}] = [c_1 P : c_2 Q : c_3 R : c_4 S],$$

into Equation (65) and solves for the coefficients $c_1, \ldots, c_4$ such that Equation (65) coincides with Equation (62). ☐

We give another description of a singular Kummer variety associated with a principally polarized abelian variety based on results in [3]. Let $\mathcal{L}$ be the ample line symmetric bundle of an abelian surface **A** defining its principal polarization and consider the rational map $\varphi_{\mathcal{L}^2} : \mathbf{A} \to \mathbb{P}^3$ associated with the line bundle $\mathcal{L}^2$. Its image $\varphi_{\mathcal{L}^2}(\mathbf{A})$ is a quartic surface in $\mathbb{P}^3$ which in projective coordinates $[w : x : y : z]$ can be written as

$$
\begin{aligned}
0 = {}& \xi_0 \left(w^4 + x^4 + y^4 + z^4\right) + \xi_4 \, wxyz + \xi_1 \left(w^2z^2 + x^2y^2\right) \\
& + \xi_2 \left(w^2x^2 + y^2z^2\right) + \xi_3 \left(w^2y^2 + x^2z^2\right),
\end{aligned}
\tag{66}
$$

with $[\xi_0 : \xi_1 : \xi_2 : \xi_3 : \xi_4] \in \mathbb{P}^4$. Any general member of the family (66) is smooth. As soon as the surface is singular at a general point, it must have sixteen singular nodal points because of its symmetry. The discriminant turns out to be a homogeneous polynomial of degree eighteen in the parameters $[\xi_0 : \xi_1 : \xi_2 : \xi_3 : \xi_4] \in \mathbb{P}^4$ and was determined in [3, Sec. 7.7 (3)]. Thus, the Kummer surfaces form an open set among the hypersurfaces in Equation (66) with parameters $[\xi_0 : \xi_1 : \xi_2 : \xi_3 : \xi_4] \in \mathbb{P}^4$, namely the ones that make the only irreducible factor of degree three in the discriminant vanish, i.e.,

$$
\xi_0 \left(16\xi_0^2 - 4\xi_1^2 - 4\xi_2^2 - 4\xi_3^3 + \xi_4^2\right) + 4\,\xi_1\xi_2\xi_3 = 0.
\tag{67}
$$

Setting $\xi_0 = 1$ and using the affine moduli $\xi_1 = -A$, $\xi_2 = -B$, $\xi_3 = -C$, $\xi_4 = 2D$, we obtain the normal form of a nodal quartic surface. We give the following:

Definition 4.35 *A Göpel-Hudson quartic is the surface in* $\mathbb{P}^3(w, x, y, z)$ *given by*

$$
\begin{aligned}
0 = {}& w^4 + x^4 + y^4 + z^4 + 2Dwxyz - A\left(w^2z^2 + x^2y^2\right) \\
& - B\left(w^2x^2 + y^2z^2\right) - C\left(w^2y^2 + x^2z^2\right),
\end{aligned}
\tag{68}
$$

where $A, B, C, D \in \mathbb{C}$ *such that*

$$
D^2 = A^2 + B^2 + C^2 + ABC - 4 .
\tag{69}
$$

Remark 4.36 The Göpel-Hudson quartic in Equation (68) is invariant under the transformations changing signs of two coordinates, generated by

$$[w : x : y : z] \to [-w : -x : y : z], \quad [w : x : y : z] \to [-w : x : -y : z],$$

and under the permutations of variables, generated by $[w : x : y : z] \to [x : w : z : y]$ and $[w : x : y : z] \to [y : z : w : x]$.

We have the following:

Lemma 4.37 *The Göpel-Hudson quartic in Equation (68) is isomorphic to the Göpel quartic in Equation (62). In particular, for generic parameters* (A, B, C, D) *satisfying Equation (69), the Göpel quartic in Equation (62) is isomorphic to the singular Kummer variety associated with a principally polarized abelian surface.*

Proof. We first introduce complex numbers $[w_0 : x_0 : y_0 : z_0] \in \mathbb{P}^3$ such that

$$A = \frac{w_0^4 - x_0^4 - y_0^4 + z_0^4}{w_0^2 z_0^2 - x_0^2 y_0^2}, \quad B = \frac{w_0^4 + x_0^4 - y_0^4 - z_0^4}{w_0^2 x_0^2 - y_0^2 z_0^2},$$

$$C = \frac{w_0^4 - x_0^4 + y_0^4 - z_0^4}{w_0^2 y_0^2 - x_0^2 z_0^2},$$

$$D = \frac{w_0 x_0 y_0 z_0 \prod_{\epsilon, \epsilon' \in \{\pm 1\}} (w_0^2 + \epsilon x_0^2 + \epsilon' y_0^2 + \epsilon \epsilon' z_0^2)}{(w_0^2 z_0^2 - x_0^2 y_0^2)(w_0^2 x_0^2 - y_0^2 z_0^2)(w_0^2 y_0^2 - x_0^2 z_0^2)}, \tag{70}$$

and

$$\alpha = 2 \frac{w_0^2 y_0^2 + x_0^2 z_0^2}{w_0^2 y_0^2 - x_0^2 z_0^2}, \quad \beta = 2 \frac{w_0^2 x_0^2 + y_0^2 z_0^2}{w_0^2 x_0^2 - y_0^2 z_0^2}, \quad \gamma = 2 \frac{w_0^2 z_0^2 + x_0^2 y_0^2}{w_0^2 z_0^2 - x_0^2 y_0^2},$$

$$\delta^2 = 16 \frac{w_0^4 x_0^4 y_0^4 z_0^4 \prod_{\epsilon, \epsilon' \in \{\pm 1\}} (w_0^2 + \epsilon x_0^2 + \epsilon' y_0^2 + \epsilon \epsilon' z_0^2)}{\left(w_0^2 y_0^2 - x_0^2 z_0^2\right)^2 \left(w_0^2 x_0^2 - y_0^2 z_0^2\right)^2 \left(w_0^2 z_0^2 - x_0^2 y_0^2\right)^2}. \tag{71}$$

Then, the linear transformation given by

$$P = w_0 w + x_0 x + y_0 y + z_0 z,$$

$$Q = w_0 w + x_0 x - y_0 y - z_0 z,$$

$$R = w_0 w - x_0 x - y_0 y + z_0 z,$$

$$S = w_0 w - x_0 x + y_0 y - z_0 z, \tag{72}$$

transforms Equation (62) into Equation (68). □

Lemma 4.38 *Using the same notation as in Equation (70), the sixteen nodes of the Göpel-Hudson quartic (68) are the points $[w : x : y : z]$ given by*

$$[w_0 : x_0 : y_0 : z_0], [-w_0 : -x_0 : y_0 : z_0], [-w_0 : x_0 : -y_0 : z_0], [-w_0 : x_0 : y_0 : -z_0],$$

$$[x_0 : w_0 : z_0 : y_0], [-x_0 : -w_0 : z_0 : y_0], [-x_0 : w_0 : -z_0 : y_0], [-x_0 : w_0 : z_0 : -y_0],$$

$$[y_0 : z_0 : w_0 : x_0], [-y_0 : -z_0 : w_0 : x_0], [-y_0 : z_0 : -w_0 : x_0], [-y_0 : z_0 : w_0 : -x_0],$$

$$[z_0 : y_0 : x_0 : w_0], [-z_0 : -y_0 : x_0 : w_0], [-z_0 : y_0 : -x_0 : w_0], [-z_0 : y_0 : x_0 : -w_0].$$

In particular, for generic parameters (A, B, C, D), no node is contained in the coordinate planes $w = 0$, $x = 0$, $y = 0$, or $z = 0$.

We also set

$$p_0^2 = w_0^2 + x_0^2 + y_0^2 + z_0^2, \quad q_0^2 = w_0^2 + x_0^2 - y_0^2 - z_0^2,$$

$$r_0^2 = w_0^2 - x_0^2 + y_0^2 - z_0^2, \quad s_0^2 = w_0^2 - x_0^2 - y_0^2 + z_0^2. \tag{73}$$

We have the following:

Lemma 4.39 *The nodes of the Göpel quartic (62) are the points* $[P : Q : R : S]$ *given by*

$$[p_0^2 : q_0^2 : r_0^2 : s_0^2], [q_0^2 : p_0^2 : s_0^2 : r_0^2], [r_0^2 : s_0^2 : r_0^2 : q_0^2], [s_0^2 : r_0^2 : q_0^2 : p_0^2],$$

$$[w_0x_0 + y_0z_0 : w_0x_0 - y_0z_0 : 0 : 0], [w_0x_0 - y_0z_0 : w_0x_0 + y_0z_0 : 0 : 0],$$

$$[0 : 0 : w_0x_0 + y_0z_0 : w_0x_0 - y_0z_0], [0 : 0 : w_0x_0 - y_0z_0 : w_0x_0 + y_0z_0],$$

$$[w_0z_0 + x_0y_0 : 0 : w_0z_0 - x_0y_0 : 0], [0 : w_0z_0 + x_0y_0 : 0 : w_0z_0 - x_0y_0],$$

$$[w_0z_0 - x_0y_0 : 0 : w_0z_0 + x_0y_0 : 0], [0 : w_0z_0 - x_0y_0 : 0 : w_0z_0 + x_0y_0],$$

$$[w_0y_0 + x_0z_0 : 0 : 0 : w_0y_0 - x_0z_0], [0 : w_0y_0 + x_0z_0 : w_0y_0 - x_0z_0 : 0],$$

$$[0 : w_0y_0 - x_0z_0 : w_0y_0 + x_0z_0 : 0], [w_0y_0 - x_0z_0 : 0 : 0 : w_0y_0 + x_0z_0].$$

4.7 (2, 2)-Polarized Kummer Surfaces

We use the Göpel and Göpel-Hudson quartics to provide an explicit model for $(2, 2)$-isogenous Kummer surfaces.

Remark 4.40 The transformation $[w : x : y : z] \to [-w : x : y : z]$ is an isomorphism between the Göpel-Hudson quartic with moduli parameters (A, B, C, D) and the one with parameters $(A, B, C, -D)$. Moreover, the two quartics coincide exactly along the coordinate planes $w = 0$, $x = 0$, $y = 0$, or $z = 0$.

The map $\pi : \mathbb{P}(w, x, y, z) \to \mathbb{P}(P, Q, R, S)$ with $P = w^2, \ldots, S = z^2$ is $8 : 1$ outside the coordinate planes. The map π induces a covering of a reducible octic surface in $\mathbb{P}^3$ given by

$$(\Psi - 2Dwxyz)(\Psi + 2Dwxyz) = 0, \tag{74}$$

with

$$\Psi = w^4 + x^4 + y^4 + z^4 - A(w^2y^2 + y^2z^2) - B(w^2z^2 + x^2y^2) - C(w^2x^2 + y^2z^2),$$

onto the Göpel quartic in Equation (62) with $\alpha = A$, $\beta = B$, $C = \gamma$, and $D = \delta$. Because of Lemma 4.37 and Lemma 4.34, we can assume that the Göpel-Hudson quartic and Göpel quartic are the singular Kummer varieties associated with two principally polarized abelian varieties, say $\mathbf{A}$ and $\mathbf{A}'$, respectively. We have the following:

Proposition 4.41 *The map* $\pi : \mathbb{P}(w, x, y, z) \to \mathbb{P}(P, Q, R, S)$ *with* $P = w^2, \ldots, S = z^2$ *restricted to the Göpel-Hudson quartic onto the Göpel quartic with* $\alpha = A$, $\beta = B$, $C = \gamma$, *and* $D = \delta$, *is induced by a* $(2, 2)$-*isogeny* $\psi : \mathbf{A} \to \mathbf{A}' \cong \hat{\mathbf{A}}$.

Proof. The rational map π is 4:1 from the Göpel-Hudson quartic onto the Göpel quartic. The map $\pi : \mathbb{P}(w, x, y, z) \to \mathbb{P}(P, Q, R, S)$ maps the sixteen nodes on the Göpel-Hudson quartic in Lemma 4.38 to four nodes on the Göpel quartic. It follows from [28] that the map π between Kummer varieties associated with two principally polarized abelian varieties $\mathbf{A}$ and $\mathbf{A}'$ is induced by an isogeny $p : \mathbf{A}' \to \mathbf{A}$ of degree four. The kernel of p is a two-dimensional isotropic subspace K of $\mathbf{A}[2]$ such that $\mathbf{A}' = \mathbf{A}/\mathsf{K}$.   □

4.7.1 An Explicit Model Using Theta Functions

We provide an explicit model for the Göpel quartic in terms of Theta functions. We have the following:

Proposition 4.42 *The surface in $\mathbb{P}^3$ given by Equation (62) is isomorphic to the singular Kummer variety $\mathbf{A}_\tau/\langle -\mathbb{I} \rangle$ where the coordinates are given by*

$$[P : Q : R : S] = \left[\theta_1(z)^2 : \theta_2(z)^2 : \theta_3(z)^2 : \theta_4(z)^2 \right] , \qquad (75)$$

and the moduli parameters are

$$\alpha = \frac{\theta_1^4 - \theta_2^4 - \theta_3^4 + \theta_4^4}{\theta_1^2 \theta_4^2 - \theta_2^2 \theta_3^2}, \qquad \beta = \frac{\theta_1^4 + \theta_2^4 - \theta_3^4 - \theta_4^4}{\theta_1^2 \theta_2^2 - \theta_3^2 \theta_4^2},$$

$$\gamma = \frac{\theta_1^4 - \theta_2^4 + \theta_3^4 - \theta_4^4}{\theta_1^2 \theta_3^2 - \theta_2^2 \theta_4^2},$$

$$\delta = \frac{\theta_1 \theta_2 \theta_3 \theta_4 \prod_{\epsilon, \epsilon' \in \{\pm 1\}} (\theta_1^2 + \epsilon \theta_2^2 + \epsilon' \theta_3^2 + \epsilon \epsilon' \theta_3^2)}{\left(\theta_1^2 \theta_2^2 - \theta_3^2 \theta_4^2 \right) \left(\theta_1^2 \theta_3^2 - \theta_2^2 \theta_4^2 \right) \left(\theta_1^2 \theta_4^2 - \theta_2^2 \theta_3^2 \right)}. \qquad (76)$$

Proof. One checks that the Rosenhain tetrahedron $\{\mathsf{T}_{256}, \mathsf{T}_{136}, \mathsf{T}_{356}, \mathsf{T}_{126}\}$, with tropes given by the Theta function $\theta_1(z)^2, \theta_2(z)^2, \theta_3(z)^2, \theta_4(z)^2$ is contained in three sets of eight tropes corresponding to the even eights Δ_{15}, Δ_{46}, Δ_{23} in Proposition 4.22. One can choose any of these sets of eight tropes to proceed. For example, n Lemma 4.8 we can choose the Rosenhain tetrahedra $\{\mathsf{T}_{256}, \mathsf{T}_{126}, \mathsf{T}_{236}, \mathsf{T}_4\}$ and $\{\mathsf{T}_{136}, \mathsf{T}_{356}, \mathsf{T}_{156}, \mathsf{T}_6\}$. Then, Equation (56) is equivalent to

$$\varphi^2 - 4 (\lambda_2 - \lambda_1)^2 (\lambda_3 - \lambda_1)^2 (\lambda_2 - 1)^2 (\lambda_3 - 1)^2 \mathsf{T}_{126} \mathsf{T}_{136} \mathsf{T}_{256} \mathsf{T}_{356} , \qquad (77)$$

with

$$\varphi = (\lambda_3 - \lambda_1)^2 (\lambda_2 - 1)^2 \mathsf{T}_{136} \mathsf{T}_{256} + (\lambda_2 - \lambda_1)^2 (\lambda_3 - \lambda_1)^2 \mathsf{T}_{126} \mathsf{T}_{236}$$
$$- (\lambda_3 - \lambda_2)^2 (\lambda_1 - 1)^2 \mathsf{T}_{156} \mathsf{T}_{236}.$$

In Proposition 4.29 we then choose the coordinates determined by the Göpel tetrahedron $\{\mathsf{T}_{256}, \mathsf{T}_{136}, \mathsf{T}_{356}, \mathsf{T}_{126}\}$. We choose coefficients $c_1, \ldots, c_4$ with

$$[\mathsf{T}_{256}, \mathsf{T}_{136}, \mathsf{T}_{356}, \mathsf{T}_{126}] = [c_1 P : c_2 Q : c_3 R : c_4 S],$$

such that Equation (77) coincides with Equation (62). It turns out that the coefficients $c_1, \ldots, c_4$ are rational functions in the squares of (even) Theta constants $\theta_1^2, \ldots, \theta_{10}^2$, and the coordinates are then given by Theta functions with non-vanishing elliptic variables $[P : Q : R : S] = [\theta_1(z)^2 : \theta_2(z)^2 : \theta_3(z)^2 : \theta_4(z)^2]$. In particular, setting

$$\left[\mathsf{T}_{256}, \mathsf{T}_{136}, \mathsf{T}_{356}, \mathsf{T}_{126} \right] = \left[\theta_2^2 \theta_3^2 \theta_4^2 \, P : \theta_1^2 \theta_3^2 \theta_4^2 \, Q : \theta_1^2 \theta_2^2 \theta_4^2 \, R : \theta_1^2 \theta_2^2 \theta_3^2 \, S \right] \tag{78}$$

changes Equation (54) into Equation (62). Using Equations (10), the transformation (78) descends to $\mathcal{A}_2(2, 4)$ with

$$\alpha = 2 \, \frac{\Theta_1^2 \Theta_3^2 + \Theta_2^2 \Theta_4^2}{\Theta_1^2 \Theta_3^2 - \Theta_2^2 \Theta_4^2}, \quad \beta = 2 \, \frac{\Theta_1^2 \Theta_2^2 + \Theta_3^2 \Theta_4^2}{\Theta_1^2 \Theta_2^2 - \Theta_3^2 \Theta_4^2}, \quad \gamma = 2 \, \frac{\Theta_1^2 \Theta_4^2 + \Theta_2^2 \Theta_3^2}{\Theta_1^2 \Theta_4^2 - \Theta_2^2 \Theta_3^2}. \tag{79}$$

Using Equations (9) the moduli parameter can be re-written

$$\alpha = \frac{\theta_1^4 - \theta_2^4 - \theta_3^4 + \theta_4^4}{\theta_1^2 \theta_4^2 - \theta_2^2 \theta_3^2}, \quad \beta = \frac{\theta_1^4 + \theta_2^4 - \theta_3^4 - \theta_4^4}{\theta_1^2 \theta_2^2 - \theta_3^2 \theta_4^2},$$

$$\gamma = \frac{\theta_1^4 - \theta_2^4 + \theta_3^4 - \theta_4^4}{\theta_1^2 \theta_3^2 - \theta_2^2 \theta_4^2}. \tag{80}$$

such that over $\mathcal{A}_2(4, 8)$ the modulus δ is given as

$$\delta = \frac{\theta_1 \theta_2 \theta_3 \theta_4 \prod_{\epsilon, \epsilon' \in \{\pm 1\}} \left(\theta_1^2 + \epsilon \theta_2^2 + \epsilon' \theta_3^2 + \epsilon \epsilon' \theta_3^2 \right)}{\left(\theta_1^2 \theta_2^2 - \theta_3^2 \theta_4^2 \right) \left(\theta_1^2 \theta_3^2 - \theta_2^2 \theta_4^2 \right) \left(\theta_1^2 \theta_4^2 - \theta_2^2 \theta_3^2 \right)}. \qquad \square$$

The Rosenhain roots of the $(2, 2)$-isogenous genus-two curve $\hat{\mathcal{C}}$ in Equation (29) can be considered coordinates of an isogenous moduli space that we denote by $\hat{\mathcal{A}}_2(2)$. We have the following:

Lemma 4.43 *The moduli* $\alpha, \beta, \gamma, \delta$ *are rational functions over* $\hat{\mathcal{A}}_2(2)$.

Proof. Equations (76) descend to rational functions on $\hat{\mathcal{A}}_2(2)$ given by

$$\alpha = 2 \, \frac{\Lambda_1 + 1}{\Lambda_1 - 1}, \quad \beta = 2 \, \frac{\Lambda_1 \Lambda_2 + \Lambda_1 \Lambda_3 - 2 \Lambda_2 \Lambda_3 - 2 \Lambda_1 + \Lambda_2 + \Lambda_3}{(\Lambda_2 - \Lambda_3)(\Lambda_1 - 1)},$$

$$\gamma = 2 \, \frac{\Lambda_3 + \Lambda_2}{\Lambda_3 - \Lambda_2}, \quad \delta = \frac{4(\Lambda_1 - \Lambda_2 \Lambda_3)}{(\Lambda_1 - 1)(\Lambda_3 - \Lambda_2)}. \tag{81}$$

$\square$

We also provide an explicit model for the Göpel-Hudson quartic in terms of Theta functions. In terms of Theta functions, the relation between Equation (68) and Equation (62) was first determined by Borchardt [4]. We have the following:

Proposition 4.44 *The surface in $\mathbb{P}^3$ given by Equation (68) is isomorphic to the singular Kummer variety $\mathbf{A}_\tau/\langle -\mathbb{I}\rangle$ where the coordinates are given by*

$$[w : x : y : z] = [\Theta_1(2z) : \Theta_2(2z) : \Theta_3(2z) : \Theta_4(2z)] , \qquad (82)$$

and the moduli parameters are

$$A = \frac{\Theta_1^4 - \Theta_2^4 - \Theta_3^4 + \Theta_4^4}{\Theta_1^2\Theta_4^2 - \Theta_2^2\Theta_3^2}, \qquad B = \frac{\Theta_1^4 + \Theta_2^4 - \Theta_3^4 - \Theta_4^4}{\Theta_1^2\Theta_2^2 - \Theta_3^2\Theta_4^2},$$

$$C = \frac{\Theta_1^4 - \Theta_2^4 + \Theta_3^4 - \Theta_4^4}{\Theta_1^2\Theta_3^2 - \Theta_2^2\Theta_4^2},$$

$$D = \frac{\Theta_1\Theta_2\Theta_3\Theta_4 \prod_{\epsilon,\epsilon' \in \{\pm 1\}}\left(\Theta_1^2 + \epsilon\Theta_2^2 + \epsilon'\Theta_3^2 + \epsilon\epsilon'\Theta_4^2\right)}{\left(\Theta_1^2\Theta_2^2 - \Theta_3^2\Theta_4^2\right)\left(\Theta_1^2\Theta_3^2 - \Theta_2^2\Theta_4^2\right)\left(\Theta_1^2\Theta_4^2 - \Theta_2^2\Theta_3^2\right)}. \qquad (83)$$

Proof. Comparing Equations (79) and (71), we find a solution in terms of Theta function given by

$$[w_0 : x_0 : y_0 : z_0] = [\Theta_1 : \Theta_2 : \Theta_3 : \Theta_4] . \qquad (84)$$

Equations (12) are equivalent to

$$\theta_1^2(z) = \Theta_1\,\Theta_1(2z) + \Theta_2\,\Theta_2(2z) + \Theta_3\,\Theta_3(2z) + \Theta_4\,\Theta_4(2z),$$

$$\theta_2^2(z) = \Theta_1\,\Theta_1(2z) + \Theta_2\,\Theta_2(2z) - \Theta_3\,\Theta_3(2z) - \Theta_4\,\Theta_4(2z),$$

$$\theta_3^2(z) = \Theta_1\,\Theta_1(2z) - \Theta_2\,\Theta_2(2z) - \Theta_3\,\Theta_3(2z) + \Theta_4\,\Theta_4(2z),$$

$$\theta_4^2(z) = \Theta_1\,\Theta_1(2z) - \Theta_2\,\Theta_2(2z) + \Theta_3\,\Theta_3(2z) - \Theta_4\,\Theta_4(2z). \qquad (85)$$

Comparing Equations (72) with Equations (85), the coordinates can be expressed in terms of Theta functions with non-vanishing elliptic arguments as

$$[w : x : y : z] = [\Theta_1(2z) : \Theta_2(2z) : \Theta_3(2z) : \Theta_4(2z)]. \qquad \square$$

It follows:

Lemma 4.45 *The moduli parameter A, B, C, D are rational functions over $\mathcal{A}_2(2)$.*

Proof. Equations (83) descend to rational functions on $\mathcal{A}_2(2)$ given by

$$A = 2\frac{\lambda_1 + 1}{\lambda_1 - 1}, \qquad B = 2\frac{\lambda_1\lambda_2 + \lambda_1\lambda_3 - 2\lambda_2\lambda_3 - 2\lambda_1 + \lambda_2 + \lambda_3}{(\lambda_2 - \lambda_3)(\lambda_1 - 1)},$$

$$C = 2\frac{\lambda_3 + \lambda_2}{\lambda_3 - \lambda_2}, \qquad D = 4\frac{\lambda_1 - \lambda_2\lambda_3}{(\lambda_2 - \lambda_3)(\lambda_1 - 1)}. \tag{86}$$

$\square$

We have proved the following:

Theorem 4.46

(1) *The singular Kummer variety associated with the Jacobian* $\mathrm{Jac}(\mathcal{C})$ *of the genus-two curve* $\mathcal{C}$ *in Rosenhain normal form given by Equation (17) is the image of* $[\Theta_1(2z) : \Theta_2(2z) : \Theta_3(2z) : \Theta_4(2z)]$ *in* $\mathbb{P}^3$ *which satisfies Equation (68) with moduli parameters given by Equations (86).*

(2) *The singular Kummer variety associated with the Jacobian* $\mathrm{Jac}(\hat{\mathcal{C}})$ *of the* $(2, 2)$*-Isogenous curve* $\hat{\mathcal{C}}$ *(cf. Lemma 3.8) in Rosenhain normal form (29) is the image of* $[\theta_1^2(z) : \theta_2^2(z) : \theta_3^2(z) : \theta_4^2(z)]$ *in* $\mathbb{P}^3$ *which satisfies Equation (62) with moduli parameters given by Equations (81).*

Proof. The proof follows from Propositions 4.41, 4.42, 4.44 and Lemmas 4.43, 4.45.

$\square$

4.8 The Rosenhain Normal Form

In [3, Prop. 10.3.2] another normal form for a nodal quartic surfaces was established. We make the following definition:

Definition 4.47 *A Rosenhain quartic is the surface in* $\mathbb{P}^3(Y_0, \ldots, Y_3)$ *given by*

$$a^2\left(Y_0^2 Y_1^2 + Y_2^2 Y_3^2\right) + b^2\left(Y_0^2 Y_2^2 + Y_1^2 Y_3^2\right) + c^2\left(Y_0^2 Y_3^2 + Y_1^2 Y_2^2\right)$$

$$+ 2ab\left(Y_0 Y_1 - Y_2 Y_3\right)\left(Y_0 Y_2 + Y_1 Y_3\right) - 2ac\left(Y_0 Y_1 + Y_2 Y_3\right)\left(Y_0 Y_3 + Y_1 Y_2\right)$$

$$+ 2bc\left(Y_0 Y_2 - Y_1 Y_3\right)\left(Y_0 Y_3 - Y_1 Y_2\right) + d^2 Y_0 Y_1 Y_2 Y_3 = 0, \tag{87}$$

with $[a : b : c : d] \in \mathbb{P}^3$.

The following was proved in [3, Prop. 10.3.2]:

Proposition 4.48 *For generic parameters* $[a : b : c : d] \in \mathbb{P}^3$, *the Rosenhain quartic in Equation (87) is isomorphic to the singular Kummer variety associated with a principally polarized abelian variety.*

We also have the following:

Lemma 4.49 *The Rosenhain quartic in Equation (87) is isomorphic to the Göpel-Hudson quartic in Equation (68).*

Proof. Using the same notation as in the proof of Lemma 4.34, we set

$$
\begin{aligned}
a &= 4\big(w_0^2 z_0^2 - x_0^2 y_0^2\big)\big(w_0 y_0 + x_0 z_0\big),\\
b &= \big(w_0^2 + x_0^2 - y_0^2 - z_0^2\big)\big(w_0^2 - x_0^2 + y_0^2 - z_0^2\big)\big(w_0 x_0 - y_0 z_0\big),\\
c &= \big(w_0^2 - x_0^2 - y_0^2 + z_0^2\big)\big(w_0^2 + x_0^2 + y_0^2 + z_0^2\big)\big(w_0 x_0 + y_0 z_0\big),
\end{aligned}
\tag{88}
$$

and a polynomial expression for d^2 of degree twelve that we do not write out explicitly. Then, the linear transformation given by

$$
\begin{aligned}
Y_0 &= w_0\, w + x_0\, x + y_0\, y + z_0\, z,\\
Y_1 &= w_0\, w + x_0\, x - y_0\, y - z_0\, z,\\
Y_2 &= z_0\, w + y_0\, x + x_0\, y + w_0\, z,\\
Y_3 &= z_0\, w + y_0\, x - x_0\, y - w_0\, z,
\end{aligned}
\tag{89}
$$

transforms Equation (87) into Equation (68).   □

We have the stronger result:

Proposition 4.50 *Every Rosenhain tetrahedron determines an isomorphism between the Cassels-Flynn quartic in Equation (37) and the Rosenhain quartic in Equation (87).*

Proof. Given a Rosenhain tetrahedron $\{T_a, T_b, T_c, T_d\}$, we use Proposition 4.29 to write the Cassels-Flynn quartic in Equation (37) in the equivalent form

$$
\big(\rho^2 T_a T_{a'} + \gamma^2 T_b T_{b'} - \beta^2 T_c T_{c'}\big)^2 - 4\gamma^2 \rho^2 T_a T_{a'} T_b T_{b'} = 0,
\tag{90}
$$

with $\beta, \gamma, \rho \in \mathbb{C}[\lambda_1, \lambda_2, \lambda_3]$. The two additional tropes $T_c T_{c'}$ are linear functions of $T_a, T_{a'}, T_b, T_{b'}$ with coefficients in $\mathbb{C}(\lambda_1, \lambda_2, \lambda_3)$. One substitutes

$$
[T_a, T_b, T_c, T_d] = [c_0 Y_0 : c_1 Y_1 : c_2 Y_2 : c_3 Y_3],
$$

into Equation (90) and solves for the coefficients $c_0, \ldots, c_3$ such that Equation (87) coincides with Equation (62).   □

Remark 4.51   Equation (87) is unchanged by the Cremona transformation

$$
[Y_0 : Y_1 : Y_2 : Y_3] \mapsto [Y_1 Y_2 Y_3 : Y_0 Y_2 Y_3 : Y_0 Y_1 Y_3 : Y_0 Y_1 Y_2].
$$

We also have the following:

Theorem 4.52 *The surface in $\mathbb{P}^3$ given by Equation (87) is isomorphic to the singular Kummer variety* $\mathbf{A}_\tau / \langle -\mathbb{I} \rangle$ *where the coordinates are given by*

$$[Y_0 : Y_1 : Y_2 : Y_3] = \left[\theta_1(z)^2 : \theta_2(z)^2 : \theta_7(z)^2 : \theta_{12}(z)^2 \right], \qquad (91)$$

and the moduli parameters are

$$a = (2\,\Theta_1\Theta_4 - 2\,\Theta_2\Theta_3)\,(2\,\Theta_1\Theta_4 + 2\,\Theta_2\Theta_3)\,(2\,\Theta_1\Theta_3 + 2\,\Theta_2\Theta_4),$$

$$b = \left(\Theta_1^2 + \Theta_2^2 - \Theta_3^2 - \Theta_4^2 \right) \left(\Theta_1^2 - \Theta_2^2 + \Theta_3^2 - \Theta_4^2 \right) (2\,\Theta_1\Theta_2 - 2\,\Theta_3\Theta_4),$$

$$c = \left(\Theta_1^2 - \Theta_2^2 - \Theta_3^2 + \Theta_4^2 \right) \left(\Theta_1^2 + \Theta_2^2 + \Theta_3^2 + \Theta_4^2 \right) (2\,\Theta_1\Theta_2 + 2\,\Theta_3\Theta_4),$$

$$d^2 = 256\,\Theta_1\Theta_2\Theta_4\Theta_3 \left(\Theta_1^2\Theta_4^2 - \Theta_2^2\Theta_3^2 \right) \left(\Theta_1^4 - \Theta_2^4 - \Theta_3^4 + \Theta_4^4 \right)$$

$$+ 8 \left(\Theta_1^2 + \Theta_4^2 \right) \left(\Theta_2^2 + \Theta_3^2 \right) \left(\Theta_1^2 + \Theta_2^2 + \Theta_3^2 + \Theta_4^2 \right)^2$$

$$\times \left(\Theta_1^2 - \Theta_2^2 - \Theta_3^2 + \Theta_4^2 \right)^2 + 8 \left(\Theta_1^2 - \Theta_4^2 \right)$$

$$\times \left(\Theta_2^2 - \Theta_3^2 \right) \left(\Theta_1^2 + \Theta_2^2 - \Theta_3^2 - \Theta_4^2 \right)^2$$

$$\times \left(\Theta_1^2 - \Theta_2^2 + \Theta_3^2 - \Theta_4^2 \right)^2 - 32 \left(\Theta_1^2\Theta_2^2 + \Theta_3^2\Theta_4^2 \right)$$

$$\times \left(\Theta_1^2 + \Theta_2^2 + \Theta_3^2 + \Theta_4^2 \right) \left(\Theta_1^2 - \Theta_2^2 + \Theta_3^2 - \Theta_4^2 \right)$$

$$\times \left(\Theta_1^2 - \Theta_2^2 - \Theta_3^2 + \Theta_4^2 \right) \left(\Theta_1^2 + \Theta_2^2 - \Theta_3^2 - \Theta_4^2 \right).$$

Proof. All tropes in Proposition 4.29 are determined in terms of the Rosenhain tetrahedron $\{\mathsf{T}_{256}, \mathsf{T}_{136}, \mathsf{T}_{246}, \mathsf{T}_2\}$, with tropes given by $\theta_1(z)^2, \theta_2(z)^2, \theta_7(z)^2, \theta_{12}(z)^2$. One checks that the Rosenhain tetrahedron is contained in three sets of eight tropes corresponding to the even eights Δ_{12}, Δ_{13} in Proposition 4.22. The proof then proceeds analogous to the proof of Proposition 4.42. □

5 Non-Principally Polarized Kummer Surfaces

In this section we consider abelian surfaces **B** with a polarization of type (d_1, d_2) given by an ample symmetric line bundle $\mathcal{N}$. We recall that by the Riemann-Roch theorem we have $\chi(\mathcal{N}) = (\mathcal{N}^2)/2 = d_1 d_2$. It follows from [3, Prop. 4.5.2] that $\mathcal{N}$ is ample if and only if $h^i(\mathbf{B}, \mathcal{N}) = 0$ for $i = 1, 2$ and $(\mathcal{N}^2) > 0$. Therefore, the line bundle defines an associated rational map $\varphi_\mathcal{N} : \mathbf{B} \to \mathbb{P}^{d_1 d_2 - 1}$. It was proven in [3, Prop. 4.1.6, Lemma 10.1.2] that the linear system $|\mathcal{N}|$ is base point free for $d_1 = 2$ and for $d_1 = 1$ and $d_2 \geq 3$, and it has exactly four base points if $(d_1, d_2) = (1, 2)$. Moreover, every polarization is induced by a principal polarization $\mathcal{L}$ on an abelian surface **A**, that is, there is an isogeny $\phi : \mathbf{B} \to \mathbf{A}$ such that $\phi^* \mathcal{L} \cong \mathcal{N}$ [3, Prop. 4.1.2].

5.1 (1, 4)-Polarized Kummer Surfaces

Let us now consider a generic abelian surface **B** with a $(1, 4)$-polarization. In [3] the following octic surface in $\mathbb{P}^3$ was considered:

Definition 5.1 *A Birkenhake-Lange octic is the surface in $\mathbb{P}^3(Z_0, \ldots, Z_3)$ given by*

$$
\begin{aligned}
a^2 &\left(Z_0^4 Z_1^4 + Z_2^4 Z_3^4\right) + b^2 \left(Z_0^4 Z_2^4 + Z_1^4 Z_3^4\right) + c^2 \left(Z_0^4 Z_3^4 + Z_1^4 Z_2^4\right) \\
&+ 2ab\left(Z_0^2 Z_1^2 - Z_2^2 Z_3^2\right)\left(Z_0^2 Z_2^2 + Z_1^2 Z_3^2\right) - 2ac\left(Z_0^2 Z_1^2 + Z_2^2 Z_3^2\right) \\
&\times \left(Z_0^2 Z_3^2 + Z_1^2 Z_2^2\right) + 2bc\left(Z_0^2 Z_2^2 - Z_1^2 Z_3^2\right)\left(Z_0^2 Z_3^2 - Z_1^2 Z_2^2\right) \\
&+ d^2 Z_0^2 Z_1^2 Z_2^2 Z_3^2 = 0,
\end{aligned}
\tag{92}
$$

with $[a : b : c : d] \in \mathbb{P}^3$.

We have the following lemma:

Lemma 5.2 *The map $\mathbb{P}(Z_0, Z_1, Z_2, Z_3) \to \mathbb{P}(Y_0, Y_1, Y_2, Y_3)$ with $Y_i = Z_i^2$ induces a covering of the octic surface in Equation (92) onto the quartic in Equation (87) which is $8 : 1$ outside the coordinate planes $Z_i = 0$.*

Proof. Along the coordinate planes the covering is $4 : 1$, and the coordinate points $P_0 = [1 : 0 : 0 : 0]$, $P_1 = [0 : 1 : 0 : 0]$, $P_2 = [0 : 0 : 1 : 0]$, $P_3 = [0 : 0 : 0 : 1]$ are of multiplicity four in the image. The octic surface has double curves along the coordinate planes. The coordinate planes $Y_i = 0$ form a Rosenhain tetrahedron. Therefore, the coordinate plane passes through six nodes on the Kummer quartic. For example, $Y_3 = 0$ passes through P_0, P_1, P_2 and three more nodes P_0', P_1', P_2'. Therefore, the preimage contains $3 + 4 \cdot 3$ nodes. $\qquad \square$

We consider the rational map $\varphi_\mathcal{N} : \mathbf{B} \to \mathbb{P}^3$ associated with the line bundle $\mathcal{N}$. The map $\varphi_\mathcal{N}$ cannot be an embedding, i.e., diffeomorphic onto its image. However, it is generically birational onto its image [3]. The following was proved in [3, Sec. 10.5]:

Corollary 5.3 [3, Prop. 10.5.7] *If $\mathcal{N}$ is the ample symmetric line bundle on an abelian surface **B** defining a polarization of type $(1, 4)$ such that the induced canonical map $\varphi_\mathcal{N} : \mathbf{B} \to \mathbb{P}^3$ is birational, then the coordinates of $\mathbb{P}^3$ can be chosen such that $\varphi_\mathcal{N}(\mathbf{B})$ is given by the Birkenhake-Lange octic in Equation (92).* $\qquad \square$

We again have the following:

Corollary 5.4 *The surface in* $\mathbb{P}^3$ *given by Equation (92) is birational to the singular Kummer variety* $\mathbf{B}/\langle -\mathbb{I} \rangle$ *where the coordinates are given by*

$$[Z_0 : Z_1 : Z_2 : Z_3] = [\theta_1(z) : \theta_2(z) : \theta_7(z) : \theta_{12}(z)]. \tag{93}$$

$\square$

5.2 (2, 4)-Polarized and (1, 2)-Polarized Kummer Surfaces

Let us now consider the generic abelian surface $\mathbf{B}$ with a $(1, 2)$-polarization. Barth studied abelian surfaces with $(1, 2)$-polarization in [1]. An excellent summary of Barth's construction was given by Garbagnati in [15, 16]. Kummer surfaces with $(1, 2)$-polarization were also discussed [27, 2]. Barth studied a projective model of the surface Kum($\mathbf{B}$) as intersection of three quadrics in $\mathbb{P}^5$, giving also the explicit equations of these quadrics. We will show how these conics arise as Mumford identities of Theta functions. We make the following:

Definition 5.5 *A Barth surface is the surface in* $\mathbb{P}^7(w, \ldots, z, X_1, \ldots, X_4)$ *given as the complete intersection of the 6 quadrics*

$$
\begin{aligned}
2p_0q_0\left(X_1^2 + X_2^2\right) &= \left(p_0^2 + q_0^2\right)\left(w^2 + x^2\right) - \left(p_0^2 - q_0^2\right)\left(y^2 + z^2\right), \\
2r_0s_0\left(X_1^2 - X_2^2\right) &= \left(r_0^2 + s_0^2\right)\left(w^2 - x^2\right) + \left(r_0^2 - s_0^2\right)\left(y^2 - z^2\right), \\
4u_0t_0X_1X_2 &= 2\left(t_0^2 + u_0^2\right)wx - 2\left(t_0^2 - u_0^2\right)yz,
\end{aligned} \tag{94}
$$

and

$$
\begin{aligned}
2p_0q_0\left(X_3^2 + X_4^2\right) &= \left(-p_0^2 + q_0^2\right)\left(w^2 + x^2\right) + \left(p_0^2 + q_0^2\right)\left(y^2 + z^2\right), \\
2r_0s_0\left(X_3^2 - X_4^2\right) &= \left(-r_0^2 + s_0^2\right)\left(w^2 - x^2\right) - \left(r_0^2 + s_0^2\right)\left(y^2 - z^2\right), \\
4u_0t_0X_3X_4 &= 2\left(t_0^2 - u_0^2\right)wx - 2\left(t_0^2 + u_0^2\right)yz,
\end{aligned} \tag{95}
$$

where $[p_0 : \cdots : u_0] \in \mathbb{P}^5$ *such that*

$$t_0^2u_0^2 = p_0^2q_0^2 - r_0^2s_0^2, \qquad t_0^4 + u_0^4 = p_0^4 + q_0^4 - r_0^4 - s_0^4. \tag{96}$$

We start by considering the polarization given by the divisor $2\mathcal{N}$ on $\mathbf{B}$ defining a polarization of type $(2, 4)$. Barth proved the following in [1, Prop. 2.1]:

Proposition 5.6 *The divisor* $2\mathcal{N}$ *is very ample on* $\mathbf{B}$ *and the linear system* $|2\mathcal{N}|$ *embeds* $\mathbf{B}$ *as a smooth surface of degree 16 in* $\mathbb{P}^7$.

Remark 5.7 It was proved in [1, Prop. 4.6] that a generic set of moduli parameters satisfies

$$\left(p_0^2 s_0^2 - q_0^2 r_0^2\right)\left(p_0^2 r_0^2 - q_0^2 s_0^2\right)\left(p_0^2 u_0^2 - q_0^2 t_0^2\right)\left(p_0^2 t_0^2 - q_0^2 u_0^2\right)$$
$$\times \left(r_0^2 u_0^2 - s_0^2 t_0^2\right)\left(r_0^2 t_0^2 - s_0^2 u_0^2\right) \neq 0.$$

We have the following:

Corollary 5.8 *For generic parameters* $[p_0 : \cdots : u_0] \in \mathbb{P}^5$, *the six quadrics defining the Barth surface in Equations (94) and (95) generate the ideal of a smooth irreducible surface of degree 16 in* $\mathbb{P}^7$ *isomorphic to* **B** *with* $(2, 4)$-*polarization given by* $2\mathcal{N}$.

Proof. The quadratic equations for the conics defining **B** in $\mathbb{P}^7$ were determined by Barth in [1, Eq. (2.10)]. The result then follows when taking suitable linear combinations of these equations using the variable transformation

$$\mathbf{x}_1 = w, \ \mathbf{x}_2 = x, \ \mathbf{x}_3 = X_1, \ \mathbf{x}_4 = X_2, \ \mathbf{x}_5 = y, \ \mathbf{x}_6 = z, \ \mathbf{x}_7 = iX_4,$$
$$\mathbf{x}_8 = -iX_3, \tag{97}$$

and identifying the moduli parameters according to

$$\lambda_1 = p_0, \ \mu_1 = q_0, \ \lambda_2 = s_0, \ \mu_2 = r_0, \ \lambda_3 = t_0, \ \mu_3 = u_0. \tag{98}$$

$\square$

With respect to the action of the involution $-\mathbb{I}$ on **B**, the space $H^0(\mathbf{B}, 2\mathcal{N})$ with $h^0(\mathbf{B}, 2\mathcal{N}) = 8$ splits into the direct sum $H^0(\mathbf{B}, 2\mathcal{N}) = H^0(\mathbf{B}, 2\mathcal{N})_+ \oplus H^0(\mathbf{B}, 2\mathcal{N})_-$ of eigenspaces of dimensions $h^0(\mathbf{B}, 2\mathcal{N})_+ = 6$ and $h^0(\mathbf{B}, 2\mathcal{N})_- = 2$, respectively. In particular, it is possible to choose the coordinates of $\mathbb{P}^7$ such that $\{X_3 = X_4 = 0\}$ is the subspace of $\mathbb{P}^7$ invariant under the action of $-\mathbb{I}$. We will denote the subspace by $\mathbb{P}^5_+ \cong \mathbb{P}H^0(\mathbf{B}, 2\mathcal{N})_+$, and the anti-invariant subspace $\{w = 0, \ldots, X_1 = 0, X_2 = 0\}$ by $\mathbb{P}^1_- \cong \mathbb{P}H^0(\mathbf{B}, 2\mathcal{N})_-$; see [1, Sec. 2]. If we consider the projection

$$\Pi : \mathbb{P}^7 \to \mathbb{P}^5_+, \quad [w : x : y : z : X_1 : X_2 : X_3 : X_4]$$
$$\mapsto [w : x : y : z : X_1 : X_2],$$

with center $\mathbb{P}^1_-$, it was proved in [1, Sec. 4.3] that $\mathbf{B} \cap \mathbb{P}^1_- = \emptyset$. Hence, the projection Π is well defined and induces a double cover from **B** onto its image. The following was proved in [1, Prop. 4.6]:

Corollary 5.9 *The three quadrics in Equations (94) generate the ideal of an irreducible surface of degree 8 in* $\mathbb{P}^5_+$ *with 16 normal singularities isomorphic to the singular Kummer variety* $\mathbf{B}/\langle -\mathbb{I} \rangle$.

Using the same argument as before, the projection

$$\pi : \mathbb{P}_+^5 \to \mathbb{P}^3, \qquad [w : x : y : z : X_1 : X_2] \mapsto [w : x : y : z], \qquad (99)$$

is well-defined and induces a double cover of $\mathbf{B}/\langle -\mathbb{I} \rangle$ onto its image. We have the following:

Proposition 5.10 *The image of the projection π is isomorphic to the singular Kummer variety $/\langle -\mathbb{I} \rangle$ associated with a principally polarized abelian variety $\mathbf{A}$.*

Proof. The statement follows by eliminating X_1, X_2 from Equations (94) and recovering a Göpel-Hudson quartic in Equation (68) with parameters

$$A = \frac{2(p_0^2 r_0^2 + q_0^2 s_0^2)}{p_0^2 r_0^2 - q_0^2 s_0^2}, \quad B = \frac{2(p_0^2 q_0^2 + r_0^2 s_0^2)}{p_0^2 q_0^2 - r_0^2 s_0^2}, \quad C = \frac{2(p_0^2 s_0^2 + q_0^2 r_0^2)}{p_0^2 s_0^2 - q_0^2 r_0^2},$$

$$D = \frac{4 p_0^2 q_0^2 r_0^2 s_0^2 (t_0^2 - u_0^2)(t_0^2 + u_0^2)}{(p_0^2 r_0^2 - q_0^2 s_0^2)(p_0^2 q_0^2 - r_0^2 s_0^2)(p_0^2 s_0^2 - q_0^2 r_0^2)}. \qquad (100)$$

These moduli parameters equal the ones in Equations (70) when using Equations (73) and (73). $\qquad \square$

Remark 5.11 Other projections are obtained by eliminating either w, x or y, z instead. These images are isomorphic Göpel-Hudson quartics and are related by the action of a projective automorphism mapping one even Göpel tetrahedron to another one with two tropes in common.

We now describe the role of the even eight in the construction above: Mehran proved in [28] that the rational double cover of a smooth Kummer surface $\text{Kum}(\mathbf{A})$ with principal polarization branched along an even eight is a Kummer surface with $(1, 2)$-polarization. In terms of the singular Kummer variety $\mathbf{B}/\langle -\mathbb{I} \rangle$ and its image – which is a singular Kummer variety $\mathbf{A}/\langle -\mathbb{I} \rangle$ associated with a principally polarized abelian variety – this can be interpreted as follows: the 16 singular points of $\mathbf{B}/\langle -\mathbb{I} \rangle$ are mapped to 8 singular points on $\mathbf{A}/\langle -\mathbb{I} \rangle$ that form the complement of the even eight.

The Kummer variety $\mathbf{B}/\langle -\mathbb{I} \rangle$ is the complete intersection of three quadrics in Equations (94) which we denote by $\mathsf{Q}_1, \mathsf{Q}_2, \mathsf{Q}_3$. The Kummer variety is contained in each quadric, or equivalently, in the hypernet of quadrics $\alpha_1 \mathsf{Q}_1 + \alpha_2 \mathsf{Q}_2 + \alpha_3 \mathsf{Q}_3$ for complex numbers $\alpha_1, \alpha_2, \alpha_3 \in \mathbb{C}$. We have the following:

Lemma 5.12 *For generic moduli parameters there are exactly four special quadrics $\mathsf{K}_1, \dots, \mathsf{K}_4$ in the hypernet where the rank equals three. The four special quadrics are given by*

$$\mathsf{K}_1 : c_{1,1}\big(w_0 x - x_0 w\big)^2 + c_{1,2}\big(y_0 z - z_0 y\big)^2 + c_{1,3}\big(X_1^{(0)} X_2 - X_2^{(0)} X_1\big)^2 = 0,$$

$$\mathsf{K}_2 : c_{2,1}\big(w_0 x + x_0 w\big)^2 + c_{2,2}\big(y_0 z + z_0 y\big)^2 + c_{2,3}\big(X_1^{(0)} X_2 + X_2^{(0)} X_1\big)^2 = 0,$$

$$\mathsf{K}_3 : c_{3,1}\big(w_0 w - x_0 x\big)^2 + c_{3,2}\big(y_0 y - z_0 z\big)^2 + c_{3,3}\big(X_1^{(0)} X_1 - X_2^{(0)} X_2\big)^2 = 0,$$

$$\mathsf{K}_4 : c_{4,1}\big(w_0 w + x_0 x\big)^2 + c_{4,2}\big(y_0 y + z_0 z\big)^2 + c_{4,3}\big(X_1^{(0)} X_1 + X_2^{(0)} X_2\big)^2 = 0,$$

$$\tag{101}$$

where $c_{i,j} \in \mathbb{C}^*$, the parameters w_0, x_0, y_0, z_0 are related to p_0, q_0, r_0, s_0 by Equations (73) and

$$X_1^{(0)\,2} X_2^{(0)\,2} = w_0^2 x_0^2 - y_0^2 z_0^2, \qquad X_2^{(0)\,4} + X_1^{(0)\,4} = w_0^4 + x_0^4 - y_0^4 - z_0^4 .$$

$$\tag{102}$$

Proof. The fact that there are exactly four sets of parameters, and thus four special quadrics $\mathsf{K}_1, \ldots, \mathsf{K}_4$ in the hypernet where the rank equals three was proven in [1, Sec. 4]; see also a similar computation outlined in [15, Sec. 2.4.2]. The four set of parameters where the rank of the quadric in the hypernet equals three are given by

$$\alpha_1 = \pm\frac{\sqrt{\big(s_0^2 u_0^2 - r_0^2 t_0^2\big)\big(s_0^2 t_0^2 - r_0^2 u_0^2\big)}}{p_0^2 s_0^2 t_0^2},$$

$$\alpha_2 = \pm\frac{\sqrt{\big(p_0^2 u_0^2 - q_0^2 t_0^2\big)\big(p_0^2 t_0^2 - q_0^2 u_0^2\big)}}{p_0^2 s_0^2 t_0^2},$$

$$\alpha_3 = \pm\frac{\sqrt{\big(p_0^2 r_0^2 - q_0^2 s_0^2\big)\big(q_0^2 r_0^2 - p_0^2 s_0^2\big)}}{p_0^2 s_0^2 t_0^2}.$$

Plugging these values into $\alpha_1 \mathsf{Q}_1 + \alpha_2 \mathsf{Q}_2 + \alpha_3 \mathsf{Q}_3$, we obtain, after a tedious computation, the four quadrics. $\qquad\square$

The singular locus of each quadric K_i in Lemma 5.12 is a plane S_i for $1 \le i \le 4$, given by

$$\mathsf{S}_1 : \quad w_0 x = x_0 w, \quad y_0 z = z_0 y, \quad X_1^{(0)} X_2 = X_2^{(0)} X_1,$$

$$\mathsf{S}_2 : \quad w_0 x = -x_0 w, \quad y_0 z = -z_0 y, \quad X_1^{(0)} X_2 = -X_2^{(0)} X_1,$$

$$\mathsf{S}_3 : \quad x_0 x = w_0 w, \quad y_0 y = z_0 z, \quad X_1^{(0)} X_1 = X_2^{(0)} X_2,$$

$$\mathsf{S}_4 : \quad x_0 x = -w_0 w, \quad y_0 y = -z_0 z, \quad X_1^{(0)} X_1 = -X_2^{(0)} X_2. \tag{103}$$

On each plane S_i with $1 \le i \le 4$, the other cones cut out, for generic moduli parameters, pencils of conics with four distinct base points [1, Lemma. 4.1]. These four points on the four different planes constitute precisely the 16 singular points $\mathbf{B}/\langle -\mathbb{I} \rangle$. We have the following:

Proposition 5.13 *The even eight of the projection* π *in Equation (99) consists of the eight nodes* $[w : x : y : z]$ *on the Göpel-Hudson quartic (68) given by*

$$[y_0 : z_0 : w_0 : x_0], [-y_0 : -z_0 : w_0 : x_0], [-y_0 : z_0 : -w_0 : x_0],$$
$$[-y_0 : z_0 : w_0 : -x_0], [z_0 : y_0 : x_0 : w_0], [-z_0 : -y_0 : x_0 : w_0],$$
$$[-z_0 : y_0 : -x_0 : w_0], [-z_0 : y_0 : x_0 : -w_0].$$

Proof. The image under the projection π is a singular Kummer variety $\mathbf{A}/\langle -\mathbb{I} \rangle$ associated with a principally polarized abelian variety realized as Göpel-Hudson quartic. We computed the 16 singular points of a Göpel-Hudson quartic in Lemma 4.38. The 16 singular points $\mathbf{B}/\langle -\mathbb{I} \rangle$ are located on the four planes S_i for $1 \le i \le 4$. Under the projection π the ratios of the first four coordinates as determined by the equations of the planes must be preserved. We computed the 16 singular points of a Göpel-Hudson quartic in Lemma 4.38. The image of the 16 singular points $\mathbf{B}/\langle -\mathbb{I} \rangle$ under the double cover π are the eight nodes on $\mathbf{A}/\langle -\mathbb{I} \rangle$ given by

$$[w_0 : x_0 : y_0 : z_0], [-w_0 : -x_0 : y_0 : z_0], [-w_0 : x_0 : -y_0 : z_0],$$
$$[-w_0 : x_0 : y_0 : -z_0], [x_0 : w_0 : z_0 : y_0], [-x_0 : -w_0 : z_0 : y_0],$$
$$[-x_0 : w_0 : -z_0 : y_0], [-x_0 : w_0 : z_0 : -y_0].$$

The even eight consists of the complimentary set of nodes. $\qquad\square$

We provide an explicit model for the Barth surface in terms of Theta functions. We have the following:

Theorem 5.14 *The surface* $\mathbf{B}$ *in* $\mathbb{P}^7$ *given by Equations (94) and (95) with moduli parameters given by*

$$[p_0 : q_0 : r_0 : s_0 : t_0 : u_0] = [\theta_1 : \theta_2 : \theta_3 : \theta_4 : \theta_8 : \theta_{10}]. \qquad (104)$$

is isomorphic to the image of six even and two odd Theta functions given by

$$[w : x : y : z : X_1 : X_2 : X_3 : X_4]$$
$$= [\Theta_1(2z) : \Theta_2(2z) : \Theta_3(2z) : \Theta_4(2z) : \Theta_8(2z) : \Theta_{10}(2z) : \Theta_{13}(2z) : \Theta_{16}(2z)].$$

In particular, the ideal of the surface $\mathbf{B}$ *in* $\mathbb{P}^7$ *is generated by the Mumford relations for the Theta function* $\Theta_1(2z), \dots, \Theta_{16}(2z)$*. Moreover, the ideal of the singular Kummer variety* $\mathbf{B}/\langle -\mathbb{I} \rangle$ *is generated by the Mumford relations for the Theta function* $\Theta_1(2z), \dots, \Theta_4(2z), \Theta_8(2z), \Theta_{10}(2z)$*. The map* $\pi :$ $\mathbf{B}/\langle -\mathbb{I} \rangle \to \mathbf{A}_\tau/\langle -\mathbb{I} \rangle$ *is a rational double cover onto the singular Kummer variety associated with the principally polarized abelian variety* $\mathbf{A}_\tau$*.*

Proof. We use a set of eight tropes consisting of the Rosenhain tetrahedra

$$\{\mathsf{T}_{256}, \mathsf{T}_{126}, \mathsf{T}_{236}, \mathsf{T}_4\} \quad \text{and} \quad \{\mathsf{T}_{136}, \mathsf{T}_{356}, \mathsf{T}_{156}, \mathsf{T}_6\}.$$

The corresponding Theta functions are

$$\{\Theta_1(2z), \Theta_4(2z), \Theta_{10}(2z), \Theta_{16}(2z)\} \quad \text{and}$$
$$\{\Theta_2(2z), \Theta_3(2z), \Theta_8(2z), \Theta_{13}(2z)\}.$$

The bi-monomial Mumford relations in Proposition 2.3, after τ is replaced by 2τ, include the equations

$$\Theta_8\Theta_{10}\Theta_8(2z)\Theta_{10}(2z) = \Theta_1\Theta_2\Theta_1(2z)\Theta_2(2z) - \Theta_3\Theta_4\Theta_3(2z)\Theta_4(2z),$$
$$\Theta_8\Theta_{10}\Theta_{13}(2z)\Theta_{16}(2z) = \Theta_3\Theta_4\Theta_1(2z)\Theta_2(2z) - \Theta_1\Theta_2\Theta_3(2z)\Theta_4(2z).$$
$$(105)$$

Using the Frobenius identities (7) we obtain

$$2\theta_8\theta_{10}\Theta_8(2z)\Theta_{10}(2z) = (\theta_8^2 + \theta_{10}^2)\Theta_1(2z)\Theta_2(2z)$$
$$- (\theta_8^2 - \theta_{10}^2)\Theta_3(2z)\Theta_4(2z),$$
$$2\theta_8\theta_{10}\Theta_{13}(2z)\Theta_{16}(2z) = (\theta_8^2 - \theta_{10}^2)\Theta_1(2z)\Theta_2(2z)$$
$$- (\theta_8^2 + \theta_{10}^2)\Theta_3(2z)\Theta_4(2z). \quad (106)$$

Similarly, the quadratic Mumford relations in Proposition 2.2 combined with Equations (9) yield

$$2\theta_1\theta_2(\Theta_8^2(2z) + \Theta_{10}^2(2z)) = (\theta_1^2 + \theta_2^2)(\Theta_1^2(2z) + \Theta_2^2(2z))$$
$$- (\theta_1^2 - \theta_2^2)(\Theta_3^2(2z) + \Theta_4^2(2z)),$$
$$2\theta_3\theta_4(\Theta_8^2(2z) - \Theta_{10}^2(2z)) = (\theta_3^2 + \theta_4^2)(\Theta_1^2(2z) - \Theta_2^2(2z))$$
$$+ (\theta_3^2 - \theta_4^2)(\Theta_3^2(2z) - \Theta_4^2(2z)),$$
$$2\theta_1\theta_2(\Theta_{13}^2(2z) + \Theta_{16}^2(2z)) = (-\theta_1^2 + \theta_2^2)(\Theta_1^2(2z) + \Theta_2^2(2z))$$
$$+ (\theta_1^2 + \theta_2^2)(\Theta_3^2(2z) + \Theta_4^2(2z)),$$
$$2\theta_3\theta_4(\Theta_{13}^2(2z) - \Theta_{16}^2(2z)) = (-\theta_3^2 + \theta_4^2)(\Theta_1^2(2z) - \Theta_2^2(2z))$$
$$- (\theta_3^2 + \theta_4^2)(\Theta_3^2(2z) - \Theta_4^2(2z)).$$

These are precisely Equations (94) and (95) for moduli parameters given by

$$[p_0 : q_0 : r_0 : s_0 : t_0 : u_0] = [\theta_1 : \theta_2 : \theta_3 : \theta_4 : \theta_8 : \theta_{10}] . \quad (107)$$

and variables given by

$$[w : x : y : z : X_1 : X_2 : X_3 : X_4]$$
$$= [\Theta_1(2z) : \Theta_2(2z) : \Theta_3(2z) : \Theta_4(2z) : \Theta_8(2z) : \Theta_{10}(2z) :$$
$$\Theta_{13}(2z) : \Theta_{16}(2z)] .$$

By eliminating X_1, X_2 one obtains the Göpel-Hudson quartic in Equation (68) with moduli parameters matching those in Equation (83). $\quad\square$

In Proposition 5.14, we used a set of eight tropes consisting of the Rosenhain tetrahedra $\{T_{256}, T_{126}, T_{236}, T_4\}$ and $\{T_{136}, T_{356}, T_{156}, T_6\}$. The corresponding Theta functions are $\{\Theta_1(2z), \Theta_4(2z), \Theta_{10}(2z), \Theta_{16}(2z)\}$ and $\{\Theta_2(2z), \Theta_3(2z), \Theta_8(2z), \Theta_{13}(2z)\}$. Only the Theta functions $\Theta_{13}(2z)$ and $\Theta_{16}(2z)$ are odd, the remaining ones are even which allowed us to identify the sub-spaces $\mathbb{P}_+^5$ and $\mathbb{P}_-^1$. This computation can be generalized. We have the following:

Proposition 5.15 *Every set of eight tropes that satisfies a quadratic relation and corresponds to an even eight, given in Proposition 4.22, determines an isomorphism between the complete intersection of the three quadrics in Equations (94) in $\mathbb{P}_+^5$ and a singular Kummer variety $\mathbf{B}/\langle -\mathbb{I} \rangle$ with $(1, 2)$-polarization.*

Proof. The construction of the Kummer variety $\mathbf{B}/\langle -\mathbb{I} \rangle$ with $(1, 2)$-polarization as projection from the Barth surface $\mathbf{B}$ in Corollary 5.9 requires a splitting of the space $H^0(\mathbf{B}, 2\mathcal{N})$ with $h^0(\mathbf{B}, 2\mathcal{N}) = 8$ into the direct sum $H^0(\mathbf{B}, 2\mathcal{N}) = H^0(\mathbf{B}, 2\mathcal{N})_+ \oplus H^0(\mathbf{B}, 2\mathcal{N})_-$ of ± 1 eigenspaces of dimensions $h^0(\mathbf{B}, 2\mathcal{N})_+ = 6$ and $h^0(\mathbf{B}, 2\mathcal{N})_- = 2$. Among the 30 sets of eight tropes given in Proposition 4.22, 15 sets correspond to even eights Δ_{ij} not containing the node p_0, the other 15 sets do contain the node p_0. The sets of eight tropes corresponding to even eights Δ_{ij} contain 6 even tropes of the form T_{kl6} and 2 odd tropes of the form T_m, whereas the other 15 sets contain 4 even and 4 odd tropes. Therefore, for the sets of eight tropes corresponding to an even eight Δ_{ij} we obtain an abelian surface $\mathbf{B}_{ij}$ with $(2, 4)$-polarization and, by eliminating the odd coordinates X_3 and X_4, a singular Kummer variety $\mathbf{B}_{ij}/\langle -\mathbb{I} \rangle$ in Corollary 5.9. $\qquad\square$

Mehran proved [28, Cor. 4.7] that there are exactly 15 distinct isomorphism classes of smooth Kummer surfaces $\mathrm{Kum}(\mathbf{B}_{ij})$ with $(1, 2)$-polarization for $1 \leq i < j < 6$ that are rational double covers of the smooth Kummer surface $\mathrm{Kum}(\mathbf{A})$ associated with a particular principally polarized abelian variety $\mathbf{A}$. It was also proved that each such double cover is induced by a two-isogeny $\phi_{ij} : \mathbf{B}_{ij} \to \mathbf{A}$ of abelian surfaces [28, Prop 2.3]. We have the following:

Theorem 5.16 *The 15 Barth surfaces obtained from the 15 sets of eight tropes that satisfy a quadratic relation and correspond to an even eight, given in Proposition 4.22, realize all distinct 15 isomorphism classes of singular Kummer varieties with $(1, 2)$-polarization covering a fixed smooth Kummer surface $\mathrm{Kum}(\mathbf{A})$ with a principal polarization.*

Proof. Each sets of eight tropes that satisfies a quadratic relation and corresponds to an even eight, given in Proposition 4.22, determines a Barth surface using Proposition 5.15. We first show that for any two different sets of eight tropes corresponding to even eights Δ_{ij} and $\Delta_{i'j'}$, the images of the singular Kummer varieties $\mathbf{B}_{ij}/\langle -\mathbb{I} \rangle$ and $\mathbf{B}_{i'j'}/\langle -\mathbb{I} \rangle$, respectively, under their respective projections π and π' are isomorphic to the same singular Kummer variety $\mathbf{A}/\langle -\mathbb{I} \rangle$ associated with a principally polarized abelian variety $\mathbf{A}$: each set of eight tropes corresponding to an even eight Δ_{ij} contains three different Göpel tetrahedra that pairwise have two tropes in common. Each Göpel tetrahedron defines a projection from $\mathbf{B}_{ij}/\langle -\mathbb{I} \rangle$ onto a Göpel-Hudson quartic in Proposition 5.10 by eliminating the complimentary pair of variables. However, all resulting quartics are isomorphic and related by the action of a projective automorphisms obtained by the composition of isomorphisms obtained in Proposition 4.29. Because of Lemma 4.23, each of the three Göpel tetrahedra also appear as projection of two other singular Kummer varieties $\mathbf{B}_{i'j'}/\langle -\mathbb{I} \rangle$. Therefore, the images obtained as projections of all the different singular Kummer varieties $\mathbf{B}_{ij}/\langle -\mathbb{I} \rangle$ are isomorphic. They realize the same singular Kummer variety $\mathbf{A}/\langle -\mathbb{I} \rangle$ associated with a principally polarized abelian variety $\mathbf{A}$ and moduli parameter given by Equations (100).

It was proved in [28, Prop. 4.2] that all distinct 15 isomorphism classes Kum($\mathbf{B}_{ij}$) are obtained by taking a double cover branched along the fifteen different even eights Δ_{ij} on the smooth Kummer surface Kum($\mathbf{A}$). One only has to consider the even eights not containing p_0 is because one obtains the exact same Kummer surface whether one takes the double cover branched along an even eight or its complement [28, Prop. 2.3]. After blowing down the exceptional divisors, the double covers are equivalent to 15 morphisms $p_{ij} : \mathbf{B}_{ij}/\langle -\mathbb{I} \rangle \to \mathbf{A}/\langle -\mathbb{I} \rangle$ which map the 16 nodes of $\mathbf{B}_{ij}$ to the eight nodes of $\mathbf{A}/\langle -\mathbb{I} \rangle$ contained in the complement of Δ_{ij}. Proposition 5.15 proves that the projection p in Equation (99) when applied to all cases in Proposition 5.15 realizes all such morphisms. $\square$

References

[1] Wolf Barth, *Abelian surfaces with* $(1, 2)$-*polarization*, Algebraic geometry, Sendai, 1985, Adv. Stud. Pure Math., vol. 10, North-Holland, Amsterdam, 1987, pp. 41–84. MR 946234

[2] F. Bastianelli, G. P. Pirola, and L. Stoppino, *Galois closure and Lagrangian varieties*, Adv. Math. **225** (2010), no. 6, 3463–3501. MR 2729013

[3] Christina Birkenhake and Herbert Lange, *Complex abelian varieties*, second ed., Grundlehren der Mathematischen Wissenschaften [Fundamental Principles of Mathematical Sciences], vol. 302, Springer-Verlag, Berlin, 2004. MR 2062673

[4] C. W. Borchardt, *über die Darstellung der Kummerschen Fläche vierter Ordnung mit sechzehn Knotenpunkten durch die Göpelsche biquadratische Relation zwischen vier Thetafunctionen mit zwei Variabeln*, J. Reine Angew. Math. **83** (1877), 234–244. MR 1579732

[5] Jean-Benoît Bost and Jean-François Mestre, *Moyenne arithmético-géométrique et périodes des courbes de genre 1 et 2*, Gaz. Math. (1988), no. 38, 36–64. MR 970659

[6] J. W. S. Cassels and E. V. Flynn, *Prolegomena to a middlebrow arithmetic of curves of genus 2*, London Mathematical Society Lecture Note Series, vol. 230, Cambridge University Press, Cambridge, 1996. MR 1406090

[7] A. Clingher, A. Malmendier, and T. Shaska, *Six line configurations and string dualities*, Comm. Math. Phys. (2019).

[8] Adrian Clingher and Charles F. Doran, *Modular invariants for lattice polarized K3 surfaces*, Michigan Math. J. **55** (2007), no. 2, 355–393. MR 2369941

[9] _____, *Lattice polarized K3 surfaces and Siegel modular forms*, Adv. Math. **231** (2012), no. 1, 172–212. MR 2935386

[10] Pierre Deligne, *La conjecture de Weil pour les surfaces K3*, Invent. Math. **15** (1972), 206–226. MR 0296076

[11] I. Dolgachev and D. Lehavi, *On isogenous principally polarized abelian surfaces*, Curves and abelian varieties, Contemp. Math., vol. 465, Amer. Math. Soc., Providence, RI, 2008, pp. 51–69. MR 2457735

[12] Igor V. Dolgachev, *Abstract configurations in algebraic geometry*, The Fano Conference, Univ. Torino, Turin, 2004, pp. 423–462. MR 2112585

[13] _____, *Classical algebraic geometry*, Cambridge University Press, Cambridge, 2012, A modern view. MR 2964027

[14] Eugene Victor Flynn, *The Jacobian and formal group of a curve of genus 2 over an arbitrary ground field*, Math. Proc. Cambridge Philos. Soc. **107** (1990), no. 3, 425–441. MR 1041476

[15] Alice Garbagnati, *Symplectic automorphisms on k3 surfaces*, Ph.D. thesis, Dipartimento di matematica, Università degli Studi di Milano, 2008.

[16] Alice Garbagnati and Alessandra Sarti, *On symplectic and non-symplectic automorphisms of K3 surfaces*, Rev. Mat. Iberoam. **29** (2013), no. 1, 135–162. MR 3010125

[17] P. Gaudry, *Fast genus 2 arithmetic based on theta functions*, J. Math. Cryptol. **1** (2007), no. 3, 243–265. MR 2372155

[18] Maria R. Gonzalez-Dorrego, *(16, 6) configurations and geometry of Kummer surfaces in $\mathbf{P}^3$*, Mem. Amer. Math. Soc. **107** (1994), no. 512, vi+101. MR 1182682

[19] R. W. H. T. Hudson, *Kummer's quartic surface*, Cambridge Mathematical Library, Cambridge University Press, Cambridge, 1990, With a foreword by W. Barth, Revised reprint of the 1905 original. MR 1097176

[20] Jun-Ichi Igusa, *On Siegel modular forms of genus two*, Amer. J. Math. **84** (1962), 175–200. MR 0141643

[21] _____, *On Siegel modular forms genus two. II*, Amer. J. Math. **86** (1964), 392–412. MR 0168805

[22] Kenji Koike, *On Jacobian Kummer surfaces*, J. Math. Soc. Japan **66** (2014), no. 3, 997–1016. MR 3238326

[23] Abhinav Kumar, *Elliptic fibrations on a generic Jacobian Kummer surface*, J. Algebraic Geom. **23** (2014), no. 4, 599–667. MR 3263663

[24] E. E. Kummer, *Über die Flächen vierten Grades, auf welchen Schaaren von Kegelschnitten liegen*, J. Reine Angew. Math. **64** (1865), 66–76. MR 1579281

[25] A. Malmendier and T. Shaska, *The Satake sextic in F-theory*, J. Geom. Phys. **120** (2017), 290–305. MR 3712162

[26] Andreas Malmendier and David R. Morrison, *K3 surfaces, modular forms, and non-geometric heterotic compactifications*, Lett. Math. Phys. **105** (2015), no. 8, 1085–1118. MR 3366121

[27] Afsaneh Mehran, *Double covers of Kummer surfaces*, Manuscripta Math. **123** (2007), no. 2, 205–235. MR 2306633

[28] ———, *Kummer surfaces associated to $(1, 2)$-polarized abelian surfaces*, Nagoya Math. J. **202** (2011), 127–143. MR 2804549

[29] David Mumford, *Tata lectures on theta. I*, Modern Birkhäuser Classics, Birkhäuser Boston, Inc., Boston, MA, 2007, With the collaboration of C. Musili, M. Nori, E. Previato and M. Stillman, Reprint of the 1983 edition. MR 2352717

[30] ———, *Abelian varieties*, Tata Institute of Fundamental Research Studies in Mathematics, vol. 5, Published for the Tata Institute of Fundamental Research, Bombay; by Hindustan Book Agency, New Delhi, 2008, With appendices by C. P. Ramanujam and Yuri Manin, Corrected reprint of the second (1974) edition. MR 2514037

[31] V. V. Nikulin, *Kummer surfaces*, Izv. Akad. Nauk SSSR Ser. Mat. **39** (1975), no. 2, 278–293, 471. MR 0429917

[32] E. Previato, T. Shaska, and G. S. Wijesiri, *Thetanulls of cyclic curves of small genus*, Albanian J. Math. **1** (2007), no. 4, 253–270. MR 2367218

[33] Fried. Jul. Richelot, *De transformatione integralium Abelianorum primi ordinis commentatio. Caput secundum. De computatione integralium Abelianorum primi ordinis*, J. Reine Angew. Math. **16** (1837), 285–341. MR 1578135

[34] E. R. Shafarevitch, *Le théorème de Torelli pour les surfaces algébriques de type K3*, (1971), 413–417. MR 0419459

[35] Tetsuji Shioda, *Classical Kummer surfaces and Mordell-Weil lattices*, Algebraic geometry, Contemp. Math., vol. 422, Amer. Math. Soc., Providence, RI, 2007, pp. 213–221. MR 2296439

Appendix A. Mumford Relations I

The three-term relations between bi-monomial combinations of Theta functions $\xi_{i,j} = \theta_i(z)\theta_j(z)$ for $1 \leq i < j \leq 16$ in Proposition 2.3 generate an ideal that is generated by the following 60 equations:

$$\theta_1\theta_2\,\xi_{3,4} - \theta_3\theta_4\,\xi_{1,2} + \theta_8\theta_{10}\,\xi_{13,16} = 0, \qquad \theta_1\theta_2\,\xi_{13,16} - \theta_3\theta_4\,\xi_{8,10} + \theta_8\theta_{10}\,\xi_{3,4} = 0,$$
$$\theta_1\theta_3\,\xi_{7,9} + \theta_2\theta_4\,\xi_{11,12} - \theta_7\theta_9\,\xi_{1,3} = 0, \qquad \theta_1\theta_3\,\xi_{11,12} + \theta_2\theta_4\,\xi_{7,9} - \theta_7\theta_9\,\xi_{2,4} = 0,$$
$$\theta_1\theta_4\,\xi_{5,6} + \theta_2\theta_3\,\xi_{14,15} - \theta_5\theta_6\,\xi_{1,4} = 0, \qquad \theta_1\theta_4\,\xi_{14,15} + \theta_2\theta_3\,\xi_{5,6} - \theta_5\theta_6\,\xi_{2,3} = 0,$$
$$\theta_1\theta_5\,\xi_{7,8} + \theta_4\theta_6\,\xi_{11,13} - \theta_7\theta_8\,\xi_{1,5} = 0, \qquad \theta_1\theta_5\,\xi_{11,13} + \theta_4\theta_6\,\xi_{7,8} - \theta_7\theta_8\,\xi_{4,6} = 0,$$
$$\theta_1\theta_6\,\xi_{4,5} - \theta_4\theta_5\,\xi_{1,6} - \theta_9\theta_{10}\,\xi_{12,16} = 0, \qquad \theta_1\theta_6\,\xi_{9,10} - \theta_4\theta_5\,\xi_{12,16} - \theta_9\theta_{10}\,\xi_{1,6} = 0,$$
$$\theta_1\theta_7\,\xi_{5,8} + \theta_3\theta_9\,\xi_{14,16} - \theta_5\theta_8\,\xi_{1,7} = 0, \qquad \theta_1\theta_7\,\xi_{14,16} + \theta_3\theta_9\,\xi_{5,8} - \theta_5\theta_8\,\xi_{3,9} = 0,$$
$$\theta_1\theta_8\,\xi_{5,7} + \theta_2\theta_{10}\,\xi_{12,15} - \theta_5\theta_7\,\xi_{1,8} = 0, \qquad \theta_1\theta_8\,\xi_{12,15} + \theta_2\theta_{10}\,\xi_{5,7} - \theta_5\theta_7\,\xi_{2,10} = 0,$$
$$\theta_1\theta_9\,\xi_{3,7} - \theta_3\theta_7\,\xi_{1,9} + \theta_6\theta_{10}\,\xi_{13,15} = 0, \qquad \theta_1\theta_9\,\xi_{6,10} - \theta_3\theta_7\,\xi_{13,15} - \theta_6\theta_{10}\,\xi_{1,9} = 0,$$

$$\theta_1\theta_{10}\,\xi_{6,9} + \theta_2\theta_8\,\xi_{11,14} - \theta_6\theta_9\,\xi_{1,10} = 0, \quad \theta_1\theta_{10}\,\xi_{11,14} + \theta_2\theta_8\,\xi_{6,9} - \theta_6\theta_9\,\xi_{2,8} = 0,$$
$$\theta_2\theta_9\,\xi_{10,14} + \theta_4\theta_7\,\xi_{5,13} - \theta_6\theta_8\,\xi_{1,11} = 0, \quad \theta_2\theta_9\,\xi_{5,13} + \theta_4\theta_7\,\xi_{10,14} - \theta_6\theta_8\,\xi_{3,12} = 0,$$
$$\theta_2\theta_7\,\xi_{8,15} - \theta_4\theta_9\,\xi_{6,16} - \theta_5\theta_{10}\,\xi_{1,12} = 0, \quad \theta_2\theta_7\,\xi_{6,16} - \theta_4\theta_9\,\xi_{8,15} + \theta_5\theta_{10}\,\xi_{3,11} = 0,$$
$$\theta_3\theta_{10}\,\xi_{2,16} + \theta_4\theta_8\,\xi_{1,13} - \theta_6\theta_7\,\xi_{5,11} = 0, \quad \theta_3\theta_{10}\,\xi_{1,13} + \theta_4\theta_8\,\xi_{2,16} - \theta_6\theta_7\,\xi_{9,15} = 0,$$
$$\theta_2\theta_6\,\xi_{4,15} - \theta_3\theta_5\,\xi_{1,14} + \theta_8\theta_9\,\xi_{7,16} = 0, \quad \theta_2\theta_6\,\xi_{10,11} + \theta_3\theta_5\,\xi_{7,16} - \theta_8\theta_9\,\xi_{1,14} = 0,$$
$$\theta_2\theta_5\,\xi_{4,14} - \theta_3\theta_6\,\xi_{1,15} + \theta_7\theta_{10}\,\xi_{9,13} = 0, \quad \theta_2\theta_5\,\xi_{9,13} - \theta_3\theta_6\,\xi_{8,12} + \theta_7\theta_{10}\,\xi_{4,14} = 0,$$
$$\theta_3\theta_8\,\xi_{2,13} + \theta_4\theta_{10}\,\xi_{1,16} - \theta_5\theta_9\,\xi_{6,12} = 0, \quad \theta_3\theta_8\,\xi_{1,16} + \theta_4\theta_{10}\,\xi_{2,13} - \theta_5\theta_9\,\xi_{7,14} = 0,$$
$$\theta_2\theta_5\,\xi_{3,6} - \theta_3\theta_6\,\xi_{2,5} - \theta_7\theta_{10}\,\xi_{11,16} = 0, \quad \theta_2\theta_5\,\xi_{11,16} - \theta_3\theta_6\,\xi_{7,10} + \theta_7\theta_{10}\,\xi_{3,6} = 0,$$
$$\theta_2\theta_6\,\xi_{8,9} + \theta_3\theta_5\,\xi_{12,13} - \theta_8\theta_9\,\xi_{2,6} = 0, \quad \theta_2\theta_6\,\xi_{12,13} + \theta_3\theta_5\,\xi_{8,9} - \theta_8\theta_9\,\xi_{3,5} = 0,$$
$$\theta_2\theta_7\,\xi_{4,9} - \theta_4\theta_9\,\xi_{2,7} - \theta_5\theta_{10}\,\xi_{13,14} = 0, \quad \theta_2\theta_7\,\xi_{13,14} - \theta_4\theta_9\,\xi_{5,10} + \theta_5\theta_{10}\,\xi_{4,9} = 0,$$
$$\theta_2\theta_9\,\xi_{4,7} - \theta_4\theta_7\,\xi_{2,9} + \theta_6\theta_8\,\xi_{15,16} = 0, \quad \theta_2\theta_9\,\xi_{15,16} + \theta_4\theta_7\,\xi_{6,8} - \theta_6\theta_8\,\xi_{4,7} = 0,$$
$$\theta_1\theta_9\,\xi_{8,14} - \theta_3\theta_7\,\xi_{5,16} - \theta_6\theta_{10}\,\xi_{2,11} = 0, \quad \theta_1\theta_9\,\xi_{5,16} - \theta_3\theta_7\,\xi_{8,14} + \theta_6\theta_{10}\,\xi_{4,12} = 0,$$
$$\theta_1\theta_7\,\xi_{10,15} + \theta_3\theta_9\,\xi_{6,13} - \theta_5\theta_8\,\xi_{2,12} = 0, \quad \theta_1\theta_7\,\xi_{6,13} + \theta_3\theta_9\,\xi_{10,15} - \theta_5\theta_8\,\xi_{4,11} = 0,$$
$$\theta_1\theta_6\,\xi_{3,15} - \theta_4\theta_5\,\xi_{2,14} - \theta_9\theta_{10}\,\xi_{7,13} = 0, \quad \theta_1\theta_6\,\xi_{8,11} - \theta_4\theta_5\,\xi_{7,13} - \theta_9\theta_{10}\,\xi_{2,14} = 0,$$
$$\theta_1\theta_5\,\xi_{3,14} - \theta_4\theta_6\,\xi_{2,15} - \theta_7\theta_8\,\xi_{9,16} = 0, \quad \theta_1\theta_5\,\xi_{9,16} + \theta_4\theta_6\,\xi_{10,12} - \theta_7\theta_8\,\xi_{3,14} = 0,$$
$$\theta_3\theta_8\,\xi_{5,9} + \theta_4\theta_{10}\,\xi_{11,15} - \theta_5\theta_9\,\xi_{3,8} = 0, \quad \theta_3\theta_8\,\xi_{11,15} + \theta_4\theta_{10}\,\xi_{5,9} - \theta_5\theta_9\,\xi_{4,10} = 0,$$
$$\theta_3\theta_{10}\,\xi_{6,7} + \theta_4\theta_8\,\xi_{12,14} - \theta_6\theta_7\,\xi_{3,10} = 0, \quad \theta_3\theta_{10}\,\xi_{12,14} + \theta_4\theta_8\,\xi_{6,7} - \theta_6\theta_7\,\xi_{4,8} = 0,$$
$$\theta_1\theta_{10}\,\xi_{4,16} + \theta_2\theta_8\,\xi_{3,13} - \theta_6\theta_9\,\xi_{5,12} = 0, \quad \theta_1\theta_{10}\,\xi_{3,13} + \theta_2\theta_8\,\xi_{4,16} - \theta_6\theta_9\,\xi_{7,15} = 0,$$
$$\theta_1\theta_8\,\xi_{4,13} + \theta_2\theta_{10}\,\xi_{3,16} - \theta_5\theta_7\,\xi_{6,11} = 0, \quad \theta_1\theta_8\,\xi_{3,16} + \theta_2\theta_{10}\,\xi_{4,13} - \theta_5\theta_7\,\xi_{9,14} = 0,$$
$$\theta_1\theta_3\,\xi_{8,16} + \theta_2\theta_4\,\xi_{10,13} - \theta_7\theta_9\,\xi_{5,14} = 0, \quad \theta_1\theta_3\,\xi_{10,13} + \theta_2\theta_4\,\xi_{8,16} - \theta_7\theta_9\,\xi_{6,15} = 0,$$
$$\theta_1\theta_2\,\xi_{5,15} - \theta_3\theta_4\,\xi_{6,14} - \theta_8\theta_{10}\,\xi_{7,12} = 0, \quad \theta_1\theta_2\,\xi_{7,12} - \theta_3\theta_4\,\xi_{9,11} - \theta_8\theta_{10}\,\xi_{5,15} = 0,$$
$$\theta_1\theta_4\,\xi_{8,13} + \theta_2\theta_3\,\xi_{10,16} - \theta_5\theta_6\,\xi_{7,11} = 0, \quad \theta_1\theta_4\,\xi_{10,16} + \theta_2\theta_3\,\xi_{8,13} - \theta_5\theta_6\,\xi_{9,12} = 0.$$

$$(108)$$

Appendix B. Mumford Relations II

In terms of the variables $t_{a,b}$ used in Proposition 4.15, the ideal generated by the Mumford relations in Equations (108) coincides with the ideal generated by the following equations whose coefficients are determined by the Rosenhain parameters of the genus-two curve $\mathcal{C}$ in Equation (17):

$t_{1,2} - t_{156,256} + t_{146,246}$	$= 0,$	$\lambda_3 t_{1,2} - t_{136,236} + t_{146,246}$	$= 0,$
$t_{1,3} - t_{156,356} + t_{146,346}$	$= 0,$	$\lambda_2 t_{1,3} - t_{126,236} + t_{146,346}$	$= 0,$
$(\lambda_2 - 1)\,t_{1,4} - t_{126,246} + t_{156,456}$	$= 0,$	$(\lambda_2 - \lambda_3)\,t_{1,4} - t_{126,246} + t_{136,346}$	$= 0,$
$\lambda_2 t_{1,5} - t_{126,256} + t_{146,456}$	$= 0,$	$\lambda_3 t_{1,5} - t_{136,356} + t_{146,456}$	$= 0,$
$(\lambda_3 - \lambda_2)\,t_{1,6} + t_{256,346} - t_{246,356}$	$= 0,$	$\lambda_2\,(\lambda_3 - 1)\,t_{1,6} + t_{256,346} - t_{236,456}$	$= 0,$
$t_{2,3} - t_{256,356} + t_{246,346}$	$= 0,$	$\lambda_1 t_{2,3} - t_{126,136} + t_{246,346}$	$= 0,$
$(\lambda_1 - 1)\,t_{2,4} - t_{126,146} + t_{256,456}$	$= 0,$	$(\lambda_3 - 1)\,t_{2,4} + t_{256,456} - t_{236,346}$	$= 0,$
$\lambda_1 t_{2,5} - t_{126,156} + t_{246,456}$	$= 0,$	$\lambda_3 t_{2,5} - t_{236,356} + t_{246,456}$	$= 0,$
$(\lambda_1 - \lambda_3)\,t_{2,6} + t_{146,356} - t_{156,346}$	$= 0,$	$\lambda_1\,(\lambda_3 - 1)\,t_{2,6} + t_{156,346} - t_{136,456}$	$= 0,$
$(\lambda_1 - 1)\,t_{3,4} - t_{136,146} + t_{356,456}$	$= 0,$	$(\lambda_2 - 1)\,t_{3,4} + t_{356,456} - t_{236,246}$	$= 0,$
$\lambda_1 t_{3,5} - t_{136,156} + t_{346,456}$	$= 0,$	$\lambda_2 t_{3,5} - t_{236,256} + t_{346,456}$	$= 0,$
$(\lambda_1 - \lambda_2)\,t_{3,6} + t_{146,256} - t_{156,246}$	$= 0,$	$\lambda_1\,(\lambda_2 - 1)\,t_{3,6} + t_{156,246} - t_{126,456}$	$= 0,$
$(\lambda_1 - \lambda_2)\,t_{4,5} - t_{146,156} + t_{246,256}$	$= 0,$	$(\lambda_1 - \lambda_3)\,t_{4,5} - t_{146,156} + t_{346,356}$	$= 0,$
$(\lambda_2 - 1)\,(\lambda_1 - \lambda_3)\,t_{4,6} + t_{156,236} - t_{126,356}$	$= 0,$	$(\lambda_2 - \lambda_3)\,(\lambda_1 - 1)\,t_{4,6} + t_{136,256} - t_{126,356}$	$= 0,$
$\lambda_2\,(\lambda_1 - \lambda_3)\,t_{5,6} + t_{146,236} - t_{126,346}$	$= 0,$	$(\lambda_2 - \lambda_3)\,\lambda_1 t_{5,6} + t_{136,246} - t_{126,346}$	$= 0,$
$t_{1,126} - \lambda_1 t_{5,256} + (\lambda_1 - 1)\,t_{4,246}$	$= 0,$	$t_{3,236} - \lambda_3 t_{5,256} + (\lambda_3 - 1)\,t_{4,246}$	$= 0,$
$t_{1,136} - \lambda_1 t_{5,356} + (\lambda_1 - 1)\,t_{4,346}$	$= 0,$	$t_{2,236} - \lambda_2 t_{5,356} + (\lambda_2 - 1)\,t_{4,346}$	$= 0,$

$$
\begin{aligned}
(\lambda_2 - 1)\, t_{1,146} - (\lambda_1 - 1)\, t_{2,246} + (\lambda_1 - \lambda_2)\, t_{5,456} &= 0, & (\lambda_2 - \lambda_3)\, t_{1,146} + (\lambda_3 - \lambda_1)\, t_{2,246} + (\lambda_1 - \lambda_2)\, t_{3,346} &= 0, \\
t_{2,126} - \lambda_2 t_{5,156} + (\lambda_2 - 1)\, t_{4,146} &= 0, & t_{3,136} - \lambda_3 t_{5,156} + (\lambda_3 - 1)\, t_{4,146} &= 0, \\
t_{6,126} + t_{5,346} - t_{4,356} &= 0, & \lambda_3 t_{6,126} + t_{3,456} - t_{4,356} &= 0, \\
t_{6,136} + t_{5,246} - t_{4,256} &= 0, & \lambda_2 t_{6,136} + t_{2,456} - t_{4,256} &= 0, \\
(\lambda_2 - 1)\, t_{6,146} + t_{2,356} - t_{5,236} &= 0, & (\lambda_3 - 1)\, t_{6,146} + t_{3,256} - t_{5,236} &= 0, \\
t_{6,236} + t_{5,146} - t_{4,156} &= 0, & \lambda_1 t_{6,236} + t_{1,456} - t_{4,156} &= 0, \\
(\lambda_1 - 1)\, t_{6,246} + t_{1,356} - t_{5,136} &= 0, & (\lambda_3 - 1)\, t_{6,246} + t_{3,156} - t_{5,136} &= 0, \\
(\lambda_1 - 1)\, t_{6,346} + t_{1,256} - t_{5,126} &= 0, & (\lambda_2 - 1)\, t_{6,346} + t_{2,156} - t_{5,126} &= 0, \\
\lambda_2 t_{6,156} + t_{2,346} - t_{4,236} &= 0, & \lambda_3 t_{6,156} + t_{3,246} - t_{4,236} &= 0, \\
\lambda_1 t_{6,256} + t_{1,346} - t_{4,136} &= 0, & \lambda_3 t_{6,256} + t_{3,146} - t_{4,136} &= 0, \\
\lambda_1 t_{6,356} + t_{1,246} - t_{4,126} &= 0, & \lambda_2 t_{6,356} + t_{2,146} - t_{4,126} &= 0, \\
(\lambda_1 - \lambda_2)\, t_{6,456} + t_{1,236} - t_{2,136} &= 0, & (\lambda_1 - \lambda_3)\, t_{6,456} + t_{1,236} - t_{3,126} &= 0, \\
\lambda_2 t_{1,156} - \lambda_1 t_{2,256} + (\lambda_1 - \lambda_2)\, t_{4,456} &= 0, & \lambda_3 t_{1,156} - \lambda_1 t_{3,356} + (\lambda_1 - \lambda_3)\, t_{4,456} &= 0.
\end{aligned}
$$

$$(109)$$

Adrian Clingher
Dept. of Mathematics and Computer Science,
University of Missouri – St. Louis,
St. Louis, MO 63121
E-mail address: clinghera@umsl.edu

Andreas Malmendier
Dept. of Mathematics and Statistics,
Utah State University,
Logan, UT 84322
E-mail address: andreas.malmendier@usu.edu

6

σ-Functions: Old and New Results

V. M. Buchstaber, V. Z. Enolski and D. V. Leykin

To Professor Emma Previato on the occasion of her 65th birthday

Abstract. We are considering multi-variable sigma functions of genus g hyperelliptic curve as a function of two groups of variables – Jacobian variables and parameters of the curve. In the theta-functional representation of sigma-function the second group arises as periods of first and second kind differentials of the curve. We develop representation of periods in terms of theta-constants; for the first kind of period, generalization of Rosenhain type formulae are obtained whilst for the second kind of period, theta-constant expressions are presented which are explicitly related to the fixed co-homology basis.

We describe a method of constructing differentiation operators for hyperelliptic analogues of ζ- and $\wp$-functions on the parameters of the hyperelliptic curve. To demonstrate this method, we give the detailed construction of these operators in the case genus 1 and 2.

1 Introduction

Our note belongs to the area in whose development Emma Previato took active part. Ever since the first publication of the present authors [14] she inspired them and gave a lot of suggestions and advice.

The aforementioned area is a construction of Abelian functions in terms of multi-variable σ-functions. Similar to the Weierstrass elliptic function, multi-variable sigma keeps its main property – it remains form-invariant under the action of symplectic groups. Abelian functions appeared to be logarithmic derivatives like $\zeta, \wp$-functions of the Weierstrass theory and similar to the standard theta-functional approach which lead to the Krichever formula for KP solutions, [51]. But the fundamental difference between sigma

175

and theta-functional theories is the following. They are both constructed by the curve and given as series in Jacobian variables, but in the first case the expansion is purely algebraic with respect to the model of the curve because its coefficients are polynomials in parameters of the defining curve equation, whilst in the second case the coefficients are transcendental being built in terms of Riemann matrix periods which are complete Abelian integrals.

In many publications, in particular see [59, 35, 17, 55, 32, 1, 11, 53, 56, 57] and references therein, it was demonstrated that the multi-variable $\wp$-function represents a language which is very suitable for discussion of completely integrable systems of KP type. In particular, very recently, using the fact that sigma is an entire function in parameters of the curve (in contrast with theta-functions) families of degenerate solutions for solitonic equations were obtained [8, 9]. In recent papers [24] and [25], an algebraic construction of a wide class of polynomial Hamiltonian integrable systems was given, and those whose solutions are given by hyperelliptic $\wp$-functions were indicated.

The revival of interest in multi-variable σ-functions is in many respects due to H. Baker's exposition of the theory of Abelian functions which goes back to K. Weierstrass and F. Klein and is well documented and developed in his remarkable monographs [4, 5]. The heart of his exposition is the representation of fundamental bi-differentials of hyperelliptic curves in algebraic form in contrast with the most well-known representation as double differentials of Riemann theta-functions developed in Fay's monograph, [41]. Recent investigations demonstrated that Baker's approach can be extended beyond hyperelliptic curves to wide classes of algebraic curves.

The multi-variable σ-function of an algebraic curve $\mathscr{C}$ is known to be represented in terms of θ-functions of the curve as a function of two groups of variables – Jacobian of the curve, Jac($\mathscr{C}$) and Riemann matrix τ. In a vast amount of recent publications properties of σ-functions as a function of the first group of variables are discussed whilst the modular part of variables and relevant objects like θ-constant representations of complete Abelian integrals are considered separately. In this paper we deal with σ as a function over both groups of variables.

Due to the pure modular part of σ-variables we consider problem of expression of complete integrals of the first and second kind in terms of theta-constants. Revival of interest in this classically known matter accords to many recent publications reconsidering such problems like Schottky problem [40], Thomae [37, 36] and Weber [58] formulae, theory of invariants and its applications [52, 31].

The paper is organized as follows. In Section 2 we consider hyperelliptic genus g curves and a complete co-homology basis of $2g$ meromorphic differentials, with g of them chosen as holomorphic ones. We discuss expressions for periods of these differentials in terms of θ-constant with half integer characteristics. Theta-constant representation of periods of holomorphic integrals is known from Rosenhain's memoir [60], where the case of genus two was elaborated. We discuss this case and generalize Rosenhain's expressions to higher genera hyperelliptic curves. Theta-constant representation of periods of the second kind is known after F. Klein [49] who presented closed formula in terms of derived even theta-constants for non-hyperelliptic genus three curves. We re-derive this formula for higher genera hyperelliptic curves.

Section 3 is devoted to the classically known problem, which was resolved in the case of elliptic curves by Frobenius and Stickelberger [44]. The general method of the solution of this problem for a wide class of so-called (n, s)- curves has been developed in [22] and represents an extension of Weierstrass's method for the derivation of a system of differential equations defining a sigma-function. All stages of this derivation are given in detail and the main result is that the sigma-function is completely defined as the solution of a system of heat conductivity equations in a nonholonomic frame. We also consider another widely known problem – description of dependence of the solutions on initial data. This problem is formulated as a description of the dependence of integrals of motion, whose levels are given as half of the curve parameters, from the remaining half of the parameters. The differential formulae obtained permit the presentation of an effective solution of this problem for Abelian functions of hyperelliptic curves. It is worth noting that because integrals of motion can be expressed in terms of periods of the second kind in this place results of Section 2 are demanded. The results obtained in Section 3 are exemplified in details by curves of genera one and two. All consideration is based on explicit uniformization of space universal bundles of the hyperelliptic Jacobian.

2 Modular Representation of Periods of Hyperelliptic Co-Homologies

Modular invariance of the Weierstrass elliptic σ-function, $\sigma = \sigma(u; g_2, g_3)$ follows from its defining in [65] in terms of recursive series in terms of variables (u, g_2, g_3). Alternatively a σ-function can be represented in terms of a Jacobi θ-function and its modular invariance follows from transformation properties of θ-functions. The last representation involves complete elliptic integrals of the first and second kind and their representations in terms of θ-constants are classically known. In this section we are studying generalizations of these representations to hyperelliptic curves of higher genera realized in the form

$$y^2 = P_{2g+1}(x) = (x - e_1) \cdots (x - e_{2g+1}) \tag{1}$$

Here $P_{2g+1}(x)$ is a monic polynomial of degree $2g + 1$, $e_i \in \mathbb{C}$ - branch points and the curve is assumed to be non-degenerate, i.e. $e_i \neq e_j$.

2.1 Problems and Methods

Representations of complete elliptic integrals of the first and second kind in terms of Jacobi θ-constants are classically known. In particular, if an elliptic curve is given in Legendre form[1]

$$y^2 = (1 - x^2)(1 - k^2 x^2) \tag{2}$$

where k is Jacobian modulus, then complete elliptic integrals of the first kind $K = K(k)$ are represented as

$$K = \int_0^1 \frac{dx}{\sqrt{(1 - x^2)(1 - k^2 x^2)}} = \frac{\pi}{2} \vartheta_3^2(0; \tau), \tag{3}$$

and $\vartheta_3 = \vartheta_3(0|\tau)$ and $\tau = \iota \frac{K'}{K}$, $K' = K(k')$, $k^2 + k'^2 = 1$.

Further, for an elliptic curve realized as Weierstrass cubic

$$y^2 = 4x^3 - g_2 x - g_3 = 4(x - e_1)(x - e_2)(x - e_3) \tag{4}$$

recall standard notation for periods of the first and second kind elliptic integrals

$$2\omega = \oint_{\mathfrak{a}} \frac{dx}{y}, \quad 2\eta = -\oint_{\mathfrak{a}} \frac{x dx}{y} \\ 2\omega' = \oint_{\mathfrak{b}} \frac{dx}{y}, \quad 2\eta' = -\oint_{\mathfrak{b}} \frac{x dx}{y} \qquad \tau = \frac{\omega'}{\omega} \tag{5}$$

and the Legendre relation for them

$$\omega\eta' - \eta\omega' = -\frac{\iota\pi}{2} \tag{6}$$

Then the following Weierstrass relation is valid

$$\eta = -\frac{1}{12\omega}\left(\frac{\vartheta_2''(0)}{\vartheta_2(0)} + \frac{\vartheta_3''(0)}{\vartheta_3(0)} + \frac{\vartheta_4''(0)}{\vartheta_4(0)}\right) \tag{7}$$

In this section we are discussing generalizations of these relations to higher genera hyperelliptic curves realized as in (1).

2.2 Definitions and Main Theorems

In this subsection we reproduce notation from H. Baker [6]. Let $\mathscr{C}$ be a genus g non-degenerate hyperelliptic curve realized as a double cover of a Riemann sphere,

$$y^2 = 4 \prod_{j=1}^{2g+1} (x - e_j) \equiv 4x^{2g+1} + \sum_{i=0}^{} \lambda_i x^i, \quad e_i \neq e_j, \ \lambda_i \in \mathbb{C} \tag{8}$$

---

[1] Here and below we punctiliously follow notations of elliptic functions theory fixed in [7]

Let $(a; b) = (a_1, \ldots, a_g; b_1, \ldots, b_g)$ be a canonic homology basis. Introduce a co-homology basis (*Baker co-homology basis*)

$$du(x, y) = (du_1(x, y), \ldots, du_g(x, y))^T, dr(x, y)$$
$$= (dr_1(x, y), \ldots, dr_g(x, y))^T$$
$$du_i(x, y) = \frac{x^{i-1}}{y}dx, \quad dr_j(x, y) = \sum_{k=j}^{2g+1-j}(k+1-j)\frac{x^k}{4y}dx, \quad i, j = 1, \ldots, g \tag{9}$$

satisfying the generalized Legendre relation,

$$\mathfrak{M}^T J \mathfrak{M} = -\frac{\iota\pi}{2}J, \quad \mathfrak{M} = \begin{pmatrix} \omega & \omega' \\ \eta & \eta' \end{pmatrix}, \quad J = \begin{pmatrix} 0_g & 1_g \\ -1_g & 0_g \end{pmatrix} \tag{10}$$

where $g \times g$ period matrices $\omega, \omega', \eta, \eta'$ are defined as

$$2\omega = \left(\oint_{a_j} du_i\right), \quad 2\omega' = \left(\oint_{b_j} du_i\right),$$
$$2\eta = -\left(\oint_{a_j} dr_i\right), \quad 2\eta' = -\left(\oint_{b_j} dr_i\right) \tag{11}$$

We also denote $dv = (dv_1, \ldots, dv_g)^T = (2\omega)^{-1}du$ the vector of normalized holomorphic differentials.

Define the Riemann matrix $\tau = \omega^{-1}\omega'$ belonging to the Siegel half-space $\mathscr{S}_g = \{\tau^T = \tau, \text{Im}\tau > 0\}$. Define the Jacobi variety of the curve $\text{Jac}(\mathscr{C}) = \mathbb{C}^g/1_g \oplus \tau$. The canonical Riemann θ-function is defined on $\text{Jac}(\mathscr{C}) \times \mathscr{S}_g$ by Fourier series

$$\theta(z; \tau) = \sum_{\kappa \in \mathbb{Z}^n} e^{\iota\pi n^T \tau n + 2\iota\pi z^T n} \tag{12}$$

We will also use θ-functions with half-integer characteristics $[\varepsilon] = \begin{bmatrix} \varepsilon'^T \\ \varepsilon'' \end{bmatrix}$, $\varepsilon_i', \varepsilon_j'' = 0$ or 1 defined as

$$\theta[\varepsilon](z; \tau) = \sum_{\kappa \in \mathbb{Z}^n} e^{\iota\pi(n+\varepsilon'/2)^T \tau(n+\varepsilon'/2)+2\iota\pi(z+\varepsilon''/2)^T(n+\varepsilon'/2)} \tag{13}$$

The characteristic is even or odd whenever $\varepsilon'^T \varepsilon'' = 0 \pmod 2$ or $1 \pmod 2$ and $\theta[\varepsilon](z; \tau)$ as function of z inherits parity of the characteristic.

Derivatives of θ-functions by arguments z_i will be denoted as

$$\theta_i[\varepsilon](z; \tau) = \frac{\partial}{\partial z_i}\theta[\varepsilon](z; \tau), \quad \theta_{i,j}[\varepsilon](z; \tau) = \frac{\partial^2}{\partial z_i \partial z_j}\theta[\varepsilon](z; \tau), \quad \text{etc.}$$

The fundamental bi-differential $\Omega(P, Q)$ is uniquely defined on the product $(P, Q) \in \mathscr{C} \times \mathscr{C}$ by the following conditions:

i Ω is symmetric, $\Omega(P, Q) = \Omega(Q, P)$

ii Ω is normalized by the condition

$$\oint_{a_i} \Omega(P, Q) = 0, \quad i = 1, \dots, g \tag{14}$$

iii Let $P = (x, y)$ and $Q = (z, w)$ have local coordinates $\xi_1 = \xi(P)$, $\xi_2 = \xi(Q)$ in the vicinity of point R, $\xi(R) = 0$, then $\Omega(P, Q)$ expands to a power series as

$$\Omega(P, Q) = \frac{d\xi_1 d\xi_2}{(\xi_1 - \xi_2)^2} + \text{homorphic 2-form} \tag{15}$$

The fundamental bi-differential can be expressed in terms of θ-functions [41]

$$\Omega(P, Q) = d_x d_z \theta \left(\int_Q^P dv + e \right), \quad P = (x, y), Q = (z, w) \tag{16}$$

where dv is a normalized holomorphic differential and e any non-singular point of the θ-divisor (θ), i.e. $\theta(e) = 0$, but not all θ-derivatives, $\partial_{z_i} \theta(z)|_{z=e}$, $i = 1, \dots, g$ vanish.

In the case of a hyperelliptic curve $\Omega(P, Q)$ can be alternatively constructed as

$$\Omega(P, Q) = \frac{1}{2} \frac{\partial}{\partial z} \frac{y + w}{y(x - z)} dx dz + dr(P)^T du(Q) + 2du^T(P) \varkappa du(Q) \tag{17}$$

where the first two terms are given as rational functions of coordinates P, Q and a necessarily symmetric matrix $\varkappa^T = \varkappa$, $\varkappa = \eta(2\omega)^{-1}$ is introduced to satisfy the normalization condition **ii**. In shorter form (18) can be rewritten as

$$\Omega(P, Q) = \frac{2yw + F(x, z)}{4(x - z)^2 yw} dx dz + 2du^T(P) \varkappa du(Q) \tag{18}$$

where $F(x, z)$ is so-called Kleinian 2-polar, given as

$$F(x, z) = \sum_{k=0}^{g} x^k z^k (2\lambda_{2k} + \lambda_{2k+1}(x + z)) \tag{19}$$

Recently algebraic representations for $\Omega(P, Q)$ similar to (18) were found in [61], [39] for wide class on algebraic curves, including (n, s)-curves [16].

The main relation underlying the theory is the Riemann formula representing Abelian integrals of the third kind as θ-quotient written in terms of the above described realization of the fundamental differential $\Omega(P, Q)$.

Theorem 2.1 (*Riemann*) *Let* $P' = (x', y')$ *and* $P'' = (x'', y'')$ *be two arbitrary distinct points on* $\mathscr{C}$ *and let* $\mathscr{D}' = \{P_1' + \cdots + P_g'\}$ *and*

$\mathscr{D}'' = \{P_1'' + \cdots + P_g''\}$ be two non-special divisors of degree g. Then the following relation is valid

$$\int_{P''}^{P'} \sum_{j=1}^{g} \int_{P_j'}^{P_j''} \left\{ \frac{2yy_i + F(x, x_i)}{4(x - x_i)^2} \frac{dx\, dx_i}{y\; y_i} + 2du(x, y)\varkappa du(x_i, y_i) \right\}$$

$$= \ln \left(\frac{\theta(\mathscr{A}(P') - \mathscr{A}(\mathscr{D}') + K_\infty)}{\theta(\mathscr{A}(P') - \mathscr{A}(\mathscr{D}'') + K_\infty)} \right) - \ln \left(\frac{\theta(\mathscr{A}(P'') - \mathscr{A}(\mathscr{D}') + K_\infty)}{\theta(\mathscr{A}(P'') - \mathscr{A}(\mathscr{D}'') + K_\infty)} \right)$$

$$(20)$$

where $\mathscr{A}(P) = \int_\infty^P dv$ is an Abel map with base point ∞, K_∞ – vector of Riemann constants with base point ∞ which is a half-period.

Introduce a multi-variable fundamental σ-function,

$$\sigma(u) = C\theta[K_\infty]((2\omega)^{-1}u)e^{u^T \varkappa u}, \qquad (21)$$

where $[K_\infty]$ is characteristic of the vector of Riemann constants, $u = \int_\infty^{P_1} du + \cdots + \int_\infty^{P_g} du$ with non-special divisor $P_1 + \cdots + P_g$. The constant C is chosen so that the expansion $\sigma(u)$ near $u \sim 0$ starts with a Schur-Weierstrass polynomial [16]. The whole expression is proved to be invariant under the action of the symplectic group $\mathrm{Sp}(2g, \mathbb{Z})$. Klein-Weierstrass multi-variable $\wp$-functions are introduced as logarithmic derivatives,

$$\wp_{i,j}(u) = -\frac{\partial^2}{\partial u_i \partial u_j}, \quad \wp_{i,j,k}(u) = -\frac{\partial^3}{\partial u_i \partial u_j \partial u_k}, \quad \text{etc.} \quad i, j, k = 1, \ldots g$$

$$(22)$$

Corollary 2.2 *For $r \neq s \in \{1, \ldots, g\}$ the following formula is valid*

$$\sum_{i,j=1}^{g} \wp_{i,j} \left(\sum_{k=1}^{g} \int_\infty^{(x_k, y_k)} du \right) x_s^{i-1} x_r^{j-1} = \frac{F(x_s, x_r) - 2y_s y_r}{4(x_s - x_r)^2} \qquad (23)$$

Corollary 2.3 *Jacobi problem of inversion of the Abel map $\mathscr{D} \to \mathscr{A}(\mathscr{D})$ with $\mathscr{D} = (x_1, y_1) + \cdots + (x_g, y_g)$ is resolved as*

$$x^g - \wp_{g,g}(u)x^{g-1} - \cdots - \wp_{g,1}(u) = 0$$

$$y_k = \wp_{g,g,g}(u)x_k^{g-1} + \cdots + \wp_{g,g,1}(u), \quad k = 1, \ldots, g$$

$$(24)$$

2.3 $\wp$-Values at Non-Singular Even Half-Periods

In this section we present a generalization of Weierstrass formulae

$$\wp(\omega) = e_1, \quad \wp(\omega + \omega') = e_2, \quad \wp(\omega') = e_3 \qquad (25)$$

to the case of a genus g hyperelliptic curve (8). To do that introduce partitions

$$\{1, \ldots, 2g+1\} = \mathscr{I}_0 \cup \mathscr{J}_0, \quad \mathscr{I}_0 \cap \mathscr{J}_0 = \emptyset$$
$$\mathscr{I}_0 = \{i_1, \ldots, i_g\}, \quad \mathscr{J}_0 = \{j_1, \ldots, j_{g+1}\} \tag{26}$$

Then any non-singular even half-period $\Omega_{\mathscr{I}}$ is given as

$$\Omega_{\mathscr{I}_0} = \int_\infty^{(e_{i_1},0)} du + \cdots + \int_\infty^{(e_{i_g},0)} du,$$
$$\mathscr{I}_0 = \{i_1, \ldots, i_g\} \subset \{1, \ldots, 2g+1\} \tag{27}$$

Denote elementary symmetric functions $s_n(\mathscr{I}_0)$, $S_n(\mathscr{J}_0)$ of order n built on branch points $\{e_{i_k}\}$, $i_k \in \mathscr{I}_0$, $\{e_{j_k}\}$, $j_k \in \mathscr{J}_0$ correspondingly. In particular,

$$s_1(\mathscr{I}_0) = e_{i_1} + \cdots + e_{i_g}, \qquad S_1(\mathscr{J}_0) = e_{j_1} + \cdots + e_{j_{g+1}}$$
$$s_2(\mathscr{I}_0) = e_{i_1}e_{i_2} + \cdots + e_{i_{g-1}}e_{i_g}, \qquad S_2(\mathscr{J}_0) = e_{j_1}e_{j_2} + \cdots + e_{j_g}e_{i_{g+1}}$$
$$\vdots \qquad\qquad\qquad \vdots$$
$$s_g(\mathscr{I}_0) = e_{i_1} \cdots e_{i_g} \qquad S_{g+1}(\mathscr{J}_0) = e_{j_1} \cdots e_{j_{g+1}} \tag{28}$$

Because of symmetry, $\wp_{p,q}(\Omega_{\mathscr{I}_0}) = \wp_{q,p}(\Omega_{\mathscr{I}_0})$ is enough to find these quantities for $p \le q \in \{1, \ldots, g\}$. The following is valid.

Proposition 2.4 (*Conjectural Proposition*) *Let an even non-singular half-period $\Omega_{\mathscr{I}_0}$ be associated to a partition $\mathscr{I}_0 \cup \mathscr{J}_0 = \{1, \ldots, 2g+1\}$. Then for all k, $j \ge k$, $k, j = 1 \ldots, g$ the following formula is valid*

$$\wp_{k,j}(\Omega_{\mathscr{I}_0})$$
$$= (-1)^{k+j} \sum_{n=1}^{k} n\big(s_{g-k+n}(\mathscr{I}_0)S_{g-j-n+1}(\mathscr{J}_0) + s_{g-j-n}(\mathscr{I}_0)S_{g+n-k+1}(\mathscr{J}_0)\big).$$
$$\tag{29}$$

Proof. The Klein formula written for even non-singular half-period $\Omega_{\mathscr{I}_0}$ leads to a linear system of equations with respect to Kleinian two-index symbols $\wp_{i,j}(\Omega_{\mathscr{I}_0})$

$$\sum_{i=1}^{g} \sum_{j=1}^{g} \wp_{i,j}(\Omega_{\mathscr{I}_0})e_{i_r}^{i-1} e_{i_s}^{j-1} = \frac{F(e_{i_r}, e_{i_s})}{4(e_{i_r} - e_{i_s})^2} \quad i_r, i_s \in \mathscr{I}_0 \tag{30}$$

To solve these equations we note that

$$\wp_{k,g}(\Omega_{\mathscr{I}_0}) = (-1)^{k+1}s_k(\mathscr{I}_0), \quad k = 1, \ldots, g \tag{31}$$

Also note that $F(e_{i_r}, e_{i_s})$ is divisible by $(e_{i_r} - e_{i_s})^2$ and

$$\frac{F(e_{i_r}, e_{i_s})}{4(e_{i_r} - e_{i_s})^2} = e_{i_r}^{g-1} e_{i_s}^{g-1} \mathfrak{S}_1 + e_{i_r}^{g-2} e_{i_s}^{g-2} \mathfrak{S}_2 + \cdots + \mathfrak{S}_{2g-1} \tag{32}$$

where $\mathfrak{S}_k$ are order k elementary symmetric functions of elements e_i $i \in \{1, \ldots, 2g+1\} - \{i_r, i_s\}$.

Let us analyse equations (30) for small genera, $g \leq 5$. One can see that plugging in the equation (32) to (30) we get non-homogeneous linear equations solvable by Kramer's rule and the solutions can be presented in the form (29).

Now suppose that (29) is valid for higher $g > 5$ where computer power is insufficient to check that by means of computer algebra. But it's possible to check (29) for arbitrary big genera numerically leading to branch points e_i, $i = 1, \ldots, i = 2g+1$ certain numeric values. Much checking confirmed (29). $\square$

2.4 Modular Representation of $\varkappa$ Matrix

Quantities $\wp_{i,j}(\Omega_{\mathscr{I}_0})$ are expressed in terms of even θ-constants as follows

$$\wp_{i,j}(\Omega_{\mathscr{I}_0}) = -2\varkappa_{i,j} - \frac{1}{\theta[\varepsilon_{\mathscr{I}_0}](0)} \partial^2_{U_i,U_j}\theta[\varepsilon_{\mathscr{I}_0}](0), \quad \forall \mathscr{I}_0,\; i,j = 1, \ldots, g. \tag{33}$$

Here $[\varepsilon_{\mathscr{I}_0}]$ is characteristic of the vector $[\Omega_{\mathscr{I}_0} + K_\infty]$, where K_∞ is a vector of Riemann constants with base point ∞ and ∂_U is the directional derivative along vector U_i, that is the ith column vector of an inverse matrix of $\mathfrak{a}$-periods, $\mathscr{A}^{-1} = (U_1, \ldots, U_g)$. The same formula is valid for all possible partitions $\mathscr{I}_0 \cup \mathscr{J}_0$, there are N_g of those, that is the number of non-singular even characteristics,

$$N_g = \binom{2g+1}{g} \tag{34}$$

Therefore one can write

$$\varkappa_{i,j} = \frac{1}{8N_g}\Lambda_{i,j} - \frac{1}{2N_g} \sum_{\text{All even non-singular } [\varepsilon]} \frac{\partial^2_{U_i,U_j}\theta[\varepsilon_{\mathscr{I}_0}](0)}{\theta[\varepsilon_{\mathscr{I}_0}](0)} \tag{35}$$

where

$$\Lambda_{i,j} = -4 \sum_{\text{All partitions } \mathscr{I}_0} \wp_{i,j}(\Omega_{\mathscr{I}_0}) \tag{36}$$

Denote by Λ_g the symmetric matrix

$$\Lambda_g = (\Lambda_{i,j})_{i,j=1,\ldots,g} \tag{37}$$

Proposition 2.5 *Entries* $\Lambda_{k,j}$ *at* $k \le j$ *to the symmetric matrix* Λ *are given by the formula*

$$\Lambda_{k,j} = \lambda_{k+j} \frac{\binom{2g+1}{g}}{\binom{2g+1}{2g+1-k-j}} \sum_{n=1}^{k} n \left[\binom{g}{g-k+n} \binom{g+1}{g-j-n+1} \right.$$
$$\left. + \binom{g}{g-j-n} \binom{g+1}{g-k+n+1} \right]$$
(38)

Proof. Execute summation in (29) and find that each $\Lambda_{k,j}$ is proportional to λ_{k+j} with integer coefficients. □

Matrix Λ_g exhibits interesting properties regarding the sum of anti-diagonal elements implemented at derivations in [33],

$$\sum_{i,j,\,i+j=k} \Lambda_{g;i,j} = \lambda_k \frac{N_g}{4g+2} \left[\frac{1}{2} k(2g+2-k) + \frac{1}{4}(2g+1)((-1)^k - 1) \right]$$
(39)

Lower genera examples of matrix Λ were given in [32], [33], but the method implemented there was unable to get expressions for Λ at big genera.

Example 2.6 *At* $g = 6$ *we get matrix*

$$\Lambda_6 = \begin{pmatrix} 792\lambda_2 & 330\lambda_3 & 120\lambda_4 & 36\lambda_5 & 8\lambda_6 & \lambda_7 \\ 330\lambda_3 & 1080\lambda_4 & 492\lambda_5 & 184\lambda_6 & 51\lambda_7 & 8\lambda_8 \\ 120\lambda_4 & 492\lambda_5 & 1200\lambda_6 & 542\lambda_7 & 184\lambda_8 & 36\lambda_9 \\ 36\lambda_5 & 184\lambda_6 & 542\lambda_7 & 1200\lambda_8 & 492\lambda_9 & 120\lambda_{10} \\ 8\lambda_6 & 51\lambda_7 & 184\lambda_8 & 492\lambda_9 & 1080\lambda_{10} & 330\lambda_{11} \\ \lambda_7 & 8\lambda_8 & 36\lambda_9 & 120\lambda_{10} & 330\lambda_{11} & 792\lambda_{12} \end{pmatrix}$$
(40)

Collecting all these together we get the following.

Proposition 2.7 *An* $\varkappa$-matrix *defining a multi-variate* σ-function *admits the following modular form representation*

$$\varkappa = \frac{1}{8N_g} \Lambda_g - \frac{1}{2N_g} (2\omega)^{-1^T}$$

$$\times \left[\sum_{N_g \text{ even } [\varepsilon]} \frac{1}{\theta[\varepsilon]} \begin{pmatrix} \theta_{1,1}[\varepsilon] & \cdots & \theta_{1,g}[\varepsilon] \\ \vdots & \cdots & \vdots \\ \theta_{1,g}[\varepsilon] & \cdots & \theta_{g,g}[\varepsilon] \end{pmatrix} \right] .(2\omega)^{-1}$$
(41)

where 2ω *is a matrix of* $\mathfrak{a}$-*periods of holomorhphic differentials and* $\theta_{i,j}[\varepsilon] = \partial^2_{z_i,z_j} \theta[\varepsilon](z)_{z=0}$.

Note that the modular form representation of period matrices η, η' follows from the above formula,

$$\eta = 2\varkappa\omega, \qquad \eta' = 2\varkappa\omega' - \iota\pi(2\omega)^{T-1} \tag{42}$$

Example 2.8 *At* $g = 2$ *for the curve* $y^2 = 4x^5 + \lambda_4 x^4 + \cdots + \lambda_0$ *the following representation of the* $\varkappa$*-matrix is valid*

$$\varkappa = \frac{1}{80}\begin{pmatrix} 4\lambda_2 & \lambda_3 \\ \lambda_3 & 4\lambda_4 \end{pmatrix} - \frac{1}{20}\sum_{10 \ even \ [\varepsilon]}\frac{1}{\theta[\varepsilon]}\begin{pmatrix} \partial^2_{U_1^2}\theta[\varepsilon] & \partial^2_{U_1,U_2}\theta[\varepsilon] \\ \partial^2_{U_1,U_2}\theta[\varepsilon] & \partial^2_{U_2^2}\theta[\varepsilon] \end{pmatrix} \tag{43}$$

with $\varkappa = \eta(2\omega)^{-1}$, $\mathscr{A}^{-1} = (2\omega)^{-1} = (U_1, U_2)$ *and directional derivatives* ∂_{U_i}, $i = 1, 2$.

A representation of the $\varkappa$ matrix of genus 2 and 3 hyperelliptic curves in terms of directional derivatives of non-singular odd constant was found in [30].

2.5 Co-Homologies of Baker and Klein

Calculation of the $\varkappa$-matrix for the hyperelliptic curve (8) were done using the co-homology basis introduced by H. Baker (9). When holomorphic differentials, $du(x, y)$, are chosen meromorphic differentials, $dr(x, y)$, can be found from the symmetry condition **I**. One can check that the symmetry condition is also fulfilled if meromorphic differentials are changed as

$$dr(x, y) \rightarrow dr(x, y) + M du(x, y), \tag{44}$$

where M is an arbitrary constant symmetric matrix $M^T = M$. One can then choose

$$M = -\frac{1}{8N_g}\Lambda_g \tag{45}$$

Then $\varkappa$ will change to

$$\varkappa = -\frac{1}{2}\frac{1}{N_g}\sum_{N_g \ even \ \left[\varepsilon_{\mathscr{I}_0}\right]}\frac{1}{\theta[\varepsilon_{\mathscr{I}_0}](0)}\left(\partial_{U_i}\partial_{U_j}\theta[\varepsilon_{\mathscr{I}_0}](0)\right)_{i,j=1,\dots,g}. \tag{46}$$

Following [33] we introduce the co-homology basis of Klein

$$du(x, y), \quad dr(x, y) - \frac{1}{8N_g}\Lambda_g du(x, y) \tag{47}$$

with constant matrix, $\Lambda_g = \Lambda_g(\lambda)$ given by (37,38). Therefore we proved

Proposition 2.9 *The* $\varkappa$*-matrix is represented in the modular form (46) in the co-homology basis (47).*

Formula (46) first appears in F. Klein ([48], [49]). It was recently revisited in a more general context by Korotkin and Shramchenko ([50]) who extended the representation for $\varkappa$ to non-hyperelliptic curves. Correspondence of this representation to the co-homology basis to the best knowledge of the authors was not earlier discussed.

Rewrite formula (46) in the equivalent form,

$$\omega^T \eta = -\frac{1}{4N_g} \sum_{N_g \text{ even } [\varepsilon]} \frac{1}{\theta[\varepsilon]} \begin{pmatrix} \theta_{1,1}[\varepsilon] & \cdots & \theta_{1,g}[\varepsilon] \\ \vdots & \cdots & \vdots \\ \theta_{1,g}[\varepsilon] & \cdots & \theta_{g,g}[\varepsilon] \end{pmatrix} \tag{48}$$

where ω, η are half-periods of holomorphic and meromorphic differentials in the Kleinian basis.

Example 2.10 *For the Weierstrass cubic $y^2 = 4x^3 - g_2 x - g_3$ (48) represents the Weierstrass relation (7).*

Example 2.11 *At $g = 2$ (48) can be written in the form*

$$\omega^T \eta = -\frac{\iota\pi}{10} \begin{pmatrix} \partial_{\tau_{1,1}} & \partial_{\tau_{1,2}} \\ \partial_{\tau_{1,2}} & \partial_{\tau_{2,2}} \end{pmatrix} \ln \chi_5 \tag{49}$$

where χ_5 is a relative invariant of weight 5,

$$\chi_5 = \prod_{10 \text{ even } [\varepsilon]} \theta[\varepsilon] \tag{50}$$

Example 2.12 *It's worth mentioning how equations of KdV flows look in both bases. For example at $g = 2$ and curve $y^2 = 4x^5 + \lambda_4 x^4 + \cdots + \lambda_0$ in the Baker basis we got [15]*

$$\wp_{2222} = 6\wp_{2,2}^2 + 4\wp_{1,2} + \lambda_4 \wp_{2,2} + \frac{1}{2}\lambda_3$$
$$\wp_{1222} = 6\wp_{2,2}\wp_{1,2} - 2\wp_{1,1} + \lambda_4 \wp_{1,2} \tag{51}$$

In the Kleinian basis the same equations change only linearly in $\wp_{i,j}$-terms

$$\wp_{2222} = 6\wp_{2,2}^2 + 4\wp_{1,2} - 47\lambda_4\wp_{2,2} + 92\lambda_4^2 - \frac{7}{2}\lambda_3$$
$$\wp_{1222} = 6\wp_{2,2}\wp_{1,2} - 2\wp_{1,1} - 23\lambda_4\wp_{1,2} - 6\lambda_3\wp_{2,2} + 23\lambda_3\lambda_4 + 8\lambda_2 \tag{52}$$

2.6 Rosenhain Modular Form Representation of First Kind Periods

Rosenhain [60] was the first to introduce θ-functions with characteristics at $g = 2$. There are 10 even and 6 odd characteristics in that case. Let us denote each of these characteristics as

$$\varepsilon_j = \begin{bmatrix} \varepsilon_j'^T \\ \varepsilon_j''^T \end{bmatrix}, \quad j = 1, \ldots 10$$

where ε_j' and ε_j'' are column 2-vectors with entries equal to 0 or 1.

Rosenhain fixed the hyperelliptic genus two curve in the form

$$y^2 = x(x - 1)(x - a_1)(x - a_2)(x - a_3)$$

and presented without proof the expression

$$\mathscr{A}^{-1} = \frac{1}{2\pi^2 Q^2} \begin{pmatrix} -P\theta_2[\delta_2] & Q\theta_2[\delta_1] \\ P\theta_1[\delta_2] & -Q\theta_1[\delta_1] \end{pmatrix} \tag{53}$$

with

$$P = \theta[\alpha_1]\theta[\alpha_2]\theta[\alpha_3], \quad Q = \theta[\beta_1]\theta[\beta_2]\theta[\beta_3]$$

and 6 even characteristics $[\alpha_{1,2,3}]$, $[\beta_{1,2,3}]$ and 2 odd $[\delta_{1,2}]$ which looks chaotic. One of the first proofs can be found in H. Weber [64]; these formulae are implemented in Bolza's dissertation [12] and [13]. Our derivations of these formulae are based on the *Second Thomae relation* [63], see [38] and [34]. To proceed we give the following definitions.

Definition 2.13 *A triplet of characteristics* $[\varepsilon_1]$, $[\varepsilon_2,]$, $[\varepsilon_3]$ *is called azygetic if*

$$\exp \iota\pi \left\{ \sum_{j=1}^{3} \varepsilon_j'^T \varepsilon_j'' + \sum_{i=1}^{3} \varepsilon_i'^T \sum_{i=1}^{3} \varepsilon_i'' \right\} = -1$$

Definition 2.14 *A sequence of* $2g + 2$ *characteristics* $[\varepsilon_1], \ldots, [\varepsilon_{2g+2}]$ *is called a* special fundamental system *if the first g characteristics are odd, the remaining are even and any triple of characteristics in it is azygetic.*

Theorem 2.15 *(Conjectural Riemann-Jacobi derivative formula) Let g odd* $[\varepsilon_1], \ldots, [\varepsilon_g]$ *and* $g + 2$ *even* $[\varepsilon_{g+1}], \ldots, [\varepsilon_{2g+2}]$ *characteristics create a special fundamental system. Then the following equality is valid*

$$\text{Det} \left. \frac{\partial(\theta[\varepsilon_1](v), \ldots, \theta[\varepsilon_g](v))}{\partial(v_1, \ldots, v_g)} \right|_{v=0} = \pm \prod_{k=1\ldots g+2} \theta[\varepsilon_{g+k}](0) \tag{54}$$

Proof. (54) proved up to $g = 5$ in [43], [45], [42]   □

Example 2.16 *Jacobi derivative formula for elliptic curve*

$$\vartheta_1'(0) = \pi \vartheta_2(0)\vartheta_3(0)\vartheta_4(0)$$

Example 2.17 *Rosenhain derivative formula for genus two curve is given without proof in the memoir* [60]*, namely, let* $[\delta_1]$ *and* $[\delta_2]$ *be any two odd characteristics from all 6 odd, then*

$$\theta_1[\delta_1]\theta_2[\delta_2] - \theta_2[\delta_1]\theta_1[\delta_2] = \pi^2 \theta[\gamma_1]\theta[\gamma_2]\theta[\gamma_3]\theta[\gamma_4] \tag{55}$$

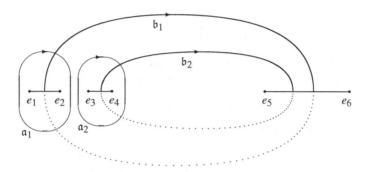

Figure 6.1 Homology basis on the Riemann surface of the curve $\mathscr{C}$ with real branching points $e_1 < e_2 < \cdots < e_6$ (upper sheet). The cuts are drawn from e_{2i-1} to e_{2i}, $i = 1, 2, 3$. The b-cycles are completed on the lower sheet (dotted lines).

where 4 even characteristics $[\gamma_1], \ldots, [\gamma_4]$ *are given as* $[\gamma_i] = [\delta_1] + [\delta_2] + [\delta_i]$, $3 \le i \le 6$. *There are 15 Rosenhain derivative formulae.*

The following geometric interpretation of the special fundamental system can be given in the case of hyperelliptic curves. Consider the genus two curve,

$$\mathscr{C}: \quad y^2 = (x - e_1) \cdots (x - e_6)$$

Denote the associated homology basis by $(\mathfrak{a}_1, \mathfrak{a}_2; \mathfrak{b}_1, \mathfrak{b}_2)$. Denote characteristics of Abelian images of branch points with base point e_6 by $\mathfrak{A}_k, k = 1, \ldots, 6$. These are half-periods given by their characteristics, $[\mathfrak{A}_k]$ with

$$\mathfrak{A}_k = \int_{(e_6,0)}^{(e_k,0)} u = \frac{1}{2}\tau\varepsilon'_k + \frac{1}{2}\varepsilon''_k, \quad k = 1, \ldots, 6 \tag{56}$$

For the homology basis drawn on the Figure we have

$$[\mathfrak{A}_1] = \frac{1}{2}\begin{bmatrix} 1 & 0 \\ 0 & 0 \end{bmatrix}, \quad [\mathfrak{A}_2] = \frac{1}{2}\begin{bmatrix} 1 & 0 \\ 1 & 0 \end{bmatrix}, \quad [\mathfrak{A}_3] = \frac{1}{2}\begin{bmatrix} 0 & 1 \\ 1 & 0 \end{bmatrix},$$

$$[\mathfrak{A}_4] = \frac{1}{2}\begin{bmatrix} 0 & 1 \\ 1 & 1 \end{bmatrix}, \quad [\mathfrak{A}_5] = \frac{1}{2}\begin{bmatrix} 0 & 0 \\ 1 & 1 \end{bmatrix}, \quad [\mathfrak{A}_6] = \frac{1}{2}\begin{bmatrix} 0 & 0 \\ 0 & 0 \end{bmatrix}$$

$$\mathfrak{A}_k = \int_\infty^{e_k} u = \frac{1}{2}\tau.\varepsilon_k + \frac{1}{2}\varepsilon'_k, \quad [\mathfrak{A}_k] = \begin{bmatrix} \varepsilon_k^T \\ \varepsilon'^T_k \end{bmatrix}$$

One can see that the set of characteristics $[\varepsilon_k] = [\mathfrak{A}_k]$ of Abelian images of branch point contains two odd $[\varepsilon_2]$ and $[\varepsilon_4]$ and the remaining four characteristics are even. One can check that the whole set of these $6 = 2g + 2$ characteristics is azygetic and therefore the set constitutes a special fundamental system. Hence one can write the Rosenhain derivative formula

$$\theta_1[\varepsilon_2]\theta_2[\varepsilon_4] - \theta_2[\varepsilon_2]\theta_1[\varepsilon_4] = \pm\pi^2\theta[\varepsilon_1]\theta[\varepsilon_3]\theta[\varepsilon_5]\theta[\varepsilon_6] \qquad (57)$$

In this way we gave a geometric interpretation of the Rosenhain derivative formula and associated set of characteristics in the case when one from even characteristic is zero. The same structure is observed for higher genera hyperelliptic curves even for $g > 5$.

Proposition 2.18 *The characteristics entering to the Rosenhain formula are described as follows. Take any of 15 Rosenhain derivative formulae say,*

$$\theta_1[p]\theta_2[q] - \theta_2[p]\theta_1[q] = \pi^2\theta[\gamma_1]\theta[\gamma_2]\theta[\gamma_3]\theta[\gamma_4]$$

Then 10 even characteristics can be grouped as

$$\underbrace{[\gamma_1], \dots, [\gamma_4]}_{4}, \quad \underbrace{[\alpha_1], [\alpha_2], [\alpha_3]}_{[\alpha_1]+[\alpha_2]+[\alpha_3]=[p]}, \quad \underbrace{[\beta_1], [\beta_2], [\beta_3]}_{[\beta_1]+[\beta_2]+[\beta_3]=[q]},$$

Then matrix of $\mathfrak{a}$-periods

$$\mathscr{A} = \frac{2Q}{PR} \begin{pmatrix} Q\theta_1[q] & Q\theta_2[q] \\ P\theta_1[p] & P\theta_2[p] \end{pmatrix}$$

with

$$P = \prod_{j=1}^{3} \theta[\alpha_j], \quad Q = \prod_{j=1}^{3} \theta[\beta_j], \quad R = \prod_{j=1}^{4} \theta[\gamma_j] \qquad (58)$$

Note, that the 15 curves are given as

$$\mathscr{C}_{p,q}: \quad y^2 = x(x-1)(x-a_1)(x-a_2)(x-a_3) \qquad (59)$$

where branch points are computed by *Bolza's formulae* [13],

$$e[\delta_j] = -\frac{\partial_{U_1}\theta[\delta_j]}{\partial_{U_2}\theta[\delta_j]}, \quad j = 1, \dots, 6 \qquad (60)$$

where ∂_{U_i} is the directional derivative along the vector U_i, $i = 1, 2$, $\mathscr{A}^{-1} = (U_1, U_2)$ and

$$e[p] = 0, \ e[q] = \infty.$$

All 15 curves $\mathscr{C}_{p,q}$ are Möbius equivalent.

2.7 Generalization of the Rosenhain Formula to Higher Genera Hyperelliptic Curve

Generalization of the Rosenhain formula to higher genera hyperelliptic curves was found in [38] and developed further in [34].

$$y^2 = \phi(x)\psi(x)$$

$$\phi(x) = \prod_{k=1}^{g}(x - e_{2k}), \quad \psi(x) = \prod_{k=1}^{g+1}(x - e_{2k-1}) \tag{61}$$

Denote by $R = \prod_{k=1}^{g+2}\theta[\gamma_k]$ monomial on the left-hand side of the Riemann-Jacobi formula (54).

Proposition 2.19 *Let a genus g hyperelliptic curve be given as in (61). Then winding vectors $(U_1, \ldots, U_g) = \mathscr{A}^{-1}$ are given by the formula*

$$U_m = \frac{\epsilon}{2\pi^g R}\text{Cofactor}\left(\left.\frac{\partial(\theta[\varepsilon_1](v), \ldots, \theta[\varepsilon_g](v))}{\partial(v_1, \ldots, v_g)}\right|_{v=0}\right)\begin{pmatrix} s_{m-1}^2 \sqrt[4]{\chi_1} \\ \vdots \\ s_{m-1}^{2g} \sqrt[4]{\chi_g} \end{pmatrix} \tag{62}$$

Here s_k^i is an order k symmetric function of elements $\{e_2, \ldots e_{2g}\}/\{e_{2i}\}$ and

$$\chi_i = \frac{\psi(e_{2i})}{\phi'(e_{2i})}, \quad i = 1, \ldots, g$$

s_k^i, χ_i are expressible in θ-constants via Thomae formulae [63].

2.8 Applications of the Above Results

Typical solution of multi-gap integration includes θ-functions $\theta(Ux + Vt + W; \tau)$ where winding vectors U, V are expressed in terms of complete holomorphic integrals and the constant W is defined by initial data. Rosenhain's formulae and their generalizations, express U, V in terms of θ-constants and gives parameters for the equation defining $\mathscr{C}$. In this way the problem of *effectivization of finite gap solutions* [28] can be solved at least for hyperelliptic curves.

Other applications of the Rosenhain formula (53) were presented in [10] where two-gap Lamé and Treibich-Verdier potentials were obtained by the reduction to elliptic functions of general Its-Matveev representations [46] of finite-gap potentials to the Schrödinger equation in terms of multi-variable θ-functions.

Another application is relevant to a computer algebra problem. In the case when, say in Maple, periods of holomorphic differentials are computed then periods of second kind differentials can be obtained by Rosenhain formula (53) and its generalization.

3 Sigma-Functions and the Problem of Differentiation of Abelian Functions

3.1 Problems and Methods

Consider the curve

$$V_\lambda = \left\{ (x, y) \in \mathbb{C}^2 : y^2 = \mathscr{C}(x; \lambda) = x^{2g+1} + \sum_{k=2}^{2g+1} \lambda_{2k} x^{2g-k+1} \right\} \quad (63)$$

where $g \geqslant 1$ and $\lambda = (\lambda_4, \ldots, \lambda_{4g+2}) \in \mathbb{C}^{2g}$ are the parameters. Set $\mathscr{D} = \{\lambda \in \mathbb{C}^{2g} : \mathscr{C}(x; \lambda)$ has multiple roots$\}$ and $\mathscr{B} = \mathbb{C}^{2g} \setminus \mathscr{D}$. For any $\lambda \in \mathscr{B}$ we obtain the affine part of a smooth projective hyperelliptic curve $\overline{V}_\lambda$ of genus g and the Jacobian variety $Jac(\overline{V}_\lambda) = \mathbb{C}^g / \Gamma_g$, where $\Gamma_g \subset \mathbb{C}^g$ is a lattice of rank $2g$ generated by the periods of the holomorphic differential on cycles of the curve V_λ.

In the general case, an *Abelian function* is a meromorphic function on a complex Abelian torus $T^g = \mathbb{C}^g / \Gamma$, where $\Gamma \subset \mathbb{C}^g$ is a lattice of rank $2g$. In other words, a meromorphic function f on $\mathbb{C}^g$ is Abelian iff $f(u) = f(u + \omega)$ for all $u = (u_1, \ldots, u_g) \in \mathbb{C}^g$ and $\omega \in \Gamma$. Abelian functions on T^g form a field $\mathscr{F} = \mathscr{F}_g$ such that:

(1) let $f \in \mathscr{F}$, then $\partial_{u_i} f \in \mathscr{F}, i = 1, \ldots, g$;
(2) let $f_1, \ldots, f_{g+1}$ be any nonconstant functions from $\mathscr{F}$, then there exists a polynomial P such that $P(f_1, \ldots, f_{g+1})(u) = 0$ for all $u \in T^g$;
(3) let $f \in \mathscr{F}$ be a nonconstant function, then any $h \in \mathscr{F}$ can be expressed rationally in terms of $(f, \partial_{u_1} f, \ldots, \partial_{u_g} f)$;
(4) there exists an entire function $\vartheta : \mathbb{C}^g \to \mathbb{C}$ such that $\partial_{u_i, u_j} \log \vartheta \in \mathscr{F}$, $i, j = 1, \ldots, g$.

For example, any elliptic function $f \in \mathscr{F}_1$ is a rational function in the Weierstrass functions $\wp(u; g_2, g_3)$ and $\partial_u \wp(u; g_2, g_3)$, where g_2 and g_3 are parameters of elliptic curve

$$V = \{(x, y) \in \mathbb{C}^2 \mid y^2 = 4x^3 - g_2 x - g_3\}.$$

It is easy to see that the function $\frac{\partial}{\partial g_2} \wp(u; g_2, g_3)$ will no longer be elliptic. This is due to the fact that the period lattice Γ is a function of the parameters g_2 and g_3. In [44] Frobenius and Stickelberger described all the differential operators L in the variables u, g_2 and g_3, such that $Lf \in \mathscr{F}_1$ for any function $f \in \mathscr{F}_1$ (see below Section 7.3).

In [21, 22] the classical problem of differentiation of Abelian functions over parameters for families of (n, s)-curves was solved. In the case of hyperelliptic curves this problem was solved more explicitly.

All genus 2 curves are hyperelliptic. We denote by $\pi : \mathscr{U}_g \to \mathscr{B}_g$ the universal bundle of Jacobian varieties $Jac(\overline{V}_\lambda)$ of hyperelliptic curves. Let us consider the mapping $\varphi \colon \mathscr{B}_g \times \mathbb{C}^g \to \mathscr{U}_g$, which defines the projection $\lambda \times \mathbb{C}^g \to \mathbb{C}^g / \Gamma_g(\lambda)$ for any $\lambda \in \mathscr{B}_g$. Let us fix the coordinates $(\lambda; u)$ in $\mathscr{B}_g \times \mathbb{C}^g \subset \mathbb{C}^{2g} \times \mathbb{C}^g$ where $u = (u_1, \ldots, u_{2g-1})$. Thus, using the mapping φ, we fixed in $\mathscr{U}_g$ the structure of the space of the bundle whose fibers J_λ are principally polarized Abelian varieties.

We denote by $F = F_g$ the field of functions on $\mathscr{U}_g$ such that for any $f \in F$ the function $\varphi^*(f)$ is meromorphic, and its restriction to the fiber J_λ is an Abelian function for any point $\lambda \in \mathscr{B}_g$.

Below, we will identify the field F with its image in the field of meromorphic functions on $\mathscr{B} \times \mathbb{C}^g$.

The following **Problem I**:

Describe the Lie algebra of differentiations of the field of meromorphic functions on $\mathscr{B}_g \times \mathbb{C}^g$, generated by the operators L, such that $Lf \in F$ for any function $f \in F$

was solved in [21, 22].

From the differential geometric point of view, Problem I is closely related to **Problem II**:

Describe the connection of the bundle $\pi \colon \mathscr{U}_g \to \mathscr{B}_g$.

The solution of Problem II leads to an important class of solutions of well-known equations of mathematical physics. In the case $g = 1$, the solution is called the Frobenius-Stikelberger connection (see [29]) and leads to solutions of the Chazy equation.

The space $\mathscr{U}_g$ is a rational variety, more precisely, there is a birational isomorphism $\varphi \colon \mathbb{C}^{3g} \to \mathscr{U}_g$. This fact was discovered by B. A. Dubrovin and S. P. Novikov in [27]. In [27], a fiber of the universal bundle is considered as a level surface of the integrals of motion of gth stationary flow of KdV system, that is, it is defined in $\mathbb{C}^{3g}$ by a system of $2g$ algebraic equations. The degree of the system grows with the growth of genus. In [14], [15] coordinates in $\mathbb{C}^{3g}$ were introduced such that a fiber is defined by $2g$ equations of degree not greater than 3.

The Dubrovin-Novikov coordinates and the coordinates from [14], [15] are the same for the universal space of genus 1 curves. But already in the case of genus 2, these coordinates differ (see [22]).

The integrals of motion of KdV systems are exactly the coefficients $\lambda_{2g+4}, \ldots, \lambda_{4g+2}$ of hyperelliptic curve V_λ, in which the coefficients $\lambda_4, \ldots, \lambda_{2g+2}$ are free parameters (see (63)). Choosing a point $z \in \mathbb{C}^{3g}$ such that the point $\varphi(z) \in \mathscr{B}_g$ is defined, one can calculate the values of the coefficients $(\lambda_4, \ldots, \lambda_{4g+2}) = \pi(z) \in \mathscr{B}_g$ substituting this point in these integrals. Thus, the solution of the Problem I of differentiation of hyperelliptic functions led to the solution of another well-known **Problem III**:

Describe the dependence of the solutions of g-th stationary flow of KdV system on the variation of the coefficients $\lambda_4, \ldots, \lambda_{4g+2}$ *of hyperelliptic curve, that is, from variation of values of the integrals of motion and parameters.*

In [47] results were obtained on Problem III, which use the fact that for a hierarchy KdV the action of polynomial vector fields on the spectral plane is given by the shift of the branch points of the hyperelliptic curve along these fields (see [62]). The deformations of the potential corresponding to this action are exactly the action of the nonisospectral symmetries of the hierarchy KdV.

Let us describe a different approach to Problem III, developed in our works. In [18] we introduced the concept of a polynomial Lie algebra over a ring of polynomials A. For brevity, we shall call them Lie A-algebras. In [21, 22] the ring of polynomials $\mathscr{P}$ in the field F was considered. This ring is generated by all logarithmic derivatives of order $k \geqslant 2$ from the hyperelliptic sigma function $\sigma(u; \lambda)$. The Lie $\mathscr{P}$-algebra $\mathscr{L} = \mathscr{L}_g$ with generators L_{2k-1}, $k = 1, \ldots, g$ and L_{2l}, $l = 0, \ldots, 2g - 1$ was constructed. The fields L_{2k-1} define isospectral symmetries, and the fields L_{2l} define nonisospectral symmetries of the hierarchy KdV. The Lie algebra $\mathscr{L}$ is isomorphic to the Lie algebra of differentiation of the ring $\mathscr{P}$ and, consequently, allows us to solve the Problem I (see property (3) of Abelian functions). The generators L_{2k-1}, $k = 1, \ldots, g$, coincide with the operators $\partial_{u_{2k-1}}$, and, consequently, commute. Thus, in the Lie $\mathscr{P}$-algebra $\mathscr{L}$ the Lie $\mathscr{P}$-subalgebra $\mathscr{L}^*$ generated by the operators L_{2k-1}, $k = 1, \ldots, g$ is defined. The generators L_{2l}, $l = 0, \ldots, 2g - 1$, are such that the Lie $\mathscr{P}$-algebra $\mathscr{L}^*$ is an ideal in the Lie $\mathscr{P}$-algebra $\mathscr{L}$.

The construction of L_{2k}, $k = 0, \ldots, 2g - 1$, is based on the following fundamental fact (see [19]):

The entire function $\psi(u; \lambda)$, satisfying the system of heat equations in a nonholonomic frame

$$\ell_{2i}\psi = H_{2i}\psi, \ i = 0, \ldots, 2g - 1,$$

under certain initial conditions (see [19]) coincides with the hyperelliptic sigma-function $\sigma(u; \lambda)$. Here ℓ_{2i} are polynomial linear first-order differential operators in the variables $\lambda = (\lambda_4, \ldots, \lambda_{4g+2})$ and H_{2i} are linear second-order differential operators in the variables $u = (u_1, \ldots, u_{2g-1})$. The methods for constructing these operators are described in [19].

The following fact was used essentially in constructing the operators ℓ_{2i}:

The Lie $\mathbb{C}[\lambda]$-algebra $\mathscr{L}_\lambda$ with generators ℓ_{2i}, $i = 0, \ldots, 2g - 1$ is isomorphic to an infinite-dimensional Lie algebra $Vect_{\mathscr{B}}$ of vector fields on $\mathbb{C}^{2g}$, that are tangent to the discriminant variety Δ. We recall that the Lie algebra $Vect_{\mathscr{B}}$ is essentially used in singularity theory and its applications (see [2]).

In the Lie $\mathscr{P}$-algebra $\mathscr{L}$ we can choose the generators L_{2k}, $k = 0, \ldots, 2g - 1$ such that for any polynomial $P(\lambda) \in \mathbb{C}[\lambda]$ and any k the formula $L_{2k}\pi^* P(\lambda) = \pi^*(\ell_{2k} P(\lambda))$ holds, where π^* is the ring homomorphism induced by the projection $\pi : \mathscr{U}_g \to \mathscr{B}_g$.

Section 3 describes the development of an approach to solving the Problem I. This approach uses:

(a) the graded set of multiplicative generators of the polynomial ring
    $$\mathscr{P} = \mathscr{P}_g;$$
(b) the description of all algebraic relations between these generators;
(c) the description of the birational isomorphism $J : \mathscr{U}_g \to \mathbb{C}^{3g}$ in terms of graded polynomial rings;
(d) the description of the polynomial projection $\pi : \mathbb{C}^{3g} \to \mathbb{C}^{2g}$, where $\mathbb{C}^{2g}$ is a space in coordinates $\lambda = (\lambda_4, \ldots, \lambda_{4g+2})$, such that for any $\lambda \in \mathscr{B}$ the space $J^{-1}\pi^{-1}(\lambda)$ is the Jacobian variety $Jac(V_\lambda)$;
(e) the construction of linear differential operators of first order

$$\widehat{H}_{2k} = \sum_{i=1}^{q} h_{2(k-i)+1}(\lambda; u)\partial_{u_{2i-1}}, \quad q = \min(k, g),$$

such that $L_{2k} = \ell_{2k} - \widehat{H}_{2k}$, $k = 0, \ldots, 2g - 1$.

Here:

- $h_{2(k-i)+1}(\lambda; u)$ are meromorphic functions on $\mathscr{B}_g \times \mathbb{C}^g$;
- $h_{2(k-i)+1}(\lambda; u)$ are homogeneous functions of degree $2(k - i) + 1$ in $\lambda = (\lambda_4, \ldots, \lambda_{4g+2})$, $\deg \lambda_{2k} = 2k$, and $u = (u_1, \ldots, u_{2g-1})$, $\deg u_{2k-1} = 1 - 2k$;
- $\partial_{u_{2l-1}} h_{2(k-i)+1}$ are homogeneous polynomials of degree $2(k + l - i)$ in the ring $\mathscr{P}_g$.

This approach was proposed in [23] and found the application in [26]. A detailed construction of the Lie algebra $\mathscr{L}_2$ is given in [23], and the Lie algebra $\mathscr{L}_3$ in [26].

General methods and results (see Section 3.2) will be demonstrated in cases $g = 1$ (see Section 3.3) and $g = 2$ (see Section 3.4).

3.2 Hyperelliptic Functions of Genus $g \geqslant 1$

For brevity, Abelian functions on the Jacobian varieties of (63) will be called *hyperelliptic functions* of genus g. In the theory and applications of these functions, that are based on the sigma-function $\sigma(u; \lambda)$ (see [4, 14, 15, 17]), the grading plays an important role. Below, the variables $u = (u_1, u_3, \ldots, u_{2g-1})$, parameters $\lambda = (\lambda_4, \ldots, \lambda_{4g+2})$ and functions are indexed in a way that clearly indicates their grading. Note that our new notations for the variables differ from the ones in [14, 15, 17] as follows

$$u_i \longleftrightarrow u_{2(g-i)+1}, \ i = 1, \ldots, g.$$

Let

$$\omega = ((2k_1 - 1) \cdot j_1, \ldots, (2k_s - 1) \cdot j_s)$$

where $1 \leqslant s \leqslant g$, $j_q > 0$, $q = 1, \ldots, s$ and $j_1 + \cdots + j_s \geqslant 2$. We draw attention to the fact that the symbol "·" in the two-component expression $(2k_q - 1) \cdot j_q$ is not a multiplication symbol. Set

$$\wp_\omega(u; \lambda) = -\partial_{u_{2k_1-1}}^{j_1} \cdots \partial_{u_{2k_s-1}}^{j_s} \ln \sigma(u; \lambda). \tag{64}$$

Thus

$$\deg \wp_\omega = (2k_1 - 1) j_1 + \cdots + (2k_s - 1) j_s.$$

Note that our ω differ from the ones in [23, 26].

Say that a multi-index ω is given in normal form if $1 \leqslant k_1 < \cdots < k_s$. According to formula (64), we can always bring the multi-index ω to a normal form using the identifications:

$$\left((2k_p - 1) \cdot j_p, (2k_q - 1) \cdot j_q\right) = \left((2k_q - 1) \cdot j_q, (2k_p - 1) \cdot j_p\right),$$
$$\left((2k_p - 1) \cdot j_p, (2k_q - 1) \cdot j_q\right) = (2k_p - 1) \cdot (j_p + j_q), \ \text{if} \ k_p = k_q.$$

In [17] (see also [14, 15]) it was proved that for $1 \leqslant i \leqslant k \leqslant g$ all algebraic relations between hyperelliptic functions of genus g follow from the relations, which in our graded notations have the form

$$\wp_{1\cdot3,(2i-1)\cdot1} = 6 \left(\wp_{1\cdot2}\wp_{1\cdot1,(2i-1)\cdot1} + \wp_{1\cdot1,(2i+1)\cdot1}\right)$$
$$- 2 \left(\wp_{3\cdot1,(2i-1)\cdot1} - \lambda_{2i+2}\delta_{i,1}\right). \tag{65}$$

Here and below, $\delta_{i,k}$ is the Kronecker symbol, $\deg \delta_{i,k} = 0$.

$$\wp_{1\cdot2,(2i-1)\cdot1}\wp_{1\cdot2,(2k-1)\cdot1} = 4 \left(\wp_{1\cdot2}\wp_{1\cdot1,(2i-1)\cdot1}\wp_{1\cdot1,(2k-1)\cdot1}\right.$$
$$+ \wp_{1\cdot1,(2k-1)\cdot1}\wp_{1\cdot1,(2i+1)\cdot1}$$
$$+ \wp_{1\cdot1,(2i-1)\cdot1}\wp_{1\cdot1,(2k+1)\cdot1} + \wp_{(2k+1)\cdot1,(2i+1)\cdot1}\right)$$
$$- 2 \left(\wp_{1\cdot1,(2i-1)\cdot1}\wp_{3\cdot1,(2k-1)\cdot1}\right.$$
$$+ \wp_{1\cdot1,(2k-1)\cdot1}\wp_{3\cdot1,(2i-1)\cdot1} + \wp_{(2k-1)\cdot1,(2i+3)\cdot1}$$
$$+ \wp_{(2i-1)\cdot1,(2k+3)\cdot1}\right) + 2 \left(\lambda_{2i+2}\wp_{1\cdot1,(2k-1)\cdot1}\delta_{i,1}\right.$$
$$+ \lambda_{2k+2}\wp_{1\cdot1,(2i-1)\cdot1}\delta_{k,1})$$
$$+ 2\lambda_{2(i+j+1)}(2\delta_{i,k} + \delta_{k,i-1} + \delta_{i,k-1}). \tag{66}$$

Corollary 3.1 *For all $g \geqslant 1$, we have the formulae:*

1. *Setting $i = 1$ in (65), we obtain*

$$\wp_{1\cdot4} = 6\wp_{1\cdot2}^2 + 4\wp_{1\cdot1,3\cdot1} + 2\lambda_4. \tag{67}$$

2. *Setting* $i = 2$ *in (65), we obtain*

$$\wp_{1\cdot3,3\cdot1} = 6(\wp_{1\cdot2}\wp_{1\cdot1,3\cdot1} + \wp_{1\cdot1,5\cdot1}) - 2\wp_{3\cdot2}. \tag{68}$$

3. *Setting* $i = k = 1$ *in (66), we obtain*

$$\wp_{1\cdot3}^2 = 4\left[\wp_{1\cdot2}^3 + (\wp_{1\cdot1,3\cdot1} + \lambda_4)\wp_{1\cdot2} + (\wp_{3\cdot2} - \wp_{1\cdot1,5\cdot1} + \lambda_6)\right]. \tag{69}$$

Theorem 3.2   1. *For any* $\omega = ((2k_1 - 1) \cdot j_1, \ldots, (2k_s - 1) \cdot j_s)$ *the hyperelliptic function* $\wp_\omega(u; \lambda)$ *is a polynomial from* $3g$ *functions* $\wp_{1\cdot j,(2k-1)\cdot1}$, $1 \leqslant j \leqslant 3$, $1 \leqslant k \leqslant g$.
Note that if $k = 1$, *we have* $\wp_{1\cdot j,1\cdot1} = \wp_{1\cdot(j+1)}$.
2. *Set* $W_\wp = \{\wp_{1\cdot j,(2k-1)\cdot1}, \ 1 \leqslant j \leqslant 3, \ 1 \leqslant k \leqslant g\}$. *The projection of the universal bundle* $\pi_g \colon \mathscr{U}_g \to \mathscr{B}_g \subset \mathbb{C}^{2g}$ *is given by the polynomials* $\lambda_{2k}(W_\wp)$, $k = 2, \ldots, 2g + 1$ *of degree at most 3 from the functions* $\wp_{1\cdot j,(2k-1)\cdot1}$.

The proof method of Theorem 3.2 will be demonstrated on the following examples:

Example 3.3   1. *Differentiating the relation (67) with respect to* u_1, *we obtain*

$$\wp_{1\cdot5} = 12\wp_{1\cdot2}\wp_{1\cdot3} + 4\wp_{1\cdot2,3\cdot1}. \tag{70}$$

2. *According to formula (67), we obtain*

$$2\lambda_4 = \wp_{1\cdot4} - 6\wp_{1\cdot2}^2 - 4\wp_{1\cdot1,3\cdot1}. \tag{71}$$

3. *According to formula (68), we obtain*

$$2\wp_{3\cdot2} = 6(\wp_{1\cdot2}\wp_{1\cdot1,3\cdot1} + \wp_{1\cdot1,5\cdot1}) - \wp_{1\cdot3,3\cdot1}. \tag{72}$$

4. *Substituting expressions for* λ_4 *(see (71)) and* $\wp_{3\cdot2}$ *(see (72)) into formula (69), we obtain an expression for the polynomial* λ_6.

The derivation of formulae (70)–(72) and the method of obtaining the polynomial λ_6 demonstrate the method of proving Theorem 3.2. Below, this method will be set out in detail in cases $g = 1$ (see Section 3.3) and $g = 2$ (see Section 3.4).

Corollary 3.4 *The operator* L *of differentiation with respect to* $u = (u_1, \ldots, u_{2g-1})$ *and* $\lambda = (\lambda_1, \ldots, \lambda_{4g+2})$ *is a derivation of the ring* $\mathscr{P}$ *if and only if* $L\wp_{1\cdot1,(2k-1)\cdot1} \in \mathscr{P}$ *for* $k = 1, \ldots, g$.

Proof. According to part 1 of Theorem 3.2, it suffices to prove that $L\wp_{1\cdot j,(2k-1)\cdot1} \in \mathscr{P}$ for $j = 2$ and 3, $k = 1, \ldots, g$. We have $L\wp_{1\cdot j,(2k-1)\cdot1} = LL_1\wp_{1\cdot(j-1),(2k-1)\cdot1} = (L_1 L + [L, L_1])\wp_{1\cdot(j-1),(2k-1)\cdot1}$. Using now that

$[L, L_1] \in \mathscr{L}^*$ and the assumption of Theorem 3.2, we complete the proof by induction. $\qquad\qquad\qquad\qquad\qquad\qquad\qquad\qquad\qquad\qquad\qquad\qquad\quad$ $\square$

Set $\mathscr{A} = \mathbb{C}[X]$, where $X = \{x_{i,2j-1}, \ 1 \leqslant i \leqslant 3, \ 1 \leqslant j \leqslant g\}$, $\deg x_{i,2j-1} = i + 2j - 1$.

Corollary 3.5 1. *The birational isomorphism* $J: \mathscr{U}_g \to \mathbb{C}^{3g}$ *is given by the polynomial isomorphism*

$$J^*: \mathscr{A} \longrightarrow \mathscr{P} \ : \ J^*X = W_{\wp}.$$

2. *There is a polynomial map*

$$p: \mathbb{C}^{3g} \longrightarrow \mathbb{C}^{2g}, \quad p(X) = \lambda,$$

such that

$$p^*\lambda_{2k} = \lambda_{2k}(X), \ k = 2, \ldots, 2g+1,$$

where $\lambda_{2k}(X)$ *are the polynomials from Theorem 3.2, item 2, obtained by substituting* $W_{\wp} \longmapsto X$.

The isomorphism J^* defines the Lie $\mathscr{A}$-algebra $\mathscr{L} = \mathscr{L}_g$ with $3g$ generators L_{2k-1}, $k = 1, \ldots, g$ and L_{2l}, $l = 0, \ldots, 2g - 1$. In terms of the coordinates $x_{i,2j-1}$, we obtain the following description of the g-th stationary flow of KdV system.

Theorem 3.6 1. *The commuting operators* L_{2k-1}, $k = 1, \ldots, g$, *define on* $\mathbb{C}^{3g}$ *a polynomial dynamical system*

$$L_{2k-1}X = G_{2k-1}(X), \ k = 1, \ldots, g, \tag{73}$$

where $G_{2k-1}(X) = \{G_{2k-1,i,2j-1}(X)\}$ *and* $G_{2k-1,i,2j-1}(X)$ *is a polynomial that uniquely defines the expression for the function* $\wp_{1\cdot i,(2j-1)\cdot 1,(2k-1)\cdot 1}$ *in the form of a polynomial from the functions* $\wp_{1\cdot i,(2q-1)\cdot 1}$.

2. *System (73) has 2g polynomial integrals* $\lambda_{2k} = \lambda_{2k}(X), \ k = 2, \ldots, 2g+1.$

3.3 Elliptic Functions

Consider the curve

$$V_\lambda = \{(x, y) \in \mathbb{C}^2 \ : \ y^2 = x^3 + \lambda_4 x + \lambda_6\}.$$

The discriminant of the family of curves V_λ is

$$\Delta = \{\lambda = (\lambda_4, \lambda_6) \in \mathbb{C}^2 \ : \ 4\lambda_4^3 + 27\lambda_6^2 = 0\}.$$

We have the universal bundle $\pi : \mathcal{U}_1 \to \mathcal{B}_1 = \mathbb{C}^2 \setminus \Delta$ and the mapping

$$\varphi : \mathcal{B}_1 \times \mathbb{C} \to \mathcal{U}_1 \; : \; \lambda \times \mathbb{C} \to \mathbb{C}/\Gamma_1(\lambda).$$

Consider the field $F = F_1$ of functions on $\mathcal{U}_1$ such that the function $\varphi^*(f)$ is meromorphic, and its restriction to the fiber $\mathbb{C}/\Gamma_1(\lambda)$ is an elliptic function for any point $\lambda \in \mathcal{B}_1$. Using the Weierstrass sigma function $\sigma(u; \lambda)$ for $\partial = \frac{\partial}{\partial u}$, we obtain

$$\zeta(u) = \partial \ln \sigma(u; \lambda) \quad \text{and} \quad \wp(u; \lambda) = -\partial \zeta(u; \lambda).$$

The ring of polynomials $\mathscr{P} = \mathscr{P}_1$ in F is generated by the elliptic functions $\wp_{1 \cdot i}$, $i \geqslant 2$. Set $\wp_{1 \cdot i} = \wp_i$. We have $\wp_2 = \wp$ and $\wp_{i+1} = \partial \wp_i = \wp_i'$. All the algebraic relations between the functions $\wp_i$ follow from the relations

$$\wp_4 = 6\wp_2^2 + 2\lambda_4 \quad \text{(see (65))}, \tag{74}$$

$$\wp_3^2 = 4[\wp_2^3 + \lambda_4 \wp_2 + \lambda_6] \quad \text{(see (66))}. \tag{75}$$

Thus, we obtain a classical result:

Theorem 3.7 1. *There is the isomorphism $\mathscr{P} \simeq \mathbb{C}[\wp, \wp', \wp'']$.*
2. *The projection $\pi : \mathcal{U}_1 \to \mathbb{C}^2$ is given by the polynomials*

$$\frac{1}{2}\wp'' - 3\wp^2 = \lambda_4, \tag{76}$$

$$\left(\frac{\wp'}{2}\right)^2 + 2\wp^3 - \frac{1}{2}\wp''\wp = \lambda_6. \tag{77}$$

Consider the linear space $\mathbb{C}^3$ with the graded coordinates x_2, x_3, x_4, $\deg x_k = k$. Set $\mathscr{A}_1 = \mathbb{C}[x_2, x_3, x_4]$.

Corollary 3.8 *The birational isomorphism $J : \mathcal{U}_1 \to \mathbb{C}^3$ is given by the ring isomorphism*

$$J^* : \mathscr{A}_1 \to \mathscr{P}_1 \; : \; J^*(x_2, x_3, x_4) = (\wp, \wp', \wp'').$$

Proof. The ring $\mathscr{P}_1$ is generated by elliptic functions $\wp_i$, $i \geqslant 2$, where $\wp_{i+1} = \wp_i'$. It follows from formula (76) that $\wp_5 = 12\wp_2\wp_3$. Hence, each function $\wp_i$ is a polynomial in $\wp_2$, $\wp_3$ and $\wp_4$ for all $i \geqslant 5$. $\qquad\square$

Corollary 3.9 1. *The operator $L_1 = \partial$ defines a polynomial dynamical system on $\mathbb{C}^3$*

$$x_2' = x_3, \quad x_3' = x_4, \quad x_4' = 12x_2x_3. \tag{78}$$

2. *The system (78) has 2 polynomial integrals*

$$\lambda_4 = \frac{1}{2}x_4 - 3x_2^2 \quad \text{and} \quad \lambda_6 = \frac{1}{4}x_3^2 + 2x_2^3 - \frac{1}{2}x_4x_2.$$

Let us consider the standard Weierstrass model of an elliptic curve

$$V_g = \{(x, y) \in \mathbb{C}^2 \ : \ y^2 = 4x^3 - g_2 x - g_3\}.$$

The discriminant of this curve has the form $\Delta(g_2, g_3) = g_2^3 - 27 g_3^2$. We have $V_\lambda = V_g$ where $g_2 = -4\lambda_4$ and $g_3 = -4\lambda_6$.

The elliptic sigma function $\sigma(u; \lambda)$ satisfies the system of equations

$$\ell_{2i}\, \sigma = H_{2i}\, \sigma, \ i = 0, 1, \tag{79}$$

where

$$\ell_0 = 4\lambda_4 \partial_{\lambda_4} + 6\lambda_6 \partial_{\lambda_6}; \quad H_0 = u\partial - 1;$$

$$\ell_2 = 6\lambda_6 \partial_{\lambda_4} - \frac{4}{3}\lambda_4^2 \partial_{\lambda_6}; \quad H_2 = \frac{1}{2}\partial^2 + \frac{1}{6}\lambda_4 u^2.$$

The operators ℓ_0, ℓ_2 and H_0, H_2, characterizing the sigma function $\sigma(u; g_2, g_3)$ of the curve V_g, were constructed in the work of Weierstrass [66]. The operators $L_i \in Der(F_1)$, $i = 0, 1$ and 2, were first found by Frobenius and Stickelberger (see [44]). Below, following work [22], we present the construction of the operators L_0 and L_2 on the basis of equations (79).

Let us construct the linear differential operators $\widehat{H}_{2i}$, $i = 0, 1$, of first order, such that $L_{2i} = \ell_{2i} - \widehat{H}_{2i}$, $i = 0, 1$, are the differentiations of the ring $\mathscr{P}_1 = \mathbb{C}[\wp, \wp', \wp'']$.

1. The formula for L_0.

We have $\ell_0 \sigma = (u\partial - 1)\sigma$. Therefore, $\ell_0 \ln \sigma = u\zeta(u) - 1$. Applying the operators ∂, ∂^2 and using the fact that operators ∂, ℓ_0 commute, we obtain:

$$\ell_0 \zeta = \zeta - u\wp, \quad \ell_0 \wp = 2\wp + u\partial\wp.$$

Setting $\widehat{H}_0 = u\partial$, we obtain $L_0 = \ell_0 - u\partial$. Consequently

$$L_0 \zeta = \zeta, \quad L_0 \wp = 2\wp.$$

2. The formula for L_2.

We have $\ell_2 \sigma = \frac{1}{2}\partial^2 \sigma - \frac{1}{6}\lambda_4 u^2 \sigma$. Therefore $\ell_2 \ln \sigma = \frac{1}{2}\frac{\partial^2 \sigma}{\sigma} - \frac{1}{6}\lambda_4 u^2$. We have $\frac{\partial^2 \sigma}{\sigma} = -\wp_2 + \zeta^2$. Thus

$$\ell_2 \ln \sigma = -\frac{1}{2}\wp_2 + \frac{1}{2}\zeta^2 - \frac{1}{6}\lambda_4 u^2. \tag{80}$$

Applying the operators ∂ and ∂^2 to (80), we obtain

$$\ell_2 \zeta = -\frac{1}{2}\wp_3 + \zeta\partial\zeta - \frac{1}{3}\lambda_4 u, \quad -\ell_2\wp_2 = -\frac{1}{2}\wp_4 + \wp_2^2 - \zeta\partial\wp_2 - \frac{1}{3}\lambda_4.$$

Setting $\widehat{H}_2 = \zeta\partial$, we obtain $L_2 = \ell_2 - \zeta\partial$. Consequently,

$$L_2 \zeta = -\frac{1}{2}\wp_3 - \frac{1}{3}\lambda_4 u, \quad L_2\wp_2 = \frac{1}{2}\wp_4 - \wp_2^2 + \frac{1}{3}\lambda_4 = \frac{2}{3}\wp_4 - 2\wp_2^2.$$

Thus, we get the following result:

Theorem 3.10 *The Lie $\mathscr{P}_1$-algebra $\mathscr{L}_1$ is generated by operators L_0, L_1 and L_2 such that*

$$[L_0,\ L_k] = kL_k,\ k = 1, 2, \qquad [L_1,\ L_2] = \wp_2 L_1, \tag{81}$$

$$L_0\wp_2 = 2\wp_2; \quad L_1\wp_2 = \wp_3; \quad L_2\wp_2 = \frac{2}{3}\wp_4 - 2\wp_2^2. \tag{82}$$

Proof. Formulas (81)–(82) completely determine the actions of the operators L_k, $k = 0, 1, 2$, on the ring $\mathscr{P}_1$ by the following inductive formula:

$$L_k\wp_{i+1} = [L_k,\ L_1]\wp_i + L_1 L_k\wp_i. \tag{83}$$

$\square$

Example 3.11 *Substituting $k = 2$ and $i = 2$ in (83), we obtain*

$$L_2\wp_3 = [L_2,\ L_1]\wp_2 + L_1 L_2\wp_2 = -5\wp_2\wp_3 + \frac{4}{3}\wp_5.$$

3.4 Hyperelliptic Functions of Genus $g = 2$

For each curve with affine part of the form

$$V_\lambda = \left\{ (x, y) \in \mathbb{C}^2 \,|\, y^2 = x^5 + \lambda_4 x^3 + \lambda_6 x^2 + \lambda_8 x + \lambda_{10} \right\},$$

one can construct a sigma-function $\sigma(u; \lambda)$ (see [14]). This function is an entire function in $u = (u_1, u_3) \in \mathbb{C}^2$ with parameters $\lambda = (\lambda_4, \lambda_6, \lambda_8, \lambda_{10}) \in \mathbb{C}^4$. It has a series expansion in u over the polynomial ring $\mathbb{Q}[\lambda_4, \lambda_6, \lambda_8, \lambda_{10}]$ in the vicinity of 0. The initial segment of the expansion has the form

$$\sigma(u; \lambda) = u_3 - \frac{1}{3} u_1^3 + \frac{1}{6} \lambda_6 u_3^3 - \frac{1}{12} \lambda_4 u_1^4 u_3 - \frac{1}{6} \lambda_6 u_1^3 u_3^2 -$$
$$- \frac{1}{6} \lambda_8 u_1^2 u_3^3 - \frac{1}{3} \lambda_{10} u_1 u_3^4 + \left(\frac{1}{60} \lambda_4 \lambda_8 + \frac{1}{120} \lambda_6^2 \right) u_3^5 + (u^7). \tag{84}$$

Here (u^k) denotes the ideal generated by monomials $u_1^i u_3^j$, $i + j = k$.

The sigma-function is an odd function in u, i.e. $\sigma(-u; \lambda) = -\sigma(u; \lambda)$.

Set

$$\nabla_\lambda = \left(\frac{\partial}{\partial \lambda_4}, \frac{\partial}{\partial \lambda_6}, \frac{\partial}{\partial \lambda_8}, \frac{\partial}{\partial \lambda_{10}} \right) \quad \text{and} \quad \partial_{u_1} = \frac{\partial}{\partial u_1}, \ \partial_{u_3} = \frac{\partial}{\partial u_3}.$$

We need the following properties of the two-dimensional sigma-function (see [15, 20] for details):

1. The following system of equations holds:

$$\ell_i \sigma = H_i \sigma, \quad i = 0, 2, 4, 6, \tag{85}$$

where $(\ell_0 \; \ell_2 \; \ell_4 \; \ell_6)^\top = T \, \nabla_\lambda$,

$$
T = \begin{pmatrix}
4\lambda_4 & 6\lambda_6 & 8\lambda_8 & 10\lambda_{10} \\
6\lambda_6 & 8\lambda_8 - \frac{12}{5}\lambda_4^2 & 10\lambda_{10} - \frac{8}{5}\lambda_4\lambda_6 & -\frac{4}{5}\lambda_4\lambda_8 \\
8\lambda_8 & 10\lambda_{10} - \frac{8}{5}\lambda_4\lambda_6 & 4\lambda_4\lambda_8 - \frac{12}{5}\lambda_6^2 & 6\lambda_4\lambda_{10} - \frac{6}{5}\lambda_6\lambda_8 \\
10\lambda_{10} & -\frac{4}{5}\lambda_4\lambda_8 & 6\lambda_4\lambda_{10} - \frac{6}{5}\lambda_6\lambda_8 & 4\lambda_6\lambda_{10} - \frac{8}{5}\lambda_8^2
\end{pmatrix}
$$

and

$$
H_0 = u_1 \partial_{u_1} + 3u_3 \partial_{u_3} - 3,
$$

$$
H_2 = \frac{1}{2} \partial_{u_1}^2 - \frac{4}{5}\lambda_4 u_3 \partial_{u_1} + u_1 \partial_{u_3} - \frac{3}{10}\lambda_4 u_1^2 + \frac{1}{10}(15\lambda_8 - 4\lambda_4^2)u_3^2,
$$

$$
H_4 = \partial_{u_1}\partial_{u_3} - \frac{6}{5}\lambda_6 u_3 \partial_{u_1} + \lambda_4 u_3 \partial_{u_3} - \frac{1}{5}\lambda_6 u_1^2 + \lambda_8 u_1 u_3
$$
$$
\qquad + \frac{1}{10}(30\lambda_{10} - 6\lambda_6\lambda_4)u_3^2 - \lambda_4,
$$

$$
H_6 = \frac{1}{2} \partial_{u_3}^2 - \frac{3}{5}\lambda_8 u_3 \partial_{u_1} - \frac{1}{10}\lambda_8 u_1^2 + 2\lambda_{10}u_1 u_3 - \frac{3}{10}\lambda_8\lambda_4 u_3^2 - \frac{1}{2}\lambda_6.
$$

2. The equation $\ell_0 \sigma = H_0\sigma$ implies that σ is a homogeneous function of degree -3 in u_1, u_3, λ_j.

3. The discriminant of the hyperelliptic curve V_λ of genus 2 is equal to $\Delta = \frac{16}{5} \det T$. It is a homogeneous polynomial in λ of degree 40. Set $\mathscr{B} = \{\lambda \in \mathbb{C}^4 : \Delta(\lambda) \neq 0\}$; then the curve V_λ is smooth for $\lambda \in \mathscr{B}$.

We have

$$
\ell_0 \Delta = 40\Delta, \quad \ell_2 \Delta = 0, \quad \ell_4 \Delta = 12\lambda_4\Delta, \quad \ell_6 \Delta = 4\lambda_6\Delta.
$$

Thus, the fields ℓ_0, ℓ_2, ℓ_4 and ℓ_6 are tangent to the variety $\{\lambda \in \mathbb{C}^4 : \Delta(\lambda) = 0\}$.

The present study is based on the following results.

Theorem 3.12 (uniqueness conditions for the two-dimensional sigma-function) *The entire function $\sigma(u; \lambda)$ is uniquely determined by the system of equations (85) and initial condition $\sigma(u; 0) = u_3 - \frac{1}{3}u_1^3$.*

We have the universal bundle $\pi : \mathscr{U}_2 \to \mathscr{B}_2 = \mathbb{C}^4 \setminus \mathscr{D}$ and the mapping

$$
\varphi : \mathscr{B}_2 \times \mathbb{C}^2 \to \mathscr{U}_2 : \lambda \times \mathbb{C}^2 \to \mathbb{C}^2 / \Gamma_2(\lambda).
$$

Consider the field $F = F_2$ of functions on $\mathscr{U}_2$ such that the function $\varphi^*(f)$ is meromorphic, and its restriction to the fiber $\mathbb{C}^2 / \Gamma_2(\lambda)$ is an hyperelliptic function for any point $\lambda \in \mathscr{B}_2$.

All the algebraic relations between the hyperelliptic functions of genus 2 follow from the relations, which in our notations have the form:

$$\wp_{1\cdot4} = 6\wp_{1\cdot2}^2 + 4\wp_{1\cdot1,3\cdot1} + 2\lambda_4, \tag{86}$$

$$\wp_{1\cdot3,3\cdot1} = 6\wp_{1\cdot2}\wp_{1\cdot1,3\cdot1} - 2\wp_{3\cdot2}, \tag{87}$$

(see (65) for $i = 1$ and $i = 2$) and

$$\wp_{1\cdot3}^2 = 4\left[\wp_{1\cdot2}^3 + (\wp_{1\cdot1,3\cdot1} + \lambda_4)\wp_{1\cdot2} + \wp_{3\cdot2} + \lambda_6\right], \tag{88}$$

$$\wp_{1\cdot3}\wp_{1\cdot2,3\cdot1} = 4\wp_{1\cdot2}^2\wp_{1\cdot1,3\cdot1} + 2\wp_{1\cdot1,3\cdot1}^2 - 2\wp_{1\cdot2}^2\wp_{3\cdot2} + 2\lambda_4\wp_{1\cdot1,3\cdot1} + 2\lambda_8, \tag{89}$$

$$\wp_{1\cdot2,3\cdot1}^2 = 4(\wp_{1\cdot2}\wp_{1\cdot1,3\cdot1}^2 - \wp_{1\cdot1,3\cdot1}\wp_{3\cdot2} + \lambda_{10}) \tag{90}$$

(see (66) for $(i, k) = (1, 1)$, $(1, 2)$ and $(2, 2)$).

Consider the linear space $\mathbb{C}^6$ with the graded coordinates $X = (x_2, x_3, x_4)$, $Y = (y_4, y_5, y_6)$, $\deg x_k = k$, $\deg y_k = k$. Set $\mathscr{A}_2 = \mathbb{C}[X, Y]$.

Theorem 3.13 1. *The birational isomorphism $J_2 \colon \mathcal{U}_2 \to \mathbb{C}^6$ is given by the isomorphism of polynomial rings*

$$J_2^* \colon \mathscr{A}_2 \longrightarrow \mathscr{P}_2 \; : \; J_2^* X = (\wp_{1\cdot2}, \wp_{1\cdot3}, \wp_{1\cdot4}),$$
$$J_2^* Y = (\wp_{1\cdot1,3\cdot1}, \wp_{1\cdot2,3\cdot1}, \wp_{1\cdot3,3\cdot1}).$$

2. *The projection $\pi_2 \colon \mathbb{C}^6 \to \mathbb{C}^4$ is given by the polynomials*

$$\lambda_4 = -3x_2^2 + \frac{1}{2}x_4 - 2y_4, \tag{91}$$

$$\lambda_6 = 2x_2^3 + \frac{1}{4}x_3^2 - \frac{1}{2}x_2x_4 - 2x_2y_4 + \frac{1}{2}y_6, \tag{92}$$

$$\lambda_8 = (4x_2^2 + y_4)y_4 - \frac{1}{2}(x_4y_4 - x_3y_5 + x_2y_6), \tag{93}$$

$$\lambda_{10} = 2x_2y_4^2 + \frac{1}{4}y_5^2 - \frac{1}{2}y_4y_6. \tag{94}$$

Proof. Using the isomorphism J_2^*, we rewrite the relations (86)–(90) in the form

$$x_4 = 6x_2^2 + 4y_4 + 2\lambda_4, \tag{95}$$

$$y_6 = 6x_2y_4 - 2\wp_{3\cdot2}, \tag{96}$$

$$x_3^2 = 4\left[x_2^3 + (y_4 + \lambda_4)x_2 + \wp_{3\cdot2} + \lambda_6\right], \tag{97}$$

$$x_3y_5 = 2\left[2x_2^2y_4 + y_4^2 - x_2^2\wp_{3\cdot2} + \lambda_4y_4 + \lambda_8\right], \tag{98}$$

$$y_5^2 = 4\left[x_2y_4^2 - y_4\wp_{3\cdot2} + \lambda_{10}\right]. \tag{99}$$

Directly from relations (95)–(99), we obtain the formula for the polynomial mapping π_2, that is, the proof of assertion 2 of the theorem.

Set $x_{i+1} = \wp_{1\cdot(i+1)}$, $y_{i+3} = \wp_{1\cdot i,3\cdot 1}$, $i \geqslant 1$. Applying the operator ∂_{u_1} to formula (95), we obtain

$$x_5 = 12x_2x_3 + 4y_5 = x_5(X, Y). \tag{100}$$

Substituting the expression for λ_4 from (95) and the expression for $\wp_{3\cdot 2}$ from (96) into the formula (97) and then applying the operator ∂_{u_1}, we obtain

$$y_7 = 4x_3y_4 + x_2(x_5 + 4y_5 - 12x_2x_3) = y_7(X, Y). \tag{101}$$

By induction from formulas (100) and (101), we obtain the polynomial formulas

$$x_{i+1} = x_{i+1}(X, Y), \qquad y_{i+3} = y_{i+3}(X, Y). \tag{102}$$

From formula (96) we obtain

$$\wp_{3\cdot 2} = 3x_2y_4 - \frac{1}{2}y_6 = z_6(X, Y), \qquad \wp_{3\cdot(i+2)} = \partial^i_{u_3}z_6(X, Y) = z_{3i+6}. \tag{103}$$

The following formulae complete the proof of assertion 1 of the theorem

$$\partial_{u_3}x_{i+1} = \partial_{u_1}\wp_{1\cdot i,3\cdot 1} = \partial_{u_1}y_{i+3}(X, Y), \tag{104}$$

$$\partial_{u_3}y_{i+3} = \partial_{u_3}\wp_{1\cdot i,3\cdot 1} = \wp_{1\cdot i,3\cdot 2} = \partial^i_{u_1}z_6(X, Y). \tag{105}$$

$\square$

In the course of the proof of Theorem 3.13, we obtained a detailed proof of Theorem 3.2 in the case $g = 2$.

Set $L_1 = \partial_{u_1}$ and $L_3 = \partial_{u_3}$. We introduce the operators $L_i \in \mathrm{Der}(F_2)$, $i = 0, 2, 4, 6$, based on the operators $\ell_i - H_i$.

Theorem 3.14 *The generators of the F_2-module $\mathrm{Der}(F_2)$ are given by the formulae*

$$L_{2k-1} = \partial_{u_{2k-1}}, \; k = 1, 2, \qquad L_{2k} = \ell_{2k} - \widehat{H}_{2k}, \; k = 0, 1, 2, 3,$$

where

$$\widehat{H}_0 = u_1\partial_{u_1} + 3u_3\partial_{u_3}, \qquad \widehat{H}_2 = \left(\zeta_1 - \frac{4}{5}\lambda_4u_3\right)\partial_{u_1} + u_1\partial_{u_3},$$

$$\widehat{H}_4 = \left(\zeta_3 - \frac{6}{5}\lambda_6u_3\right)\partial_{u_1} + (\zeta_1 + \lambda_4u_3)\partial_{u_3}, \qquad \widehat{H}_6 = -\frac{3}{5}\lambda_8u_3\partial_{u_1} + \zeta_3\partial_{u_3}.$$

Proof. We will use the methods of [22] to obtain the explicit form of operators L_i and to describe their action on the ring $\mathscr{P}_2$. Note here that this theorem corrects misprints made in [22, 23].

We have $L_1 = \partial_{u_1} \in \text{Der}(F_2)$ and $L_3 = \partial_{u_3} \in \text{Der}(F_2)$.

Below we use the fact that $[\partial_{u_k}, \ell_q] = 0$ for $k = 1, 3$ and $q = 0, 2, 4, 6$.

1). Derivation of the formula for L_0.

Using (85), we have $\ell_0 \sigma = H_0 \sigma = (u_1 \partial_{u_1} + 3u_3 \partial_{u_3} - 3)\sigma$. Therefore

$$\ell_0 \ln \sigma = u_1 \partial_{u_1} \ln \sigma + 3u_3 \partial_{u_3} \ln \sigma - 3. \tag{106}$$

Applying the operators ∂_{u_1} and ∂_{u_3} to (106), we obtain

$$\ell_0 \zeta_1 = \zeta_1 - u_1 \wp_{1 \cdot 2} - 3u_3 \wp_{1 \cdot 1, 3 \cdot 1}, \tag{107}$$

$$\ell_0 \zeta_3 = 3\zeta_3 - u_1 \wp_{1 \cdot 1, 3 \cdot 1} - 3u_3 \wp_{3 \cdot 2}. \tag{108}$$

We apply the operator ∂_{u_1} to (107) to obtain

$$-\ell_0 \wp_{1 \cdot 2} = -2\wp_{1 \cdot 2} - u_1 \wp_{1 \cdot 3} - 3u_3 \wp_{1 \cdot 2, 3 \cdot 1}.$$

Therefore

$$(\ell_0 - u_1 \partial_{u_1} - 3u_3 \partial_{u_3})\wp_{1 \cdot 2} = 2\wp_{1 \cdot 2}.$$

Applying the operator ∂_{u_1} to (108), we obtain

$$-\ell_0 \wp_{1 \cdot 1, 3 \cdot 1} = -\wp_{1 \cdot 1, 3 \cdot 1} - u_1 \wp_{1 \cdot 2, 3 \cdot 1} - 3\wp_{1 \cdot 1, 3 \cdot 1} - 3u_3 \wp_{1 \cdot 1, 3 \cdot 2}.$$

Therefore,

$$(\ell_0 - u_1 \partial_{u_1} - 3u_3 \partial_{u_3})\wp_{1 \cdot 1, 3 \cdot 1} = 4\wp_{1 \cdot 1, 3 \cdot 1}.$$

Thus, we have proved that

$$L_0 = \ell_0 - u_1 \partial_{u_1} - 3u_3 \partial_{u_3} \in \text{Der}(F_2).$$

2). Derivation of the formula for L_2.

Using (85), we have

$$\ell_2 \sigma = H_2 \sigma = \left(\frac{1}{2} \partial_{u_1}^2 - \frac{4}{5} \lambda_4 u_3 \partial_{u_1} + u_1 \partial_{u_3} + w_2 \right) \sigma$$

where

$$w_2 = w_2(u_1, u_3) = -\frac{3}{10} \lambda_4 u_1^2 + \frac{1}{10} (15\lambda_8 - 4\lambda_4^2) u_3^2.$$

Therefore

$$\ell_2 \ln \sigma = \frac{1}{2} \frac{\partial_{u_1}^2 \sigma}{\sigma} - \frac{4}{5} \lambda_4 u_3 \partial_{u_1} \ln \sigma + u_1 \partial_{u_3} \ln \sigma + w_2.$$

It holds that

$$\frac{\partial_{u_1}^2 \sigma}{\sigma} = -\wp_{1 \cdot 2, 0} + \zeta_1^2.$$

We get

$$\ell_2 \ln \sigma = -\frac{1}{2}\wp_{1\cdot 2} + \frac{1}{2}\zeta_1^2 - \frac{4}{5}\lambda_4 u_3 \zeta_1 + u_1 \zeta_3 + w_2. \tag{109}$$

Applying the operators ∂_{u_1} and ∂_{u_3} to (109), we obtain

$$\ell_2\zeta_1 = -\frac{1}{2}\wp_{1\cdot 3} - \zeta_1\wp_{1\cdot 2} + \frac{4}{5}\lambda_4 u_3\wp_{1\cdot 2} + \zeta_3 - u_1\wp_{1\cdot 1,3\cdot 1} + \partial_{u_1} w_2,$$

$$\ell_2\zeta_3 = -\frac{1}{2}\wp_{1\cdot 2,3\cdot 1} - \zeta_1\wp_{1\cdot 1,3\cdot 1} - \frac{4}{5}\lambda_4\zeta_1 + \frac{4}{5}\lambda_4 u_3\wp_{1\cdot 1,3\cdot 1} - u_1\wp_{3\cdot 2} + \partial_{u_3} w_2.$$

Applying the operator ∂_{u_1} again, we obtain

$$-\ell_2\wp_{1\cdot 2} = -\frac{1}{2}\wp_{1\cdot 4} + \wp_{1\cdot 2}^2 - \zeta_1\wp_{1\cdot 3} + \frac{4}{5}\lambda_4 u_3\wp_{1\cdot 3} - 2\wp_{1\cdot 1,3\cdot 1}$$
$$- u_1\wp_{1\cdot 2,3\cdot 1} + \partial_{u_1}^2 w_2,$$

$$-\ell_2\wp_{1\cdot 1,3\cdot 1} = -\frac{1}{2}\wp_{1\cdot 3,3\cdot 1} + \wp_{1\cdot 2}\wp_{1\cdot 1,3\cdot 1} - \zeta_1\wp_{1\cdot 2,3\cdot 1} + \frac{4}{5}\lambda_4\wp_{1\cdot 2}$$
$$+ \frac{4}{5}\lambda_4 u_3\wp_{1\cdot 2,3\cdot 1} - \wp_{3\cdot 2} - u_1\wp_{1\cdot 1,3\cdot 2} + \partial_{u_1}\partial_{u_3} w_2.$$

Thus, we have proved that

$$L_2 = \left(\ell_2 - \zeta_1\partial_{u_1} - u_1\partial_{u_3} + \frac{4}{5}\lambda_4 u_3\partial_{u_1}\right) \in \mathrm{Der}(F_2).$$

We have $\partial_{u_1}^2 w_2 = -\frac{3}{5}\lambda_4$ and $\partial_{u_1}\partial_{u_3} w_2 = 0$.

3). Derivation of the formula for L_4.

Using (85), we have

$$\ell_4\sigma = H_4\sigma = \left(\partial_{u_1}\partial_{u_3} - \frac{6}{5}\lambda_6 u_3\partial_{u_1} + \lambda_4 u_3\partial_{u_3} + w_4\right)\sigma$$

where

$$w_4 = -\frac{1}{5}\lambda_6 u_1^2 + \lambda_8 u_1 u_3 + \frac{1}{10}(30\lambda_{10} - 6\lambda_6\lambda_4)u_3^2 - \lambda_4.$$

Therefore,

$$\ell_4 \ln \sigma = \frac{\partial_{u_1}\partial_{u_3}\sigma}{\sigma} - \frac{6}{5}\lambda_6 u_3\partial_{u_1}\ln\sigma + \lambda_4 u_3\partial_{u_3}\ln\sigma + w_4.$$

It holds that

$$\frac{\partial_{u_1}\partial_{u_3}\sigma}{\sigma} = -\wp_{1\cdot 1,3\cdot 1} + \zeta_1\zeta_3.$$

We obtain

$$\ell_4 \ln \sigma = -\wp_{1\cdot 1,3\cdot 1} + \zeta_1\zeta_3 - \frac{6}{5}\lambda_6 u_3\zeta_1 + \lambda_4 u_3\zeta_3 + w_4. \tag{110}$$

Applying the operators ∂_{u_1} and ∂_{u_3} to (110), we obtain

$$\ell_4\zeta_1 = -\wp_{1\cdot2,3\cdot1} - \wp_{1\cdot2}\zeta_3 - \zeta_1\wp_{1\cdot1,3\cdot1} + \frac{6}{5}\lambda_6 u_3\wp_{1\cdot2} - \lambda_4 u_3\wp_{1\cdot1,3\cdot1} + \partial_{u_1}w_4,$$

$$\ell_4\zeta_3 = -\wp_{1\cdot1,3\cdot2} - \wp_{1\cdot1,3\cdot1}\zeta_3 - \zeta_1\wp_{3\cdot2} - \frac{6}{5}\lambda_6\zeta_1 + \frac{6}{5}\lambda_6 u_3\wp_{1\cdot1,3\cdot1}$$

$$+ \lambda_4\zeta_3 - \lambda_4 u_3\wp_{3\cdot2} + \partial_{u_3}w_4.$$

Applying the operator ∂_{u_1} again, we obtain

$$-\ell_4\wp_{1\cdot2} = -\wp_{1\cdot3,3\cdot1} - \wp_{1\cdot3}\zeta_3 + \wp_{1\cdot2}\wp_{1\cdot1,3\cdot1} + \wp_{1\cdot2}\wp_{1\cdot1,3\cdot1} - \zeta_1\wp_{1\cdot2,3\cdot1}$$

$$+ \frac{6}{5}\lambda_6 u_3\wp_{1\cdot3} - \lambda_4 u_3\wp_{1\cdot2,3\cdot1} + \partial_{u_1}^2 w_4,$$

$$-\ell_4\wp_{1\cdot1,3\cdot1} = -\wp_{1\cdot2,3\cdot2} - \wp_{1\cdot2,3\cdot1}\zeta_3 + \wp_{1\cdot1,3\cdot1}^2 + \wp_{1\cdot2}\wp_{3\cdot2} - \zeta_1\wp_{1\cdot1,3\cdot2}$$

$$+ \frac{6}{5}\lambda_6\wp_{1\cdot2} + \frac{6}{5}\lambda_6 u_3\wp_{1\cdot2,3\cdot1} - \lambda_4\wp_{1\cdot1,3\cdot1} - \lambda_4 u_3\wp_{1\cdot1,3\cdot2}$$

$$+ \partial_{u_1}\partial_{u_3} w_4.$$

Therefore, we have proved that

$$L_4 = \left(\ell_4 - \zeta_3\partial_{u_1} - \zeta_1\partial_{u_3} + \frac{6}{5}\lambda_6 u_3\partial_{u_1} - \lambda_4 u_3\partial_{u_3}\right) \in \text{Der}(F_2).$$

We have $\partial_{u_1}^2 w_4 = -\frac{2}{5}\lambda_6$ and $\partial_{u_1}\partial_{u_3} w_4 = \lambda_8$.

4). <u>Derivation of the formula for L_6.</u>

Using (85), we have

$$\ell_6\sigma = H_6\sigma = \left(\frac{1}{2}\partial_{u_3}^2 - \frac{3}{5}\lambda_8 u_3\partial_{u_1} + w_6\right)\sigma$$

where

$$w_6 = -\frac{1}{10}\lambda_8 u_1^2 + 2\lambda_{10}u_1 u_3 - \frac{3}{10}\lambda_8\lambda_4 u_3^2 - \frac{1}{2}\lambda_6.$$

Therefore,

$$\ell_6\ln\sigma = \frac{1}{2}\frac{\partial_{u_3}^2\sigma}{\sigma} - \frac{3}{5}\lambda_8 u_3\partial_{u_1}\ln\sigma + w_6.$$

We obtain

$$\ell_6\ln\sigma = -\frac{1}{2}\wp_{3\cdot2} + \frac{1}{2}\zeta_3^2 - \frac{3}{5}\lambda_8 u_3\zeta_1 + w_6. \tag{111}$$

Applying the operators ∂_{u_1} and ∂_{u_3} to (111), we obtain

$$\ell_6\zeta_1 = -\frac{1}{2}\wp_{1\cdot1,3\cdot2} - \zeta_3\wp_{1\cdot1,3\cdot1} + \frac{3}{5}\lambda_8 u_3\wp_{1\cdot2} + \partial_{u_1}w_6,$$

$$\ell_6\zeta_3 = -\frac{1}{2}\wp_{3\cdot3} - \zeta_3\wp_{3\cdot2} - \frac{3}{5}\lambda_8\zeta_1 + \frac{3}{5}\lambda_8 u_3\wp_{1\cdot1,3\cdot1} + \partial_{u_3}w_6.$$

Applying the operator ∂_{u_1} again, we obtain

$$-\ell_6\wp_{1\cdot2} = -\frac{1}{2}\wp_{1\cdot2,3\cdot2} + \wp^2_{1\cdot1,3\cdot1} - \zeta_3\wp_{1\cdot2,3\cdot1} + \frac{3}{5}\lambda_8 u_3\wp_{1\cdot3} + \partial^2_{u_1} w_6,$$

$$-\ell_6\wp_{1\cdot1,3\cdot1} = -\frac{1}{2}\wp_{1\cdot1,3\cdot3} + \wp_{1\cdot1,3\cdot1}\wp_{3\cdot2} - \zeta_3\wp_{1\cdot1,3\cdot2} + \frac{3}{5}\lambda_8\wp_{1\cdot2}$$

$$+ \frac{3}{5}\lambda_8 u_3\wp_{1\cdot2,3\cdot1} + \partial_{u_1}\partial_{u_3} w_6.$$

Therefore, we have proved that

$$L_6 = \left(\ell_6 - \zeta_3\partial_{u_3} + \frac{3}{5}\lambda_8 u_3\partial_{u_1}\right) \in \mathrm{Der}(F_2).$$

We have $\partial^2_{u_1} w_6 = -\frac{1}{5}\lambda_8$ and $\partial_{u_1}\partial_{u_3} w_6 = 2\lambda_{10}$. This completes the proof. $\square$

The description of commutation relations in the differential algebra of Abelian functions of genus 2 was given in [22, 23], see also [17]. We obtain this result directly from Theorem 3.14 and correct some misprints made in [22, 23]. To simplify the calculations, we use the following results:

Lemma 3.15 *The following commutation relations hold for* ℓ_k:

$$[\partial_{u_1}, \ell_k] = 0, \quad k = 0, 2, 4, 6, \quad [\partial_{u_3}, \ell_k] = 0, \quad k = 0, 2, 4, 6,$$

$$[\ell_0, \ell_k] = k\ell_k, \quad k = 2, 4, 6, \quad [\ell_2, \ell_4] = \frac{8}{5}\lambda_6\ell_0 - \frac{8}{5}\lambda_4\ell_2 + 2\ell_6,$$

$$[\ell_2, \ell_6] = \frac{4}{5}\lambda_8\ell_0 - \frac{4}{5}\lambda_4\ell_4, \quad [\ell_4, \ell_6] = -2\lambda_{10}\ell_0 + \frac{6}{5}\lambda_8\ell_2 - \frac{6}{5}\lambda_6\ell_4 + 2\lambda_4\ell_6.$$

Proof. These relations follow directly from (85). $\square$

Lemma 3.16 *The operators* L_i, $i = 0, 1, 2, 3, 4, 6$, *act on* $-\zeta_1$ *and* $-\zeta_3$ *according to the formulae*

$$L_0(-\zeta_1) = -\zeta_1, \qquad\qquad L_0(-\zeta_3) = -3\zeta_3,$$

$$L_1(-\zeta_1) = \wp_{1\cdot2}, \qquad\qquad L_1(-\zeta_3) = \wp_{1\cdot1,3\cdot1},$$

$$L_2(-\zeta_1) = \frac{1}{2}\wp_{1\cdot3} - \zeta_3 + \frac{3}{5}\lambda_4 u_1, \qquad L_2(-\zeta_3) = \frac{1}{2}\wp_{1\cdot2,3\cdot1} + \frac{4}{5}\lambda_4\zeta_1$$

$$+ \left(\frac{4}{5}\lambda_4^2 - 3\lambda_8\right)u_3,$$

$$L_3(-\zeta_1) = \wp_{1\cdot1,3\cdot1}, \qquad\qquad L_3(-\zeta_3) = \wp_{3\cdot2},$$

$$L_4(-\zeta_1) = \wp_{1\cdot2,3\cdot1} + \frac{2}{5}\lambda_6 u_1 - \lambda_8 u_3, \qquad L_4(-\zeta_3) = \wp_{1\cdot1,3\cdot2} + \frac{6}{5}\lambda_6 \zeta_1$$
$$- \lambda_4 \zeta_3 - \lambda_8 u_1 +$$
$$+ 6\left(\frac{1}{5}\lambda_4\lambda_6 - \lambda_{10}\right)u_3,$$

$$L_6(-\zeta_1) = \frac{1}{2}\wp_{1\cdot1,3\cdot2} + \frac{1}{5}\lambda_8 u_1 - 2\lambda_{10} u_3, \quad L_6(-\zeta_3) = \frac{1}{2}\wp_{3\cdot3} + \frac{3}{5}\lambda_8 \zeta_1$$
$$- 2\lambda_{10} u_1 + \frac{3}{5}\lambda_4\lambda_8 u_3.$$

Proof. For the operators L_1, L_3 this result follows from definitions. For the operators L_0, L_2, L_4 and L_6 this result follows from the proof of Theorem 3.14.   □

The following theorem completes the description of the action of generators of the Lie $\mathscr{P}_2$-algebra $\mathscr{L}_2$ on the ring of polynomials $\mathscr{P}_2$.

Theorem 3.17 *The operators* L_i, $i = 0, 1, 2, 3, 4, 6$ *act on* $\wp_{1\cdot2}$ *and* $\wp_{1\cdot1,3\cdot1}$ *according to the formulae*

$$L_0\wp_{1\cdot2} = 2\wp_{1\cdot2}, \qquad L_0\wp_{1\cdot1,3\cdot1} = 4\wp_{1\cdot1,3\cdot1},$$

$$L_1\wp_{1\cdot2} = \wp_{1\cdot3}, \qquad L_1\wp_{1\cdot1,3\cdot1} = \wp_{1\cdot2,3\cdot1},$$

$$L_2\wp_{1\cdot2} = \frac{1}{2}\wp_{1\cdot4} - \wp_{1\cdot2}^2 + 2\wp_{1\cdot1,3\cdot1} + \frac{3}{5}\lambda_4,$$

$$L_2\wp_{1\cdot1,3\cdot1} = \frac{1}{2}\wp_{1\cdot3,3\cdot1} - \wp_{1\cdot2}\wp_{1\cdot1,3\cdot1} - \frac{4}{5}\lambda_4\wp_{1\cdot2} + \wp_{3\cdot2},$$

$$L_4\wp_{1\cdot2} = \wp_{1\cdot3,3\cdot1} - 2\wp_{1\cdot2}\wp_{1\cdot1,3\cdot1} + \frac{2}{5}\lambda_6,$$

$$L_4\wp_{1\cdot1,3\cdot1} = \wp_{1\cdot2,3\cdot2} - \wp_{1\cdot1,3\cdot1}^2 - \wp_{1\cdot2}\wp_{3\cdot2} - \frac{6}{5}\lambda_6\wp_{1\cdot2} + \lambda_4\wp_{1\cdot1,3\cdot1} - \lambda_8,$$

$$L_6\wp_{1\cdot2} = \frac{1}{2}\wp_{1\cdot2,3\cdot2} - \wp_{1\cdot1,3\cdot1}^2 + \frac{1}{5}\lambda_8,$$

$$L_6\wp_{1\cdot1,3\cdot1} = \frac{1}{2}\wp_{1\cdot1,3\cdot3} - \wp_{1\cdot1,3\cdot1}\wp_{3\cdot2} - \frac{3}{5}\lambda_8\wp_{1\cdot2} - 2\lambda_{10}.$$

Note, that in these formulae the parameters λ_{2k}, $k = 4, \ldots, 10$, are considered as polynomials $\lambda_{2k}(W_\wp)$ (see 91–94).

Proof. See the derivation of the formulae for the operators L_{2k}, $k = 0, 2, 3, 4$, in the proof of Theorem 3.14.   □

The following result is based on the formulae of Theorem 3.14.

Theorem 3.18 *The commutation relations in the Lie F_2-algebra* $\mathrm{Der}(F_2)$ *of derivations of the field F_2 have the form*

$$[L_0, L_k] = kL_k, \quad k = 1, 2, 3, 4, 6; \qquad [L_1, L_2] = \wp_{1 \cdot 2} L_1 - L_3;$$

$$[L_1, L_3] = 0; \qquad\qquad\qquad\qquad [L_1, L_4] = \wp_{1 \cdot 1, 3 \cdot 1} L_1$$
$$\qquad\qquad\qquad\qquad\qquad\qquad\qquad + \wp_{1 \cdot 2} L_3;$$

$$[L_1, L_6] = \wp_{1 \cdot 1, 3 \cdot 1} L_3; \qquad\qquad [L_3, L_2] = \left(\wp_{1 \cdot 1, 3 \cdot 1} + \frac{4}{5}\lambda_4\right) L_1;$$

$$[L_3, L_4] = \left(\wp_{3 \cdot 2} + \frac{6}{5}\lambda_6\right) L_1 \qquad [L_3, L_6] = \frac{3}{5}\lambda_8 L_1 + \wp_{3 \cdot 2} L_3;$$
$$\qquad\qquad + \left(\wp_{1 \cdot 1, 3 \cdot 1} - \lambda_4\right) L_3;$$

$$[L_2, L_4] = \frac{8}{5}\lambda_6 L_0 - \frac{1}{2}\wp_{1 \cdot 2, 3 \cdot 1} L_1 - \frac{8}{5}\lambda_4 L_2 + \frac{1}{2}\wp_{1 \cdot 3} L_3 + 2L_6;$$

$$[L_2, L_6] = \frac{4}{5}\lambda_8 L_0 - \frac{1}{2}\wp_{1 \cdot 1, 3 \cdot 2} L_1 + \frac{1}{2}\wp_{1 \cdot 2, 3 \cdot 1} L_3 - \frac{4}{5}\lambda_4 L_4;$$

$$[L_4, L_6] = -2\lambda_{10} L_0 - \frac{1}{2}\wp_{3 \cdot 3} L_1 + \frac{6}{5}\lambda_8 L_2 + \frac{1}{2}\wp_{1 \cdot 1, 3 \cdot 2} L_3 - \frac{6}{5}\lambda_6 L_4 + 2\lambda_4 L_6.$$

Proof. Due to linearity, the relation $[L_0, L_k] = kL_k$, $k = 1, 2, 3, 4, 6$, can be checked independently for every summand in the expression for L_k.

The expressions for $[L_m, L_n]$, where m or n is equal to 1 or 3, can be obtained by simple calculations using Theorem 3.14.

It remains to prove the commutation relations among L_2, L_4 and L_6. We express $[L_m, L_n]$, where $m < n$ and $m, n = 2, 4, 6$, in the form

$$[L_m, L_n] = a_{m,n,0} L_0 + a_{m,n,-1} L_1 + a_{m,n,-2} L_2 + a_{m,n,-3} L_3$$
$$+ a_{m,n,-4} L_4 + a_{m,n,-6} L_6.$$

We have $\deg a_{i,j,-k} = i + j - k$. Applying both sides of this equation to λ_k and using the explicit expressions for L_k, we get

$$[\ell_m, \ell_n]\lambda_k = (a_{m,n,0}\ell_0 + a_{m,n,-2}\ell_2 + a_{m,n,-4}\ell_4 + a_{m,n,-6}\ell_6)\lambda_k.$$

This formula and Lemma 3.15 yield the values of the coefficients $a_{m,n,-k}$, $k = 0, 2, 4, 6$:

$$[L_2, L_4] = \frac{8}{5}\lambda_6 L_0 + a_{2,4,-1} L_1 - \frac{8}{5}\lambda_4 L_2 + a_{2,4,-3} L_3 + 2L_6; \qquad (112)$$

$$[L_2, L_6] = \frac{4}{5}\lambda_8 L_0 + a_{2,6,-1} L_1 + a_{2,6,-3} L_3 - \frac{4}{5}\lambda_4 L_4; \qquad (113)$$

$$[L_4, L_6] = -2\lambda_{10} L_0 + a_{4,6,-1} L_1 + \frac{6}{5}\lambda_8 L_2 + a_{4,6,-3} L_3 - \frac{6}{5}\lambda_6 L_4 + 2\lambda_4 L_6.$$
$$\qquad\qquad (114)$$

In subsequent calculations we compare the actions of the left- and right-hand sides of the expressions (112)–(114) on the coordinates u_1 and u_3. To this end we use the expressions (85), Theorem 3.14 and Lemma 3.16.

We present the calculation of the coefficient $a_{2,4,-1}$. The left-hand side of (112) gives

$$[L_2, L_4]u_1 = L_2(-\zeta_3 + \frac{6}{5}\lambda_6 u_3) - L_4(-\zeta_1 + \frac{4}{5}\lambda_4 u_3)$$

$$= L_2(-\zeta_3) + \frac{6}{5}\ell_2(\lambda_6)u_3 - \frac{6}{5}\lambda_6 u_1 - L_4(-\zeta_1) - \frac{4}{5}\ell_4(\lambda_4)u_3$$

$$- \frac{4}{5}\lambda_4(-\zeta_1 - \lambda_4 u_3)$$

$$= -\frac{1}{2}\wp_{1\cdot2,3\cdot1} + \frac{8}{5}\lambda_4\zeta_1 - \frac{8}{5}\lambda_6 u_1 + \frac{2}{5}\left(3\lambda_8 - \frac{16}{5}\lambda_4^2\right)u_3.$$

The right-hand side of (112) gives

$$[L_2, L_4]u_1 = a_{2,4,-1} + \frac{8}{5}\lambda_4\zeta_1 - \frac{8}{5}\lambda_6 u_1 + \frac{2}{5}\left(3\lambda_8 - \frac{16}{5}\lambda_4^2\right)u_3.$$

By equating them, we obtain $a_{2,4,-1} = -\frac{1}{2}\wp_{1\cdot2,3\cdot1}$.

The coefficients $a_{2,4,-3}$, $a_{2,6,-1}$, $a_{2,6,-3}$, $a_{4,6,-1}$ and $a_{4,6,-3}$ are calculated in a similar way. □

References

[1] T. Ayano and A. Nakayashiki, *On Addition Formulae for Sigma Functions of Telescopic Curves*, Symmetry, Integrability and Geometry: Methods and Applications (SIGMA) **9** (2013), 046, 14 pages.

[2] V. I. Arnold, *Singularities of Caustics and Wave Fronts*, Kluwer Academic Publishers Group, Dordrecht, 1990.

[3] H. F. Baker, *Abel's theorem and the allied theory of theta functions*, Cambridge University Press, Cambridge, (1897), Reprinted in 1995.

[4] ———, *On the hyperelliptic sigma functions*, Amer. Journ. Math. **20** (1898), 301–384.

[5] ———, *On a system of differential equations leading to periodic functions*, Acta Math. **27** (1903), 135–156.

[6] ———, *Multiply Periodic Functions*, Cambridge University Press, Cambridge, 1907.

[7] H. Bateman and A. Erdélyi, *Higher Transcendental Functions*, Vol. 3, New York, Toronto, London McGraw-Hill Book Company, 1955.

[8] J. Bernatska and D. Leykin, *On degenerate sigma-functions in genus two*, Glasgow Math. J. **60** (2018), arXiv:1509.01490.

[9] J. Bernatska, V. Enolski and A. Nakayashiki, *Sato Grassmannian and Degenerate Sigma Function*, (2018), arXiv: 1810.01224 [nlin.SI].

[10] E. D. Belokolos and V. Z. Enol'skii, *Reduction of Theta Functions and Elliptic Finite-Gap Potentials*, Acta Appl. Math. **36** (1994), 87–117.

[11] H. W. Braden, V. Z. Enolski and Yu N. Fedorov, *Dynamics on strata of trigonal Jacobians and some integrable problems of rigid body motion*, Nonlinearity **26** (2013), 1865–1889.

[12] O. Bolza, *Reduction hyperelliptischer Integrale erster Ordnung und erster Gattung auf elliptische insbesondere über die Reduction durch eine Transformation vierten Grades*, Inaugural-Dissertation; Georg-August Universit at zu Göttingen, 1885.

[13] O. Bolza, *Ueber die Reduction hyperelliptischer Integrale erster Ordnung und erster Gattung auf elliptische durch eine Transformation vierten Grades*, Math. Ann. **XXVIII** (1886), 447.

[14] V. M. Buchstaber, V. Z. Enolskii and D. V. Leikin, *Hyperelliptic Kleinian functions and applications*, "Solitons, Geometry and Topology: On the Crossroad", Adv. Math. Sci., AMS Transl., 179:2, Providence, RI, 1997, 1–33.

[15] V. M. Buchstaber, V. Z. Enolskii and D. V. Leykin, *Kleinian functions, hyperelliptic Jacobians and applications*, Rev. Math. and Math. Phys., vol. 10:2, Gordon and Breach, 1997, 3–120.

[16] V. M. Bukhshtaber, D. V. Leikin and V. Z. Enol'skii, *Rational analogues of Abelian functions*, Funkts. Anal. Prilozhen. **33**, No. 2, 1–15 (1999).

[17] V. M. Buchstaber, V. Z. Enolskii and D. V. Leikin, *Multi-Dimensional Sigma-Functions*, arXiv: 1208.0990, 2012, 267 pp.

[18] V. M. Buchstaber and D. V. Leykin, *Polynomial Lie algebras*, Functional Anal. Appl. **36**:4 (2002), 267–280.

[19] V. M. Buchstaber and D. V. Leykin, *Heat Equations in a Nonholonomic Frame*, Funct. Anal. Appl. **38**:2 (2004), 88–101.

[20] V. M. Buchstaber and D. V. Leikin, *Addition Laws on Jacobian Varieties of Plane Algebraic Curves*, Proc. Steklov Inst. Math. **251** (2005), 49–120.

[21] V. M. Buchstaber and D. V. Leikin, *Differentiation of Abelian functions with respect to parameters*, Russian Math. Surveys **62**:4 (2007), 787–789.

[22] V. M. Buchstaber and D. V. Leikin, *Solution of the Problem of Differentiation of Abelian Functions over Parameters for Families of (n, s)-Curves*, Funct. Anal. Appl. **42**:4 (2008), 268–278.

[23] V. M. Buchstaber, *Polynomial dynamical systems and the Kortewegde Vries equation*, Proc. Steklov Inst. Math. **294** (2016), 176–200.

[24] V. M. Buchstaber and A. V. Mikhailov, *The space of symmetric squares of hyperelliptic curves and integrable Hamiltonian polynomial systems on R^4*, arXiv: 1710.00866 v1 [nlin.SI] 2 Oct 2017, 24 pages.

[25] V. M. Buchstaber and A. V. Mikhailov, *Polynomial Hamiltonian integrable systems on symmetric powers of plane curves*, Russian Math. Surveys **73**:6(439) (2018), 1122–1124.

[26] E. Yu. Bunkova, *Differentiation of genus 3 hyperelliptic functions.*, European Journal of Mathematics, **4**:1 (2018), 93-112; arXiv: 1703.03947.

[27] B. A. Dubrovin and S. P. Novikov, *A periodic problem for the Korteweg-de Vries and Sturm-Liouville equations. Their connection with algebraic geometry.*, Dokl. Akad. Nauk SSSR **219**:3 (1974), 531–534.

[28] B. A. Dubrovin, *Theta functions and nonlinear equations*, Russ. Math. Surveys **36** (1981), 11–80.

[29] B. A. Dubrovin, *Geometry of 2D Topological Field Theories*, in Integrable Systems and Quantum Groups (Springer, Berlin, 1996), Lect. Notes in Math. v. 1620, 120–348; arXiv:hep-th/9407018.

[30] J. C. Eilbeck, K. Eilers and V. Z. Enolski, *Periods of second kind differentials of (n, s)-curves*, Trans. Moscow Math. Soc. **74**:2 (2013), 245–260.

[31] V. Z. Enolski and Yu. N. Fedorov, *Algebraic description of Jacobians isogeneous to certain Prym varieties with polarization (1,2)*, Experimental Mathematics **25**, 1–32, 2016, Preprint : arXiv: 1120.744 [nlin.SI], [math.AG] 22 Nov. 2014. 52 pp.

[32] V. Enolski, B. Hartmann Betti, V. Kagramanova, J.Kunz Jutta, C. Lämmerzahl and P. Sirimachan, *Inversion of a general hyperelliptic integral and particle motion in Hořava-Lifshitz black hole space-times*, J.Math.Phys. **53** (2012), 012504; arXiv:1106.2408v1 [gr-qc].

[33] K. Eilers, *Modular form representation for periods of hyperelliptic integrals*, SIGMA **12** (2016), 060, 13 pages.

[34] K. Eilers, *Rosenhain-Thomae formulae for higher genera hyperelliptic curves*, Journ. Nonlin. Math. Phys. **25** (2018), 85–105.

[35] J. C. Eilbeck, V. Z. Enolski, S. Matsutani, Y. Ônishi and E. Previato, *Abelian functions for trigonal curves of genus of genus three*, Int. Math. Res. Not. (IMRN) **2007** (2007): rnm 140–168.

[36] A. Eisenmann and H. M. Farkas, *An elementary proof of Thomaes formulae*, Online J. Anal. Comb., (3) Issue 3, (2008) 14p.

[37] V. Z. Enolski and T. Grava, *Thomae type formulae for singular Z_N curves*, Letters in Mathematical Physics (LMP) **76**:2 (2006), 187–214.

[38] V. Enolskii and P. Richter, *Periods of hyperelliptic integrals expressed in terms of θ-constants by means of Thomae formulae*, Phil. Trans. R. Soc. A **366** (2008), 1005–1024.

[39] B. Eynard, *Notes about a combinatorial expression of the fundamental second kind differential on an algebraic curve*, arXiv: 1805.07247 [math-ph], 2018.

[40] H. Farkas, S. Grushevsky and R. Salvati Manni. *An explicit solution to the weak Schottky problem*, https://arxiv.org/abs/1710.02938, 2017.

[41] J. D. Fay, *Theta functions on Riemann surfaces*, Lectures Notes in Mathematics (Berlin), vol. 352, Springer, 1973.

[42] J. Fay, *On the Riemann-Jacobi formula*, Nachr. Akad. Wiss. Gottingen Math.-Phys. Kl. II 1979, no. 5, 61–73.

[43] G. Frobenius, *Uber die constanten Factoren der Thetareihen*, J. Reine Angew. Math.**98** (1885), 244–265.

[44] F. G. Frobenius and L. Stickelberger, *Über die Differentiation der elliptischen Functionen nach den Perioden und Invarianten*, J. Reine Angew. Math. **92** (1882), 311–337.

[45] J.-I. Igusa, *On Jacobis derivative formula and its generalizations*, Amer. J. Math. **102** (1980), 409–446.

[46] A. R. Its and V. B. Matveev, *Schrödinger operators with finite-gap spectrum and N-soliton solutions of the Korteweg–de Vries equation*, TMF **23**(1) (1975), 51-68; Theoret. and Math. Phys. **23**:1 (1975), 343–355.

[47] P. G. Grinevich and A. Yu. Orlov, *Virasoro action on Riemann surfaces, Grassmanians,* det $\bar{\partial}$ *and Segal–Wilson τ-function*, in Problem of modern quantum field theory, Springer-Verlag, 1989, 86–106.

[48] F. Klein, *Ueber hyperelliptische Sigmafunctionen*, Math. Ann. **27** (1886), 431–464.

[49] F. Klein, *Ueber hyperelliptische Sigmafunctionen - Zweite Abhandlung*, Math. Ann. **32** (1888), 351–380.

[50] D. Korotkin and V. Shramchenko, *On higher genus Weierstrass sigma-functions*, Physica D (2012).

[51] I. M. Krichever, *Methods of algebraic geometry in the theory of nonlinear equations*, Russ. Math. Surv. **32** (1977), 185–213.

[52] V. Krishnamoorthy, T. Shaska and H. Vlklein, *Invariants of Binary Forms*, In Progress in Galois Theory, pp. 101–122, Dev. Math., 12, New York: Springer, 2005.

[53] S. Matsutani and E. Previato, *Jacobi inversion on strata of the Jacobian of the C_{rs} curve $y^r = f(x)$, I,II*, J. Math. Soc. Japan **60**, No. 4 (2008), 1009–1044. **66**, No. 2 (2014), 647–692.

[54] A. Nakayashiki, *Algebraic Expression of Sigma Functions of (n, s) Curves*, Asian J.Math. **14**:2 (2010), 174–211; arXiv:0803.2083, 2008.

[55] A. Nakayashiki and K. Yori, *Derivatives of Schur, tau and sigma functions, on Abel-Jacobi images*, In Symmetries, Integrable Systems and Representations, K.Iohara, S. Morier-Genoud, B. Remy (eds.), Springer, 2012, 429–462.

[56] A. Nakayashiki, *Tau function approach to theta functions*, Int. Math. Res. Not. (IMRN) **2016-17** (2016), 5202–5248.

[57] A. Nakayashiki, *Degeneration of trigonal curves and solutions of the KP-hierarchy*, Nonlinearity **31** (2018), 3567–3590.

[58] E. Nart and C. Ritzenthaler, *A new proof of a Thomae-like formula for non hyperelliptic genus 3 curves*, In Arithmetic, geometry, cryptography and coding theory, volume 686 of Contemporary Mathematics, pages 137–155. American Mathematical Society, 2017

[59] Y. Ônishi, *Determinant expressions for hyperelliptic functions, with an Appendix by Shigeki Matsutani: Connection of the formula of Cantor and Brioschi-Kiepert type*, Proc. Edinburgh Math. Soc. **48** (2005), 705–742.

[60] G. Rosenhain, *Abhandlung über die Functionen zweier Variablen mit fier Perioden welche die Inversion sind der ultra-elliptische Integrale erster Klasse*, Translation to German from Latin manuscript published in 1851. Ostwald Klassiker der Exacten Wissenschaften, Nr. **65**, pp 1–96, Leipzig, 1895, Verlag von Wilhelm Engelmann.

[61] J. Suzuki, *Kleins Fundamental 2-Form of Second Kind for the C_{ab} Curves*, SIGMA **13** (2017), 017, 13 pages.

[62] M. Schiffer and D. C Spencer, *Functionals of Finite Riemann Surfaces*, Paperback May 21, 2014; originally published as v.16 of the Princeton Mathematical Series, by Princeton University Press, Princeton, New Jersey, 1954.

[63] J. Thomae, *Beitrag zur Bestimmung (0, ..., 0) durch die Klassenmodulen algebraischer Funktionen*, Journ. reine angew. Math. **71** (1870), 201–222.

[64] H. Weber, *Anwendung der Thetafunctionen sweier Veräderlicher auf die Theorie der Bewegung eines festen Körpers in einer Flüssigkeit*, Math. Ann. **14** (1879), 173–206.

[65] K. Weierstrass, *Formeln und Lehrsätze der elliptiechen Functionen bearbeitet und herausgegeben von H.A.Schwarz*, Göttingen, 1885.

[66] K. Weierstraß, *Zur Theorie der elliptischen Funktionen*, Mathematische Werke, Vol. 2, Teubner, Berlin, 1894, 245–255.

V. M. Buchstaber
Steklov Mathematical Institute, Moscow
E-mail address: buchstab@mi-ras.ru

V. Z. Enolski
National University of Kyiv-Mohyla Academy, KIEV
E-mail address: venolski@gmail.com,enolsky@ukma.edu.ua

D. V. Leykin
Institute of Magnetism, NASU, UKRAINE
E-mail address: dmitry.leykin@gmail.com

7

Bergman Tau-Function: From Einstein Equations and Dubrovin-Frobenius Manifolds to Geometry of Moduli Spaces

Dmitry Korotkin

Dedicated to the 65th birthday of Emma Previato

Abstract. We review the role played by tau-functions of special type - called *Bergman* tau-functions in various areas: theory of isomonodromic deformations, solutions of Einstein's equations, theory of Dubrovin-Frobenius manifolds, geometry of moduli spaces and spectral theory of Riemann surfaces. These tau-functions are natural generalizations of Dedekind's eta-function to higher genus. Study of their properties allows to get an explicit form of Einstein's metrics, obtain new relations in Picard groups of various moduli spaces and derive holomorphic factorization formulas of determinants of Laplacians in flat singular metrics on Riemann surfaces, among other things.

1 Introduction

There exist several alternative definitions of tau-function of integrable systems. One of them is based on Hirota's equation [34]; the analytical theory of these tau-functions (which in particular include the tau-functions of the KP hierarchy) was essentially developed in the paper by Segal and Wilson [70]. Alternatively the tau-function can be defined as the partition function of some integrable quantum model. It was this definition which led Jimbo, Miwa and their collaborators to the notion of the isomonodromic tau-function of Schlesinger system [68] based on the theory of holonomic quantum fields. The divisor of zeros of the Jimbo-Miwa tau-function was then shown by Malgrange [61] to play the main role in the solvability of matrix Riemann-Hilbert problems.

To describe the class of tau-functions discussed in this paper we consider a Riemann surface $\mathcal{R}$ of genus g with some choice of canonical basis of cycles (the Torelli marking) and introduce the canonical bimeromorphic differential

215

on $\mathcal{R}$ by the formula $B(x, y) = d_x d_y \ln E(x, y)$ where $E(x, y)$ is the prime-form. This bidifferential has pole of second order on the diagonal with biresidue 1. Let v be some (holomorphic or meromorphic) abelian differential on $\mathcal{R}$. Using v one can regularize $B(x, y)$ near diagonal as follows:

$$B_{reg}^v(x, x) = \left(B(x, y) - \frac{v(x)v(y)}{(\int_x^y v)^2} \right) \bigg|_{y=x}.$$

Although Schlesinger systems are in general not explicitly solvable, they admit explicit solutions corresponding to special monodromy groups. The first class of examples of such explicit solutions arises in the theory of Dubrovin-Frobenius manifolds; corresponding monodromy groups don't have any continuous parameters. Most ingredients of Dubrovin-Frobenius manifold structures on Hurwitz spaces allow an explicit description [16, 17]. In partic- ular, their isomonodromic tau-function (which coincides with genus one free energy up to an auxiliary factor) satisfies the following system of equations with respect to critical values $z_j = f(x_j)$ (x_j is a critical point of f which we assume to be simple) of the meromorphic function f [48]:

$$\frac{\partial \ln \tau_B(\mathcal{R}, df)}{\partial z_j} = -\text{res}\bigg|_{x_k} \frac{B_{reg}^{df}(x, x)}{df(x)}. \tag{1}$$

We call τ_B the *Bergman tau-function* on the Hurwitz space since the main contribution to the right hand side of this equation is given by the *Bergman projective connection*. In the framework of conformal field theory $\tau_B(\mathcal{R}, df)$ has the meaning of chiral partition function of free bosons on a Riemann surface with flat metric of infinite volume $|df|^2$ [41, 71]. An explicit formula for τ_B was found in [44].

Another class of explicitly solvable Riemann-Hilbert problems corresponds to monodromy groups with quasi-permutation monodromies; such RH prob- lems were solved in [59] in terms of Szegö kernel on branched coverings of the Riemann sphere (the 2×2 case was treated earlier in [39, 14]). The corresponding Jimbo-Miwa tau-function is a product of three factors:

$$\tau_{JM} = \tau_0 \, \tau_B^{-1/2} \, \Theta \begin{bmatrix} \mathbf{p} \\ \mathbf{q} \end{bmatrix} (Q) \tag{2}$$

each of which has an independent meaning. The last factor is the theta-function with characteristics $\mathbf{p}, \mathbf{q} \in \mathbb{C}^g$; the Malgrange divisor is defined by vanishing of this theta-function. The factor τ_0 satisfies a system of equations

$$\frac{\partial \ln \tau_0}{\partial z_m} = \frac{1}{2} \text{res} \, |_{z=z_m} \sum_{j=1}^n \frac{W^2(z^{(j)})}{dz}; \qquad W(x) = \sum_{m=1}^M \sum_{j=1}^n r_m^{(j)} d_x \ln E(x, z_m^{(j)})$$

(here n is the degree of the function f, $r_m^{(j)}$ are constants which, together with $\mathbf{p}, \mathbf{q}$ define monodromy matrices; $z_m^{(j)}$ is the point on the jth sheet projecting to the branch point $z_m \in \mathbb{C}$) which resembles equations from Seiberg-Witten theory (this analogy was recently observed in [28]); for a special class of coverings this factor can be expressed via a product of prime-forms [28] as follows:

$$\tau_0^2 = \prod_{z_m^{(j)} \neq z_k^{(i)}} E(z_m^{(j)}, z_k^{(i)})^{r_m^{(j)} r_k^{(i)}}.$$

We give an alternative proof of this formula for another characteristic special case of Riemann-Hilbert problems in this paper. The vector Q from (2) is given by $Q = \sum_{j,m} r_m^{(j)} \mathcal{A}_{x_0}(z_m^{(j)})$ where $\mathcal{A}_{x_0}$ is the Abel map with basepoint x_0. The second factor in (2) is the $-1/2$ power of the Bergman tau-function (1).

The first applications of tau-functions (1), (2) we consider here are in the area of explicit solutions of Einstein's equations. The first case is self-dual Einstein manifolds with $SU(2)$ symmetry which are equivalent to a special case of 2×2 fuchsian Schlesinger system with four singularities. The metric on such manifolds is written in the form

$$g = F \left\{ d\mu^2 + \frac{\sigma_1^2}{W_1^2} + \frac{\sigma_2^2}{W_2^2} + \frac{\sigma_3^2}{W_3^2} \right\}$$

where the 1-forms σ_i satisfy $d\sigma_i = \sigma_j \wedge \sigma_k$ where (i, j, k) is any cyclic permutation of $(1, 2, 3)$; μ is the "euclidean time" variable and functions W_j and F depend only on μ. In the most non-trivial case of such manifolds the self-duality conditions are equivalent to the following system of equations:

$$\frac{dW_j}{d\mu} = -W_k W_l + 2W_j \frac{d}{d\mu} \ln(\vartheta_{k+1}\vartheta_{l+1}), \tag{3}$$

(where (i, j, k) is an arbitrary permutation of $(1, 23)$ and $\theta_j, j = 2, 3, 4$ are theta-constants with module $i\mu$) which is equivalent to a 4-point 2×2 Schlesinger system. The conformal factor F can be chosen such that the Einstein's equations (with cosmological constant Λ) hold for the metric g when an integral of motion of the system has a special value corresponding to an explicitly solvable case of (3). The remarkably simple explicit formulas for W_j can be found from the tau-function $\tau_{JM} = \theta[p, q]/(\theta_2\theta_4)$ (where $p, q \in \mathbb{C}$ satisfy appropriate reality conditions) [39]:

$$W_1 = -\frac{i}{2}\vartheta_3\vartheta_4 \frac{\frac{d}{dq}\vartheta[p, q + \frac{1}{2}]}{e^{\pi i p}\vartheta[p, q]}, \qquad W_2 = \frac{i}{2}\vartheta_2\vartheta_4 \frac{\frac{d}{dq}\vartheta[p + \frac{1}{2}, q + \frac{1}{2}]}{e^{\pi i p}\vartheta[p, q]},$$

$$W_3 = -\frac{1}{2}\vartheta_2\vartheta_3 \frac{\frac{d}{dq}\vartheta[p + \frac{1}{2}, q]}{\vartheta[p, q]}.$$

The corresponding conformal factor F is given by the formula

$$F = \frac{2}{\pi \Lambda} \frac{W_1 W_2 W_3}{\left(\frac{d}{dq} \ln \vartheta [p,q]\right)^2}.$$

Such metrics have remarkable modular properties which were exploited in various contexts (see [63, 23]).

Another appearance of the tau-function (2) is in the theory of stationary axially symmetric Einstein's equation (in this case the spacetime has physical signature $3 + 1$ and there is no cosmological constant). The non-trivial part of Einstein's equations is then encoded in the Ernst equation

$$(\mathcal{E} + \overline{\mathcal{E}})(\mathcal{E}_{\rho\rho} + \rho^{-1}\mathcal{E}_\rho + \mathcal{E}_{zz}) = 2(\mathcal{E}_z^2 + \mathcal{E}_\rho^2)$$

where $\mathcal{E}$ is a complex-valued function of variables z and ρ. The Ernst equation admits a class of solutions associated to hyperelliptic Riemann surfaces of the form

$$y^2 = (w - \xi)(w - \bar{\xi}) \prod_{j=1}^{2g} (w - w_j)$$

where two branch points $\xi = z + i\rho$ and $\bar{\xi}$ depend on space time variables. The corresponding solution of the Ernst equation depending on two constant vectors $\mathbf{p}, \mathbf{q} \in \mathbb{C}^g$ (which satisfy certain reality conditions) has the form [51]

$$\mathcal{E}(\xi, \bar{\xi}) = \frac{\Theta \begin{bmatrix} \mathbf{p} \\ \mathbf{q} \end{bmatrix} (\mathcal{A}(\infty^1) - \mathcal{A}(\xi))}{\Theta \begin{bmatrix} \mathbf{p} \\ \mathbf{q} \end{bmatrix} (\mathcal{A}(\infty^2) - \mathcal{A}(\xi))}.$$

Such solutions can be applied to solve various physically meaningful boundary value problems for Ernst equation [40].

For a given Ernst potential coefficients of the corresponding metric can be found in quadratures. However, having an explicit formula for the tau-function (2) in the hyperelliptic case one can find these coefficients explicitly, in particular, the so-called *conformal factor* is given by the formula [53]

$$e^{2k} = \frac{\Theta \begin{bmatrix} \mathbf{p} \\ \mathbf{q} \end{bmatrix} (0) \Theta \begin{bmatrix} \mathbf{p} \\ \mathbf{q} \end{bmatrix} (\frac{1}{2}\mathbf{e})}{\Theta(0)\Theta(\frac{1}{2}\mathbf{e})}$$

where $\mathbf{e} = (1, \ldots, 1)$

The key property of the Bergman tau-function which makes it useful in geometry of moduli spaces is its transformation law under the change of Torelli marking of C:

$$\tau_B(\mathcal{R}, v) \rightarrow \epsilon \det(C\Omega + D)\tau_B(\mathcal{R}, v) \tag{4}$$

where Ω is the period matrix of $\mathcal{R}$ and $\begin{pmatrix} C & D \\ B & A \end{pmatrix}$ is an $Sp(2g, \mathbb{Z})$ transformation of the canonical basis of cycles on $\mathcal{R}$. and where ϵ is a root of unity of degree which depends on multiplicities of zeros of the differential v. The property (4) allows to interpret the Bergman tau-function τ_B as a section of the determinant of the Hodge vector bundle (or a higher genus version of the Dedekind eta function) and can be used to study the geometry of moduli spaces.

The list of moduli spaces where this strategy was successfully applied is quite long. Studying analytical properties of Bergman tau-function on Hurwitz spaces and their compactifications, called spaces of admissible covers gives an expression for the Hodge class λ on the space of admissible covers of degree n and genus g [46]. An alternative algebro-geometric proof of this relation was obtained in [29] and further used to compute classes of certain divisors within $\mathcal{M}_g$ in [30].

Applying a similar strategy to the moduli space $\overline{\mathcal{H}}_g$ of stable abelian differentials on Deligne-Mumford stable Riemann surfaces of genus g we get the following relation in the rational Picard group $\mathrm{Pic}(P\overline{\mathcal{H}}_g) \otimes \mathbb{Q}$:

$$\lambda = \frac{g-1}{4}\varphi + \frac{1}{24}\delta_{\mathrm{deg}} + \frac{1}{12}\delta_0 + \frac{1}{8}\sum_{j=1}^{[g/2]}\delta_j$$

where λ is the Hodge class, φ is the first Chern class of the line bundle associated to the projection $\overline{\mathcal{H}}_g \to P\overline{\mathcal{H}}_g$, δ_{deg} is the class of the divisor of abelian differentials with multiple zeros and δ_j are classes of Deligne-Mumford boundary divisors.

Further generalization to moduli spaces $\mathcal{Q}_g$ of holomorphic quadratic differentials looks as follows. For a pair $(\mathcal{R}, Q)$ where $\mathcal{R}$ is a Riemann surface of genus g and Q is a holomorphic quadratic differential with simple zeros on $\mathcal{R}$ we define the canonical cover $\widehat{\mathcal{R}}$ by $v^2 = Q$; the genus of $\widehat{\mathcal{R}}$ equals $4g - 3$. Then we define two vector bundles over $\mathcal{Q}_g$: one is the Hodge vector bundle whose fiber has rank g; the fiber is the space of abelian differentials on $\widehat{\mathcal{R}}$ invariant under the natural involution (this fiber can be identified with the space of abelian differentials on $\mathcal{R}$). The second is the Prym vector bundle whose fiber has rank $3g - 3$; the fiber is the space of abelian differentials on $\widehat{\mathcal{R}}$ which are skew symmetric under the natural involution. The fiber of the Prym vector bundle turns out to be isomorphic to the space of holomorphic quadratic differentials on $\mathcal{R}$. One can define then two tau-functions: the *Hodge* tau-function $\tau_+(\mathcal{R}, Q)$ and the *Prym* tau-function $\tau_-(\mathcal{R}, Q)$; analysis of their analytical properties allows to express the Hodge class and the class λ_2 (the first Chern class of the vector bundle of quadratic differentials) via

ψ-classes, the class δ_{deg} of quadratic differentials with multiple zeros and the Deligne-Mumford boundary divisor classes. Further elimination of the classes ψ and δ_{deg} from these expressions leads to the famous Mumford's formula in the rational Picard group of $\overline{\mathcal{M}}_g$

$$\lambda_2 - 13\lambda = \delta_{DM}.$$

Generalizing this scheme to spaces of meromorphic quadratic differentials with n second order poles the formalism of Prym and Hodge tau-functions reproduces another classical relation which holds in the Picard group of $\overline{\mathcal{M}}_{g,n}$:

$$\lambda_2^{(n)} - 13\lambda + \sum_{j=1}^{n} \psi_j = -\delta_{DM}. \tag{5}$$

Here $\lambda_2^{(n)}$ is the first Chern class of the vector bundle of quadratic differentials with first order poles at the punctures; ψ_j is the class of line bundle whose fiber is the cotangent space to C at the jth marked point.

The tau-functions turn out to be a useful tool also in the real analytic context of combinatorial model of $\mathcal{M}_{g,n}$ based on Strebel differentials. This combinatorial model, denoted by $\mathcal{M}_{g,n}[\mathbf{p}]$, can be considered as a real slice of the moduli space $\mathcal{Q}_{g,n}[\mathbf{p}]$ of meromorphic quadratic differentials on $\mathcal{R}$ with second order poles at n marked points and biresidues given by $-p_j^2/4\pi^2$. The real slice is defined by the condition that all periods of v on the canonical cover $v^2 = Q$ are real. Such combinatorial model is a union of strata labelled by the set of multiplicities of zeros of Q; each stratum consists of several topologically trivial cells while the facets between the cells belong to strata of lower dimension.

Strata corresponding to odd multiplicities of all zeros of Q turn out to be cycles; they are named after Witten and Kontsevich. Combinations of these strata are known to be Poincaré dual to tautological classes [1, 65]. In real codimension 2 there are two such cycles: the Witten cycle W_5 whose largest stratum corresponds to differentials Q with one triple zero while other zeros are simple, and Kontsevich boundary $W_{1,1}$ whose largest stratum corresponds to quadratic differentials with two simple poles at points obtained by resolution of the node of a stable Riemann surface with one node. Arguments of Hodge and Prym tau-functions $\tau_{\pm}$ then give sections of certain circle bundles over cells in the largest stratum of $\mathcal{M}_{g,n}[\mathbf{p}]$ which can be continuously propagated from cell to cell but have monodromies around W_5 and $W_{1,1}$. Calculation of these monodromies leads to the following two relations [10]:

$$12\kappa_1 = W_5 + W_{1,1}$$

where κ_1 is the first kappa-class (this relation is an analog of the relation derived by Penner in the framework of combinatorial model based on hyperbolic geometry [67]) and

$$\lambda_2^{(n)} - 13\lambda - \sum_{i=1}^{n} \psi_i = -W_{1,1}$$

which is a combinatorial version of the complex analytic formula (5).

Tau-functions on moduli spaces of N-differentials were used in [55] to compute classes of determinants of Prym-Tyurin vector bundles over these moduli spaces. In [60] these results were applied to find the class of the universal Hitchin's discriminant in the Picard group of the universal moduli space of Hitchin's spectral covers.

The equation of a general $GL(n)$ spectral cover $\widehat{\mathcal{R}}$ over a Riemann surface $\mathcal{R}$ of genus g is given by

$$P_n(v) = v^n + Q_{n-1} v^{n-1} + \cdots + Q_1 v + Q_0 = 0 \qquad (6)$$

where Q_j is a holomorphic $(n - j)$ - differential over $\mathcal{R}$. The branch points of $\widehat{\mathcal{R}}$ are zeros of discriminant W of equation (6) which is a holomorphic $n(n-1)$-differential; thus the total number of branch points equals $n(n - 1)(2g - 2)$ and the genus of $\widehat{\mathcal{R}}$ equals $\hat{g} = n^2(g - 1) + 1$. Denote the moduli space of covers (6) when both the base $\mathcal{R}$ and coefficients Q_j are allowed to vary by $\mathcal{M}_H$; this is the universal space of Hitchin's spectral covers. We denote by $\mathcal{M}_H^{\mathcal{R}}$ the space of spectral covers for fixed base $\mathcal{R}$. The universal discriminant $D_H \subset \mathcal{M}_H$ consists of covers with coinciding branch points; (D_H consists of three components corresponding to different structures of ramification points corresponding to the double branch point). The class of divisor D_H in the space $P\mathcal{M}_H$ can be computed using the formula for the Hodge class on the moduli space of holomorphic $n(n - 1)$-differentials [55]:

$$\frac{1}{n(n-1)}[D_H] = (n^2 - n + 1)(12\lambda - \delta) - 2(g - 1)(2n^2 - 2n + 1)\varphi$$

where λ is (the pullback of) the Hodge class on $\overline{\mathcal{M}}_g$, φ is the tautological class of the line bundle arising from projectivization $\overline{\mathcal{M}}_H \rightarrow P\overline{\mathcal{M}}_H$, and δ is the (pullback to $P\mathcal{M}_H$) of the class of DM boundary of $\overline{\mathcal{M}}_g$.

We mention also an interesting fact about the space of spectral covers with fixed base $\mathcal{M}_H^{\mathcal{R}}$. The set of natural coordinates on $\mathcal{M}_H^{\mathcal{R}}$ is given by a-periods A_j of the abelian differential v on $\widehat{\mathcal{R}}$. The variation of the period matrix $\widehat{\Omega}$ of $\widehat{\mathcal{R}}$ with respect to these coordinates can be deduced from variational formulas (258) on the moduli space of abelian differentials on Riemann surfaces of genus $\hat{g}$. Applying the elementary chain rule to the formulas (258) one gets certain sum of residues over ramification points $\hat{x}_r$ of $\widehat{\mathcal{R}}$ corresponding to branch points $x_r \in \mathcal{R}$ [11] (assuming that all of these branch points are simple)

$$\frac{d\widehat{\Omega}_{lk}}{dA_j} = -2\pi i \sum_{ramification \ points \ \hat{x}_r} \mathrm{res}|_{\hat{x}_r} \frac{v_l v_k v_r}{d\xi \, d(v/d\xi)} \tag{7}$$

where ξ is an arbitrary local parameter on $\mathcal{R}$ near branch point x_r; $\{v_j\}_{j=1}^{\hat{g}}$ are normalized abelian differentials on $\widehat{\mathcal{R}}$. The formula (7) provides an explicit coordinate realization of the Donagi-Markman cubic [15]

The Bergman tau-functions have also an application to the spectral theory of Riemann surfaces with flat singular metrics and special holonomy: they allow to calculate the determinant of Laplace operator in such metrics. An example of such formula is given by [45]

$$\mathrm{det}\Delta^{C,|v|^2} = const \, (\mathrm{det}\Im\Omega) \, \mathrm{Area}(\mathcal{R}, |v|^2) \, |\tau(\mathcal{R}, v)|^2$$

where v is a holomorphic abelian differential on $\mathcal{R}$. This formula as well as its generalization to spaces of quadratic differentials turn out to be useful in the theory of the Teichmüller flow [18].

We did not cover all aspects of the theory and applications of the Bergman tau-function. Among other applications of this subject are the theory of random matrices [21], the Chern-Simons theory [64] and the theory of topological recursion [7].

The goal of this paper is to to summarize these links and establish a few new facts along the way. We start from describing the role of Bergman tau-function on Hurwitz spaces in the theory of Dubrovin-Frobenius manifolds in section 2. This section is based on [17, 48, 44]. Here we introduce the Schlesinger system and Jimbo-Miwa tau-function and present an explicit formula for the Bergman tau-function on Hurwitz spaces. In section 3, following [59], we describe solutions of an arbitrary Riemann-Hilbert problem with quasi-permutation monodromies and give a detailed calculation of its tau-function. In section 4, following [6], we show how to use the most elementary version of this tau-function to get elementary expressions for metric coefficients of $SU(2)$ invariant self-dual Einstein manifolds. In section 5 we show how to use the solutions of 2×2 Riemann-Hilbert problems with quasi-permutation monodromies to get solutions of Einstein equations with two Killing vectors in terms of hyperelliptic theta-functions and express metric coefficients in terms of corresponding theta-functions [51, 52, 53]. In section 6 we describe applications of Bergman tau-function to geometry of moduli spaces of abelian, quadratic and N-differentials. In section 6.1, following [46] we compute the Hodge class on the space of admissible covers in terms of boundary divisors. In section 6.2 which is based on [57] we apply the formalism to compute the Hodge class on spaces of abelian differentials. In section 6.3 we extend the formalism to spaces of quadratic differentials by defining the Hodge and Prym tau-functions [58]. We show in particular how Mumford's relations between Hodge class and the class of the determinant of the vector bundle of quadratic differentials can be derived by analyzing the analytical properties of tau-functions. In section 6.4 we introduce the Hodge and Prym tau-functions on an arbitrary stratum of the space of holomorphic quadratic differentials. In section 6.5 we extend this formalism to spaces of meromorphic quadratic differentials with second order poles and get the analog of Mumford's relations for $\overline{\mathcal{M}}_{g,n}$. In section 7 (following [10]) we consider the flat combinatorial model of $\mathcal{M}_{g,n}$ based on Jenkins-Strebel differentials. Arguments of Hodge and Prym tau-functions give sections of circle line bundles which are combinations of tautological classes; computation of monodromy of these arguments around Witten's cycle and Kontsevich's boundary of the combinatorial model gives combinatorial analogs of Mumford's relations in the Picard group of $\overline{\mathcal{M}}_{g,n}$. In section 8 we briefly describe the application of tau-functions to spaces of holomorphic N-differentials and spin moduli spaces following [55, 4]. In particular we show how the analytical properties of tau-functions on spin moduli spaces imply the Farkas-Verra formula for the divisor of degenerate odd spinors. In section 9 we use tau-functions to study moduli spaces of Hitchin's spectral covers following [60, 11]. In section 9.1 we determine the class of the universal discriminant locus in the space of all spectral curves i.e. the locus where not all branch points of the spectral cover are simple. In section 9.2 we, following [11], derive variational formulas for the period matrix on the moduli space of spectral covers with fixed base from variational formulas on moduli spaces of abelian differentials [45], establishing the link with the Donagi-Markman cubic. In section 10 we summarize the links between tau-functions, determinants of Laplace operator on Riemann surfaces with flat singular metrics and the sum of Lyapunov exponents of the Teichmüller flow on spaces of abelian and quadratic differentials. In the Appendix (section Appendix A) we describe canonical objects associated to Riemann surfaces which are used in the main text as well as variational formulas on Hurwitz spaces and spaces of abelian and N-differentials.

Acknowledgements. The author thanks Marco Bertola, Alexey Kokotov and Peter Zograf for numerous illuminating discussions. This work was supported in part by the Natural Sciences and Engineering Research Council of Canada grant RGPIN/3827-2015.

2 Tau-Function of Dubrovin-Frobenius Manifold Structure on Hurwitz Spaces

2.1 Schlesinger System and Jimbo-Miwa Tau-Function

Consider the system of linear differential equations with the initial condition:

$$\frac{d\Psi}{dz} = A(z)\Psi , \qquad \Psi(z_0) = I \tag{8}$$

where $A(z)$ is an $N \times N$ matrix whose entries are meromorphic functions on $\mathbb{C}P^1$. The system (8) is called Fuchsian if the matrix $A(z)$ has only simple poles, i.e. $A(z) = C + \sum_{i=1}^M \frac{A_i}{z-z_i}$. We assume that $z = \infty$ is not a singular point of (8) i.e. $\sum_{i=1}^M A_i = 0$. The solution $\Psi(z)$ of (8) is single-valued on the universal cover of $\mathbb{C}P^1 \setminus \{z_i\}_{i=1}^M$. If one starts at a point z_0 on some sheet of the universal cover, and analytically continues Ψ along a loop $\gamma \in \pi_1(\mathbb{C}P^1 \setminus \{z_i\}_{i=1}^M)$ one gets a new solution, Ψ_γ, of the same system; therefore, Ψ_γ is related to Ψ by a right multiplier, M_γ, which is called the monodromy matrix: $\Psi_\gamma = \Psi M_\gamma$. In this way, one gets an anti-homomorphism from the fundamental group $\pi_1(\mathbb{C}P^1 \setminus \{z_i\}_{i=1}^M)$ to $GL(N)$. The image of this anti-homomorphism is called the monodromy group of the system (8). The monodromy matrices and the positions of singularities form the set of monodromy data of the equation (8). The Riemann-Hilbert (or inverse monodromy) problem is the problem of finding a matrix-valued function Ψ [and, therefore, also the coefficients A_j] knowing the monodromy data. One can vary the positions of singularities z_j such that the monodromy matrices remain unchanged. Such a deformation (called the isomonodromic deformation) generically implies the following set of non-linear differential *Schlesinger equations* for A_j as functions of $\{z_k\}$:

$$\frac{\partial A_j}{\partial z_k} = \frac{[A_j, A_k]}{z_j - z_k} - \frac{[A_j, A_k]}{z_0 - z_k} , \qquad j \neq k; \tag{9}$$

$$\frac{\partial A_k}{\partial z_k} = -\sum_{j \neq k} \left(\frac{[A_j, A_k]}{z_j - z_k} - \frac{[A_j, A_k]}{z_j - z_0} \right). \tag{10}$$

In terms of a solution of the Schlesinger system the Jimbo-Miwa tau-function is defined by the system of equations [68]

$$\frac{\partial}{\partial z_j} \ln \tau = H_j := \frac{1}{2} \text{res}|_{z=z_j} \frac{\text{tr} \left(d\Psi \, \Psi^{-1} \right)^2}{dz}. \tag{11}$$

According to Jimbo-Miwa and Malgrange [61], the isomonodromic tau-function is a holomorphic section of a holomorphic line bundle over the space $\mathbb{C}^n \setminus \{z_k = z_j\}$. The divisor (τ) of its zeros (the *Malgrange divisor*) has an important meaning: if $\{z_j\}_{j=1}^n \in (\tau)$ then the Riemann-Hilbert problem with the given set of monodromy data does not have a solution. Moreover, the solution $\{A_j\}_{j=1}^n$ of the Schlesinger system is singular on (τ).

2.2 Bergman Tau-Function on Hurwitz Space

Here we discuss the appearance of the Bergman tau-function in the theory of Dubrovin-Frobenius manifolds [16, 17]. The class of Dubrovin-Frobenius manifolds which admits a complete analytical description is provided by Hurwitz spaces. Restricting ourselves to the semi-simple manifolds consider a Hurwitz space $H_{g,n}$; a point of $H_{g,n}$ is a pair $(\mathcal{R}, f)$ where $\mathcal{R}$ is a Riemann surface of genus g and f is a function of degree n on $\mathcal{R}$ with simple poles and simple critical points. To each Dubrovin-Frobenius manifold one can associate a natural Riemann-Hilbert problem whose solution, together with associated solutions of the Schlesinger system, encodes all essential ingredients of the manifold - from flat coordinates to the prepotential. Matrix entries of the solution $\{A_j\}$ of the Schlesinger system are given by the rotation coefficients of a flat metric associated to each Dubrovin-Frobenius manifold, see eq (3.70) of [16]. Therefore, the rotation coefficients define also the Jimbo-Miwa isomonodromic tau-function which in turn gives the genus one contribution (the G-function) to the free energy (see Th.3.2 of [17]). Referring for details to [16, 17] we will only present formulas which are necessary to make this section reasonably self-contained.

The Jimbo-Miwa tau-function of a Dubrovin-Frobenius manifold, which in the present context is nothing but the Bergman tau-function on the Hurwitz space, is defined by the system of equations (see Eq. (3.40) of [17])

$$\frac{\partial \ln \tau_B}{\partial z_j} = \frac{1}{2} \sum_{i \neq j} \frac{\beta_{ij}^2}{z_j - z_i} \tag{12}$$

where β_{ij} are the rotation coefficients of the Darboux-Egoroff metric associated to the Frobenius manifold and z_j are the canonical coordinates on the manifold.

For semi-simple Frobenius structures on the Hurwits space $H_{g,n}$ the canonical coordinates are given by the critical values of the function f: $z_i = f(x_i)$ where x_i are critical points of the function f $(df(x_i) = 0)$. Equivalently, the critical values z_i are called the branch points of the corresponding n-sheeted branched cover while for the critical points x_i we reserve the term "ramification points". The natural local coordinates (also called "distinguished") on $\mathcal{R}$ near x_j are given by $\zeta_j(x) = [f(x) - z_j]^{1/2}$. The rotation coefficients for the Frobenius structures on $H_{g,n}$ are expressed in terms of the canonical bidifferential $B(x, y)$ as follows [43]:

$$\beta_{jk} = \frac{1}{2} \frac{B(x, y)}{d\zeta_j(x)d\zeta_k(y)}\Big|_{x=x_j,\, y=x_k}. \tag{13}$$

Then the equations (12) can be alternatively written as follows:

$$\frac{\partial \ln \tau_B}{\partial z_j} = -\text{res}\Big|_{x_k} \frac{B_{reg}(x, x)}{df(x)} \tag{14}$$

where $B_{reg}(x, x)$ is the meromorphic quadratic differential which equals to the constant term of $B(x, y)$ on the diagonal:

$$B_{reg}(x, x) = \left(B(x, y) - \frac{df(x)df(y)}{(f(x - f(y))^2} \right)\Big|_{x=y}. \tag{15}$$

The differential $B_{reg}(x, x)$ can also be represented in terms of the Bergman projective connection S_B and the projective connection $S_{df}(\cdot) = \{f, \cdot\}$ as follows:

$$B_{reg}(x, x) = \frac{1}{6}(S_B - S_{df}).$$

Let us introduce the divisor of the differential (df):

$$(df) = \sum k_j x_j \tag{16}$$

where $k_j = 1$ if x_j is a (simple) ramification point. If x_j is a pole of f then the order of the pole of f at x_j equals $-(k_j + 1)$. Near pole x_j the distinguished local parameters are given by

$$\zeta_j(x) = f(x)^{1/(k_j+1)}. \tag{17}$$

Near zeros x_j of df the distinguished local coordinates are defined by

$$\zeta_j(x) = (f(x) - f(x_j))^{1/(k_j+1)}. \tag{18}$$

Introduce also the following notations:

$$E(x, x_j) = \lim_{y \to x_j} E(x, y)\sqrt{d\zeta_i(y)} \tag{19}$$

and

$$E(x_i, x_j) = \lim_{x \to x_j, y \to x_i} E(x, y)\sqrt{d\zeta_i(y)}\sqrt{d\zeta_j(x)}. \tag{20}$$

The solution of the system (14) is given by the following theorem:

Proposition 2.1 [44] *Let* $(df) = \sum k_i q_i$ *be the divisor of the differential* df. *Then the solution* τ_B *of the system (14) is given by the following expression:*

$$\tau_B = C^{2/3}(x)\left(\frac{df(x)}{\prod_i E^{k_i}(x, q_i)} \right)^{(g-1)/3} \left(\prod_{i<j} E(q_i, q_j)^{\frac{k_i k_j}{6}} \right) e^{\frac{\pi i}{6}\langle s, \Omega s\rangle - \frac{2\pi i}{3}\langle s, K^x\rangle}. \tag{21}$$

where K^x is the vector of Riemann constants. Here the vector $\mathbf{s}$ is defined by relation

$$\mathcal{A}_x((df)) + 2K^x + \Omega\mathbf{s} + \mathbf{r} = 0 \qquad (22)$$

where $\mathbf{s}, \mathbf{r} \in \mathbb{Z}^g$.

For the space of hyperelliptic coverings given by $w^2 = \prod_{i=1}^{2g+2}(z - z_i)$ the following simpler expression for the Bergman tau-function was found earlier in [39]:

$$\tau_B(\{z_m\}) = \det A \prod_{m<n}(z_m - z_n)^{\frac{1}{4}} \qquad (23)$$

where $A_{ij} = \int_{a_i} z^{j-1}dz/w$ is the matrix of a-periods of the non-normalized holomorphic differentials on $\mathcal{R}$. The expression (23) appeared first as the correlation function of the Ashkin-Teller model [74].

The function τ_B is simultaneously the Jimbo-Miwa tau-function of two Riemann-Hilbert problems whose solutions are related by the Laplace transform. One of them if non-Fuchsian; it's solution was implicitly given in [16] and later given in the complete form in [69]. Another Riemann-Hilbert problem is Fuchsian; its monodromy group and solution were described in [56]. Monodromy matrices of such monodromy group always have integer values; therefore, they do not contain any parameters and represent isolated points in the monodromy manifold. The function τ_B does not vanish unless two critical values of the function f coincide; therefore, the Malgrange divisor of the Dubrovin's Riemann-Hilbert problems is empty.

There exists another class of monodromy groups which can be associated to the branch coverings: these are the monodromy groups with quasi-permutation monodromies [59]. These monodromy groups depend on parameters, and the Bergman tau-function gives only one of the contributions to the Jimbo-Miwa tau-function.

3 Riemann-Hilbert Problem for Quasi-Permutation Monodromy Groups

Riemann-Hilbert problems with the quasi-permutation monodromy groups can be solved as follows [59]. Suppose we are given a $GL(n)$ representation of $\pi_1(\mathbb{C}P^1 \setminus \{z_j = z_k\}, z_0)$ such that all monodromy matrices have exactly one non-vanishing entry in each column and in each row. Then for any monodromy matrix M_γ ($\gamma \in \pi_1(\mathbb{C}P^1 \setminus \{z_j = z_k\}, z_0)$) one can construct an associate permutation matrix M_γ^0 by replacing all non-vanishing entries of M_γ by 1. In this way we get a *permutation* representation of $\pi_1(\mathbb{C}P^1 \setminus \{z_j = z_k\}, z_0)$, which, according to Riemann's theorem, defines an n-sheeted branched covering of $\mathbb{C}P^1$. This gives the pair $(\mathcal{R}, f)$ where $\mathcal{R}$ is the Riemann surface of a genus g (the genus is determined by the monodromy group) and f is the meromorphic function on $\mathcal{R}$ of degree n.

Assume that the monodromy representations can not be decomposed into direct sum of two other quasi-permutation representations. Then the Riemann surface $\mathcal{R}$ is connected. Denote by $z^{(j)}$ the point on jth sheet (under some enumeration of sheets) of $(\mathcal{R}, f)$ having projection z to the base; $\{z^{(j)}\}_{j=1}^{n} = f^{-1}(z)$ if z is not a branch point.

The solution of an arbitrary RH problem with quasi-permutation monodromies constructed in [59] depends on the following set of parameters:

- Two vectors $\mathbf{p}, \mathbf{q} \in \mathbb{C}^g$.
- Constants $r_m^{(j)} \in \mathbb{C}$ assigned to each point $z_m^{(j)}$; we assume that the constants $r_m^{(j)}$ and $r_m^{(j')}$ coincide if $z_m^{(j)} = z_m^{(j')}$ i.e. if $z_m^{(j)}$ is a ramification point. These constants are assumed to satisfy the relation

$$\sum_{m=1}^{M} \sum_{j=1}^{n} r_m^{(j)} = 0. \tag{24}$$

Therefore, there are $Mn - 2g - 2n + 1$ independent parameters among the constants $r_m^{(j)}$.

Altogether we have $Mn - 2n + 1$ independent constants $\mathbf{p}, \mathbf{q}$ and $r_m^{(j)}$; this number coincides with the number of non-trivial parameters carried by the quasi-permutation monodromy matrices $M_1, \ldots, M_M$.

Denote by $\widehat{S}(x, y)$ the modified Szegö kernel given by the following formula (we assume that x and y lie inside of the fundamental polygon $\tilde{\mathcal{R}}$ of $\mathcal{R}$):

$$\widehat{S}(x, y) := \frac{\Theta\left[\begin{smallmatrix}\mathbf{p}\\\mathbf{q}\end{smallmatrix}\right] (\mathcal{A}(x) - \mathcal{A}(y) + Q)}{\Theta\left[\begin{smallmatrix}\mathbf{p}\\\mathbf{q}\end{smallmatrix}\right] (Q) E(x, y)} \prod_{m=1}^{M} \prod_{l=1}^{n} \left[\frac{E(x, z_m^{(l)})}{E(y, z_m^{(l)})}\right]^{r_m^{(l)}} \tag{25}$$

where

$$Q := \sum_{m=1}^{M} \sum_{j=1}^{n} r_m^{(j)} \mathcal{A}_{x_0}(z_m^{(j)}).$$

The vector Q does not depend on the choice of initial point x_0 of the Abel map due to assumption (24). The kernel (27) is well-defined if $\Theta\left[\begin{smallmatrix}\mathbf{p}\\\mathbf{q}\end{smallmatrix}\right] (Q) \neq 0$.

Define

$$\psi(x, y) = \widehat{S}(x, y) E_0(f(x), f(y)) \tag{26}$$

where

$$E_0(z, w) = \frac{z - w}{\sqrt{dz\,dw}}$$

is the prime form on $\mathbb{C}P^1$.

According to [59] the solution of the RH problem is given by the analytical continuation (in z) of the matrix function

$$\Psi_{kj}(z_0, z) = \psi(z^{(j)}, z_0^{(k)}) \tag{27}$$

from a neighbourhood of the normalization point $z \sim z_0$.

The function Ψ satisfies the condition $\Psi(z_0, z_0) = I$. Moreover, the Fay identity (246) allows to compute its determinant:

$$\det\Psi = \prod_{m=1}^{M} \prod_{j,k=1}^{N} \left[\frac{E(z^{(j)}, z_m^{(k)})}{E(z_0^{(j)}, z_m^{(k)})} \right]^{r_m^{(k)}}.$$

In the proof of this expression for $\det\Psi$ one uses the elementary fact that $\sum_{p \in f^{-1}(z)} \frac{v}{df}(p) = 0$ for any holomorphic differential v on $\mathcal{R}$, see [59].

3.1 Jimbo-Miwa Tau-Function

The computation of the Jimbo-Miwa tau-function (11) corresponding to the solution (27) starts from transforming $\operatorname{tr}\left(\Psi_z \Psi^{-1}\right)^2$ to a suitable form. Since this expression is independent of the choice of the normalization point z_0 one can take the limit $z_0 \to z$ in the formulas (27), (25):

$$\Psi_{kj}(z, z_0) = (z_0 - z)\frac{\widehat{S}(z^{(j)}, z^{(k)})}{dz} + O((z_0 - z)^2), \qquad k \neq j \tag{28}$$

$$\Psi_{jj}(z, z_0) = 1 + (z_0 - z)\frac{W_1(z^{(j)}) - W_2(z^{(j)})}{dz} \tag{29}$$

where $W_1(x)$ is the linear combination of the basic holomorphic differentials on $\mathcal{R}$:

$$W_1(x) = \frac{1}{\Theta\left[\begin{smallmatrix} p \\ q \end{smallmatrix}\right](Q)} \sum_{\alpha=1}^{g} \partial_{z_\alpha}\{\Theta\left[\begin{smallmatrix} p \\ q \end{smallmatrix}\right](Q)\} v_\alpha(x) \tag{30}$$

and $W_2(x)$ is the following meromorphic differential with simple poles at the points $z_m^{(j)}$ and residues $r_m^{(j)}$:

$$W_2(x) = \sum_{m=1}^{M} \sum_{j=1}^{n} r_m^{(j)} d_x \ln E(x, z_m^{(j)}). \tag{31}$$

Then

$$\operatorname{tr}\left(\Psi_z \Psi^{-1}\right)^2 (dz)^2 = 2 \sum_{j<k} \widehat{S}(z^{(j)}, z^{(k)}) \widehat{S}(z^{(k)}, z^{(j)})$$

$$+ \sum_{j=1}^{n} \left(W_1(z^{(j)}) - W_2(z^{(j)}) \right)^2.$$

Due to (245) we have

$$\widehat{S}(x, y)\widehat{S}(y, x) = -B(x, y) - \sum_{\alpha,\beta=1}^{g} \partial_{\alpha\beta}^2 \left\{ \ln \Theta \begin{bmatrix} \mathbf{p} \\ \mathbf{q} \end{bmatrix}(Q) \right\} v_\alpha(x)v_\beta(y).$$

Furthermore, since $W_1(x)$ is holomorphic on $\mathcal{R}$, we have $\sum_{j=1}^{N} W_1(z^{(j)}) = 0$ and

$$\sum_{j=1}^{n} \{W_1(z^{(j)})\}^2$$

$$= -2 \sum_{\substack{j,k=1 \\ j<k}}^{N} \sum_{\alpha,\beta=1}^{g} \partial_\alpha \left\{ \ln \Theta \begin{bmatrix} \mathbf{p} \\ \mathbf{q} \end{bmatrix}(Q) \right\} \partial_\beta \left\{ \ln \Theta \begin{bmatrix} \mathbf{p} \\ \mathbf{q} \end{bmatrix}(Q) \right\} v_\alpha(z^{(j)})v_\beta(z^{(k)}).$$

Therefore,

$$\frac{1}{2}\operatorname{tr}\left(\Psi_z \Psi^{-1}\right)^2 (dz)^2 = -\sum_{j<k} B(z^{(j)}, z^{(k)}) + \frac{1}{2} \sum_{j=1}^{n} W_2^2(z^{(j)})$$

$$- \frac{1}{\Theta \begin{bmatrix} \mathbf{p} \\ \mathbf{q} \end{bmatrix}(Q)} \sum_{j<k} \sum_{\alpha,\beta} \partial_{\alpha\beta}^2 \left\{ \Theta \begin{bmatrix} \mathbf{p} \\ \mathbf{q} \end{bmatrix}(Q) \right\} v_\alpha(z^{(j)})v_\beta(z^{(k)})$$

$$- \frac{1}{\Theta \begin{bmatrix} \mathbf{p} \\ \mathbf{q} \end{bmatrix}(Q)} \sum_{\alpha} \partial_\alpha \left\{ \Theta \begin{bmatrix} \mathbf{p} \\ \mathbf{q} \end{bmatrix}(Q) \right\} \sum_{m} \sum_{j} r_m^{(j)} v_\alpha(z^{(j)}) d_x \ln E(x, z_m^{(j)}).$$

$$(32)$$

Using the heat equation for the theta-function (232) and the variational formula (267), we see that the contribution of the last two terms in (32) to the residue at z_m is given by $\partial_{z_m} \Theta \begin{bmatrix} \mathbf{p} \\ \mathbf{q} \end{bmatrix}(Q)$ and, therefore,

$$\frac{\partial \ln \tau_{JM}}{\partial z_m} = \frac{1}{2} \operatorname{res}|_{z=z_m} \left\{ \operatorname{tr}\left(\Psi_z \Psi^{-1}\right)^2 dz \right\}$$

$$= -\operatorname{res}|_{z=z_m} \sum_{j<k} \frac{B(z^{(j)}, z^{(k)})}{dz}$$

$$+ \frac{1}{2} \operatorname{res}|_{z=z_m} \sum_{j=1}^{n} \frac{W_2^2(z^{(j)})}{dz} + \partial_{z_m} \Theta \begin{bmatrix} \mathbf{p} \\ \mathbf{q} \end{bmatrix}(Q). \qquad (33)$$

We get the following proposition [59]:

Proposition 3.1 *The tau-function τ_{JM} is given by*

$$\tau_{JM} = \tau_0 \, \tau_B^{-1/2} \, \Theta \begin{bmatrix} \mathbf{p} \\ \mathbf{q} \end{bmatrix} (Q) \tag{34}$$

where τ_B is the Bergman tau-function defined by the system (14) and given by the formula (21).

The factor τ_0 satisfies the system of equations

$$\frac{\partial \ln \tau_0}{\partial z_m} = \frac{1}{2} \operatorname{res} |_{z=z_m} \sum_{j=1}^{n} \frac{W_2^2(z^{(j)})}{dz} \tag{35}$$

where the Abelian differential of third kind W_2 is given by (31).

Proof. The theorem follows from (33) and the identity proven in [43]:

$$2 \operatorname{res} |_{z=z_m} \frac{1}{dz} \left\{ \sum_{j<k} B(z^{(j)}, z^{(k)}) \right\} = \frac{1}{6} \operatorname{res}|_{z=z_m} \sum_{p \in f^{-1}(z)} \frac{S_B - S_{df}}{df} \tag{36}$$

where S_B is the Bergman projective connection and $f(x) = z$ is the meromorphic function defining the covering $\mathcal{R}$. The right hand side of (36) coincides with the logarithmic derivative of the Bergman tau-function (21). $\qquad \square$

3.2 Function τ_0

The factor τ_0 in (34) is given by the following theorem

Proposition 3.2 *The solution of the system of equations*

$$\frac{\partial \ln \tau_0}{\partial z_m} = \frac{1}{2} \operatorname{res} |_{z=z_m} \sum_{j=1}^{n} \frac{W_2^2(z^{(j)})}{dz}. \tag{37}$$

with

$$W_2(x) = \sum_{m=1}^{M} \sum_{j=1}^{n} r_m^{(j)} d_x \ln E(x, z_m^{(j)}) \tag{38}$$

is given by

$$\tau_0^2 = \prod_{z_m^{(j)} \neq z_k^{(i)}} E(z_m^{(j)}, z_k^{(i)})^{r_m^{(j)} r_k^{(i)}} \tag{39}$$

where the prime-forms whose arguments coincide with $z_m^{(j)}$ are computed with respect to the distinguished local parameters as in (20) (notice that all the terms in the r.h.s. of (39) enter twice; we use this convention since there is no natural ordering of the points $z_m^{(j)}$).

Proof. The proof is based on the variational formula for the prime-form (266); to avoid unnecessary technicalities we give the proof in the special situation which, however, reflects all the characteristic features of the general case. Namely, assume that z_1 is not a critical value (this means that the corresponding monodromy matrix M_1 is diagonal) while all residues corresponding to other z_m's equal zero. In this case (39) gives

$$\tau_0^2 = \prod_{i \neq j} E(z_1^{(i)}, z_1^{(j)})^{r_1^{(i)} r_1^{(j)}} \tag{40}$$

and

$$W_2 = \sum_{i=1}^{n} r_1^{(i)} d \ln E(\lambda, z_1^{(i)}) \tag{41}$$

with $\sum_{i=1}^{n} r_1^{(i)} = 0$.

Let us first differentiate (40) with respect to some branch point z_2 using (266):

$$\frac{\partial}{\partial z_2} \ln \prod_{i \neq j} E(z_1^{(i)}, z_1^{(j)})^{r_1^{(i)} r_1^{(j)}} \tag{42}$$

$$= -\frac{1}{2} \sum_{p \in f^{-1}(z_2)} \text{res} \Big|_{t=p} (z_2) \frac{1}{df(t)} \sum_{i \neq j} r_1^{(i)} r_1^{(j)} (d \ln E(t, z_1^{(i)}) - d \ln E(t, z_1^{(j)}))^2 \tag{43}$$

$$= -\frac{1}{2} \sum_{p \in \pi^{-1}(z_2)} \text{res} \Big|_{t=p} \frac{1}{df(t)} \sum_{i \neq j} r_1^{(i)} r_1^{(j)}$$

$$\times \Big[(d \ln E(t, z_1^{(i)}))^2 + (d \ln E(t, z_1^{(j)}))^2 - 2d \ln E(t, z_1^{(i)})) d \ln E(t, z_1^{(j)}) \Big].$$

In the first term of the last expression we perform the summation over j using $\sum_{i=1}^{n} r_1^{(i)} = 0$ and in the second term the summation is performed over i. The results equal to each other which gives an extra factor of -2. Summing only over $i < j$ in the last sum gives the following expression for (42):

$$\sum_{p\in\pi^{-1}(z_2)} \text{res}\Big|_{t=p} \frac{1}{df(t)}\left[\sum_i (r_1^{(i)})^2 (d\ln E(t, z_1^{(i)}))^2\right.$$

$$\left.+2\sum_{i<j} r_1^{(i)} r_1^{(j)} d\ln E(t, z_1^{(i)})) d\ln E(t, z_1^{(j)}))\right]$$

$$= \sum_{p\in\pi^{-1}(z_2)} \text{res}\Big|_{t=z_2^{(k)}} \frac{1}{df(x)}\left[\sum_i r_1^{(i)} d\ln E(t, z_1^{(i)})\right]^2 \tag{44}$$

which coincides with the r.h.s. of (37).

Consider now the derivative with respect to z_1. The moduli of the Riemann surface $\mathcal{R}$ remain unchanged but each prime form has to be differentiated with respect to its arguments. Then

$$\frac{\partial}{\partial z_1} \ln \prod_{i\neq j} E(z_1^{(i)}, z_1^{(j)})^{r_1^{(i)} r_1^{(j)}} = 2\sum_{i\neq j} r_1^{(i)} r_1^{(j)} \frac{d_x \ln E(x, y)}{df(x)}\Big|_{x=z_1^{(i)}, y=z_1^{(j)}}. \tag{45}$$

On the other hand, the r.h.s. of (37) gives

$$\sum_{k=1}^{n} \text{res}\Big|_{x=z_1^{(k)}} \frac{1}{df(x)}\left[\sum_i r_1^{(i)} d\ln E(x, z_1^{(i)})\right]^2$$

$$= \text{res}\Big|_{z=z_1} \frac{1}{dz} \sum_k \sum_i (r_1^{(i)})^2 (d\ln E(z^{(k)}, z_1^{(i)}))^2$$

$$+ \text{res}\Big|_{z=z_1} \frac{1}{dz} \sum_k \sum_{i\neq j} r_1^{(i)} r_1^{(j)} d\ln E(z^{(k)}, z_1^{(i)}) d\ln E(z^{(k)}, z_1^{(j)}). \tag{46}$$

In the first sum in (46) we can interchange the order of summation and sum over k first; this gives the meromorphic differential on the base:

$$\frac{1}{dz} \sum_k (d\ln E(z^{(k)}, z_1^{(i)}))^2 = \frac{dz}{(z-z_1)^2} \tag{47}$$

which does not have residue at z_1.

The second sum contributes when either $k = i$ or $k = j$; due to the symmetry we get

$$2\sum_{i\neq j} r_1^{(i)} r_1^{(j)} d\ln E(z_1^{(j)}, z_1^{(i)})$$

computed in the local parameter $z - z_1$, which coincides with (45). $\qquad\square$

Factors τ_0 and $\tau_B^{-1/2}$ in the formula (34) for the tau-function never vanish in $\mathbb{C}^M \setminus \{z_j = z_k\}$. Therefore, the Malgrange divisor of the Riemann-Hilbert problems with quasi-permutation monodromies is defined by the equation

$$\Theta \begin{bmatrix} \mathbf{p} \\ \mathbf{q} \end{bmatrix} \left(\sum_{m=1}^{M} \sum_{j=1}^{n} r_m^{(j)} \mathcal{A}_{x_0}(z_m^{(j)}) \right) = 0.$$

Remark 3.3   The formula for τ_0 proposed in [59] was incorrect. The expression (39) was first given in [28] for the case of coverings with special monodromy groups (the monodromy matrices from [28] were split in pairs such that $M_{2k} M_{2k-1} = I$); the tau-function τ_0 was named in [28] the "Seiberg-Witten" tau-function although we are not aware of the logic behind this terminology. The full expression (34) was interpreted in [28] as conformal block of a conformal field theory in the spirit of recent works on the link between conformal blocks and isomonodromic tau-functions (see for example [27]).

4 Self-Dual $SU(2)$ Invariant Einstein Manifolds

The complete description of $SU(2)$ invariant self-dual Einstein metrics metrics of this type was given by Hitchin [37] in terms of special solutions of Painlevé 6 equation. The analysis presented in [37] essentially relied on the previous works [72, 73]. The final formulas derived in [37] were rather complicated; the elementary formulas for these metrics were obtained later in [6] using the expression for the Bergman tau-function on the space of elliptic curves derived in [39]. In this section we summarize the approach of [6].

The $SU(2)$-invariant self-dual Einstein metric can be written in the following form [72, 73]

$$g = F \left\{ d\mu^2 + \frac{\sigma_1^2}{W_1^2} + \frac{\sigma_2^2}{W_2^2} + \frac{\sigma_3^2}{W_3^2} \right\} \tag{48}$$

where the 1-forms σ_j satisfy

$$d\sigma_1 = \sigma_2 \wedge \sigma_3 , \quad d\sigma_2 = \sigma_3 \wedge \sigma_1 , \quad d\sigma_3 = \sigma_1 \wedge \sigma_2, \tag{49}$$

and the functions W_j depend only on the "euclidean time" μ. In terms of the new variables $A_j(\mu)$ (the connection coefficients) defined by equations

$$\frac{dW_j}{d\mu} = -W_k W_l + W_j(A_k + A_k), \tag{50}$$

where (j, k, l) is an arbitrary permutation of the indexes $(1, 2, 3)$ the condition of the self-duality of the metric is given by the classical Halphen system:

$$\frac{dA_j}{d\mu} = -A_k A_l + A_j(A_k + A_k). \tag{51}$$

Any solution of (51) can be substituted into the system (50), which defines then metric coefficients W_j.

The full system (50), (51) is invariant with respect to an $SL(2, \mathbb{R})$ Möbius transformations:

$$\mu \to \tilde{\mu} = \frac{a\mu + b}{c\mu + d}, \qquad ad - bc = 1, \tag{52}$$

$$W_j \to \tilde{W}_j = (c\mu + d)^2 W_j, \tag{53}$$

$$A_j \to \tilde{A}_j = (c\mu + d)^2 A_j + c(c\mu + d)^2. \tag{54}$$

The system (51) has several interesting special cases. When one chooses the trivial solution $A_j = 0$ of the system (51), the corresponding system (50) reduces to the equations of the Euler top. If two of the functions A_j vanish, the remaining one must be constant. Another special case is when for all j one chooses $W_j = A_j$. We refer to [37] for the list of references where these special cases were discussed.

The system (50), corresponding to the general solution of the system (51), was related in the papers [73, 37] to the four-point Schlesinger system. Moreover, it turned out that the conformal factor F can be chosen to make the metric (48) satisfy the Einstein equation exactly in the case when the system (50) can be solved in terms of elliptic functions.

It is well-known that the functions

$$A_1 = 2\frac{d}{d\mu} \ln \vartheta_2 , \qquad A_2 = 2\frac{d}{d\mu} \ln \vartheta_3 , \qquad A_3 = 2\frac{d}{d\mu} \ln \vartheta_4 \tag{55}$$

where

$$\vartheta_2 \equiv \vartheta\left[\frac{1}{2}, 0\right](0, i\mu) , \qquad \vartheta_3 \equiv \vartheta[0, 0](0, i\mu) , \qquad \vartheta_4 \equiv \vartheta\left[0, \frac{1}{2}\right](0, i\mu)$$

are standard theta-constants, solve the Halphen system (51). The general solution of (51) may be obtained applying the Möbius transformations (52), (54) to solution (55). Therefore, it is sufficient to solve the system (50), where the functions A_j are given by (55). Then the general solution of the system (50) can be obtained by the Möbius transformation from the general solution of the system

$$\frac{dW_j}{d\mu} = -W_k W_l + 2W_j \frac{d}{d\mu} \ln(\vartheta_{k+1}\vartheta_{l+1}). \tag{56}$$

In general, the system (56) can not be solved in elementary or special functions. However, such solution is possible if W_i satisfy the following additional condition (which is nothing but the fixing of the value of an integral of motion of the system (56)):

$$\vartheta_2^4 W_1^2 - \vartheta_3^4 W_2^2 + \vartheta_4^4 W_3^2 = \frac{\pi^2}{4} \vartheta_2^4 \vartheta_3^4 \vartheta_4^4. \tag{57}$$

The relationship of the system (56) to isomonodromic deformations and the tau-function is most naturally seen in terms of variables

$$\Omega_1 = -\frac{W_2}{\pi \vartheta_2^2 \vartheta_4^2}, \qquad \Omega_2 = -\frac{W_3}{\pi \vartheta_2^2 \vartheta_3^2}, \qquad \Omega_3 = -\frac{W_1}{\pi \vartheta_3^2 \vartheta_4^2} \tag{58}$$

which were used in [73, 37]. Then the relation (57) takes the simple form

$$-\Omega_1^2 + \Omega_2^2 + \Omega_3^2 = \frac{1}{4}. \tag{59}$$

To rewrite the system (56) in terms of Ω_j's one also makes the change of the independent variable from μ to x such that the period of the elliptic curve

$$v^2 = \lambda(\lambda - 1)(\lambda - x). \tag{60}$$

equals $i\mu$. It is assumed that $0 < x < 1$, the a-cycles goes around $[0, x]$ and b-cycle goes around $[x, 1]$. Then the system (56) is equivalent to

$$\frac{d\Omega_1}{dx} = -\frac{\Omega_2 \Omega_3}{x(1-x)}, \qquad \frac{d\Omega_2}{dx} = -\frac{\Omega_3 \Omega_1}{x}, \qquad \frac{d\Omega_3}{dx} = -\frac{\Omega_1 \Omega_2}{1-x}. \tag{61}$$

To derive equations (61) from (56) one needs to use the elementary version of the variational formula (264):

$$\frac{d\mu}{dx} = \frac{\pi}{4K^2 x(x-1)}. \tag{62}$$

where K is the full elliptic integral

$$K = \frac{1}{2} \int_0^x \frac{d\lambda}{\sqrt{\lambda(\lambda - 1)(\lambda - x)}} \tag{63}$$

and the standard expressions for theta-constants are given by

$$\vartheta_2^4 = \frac{4}{\pi^2} K^2 x, \qquad \vartheta_3^4 = \frac{4}{\pi^2} K^2, \qquad \vartheta_4^4 = \frac{4}{\pi^2} K^2 (1-x). \tag{64}$$

The metric (48) in terms of Ω_j's can be written as follows

$$g = \tilde{F} \left\{ \frac{dx^2}{x(1-x)} + \frac{\sigma_1^2}{\Omega_1^2} + \frac{(1-x)\sigma_2^2}{\Omega_2^2} + \frac{x\sigma_3^2}{\Omega_3^2} \right\}, \tag{65}$$

where the conformal factors $\tilde{F}$ and F are related by $\tilde{F} = \frac{1}{\vartheta_2^4 \vartheta_4^4} F$.

Let variables Ω_j solve the system (61), and satisfy the condition (59). Then the metric (65) satisfies the Einstein equations with the cosmological constant Λ if the conformal factor $\tilde{F}$ is given by the complicated explicit expression found in [73].

The system (61) arises in the context of isomonodromic deformations of the equation

$$\frac{d\Psi}{d\lambda} = \left(\frac{A^0}{\lambda} + \frac{A^1}{\lambda - 1} + \frac{A^x}{\lambda - x} \right) \Psi ; \tag{66}$$

the isomonodromy condition (assuming the normalization $\Psi(\infty) = I$) implies the equation

$$\frac{d\Psi}{dx} = -\frac{A^x}{\lambda - x} \Psi. \tag{67}$$

The compatibility condition of equations (66) and (67) is given by the Schlesinger system

$$\frac{dA^0}{dx} = \frac{[A^x, A^0]}{x} , \quad \frac{dA^1}{dx} = \frac{[A^x, A^1]}{x - 1} , \quad \frac{dA^x}{dx} = -\frac{[A^x, A^0]}{x} - \frac{[A^x, A^1]}{x - 1}. \tag{68}$$

Eigenvalues of the matrices A^0, A^x and A^1 are integrals of motion of the Schlesinger system. If one chooses

$$\text{tr}(A^0)^2 = \text{tr}(A^1)^2 = \text{tr}(A^x)^2 = \frac{1}{8}, \tag{69}$$

then the formulas

$$\Omega_1^2 = -\left(\frac{1}{8} + \text{tr}A^0 A^1 \right) , \quad \Omega_2^2 = \frac{1}{8} + \text{tr}A^1 A^x , \quad \Omega_3^2 = \frac{1}{8} + \text{tr}A^0 A^x \tag{70}$$

give the solution of the system (61), (59) (see [37]).

As a corollary of the conditions (69), the eigenvalues of all monodromies M^0, M^1 and M^x equal to $\pm i$. Such sets of monodromy matrices can be completely classified [37]: up to a simultaneous constant similarity transformation they are either given by the triple

$$
M_0 = \begin{pmatrix} 0 & -ie^{-2\pi iq} \\ -ie^{2\pi iq} & 0 \end{pmatrix} ; \quad M_1 = \begin{pmatrix} 0 & ie^{-2\pi i(p+q)} \\ ie^{2\pi i(p+q)} & 0 \end{pmatrix} ;
$$

$$
M_x = \begin{pmatrix} 0 & -ie^{-2\pi ip} \\ -ie^{2\pi ip} & 0 \end{pmatrix} \quad p, q \in \mathbb{C}, \tag{71}
$$

or the triple

$$
M_0 = \begin{pmatrix} -i & q_0 \\ 0 & i \end{pmatrix} ; \quad M_1 = \begin{pmatrix} -i & -i + q_0 \\ 0 & i \end{pmatrix} ;
$$

$$
M_x = \begin{pmatrix} -i & -i \\ 0 & i \end{pmatrix} , \quad q_0 \in \mathbb{C}. \tag{72}
$$

In [37] the system (61), (59) was solved in terms of elliptic functions (independently in the cases (71) and (72)) using the link between $\{\Omega_j\}$ and the solution $y(x)$ of the Painlevé 6 equation with coefficients $\left(\frac{1}{8}, -\frac{1}{8}, \frac{1}{8}, \frac{3}{8} \right)$, which is known to be equivalent to the system (68).

If one then directly expresses $\{\Omega_j\}$ in terms of elliptic functions then the resulting expressions turn out to be very cumbersome (see [37]).

Simple formulas for variables Ω_j and W_j were derived in [6] using the expression for the Jimbo-Miwa tau-function of the Schlesinger system (68) obtained in [39]. This tau-function is defined by equation

$$
\frac{d}{dx} \ln \tau(x) = \frac{\mathrm{tr} A^0 A^x}{x} + \frac{\mathrm{tr} A^1 A^x}{x-1}. \tag{73}
$$

The solution of the system (61) can be expressed in terms of the τ-function (73) as follows:

$$
\Omega_1^2 = \frac{d}{dx} \left\{ x(x-1) \frac{d}{dx} \ln \tau(x) \right\}, \tag{74}
$$

$$
\Omega_2^2 = (1-x) \frac{d}{dx} \left\{ x(x-1) \frac{d}{dx} \ln \tau(x) \right\} + x(x-1) \frac{d}{dx} \ln \tau(x) + \frac{1}{8}, \tag{75}
$$

$$
\Omega_3^2 = x \frac{d}{dx} \left\{ x(x-1) \frac{d}{dx} \ln \tau(x) \right\} - x(x-1) \frac{d}{dx} \ln \tau(x) + \frac{1}{8} \tag{76}
$$

solve (61), (59).

The Jimbo-Miwa τ-function of the Schlesinger system (68), corresponding to monodromy matrices (71), is given by the formulas (23) and (34) where $\tau_0 = 1$ and

$$
\tau_B = K[x(1-x)]^{1/4}.
$$

(In present case det A from (23) equals 4 times the elliptic integral K (63)). Therefore,

$$\tau(x) = \frac{\vartheta[p,q]}{K^{1/2}[x(1-x)]^{1/8}} \tag{77}$$

If, moreover, we make use of the Thomae formulas (64) we get (up to an unessential constant)

$$\tau(x) = \frac{\vartheta[p,q]}{\sqrt{\vartheta_2\vartheta_4}}. \tag{78}$$

Now a long but straightforward computation leads to the following theorem.

Theorem 4.1 ([6]) *The general two-parametric family of solutions of the system (56), satisfying condition (57), is given by the following formulas:*

$$W_1 = -\frac{i}{2}\vartheta_3\vartheta_4\frac{\frac{d}{dq}\vartheta[p,q+\frac{1}{2}]}{e^{\pi i p}\vartheta[p,q]}, \qquad W_2 = \frac{i}{2}\vartheta_2\vartheta_4\frac{\frac{d}{dq}\vartheta[p+\frac{1}{2},q+\frac{1}{2}]}{e^{\pi i p}\vartheta[p,q]},$$

$$W_3 = -\frac{1}{2}\vartheta_2\vartheta_3\frac{\frac{d}{dq}\vartheta[p+\frac{1}{2},q]}{\vartheta[p,q]} \tag{79}$$

where $\vartheta[p,q]$ denotes the theta-function with characteristics of vanishing first argument $\vartheta[p,q](0,i\mu)$; $p,q \in \mathbb{C}$.

The corresponding metric (48) is real and satisfies Einstein's equations with the negative cosmological constant Λ if

$$p \in \mathbb{R}, \qquad \Re q = \frac{1}{2} \tag{80}$$

and the conformal factor F is given by the following formula:

$$F = \frac{2}{\pi\Lambda}\frac{W_1 W_2 W_3}{\left(\frac{d}{dq}\ln\vartheta[p,q]\right)^2}. \tag{81}$$

The metric (48) is real and satisfies Einstein's equations with the positive cosmological constant Λ if

$$q \in \mathbb{R}, \qquad \Re p = \frac{1}{2}, \tag{82}$$

and the conformal factor is given by the same formula (81).

There exists also the additional one-parametric family of solutions of the system (56):

$$W_1 = \frac{1}{\mu+q_0}+2\frac{d}{d\mu}\ln\vartheta_2, \qquad W_2 = \frac{1}{\mu+q_0}+2\frac{d}{d\mu}\ln\vartheta_3,$$

$$W_3 = \frac{1}{\mu+q_0}+2\frac{d}{d\mu}\ln\vartheta_4 \tag{83}$$

($q_0 \in \mathbb{R}$), *which defines manifolds with the vanishing cosmological constant* $\Lambda = 0$ *if the conformal factor* F *is defined by the formula*

$$F = C(\mu + q_0)^2 W_1 W_2 W_3, \qquad (84)$$

where $C > 0$ *is an arbitrary constant.*

The metric coefficients (79) and (83) have nice modular properties which in turn translate into modular properties of objects arising in the spectral geometry of such Einstein's manifolds [23]. These modular properties were analyzed from cosmological perspective in [63].

5 Einstein Equations with Two Killing Symmetries

Here we give another example of appearance of the Schlesinger equations and the Jimbo-Miwa tau-function in the theory of Einstein equations. Our presentation is based on the papers [51, 52, 53]. Unlike the previous example here we consider the Einstein equations without cosmological constant in the physical $3 + 1$ signature. The physical context depends on the choice of two commuting Killing vectors: if both Killing vectors are space-like one gets either the Einstein-Rosen waves with two polarizations or the Gowdy model. If one of these Killing vectors is space-like and another one is time-like, the resulting spacetimes are stationary and axially symmetric. It is the latter case which was considered in the original paper [51] although the solutions found there can be easily extended to two other cases.

The metric on the stationary axially symmetric manifold can be written in the standard form

$$ds^2 = f^{-1}[e^{2k}(dz^2 + d\rho^2) + \rho^2 d\varphi^2] - f(dt + Fd\varphi)^2 \qquad (85)$$

The metric coefficients k, A and f are functions of two coordinates (ρ, z) only. The most non-trivial part of the Einstein equations (called the *Ernst equation*) can be written in terms of only one complex-valued function $\mathcal{E}(\rho, z)$ (the *Ernst potential*):

$$(\mathcal{E} + \overline{\mathcal{E}})(\mathcal{E}_{\rho\rho} + \rho^{-1}\mathcal{E}_\rho + \mathcal{E}_{zz}) = 2(\mathcal{E}_z^2 + \mathcal{E}_\rho^2). \qquad (86)$$

The metric coefficients of (85) can be restored from the Ernst potential via equations

$$f = \Re\mathcal{E} \,, \qquad \frac{\partial F}{\partial \xi} = 2\rho\frac{(\mathcal{E} - \overline{\mathcal{E}})_\xi}{(\mathcal{E} + \overline{\mathcal{E}})^2} \qquad \frac{\partial k}{\partial \xi} = 2i\rho\frac{\mathcal{E}_\xi \overline{\mathcal{E}}_\xi}{(\mathcal{E} + \overline{\mathcal{E}})^2} \qquad (87)$$

where $\xi = z + i\rho$.

The Ernst equation is integrable in the sense of the theory of integrable systems. Being nonlinear itself, it can be represented as the compatibility condition of the associated linear system found in [62, 5]:

$$\Psi_\xi = \frac{G_\xi G^{-1}}{1 - \gamma} \Psi , \qquad \Psi_{\bar\xi} = \frac{G_{\bar\xi} G^{-1}}{1 - \gamma} \Psi \tag{88}$$

where

$$G = \begin{pmatrix} 2 & i(\mathcal{E} - \bar{\mathcal{E}}) \\ i(\mathcal{E} - \bar{\mathcal{E}}) & 2\mathcal{E}\bar{\mathcal{E}} \end{pmatrix} ; \tag{89}$$

$$\gamma = \frac{2}{\xi - \bar\xi} \left\{ w - \frac{\xi + \bar\xi}{2} + \sqrt{(w - \xi)(w - \bar\xi)} \right\} \tag{90}$$

is the "variable spectral parameter" while $w \in \mathbb{C}$ is the constant spectral parameter; $\Psi(x, \rho, w)$ is the 2×2 matrix-valued function.

The link between the Fuchsian Schlesinger system and the Ernst equation was established in [52]. Namely, let a solution Ψ of the system (8) with simple poles (and z replaced by γ) satisfy two additional conditions:

$$\Psi^t(1/\gamma)\Psi^{-1}(0)\Psi(\gamma) = I , \tag{91}$$

$$\Psi(-\bar\gamma) = \overline{\Psi}(\gamma). \tag{92}$$

Let matrices $\{A_j\}$ be the corresponding solution of the Schlesinger system (9), (10) with $z_0 = \infty$ and let $z_j = \gamma(w_j, \xi, \bar\xi)$ where the function γ is given by (90). Then the matrix $G = \Psi(\gamma = 0, \xi, \bar\xi)$ has the structure (89) and the corresponding function $\mathcal{E}$ satisfies the Ernst equation (86).

Moreover, the metric coefficient k is related to the Jimbo-Miwa tau-function of the Schlesinger system by the following formula:

$$e^{2k} = C \prod_j \left\{ \frac{\partial \gamma_j}{\partial w_j} \right\}^{\mathrm{tr}A_j^2/2} \tau_{JM} \tag{93}$$

where C is a constant. We notice that $\mathrm{tr}A_j^2$ are also constants of motion of the Schlesinger system which can be found directly from the monodromy matrices.

This observation means that knowing a solution of Riemann-Hilbert problem with some set of monodromy matrices one can always find related solution of the Ernst equation. In particular, solutions of RH problem with quasi-permutation monodromies described above give rise to algebro-geometric solutions of Ernst equation found in [51].

Define the hyperelliptic curve $\mathcal{R}$ of genus g by

$$y^2 = (w - \xi)(w - \bar\xi) \prod_{j=1}^{2g} (w - w_j) \tag{94}$$

where $w_j \in \mathbb{C}$. The curve (94) is assumed to be real i.e. the branch points w_j are either real or form conjugated pairs. The branch points ξ and $\bar{\xi}$ of the curve $\mathcal{R}$ depend on the space variables (x, ρ). Denote the Abel map on $\mathcal{R}$ by $\mathcal{A}$ and introduce two constant vectors $\mathbf{p}, \mathbf{q} \in \mathbb{C}^g$ which satisfy appropriate reality conditions [51, 53]. These vectors define the monodromy group with off-diagonal 2×2 monodromy matrices, see [39].

The the solution of (94) is given by

$$\mathcal{E}(\xi, \bar{\xi}) = \frac{\Theta \begin{bmatrix} \mathbf{p} \\ \mathbf{q} \end{bmatrix} (\mathcal{A}(\infty^1) - \mathcal{A}(\xi))}{\Theta \begin{bmatrix} \mathbf{p} \\ \mathbf{q} \end{bmatrix} (\mathcal{A}(\infty^2) - \mathcal{A}(\xi))} \tag{95}$$

where ∞^1 and ∞^2 are points at infinity on different sheets of $\mathcal{R}$.

The expression (34) for the tau-function of 2×2 Riemann-Hilbert problem with quasi-permutation monodromies, combined with the link (93) between the conformal factor and the Jimbo-Miwa tau-function, leads then to the following simple formula for the conformal factor e^{2k} defined by equations (87):

$$e^{2k} = \frac{\Theta \begin{bmatrix} \mathbf{p} \\ \mathbf{q} \end{bmatrix} (0) \Theta \begin{bmatrix} \mathbf{p} \\ \mathbf{q} \end{bmatrix} (\frac{1}{2}\mathbf{e})}{\Theta(0)\Theta(\frac{1}{2}\mathbf{e})} \tag{96}$$

where $\mathbf{e} = (1, \ldots, 1)$.

The theta-functional solutions (95) in genus 2 turned out to have a physical significance: they arise in the problem of analytical description of rigidly rotating infinitely thin dust disk (see [40] and references therein for details).

6 Bergman Tau-Function and Moduli Spaces

6.1 Hodge Class on Spaces of Admissible Covers

The Bergman tau-function τ_B (21) turns out to be an efficient tool in the study of geometry of moduli spaces. In particular, the close look at the properties of τ_B allows to interpret it as the natural higher genus analog of the Dedekind's eta-function.

Consider the Hurwitz space $\mathcal{H}_{g,n}$ of pairs $(\mathcal{R}, f)$ where $\mathcal{R}$ is a Riemann surface of genus g and f is a meromorphic function of degree n on $\mathcal{R}$ such that all poles and critical points of f are simple. Then the number of critical points equals to $M = 2g + 2n - 2$. Consider the quotient $\mathcal{H}_{g,n}/\sim$ where $\sim$ is the equivalence relation between pairs $(\mathcal{R}, f)$ and $(\mathcal{R}, (af + b)/(cf + d))$ for an arbitrary $GL(2, \mathbb{C})$ matrix $\begin{pmatrix} a & b \\ c & d \end{pmatrix}$. There exists the compactification of the space $\mathcal{H}_{g,n}/\sim$, called the space of *admissible covers* [33], which we denote by $\mathcal{H}_{g,n}^{adm}$. Study of analytical properties of the Bergman tau-function (21) near various boundary components of the space $\mathcal{H}_{g,n}^{adm}$ allows to get a new relation in the Picard group of this space [46].[1]

The first key fact about τ_B is its transformation law under a change of Torelli marking of $\mathcal{R}$ which is given by the following proposition.

Proposition 6.1 *Under the symplectic change of canonical basis of cycles* (a_i, b_i) *on $\mathcal{R}$ given by*

$$\begin{pmatrix} \tilde{b}_i \\ \tilde{a}_i \end{pmatrix} = \begin{pmatrix} A & B \\ C & D \end{pmatrix} \begin{pmatrix} b_i \\ a_i \end{pmatrix} \tag{97}$$

the function τ_B transforms as follows:

$$\tau_B(\mathcal{R}, f) \to \epsilon \det(C\Omega + D)\tau_B(\mathcal{R}, f) \tag{98}$$

where $\epsilon^{24} = 1$.

This proposition can be easily deduced from the explicit formula (21) (equations (14) give this relation up to a multiplicative constant; to prove that this constant is a 24th root of unity one needs to use the explicit formula (21)).

Another important property of τ_B is given by the following

Proposition 6.2 ([46]) *Let* $V(z_1, \ldots, z_M) = \prod_{i<j}(z_i - z_j)$ *be the Vandermonde determinant of the critical values $z_1, \ldots, z_M$ of f. Then*

$$\eta = \tau_B^{24(M-1)} V^{-6} \tag{99}$$

is invariant under any Möbius transformation $f \to (af + b)/(cf + d)$.

Propositions 6.1 and 6.2 imply that η is the section of the determinant of Hodge vector bundle over $\mathcal{H}_{g,n}^{adm}$. This section is singular on boundary divisors $\Delta_\mu^{(k)}$ of $\mathcal{H}_{g,n}^{adm}$ which are parametrized by the following data [33]: $\mu = [n_1, \ldots, n_r]$ is a partition of $1, \ldots, n$ and $(k, M - k)$ is a partition of the set of critical values. A generic nodal curve from $\Delta_\mu^{(k)}$ is obtained as follows. Consider the degeneration of the base $\mathbb{C}P^1$ into the union of two Riemann spheres such that one of them carries k critical values and another one carries the remaining $M - k$ critical values (such stable curves form the divisor D_k in $\mathcal{M}_{0,M}$). Then μ is the partition corresponding to S_n monodromy of the covering around the nodal point formed on the base. The number r of cycles in the permutation corresponding to the nodal point on the base equals to the number of nodes of the coverings forming $\Delta_\mu^{(k)}$. The union of $\Delta_\mu^{(k)}$ over all μ for a given k is the pre-image of the divisor D_k under the inverse of the

---

[1] The tau-function (2.9) used in [46] equals to 24th power of the Bergman tau-function on a Hurwitz space used in this paper

branching map from the Hurwitz space to the set of critical values of the function f i.e. to $\mathcal{M}_{0,M}$.

By studying the asymptotics of τ_B and η near $\Delta_\mu^{(k)}$ and using (97) we get the following

Theorem 6.3 ([46]) *The Hodge class* $\lambda \in \mathrm{Pic}(\overline{\mathcal{H}}_{g,n}) \otimes \mathbb{Q}$ *can be expressed as follows in terms of the classes of boundary divisors:*

$$\lambda = \sum_{k=2}^{g+n-1} \sum_{\mu=[n_1,\dots,n_r]} \prod_{i=1}^{r} n_i \left(\frac{k(M-k)}{8(M-1)} - \frac{1}{12} \sum_{i=1}^{r} \frac{n_i^2-1}{n_i} \right) \delta_\mu^{(k)} \qquad (100)$$

where $\delta_\mu^{(k)}$ *is the class of divisor* $\Delta_\mu^{(k)}$.

In the case $n = 2$, i.e. for the moduli spaces of hyperelliptic curves of given genus $(M-2)/2$ the formula (100) was found earlier by Cornalba and Harris in [13]:

$$\lambda = \sum_{i=1}^{[(g+1)/2]} \frac{i(g+1-i)}{4g+2} \delta_{[1^2]}^{(2i)} + \sum_{j=1}^{[g/2]} \frac{j(g-j)}{2g+1} \delta_{[2]}^{(2j+1)}. \qquad (101)$$

The formula (100) got later an alternative proof using algebro-geometric techniques and then used to compute the classes of the Mumford divisors in $\mathcal{M}_g$ by van der Geer and Kouvidakis in [29, 30, 31].

The theory of the Bergman tau-function on Hurwitz spaces can be naturally extended to other moduli spaces: spaces of holomorphic Abelian, quadratic and n-differential, as well as to moduli spaces of spin curves. In all of these cases it allows to derive old and new facts about the geometry of these spaces.

6.2 Spaces of Holomorphic Abelian Differentials

Denote by $\mathcal{H}_g$ the Hodge bundle over $\mathcal{M}_g$ and by $\overline{\mathcal{H}}_g$ its extension to $\overline{\mathcal{M}}_g$.

The space $\mathcal{H}_g$ is the union of strata $\mathcal{H}_g(k_1, \dots, k_n)$ such that $k_1 + \cdots + k_n = 2g - 2$; an element of $\mathcal{H}_g(k_1, \dots, k_n)$ is the pair $(\mathcal{R}, v)$, where $\mathcal{R}$ is a Riemann surface of genus g and v is a holomorphic abelian differential on $\mathcal{R}$ with zeros $x_1, \dots, x_n$ of multiplicities $(k_1, \dots, k_n)$. Then the tau-function $\tau_B(\mathcal{R}, v)$ on $\mathcal{H}_g(k_1, \dots, k_N)$ is defined by the formula which looks exactly as (21) but where the divisor of the exact meromorphic differential df is replaced by the divisor of the holomorphic differential v: $(v) = \sum_{j=1}^{n} k_j q_j = \sum_{j=1}^{n} k_j x_j$.

To write down the system of equations for $\tau_B(\mathcal{R}, v)$ similar to (14) we introduce the system of local *homological* coordinates on $\mathcal{H}_g(k_1, \dots, k_n)$ given by integrals of v over a basis of the relative homology group $H_1(\mathcal{R}, \{x_j\}_{j=1}^{n})$; the number $2g + n - 1$ of cycles in this basis coincides with the dimension of $\mathcal{H}_g(k_1, \dots, k_n)$. The set of independent cycles in this homology group can be chosen as follows (this is a special case of the basis (254) in the case of meromorphic differentials):

$$\{s_j\}_{j=1}^{2g+n-1} = (a_i, b_i, l_j), \qquad i = 1, \ldots, g, \quad j = 1, \ldots, n-1 \qquad (102)$$

where l_j is a contour connecting x_1 with x_{j+1}. The homology group $H_1(\mathcal{R} \setminus \{x_j\}_{j=1}^n)$ is dual to $H_1(\mathcal{R}, \{x_j\}_{j=1}^n)$. The basis in the dual group which is dual to the basis (102) in $H_1(\mathcal{R}, \{x_j\}_{j=1}^n)$ is given by (this is the special case of (256)):

$$\{s_j^*\}_{j=1}^{2g+n-1} = (-b_i, a_i, \tilde{c}_j) \qquad (103)$$

where $\tilde{c}_j$ is a small counter-clockwise contour around x_{j+1}; $s_j^* \circ s_k = \delta_{jk}$.

The equations for τ_B on the stratum $\mathcal{H}(k_1, \ldots, k_n)$ are given by the following analog of (14):

$$\frac{\partial \ln \tau_B}{\partial (\int_{s_j} v)} = \int_{s_j^*} \frac{B_{reg}(x,x)}{v(x)}. \qquad (104)$$

The main properties of τ_B on $\mathcal{H}_g(k_1, \ldots, k_n)$ are the following. First, τ_B is holomorphic and non-vanishing as long as the curve $\mathcal{R}$ is non-degenerate i.e. $\mathcal{R}$ does not approach the boundary of $\mathcal{M}_g$ and the zeros of v do not change their multiplicities (i.e. do not merge). Second, as well as in the case of Hurwitz spaces, τ_B transforms as follows under the change of the Torelli marking of $\mathcal{R}$ given by (97):

$$\tau_B(\mathcal{R}, f) \to \epsilon \det(C\Omega + D)\tau_B(\mathcal{R}, f) \qquad (105)$$

where ϵ is a root of unity of degree $12N$ and N is the least common multiple of $k_1 + 1, \ldots, k_n + 1$ (see (4.27) of [38]).

Third, τ_B transforms as follows under the rescaling of v:

$$\tau_B(\mathcal{R}, \kappa v) = \tau_B(\mathcal{R}, v)\kappa^{\frac{1}{12}\sum_{i=1}^n \frac{k_i(k_i+2)}{k_i+1}}. \qquad (106)$$

These properties can not be used to get relations in the Picard group of compactification of any stratum $\mathcal{H}(k_1, \ldots, k_n)$ since the geometric structure of such compactification is not well understood yet. However, when all zeros of v are simple such compactification coincides with $P\overline{\mathcal{H}}_g$. In this case one can prove the following theorem.

Theorem 6.4 ([57]) *In the rational Picard group of* $\mathrm{Pic}(P\overline{\mathcal{H}}_g) \otimes \mathbb{Q}$ *the following relation holds:*

$$\lambda = \frac{g-1}{4}\varphi + \frac{1}{24}\delta_{\deg} + \frac{1}{12}\delta_0 + \frac{1}{8}\sum_{j=1}^{[g/2]} \delta_j. \qquad (107)$$

Here λ is the pullback of the Hodge class from $\overline{\mathcal{M}}_g$ to $\overline{\mathcal{H}}_g$, φ is the first Chern class of the line bundle associated to the projection $\overline{\mathcal{H}}_g \rightarrow P\overline{\mathcal{H}}_g$, $\delta_{\deg}$ is the class of the divisor of abelian differentials with multiple zeros, and δ_j, $j = 0, \ldots, [g/2]$, are the pullbacks of the classes of the Deligne-Mumford boundary divisors from $\overline{\mathcal{M}}_g$ to $P\overline{\mathcal{H}}_g$.

6.3 Spaces of Quadratic Differentials: Hodge and Prym Tau-Functions

Denote by $\mathcal{Q}_g$ the moduli space of pairs $(\mathcal{R}, Q)$ where $\mathcal{R}$ is a Riemann surface of genus g and Q is a holomorphic quadratic differential on $\mathcal{R}$. The canonical covering $\widehat{\mathcal{R}}$ of $\mathcal{R}$ is defined by the equation

$$v^2 = Q \tag{108}$$

in $T^*\mathcal{R}$. Introduce a subspace $\mathcal{Q}_g^0$ of $\mathcal{Q}_g$ corresponding to differentials Q with $4g - 4$ simple zeros at $x_1, \ldots, x_{4g-4}$; we have $\dim \mathcal{Q}_g^0 = \dim \mathcal{Q}_g = 6g - 6$. For a point $(\mathcal{R}, Q) \in \mathcal{Q}_g^0$ denote zeros of Q by $x_1, \ldots, x_{4g-4}$; the genus of the canonical cover $\widehat{\mathcal{R}}$ equals $4g - 3$. The zeros x_j are branch points of the covering $\pi : \widehat{\mathcal{R}} \rightarrow \mathcal{R}$. Denote the involution on $\widehat{\mathcal{R}}$ interchanging the sheets by μ. The linear operator μ_* acting in $H_1(\widehat{\mathcal{R}}, \mathbb{R})$ has eigenvalues ± 1; denote the corresponding eigenspaces by H_+ and H_-. We have $\dim H_+ = 2g$ (H_+ can be identified with $H_1(\mathcal{R}, \mathbb{R})$) and $\dim H_- = 2g_-$ with $g_- = 3g - 3$. The system of homological coordinates on $\mathcal{Q}_g^0$ is given by choosing a symplectic basis $\{s_j\}_{j=1}^{2g_-} = \{a_i^-, b_i^-\}_{i=1}^{g_-}$ in H_- and integrating v over this basis:

$$\mathcal{P}_{s_i} = \int_{s_i} v \tag{109}$$

Similarly, we decompose the space $H^{(1,0)}(\widehat{\mathcal{R}})$ of holomorphic differentials on $\widehat{\mathcal{R}}$ into ± 1 eigenspaces of μ^*: $H^{(1,0)}(\widehat{\mathcal{R}}) = H^+ \oplus H^-$ where elements of H^- are called *Prym differentials*.

Denote the canonical bidifferentials on $\mathcal{R}$ and $\widehat{\mathcal{R}}$ by B and $\widehat{B}$, respectively and define

$$B_{reg}(x, x) = \left(B(x, y) - \frac{v(x)v(y)}{(\int_x^y v)^2} \right) \Bigg|_{y=x}. \tag{110}$$

($B_{reg}(x, x)$ is the meromorphic quadratic differential on $\mathcal{R}$ depending on the choice of the Torelli marking on $\mathcal{R}$; we can also pull it back to $\widehat{\mathcal{R}}$ and denote by the same letter, slightly abusing the notations) and

$$\widehat{B}_{reg}(x, x) = \left(\widehat{B}(x, y) - \frac{v(x)v(y)}{(\int_x^y v)^2} \right) \Bigg|_{y=x} \tag{111}$$

which is the meromorphic quadratic differential on $\widehat{\mathcal{R}}$ depending on the choice of the Torelli marking on $\widehat{\mathcal{R}}$.

Equivalently,

$$B_{reg}(x, x) = \frac{1}{6}(S_B - S_v), \qquad \widehat{B}_{reg}(x, x) = \frac{1}{6}(\widehat{S}_B - S_v) \qquad (112)$$

where S_B and $\widehat{S}_B$ are the Bergman projective connections on $\mathcal{R}$ and $\widehat{\mathcal{R}}$, respectively and $S_v = \{\int^x v, \cdot\}$ is the projective connection (defined on $\widehat{\mathcal{R}}$ and the invariant under μ, thus also giving the projective connection on $\mathcal{R}$) defined by differential v.

Now we define two Bergman tau-functions, $\tau(\mathcal{R}, Q)$ and $\widehat{\tau}(\mathcal{R}, Q)$, by equations

$$\frac{\partial \ln \tau}{\partial \int_{s_i} v} = \int_{s_i^*} \frac{B_{reg}(x, x)}{v(x)}, \qquad (113)$$

$$\frac{\partial \ln \widehat{\tau}}{\partial \int_{s_i} v} = \int_{s_i^*} \frac{\widehat{B}_{reg}(x, x)}{v(x)}. \qquad (114)$$

Let us assume now that the Torelli markings used to define B and $\widehat{B}$ are compatible i.e. we choose the canonical basis in $H_1(\widehat{\mathcal{R}}, \mathbb{Z})$ as follows:

$$\{a_j, a_j^\mu, \tilde{a}_k, b_j, b_j^\mu, \tilde{b}_k\} \qquad (115)$$

such that

$$\mu_* a_j = a_j^\mu, \qquad \mu_* b_j = b_j^\mu, \qquad \mu_* \tilde{a}_k + \tilde{a}_k = \mu_* \tilde{b}_k + \tilde{b}_k = 0 \qquad (116)$$

where $(a_i, b_i, a_i^\mu, b_i^\mu)$ is the pullback of the canonical basis of cycles (a_i, b_i) from $\mathcal{R}$ to $\widehat{\mathcal{R}}$.

Denote by

$$(\hat{v}_j, \hat{v}_j^\mu, \hat{w}_k)$$

the corresponding basis of normalized abelian differentials on $\widehat{\mathcal{R}}$. The differentials $v_j^+ = \hat{v}_j + \hat{v}_j^\mu$ for $j = 1, \ldots, g$ form a basis in the space H^+; these differentials are naturally identified with the normalized holomorphic differentials v_j on $\mathcal{R}$. The basis in H_- is given by $g_- = 3g - 3$ Prym differentials v_l^-:

$$\{\hat{v}_l - \hat{v}_l^\mu, \hat{w}_k\}, \qquad l = 1, \ldots, g, \quad k = 1, \ldots, 2g - 3. \qquad (117)$$

The classes in $H_1(\widehat{\mathcal{R}}, \mathbb{R})$ given by

$$a_j^+ = \frac{1}{2}(a_j + a_j^\mu), \qquad b_j^+ = b_j + b_j^\mu \qquad (118)$$

with the intersection index $a_j^+ \circ b_k^+ = \frac{1}{2}\delta_{jk}$, form the basis in H_+ whereas the classes

$$a_l^- = \frac{1}{2}(a_l - a_l^\mu), \qquad b_l^- = b_l - b_l^\mu \qquad l = 1, \ldots, g \qquad (119)$$

$$a_l^- = \tilde{a}_{l-g}, \qquad b_l^- = \tilde{b}_{l-g} \qquad l = g+1, \ldots, g_- \qquad (120)$$

form the basis in H_- with the intersection indeces $a_j^- \circ b_k^- = \delta_{jk}$.

The period matrix Ω^+ of differentials v_j^+ is equal to twice the period matrix of $\mathcal{R}$:

$$\Omega_{jk}^+ = \int_{b_k^+} v_j^+ = 2\Omega_{jk}. \qquad (121)$$

Integrating the Prym differentials (117) over the cycles (119), (120) we get the matrix Ω^- which is equal to 2Π where Π is the Prym matrix defined in [25], p. 86:

$$\Omega_{jk}^- = \int_{b_k^-} v_j^- = 2\Pi_{jk}. \qquad (122)$$

Then the period matrix $\widehat{\Omega}$ of $\widehat{\mathcal{R}}$ is given by

$$\widehat{\Omega} = T^{-1} \begin{pmatrix} \Omega^+ & 0 \\ 0 & \Omega^- \end{pmatrix} T^{-1} = 2T^{-1} \begin{pmatrix} \Omega & 0 \\ 0 & \Pi \end{pmatrix} T^{-1} \qquad (123)$$

where

$$T = \begin{pmatrix} I_g & I_g & 0 \\ I_g & -I_g & 0 \\ 0 & 0 & I_{g_- - g} \end{pmatrix}. \qquad (124)$$

Under this choice of Torelli marking of $\mathcal{R}$ and $\widehat{\mathcal{R}}$ we introduce (following [47, 10]) the *Hodge tau-function*

$$\tau_+(\mathcal{R}, Q) = \tau(\mathcal{R}, Q) \qquad (125)$$

and *the Prym tau-function*

$$\tau_-(\mathcal{R}, Q) = \frac{\widehat{\tau}(\mathcal{R}, Q)}{\tau(\mathcal{R}, Q)}. \qquad (126)$$

The explicit formula for $\tau_+(\mathcal{R}, Q)$ is given by (21) with only a slight modification: the divisor $\sum k_i q_i$ should be formally replaced by $\sum_{i=1}^{4g-4} \frac{1}{2} x_i$; the distnguished local parameters near x_i should be chosen as $[\int_{x_i}^x v]^{2/3}$ and the vectors $\mathbf{s}, \mathbf{r} \in \mathbb{Z}^g/2$ should be defined by the following analog of (22):

$$\frac{1}{2}\mathcal{A}_x((Q)) + 2K^x + \Omega\mathbf{s} + \mathbf{r} = 0. \qquad (127)$$

The tau-function $\widehat{\tau}(\mathcal{R}, Q)$ is also given by the formula (21), but this time applied to the pair $(\widehat{\mathcal{R}}, v)$ i.e. it coincides with the tau-function on the space of holomorphic Abelian differentials $\tau(\widehat{\mathcal{R}}, v)$ as defined by (21) with $\mathcal{R}$ replaced by $\widehat{\mathcal{R}}$ and the differential df replaced by v; the divisor (v) on $\widehat{\mathcal{R}}$ is given by $\sum_{i=1}^{4g-4} 2x_i)$.

The key property of the tau-functions $\tau_{\pm}$ is their transformation under the change of the symplectic bases in H_+ and H_-. Namely, for a symplectic transformation σ in $H_1(\widehat{\mathcal{R}})$ which is commuting with μ_* and acting by matrices

$$\sigma_{\pm} = \begin{pmatrix} A_{\pm} & B_{\pm} \\ C_{\pm} & D_{\pm} \end{pmatrix} \tag{128}$$

in $H_{\pm}$ the tau-functions $\tau_{\pm}$ transform as follows:

$$\frac{\tau_{\pm}^{\sigma}}{\tau_{\pm}} = \gamma_{\pm}(\sigma)\det(C_{\pm}\Omega_{\pm} + D_{\pm}) \tag{129}$$

where $\gamma_{\pm}^{48}(\sigma) = 1$.

Another important property is the behaviour of $\tau_{\pm}$ under the rescaling of Q:

$$\tau_+(\mathcal{R}, \kappa Q) = \kappa^{\frac{5(g-1)}{36}} \tau_+(\mathcal{R}, \kappa Q) \,, \qquad \tau_-(\mathcal{R}, \kappa Q) = \kappa^{\frac{11(g-1)}{36}} \tau_-(\mathcal{R}, \kappa Q). \tag{130}$$

Both tau-functions, $\tau_{\pm}$, are holomorphic and non-vanishing on $\mathcal{Q}_g^0$. Denote by $D_{\deg}$ the divisor in $\overline{\mathcal{H}}_g$ containing differentials Q with multiple zeros. The asymptotics of $\tau_{\pm}$ near $D_{\deg}$ was computed in [58]. Namely, let two zeros of Q, say x_1 and x_2, merge. It is easy to show (Lemma 8 of [58]) that the transversal local coordinate on $\overline{\mathcal{H}}_g$ in a neighbourhood of $D_{\deg}$ can be chosen to be $t_{\deg} = \int_{x_1}^{x_2} v$, and near $D_{\deg}$ the tau-functions have the following asymptotics:

$$\tau_+ \sim t_{\deg}^{1/72}(const + o(1)) \,, \qquad \tau_- \sim t_{\deg}^{13/72}(const + o(1)). \tag{131}$$

The transversal local coordinate t on $\mathcal{H}_g$ near components of Deligne-Mumford boundary can be expressed in terms of the periods of v along the corresponding vanishing cycle a and its dual cycle b as $t = \exp\{2\pi i \int_b v / \int_a v\}$. The asymptotics of $\tau_{\pm}$ in terms of t can also be computed (Sec 5.2 and 5.3 of [58]) to give $\tau_{\pm} \sim t^{1/12}(1 + o(1))$.

These analytical facts about $\tau_{\pm}$ allow to relate different classes in $Pic(P\overline{\mathcal{Q}}_g, \mathbb{Q})$. Namely, denote by $\lambda = \lambda_H$ the first Chern class of the Hodge vector bundle and by Λ_P the first Chern class of the Prym vector bundle with the fiber H^-. Denote also by φ the first Chern class of the tautological line bundle L associated to the projection $P\overline{\mathcal{Q}}_g \to \overline{\mathcal{Q}}_g$.

Then the analytical properties of $\tau_\pm$ imply the following expressions for the classes λ and λ_P in $Pic(P\overline{\mathcal{Q}}_g, \mathbb{Q})$ (Th.1 of [58]):

$$\lambda = \frac{5(g-1)}{36}\varphi + \frac{1}{72}\delta_{\deg} + \frac{1}{12}\delta_{DM} \tag{132}$$

$$\lambda_P = \frac{11(g-1)}{36}\varphi + \frac{13}{72}\delta_{\deg} + \frac{1}{12}\delta_{DM} \tag{133}$$

where δ_{DM} is the class of pullback of the Deligne-Mumford boundary to $P\overline{\mathcal{Q}}_g$. Excluding the class $\delta_{\deg}$ from (132), (133), we get

$$\lambda_P - 13\lambda = -\sum_{i=0}^{[g/2]}\delta_i - \frac{3g-3}{2}\varphi. \tag{134}$$

For each pair $(\mathcal{R}, Q)$ the vector space H^- of Prym differentials is closely related to the space $\Lambda^{(2)}$ of holomorphic quadratic differentials on $\mathcal{R}$. Namely, for each quadratic differential $\tilde{Q} \in \Lambda^{(2)}$ the ratio $\pi_* \tilde{Q}/v$ in a Prym differential. This relation implies the following equality between the determinant classes: $\lambda_P = p^*\lambda_2 - \frac{3g-3}{2}\psi$ where λ_2 is c_1 of the determinant line bundle of the vector bundle of quadratic differentials and p is the natural projection of $P\overline{\mathcal{Q}}_g$ to $\mathcal{M}_g$. In this formula $3g - 3$ is the dimension of the fiber and $1/2$ appears since multiplication of Q with κ corresponds to transformation $v \to \kappa^{1/2}v$ of v. Now (134) leads to the celebrated Mumford formula

$$\lambda_2 - 13\lambda = \delta_{DM}. \tag{135}$$

6.4 Tau-Functions on Higher Strata of $\mathcal{Q}_g$

The construction of the previous section can be extended to an arbitrary stratum of the space $\mathcal{Q}_g$ with fixed multiplicities of zeros of Q. Namely, consider a stratum $\mathcal{Q}_g^{k,l}$ of the space $\mathcal{Q}_g$ such that the differential Q has m_{odd} zeros of odd multiplicity and m_{even} zeros of even multiplicity. Then the divisor (Q) has the form

$$(Q) = \sum_{i=1}^{m} d_i q_i \equiv \sum_{i=1}^{m_{odd}} (2k_i + 1)x_i + \sum_{i=m_{odd}+1}^{m} 2l_i x_i \tag{136}$$

where $k_i, l_i \geq 0$, $m = m_{odd} + m_{even}$.

The canonical covering $v^2 = Q$ (172) is branched at zeros of odd multiplicity $\{x_i\}_{i=1}^{m_{odd}}$. Note that m_{odd} is always even because $\deg(Q) = 4g - 4$ is even. Therefore the genus of $\widehat{\mathcal{R}}$ equals

$$\hat{g} = 2g + \frac{m_{odd}}{2} - 1. \tag{137}$$

Each zero of even multiplicity $\{x_i\}_{i=m_{odd}+1}^m$ has two pre-images on $\widehat{\mathcal{R}}$ which we denote by $\hat{x}_i$ and $\hat{x}_i^\mu$. The pre-images of poles and zeros of odd multiplicity $\{x_i\}$ on $\widehat{\mathcal{R}}$ are branch points (as before, we continue to denote them by the same letters, omitting the hat).

Decomposing the space of holomorphic differentials into invariant subspaces of μ^*, $H^{(1,0)}(\widehat{\mathcal{R}}) = H^+ \oplus H^-$, and computing their dimensions, we get dim $H^+ = g$ and dim $H^- = g_- = g + \frac{m_{odd}}{2} - 1$. The space H_+ is the fiber of the Hodge vector bundle Λ_H over $\mathcal{Q}_g^{\mathbf{k,l}}$ while the space H^- is the fiber of the Prym vector bundle Λ_P over $\mathcal{Q}_g^{\mathbf{k,l}}$.

$$H^{(1,0)}(\widehat{\mathcal{R}}) = H^+ \oplus H^-, \qquad \dim H^+ = g, \qquad \dim H^- = g_- \qquad (138)$$

where

$$g_- = g + \frac{m_{odd}}{2} - 1. \qquad (139)$$

To introduce homological coordinates on the stratum $\mathcal{Q}_g^{\mathbf{k,l}}$ decompose the relative homology group

$$H_1(\widehat{\mathcal{R}}, \{\hat{x}_j, \hat{x}_j^\mu\}_{j=m_{odd}+1}^m) \qquad \text{as} \qquad H_+ \oplus H_- \qquad (140)$$

where H_+ can be identified with $H_1(\mathcal{R}, \{x_j\}_{j=m_{odd}+1}^m)$. We have for $m_{even} \geq 1$

$$\dim H_1(\widehat{\mathcal{R}}, \{\hat{x}_j, \hat{x}_j^\mu\}_{j=m_{odd}+1}^m) = 2\hat{g} + 2m_{even} - 1,$$

$$\dim H_+ = 2g + m_{even} - 1, \qquad \dim H_- = 2g_- + m_{even}.$$

(for $m_{even} = 0$ we have $\dim H_+ = 2g$, $\dim H_- = 2g_-$ and $g_- = g + m/2 - 1$).

Similarly to the case of all simple zeros we have the equality $\dim \mathcal{Q}_g^{\mathbf{k,l}} = \dim H_-$, and, therefore, choosing some set of independent cycles $\{s_i\}_{i=1}^{\dim H_-}$ in H_- one can use the integrals $\int_{s_i} v$ as local coordinates on the stratum $\mathcal{Q}_g^{\mathbf{k,l}}$.

The homology group dual to (140) is

$$H_1(\widehat{\mathcal{R}} \setminus \{\hat{x}_j, \hat{x}_j^\mu\}_{j=m_{odd}+1}^m) = H_+^* \otimes H_-^*; \qquad (141)$$

its symmetric part H_+^* can be identified with $H_1(\mathcal{R} \setminus \{x_j\}_{j=m_{odd}+1}^m)$.

The skew-symmetric part H_-^* of the group (141) is dual to H_-, and we denote by $\{s_i^*\}$ the basis in H_-^* which is dual to the basis $\{s_i\}$ in H_-.

Now the tau-functions $\tau_+ = \tau(\mathcal{R}, Q)$ and $\hat{\tau}(\mathcal{R}, Q)$ on the space $\mathcal{Q}_g^{\mathbf{k,l}}$ formally coincide with (113) and (114) assuming that the Torelli markings of $\mathcal{R}$ and $\widehat{\mathcal{R}}$ agree.

As well as in the case of simple zeros, we introduce the Prym tau-function via the ratio $\tau_-(\mathcal{R}, Q) = \hat{\tau}(\mathcal{R}, Q)/\tau(\mathcal{R}, Q)$.

The explicit formulas for τ and $\hat{\tau}$ can formally be written in the same form as (21) under an appropriate identification of the corresponding divisors.

Namely, $\tau_+ = \tau(\mathcal{R}, Q)$ is then expressed in terms of the divisor $(Q) = \sum d_i q_i$ as follows:

$$\tau(\mathcal{R}, Q) = C^{2/3}(x) \left(\frac{Q(x)}{\prod_{i=1}^{M} E^{d_i}(x, q_i)} \right)^{(g-1)/6}$$

$$\times \prod_{i<j} E(q_i, q_j)^{\frac{d_i d_j}{24}} e^{\frac{\pi i}{6} \langle s, \Omega s \rangle - \frac{2\pi i}{3} \langle s, K^x \rangle}, \qquad (142)$$

where the vectors $\mathbf{s}, \mathbf{r} \in \mathbb{Z}^g / 2$ are defined by the equations

$$\frac{1}{2} \mathcal{A}_x((Q)) + 2K^x + \Omega \mathbf{s} + \mathbf{r} = 0.$$

Near zeros x_j of odd multiplicities the distinguished local coordinates are given by

$$\zeta_j(x) = \left[\int_{x_j}^{x} v \right]^{2/(2k_j+3)}, \qquad j = 1, \ldots, m_{odd} \qquad (143)$$

and near zeros of even multiplicity by

$$\zeta_j(x) = \left[\int_{x_j}^{x} v \right]^{1/(l_j+1)}, \qquad j = m_{odd} + 1, \ldots, m. \qquad (144)$$

To give the explicit expression for $\hat{\tau}(\mathcal{R}, Q)$ we introduce the following notation for the divisor (v) on $\widehat{\mathcal{R}}$:

$$(v) = \sum \hat{d}_i \hat{q}_i = \sum_{i=1}^{m_{odd}} (2k_i + 2)x_i + \sum_{i=m_{odd}+1}^{m} l_i(\hat{x}_i + \hat{x}_i^\mu). \qquad (145)$$

Let $\widehat{E}$ and $\widehat{C}$ be the prime-form and the multidifferential (247) associated to $\widehat{\mathcal{R}}$. Then

$$\hat{\tau}(\mathcal{R}, Q) = \widehat{C}^{2/3}(x) \left(\frac{v(x)}{\prod_{i=1}^{\hat{m}} \widehat{E}^{\hat{d}_i}(x, \hat{q}_i)} \right)^{(\hat{g}-1)/3} \left(\prod_{i<j} \widehat{E}(\hat{q}_i, \hat{q}_j)^{\hat{d}_i \hat{d}_j} \right)^{1/6}$$

$$\times e^{-\frac{\pi}{6} \langle \widehat{\Omega} \hat{\mathbf{r}}, \hat{\mathbf{s}} \rangle - \frac{2\pi i}{3} \langle \hat{\mathbf{r}}, \widehat{K}^x \rangle} \qquad (146)$$

where the vectors $\hat{\mathbf{r}}, \hat{\mathbf{s}} \in \mathbb{Z}^{\hat{g}}$ are defined by the relation

$$\widehat{\mathcal{A}}_x((v)) + 2\widehat{K}^x + \widehat{\Omega} \hat{\mathbf{r}} + \hat{\mathbf{s}} = 0 \qquad (147)$$

and $\widehat{\mathcal{A}}_x$ is the Abel map on $\widehat{\mathcal{R}}$ with the basepoint x.

The distinguished local coordinates on $\widehat{\mathcal{R}}$ near zeros of v are given by

$$\hat{\zeta}_i = \zeta_i^{1/2} = \left[\int_{x_i}^{x} v \right]^{1/(2k_i+3)} \qquad (148)$$

for $i = 1, \ldots, m_{odd}$ and

$$\widehat{\zeta}_i = \zeta_i = \left[\int_{x_i}^x v \right]^{1/(l_i+1)} \tag{149}$$

for $i = m_{odd} + 1, \ldots, m$.

The tau-functions $\tau_\pm$ on $\mathcal{Q}_g^{\mathbf{k},\mathbf{l}}$ transform according to (129) with $\gamma_\pm^{48\alpha} = 1$ where

$$\alpha = LCM(d_1 + 2, \ldots, d_m + 2).$$

Under a rescaling of Q the functions $\tau_\pm$ transform as $\tau_\pm(\epsilon Q, \mathcal{R}) = \epsilon^{\kappa_\pm} \tau_\pm(Q, \mathcal{R})$ with the homogeneity coefficients given by

$$\kappa_+ = \frac{1}{48} \sum_{i=1}^{\mathbf{M}} \frac{d_i(d_i + 4)}{d_i + 2}, \tag{150}$$

$$\kappa_- = \kappa_+ + \frac{1}{8} \sum_{i=1}^{m_{odd}} \frac{1}{d_i + 2}. \tag{151}$$

The direct use of $\tau_\pm$ to study relations between various classes in the Picard group of compactification of $\mathcal{Q}_g^{\mathbf{k},\mathbf{l}}$ for arbitrary $\{\mathbf{k}, \mathbf{l}\}$ is problematic since the structure of such compactification is not well understood until now.

The explicit formulas for $\tau_\pm$ as well as their homogeneity coefficients turn out to be useful in the problems of the holomorphic factorization of the determinant of Laplacian in flat metrics over Riemann surfaces and also in the theory of Teichmüller flow on moduli spaces, as we discuss in section 10 below.

6.5 Quadratic Differentials with Poles of Second Order and Classes on $\overline{\mathcal{M}}_{g,n}$

Here we show that the tau-functions $\tau_\pm$ on spaces of quadratic differentials with second order poles can be used to get the generalization of the Mumford relation (135) to moduli spaces $\mathcal{M}_{g,n}$.

Let Q be a meromorphic quadratic differential on a curve $\mathcal{R}$ of genus g such that all poles $z_1, \ldots, z_n$ of Q are of second order. We now write the divisor (Q) in the form

$$(Q) = \sum_{i=1}^{m+n} d_i q_i \equiv \sum_{i=1}^{m_{odd}} (2k_i + 1)x_i + \sum_{i=m_{odd}+1}^{m} 2l_i x_i - \sum_{i=1}^{n} 2z_i \tag{152}$$

where $k_i \in \mathbb{Z}, l_i \in \mathbb{N}, m = m_{odd} + m_{even}$.

Let us fix n non-vanishing numbers $p_i \in \mathbb{C}$ and assume that the singular part of Q at z_i looks as follows:

$$Q(\zeta) = -\frac{p_i^2}{4\pi^2} \frac{(d\zeta)^2}{\zeta^2}. \tag{153}$$

The moduli space of pairs $(\mathcal{R}, Q)$ where the divisor of Q has form (152) will be denoted by $\mathcal{Q}_{g,n}^{\mathbf{k},\mathbf{l}}$. The subspace of $\mathcal{Q}_{g,n}^{\mathbf{k},\mathbf{l}}$ such that singular parts of Q near x_i are given by (153) will be denoted by $\mathcal{Q}_{g,n}^{\mathbf{k},\mathbf{l}}[\mathbf{p}]$.

The divisor of Abelian differential of third kind v on the canonical cover $v^2 = Q$ now has the form

$$(v) = \sum_{i=1}^{\widehat{m}} \hat{d}_i \hat{q}_i \equiv \sum_{i=1}^{m_{odd}} (2k_i + 2)x_i + \sum_{i=m_{odd}+1}^{m} l_i(\hat{x}_i + \hat{x}_i^{\mu}) - \sum_{i=1}^{n} (\hat{z}_i + \hat{z}_i^{\mu}) \tag{154}$$

where $\widehat{m} = m_{odd} + 2m_{even} + 2n$.

The genus of $\widehat{\mathcal{R}}$ is agian given by $\hat{g} = g + g_-$ where $g_- = g + m_{odd}/2 - 1$. The distinguished local coordinates on $\mathcal{R}$ and $\widehat{\mathcal{R}}$ near zeros x_i and $\hat{x}_i$ are given by the same formulas (143), (144), (148), (149) as in the case $n = 0$. To define distinguished local coordinates near z_i (both on $\mathcal{R}$ and $\widehat{\mathcal{R}}$) we fix some zero, say, x_i and choose a set of cuts $\gamma_1, \ldots, \gamma_n$ within the fundamental polygon $\mathcal{R}_0$ of $\mathcal{R}$ connecting x_1 with $z_1, \ldots, z_n$. Now we define the local coordinates near z_i and z_i^{μ} by

$$\hat{\xi}_i^{\pm}(x) = \exp\left\{ \frac{\pm 2\pi i}{p_i} \int_{x_1}^{x} v \right\} \tag{155}$$

where the integration paths are chosen not to cross the cuts γ_i.

Now tau-functions τ and $\hat{\tau}$ are defined by formally the same expressions (142), (146) on the whole space $\mathcal{Q}_{g,n}^{\mathbf{k},\mathbf{l}}$ as well as on each subspace $\mathcal{Q}_{g,n}^{\mathbf{k},\mathbf{l}}[\mathbf{p}]$.

However, besides Torelli markings of $\mathcal{R}$ and $\widehat{\mathcal{R}}$ they now depend on the choice of contours γ_i through the distinguished coordinates near z_i and $\hat{z}_i$. On the other hand, the (48α)th power of the expressions

$$\tau_{\pm}(\mathcal{R}, Q) \left(\prod_{i=1}^{n} d\xi_i(z_i) \right)^{1/12}, \tag{156}$$

where

$$\alpha = LCM(d_1 + 2, \ldots, d_{\mathbf{M}} + 2) \tag{157}$$

is independent of the choice of local parameters ξ_i near z_i.

Transformations of $\tau_{\pm}$ under a change of canonical bases in H_+ and H_-, respectively, are again given by (129).

Therefore, on each space $\mathcal{Q}_{g,n}^{k,l}[\mathbf{p}]$ the powers $\tau_{\pm}^{48\alpha}$ are sections of the line bundles $(\det \Lambda_{\pm}) \prod_{j=1}^{n} L_j^{1/12}$ where L_i is the line bundle associated to the marked point z_i (the fiber of L_i is the cotangent space to $\mathcal{R}$ at z_i).

Consider now the largest stratum $\mathcal{Q}_{g,n}^0[\mathbf{p}]$ which corresponds to differentials with all simple zeros $x_1, \ldots, x_{4g-4+2n}$. We have $\dim \mathcal{Q}_{g,n}^0[\mathbf{p}] = \dim \mathcal{Q}_{g,n}[\mathbf{p}] = 6g - 6 + 2n$ which coincides with dimension of $T^*\mathcal{M}_{g,n}$. Actually, the space $\mathcal{Q}_{g,n}[\mathbf{p}]$ can be identified with $T^*\mathcal{M}_{g,n}$ if on each Riemann surface one chooses a quadratic differential Q_0 with given singular part which holomorphically depends on moduli. Than any other differential with the same singular parts can be obtained by adding to it a quadratic differential with simple poles at z_i i.e. a cotangent vector to $\mathcal{M}_{g,n}$. Thus $\mathcal{Q}_{g,n}[\mathbf{p}]$ is the affine space modelled on the vector space $T^*\mathcal{M}_{g,n}$.

For a point $(\mathcal{R}, Q) \in \mathcal{Q}_{g,n}^0[\mathbf{p}]$ we have $\hat{g} = 4g - 3 + n$. Let $D_{\deg} = \mathcal{Q}_{g,n}[\mathbf{p}] \setminus \mathcal{Q}_{g,n}^0[\mathbf{p}]$ be the divisor corresponding to differentials with multiple zeros. Consider the partial compactification $\tilde{\mathcal{Q}}_{g,n}[\mathbf{p}]$ of $\mathcal{Q}_{g,n}[\mathbf{p}]$ obtained by the natural extension of fibers of the affine bundle $\mathcal{Q}_{g,n}[\mathbf{p}]$ to the Deligne-Mumford boundary of $\mathcal{M}_{g,n}$; denote this divisor in $\tilde{\mathcal{Q}}_{g,n}[\mathbf{p}]$ by D_{DM}. Introduce the classes $\psi_i = c_1(L_i)$, the Hodge class $\lambda = c_1(\det \Lambda_+)$ and the Prym class $\lambda_P = c_1(\det \Lambda_-)$ in the Picard group of $\mathcal{Q}_{g,n}^{k,l}[\mathbf{p}]$.

Asymptotics of $\tau_{\pm}$ near $D_{\deg}$ and D_{DM} are the same as in the case of holomorphic quadratic differentials. This implies the following formulas relating the Hodge and Prym classes with classes ψ_i and classes of divisors $D_{\deg}$ and D_{DM} in $Pic(\tilde{\mathcal{Q}}_{g,n}[\mathbf{p}], \mathbb{Q})$:

$$\lambda + \frac{1}{12} \sum_{i=1}^{n} \psi_i = \frac{1}{72} \delta_{\deg} + \frac{1}{12} \delta_{DM}, \tag{158}$$

$$\lambda_P + \frac{1}{12} \sum_{i=1}^{n} \psi_i = \frac{13}{72} \delta_{\deg} + \frac{1}{12} \delta_{DM}. \tag{159}$$

One difference between these formulas and the formulas (132), (133) is the absence of the class φ which appears in (132), (133) due to projectivization of the underlying moduli space. Another difference is the presence of the classes ψ_i in (158), (159).

Linear combinations of (158) and (159) gives the following relations between the classes in $Pic(\tilde{\mathcal{Q}}_{g,n}[\mathbf{p}], \mathbb{Q})$:

$$\lambda_P - 13\lambda = \sum_{i=1}^{n} \psi_i - \delta_{DM} \tag{160}$$

and

$$\lambda_P - \lambda = \frac{1}{6} \delta_{\deg}. \tag{161}$$

Observe now that for each point of $\mathcal{Q}^0_{g,n}[\mathbf{p}]$ the fiber H^- of the Prym bundle can be identified with the space of quadratic differentials on $\mathcal{R}$ with at most simple poles at the marked points. The correspondence is given by the relation

$$\pi^* Q_1 = uv \qquad (162)$$

where $u \in H^-$ is a holomorphic Prym differential on $\widehat{\mathcal{R}}$ and Q_1 is a quadratic differentials on $\mathcal{R}$ with at most simple poles at $z_1, \ldots, z_n$. Since both u and v change their sign under the interchange of the sheets of $\widehat{\mathcal{R}}$ the product uv can be identified with a quadratic differential on $\mathcal{R}$. Such correspondence naturally extends to the space $\tilde{\mathcal{Q}}_{g,n}[\mathbf{p}]$ (see [55]).

Denoting the (pullback from $\overline{\mathcal{M}}_{g,n}$) of the first Chern class of the vector bundle of quadratic differentials with simple poles by $\lambda_2^{(n)}$ we get the equality $\lambda_P = \lambda_2^{(n)}$ and the relations (160), (161) imply

$$\lambda_2^{(n)} - 13\lambda + \sum_{i=1}^{n} \psi_i = -\delta_{DM} \qquad (163)$$

and

$$\kappa_1 = \frac{1}{6}\delta_{\text{deg}} \qquad (164)$$

where $\kappa_1 = \lambda_2^{(n)} - \lambda$ is the pullback of the first kappa-class from $\overline{\mathcal{M}}_{g,n}$ to $\tilde{\mathcal{Q}}_{g,n}[\mathbf{p}]$.

The relation (163) on the affine bundle over $\overline{\mathcal{M}}_{g,n}$ implies the identical relation for the corresponding classes on $\overline{\mathcal{M}}_{g,n}$. This reproduces the classical extension of the Mumford relation (135) to $\overline{\mathcal{M}}_{g,n}$ (see the formula (7.8) of [2]).

The relation (164) is the analog of the Penner expression for the class κ_1 in the hyperbolic combinatorial model [67]; however, unlike the Penner relation, our formula (164) does not have any contribution from hte Deligne-Mumford boundary. We notice that the divisor δ_{deg} in (164) essentially depends on $[\mathbf{p}]$; this situation is parallel to the analog of the Penner formula which is written in the framework of the combinatorial model of $\mathcal{M}_{g,n}$ based on Jenkins-Strebel differentials.

7 Flat Combinatorial Model of $\mathcal{M}_{g,n}$: Tau-Functions and Witten's Classes

The formalism of the previous section can be further applied to the combinatorial model of $\mathcal{M}_{g,n}$ which is based on the Jenkins-Strebel (JS) differentials [10]. This combinatorial model is based on the fact that for a given Riemann surface $\mathcal{R}$ with n marked points $z_1, \ldots, z_n$ and for a given vector $\mathbf{p} \in \mathbb{R}^n_+$

there exists the unique quadratic differential Q with the singular part at z_i of the form (153) which has all real periods of $v = \sqrt{Q}$ on $\widehat{\mathcal{R}}$ (this means that all periods of v over cycles from H_- are real, since all integrals of v over cycles from H_+ vanish). Riemann surfaces with n marked points which are equipped with Jenkins-Strebel differentials form the *flat* combinatorial model $\mathcal{M}_{g,n}[\mathbf{p}]$ of $\mathcal{M}_{g,n}$. This model was developed starting from ideas of Harer, Mumford and Thurston; the modern exposition can be found in [65]). The flat combinatorial model of $\mathcal{M}_{g,n}$ was used in Kontsevich's proof [49] of the Witten conjecture [76]. Another combinatorial model of $\mathcal{M}_{g,n}$ which is based on hyperbolic geometry of $\mathcal{R}$ was proposed in [67].

The cells of $\mathcal{M}_{g,n}[\mathbf{p}]$ with $\mathbf{p} = (p_1, \ldots, p_n) \in \mathbb{R}_+^n$ are labeled by ribbon graphs of given topology on a Riemann surface $\mathcal{R}$ of genus g punctured at n points. The vertices of the ribbon graph are zeros of Q while each face contains one pole of Q.

The quadratic residues at the poles are given by $-p_i^2/4\pi^2$, where $p_1, \ldots, p_n$ are the perimeters of the faces. The lengths of the edges in the metric $|Q|$ are used as coordinates on a given cell; these lengths are equal to (half-integer combinations of) the integrals of v over cycles in H_-. The union of cells with given multiplicities $k_1, \ldots, k_m$ of zeros of Q forms the stratum of $\mathcal{M}_{g,n}[\mathbf{p}]$ labeled by the vector $\mathbf{k}$. The valencies of vertices of the ribbon graph are equal to $k_j + 2$. The special role is played by the strata corresponding to all odd k_i; these strata turn out to be cycles, called *Kontsevich-Witten cycles* [65]. The compactification $\overline{\mathcal{M}}_{g,n}[\mathbf{p}]$ is obtained by adding the *Kontsevich boundary* $W_{1,1}$ to $\mathcal{M}_{g,n}[\mathbf{p}]$. Cells of $W_{1,1}$ of the highest dimension correspond to JS differentials with two simple poles; corresponding ribbon graphs have two univalent vertices while all other vertices are three-valent.

In the main stratum W of $\mathcal{M}_{g,n}[\mathbf{p}]$ all zeros of Q are simple, so that the ribbon graph has only three-valent vertices. The boundaries of real co-dimension 1 of the cells of W (the *facets*) correspond to JS differentials Q with one double zero, while all other zeros remain simple; the corresponding ribbon graphs have one 4-valent vertex while all other vertices are three-valent. The union of cells of W and their facets will be denoted $\tilde{W}$. The complement, $\overline{\mathcal{M}}_{g,n}[\mathbf{p}] \setminus \tilde{W}$, contains cells of (real) co-dimension 2 and higher.

In the (real) co-dimension 2 there exist two special sub-complexes of $\overline{\mathcal{M}}_{g,n}[\mathbf{p}]$ which are also cycles. The first one is the "Witten's cycle" W_5 which corresponds to ribbon graphs with at least one 5-valent vertex; in the cells of highest dimension of W_5 the Strebel differential Q has one zero of order 3 while all other zeros are simple. Ribbon graphs corresponding to cells of W_5 are obtained from three-valent ribbon graphs by degeneration of two edges having one common vertex. The second one is the Kontsevich's boundary $W_{1,1}$; ribbon graphs corresponding to cells of $W_{1,1}$ are obtained from three-valent graphs by degeneration of two edges having two common vertices.

The cycle $W_{1,1}$ is the sum of several sub-cycles that correspond to different topological types of the boundary of $\mathcal{M}_{g,n}$. While $\mathcal{M}_{g,n}[\mathbf{p}]$ is in one-to-one correspondence with $\mathcal{M}_{g,n}$, this isomorphism does not extend to the boundary: some components of the Deligne-Mumford boundary $\overline{\mathcal{M}}_{g,n} \setminus \mathcal{M}_{g,n}$ (namely, the components corresponding to homologically trivial vanishing cycles which leave all punctures in one of the connected components of the stable curve) are not represented by cells of highest dimension of $W_{1,1}$.

The orientation on $\mathcal{M}_{g,n}[\mathbf{p}]$ is induced by the symplectic form $\sum_{i=1}^{n} p_i^2 \psi_1$, where ψ_i is the 2-form representing the corresponding tautological class and which can be written in terms of lengths of the edges [49]. This symplectic form on each cell of a Kontsevich-Witten cycle can be written as $\sum_{k=1}^{g-} dA_k \wedge dB_k$ where A_k and B_k are periods of v over a set of cycles (a_k^-, b_k^-) in H_- satisfying $a_j^- \circ b_k^- = \delta_{jk}$ [10] ((a_k^-, b_k^-) do not form a basis in H_- since H_- also contains combinations of small circles around z_i and z_i^μ, but the integrals of v along these circles are combinations of p_i, which are constant in a given combinatorial model).

The real-analytic complex $\mathcal{M}_{g,n}[\mathbf{p}]$ is nothing but the real slice (i.e. all homological coordinates including the perimeters p_j are real) of the complex-analytic space $\mathcal{Q}_{g,n}[\mathbf{p}]$ considered in previous section. In particular, the stratum W is the real slice of the space $\mathcal{Q}_{g,n}^0[\mathbf{p}]$.

Therefore, the Hodge and Prym vector bundle, as well as tau-functions $\tau_\pm$ can be defined over $\mathcal{M}_{g,n}[\mathbf{p}]$ and $\overline{\mathcal{M}}_{g,n}[\mathbf{p}]$ by restriction of constructions of the previous section to the real slice. However, then $\tau_\pm$ become only real-analytic on each cell of $\mathcal{M}_{g,n}[\mathbf{p}]$. Defining $\tau_\pm$ on each cell of W one can verify that $|\tau_\pm|$ diverge at the facets betwen cells, but $\varphi_\pm = \arg\tau_\pm$ vary continuously and define sections of $U(1)$ (circle) bundles which are combinations of $U(1)$ bundles related to $\det\Lambda_P$, $\det\Lambda_H$ and classes ψ_i.

Further (rather technically non-trivial) computation of the increments of $\arg\tau_\pm$ around Witten's cycle W_5 and the Kontsevich's boundary $W_{1,1}$ within $\tilde{W}$ leads to the following relations which should be understood as equivalences between classes and their Poincare dual cycles [10]:

$$\lambda + \frac{1}{12}\sum \psi_i = \frac{1}{144}W_5 + \frac{13}{144}W_{1,1}, \qquad (165)$$

$$\lambda_2^{(n)} + \frac{1}{12}\sum \psi_i = \frac{13}{144}W_5 + \frac{25}{144}W_{1,1}. \qquad (166)$$

The derivation of these formulas also uses the isomorphism between the Prym vector bundle Λ_P and the vector bundle $\Lambda_2^{(n)}$ of meromorphic quadratic differentials with simple poles at the marked points over $\overline{\mathcal{M}}_{g,n}[\mathbf{p}]$.

Taking into account that $\kappa_1 = \lambda_2^{(n)} - \lambda$ one can express the class κ_1 via W_5 and $W_{1,1}$:

$$12\kappa_1 = W_5 + W_{1,1} \tag{167}$$

reproducing the Arbarello-Cornalba formula [1].

Eliminating the Witten cycle W_5 from (165), (166) we get to the following "combinatorial" version of the Mumford relation (163):

$$\lambda_2^{(n)} - 13\lambda - \sum_{i=1}^{n} \psi_i = -W_{1,1}. \tag{168}$$

8 Other Examples: Moduli of N-Differentials and Spin Moduli Spaces

Here we briefly outline the applications of the formalism of the Bergman tau-function to two other problems: the computation of the class of degenerate n-differentials in the vector bundle of holomorphic N-differentials over $\mathcal{M}_g$ [55] and the computation of the class of degenerate odd spinors in the spin moduli space [4].

8.1 N-Differentials

Denote by $M_g^{(N)}$ the moduli space of pairs $(\mathcal{R}, W)$ where $\mathcal{R}$ is a Riemann surface of genus g and W is a holomorphic N-differential on $\mathcal{R}$; $\dim M_g^{(N)} = (2N + 2)(g - 1)$. Projectivisation of this space admits the natural compactification $PM_g^{(N)}$, see [55]. The pair $(\mathcal{R}, W)$ defines an N-sheeted covering $\widehat{\mathcal{R}}$ of $\mathcal{R}$ via equation

$$v^N = W. \tag{169}$$

The covering (169) possesses $\mathbb{Z}_N$ symmetry; denote the natural automorphism of order N on $\widehat{\mathcal{R}}$ by μ. Then the homology space $H_1(\widehat{\mathcal{R}}, \mathbb{C})$ can be represented as the direct sum of N invariant subspaces of μ_*: $\mathcal{H}_0$ (which can be identified with $H_1(\mathcal{R})$, so $\dim \mathcal{H}_0 = 2g$), $\mathcal{H}_1, \ldots, \mathcal{H}_{N-1}$. If all zeros of W are simple then the genus of $\widehat{\mathcal{R}}$ equals $\hat{g} = n^2(g - 1) + 1$. Then also $\dim \mathcal{H}_k = (2N + 2)(g - 1)$ for $k = 1, \ldots, N - 1$ which coincides with $\dim M_g^{(N)}$. Integrals of v over a basis in $\mathcal{H}_1$ can be used as local homological coordinates on each stratum of $M_g^{(N)}$ (integrals of v over cycles from other $\mathcal{H}_j$ vanish).

The Bergman tau-function on the stratum of generic N-differentials i.e. differentials with simple zeros can formally be defined by the formula (21) where the multiplicities k_i are formally replaced by $1/N$ and points q_i are replaced by zeros $x_1, \ldots, x_{N(2g-2)}$ of W. The same substitution should be done in expressions for distinguished local coordinates (18).

In analogy to the $N = 2$ case, computing the asymptotics of the tau-function near Deligne-Mumford boundary and the subspace D_{deg} of differentials with multiple zeros we get the following relation in $Pic(PM_g^{(N)}, \mathbb{Q})$ [55]:

$$\lambda = \frac{(g-1)(2N+1)}{6N(N+1)} \varphi + \frac{1}{12N(N+1)} \delta_{\mathrm{deg}} + \frac{1}{12} \delta_{DM} \qquad (170)$$

where, as before, φ is the class of the line bundle arising from projectivization, λ is the (pullback from $\mathcal{M}_g$ of) the Hodge class, δ_{deg} is the class of D_{deg}.

The formula (170) will be used below to compute the class of the universal Hitchin discriminant in the universal moduli space of Hitchin spectral covers.

8.2 Spin Moduli Spaces and Farkas-Verra Formula

The spin moduli space $\mathcal{S}_g$ is the space of pairs $(\mathcal{R}, \chi)$ where $\mathcal{R}$ is a Riemann surface of genus g and χ is a spin line bundle over $\mathcal{R}$. This space has two connected components, $\mathcal{S}_g^+$ and $\mathcal{S}_g^-$, corresponding to even and odd χ, respectively, which are finite-sheeted coverings of $\mathcal{M}_g$. We refer to [24] for more detailed description of these spaces and references. The formalism of the Bergman tau-function, being applied to the space $\overline{\mathcal{S}}_g^-$ [4], allows to reproduce the formula for the class of degenerate odd spinors originally derived by Farkas and Verra in Th.0.5 of [24] (this theorem was the main tool in determining the type of spaces $\overline{\mathcal{S}}_g^-$ for $g > 11$).

Here we briefly explain the construction of [4]. An odd spin line bundle for the generic curve $\mathcal{R}$ has exactly one holomorphic section (the co-dimension of the subspace of $\mathcal{M}_g$ where there exist 3 or more holomorphic sections equals two), and this section (for $\mathcal{R} \in \mathcal{M}_g$) is nothing but the spinor h (234) which enters the definition of the prime form.

The divisor $\mathcal{Z}_g$ of [24] corresponds to the spinors h with one zero of order 2 while other zeros remain simple. To compute the class of this divisor via the tau-function formalism consider the Bergman tau-function on the stratum $\mathcal{H}_g(2, \ldots, 2)$ of Abelian differentials (the Abelian differential v is taken to be h^2). Then $\mathcal{Z}_g$ is a subspace of the boundary component of $\mathcal{H}_g(2, \ldots, 2)$ where two zeros merge, i.e. $\mathcal{Z}_g \subset \mathcal{H}_g(4, 2, \ldots, 2)$. The computation of the asymptotics of $\tau_B(\mathcal{R}, h^2)$ near $\mathcal{Z}_g$ and boundary components of $\mathcal{S}_g^-$ which correspond to DM boundary of $\mathcal{M}_g$ gives, as shown in [4], the following expression of the class of $\mathcal{Z}_g$ in terms of the Hodge class λ and boundary divisors. This expression reproduces the original Farkas-Verra formula of [24]:

$$[\mathcal{Z}_g] = (g+8)\lambda - \frac{g+2}{4}[A_0] - 2[B_0] - \sum_{j=1}^{[g/2]} 2(g-j)[A_j] - \sum_{j=1}^{[g/2]} 2j[B_j]$$

(171)

where for each j the union of divisors $A_j \cup B_j$ in the formula (171) is the pullback of the component D_j of the Deligne-Mumford boundary. For $j = 1, \ldots, [g/2]$ the divisor A_j corresponds to a reducible curve such that on the component of genus j the spin line bundle is odd. The divisor B_j corresponds to reducible curve such that on component of genus j the spin line bundle is even. The definition of the divisors A_0 and B_0 whose union forms the pullback of the irreducible Deligne-Mumford boundary components Δ_0 is slightly more subtle, and we refer to [24, 4] for more details.

9 Moduli Spaces of Hitchin's Spectral Covers

Hitchin in [35, 36] proposed the dimensional reduction of the self-dual Yang-Mills equations by splitting the 4-dimensional space into the product of a Riemann surface $\mathcal{R}$ and the real plane $\mathbb{R}^2$; the gauge fields are assumed to be independent of coordinates on $\mathbb{R}^2$. As a result of such dimensional reduction, one arrives at the class of the finite-dimensional completely integrable systems, called *Hitchin's systems*. Here we refer to Atiyah's book [3] (Sect. 6.3) for an introduction to the topic and to the original papers of Hitchin [35, 36]. In the paper by van Greemen and Previato [32] the detailed treatment of genus 2 case was given; [15] contains the detailed review of the subject. Hitchin systems provide the most general class of integrable systems associated to Riemann surfaces of an arbitrary genus.

The Hamiltonians of the Hitchin system are encoded in the so-called *spectral cover* $\widehat{\mathcal{R}}$ which is the n-sheeted cover of $\mathcal{R}$ defined by the following equation in $T^*\mathcal{R}$:

$$\widehat{\mathcal{R}} = \{(x, v) \in \mathcal{R} \times T_x^*\mathcal{R} \mid P_n(v) = 0\} \tag{172}$$

where

$$P_n(v) = v^n + Q_{n-1}v^{n-1} + \cdots + Q_1 v + Q_0,$$

Q_k is a holomorphic $n-k$-differential on $\mathcal{R}$, and v is a holomorphic 1-form on $\widehat{\mathcal{R}}$. In the framework of [35] the equation (172) is given by the characteristic polynomial $P_n(V) = \det(\Phi - vI)$ of the *Higgs field* Φ on $\mathcal{R}$ (Φ is a section of $ad_\chi \otimes K$ where χ is a vector bundle and K is the canonical line bundle over $\mathcal{R}$).

For the case of $GL(n)$ Hitchin's systems all differentials Q_k from (172) are arbitrary; in the case of $SL(n)$ systems $Q_1 = 0$.

The branch points of the cover $\widehat{\mathcal{R}}$ are zeros of the discriminant W of $P_n(v)$ which is the holomorphic $n(n-1)$-differential on $\mathcal{R}$. Thus, the number m of zeros of W, counted with multiplicities, equals

$$m = n(n-1)(2g-2), \tag{173}$$

and the Riemann-Hurwitz formula gives the genus $\hat{g}$ of $\widehat{\mathcal{R}}$:

$$\hat{g} = n^2(g-1)+1 \tag{174}$$

so that $m = 2(\hat{g}-1-n(g-1))$. When all zeros of W are simple, all branch points of the covering $\pi : \widehat{\mathcal{R}} \to \mathcal{R}$ are also simple.

Let $\mathcal{R}$ be a smooth curve of genus g and denote by $\mathcal{M}_H^{\mathcal{R}}$ the moduli space of $GL(n)$ Hitchin spectral covers of the form (172). Then

$$\mathcal{M}_H^{\mathcal{R}} = \bigoplus_{j=1}^{n} H^0(\mathcal{R}, K_{\mathcal{R}}^{\otimes j}) \tag{175}$$

where $K_{\mathcal{R}} = T^*\mathcal{R}$ is the canonical line bundle on $\mathcal{R}$, and

$$\dim \mathcal{M}_H^{\mathcal{R}} = \hat{g} = n^2(g-1)+1 \tag{176}$$

(recall that $\dim H^0(\mathcal{R}, K) = g$ and $\dim H^0(\mathcal{R}, K^j) = (2j-1)(g-1)$ for $j \geq 2$).

There is the natural coordinate system on $\mathcal{M}^{\mathcal{R}}$ given by the a-periods of v on $\widehat{\mathcal{R}}$

$$A_j = \oint_{\hat{a}_j} v \tag{177}$$

where $\{\hat{a}_j, \hat{b}_j\}_{j=1}^{\hat{g}}$ is a canonical symplectic basis in $H_1(\widehat{\mathcal{R}}, \mathbb{Z})$.

9.1 Class of Universal Hitchin's Discriminant Locus

The *Hitchin's discriminant locus* $D_H^{\mathcal{R}} \subset \mathcal{M}^{\mathcal{R}}$ is defined by the condition that the discriminant W of $P_n(v)$ has multiple zeroes i.e. $\widehat{\mathcal{R}}$ has non-simple branch points.

Denote by $\mathcal{M}_H$ the space of all $GL(n)$ spectral covers for a fixed genus of $\mathcal{R}$ (the base $\mathcal{R}$ is allowed to vary);

$$\dim \mathcal{M}_H = \dim \mathcal{M}_H^{\mathcal{R}} + 3g - 3 = (n^2+3)(g-1)+1. \tag{178}$$

For the precise definition of the space $\mathcal{M}_H$ and the compactification of its projectivization (with respect to the equivalence under rescaling $Q_j \mapsto \epsilon^j Q_j$, $\epsilon \in \mathbb{C}^*$, $j = 1, \ldots, n$.) $P\overline{\mathcal{M}}_H$ we refer to [60]. The divisor $D_H \subset P\overline{\mathcal{M}}_H$ is called the *universal Hitchin discriminant locus*; it's the union of (projectivizations) of $D_H^{\mathcal{R}}$ over all stable curves $\mathcal{R} \in \overline{\mathcal{M}}_g$.

Two zeros of W can coalesce in three different ways. First, the double branch point can form on $\widehat{\mathcal{R}}$; this component of D_H is called the *caustic* and denoted by $D_H^{(c)}$. Second, two simple branch points with the same projection to $\mathcal{R}$ can form; this component of D_H is called the *Maxwell stratum* and denoted by $D_H^{(m)}$. Finally, the node can form i.e. $\widehat{\mathcal{R}}$ gets to the boundary of $\mathcal{M}_{\hat{g}}$. This component of D_H is called *the boundary* and denoted by $D_H^{(b)}$. The simple local analysis [60] shows that the divisor class of D_H equals

$$[D_H] = [D_H^{(b)}] + 2[D_H^{(m)}] + 3[D_H^{(c)}]. \tag{179}$$

Consider the following natural bundles over the moduli space $P\overline{\mathcal{M}_H}$:

- The Hodge vector bundle $\Lambda \to \overline{M}_g$ (the fiber of Λ over a smooth $\mathcal{R}$ is the g-dimensional vector space of holomorphic 1-forms on $\mathcal{R}$). This bundle naturally lifts to $P\overline{\mathcal{M}_H}$, and we put $\lambda = c_1(\Lambda)$.
- The Hodge vector bundle $\hat{\mathcal{L}} \to \overline{M}_{\hat{g}}$ (the fiber of $\hat{\mathcal{L}}$ over a smooth spectral cover $\widehat{\mathcal{R}}$ is the $\hat{g}$-dimensional vector space of holomorphic 1-forms on $\widehat{\mathcal{R}}$). This bundle also lifts to $P\overline{\mathcal{M}_H}$, and, similarly, we put $\hat{\lambda} = c_1(\hat{\mathcal{L}})$.
- There is the natural action of $\mathbb{C}^*$ on $\mathcal{M}_H$ by

$$Q_k \mapsto \mu^k Q_k, \qquad \mu \in \mathbb{C}^*. \tag{180}$$

Denote by $P\mathcal{M}_H$ the projectivization of $\mathcal{M}_H$ with respect to the action (180). Let L be the tautological line bundle of the natural projection $\mathcal{M}_H \to P\mathcal{M}_H$. The bundle L extends to $P\overline{\mathcal{M}_H}$, and we put $\varphi = c_1(L)$.

The space $\mathcal{M}_H$ can be naturally embedded into the moduli space of holomorphic $n(n-1)$ differentials on Riemann surfaces of genus g: the $n(n-1)$ - differential corresponding to a point of $\mathcal{M}_H$ is the discriminant W. This allows to define the tau-function τ_B on $\mathcal{M}_H$ by the pullback of the tau-function $\tau(\mathcal{R}, W)$ from the moduli space of N-differentials for $N = n(n-1)$ (notice that the dimension of $\mathcal{M}_H$ which equals to $3g - 3 + n^2(g-1) - 1 = (n^2 + 3)(g - 1) + 1$ is smaller than the dimension of the moduli space of $n(n-1)$ differentials which equals to $(2n(n-1) - 1)(g-1) + 3g - 3 = (2n^2 - 2n + 3)(g-1))$.

As before, computing the asymptotics of the tau-function near the different components of D_H we conclude that the class D_H of the universal Hitchin discriminant (179) is expressed in terms of the standard generators of $\mathrm{Pic}(P\overline{\mathcal{M}_H}) \otimes \mathbb{Q}$ as follows:

$$\frac{1}{n(n-1)}[D_H] = (n^2 - n + 1)(12\lambda - \delta) - 2(g-1)(2n^2 - 2n + 1)\varphi. \tag{181}$$

Here λ is (the pullback of) the Hodge class on $\overline{M}_g$, φ is the tautological class of the projectivization $\overline{\mathcal{M}}_H \to P\overline{\mathcal{M}}_H$, and δ is the pullback of the class of the $\overline{\mathcal{M}}_g \setminus \mathcal{M}_g$ to $P\overline{\mathcal{M}}_H$.

9.2 Variation of the Period Matrix

To conclude this section we show how the variational formulas on the space $\mathcal{M}_{\mathcal{R}}$ can be deduced from the variational formulas (Th.A.3) on strata of the moduli space of Abelian differentials found in [45]. This section is based on the paper [11] to which we refer for more details.

We shall discuss here only variations of the period matrix $\widehat{\Omega}$ of $\widehat{\mathcal{R}}$ on the moduli space $\mathcal{M}_H^{\mathcal{R}}$ with a fixed base $\mathcal{R}$. We get these formulas by the pullback of the variational formulas on the space $H_{\hat{g}}(1, \ldots, 1)$ of abelian differentials on Riemann surfaces of genus $\hat{g}$. The map of $\mathcal{M}_H^{\mathcal{R}}$ to $H_{\hat{g}}(1, \ldots, 1)$ is defined by assigning to a point of $\mathcal{M}_H^{\mathcal{R}}$ the pair $(\widehat{\mathcal{R}}, v)$; for a generic point of $\mathcal{M}_H^{\mathcal{R}}$ all zeros of v are simple. Assume that the branch points of $\widehat{\mathcal{R}}$ i.e. the zeros of W are also simple. The total number of zeros of v is $2\hat{g} - 2 = 2n^2(g - 1)$; in a general position $(v) = D_{br} + D_0$ where D_{br} is the divisor of branch points of $\widehat{\mathcal{R}}$; the projection of D_{br} on $\mathcal{R}$ coincides with the divisor of discriminant W: $\pi(D_{br}) = (W)$; $\deg D_{br} = n(n - 1)(2g - 2)$. The projection of D_0 on $\mathcal{R}$ coincides with the divisor of the n-differential Q_0: $\pi(D_0) = (Q_0)$; $\deg D_0 = n(2g - 2)$ i.e. $\deg D_{br} + \deg D_0 = \deg(v)$ as expected. Let

$$D_{Br} = \{x_i\}_{i=1}^{\deg D_{br}} , \qquad D_0 = \{x_i\}_{\deg D_{br}+1}^{\deg(v)} .$$

The homological coordinates on $H_{\hat{g}}(1, \ldots, 1)$ are given by (255):

$$A_j = \int_{\hat{a}_j} v , \qquad B_j = \int_{\hat{b}_j} v , \qquad j = 1, \ldots, \hat{g},$$

$$z_k = \int_{x_{2\hat{g}-2}}^{x_k} , \qquad k = 1, \ldots, 2\hat{g} - 3.$$

On the submanifold $\mathcal{M}_H^{\mathcal{R}}$ one can choose A_j to be the set of independent coordinates; then all remaining coordinates B_j and z_j become A_j-dependent. Their dependence on A_j's can be computed taking into account that (see [11] for the proof)

$$\frac{\partial v(x)}{\partial A_j} = v_j(x) \qquad (182)$$

where v_j is a holomorphic differential on $\widehat{\mathcal{R}}$ normalized by $\int_{\hat{a}_k} v_j = \delta_{jk}$ and the projection of x on the base $\mathcal{R}$ is assumed to be fixed under differentiation. Then we have on $\mathcal{M}_H^{\mathcal{R}}$

$$\frac{\partial B_k}{\partial A_j} = \widehat{\Omega}_{jk}, \qquad j, k = 1, \ldots, \hat{g} \tag{183}$$

and, since v vanishes at the endpoints of contours $[x_{2\hat{g}-2}, x_k]$ for $k = \deg D_{br} + 1, \ldots, 2\hat{g} - 3$,

$$\frac{\partial z_k}{\partial A_j} = \int_{x_{2\hat{g}-2}}^{x_k} v_j, \qquad k = \deg D_{br} + 1, \ldots, 2\hat{g} - 3. \tag{184}$$

The variational formulas for z_k, $k = 1, \ldots, \deg D_{br}$ are more subtle.

Let $x_k \in \widehat{\mathcal{R}}$ be a ramification point of $\widehat{\mathcal{R}}$ and $\xi_k = \xi(\pi(x_k)) \in \mathcal{R}$ be the corresponding critical value in some local coordinate ξ on $\mathcal{R}$. We assume that ξ remains fixed under a deformation of $\widehat{\mathcal{R}}$; let $\zeta = \xi - \xi_k$ be a coordinate on $\mathcal{R}$ vanishing at $\pi(x_k)$ (the coordinate ζ deforms when $\widehat{\mathcal{R}}$ varies). The local coordinate on $\widehat{\mathcal{R}}$ near x_k can then be chosen to be $\widehat{\zeta}(x) = \zeta^{1/2}$.

Then the differentiation of the endpoint of the integration contour with respect to A_j also gives the contribution to $\partial z_k / \partial A_j$ and we get [11]

$$\frac{\partial z_k}{\partial A_j} = \int_{x_{2\hat{g}-2}}^{x_k} v_j + \frac{\partial \xi_k}{\partial A_j} \frac{v}{d\xi}(\xi_k) \tag{185}$$

for $k = 1, \ldots, \deg D_{br}$.

To compute the derivative $\partial \xi_k / \partial A_j$ we write $v(\xi)$ near ξ_k in the form

$$v = (a + b\sqrt{\xi - \xi_k} + \ldots) d\xi \tag{186}$$

(recall that v has simple zero in the local parameter $\sqrt{\xi - \xi_k}$). Then

$$b = \frac{d(v/d\xi)}{d\sqrt{\xi - \xi_k}}\bigg|_{\xi = \xi_k} \tag{187}$$

and

$$v_j = \frac{\partial v}{\partial A_j} = \left(a_{A_j} - \frac{(\xi_k)_{A_j}}{2\sqrt{\xi - \xi_k}} b + \ldots \right) d\xi. \tag{188}$$

Therefore,

$$-\frac{v_j}{d\sqrt{\xi - \xi_k}}\bigg|_{\xi = \xi_k} = b\frac{\partial \xi_k}{\partial A_j} \tag{189}$$

and,

$$\frac{\partial \xi_k}{\partial A_j} = -\frac{v_j/d\widehat{\zeta}_k}{(v/d\xi)_{\widehat{\zeta}_k}}(x_k). \tag{190}$$

Now (185) takes the form

$$\frac{\partial z_k}{\partial A_j} = \int_{x_{2\hat{g}-2}}^{x_k} v_j - \frac{v_j/d\widehat{\zeta}_k}{[\ln(v/d\xi)]_{\widehat{\zeta}_k}}(x_k) \tag{191}$$

for $k = 1, \ldots, \deg D_{br}$.

Now, applying the chain rule we get on $\mathcal{M}_H^{\mathcal{R}}$

$$\frac{d\widehat{\Omega}_{lk}}{dA_j} = \frac{\partial\widehat{\Omega}_{lk}}{\partial A_j} + \sum_{s=1}^{\hat{g}}\frac{\partial\widehat{\Omega}_{lk}}{\partial B_s}\frac{\partial B_s}{\partial A_j} + \sum_{r=1}^{2\hat{g}-3}\frac{\partial\widehat{\Omega}_{lk}}{\partial z_r}\frac{\partial z_r}{\partial A_j}. \tag{192}$$

Using the variational formulas (258)

$$\frac{\partial\widehat{\Omega}_{lk}}{\partial A_j} = -\int_{\hat{b}_j}\frac{v_l v_k}{v}, \qquad \frac{\partial\widehat{\Omega}_{lk}}{\partial B_j} = \int_{\hat{a}_j}\frac{v_l v_k}{v}, \qquad \frac{\partial\widehat{\Omega}_{lk}}{\partial z_r} = 2\pi i \left. \text{res}\right|_{x_r}\frac{v_l v_k}{v} \tag{193}$$

for $r = 1\ldots, 2\hat{g}-3$, we can transform (192) by applying the Riemann bilinear identity to the pair $(v_j, \frac{v_l v_k}{v})$ to get the sum of residues only at the branch points of $\widehat{\mathcal{R}}$. The result looks as follows (see [11] for details):

$$\frac{d\widehat{\Omega}_{lk}}{dA_j} = -2\pi i \sum_{\substack{branch\ points\ x_r}} \frac{v_j}{d\ln(v/d\xi)}(x_r) \left.\text{res}\right|_{x_r}\frac{v_\alpha v_\beta}{v} \tag{194}$$

or, equivalently, since the differential under the residue has the first order poles at x_r, we get the more familiar form of this variational formula which makes the symmetry between (l, k, r) manifest:

Proposition 9.1

$$\frac{d\widehat{\Omega}_{lk}}{dA_j} = -2\pi i \sum_{\substack{branch\ points\ x_r}} \left.\text{res}\right|_{x_r}\frac{v_l v_k v_r}{d\xi\, d(v/d\xi)} \tag{195}$$

where ξ is a local coordinate on $\mathcal{R}$ near x_r.

Notice that the right-hand side of (195) is independent of the choice of the local coordinates coordinates near x_r. The formula (195) is compatible with variational formula for $B(x, y)$ derived in [7]. It provides the explicit version of the variation of moduli based on the Donagi-Markman cubic [15].

10 Tau-Functions, Determinants of Flat Laplacians and Teichmüller Flow

10.1 Determinant of Laplace Operator in Flat Metrics with Trivial and $\mathbb{Z}_2$ Holonomy

Any metric with trivial holonomy and finite volume on a Riemann surface $\mathcal{R}$ can be represented in the form $|v|^2$ where v is a holomorphic abelian differential on $\mathcal{R}$. Denote zeros of v by $x_1, \ldots, x_N$ and their multiplicities by $m_1, \ldots, m_N$. Since the metric has conical singularities, the definition of the Laplace operator Δ is ambiguous; the most natural is the Friedrich's self-adjoint extension (see [45]). The spectrum of Δ can then be shown to be discrete and the determinant of Laplacian can then be defined via the usual ζ-function regularization:

$$\det\Delta = e^{-\zeta'(0)}, \qquad \zeta(s) = \sum_{n=1}^{\infty} \lambda_n^s$$

where $\lambda_1, \lambda_2, \ldots$ are eigenvalues of Δ.

Then the determinant of Laplacian can be expressed as follows [45]

$$\det\Delta^{C,|v|^2} = const\,(\det\Im\Omega)\,\mathrm{Area}(\mathcal{R}, |v|^2)\,|\tau(\mathcal{R}, v)|^2 \qquad (196)$$

where $\tau(\mathcal{R}, v)$ is the tau-function on the connected component of the moduli space $\mathcal{H}(m_1, \ldots, m_N)$; Ω is the period matrix of $\mathcal{R}$ and

$$\mathrm{Area}(\mathcal{R}, |v|^2) = \Im \sum_{j=1}^{g} A_j \overline{B}_j. \qquad (197)$$

The constant in (196) depends on the partition $(m_1, \ldots, m_N)$ and may depend also on the connected component of the moduli space $\mathcal{H}(m_1, \ldots, m_N)$.

The formula (196) is the natural generalization of the first Kronecker limit formula which holds in genus one. Namely, on the torus with the flat coordinate z and the periods (A, B)

$$\det\Delta^{|dz|^2} = const\,\Im(B/A)\Im(A\overline{B})|\eta(B/A)|^4$$

where the constant is explicitly computable.

Similarly, an arbitrary metric on $\mathcal{R}$ with $\mathbb{Z}_2$ holonomy and finite volume has the form $|Q|$ where Q is a holomorphic quadratic differential on $\mathcal{R}$. Denote multiplicities of zeros of Q by $k_1, \ldots, k_M$ and define the canonical cover $\widehat{\mathcal{R}}$ by the equation

$$v^2 = Q$$

whose genus equals $\hat{g} = 2g + m_{odd}/2 + 1$ where m_{odd} is the number of zeros of Q of odd multiplicity. Then v is a holomorphic abelian differential on $\widehat{\mathcal{R}}$; multiplicities of zeros of v are determined by $k_1, \ldots, k_M$.

Introduce the Hodge tau-function $\tau(\mathcal{R}, Q)$ and the tau-function $\widehat{\tau}(\widehat{\mathcal{R}}, v)$; define also the Prym tau-function via the ratio (126)

$$\tau_-(\mathcal{R}, Q) = \frac{\tau(\widehat{\mathcal{R}}, v)}{\tau(\mathcal{R}, Q)}. \qquad (198)$$

The homogeneity coefficient of τ_- is the difference between the homogeneity coefficients of $\tau(\widehat{\mathcal{R}}, v)$ and $\tau(\mathcal{R}, Q)$:

$$\kappa_- = \kappa_+ + \frac{1}{8} \sum_{k_j \text{ even}} \frac{1}{k_j + 2}. \tag{199}$$

The formula for the determinant of Laplacian on $\mathcal{R}$ in the metric $|Q|$ is given by the formula which looks like a direct analog of (196):

$$\det\Delta^{\mathcal{R},|Q|} = const \, (\det\Im\Omega) \, \text{Area}(\mathcal{R}, |Q|) |\tau(\mathcal{R}, Q)|^2 \tag{200}$$

where the area is expressed in terms of the periods $(A_i, B_i)_{i=1}^{g-}$ of v:

$$\text{Area}(\mathcal{R}) = \Im \sum_{j=1}^{g-} A_i \overline{B}_i. \tag{201}$$

The constant in the r.h.s. of (200) depends on the connected component of the space $\mathcal{Q}(k_1, \ldots, k_M)$.

On the other hand, applying the formula (196) to the canonical cover $\widehat{\mathcal{R}}$ equipped with the abelian differential $v = \sqrt{Q}$ we have the following formula for the determinant of Laplacian on $\widehat{\mathcal{R}}$:

$$\det\Delta^{\widehat{\mathcal{R}},|Q|} = const \, (\det\Im\widehat{\Omega}) \, \text{Area}(\widehat{\mathcal{R}}, |Q|) \, |\tau(\widehat{\mathcal{R}}, v)|^2 \tag{202}$$

where

$$\text{Area}(\widehat{\mathcal{R}}, |Q|) = 2\text{Area}(\mathcal{R}, |Q|); \tag{203}$$

and the constant in (202) may be different from the constant in (200), but may depend on the connected component in the moduli space $\mathcal{Q}(k_1, \ldots, k_M)$ as well.

Take the ratio of (202) and (200) and define the *Prym Laplacian* $\Delta_-^{\widehat{\mathcal{R}},Q}$ which is the Laplacian on $\widehat{\mathcal{R}}$ with metric $|Q|$, restricted to the subspace of functions which are skew-symmetric under the involution interchanging the sheets of $\widehat{\mathcal{R}}$. Then the determinant of $\Delta_-^{\widehat{\mathcal{R}},Q}$ is naturally defined by

$$\det\Delta_-^{\widehat{\mathcal{R}},|Q|} = \frac{\det\Delta^{\widehat{\mathcal{R}},|Q|}}{\det\Delta^{\mathcal{R},|Q|}}. \tag{204}$$

The determinant of Prym Laplacian can then be expressed as follows in terms of Prym tau-function τ_-:

$$\det\Delta_-^{\widehat{\mathcal{R}},|Q|} = const \, (\det\Im\Pi) \, |\tau_-(\widehat{\mathcal{R}}, Q)|^2 \tag{205}$$

where $\tau_-(\widehat{\mathcal{R}}, Q)$ is the Prym tau-function (198) and Π is the Prym matrix of $\widehat{\mathcal{R}}$ defined by (123). To derive (205) from (200) and (202) one needs to use the relation (203) between the areas of $\mathcal{R}$ and $\widehat{\mathcal{R}}$ and the relation

$$\frac{\det \Im \widehat{\Omega}}{\det \Im \Omega} = \det \Im \Pi \tag{206}$$

which follows from the definition of the Prym matrix [25].

10.2 Tau-Functions in the Theory of Teichmüller Flow

Tau functions on moduli spaces of abelian and holomorphic differentials naturally appear in the theory of the Teichmüller flow. We are going to briefly describe the context following [18, 19].

10.2.1 Lyapunov Exponents of Hogde Bundle

Consider any stratum $\mathcal{H}_g(k_1, \ldots, k_n)$ of the moduli space of holomorphic differentials. The Teichmüller geodesic flow on $\mathcal{H}_g(k_1, \ldots, k_n)$ is defined by the action of diagonal matrices of the form $\begin{pmatrix} e^s & 0 \\ 0 & e^{-s} \end{pmatrix}$ in the plane of the flat coordinate z which is defined as the abelian integral of the holomorphic differential v ($(\mathcal{R}, v)$ is a point of this stratum).

An arbitrary $SL(2, \mathbb{R})$ linear transformation of z-coordinate is also acting on the stratum $\mathcal{H}_g(k_1, \ldots, k_n)$; this defines foliation of the stratum into the orbits of $SL(2, \mathbb{R})$. Then projectivisation $P\mathcal{H}_g(k_1, \ldots, k_n)$ is foliated into the union of Teichmüller disks which are isomorphic to $SL(2, \mathbb{R})/SO(2)$ i.e. to the upper half-plane.

The $SL(2, \mathbb{R})/SO(2)$ matrix can be parametrized by a complex parameter t as follows:

$$M_t = \frac{1}{t + \bar{t}} \begin{pmatrix} 2 & i(t - \bar{t}) \\ i(t - \bar{t}) & 2t\bar{t} \end{pmatrix}. \tag{207}$$

On the Teichmüller disk parametrized by $t \in \mathbb{C}$ one defines the hyperbolic "Teichmüller" Laplacian [18] which is the hyperbolic Laplacian in the metric of curvature -4):

$$\Delta^T = 16(\Re t)^2 \partial^2_{t\bar{t}}.$$

For the definition of the Lyapunov exponents of the Teichmüller flow we give the quote from the section 1.4 of [18]:

"Informally, the Lyapunov exponents of a vector bundle endowed with a connection can be viewed as logarithms of mean eigenvalues of monodromy of the vector bundle along a flow on the base.

In the case of the Hodge bundle, we take a fiber of $H^1_\mathbb{R}$ and pull it along a Teichmüller geodesic on the moduli space. We wait till the geodesic winds a lot and comes close to the initial point and then compute the resulting monodromy matrix $A(t)$. Finally, we compute logarithms of eigenvalues of $A^T A$, and normalize them by twice the length t of the geodesic."

In this way one gets a set of $2g$ real numbers which is symmetric with respect to the origin; the largest of them equals $\lambda_1 = 1$ and the other $g - 1$ are ordered in such a way that $\lambda_1 \geq \lambda_2 \leq \cdots \geq \lambda_g$. The sum $\lambda_1 + \cdots + \lambda_g$ turns out to have many nice properties (in particular, it is always rational). It also has a deep geometrical meaning as the integral over a connected component of the stratum of the moduli space (this formula was proposed in [49] and proved by [20] in more general setting when the connected component of the stratum is replaced by any subspace of the moduli space invariant under the action of the Teichmüller flow, see [18] for the complete history and references). This relation looks as follows

$$\lambda_1 + \cdots + \lambda_g = -\frac{1}{4} \int_{\mathcal{M}_1} \Delta^T \ln\{\det \Im\Omega\} \, dv_1 \tag{208}$$

where $\mathcal{M}_1$ is any connected component (there could be up to 3 of them, see [50]) of $\mathcal{H}_g(k_1, \ldots, k_n)$ and Ω is the period matrix of $\mathcal{R}$; v_1 is the natural volume form (defined in the homological coordinates) on the component $\mathcal{M}_1$ normalized such that the total volume of $\mathcal{M}_1$ is 1.

The formula (208) can be rewritten as the sum of two geometrically important parts by representing the expression $\det\Im\Omega$ as the ratio of two "more complicated" expressions using the holomorphic factorization formula (196):

$$\det\Im\Omega = \frac{\det\Delta^{\mathcal{R},|v|^2}}{|\tau(\mathcal{R}, v)|^2 \, \mathrm{Area}(\mathcal{R}, |v|^2)}. \tag{209}$$

Then, since $SL(2)$ transformation preserves the area of $\mathcal{R}$, we get

$$\Delta^{(h)} \ln \det\Im\Omega = \Delta^{(h)}\{\ln \det\Delta^{|v|^2}\} - \Delta^{(h)}\{\ln(|\tau(\mathcal{R}, v)|^2)\}. \tag{210}$$

The last term of (210) can be explicitly evaluated using the homogeneity property (106) of the tau-function $\tau(\mathcal{R}, v)$.

Namely, the $SL(2)$ transformation (207) acts on each homological coordinate z_j in the same way as it acts on the flat coordinate z:

$$\begin{pmatrix} \Re z'_j \\ \Im z'_j \end{pmatrix} = M_t \begin{pmatrix} \Re z_j \\ \Im z_j \end{pmatrix}, \tag{211}$$

or

$$z_i^t = \frac{1}{t + \bar{t}} \left\{ (1 + t\bar{t}) z_i + (1 - t + \bar{t} - t\bar{t}) \bar{z}_i \right\}, \tag{212}$$

$$\bar{z}_i^t = \frac{1}{t + \bar{t}} \left\{ (1 + t\bar{t}) \bar{z}_i + (1 + t - \bar{t} - t\bar{t}) z_i \right\}. \tag{213}$$

For any function $f(\{z_j, \bar{z}_j\})$ the Teichmüller Laplacian is defined by

$$\Delta^T f = 4(t + \bar{t})^2 \, \partial_{t\bar{t}}^2 \{ f(\{z_j(t, \bar{t}), \bar{z}_j(t, \bar{t})\}) \} \big|_{t=1}.$$

We have

$$f_{t\bar{t}} = \sum_{i,j} \left(f_{z_i z_j} z_{i\,t} z_{j\,\bar{t}} + f_{\bar{z}_i \bar{z}_j} \bar{z}_{i\,t} \bar{z}_{j\,\bar{t}} + f_{z_i \bar{z}_j} z_{i\,t} \bar{z}_{j\,\bar{t}} \right) + \sum_i \left(f_{z_i} z_{i\,t\bar{t}} + f_{\bar{z}_i} \bar{z}_{i\,t\bar{t}} \right).$$

$$\tag{214}$$

From (212) it follows that

$$\frac{\partial z_i}{\partial \bar{t}} \Big|_{t=1} = \frac{\partial \bar{z}_i}{\partial t} \Big|_{t=1} = 0.$$

Moreover, since τ is a holomorphic function of $\{z_i\}$, we have $(\ln |\tau|^2)_{z_i \bar{z}_j} = 0$, and, therefore, all terms in the double sum in (214) vanish if $f = \ln |\tau|^2$.

The remaining terms we compute using (212):

$$\frac{\partial^2 z_i}{\partial t \, \partial \bar{t}} \Big|_{t=1} = \frac{z_i}{8}, \qquad \frac{\partial \bar{z}_i}{\partial t \, \partial \bar{t}} \Big|_{t=1} = \frac{\bar{z}_i}{8} \tag{215}$$

and, therefore (we need to multiply by 16 to take care of the factor $4(t + \bar{t})^2$)

$$\Delta^T \ln |\tau|^2 = 2 \sum_i \left(z_i \partial_{z_i} \right) \ln \tau + 2 \left(\sum_i \bar{z}_i \partial_{\bar{z}_i} \right) \ln \bar{\tau}. \tag{216}$$

Thus, taking into account the homogeneity (106) of the tau-function, we get the following simple formula:

$$\Delta^T \ln |\tau|^2 = \frac{1}{3} \sum_{i=1}^N \frac{m_i (m_i + 2)}{m_i + 1}. \tag{217}$$

Then the Kontsevich-Zorich-Forni formula (208), being combined with (210), gives the expression which was first derived in [18]:

$$\lambda_1 + \cdots + \lambda_g = \frac{1}{12} \sum_{i=1}^N \frac{m_i (m_i + 2)}{m_i + 1} - \frac{1}{4} \int_{\mathcal{M}_1} \Delta^T \ln \det \Delta^{|v|^2}. \tag{218}$$

The last integral turns out also to have an important geometrical meaning: according to [18] there is the relation

$$\int_{\mathcal{M}_1} \Delta^T \ln \det \Delta^{|v|^2} = -\frac{4\pi^2}{3} c_{area}(\mathcal{M}_1) \tag{219}$$

where $c_{area}(\mathcal{M}_1)$ is the Siegel-Veech constant of $\mathcal{M}_1$.

10.2.2 Lyapunov Exponents of Hodge and Prym Bundles on Spaces
of Quadratic Differentials

The Teichmüller flow preserves also each stratum $\mathcal{Q}_g(k_1, \ldots, k_M)$ of the space of quadratic differentials. There are two natural vector bundles over such stratum or any of its subspaces which are invariant under the flow: the Hodge bundle Λ_H and the Prym bundle Λ_P. The fiber of the latter is the space H^- of Prym differentials on the canonical cover $\widehat{\mathcal{R}}$ (we denote $\dim H^- = 2g_-$ as before). The fiber of their direct sum

$$\widehat{\Lambda}_H = \Lambda_H \otimes \Lambda_P$$

is the space of holomorphic abelian differentials on $\widehat{\mathcal{R}}$.

Introduce now the Hodge Lyapunov exponents $\lambda_1^+, \ldots, \lambda_g^+$, Prym Lyapunov exponents $\lambda_1^-, \ldots, \lambda_{g_-}^-$ and Lyapunov exponents $\hat{\lambda}_1, \ldots, \hat{\lambda}_{\hat{g}}$ of the bundle $\widehat{\Lambda}_H$; clearly

$$\hat{\lambda}_1 + \cdots + \hat{\lambda}_{\hat{g}} = (\lambda_1^+ + \cdots + \lambda_g^+) + (\lambda_1^- + \cdots + \lambda_{g_-}^-). \tag{220}$$

Denote again by $\mathcal{M}_1$ a subspace of $\mathcal{Q}_g(k_1, \ldots, k_M)$ invariant under the action of the Teichmüller flow. Then the Kontsevich-Zorich-Forni theorem implies

$$\lambda_1^+ + \cdots + \lambda_g^+ = -\frac{1}{4} \int_{\mathcal{M}_1} \Delta^T \ln\{\det \Im \Omega\} \, d\nu_1 \tag{221}$$

or, equivalently, using (200),

$$\lambda_1^+ + \cdots + \lambda_{g+}^+ = \frac{1}{12} \sum_{i=1}^{N} \frac{k_i(k_i+4)}{k_i+2} - \frac{1}{4} \int_{\mathcal{M}_1} \Delta^T \ln \det \Delta^{\mathcal{R},|\mathcal{Q}|}. \tag{222}$$

Applying the Kontsevich-Zorich-Forni theorem to the Hodge bundle over the moduli space of covers $\widehat{\mathcal{R}}$ (which is an invariant subspace of $\mathcal{H}_g(\{\widehat{k}_i\})$ where $(v) = \sum \widehat{k}_j x_j$ on $\widehat{\mathcal{R}}$) we get

$$\hat{\lambda}_1 + \cdots + \hat{\lambda}_{\hat{g}} = -\frac{1}{4} \int_{\mathcal{M}_1} \Delta^T \ln\{\det \Im \widehat{\Omega}\} \, d\nu_1 \tag{223}$$

where $\widehat{\Omega}$ is the period matrix of $\widehat{\mathcal{R}}$. Due to (205) we get from (220) the following formula for the sum of Prym Lyapunov exponents:

$$\lambda_1^- + \cdots + \lambda_{g_-}^- = -\frac{1}{4} \int_{\mathcal{M}_1} \Delta^T \ln\{\det \Im \Pi\} \, d\nu_1. \tag{224}$$

Using (205), we can express this sum via the determinant of Prym Laplacian and the Prym tau-function:

$$\lambda_1^- + \cdots + \lambda_{g_-}^- = -\frac{1}{4}\int_{\mathcal{M}_1} \Delta^T \ln\det\Delta_-^{\widehat{\mathcal{R}},|\mathcal{Q}|}\, dv_1$$

$$+ \frac{1}{4}\int_{\mathcal{M}_1} \Delta^T \ln|\tau_-(\widehat{\mathcal{R}}, \mathcal{Q})|^2\, dv_1. \tag{225}$$

Similarly to (217), using the expression (199) for the homogeneity coefficient of τ_-, we get

$$\lambda_1^- + \cdots + \lambda_{g_-}^- = \frac{1}{24}\sum_{j=1}^{M}\frac{k_j(k_j+4)}{k_j+2} + \frac{1}{4}\sum_{k_j \text{ even}}\frac{1}{k_j+2}$$

$$- \frac{1}{4}\int_{\mathcal{M}_1}\Delta^T \ln\det\Delta_-^{\widehat{\mathcal{R}},|\mathcal{Q}|}\, dv_1. \tag{226}$$

It was proved in [18] (Lemma 1.1) that the Siegel-Veech constant of $\widehat{\mathcal{R}}$ and $\mathcal{R}$ are related by the factor of 2 which implies the somewhat unexpected identity

$$\int_{\mathcal{M}_1}\Delta^T \ln\det\Delta_-^{\widehat{\mathcal{R}},|\mathcal{Q}|}\, dv_1 = \int_{\mathcal{M}_1}\Delta^T \ln\det\Delta_-^{\mathcal{R},|\mathcal{Q}|}\, dv_1. \tag{227}$$

In turn, this gives the link between the sums of Hodge and Prym Lyapunov's exponents which was originally proved in [18]:

$$\lambda_1^- + \cdots + \lambda_{g_-}^- = \lambda_1^+ + \cdots + \lambda_g^+ + \frac{1}{4}\sum_{k_j \text{ even}}\frac{1}{k_j+2}. \tag{228}$$

Concluding, we see that the explicit terms in the sum of the Lyapunov exponents of both Hodge and Prym vector bundles have the meaning of homogeneity coefficients of the tau-functions τ_+ and τ_-, respectively. This observation provides a non-trivial link between the world of integrable dynamical systems and the world of ergodic dynamical systems where the theory of Teichmüller flows belongs.

Appendix A. Canonical Bidifferential and Szegö Kernel on Riemann Surfaces. Variational Formulas

Here we summarize a few facts from the theory of Riemann surfaces and their variations which are used in the main text.

On a compact Riemann surface $\mathcal{R}$ of genus g introduce a canonical basis of cycles (a_α, b_α) in $H_1(\mathcal{R}, \mathbb{Z})$. Denote by $B(x, y)$ for $x, y \in \mathcal{R}$ the canonical bidifferential, which is the symmetric bimeromorphic differential on $\mathcal{R}$ having quadratic pole with biresidue 1 on the diagonal and vanishing a-periods.

The bidifferential B is expressed via the prime-form $E(x, y)$ as follows: $B(x, y) = d_x d_y \ln E(x, y)$. Consider the basis of holomorphic differentials v_α on $\mathcal{R}$ normalized as follows:

$$\oint_{a_\alpha} v_\beta = \delta_{\alpha\beta}. \tag{229}$$

The period matrix Ω of $\mathcal{R}$ is given by

$$\Omega_{\alpha\beta} = \oint_{b_\alpha} v_\beta. \tag{230}$$

The Abel map with the base-point $x_0 \in \mathcal{R}$ is defined by

$$\mathcal{A}_\alpha(x) = \int_{x_0}^{x} v_\alpha. \tag{231}$$

Introduce the theta-function with characteristics $\Theta \begin{bmatrix} \mathbf{p} \\ \mathbf{q} \end{bmatrix} (\mathbf{z}|\Omega)$, where $\mathbf{p}, \mathbf{q} \in \mathbb{C}^g$ are vectors of characteristics; $\mathbf{z} \in \mathbb{C}^g$ is the argument. The theta-function satisfies the heat equation:

$$\frac{\partial^2 \Theta \begin{bmatrix} \mathbf{p} \\ \mathbf{q} \end{bmatrix} (\mathbf{z})}{\partial z_\alpha \partial z_\beta} = 4\pi i \cdot \frac{\partial \Theta \begin{bmatrix} \mathbf{p} \\ \mathbf{q} \end{bmatrix} (\mathbf{z})}{\partial \Omega_{\alpha\beta}}. \tag{232}$$

Let us consider some non-singular odd half-integer characteristic $[\mathbf{p}^*, \mathbf{q}^*]$. The prime-form $E(x, y)$ is defined as follows:

$$E(x, y) = \frac{\Theta \begin{bmatrix} \mathbf{p}^* \\ \mathbf{q}^* \end{bmatrix} (\mathcal{A}(x) - \mathcal{A}(y))}{h(x)h(y)} \tag{233}$$

where the square of a section $h(x)$ of a spinor bundle over $\mathcal{R}$ is given by the following expression:

$$h^2(x) = \sum_{\alpha=1}^{g} \partial_{z_\alpha} \left\{ \Theta \begin{bmatrix} \mathbf{p}^* \\ \mathbf{q}^* \end{bmatrix} (0) \right\} v_j(x). \tag{234}$$

Then $h(x)$ itself is a section of the spinor bundle corresponding to the characteristic $\begin{bmatrix} \mathbf{p}^* \\ \mathbf{q}^* \end{bmatrix}$. The automorphy factors of the prime-form along all cycles a_α are trivial; the automorphy factor along cycle b_α equals to $\exp\{-\pi i \Omega_{\alpha\alpha} - 2\pi i (\mathcal{A}_\alpha(x) - \mathcal{A}_\alpha(y))\}$. The prime-form is independent of the choice of characteristic $\begin{bmatrix} \mathbf{p}^* \\ \mathbf{q}^* \end{bmatrix}$. It has also the following local behaviour as $x \to y$:

$$E(x, y) = \frac{\xi(x) - \xi(y)}{\sqrt{d\xi(x)}\sqrt{d\xi(y)}}(1 + o(1)) \tag{235}$$

where $\xi(x)$ is a local parameter.

Choosing some local coordinate ξ near the diagonal $\{x = y\} \subset \mathcal{R} \times \mathcal{R}$, we have the following expansion of $B(x, y)$ as $y \to x$:

$$B(x, y) = \left(\frac{1}{(\xi(x) - \xi(y))^2} + \frac{S_B(\xi(x))}{6} + O((\xi(x) - \xi(y))^2) \right) d\xi(x)d\xi(y)$$

(236)

where S_B is a projective connection on $\mathcal{R}$ called the *Bergman projective connection*.

If two canonical bases of cycles on $\mathcal{R}$, $\{a'_\alpha, b'_\alpha\}_{\alpha=1}^{g}$ and $\{a_\alpha, b_\alpha\}_{\alpha=1}^{g}$ are related by a matrix

$$\sigma = \begin{pmatrix} d & c \\ b & a \end{pmatrix} \in Sp(2g, \mathbb{Z}),$$

(237)

then the corresponding canonical bidifferentials are related as follows (see item 4 on page 21 of [25]):

$$B^\sigma (x, y) = B(x, y) - 2\pi i \sum_{\alpha, \beta = 1}^{g} [(c\Omega + d)^{-1}c]_{\alpha\beta} \, v_\alpha(x)v_\beta(y).$$

(238)

The period matrix Ω^σ, corresponding to the new canonical basis of cycles, is related to Ω as follows:

$$\Omega^\sigma = (a\Omega + b)(c\Omega + d)^{-1}.$$

(239)

For any two vectors $\mathbf{p}, \mathbf{q} \in \mathbb{C}^g$ such that $\Theta \begin{bmatrix} \mathbf{p} \\ \mathbf{q} \end{bmatrix} (0) \neq 0$ the Szegö kernel is defined by the formula

$$S_{pq}(x, y) = \frac{1}{\Theta \begin{bmatrix} \mathbf{p} \\ \mathbf{q} \end{bmatrix} (0)} \frac{\Theta \begin{bmatrix} \mathbf{p} \\ \mathbf{q} \end{bmatrix} (\mathcal{A}(x) - \mathcal{A}(y))}{E(x, y)}.$$

(240)

Near the diagonal, when $y \to x$, $S_{p,q}$ behaves as follows

$$S_{pq}(x, y) = \left(\frac{1}{\xi(x) - \xi(y)} + a_0(x) + O(\xi(x) - \xi(y)) \right) \sqrt{d\xi(x)}\sqrt{d\xi(y)}$$

(241)

where $\xi(x)$ is a local coordinate and coefficient a_0 is given by ([26], p.29)

$$a_0(x) = \frac{1}{d\xi(x)} \sum_{\alpha=1}^{g} \partial_\alpha \{\ln \Theta \begin{bmatrix} \mathbf{p} \\ \mathbf{q} \end{bmatrix} (0)\} v_\alpha(x).$$

(242)

Proposition A.1 (Prop. 1 of [38]) *The following variational formulas for the Szegö kernel with respect to components of characteristic vectors p and q hold:*

$$\frac{d}{dp_\alpha} S_{pq}(x, y) = - \oint_{b_\alpha} S_{pq}(x, t) S_{pq}(t, y), \qquad (243)$$

$$\frac{d}{dq_\alpha} S_{pq}(x, y) = - \oint_{a_\alpha} S_{pq}(x, t) S_{pq}(t, y). \qquad (244)$$

The Szegö kernel is related to $B(x, y)$ by the following relation [25]:

$$S_{pq}(x, y) S_{pq}(y, x) = -B(x, y) - \sum_{\alpha,\beta=1}^{g} \partial_\alpha \partial_\beta \ln \vartheta_{pq}(0)\, v_\alpha(x) v_\beta(y). \qquad (245)$$

For any two sets $x_1, \ldots, x_n$ and $y_1, \ldots, y_n$ of points on the Riemann surface Fay discovered the following identity (see [25], p.33):

$$\det\{S(x_j, y_k)\} = \frac{\Theta\begin{bmatrix} \mathbf{p} \\ \mathbf{q} \end{bmatrix}\left(\sum_{j=1}^{n}(\mathcal{A}(x_j) - \mathcal{A}(y_j))\right)}{\Theta\begin{bmatrix} \mathbf{p} \\ \mathbf{q} \end{bmatrix}(0)} \frac{\prod_{j<k} E(x_j, x_k) E(y_k, y_j)}{\prod_{j,k} E(x_j, y_k)}. \qquad (246)$$

In particular, for $n = 2$ (246) is called *Fay trisecant identity*.

Let us also define

$$\mathcal{C}(x) = \frac{1}{\mathcal{W}(x)} \sum_{\alpha_1,\ldots,\alpha_g=1}^{g} \partial^g_{\alpha_1,\ldots\alpha_g} \Theta(K^x) v_{\alpha_1} \ldots v_{\alpha_g}(x) \qquad (247)$$

where

$$\mathcal{W}(x) := \det_{1\leq\alpha,\beta\leq g} ||v_\beta^{(\alpha-1)}(x)|| \qquad (248)$$

is the Wronskian determinant of the basic holomorphic differentials, and K^x is the vector of Riemann constants with initial point x. The expression (247) is a multi-valued $n(1 - n)/2$-differential on $\mathcal{R}$ which does not have any zeros or poles [26].

In the case of genus 1 the x-dependence in (247) drops out and $\mathcal{C}(x)$ turns into $\vartheta'((\Omega + 1)/2)$.

Let us now consider some Abelian differential v on $\mathcal{R}$; it could be a differential of either first, second or third kind on; it can be also a meromorphic differential of mixed type (i.e having both: poles of higher order and residues).

Introduce the meromorphic differential Q_v, which is given by the zero order term of the asymptotics of the Bergman bidifferential on the diagonal:

$$Q_v = \frac{B_{reg}(x, x)}{v(x)}$$

where the regularization is done using the differential v, i.e.,

$$Q_v(x) = \frac{1}{v(x)} \left(B(x, y) - \frac{v(x)v(y)}{(\int_x^y v)^2} \right) \bigg|_{y=x}. \tag{249}$$

Denote by $\{\cdot, \cdot\}$ the Schwarzian derivative and define the following meromorphic projective connection associated with the differential v:

$$S_v := \left\{ \int^x v, \xi(x) \right\} \equiv \frac{v''}{v} - \frac{3}{2} \left(\frac{v'}{v} \right)^2 \tag{250}$$

where prime denotes the derivative with respect to the local coordinate ξ.

Since S_v is the meromorphic projective connection on $\mathcal{R}$, the difference $S_B - S_v$ is the meromorphic quadratic differential. Dividing $S_B - S_v$ by $6v$ we also get the differential Q_v:

$$Q_v := \frac{1}{6} \frac{S_B - S_v}{v}. \tag{251}$$

The meromorphic Abelian differential Q_v, constructed from the (holomorphic or meromorphic) Abelian differential v plays the key role in the construction of the Bergman tau-function. Notice that Q_v is determined by v and depends on Torelli marking of $\mathcal{R}$.

Remark A.2 Using the definition of the Schwarzian derivative (250) it is easy to verify that at the pole x_i ($i = 1, \ldots, n$) of order $-k_i$ the differential Q_v has zero of order $k_i - 2$ for poles of order 3 and higher (i.e. when $k_i \leq -3$). At a pole of v of second order, i.e. when $k_i = -2$, the differential Q_v has zero of order at least 1. Finally, at a simple pole of v with residue r, the differential Q_v also has the simple pole, and its residue equals to $-1/(12r)$. At a zero x_{n+i} of order k_{n+i}, the differential Q_v has the pole of order $k_{n+i} + 2$.

As a corollary of (238), Q_v transforms as follows under a symplectic transformation σ:

$$Q_v^\sigma(x) = Q_v(x) - 12\pi i \sum_{\alpha, \beta = 1}^{g} [(c\Omega + d)^{-1}c]_{\alpha\beta} \frac{v_\alpha(x)v_\beta(x)}{v(x)}. \tag{252}$$

A.1 Spaces of Meromorphic Abelian Differentials

Denote by $\mathcal{H}_g(\mathbf{k}_{n+m})$ the space of pairs $(\mathcal{R}, v)$ where $\mathcal{R}$ is a Riemann surface of genus g and v is a meromorphic differentials on $\mathcal{R}$ such that the divisor of v is given by $(v) = \sum_{i=1}^{m+n} k_i x_i$ with $k_1, \ldots, k_n < 0$ and $k_{n+1}, \ldots, k_{n+m} > 0$.

Considering the homology group of $\mathcal{R}$ punctured at poles of v, relative to the set of zeros of v, which is denoted by

$$H_1 \left(\mathcal{R} \setminus \{x_i\}_{i=1}^n; \{x_i\}_{i=n+1}^{n+m} \right), \tag{253}$$

we can choose its set of generators $\{s_i\}_{i=1}^{2g+m+n-2}$ as follows:

$$s_\alpha = a_\alpha, \quad s_{\alpha+g} = b_\alpha \quad \alpha = 1, \ldots, g; \quad s_{2g+k} = c_{k+1}, \quad k = 1, \ldots, n-1,$$

$$s_{2g+n-1+k} = l_{n+1+k}, \quad k = 1, \ldots, m-1 \tag{254}$$

where $c_2, \ldots, c_n$ are small contours around the poles $x_2, \ldots, x_n$; $l_{n+2}, \ldots, l_{n+m}$ are contours connecting the "first" zero x_{n+1} with the other zeros $x_{n+2}, \ldots, x_{n+m}$.

The homological coordinates on $\mathcal{H}_g(\mathbf{k}_{n+m})$ are defined as integrals of v over $\{s_i\}$:

$$z_i = \int_{s_i} v, \quad i = 1, \ldots, 2g + n + m - 2. \tag{255}$$

The homology group dual to (253) is the homology group of $\mathcal{R}$, punctured at *zeros* of v, relative to the set of *poles* of v; it is denoted by

$$H_1\left(\mathcal{R} \setminus \{x_i\}_{i=n+1}^{n+m}; \{x_i\}_{i=1}^{n}\right). \tag{256}$$

The set of generators $\{s_i^*\}_{i=1}^{2g+n+m-2}$ of the group (256) dual to (254) is given by

$$s_\alpha^* = -b_\alpha, \quad s_{\alpha+g}^* = a_\alpha \quad \alpha = 1, \ldots, g; \quad s_{2g+k}^* = -\tilde{l}_{k+1},$$

$$k = 1, \ldots, n-1, \quad s_{2g+n-1+k}^* = \tilde{c}_{n+1+k}, \quad k = 1, \ldots, m-1 \tag{257}$$

where $\tilde{l}_2, \ldots, \tilde{l}_n$ are contours connecting the "first" pole x_1 with other poles $x_2, \ldots, x_n$, respectively; $\tilde{c}_{n+2}, \ldots, \tilde{c}_{m+n}$ are small circles around the zeros $x_{n+2}, \ldots, x_{n+m}$.

The variational formulas for the period matrix, the basic holomorphic differentials, the Bergman bidifferential and the differential Q_v formally look analogous to the case of the space of holomorphic differentials (see Theorem 3 of [45]).

To write down these formulas we introduce on $\mathcal{R}$ a system of cuts homologous to a- and b-cycles to get the fundamental polygon $\mathcal{R}_0$; inside of $\mathcal{R}_0$ we also introduce branch cuts connecting poles of v with non-vanishing residues; these branch cuts are assumed to start at x_1, i.e. they connect x_1 with $x_2, \ldots, x_n$; denote them by the same letters $\tilde{l}_2, \ldots, \tilde{l}_n$ as their homology classes in (257). In this way we get the domain $\tilde{\mathcal{R}}_0$ where the Abelian integral $z(x) = \int_{x_{n+1}}^{x} v$ is single-valued.

Now we are in a position to formulate the following theorem, which gives variational formulas on $\mathcal{H}_g(\mathbf{k}_{n+m})$ with respect to homological coordinates $\mathcal{P}_{s_i} = \int_{s_i} v$.

Theorem A.3 [38] *The variational formulas for period matrix Ω, the normalized holomorphic differentials u_α, the Bergman bidifferential $B(x, y)$ and the* *differential Q_v on the moduli space $\mathcal{H}_g(\mathbf{k}_{n+m})$ look as follows (in the cases of v_α, $B(x, y)$ and $Q_v(x)$ the coordinates $z(x)$ and $z(y)$ remain fixed under differentiation):*

$$\frac{\partial \Omega_{\alpha\beta}}{\partial \mathcal{P}_{s_i}} = \int_{s_i^*} \frac{v_\alpha v_\beta}{v}, \tag{258}$$

$$\frac{\partial v_\alpha(x)}{\partial \mathcal{P}_{s_i}}\bigg|_{z(x)=const} = \frac{1}{2\pi i} \int_{t \in s_i^*} \frac{v_\alpha(t) B(x, t)}{v(t)}, \tag{259}$$

$$\frac{\partial}{\partial \mathcal{P}_{s_i}} \ln(E(x, y)\sqrt{v(x)}\sqrt{v(y)})\bigg|_{z(x),z(y)=const} = -\frac{1}{4\pi i} \int_{t \in s_i^*} \frac{1}{v(t)} \left[d_t \ln \frac{E(x, t)}{E(y, t)} \right]^2, \tag{260}$$

$$\frac{\partial B(x, y)}{\partial \mathcal{P}_{s_i}}\bigg|_{z(x),z(y)=const} = \frac{1}{2\pi i} \int_{t \in s_i^*} \frac{B(x, t) B(y, t)}{v(t)}, \tag{261}$$

$$\frac{\partial Q_v(x)}{\partial \mathcal{P}_{s_i}}\bigg|_{z(x)=const} = \frac{1}{2\pi i} \int_{t \in s_i^*} \frac{B^2(x, t)}{v(t)}, \tag{262}$$

$$\frac{\partial}{\partial \mathcal{P}_{s_i}} S_{pq}(x, y)\bigg|_{z(x),z(y)} = \frac{1}{4} \int_{t \in s_i^*} \frac{W_t[S_{pq}(x, t), S_{pq}(t, y)]}{v(t)} \tag{263}$$

where W_t denotes the Wronskian with respect to the variable t.

The variational formulas for v_α, B and Q_v with respect to the relative periods were first given in [59]. The proof of variational formulas with respect to coordinates $\int_{a_\alpha} v$, $\int_{b_\alpha} v$ and $\int_{x_{n+1}}^{x_{n+j}} v$ is given by Theorem 3 of [45]. The proof of variational formulas with respect to residues r_j of v is given in Theorem 1 of [38].

We notice that the Wronskian $W_t[S_{pq}(x, t), S_{pq}(t, y)]$ is a quadratic differential with respect to t; dividing it by $v(t)$ we get a meromorphic differential on $\mathcal{R}$ (with respect to t).

A.2 Hurwitz Spaces

Let $\mu_1, \dots, \mu_M$ be M partitions of n. Denote by $Hur_{g,n}(\mu_1, \dots, \mu_M)$ the Hurwitz space of pairs $(\mathcal{R}, f)$ where $\mathcal{R}$ is a Riemann surface of genus g and f is a meromorphic function on $\mathcal{R}$ with simple poles and M critical values such that the multiplicities of critical points corresponding to kth critical value are determined by the partition μ_k (such that the simple critical point corresponds to entry 2 of the partition; entry 1 of μ_k corresponds to a regular point of df). The genus g is expressed via the numbers of parts $p_1, \dots, p_M$ of the partitions via the Riemann-Hurwitz formula: $g = \sum_{i=1}^{M}(n - p_i)/2 - n + 1$.

The Hurwitz space $Hur_{g,n}(\mu_1, \ldots, \mu_M)$ is the subspace of the space $\mathcal{H}_g(\mathbf{k}_{n+m})$ with $k_1 = \cdots = k_n = -2$: on the Hurwitz space the differential v is exact: $v = df$. It means in particular that all periods of v over cycles a_α, b_α and c_{k+1} in (254) vanish. Moreover, periods over contours l_{n+1+k} are dependent if there is more than one critical point of f corresponding to the same critical value. In fact, the periods of v over the cycles (254) reduces to the set of critical values (the *branch points*) of the function f, which we denote by $z_1, \ldots, z_M$.

Then theorem (A.3) implies the following corollary

Corollary A.4 ([59, 42]) *The variational formulas on $Hur_{g,n}(\mu_1, \ldots, \mu_M)$ with respect to the critical values $z_1, \ldots, z_M$ of the function f look as follows*

$$\frac{\partial \Omega_{\alpha\beta}}{\partial z_i} = 2\pi i \sum_{p \in f^{-1}(z_i)} \operatorname{res}|_{t=p} \frac{v_\alpha v_\beta}{df}, \tag{264}$$

$$\frac{\partial v_\alpha(x)}{\partial z_i}\bigg|_{z(x)=const} = 2\pi i \sum_{p \in f^{-1}(z_i)} \operatorname{res}|_{t=p} \frac{v_\alpha(t) B(x, t)}{df(t)}, \tag{265}$$

$$\frac{\partial}{\partial z_i} \ln(E(x, y)\sqrt{df(x)}\sqrt{df(y)})\bigg|_{z(x),z(y)=const}$$
$$= -\frac{1}{2} \sum_{p \in f^{-1}(z_i)} \operatorname{res}|_{t=p} \frac{1}{df(t)} \left[d_t \ln \frac{E(x, t)}{E(y, t)} \right]^2, \tag{266}$$

$$\frac{\partial B(x, y)}{\partial z_i}\bigg|_{z(x),z(y)=const} = \sum_{p \in f^{-1}(z_i)} \operatorname{res}|_{t=p} \frac{B(x, t)B(y, t)}{df(t)}, \tag{267}$$

$$\frac{\partial Q_v(x)}{\partial z_i}\bigg|_{z(x)=const} = \sum_{p \in f^{-1}(z_i)} \operatorname{res}|_{t=p} \frac{B^2(x, t)}{df(t)}, \tag{268}$$

$$\frac{\partial}{\partial z_i} S_{pq}(x, y)\bigg|_{z(x),z(y)} = \frac{1}{2} \sum_{p \in f^{-1}(z_i)} \operatorname{res}|_{t=p} \frac{W_t[S_{pq}(x, t), \ S_{pq}(t, y)]}{df(t)}. \tag{269}$$

A.3 Spaces of Holomorphic Quadratic and n-Differentials

The moduli spaces of quadratic differentials (both holomorphic and meromorphic) can also be considered as subspaces of the moduli spaces of abelian differentials, and corresponding variational formulas can be obtained by reduction of the variational formulas from Th.A.3. Here we consider the simplest and the most important case of the space $\mathcal{Q}_g^0$ which is the space of pairs $(\mathcal{R}, Q)$ where $\mathcal{R}$ is a Riemann surface of genus g and Q is a holomorphic differential on $\mathcal{R}$ with $4g - 4$ simple zeros (we refer [58, 8, 10] for the general case); $\dim \mathcal{Q}_g^0 = 6g - 6$.

The periods $\{A_i, B_i\}_{i=1}^{3g-3}$ of the holomorphic Abelian differential v over a basis (a_α^-, b_α^-) in odd homologies H_- of the canonical cover $\widehat{\mathcal{R}}$ defined by $v^2 = Q$ can be used as local coordinates on $\mathcal{Q}_g^0$. Since v has on $\widehat{\mathcal{R}}$ zeros of order 2 at all branch points $\{x_i\}_{i=1}^{4g-4}$, the space $\mathcal{Q}_g^0$ can be considered as a subspace of the moduli space of Abelian differentials with all double zeros on Riemann surfaces of genus $\hat{g}$. Therefore variational formulas on $\mathcal{Q}_g^0$ can be obtained from Th.A.3 (see Prop. 3.1 of [8]) via an elementary chain rule. Let us consider the fundamental polygon $\widehat{\mathcal{R}}_0$ which is invariant under the involution μ and such that x_1 is at the vertex of $\widehat{\mathcal{R}}_0$. Define the coordinate on $\widehat{\mathcal{R}}_0$ by $z(x) = \int_{x_1}^x v$. Then we have

Corollary A.5 *The variational formulas on the moduli space $\mathcal{Q}_g^0$ look as follows (in the cases of v_α, $B(x, y)$ and $Q_v(x)$ the coordinates $z(x)$ and $z(y)$ remain fixed under differentiation):*

$$\frac{\partial \Omega_{\alpha\beta}}{\partial \mathcal{P}_{s_i}} = \frac{1}{2} \int_{s_i^*} \frac{v_\alpha v_\beta}{v}, \tag{270}$$

$$\left. \frac{\partial v_\alpha(x)}{\partial \mathcal{P}_{s_i}} \right|_{z(x)=const} = \frac{1}{4\pi i} \int_{t \in s_i^*} \frac{v_\alpha(t) B(x, t)}{v(t)}, \tag{271}$$

$$\left. \frac{\partial}{\partial \mathcal{P}_{s_i}} \ln(E(x, y)\sqrt{v(x}\sqrt{v(y)})) \right|_{z(x),z(y)=const} = -\frac{1}{8\pi i} \int_{t \in s_i^*} \frac{1}{v(t)} \left[d_t \ln \frac{E(x, t)}{E(y, t)} \right]^2, \tag{272}$$

$$\left. \frac{\partial B(x, y)}{\partial \mathcal{P}_{s_i}} \right|_{z(x),z(y)=const} = \frac{1}{4\pi i} \int_{t \in s_i^*} \frac{B(x, t) B(y, t)}{v(t)}, \tag{273}$$

$$\left. \frac{\partial Q_v(x)}{\partial \mathcal{P}_{s_i}} \right|_{z(x)=const} = \frac{1}{4\pi i} \int_{t \in s_i^*} \frac{B^2(x, t)}{v(t)}. \tag{274}$$

The variational formulas for $B(x, y)$ used in [58] differ from (273) by the factor of 2 due to the different normalization of homological coordinates.

Spaces of N-differentials.

Denote by $M_g^{(N)}$ the moduli space of pairs $(\mathcal{R}, W)$ where $\mathcal{R}$ is a Riemann surface of genus g and W is a holomorphic N-differential on $\mathcal{R}$; $\dim M_g^{(N)} = (2N + 2)(g - 1)$. Consider the N-sheeted cover $\widehat{\mathcal{R}}$ defined by $v^N = W$; $\widehat{\mathcal{R}}$ possesses the natural $\mathbb{Z}_N$ symmetry; denote by μ the degree N isomorphizm acting on v as $\mu^* v = \epsilon v$ (here $\epsilon = e^{2\pi i/N}$).

Denote by $\mathfrak{M}_g^{(N),0}$ the subspace of $M_g^{(N)}$ defined by the condition that all the zeros of W are simple. Then the genus of $\widehat{\mathcal{R}}$ equals $\hat{g} = N^2(g - 1) + 1$. Then the homology group of $\widehat{\mathcal{R}}$ can be decomposed as $H_1(\widehat{\mathcal{R}}, \mathbb{C}) = \oplus_{k=0}^{N-1} \mathcal{H}_k$ where $\mathcal{H}_k$ is the kth invariant subspace of μ_*: for any $s \in \mathcal{H}_k$ we have $\mu_* s = \epsilon^k s$. Then $\dim \mathcal{H}_0 = g$ and $\dim \mathcal{H}_k = (2N+2)(g-1)$ for $k = 1, \ldots, N-1$.

Denote by $\{s_j\}_{j=1}^{(2N+2)(g-1)}$ a basis in $\mathcal{H}_1$; then $\mathcal{P}_j = \int_{s_j} v$ can be used as local coordinates on $\mathfrak{M}_g^{(N),0}$; the integrals of v over other cycles from other $\mathcal{H}_k$'s vanish. Let $\{s_j^*\}_{j=1}^{(2N+2)(g-1)}$ be the dual basis in $\mathcal{H}_{N-k}$ satisfying $s_k^* \circ s_j = \delta_{jk}$.

Denote the fundamental polygon of $\widehat{\mathcal{R}}$ by $\widehat{\mathcal{R}}_0$ and define the coordinate on $\widehat{\mathcal{R}}_0$ by $z(x) = \int_{x_1}^{x} v$. Then we have

Corollary A.6 *The variational formulas on the moduli space $\mathfrak{M}_g^{(N),0}$ look as follows (in the cases of v_α, $B(x, y)$ and $Q_v(x)$ the coordinates $z(x)$ and $z(y)$ remain fixed under differentiation):*

$$\frac{\partial \Omega_{\alpha\beta}}{\partial \mathcal{P}_{s_i}} = \frac{1}{N} \int_{s_i^*} \frac{v_\alpha v_\beta}{v}, \tag{275}$$

$$\frac{\partial v_\alpha(x)}{\partial \mathcal{P}_{s_i}}\bigg|_{z(x)=const} = \frac{1}{2N\pi i} \int_{t \in s_i^*} \frac{v_\alpha(t) B(x, t)}{v(t)}, \tag{276}$$

$$\frac{\partial \ln E(x, y)}{\partial \mathcal{P}_{s_i}}\bigg|_{z(x),z(y)=const} = -\frac{1}{4N\pi i} \int_{t \in s_i^*} \frac{1}{v(t)} \left[d_t \ln \frac{E(x, t)}{E(y, t)} \right]^2, \tag{277}$$

$$\frac{\partial B(x, y)}{\partial \mathcal{P}_{s_i}}\bigg|_{z(x),z(y)=const} = \frac{1}{2N\pi i} \int_{t \in s_i^*} \frac{B(x, t) B(y, t)}{v(t)}, \tag{278}$$

$$\frac{\partial Q_v(x)}{\partial \mathcal{P}_{s_i}}\bigg|_{z(x)=const} = \frac{1}{2N\pi i} \int_{t \in s_i^*} \frac{B^2(x, t)}{v(t)}. \tag{279}$$

Proof. We give the short derivation of (278) from the variational formula (261) on the stratum $\mathcal{H}_{\hat{g}}(N, \ldots, N)$ of the moduli space of holomorphic 1-differentials of genus $\hat{g} = N^2(g-1) + 1$, similarly to Lemma 5 of [58] and Proposition 3.2 of [8].

Namely, consider a point $(\widehat{\mathcal{R}}, v)$ of the space $\mathcal{H}_{\hat{g}}(N, \ldots, N)$. Denote the zeros of v on $\mathcal{R}$ by $x_1, \ldots, x_{N(2g-2)}$; the same notation will be used for the zeros of v on $\widehat{\mathcal{R}}$. The canonical bidifferential on $\widehat{\mathcal{R}}$ will be denoted by $\widehat{B}(x, y)$. The homological coordinates on $\mathcal{H}_{\hat{g}}(N, \ldots, N)$ are given by $\int_{s_i} v$ where $\{s_i\} = \left\{ \{a_j, b_j\}_{j=1}^{\hat{g}}, \{l_i\}_{i=1}^{N(2g-2)-1} \right\}$ where the path l_i connects $x_{N(2g-2)}$ with x_i. Consider the flat coordinate on $\widehat{\mathcal{R}}$ given by $z(x) = \int_{x_{N(2g-2)}}^{x} v$.

Under any choice of Torelli marking on $\widehat{\mathcal{R}}$ used to define $\widehat{B}$ the variational formulas on the space $\mathcal{H}_{\hat{g}}(N, \ldots, N)$ look as follows (261):

$$\frac{\partial \widehat{B}(x, y)}{\partial \int_{s_i} v}\bigg|_{z(x),z(y)} = \frac{1}{2\pi\sqrt{-1}} \int_{t \in s_i^*} \frac{\widehat{B}(x, t) \widehat{B}(t, y)}{v(t)}. \tag{280}$$

The cycles s_i^* form the basis in $H_1\left(\widehat{\mathcal{R}} \setminus \{x_i\}_{i=1}^{N(2g-2)-1}\right)$ which is dual to the basis $\{s_i\}$ ($s_i^* \circ s_j = \delta_{ij}$); we have $\{s_i\} = \{-b_j, a_j\}_{j=1}^{\hat{g}}$, $\{r_i\}_{i=1}^{N(2g-2)-1}$ where r_i is the small positively oriented cycle around x_i.

Let now the curve $\widehat{\mathcal{R}}$ be defined by the equation $v^N = W$. We shall require the following correspondence between Torelli markings of $\widehat{\mathcal{R}}$ and $\mathcal{R}$.

Choosing a Torelli marking of $\mathcal{R}$ consider the set of g contours on $\mathcal{R}$ representing a-cycles $(a_1, \ldots, a_g)$. Consider their lifts to $\widehat{\mathcal{R}}$ (each $f_*[a_j]$ is a system of non-intersecting closed loops on $\widehat{\mathcal{R}}$) and require the Torelli marking of $\widehat{\mathcal{R}}$ to be chosen such that each cycle from $f_*[a_j]$ belongs to the Lagrangian subspace in $H_1(\widehat{\mathcal{R}}, \mathbb{Z})$ generated by the cycles $a_1, \ldots, a_{\hat{g}}$. Under such agreement between the Torelli markings we have the identity

$$f_x^* f_y^* B(x, y) = \widehat{B}(x, y) + \mu_y^* \widehat{B}(x, y) + \cdots + (\mu_y^*)^{n-1} \widehat{B}(x, y) \qquad (281)$$

(here $f : \widehat{\mathcal{R}} \to \mathcal{R}$ is the natural projection and μ is the $\mathbb{Z}_N$ automorphism of $\widehat{\mathcal{R}}$) which can be verified by comparing the singularity structure and normalization of both sides. Further averaging of (281) over the action of μ^* on x we get

$$f_x^* f_y^* B(x, y) = \frac{1}{N} \sum_{l,k=1}^{N-1} (\mu_x^*)^k (\mu_y^*)^l \widehat{B}(x, y). \qquad (282)$$

Now choosing $s \in \mathcal{H}_1$ and the restricting variational formulas (280) to $M_g^{(N)}$ we get

$$\frac{\partial \widehat{B}(x, y)}{\partial (\int_s v)} = \frac{1}{2\pi\sqrt{-1}} \sum_{k=1}^{\hat{g}} \left[-\frac{\partial(\int_{\hat{a}_k} v)}{\partial(\int_s v)} \int_{t \in \hat{b}_k} \frac{\widehat{B}(x, t)\widehat{B}(t, y)}{v(t)} \right.$$

$$\left. + \frac{\partial(\int_{\hat{b}_k} v)}{\partial(\int_s v)} \int_{t \in \hat{a}_k} \frac{\widehat{B}(x, t)\widehat{B}(t, y)}{v(t)} \right]$$

$$+ \frac{1}{2\pi\sqrt{-1}} \sum_{j=1}^{(N-1)(2g-2)-1} \frac{\partial z_j}{\partial(\int_s v)} \int_{t \in r_j} \frac{\widehat{B}(x, t)\widehat{B}(t, y)}{v(t)}. \qquad (283)$$

Furthermore, averaging both sides of (283) as in (282) we get

$$\frac{\partial B(x, y)}{\partial (\int_s v)} = \frac{1}{2\pi\sqrt{-1}N} \sum_{k=1}^{\hat{g}} \left[-\frac{\partial(\int_{\hat{a}_k} v)}{\partial(\int_s v)} \int_{t \in \hat{b}_k} \frac{f_t^*[B(x, t)B(t, y)]}{v(t)} \right.$$

$$\left. + \frac{\partial(\int_{\hat{b}_k} v)}{\partial(\int_s v)} \int_{t \in \hat{a}_k} \frac{f_t^*[B(x, t)B(t, y)]}{v(t)} \right]$$

$$+ \frac{1}{2\pi\sqrt{-1}} \sum_{j=1}^{(N-1)(2g-2)-1} \frac{\partial z_j}{\partial(\int_s v)} \int_{t \in r_j} \frac{f_t^*[B(x, t)B(t, y)]}{v(t)}.$$

$$(284)$$

Each integral in the sum over j in (284) vanishes since the contour r_j is invariant under the action of μ_* while the integrand gets multiplied with ρ^{-1} under the action of μ^*.

The first sum in (284) can be computed by introducing a symplectic basis $\{\hat{s}_j\}_{j=1}^{2\hat{g}}$ in $H_1(\widehat{\mathcal{R}}, \mathbb{R})$ such that each $\hat{s}_j$ belongs to some of eigenspaces of μ_*. Since the transformation from the basis $(\hat{a}_j, \hat{b}_j)$ to the basis $\{\hat{s}_j\}$ is $Sp(2\hat{g}, \mathbb{R})$ this sum equals to the right hand side of (278). □

References

[1] E.Arbarello and M.Cornalba, *Combinatorial and algebro-geometric cohomology classes on the moduli spaces of curves*, J. Algebraic Geom. 5 (1996), no. 4, 705749

[2] Arbarello E., Cornalba M., Griffiths P., *Geometry of algebraic curves*, Vol.2, Springer (2011)

[3] Atiyah, M., *Geometry and Physics of knots*, Cambridge University Press (1990)

[4] Basok,M., *Tau function and moduli of spin curves*, IMRN, **2015** Iss. 20, 1009510117 (2015)

[5] Belinskij V.A., Zakharov, V.E., *Integration of Einstein's equations by the inverse scattering problem and explicit soliton solutions*, JETP **48**, 985 (1978)

[6] Babich, M., Korotkin, D., Self-dual $SU(2)$-invariant Einstein manifolds and modular dependence of theta-functions, *Lett.Math.Phys.* **46** (1998), 323–337

[7] Baraglia,D., Zhenxi Huang, *Special Kähler geometry of the Hitchin system and topological recursion* arXiv:1707.04975

[8] Bertola,M., Korotkin,D., Norton, C., *Symplectic geometry of the moduli space of projective structures in homological coordinates*, Invent.Math. **210**, Issue 3, p. 759–814 (2017)

[9] Bertola, M., Korotkin, D., *Discriminant circle bundles over local models of Strebel graphs and Boutroux curves*, Theoretical and Mathematical Physics, to appear

[10] Bertola, M., Korotkin, D., *Hodge and Prym tau-functions, Jenkins-Strebel differentials and combinatorial model of* $\mathcal{M}_{g,n}$, arXiv:1804.02495

[11] Bertola, M., Korotkin, D., *Spaces of abelian differentials and Hitchin's spectral covers*, arXiv: 1812.05789

[12] Bolibruch, A., *The Riemann-Hilbert problem*, Russ. Math. Surveys **45** (1990), 1–58

[13] Cornalba, M., Harris, J., *Divisor classes associated to families of stable varieties, with applications to the moduli space of curves*, Ann. Sci. Ecole Norm. Sup. (4) **21** (1988), 455-475

[14] Deift, P., Its, A., Kapaev, A., Zhou, X., *On the algebro-geometric integration of the Schlesinger equations*, Commun. Math. Phys. **203** (1999), 613–633

[15] Donagi, R, Markman, E., *Spectral covers, algebraically completely integrable, Hamiltonian systems, and moduli of bundles*, Integrable systems and quantum groups (Montecatini Terme, 1993) , Lecture Notes in Math. Vol. 1620, pp. 1119, Springer, Berlin (1996)

[16] Dubrovin, B., *Geometry of 2D topological field theories*, Integrable systems and quantum groups (Montecatini Terme, 1993), 120348, Lecture Notes in Math., 1620, Springer, Berlin, (1996)

[17] Dubrovin, B., *Painlevé transcendents in two-dimensional topological field theory*, The Painlevé property, 287–412, CRM Ser. Math. Phys., Springer, New York (1999)

[18] Eskin, A., Kontsevich, M., Zorich, A., *Sum of Lyapunov exponents of the Hodge bundle with respect to the Teichmüller geodesic flow*, Publ. Math. Inst. Hautes Études Sci. **120** 207333 (2014)

[19] Forni, G., *On the Lyapunov exponents of the Kontsevich-Zorich cocycle*, in B. Hasselblatt and A. Katok (eds.) Handbook of Dynamical Systems, vol. 1B, pp. 549580, Elsevier, Amsterdam, 2006.

[20] Forni, G., *Deviation of ergodic averages for area-preserving flows on surfaces of higher genus*, Ann. Math., **155**, 1–103 (2002)

[21] Eynard, B.; Kokotov, A.; Korotkin, D. *Genus one contribution to free energy in Hermitian two-matrix model.*, Nuclear Phys. **B 694** no. 3, 443–472 (2004)

[22] Faddeev, L.D.; Takhtajan, L.A. *Hamiltonian methods in the theory of solitons*, Translated from the 1986 Russian original by Alexey G. Reyman, Classics in Mathematics. Springer, Berlin, 592 pp (2007)

[23] Fan,W., Fathizadeh, F., Marcolli, M., *Modular forms in the spectral action of Bianchi IX gravitational instantons* arXiv:1511.05321

[24] Farkas,G., Verre,V., The geometry of the moduli space of odd spin curve, Annals of Math., **180** 927–970 (2014)

[25] Fay, J., *Theta Functions on Riemann Surfaces*, Lect.Notes in Math., **352**, Springer, Berlin, 1973

[26] Fay, J., *Kernel functions, Analytic torsion and Moduli spaces*, Memoirs of the American Mathematical Society, **96** No.464 (1992), 1–123

[27] Gamayun, O.; Iorgov, N.; Lisovyy, O. *Conformal field theory of Painlevé VI.* J. High Energy Phys. 2012, no. 10, 038, 24 pp

[28] Gavrylenko, P., Marshakov, A., *Exact conformal blocks for the W-algebras, twist fields and isomonodromic deformations*, J. High Energy Phys. no. 2, **181**, 31 pp (2016)

[29] van der Geer, G., Kouvidakis, *The Hodge bundle on Hurwitz spaces*, Pure Appl. Math. Q. **7**, no. 4, Special Issue: In memory of Eckart Viehweg, 1297–1307, (2011)

[30] van der Geer, G., Kouvidakis, A., *The class of a Hurwitz divisor on the moduli of curves of even genus*. Asian J. Math. 16 (2012), no. 4, 787–805

[31] van der Geer, G., Kouvidakis, A. *The cycle classes of divisorial Maroni loci*, IMRN 2017, no. 11, 3463–3509

[32] van Geemen, B., Previato, E., *On the Hitchin System*, Duke Math. J. **85** 659–683 (1996)

[33] Harris, J., Mumford,D., *On the Kodaira dimension of the moduli space of curves*, Invent. Math. **67** (1982) 23–86

[34] Hirota, R., *The direct method in soliton theory*, Cambridge Tracts in Mathematics, 155. Cambridge University Press, Cambridge, 2004

[35] Hitchin, N. J. *The self-duality equations on a Riemann surface.* Proc. London Math. Soc. (3) **55** no. 1, 59–126 (1987)

[36] Hitchin, N. J., *Stable bundles and integrable systems*, Duke Math. J. **54**, no. 1, 91–114 (1987)

[37] Hitchin, N., *Twistor spaces, Einstein metrics and isomonodromic deformations*, J. Diff. Geom. **42** (1995), 30–112

[38] Kalla, C., Korotkin, D., *Baker-Akhiezer spinor kernel and tau-functions on moduli spaces of meromorphic differentials*, Comm. Math. Phys. **331**, no. 3, 1191–1235 (2014)

[39] Kitaev, A., Korotkin, D., *On solutions of Schlesinger equations in terms of theta-functions*, IMRN **17** (1998), 877–905.

[40] Klein, C., Richter, O., *Ernst Equation and Riemann Surfaces. Analytical and Numerical Methods*, Lecture Notes in Physics, **685**

[41] Knizhnik, V.G., Multiloop amplitudes in the theory of quantum strings and complex geometry, *Sov.Phys.Uspekhi.* **32** (1989), 945–971

[42] Kokotov, A.; Korotkin, D., *Tau-functions on Hurwitz spaces*, Math. Phys. Anal. Geom. **7** no. 1, 47-96 (2004)

[43] Kokotov, A.; Korotkin, D., *On G-function of Frobenius manifolds related to Hurwitz spaces*, Int. Math. Res. Not. 2004, no. 7, 343–360

[44] Kokotov, A.; Korotkin, D. *Isomonodromic tau-function of Hurwitz Frobenius manifolds and its applications*, Int. Math. Res. Not. 2006, Art. ID 18746, 34 pp

[45] Kokotov, A., Korotkin, D., *Tau-functions on spaces of Abelian differentials and higher genus generalization of Ray-Singer formula*, J. Diff. Geom. **82** (2009), 35–100.

[46] Kokotov, A.; Korotkin, D.; Zograf, P. *Isomonodromic tau-function on the space of admissible covers*, Adv. Math. **227**, no. 1, 586–600 (2011)

[47] Kokotov, A., Korotkin, D., *Tau-functions on spaces of Abelian and quadratic differentials and determinants of Laplacians in Strebel metrics of finite volume*, Preprint 46/2004 of Max-Planck Institute for Mathematics in Science, Leipzig (2004)

[48] Kokotov, A., Korotkin, D., *On G-function of Frobenius manifolds associated to Hurwitz spaces*, Int. Math. Res. Not. 2004, no. 7, 343–360 (2004)

[49] Kontsevich, M., *Intersection theory on the moduli space of curves and the matrix Airy function.* Comm. Math. Phys. **147** (1992), no. 1, 1–23

[50] Kontsevitch, M., Zorich, A., *Lyapunov exponents and Hodge theory*, arXiv hep-th/9701164

[51] Korotkin, D., *Finite-gap solutions of stationary axisymmetric Einstein equations in vacuum*, Theoretical and Mathematical Physics, **77** 1018–1031 (1989)

[52] Korotkin, D., and Nicolai, H., *Separation of variables and Hamiltonian formulation of the Ernst equation*, Physical Review Letters **74**, 1272–1275 (1995)

[53] D.Korotkin, V.Matveev, *Theta function solutions of the Schlesinger system and the Ernst equation*, Functional Analysis and Its Applications, **34** 252–264 (2000)

[54] Korotkin, D., Samtleben, H., *Quantization of coset-space σ-model coupled to 2D gravity*, Communications in Mathematical Physics, **190** 411–457 (1997)

[55] Korotkin, D., Sauvaget, A., Zograf, P., *Tau functions, Prym-Tyurin classes and loci of degenerate differentials*, arXiv:1710.01239

[56] Korotkin, D., Shramchenko, V., *Riemann-Hilbert problem for Hurwitz Frobenius manifolds: regular singularities* Journal-ref: J. Reine Angew.Math., issue 661, 125–187 (2011)

[57] Korotkin, D., Zograf, P., *Tau function and moduli of differentials*, Math. Res. Lett. **18** no. 3, 447458 (2011)

[58] Korotkin,D., Zograf, P., *Tau function and the Prym class.* Algebraic and geometric aspects of integrable systems and random matrices, 241–261, Contemp. Math., 593, Amer. Math. Soc., Providence, RI, 2013

[59] Korotkin, D. *Solution of matrix Riemann-Hilbert problems with quasipermutation monodromy matrices*, Math. Ann. **329** no. 2, 335–364 (2004)

[60] D.Korotkin, P.Zograf, *Tau functions, Hodge classes and discriminant loci on moduli spaces of Hitchin's spectral covers*, J. of Math.Phys, (L.D.Faddeev memorial volume), ed. by A.Its and N.Reshetikhin, **59**, 091412 (2018), doi: 10.1063/1.5038650

[61] Malgrange, B., Sur les Déformation Isomonodromiques, in *Mathématique et Physique (E.N.S. Séminaire 1979-1982)*, p.401–426, Birkhäuser, Boston, 1983

[62] Maison D. *Are the stationary, axially symmetric Einstein equations completely integrable?* Phys. Rev. Lett. **41**, 521 (1978)

[63] Manin, Y., Marcolli,M., *Symbolic dynamics, modular curves, and Bianchi IX cosmologies*, arXiv:1504.04005.

[64] McIntyre, A., Park, Jinsung, *Tau function and Chern-Simons invariant*, Adv. Math. **262** 158 (2014)

[65] G.Mondello, *Riemann surfaces, ribbon graphs and combinatorial classes*, Handbook of Teichmüller theory. Vol. II, 151–215, IRMA Lect. Math. Theor. Phys., 13, Eur. Math. Soc., Zürich, 2009

[66] Nekrasov, N., Rosly, A., Shatashvili, S., *Darboux coordinates, Yang-Yang functional, and gauge theory*, arXiv: 1103.3919 [hep-th], Nuclear Phys. B Proc. Suppl. **216**, 69–93 (2011)

[67] R.Penner, *The Poincaré dual of the Weil-Petersson Kähler two-form*, Comm. Anal. Geom. **1** (1993), no. 1, 43-69

[68] Sato,M., Miwa,T., Jimbo, M., *Holonomic quantum fields II. The Riemann-Hilbert problem*, Publ. RIMS, Kyoto Univ., **15** (1979), 201–278

[69] Shramchenko, V., *Riemann-Hilbert problem associated to Frobenius manifold structures on Hurwitz spaces: Irregular singularity*, Duke Math. J. **144** No. 1 1–52 (2008)

[70] Segal, G., Wilson, G., *Loop groups and equations of KdV type*, Inst. Hautes udes Sci. Publ. Math. No. 61 (1985), 565

[71] Sonoda, H., *Functional determinants on punctured Riemann surfaces and their application to string theory*, Nuclear Phys., **B294** (1987), 157–192

[72] Tod,K.P., *A comment on a paper of Pederson and Poon*, Class.Quant.Grav, **8** (1991) 1049

[73] Tod,K.P., *Self-dual Einstein metrics from the Painlevé VI equation* ,Phys.Lett.A **190** (1994) 221–224

[74] Zamolodchikov, Al.B., Conformal scalar field on the hyperelliptic curve and critical Ashkin-Teller multipoint correlation functions, *Nucl.Phys.* **B285** (1986), 481–503

[75] Zvonkine, D., *Strebel differentials on stable curves and Kontsevich's proof of Witten's conjecture*, arXiv:math/0209071

[76] E.Witten, *Two-dimensional gravity and intersection theory on moduli spaces*, Surveys in Differential Geometry, **1**, 243–310 (1991)

Dmitry Korotkin
Department of Mathematics and Statistics,
Concordia University
1455 de Maisonneuve W., Montréal, Québec,
Canada H3G 1M8

8

The Rigid Body Dynamics in an Ideal Fluid: Clebsch Top and Kummer Surfaces

Jean-Pierre Françoise and Daisuke Tarama

Dedicated to Emma Previato

Abstract. This is an expository presentation of a completely integrable Hamiltonian system of the Clebsch top under a special condition introduced by Weber. After a brief account of the geometric setting of the system, the structure of the Poisson commuting first integrals is discussed following the methods by Magri and Skrypnyk. Introducing supplementary coordinates, a geometric connection to Kummer surfaces, which form a typical class of K3 surfaces, is mentioned and also the system is linearized on the Jacobian of a hyperelliptic curve of genus two determined by the system. Further some special solutions contained in certain vector subspaces are discussed. Finally, an explicit computation of the action variables is introduced.

1 Introduction

This article explains a class of completely integrable Hamiltonian systems describing the rotational motion of a rigid body in an ideal fluid, called a Clebsch top. Further, some new results are exhibited about the computation of action variables of the system.

In analytical mechanics, the Hamiltonian systems of rigid bodies are typical fundamental problems, among which one can find completely integrable systems in the sense of Liouville, such as Euler, Lagrange, Kowalevski, Chaplygin, Clebsch, and Steklov tops. These completely integrable systems are studied from the viewpoint of geometric mechanics, dynamical systems, (differential) topology, and algebraic geometry. A nice survey on the former

Key words Algebraic curves and their Jacobian varieties, Integrable Systems, Clebsch top, Kummer surfaces

MSC(2010): 14H40, 14H70, 14J28, 37J35, 70E40

four systems (Euler, Lagrange, Kowalevski, Chaplygin tops) of completely integrable heavy rigid body is given in [5]. In particular, the authors have studied some elliptic fibrations arising from free rigid bodies, Euler tops, with connections to Birkhoff normal forms; see [24, 29, 11].

As for the Clebsch top, an integrable case of rigid bodies in an ideal fluid, classical studies about the integration of the system were made in [17]. In [17], Kötter has given the integration of the flows on a certain Jacobian variety. One of the achievements of [17] was to identify the intersection of the quadrics with the manifold of lines tangent to two different confocal ellipsoids. Modern studies on this topic are found in [8, 19].

Before [17], Weber has also worked on the same system under a special condition about the parameters in [31], as is also explained briefly in [3]. The same type of integration method was applied for Frahm-Manakov top in [27] by Schottky. Modern studies on this type of systems can be found in [1, 2, 12].

One of the interesting aspects of the Clebsch top under Weber's condition is its relation to Kummer surfaces, which form a typical class of K3 surfaces. In complex algebraic geometry, it is known that a Kummer surface is characterized as a quartic surface with the maximal number (i.e. 16) of double points. It is usually defined as the quotient of an Abelian surface by a special involution; see [15]. Recall that describing the Abelian surface as the quotient $\mathbb{C}^2/\Lambda$ of $\mathbb{C}^2$ by a lattice Λ, group-theoretically isomorphic to $\mathbb{Z}^4$, the involution is induced from the one of $\mathbb{C}^2$ defined through $z \mapsto -z$ for all $z \in \mathbb{C}^2$; see [6]. As to the real aspects of Kummer surfaces, one can look at the beautiful book [9] edited by G. Fischer.

In the case of the Clebsch top, the Abelian surface appears as the Jacobian variety of a hyperelliptic curve of genus two. The equations of motion for the Clebsch top allow to determine this hyperelliptic curve and the Hamiltonian flow is linearized on the Jacobian variety.

The rigid body in an ideal fluid, described by the Kirchhoff equations, is also studied in [14] from the viewpoint of dynamical systems theory. In [14], there are given some interesting special solutions which seem similar to the ones appearing in the case of free rigid body dynamics for the Euler top.

An important problem about completely integrable systems in the sense of Liouville is to describe the action-angle coordinates whose existence is guaranteed by Liouville-Arnol'd(-Mineur-Jost) Theorem; see [4]. The problem is also closely related to Birkhoff normal forms, which was studied by the authors in the cases of the Euler top; see [29, 11].

In the present paper, the geometric setting of the Kirchhoff equations is explained and the complete integrability is considered following the methods by [19]. The connection to a Kummer surface is observed directly, but, introducing two supplementary coordinates, the Kummer surface is also connected to the Jacobian variety of a hyperelliptic curve of genus two, with which the integration of the system is carried out. Two important aspects of the system from the viewpoint of dynamical systems theory are also discussed: some special solutions contained in invariant vector subspaces of dimension three and the computation of action-angle coordinates.

It should be pointed out that there are many studies on the relation between integrable systems and K3 surfaces from general point of view as can be found e.g. in [7, 20, 22]. However, such connection can be found more concretely in the case of the Clebsch top under Weber's condition. In this sense, the present paper has the same type of intention as [23], where the relation between the Euler top and some Kummer surfaces is discussed.

The structure of the present paper is as follows. Section 2 gives a brief explanation on the Kirchhoff equations and the Clebsch condition. The Kirchhoff equations are Hamilton's equations with respect to the Lie-Poisson bracket of $\mathfrak{se}(3)^* = \left(\mathfrak{so}(3) \ltimes \mathbb{R}^3\right)^* \equiv \mathbb{R}^3 \times \mathbb{R}^3 = \mathbb{R}^6$, which has two quadratic Casimir functions C_1 and C_2. In fact, the Kirchhoff equations can be restricted to the symplectic leaf, defined as the intersection of level hypersurfaces of C_1 and C_2. In the Clebsch case, there exists an additional constants of motion other than the Hamiltonian and hence the restricted system is completely integrable in the sense of Liouville with two degrees of freedom. In the present paper, we mainly focus on the case of Weber where $C_1 = 0$ and $C_2 = 1$, with which he was able to integrate the Kirchhoff equations by using his methods of theta functions in [30]. Note that the condition $C_1 = 0$ is a strong simplification, while $C_2 = 1$ can be assumed without loss of generality because it is only a convenient scaling.

In Section 3, a change of parameters for the Kirchhoff equations in the Clebsch case is introduced following the methods of [19], which allows to describe symmetrically the two first integrals other than the Casimir functions C_1 and C_2, as is observed through the functions C_3 and C_4. The intersection of the four quadrics $C_1 = 0, C_2 = 1, C_3 = c_3, C_4 = c_4$ defines, generically, a natural eightfold covering of a Kummer's quartic surface. Further, one introduces the two supplementary coordinates (x_1, x_2) with which the coordinates of the intersection of the four quadrics $C_1 = 0, C_2 = 1, C_3 = c_3, C_4 = c_4$ are clearly described. These coordinates are also useful to linearize the Hamiltonian flows.

In section 4, a family of invariant subspaces of dimension 3 on which solutions are elliptic curves is discussed; see also [14]. Such solutions are contained in the discriminant locus of the family of hyperelliptic curves. We miss the special solution with $p = 0$ where the motion is equivalent to the Euler case of the rigid body, since this is incompatible with Weber's condition $C_2 = 1$.

Section 5 provides an explicit computation of the actions following the method of [10]. In Section 6, the conclusion and the future perspectives are mentioned.

2 Kirchhoff Equations and Clebsch Condition

Following [14], we consider the Kirchhoff equations

$$
\begin{cases}
\dfrac{dK_1}{dt} = \left(\dfrac{1}{I_3} - \dfrac{1}{I_2}\right) K_2 K_3 + \left(\dfrac{1}{m_3} - \dfrac{1}{m_2}\right) p_2 p_3, \\[2mm]
\dfrac{dK_2}{dt} = \left(\dfrac{1}{I_1} - \dfrac{1}{I_3}\right) K_3 K_1 + \left(\dfrac{1}{m_1} - \dfrac{1}{m_3}\right) p_3 p_1, \\[2mm]
\dfrac{dK_3}{dt} = \left(\dfrac{1}{I_2} - \dfrac{1}{I_1}\right) K_1 K_2 + \left(\dfrac{1}{m_2} - \dfrac{1}{m_1}\right) p_1 p_2,
\end{cases}
\tag{1}
$$

$$
\begin{cases}
\dfrac{dp_1}{dt} = \dfrac{1}{I_3} p_2 K_3 - \dfrac{1}{I_2} p_3 K_2, \\[2mm]
\dfrac{dp_2}{dt} = \dfrac{1}{I_1} p_3 K_1 - \dfrac{1}{I_3} p_1 K_3, \\[2mm]
\dfrac{dp_3}{dt} = \dfrac{1}{I_2} p_1 K_2 - \dfrac{1}{I_1} p_2 K_1,
\end{cases}
\tag{2}
$$

which describe the rotational motion of an ellipsoidal rigid body in an ideal fluid, where the center of buoyancy and that of gravity for the rigid body are assumed to coincide. In (1), (2), $(\boldsymbol{K}, \boldsymbol{p}) = (K_1, K_2, K_3, p_1, p_2, p_3) \in \mathbb{R}^3 \times \mathbb{R}^3 \equiv \mathbb{R}^6$ are the phase space coordinates and $I_1, I_2, I_3, m_1, m_2, m_3 \in \mathbb{R}$ are parameters of the dynamics, which do not depend on the time t. The Kirchhoff equations (1), (2) are Hamilton's equation for the Hamiltonian $H(\boldsymbol{K}, \boldsymbol{p}) = \frac{1}{2}\left(\sum_{\alpha=1}^{3} \frac{K_\alpha^2}{I_\alpha} + \sum_{\alpha=1}^{3} \frac{p_\alpha^2}{m_\alpha}\right)$ with respect to the Lie-Poisson bracket

$$
\{F, G\}(\boldsymbol{K}, \boldsymbol{p}) = \langle \boldsymbol{K}, \nabla_K F \times \nabla_K G \rangle + \langle \boldsymbol{p}, \nabla_K F \times \nabla_p G - \nabla_K G \times \nabla_p F \rangle
$$

on $\mathfrak{se}(3)^* = \left(\mathfrak{so}(3) \ltimes \mathbb{R}^3\right)^* \equiv \mathbb{R}^3 \times \mathbb{R}^3 = \mathbb{R}^6$, where F, G are differentiable functions on $\mathbb{R}^6$, $\langle \cdot, \cdot \rangle$ is the standard Euclidean inner product on $\mathbb{R}^3$, and $\nabla_p F = \left(\dfrac{\partial F}{\partial p_1}, \dfrac{\partial F}{\partial p_2}, \dfrac{\partial F}{\partial p_3}\right)$, $\nabla_K F = \left(\dfrac{\partial F}{\partial K_1}, \dfrac{\partial F}{\partial K_2}, \dfrac{\partial F}{\partial K_3}\right)$. Note that the Hamiltonian vector field Ξ_F for an arbitrary Hamiltonian F is defined through $\Xi_F[G] = \{F, G\}$ for all differentiable functions G on $\mathbb{R}^6$ and can be written as

$$
(\Xi_F)_{(K,p)} = \left(\boldsymbol{K} \times \nabla_K F + \boldsymbol{p} \times \nabla_p F, \boldsymbol{p} \times \nabla_K F\right).
$$

The two functions $C_1(\boldsymbol{K}, \boldsymbol{p}) = \langle \boldsymbol{K}, \boldsymbol{p} \rangle = K_1 p_1 + K_2 p_2 + K_3 p_3$, $C_2(\boldsymbol{K}, \boldsymbol{p}) = \langle \boldsymbol{p}, \boldsymbol{p} \rangle = p_1^2 + p_2^2 + p_3^2$ are Casimir functions, i.e. $\{C_1, G\} \equiv 0$, $\{C_2, G\} \equiv 0$

for an arbitrary differentiable function G on $\mathbb{R}^6$ and hence they are constant of motion, as well as the Hamiltonian H.

We now focus on the Clebsch case where the parameters $I_1, I_2, I_3, m_1, m_2, m_3$ satisfy

$$m_1 I_1 (m_2 - m_3) + m_2 I_2 (m_3 - m_1) + m_3 I_3 (m_1 - m_2) = 0,$$

$$\Longleftrightarrow \frac{I_2 - I_3}{m_1} + \frac{I_3 - I_1}{m_2} + \frac{I_1 - I_2}{m_3} = 0,$$

$$\Longleftrightarrow \exists v, v' \in \mathbb{R} \text{ s.t. } \forall \alpha = 1, 2, 3, \ \frac{1}{m_\alpha} = v + \frac{v' I_\alpha}{I_1 I_2 I_3}. \tag{3}$$

Under this condition, the system (1), (2) admits the fourth constant of motion

$$L(K, p) = -\left(\frac{p_1^2}{m_2 m_3} + \frac{p_2^2}{m_3 m_1} + \frac{p_3^2}{m_1 m_2} \right) + \frac{K_1^2}{m_1 I_1} + \frac{K_2^2}{m_2 I_2} + \frac{K_3^2}{m_3 I_3}.$$

The four constants of motion H, L, C_1, C_2 Poisson commute and hence they define a completely integrable system in the sense of Liouville on each generic coadjoint orbit (symplectic leaf) in $\mathfrak{se}(3)^* \equiv \mathbb{R}^6$ defined as the intersection of the level hypersurfaces of C_1 and C_2. (See [4, 21] for the generalities of completely integrable systems and the Lie-Poisson brackets, although some of their notations do not coincide with ours.)

It is known that a coadjoint orbit V is diffeomorphic either to a point, to the two-dimensional sphere S^2, or to its cotangent bundle $T^* S^2$. In the following, we assume that the orbit V^4 is defined as the common level set $C_1 = 0, C_2 = 1$, in which case V^4 is symplectomorphic to the cotangent bundle $T^* S^2$ of the two-dimensional unit sphere; see e.g. [26, §7.5]. On the coadjoint orbit V^4 equipped with the orbit symplectic form, we can define the "momentum mapping" $J : V^4 \ni (K, p) \mapsto (H(K, p), L(K, p)) \in \mathbb{R}^2$. (Later, we use another description of first integrals C_3 and C_4, but the "momentum mapping" J is equivalent to the one J' defined as $J'(K, p) := (C_3(K, p), C_4(K, p))$, since C_3 and C_4 are linear transformations H, L, and C_2.)

Clearly, the level hypersurfaces of the function H are ellipsoids and hence it is compact. This means that the fiber $J^{-1}(c)$ of the "momentum mapping" J for the value $c \in \mathbb{R}^2$ is compact and hence we are able to use Liouville-Arnol'd(-Mineur-Jost) Theorem [4] which implies that regular fibers are (finite disjoint union of) real two-dimensional torus (tori) and around these regular fibers we can take the action-angle coordinates.

In what follows, we mainly work on the complexified system except for Section 5. The variables and parameters of the Kirchhoff equations are naturally extended to complex analytic ordinary differential equations on $\mathbb{C}^6$ from the original ones on $\mathbb{R}^6$ and the real quadratic polynomial first integrals C_1, C_2, H, L are also automatically extended to complex quadratic polynomials.

3 The Algebraic Linearization

In this section, we describe the linearization of the Kirchhoff equations (2), (1) with the Clebsch condition (3).

Following the lines of the presentation of [19], we introduce the new parameters (j_1, j_2, j_3) through which the original parameters $(I_1, I_2, I_3, m_1, m_2, m_3)$ with the Clebsch condition (3) are automatically obtained. The advantage of introducing these new parameters is symmetric descriptions of the first integrals for the Kirchhoff equations (1), (2) with the Clebsch condition (3) and the linearization for the flow; see (4) and Theorem 3.4 below.

We consider the intersections of the four quadrics defined as the level hypersurfaces of quadratic functions

$$
\begin{cases}
C_1 = \sum_{\alpha=1}^{3} K_\alpha p_\alpha, & C_2 = \sum_{\alpha=1}^{3} p_\alpha^2, \\
C_3 = \sum_{\alpha=1}^{3} \left\{ K_\alpha^2 + (j_1 + j_2 + j_3 - j_\alpha) p_\alpha^2 \right\}, & C_4 = \sum_{\alpha=1}^{3} \left(j_\alpha K_\alpha^2 + \frac{j_1 j_2 j_3}{j_\alpha} p_\alpha^2 \right),
\end{cases}
\tag{4}
$$

in $(K_1, K_2, K_3, p_1, p_2, p_3) \in \mathbb{C}^6$. In what follows, the parameters $j_1, j_2, j_3 \in \mathbb{C}$ are fixed to recover the Kirchhoff equations (1), (2).

Lemma 3.1 *The four functions C_1, C_2, C_3, C_4 Poisson commute with respect to the Lie-Poisson bracket $\{\cdot, \cdot\}$.*

This is a result of straightforward computations of $\{C_3, C_4\} = 0$. Recall that C_1 and C_2 are Casimir functions for the Lie-Poisson bracket $\{\cdot, \cdot\}$.

We take the Hamiltonian

$$
\lambda C_3 + \lambda' C_4 = \sum_{\alpha=1}^{3} \left(n_\alpha K_\alpha^2 + n_\alpha' p_\alpha^2 \right),
$$

a member of the linear pencil generated by C_3 and C_4, as well as its Hamiltonian vector field $\Xi_{\lambda C_3 + \lambda' C_4} = \lambda \Xi_{C_3} + \lambda' \Xi_{C_4}$ with respect to the Lie-Poisson bracket $\{\cdot, \cdot\}$. Here, we put

$$
n_\alpha = \lambda + \lambda' j_\alpha, \qquad n_\alpha' = \lambda (j_1 + j_2 + j_3 - j_\alpha) + \lambda' \frac{j_1 j_2 j_3}{j_\alpha}.
$$

As is pointed out in [19], we recover the original Kirchhoff equations (1), (2), when we choose the parameters λ, λ' correctly. In this case, we have $n_\alpha = 1/2I_\alpha$ and $n'_\alpha = 1/2m_\alpha$. In other words, given the parameters μ, μ' in (3), the parameters λ, λ' are given as $\lambda = \pm\sqrt{\frac{2\lambda''-\nu}{4\nu'}}, \lambda' = \frac{1}{2\nu'}$, where λ'' is an arbitrary parameter. Note that if $I_\alpha, m_\alpha, \alpha = 1, 2, 3$, are real, then we can assume that $\lambda, \lambda', j_\alpha, \alpha = 1, 2, 3$, are real.

The family of commuting Hamiltonians leaves invariant the level sets of the functions C_1, C_2, C_3, C_4, which are the symplectic leaves of the Poisson space $\left(\mathbb{C}^6, \{\cdot, \cdot\}\right)$

3.1 Intersection of the Three Quadrics $C_2 = 1, C_3 = c_3, C_4 = c_4$

We describe the intersection of the three quadrics $C_2 = 1, C_3 = c_3, C_4 = c_4$. We first observe that the three equations are linear in p_1^2, p_2^2, p_3^2. We introduce the matrix

$$A = \begin{pmatrix} 1 & 1 & 1 \\ j_2 + j_3 & j_1 + j_3 & j_1 + j_2 \\ j_2 j_3 & j_3 j_1 & j_1 j_2 \end{pmatrix}$$

and write the conditions of intersection of the three quadrics as:

$$\begin{pmatrix} 1 & 1 & 1 \\ j_2 + j_3 & j_1 + j_3 & j_1 + j_2 \\ j_2 j_3 & j_3 j_1 & j_1 j_2 \end{pmatrix} \begin{pmatrix} p_1^2 \\ p_2^2 \\ p_3^2 \end{pmatrix} = \begin{pmatrix} 1 \\ -(K_1^2 + K_2^2 + K_3^2) + c_3 \\ -(j_1 K_1^2 + j_2 K_2^2 + j_3 K_3^2) + c_4 \end{pmatrix}. \tag{5}$$

We further assume the genericity condition

$$\det A = j_1 j_2 (j_1 - j_2) + j_2 j_3 (j_2 - j_3) + j_1 j_3 (j_3 - j_1)$$
$$= -(j_1 - j_2)(j_2 - j_3)(j_3 - j_1) \neq 0.$$

Under this condition, we have

$$A^{-1} = -\frac{1}{\Delta} \begin{pmatrix} j_1^2(j_2 - j_3) & j_1(j_3 - j_2) & j_2 - j_3 \\ j_2^2(j_3 - j_1) & j_2(j_1 - j_3) & j_3 - j_1 \\ j_3^2(j_1 - j_2) & j_3(j_2 - j_1) & j_1 - j_2 \end{pmatrix}, \tag{6}$$

where we set $\Delta := (j_1 - j_2)(j_2 - j_3)(j_3 - j_1)$, and hence the solution to (5) is written as

$$p_1^2 = l - \frac{1}{\Delta}\left\{(j_1 - j_2)(j_2 - j_3)K_2^2 - (j_2 - j_3)(j_3 - j_1)K_3^2\right\},$$

$$p_2^2 = m - \frac{1}{\Delta}\left\{(j_2 - j_3)(j_3 - j_1)K_3^2 - (j_3 - j_1)(j_1 - j_2)K_1^2\right\},$$

$$p_3^2 = n - \frac{1}{\Delta}\left\{(j_3 - j_1)(j_1 - j_2)K_1^2 - (j_1 - j_2)(j_2 - j_3)K_2^2\right\}, \qquad (7)$$

where (l, m, n) are functions of the parameters (j_1, j_2, j_3) and of the (c_3, c_4), such that $l + m + n = 1$.

3.2 The Associated Kummer Surface

From the equation and the vanishing condition of the first Casimir, we get

$$\sqrt{l - \frac{1}{\Delta}\left\{(j_1 - j_2)(j_2 - j_3)K_2^2 - (j_2 - j_3)(j_3 - j_1)K_3^2\right\}}K_1$$

$$+ \sqrt{m - \frac{1}{\Delta}\left\{(j_2 - j_3)(j_3 - j_1)K_3^2 - (j_3 - j_1)(j_1 - j_2)K_1^2\right\}}K_2$$

$$+ \sqrt{n - \frac{1}{\Delta}\left\{(j_3 - j_1)(j_1 - j_2)K_1^2 - (j_1 - j_2)(j_2 - j_3)K_2^2\right\}}K_3 = 0. \qquad (8)$$

This equation represents an eight-fold covering of a Kummer's quartic surface as in [15, §54] and [31, p.341], given by the correspondence $(K_1, K_2, K_3) \mapsto (K_1^2, K_2^2, K_3^2)$. We homogenize (8) by setting $\left(K_1^2 = X_1/X_4, K_2^2 = X_2/X_4, K_3^2 = X_3/X_4\right)$ and by using the parameters

$$d_1 = -\frac{1}{\Delta}(j_3 - j_1)(j_1 - j_2) = \frac{1}{j_3 - j_2},$$

$$d_2 = -\frac{1}{\Delta}(j_1 - j_2)(j_2 - j_3) = \frac{1}{j_1 - j_3},$$

$$d_3 = -\frac{1}{\Delta}(j_2 - j_3)(j_3 - j_1) = \frac{1}{j_2 - j_1}.$$

As a result, we have

$$X_1^2(lX_4 + d_2X_2 - d_3X_3)^2 + X_2^2(mX_4 + d_3X_3 - d_1X_1)^2$$
$$+ X_3^2(nX_4 + d_1X_1 - d_2X_2)^2$$
$$- 2X_1X_2(lX_4 + d_2X_2 - d_3X_3)(mX_4 + d_3X_3 - d_1X_1)$$
$$- 2X_1X_3(lX_4 + d_2X_2 - d_3X_3)(nX_4 + d_1X_1 - d_2X_2)$$
$$- 2X_2X_3(mX_4 + d_3X_3 - d_1X_1)(nX_4 + d_1X_1 - d_2X_2) = 0, \qquad (9)$$

where $(X_1 : X_2 : X_3 : X_4)$ are regarded as homogeneous coordinates in $\mathbb{CP}^3$. Note that we have the relation $1/d_1 + 1/d_2 + 1/d_3 = 0$. As is described in [15, §54], the quartic surface defined through (8) can be shown to be a Kummer surface. Here, we give a brief justification to this.

Theorem 3.2 *The quartic surface defined through (9) in $\mathbb{CP}^3$ has 16 double points and hence it is a Kummer surface.*

Proof. The maximal number of double points in $\mathbb{CP}^3$ is 16; see [18, 25]. Thus, we only need to count the number of double points for the surface defined through (9). Setting $U_1 = lX_4 + d_2X_2 - d_3X_3$, $U_2 = mX_4 + d_3X_3 - d_1X_1$, $U_3 = nX_4 + d_1X_1 - d_2X_2$, we can write down (9) as

$$X_1^2U_1^2 + X_2^2U_2^2 + X_3^2U_3^2 - 2X_1X_2U_1U_2 - 2X_2X_3U_2U_3 - 2X_3X_1U_3U_1 = 0. \tag{10}$$

The singular points of the surface appear at the following 14 points:
(1) $(X_1 : X_2 : X_3 : X_4) = (1:0:0:0), (0:1:0:0), (0:0:1:0),$ $(0:0:0:1)$; (2) two points defined by $X_i = U_i = 0$, $i = 1, 2, 3$; (3) $X_\alpha = U_\beta = U_\gamma = 0$, $\alpha = 1, 2, 3$, $\{\alpha, \beta, \gamma\} = \{1, 2, 3\}$; and (4) $U_1 = U_2 = U_3 = 0$.

We find double points of the surface on each of these lines. Recall that a normal form of double points is given as $z_1 = z_2 = z_3 = 0$ with the equation $z_1^2 + z_2z_3 = 0$; see [6].

(1) Consider the point $(X_1 : X_2 : X_3 : X_4) = (0:0:0:1)$, where the linear functions U_1/X_4, U_2/X_4, U_3/X_4 are non-zero. These functions are regarded as units in the local ring the polynomials. The equation (10) of the surface is written as

$$\left(\frac{U_1}{X_4}\frac{X_1}{X_4} - \frac{U_2}{X_4}\frac{X_2}{X_4}\right)^2 - \frac{U_3}{X_4}\frac{X_3}{X_4}\left(2\frac{U_1}{X_4}\frac{X_1}{X_4} + 2\frac{U_2}{X_4}\frac{X_2}{X_4} - \frac{U_3}{X_4}\frac{X_3}{X_4}\right) = 0, \tag{11}$$

which apparently has a double point at $X_1/X_4 = X_2/X_4 = X_3/X_4 = 0$ since three polynomials $U_1/X_4\,X_1/X_4 - U_2/X_4\,X_2/X_4$, $U_3/X_4\,X_3/X_4$, $2U_1/X_4\,X_1/X_4 + 2U_2/X_4\,X_2/X_4 - U_3/X_4\,X_3/X_4$ are of weight one.

Concerning the point $(X_1 : X_2 : X_3 : X_4) = (0:0:1:0)$, we see that $U_1 \neq 0$, $U_2 \neq 0$, $U_3 = 0$ and hence the polynomials $U_1/X_3\,X_1/X_3 - U_2/X_3\,X_2/X_3$, U_3/X_3, $2U_1/X_3\,X_1/X_3 + 2\,U_2/X_3\,X_2/X_3 - U_3/X_3$ are of weight one. Thus, the transformation

$$\left(\frac{U_1}{X_3}\frac{X_1}{X_3} - \frac{U_2}{X_3}\frac{X_2}{X_3}\right)^2 - \frac{U_3}{X_3}\left(2\frac{U_1}{X_3}\frac{X_1}{X_3} + 2\frac{U_2}{X_3}\frac{X_2}{X_3} - \frac{U_3}{X_3}\right) = 0 \quad (12)$$

of (10) indicates that $(X_1 : X_2 : X_3 : X_4) = (0 : 0 : 1 : 0)$ is a double point of the surface. The argument can also be applied to the other two points $(X_1 : X_2 : X_3 : X_4) = (0 : 1 : 0 : 0)$ and $(X_1 : X_2 : X_3 : X_4) = (1 : 0 : 0 : 0)$. We have in total four double points of this type.

(2) Consider the points of the surface on the line $X_3 = U_3 = 0$. Here, we can assume $X_4 \neq 0$, since if $X_4 = 0$, we have $X_1 X_2 = 0$ by (10) and hence it reduces to the case (1). Under this assumption, we have $U_1 X_1 - U_2 X_2 = 0$ by (10), which consists of two distinct points, as the condition is reduced to the quadratic equation in $(X_1 : X_2)$: $ld_1 X_1^2 - \{(m + n)d_1 + (n + l)d_2\} X_1 X_2 + md_2 X_2^2 = 0$. Around each of these two points, the polynomials U_3/X_4, X_3/X_4, $U_1/X_4 \, X_1/X_4 - U_2/X_4 \, X_2/X_4$ are of weight one, while $2U_1/X_4 \, X_1/X_4 + 2U_2/X_4 \, X_2/X_4 - U_3/X_4 \, X_3/X_4$ is a unit. Thus, the transformation (11) of (10) indicates that each of the two points is double point. The same method can be applied to the points where $X_1 = U_1 = 0$ and $X_2 = U_2 = 0$. The total number of these double points is six.

(3) Consider the points with the condition $U_1 = U_2 = X_3 = 0$, with which we have $U_3 \neq 0$, $X_1 \neq 0$, $X_2 \neq 0$, $X_4 \neq 0$. Around this points, the polynomials $U_1/X_4 \, X_1/X_4 - U_2/X_4 \, X_2/X_4$, $2U_1/X_4 \, X_1/X_4 + 2U_2/X_4 \, X_2/X_4 - U_3/X_4 \, X_3/X_4$, X_3/X_4 are of weight one, while U_3/X_4 is a unit. Thus, from the transformation (11) of (10), we see that the point is a double point. The method is applied also to the points $U_2 = U_3 = X_1 = 0$, $U_3 = U_1 = X_2 = 0$. In total, we have three double points.

(4) At the point $U_1 = U_2 = U_3 = 0$, we have $0 = U_1 + U_2 + U_3 = (l + m + n) X_4 = X_4$ and hence we obtain $(X_1 : X_2 : X_3 : X_4) = (1/d_1 : 1/d_2 \, 1/d_3 : 0)$. In particular, X_1, X_2, X_3 are non-zero at this point. The functions $U_1/X_3 \, X_1/X_3 - U_2/X_3 \, X_2/X_3$, U_3/X_3, $2 U_1/X_3 \, X_1/X_3 + 2 U_2/X_3 \, X_2/X_3 - U_3/X_3$ are polynomials of weight one. So, again, we have a double point because of the transformation (12) of (10).

The other two points can be detected as in [15, §55]. □

Note that we obtain Kummer surfaces depending both on the parameters (j_1, j_2, j_3) and on the values of the two first integrals (C_3, C_4).

To sum up, the 16 double points on the surface are described (9) as follows:

(1) The origin in the affine space $[0 : 0 : 0 : 1]$.
(2) Four points at infinity ($X_4 = 0$) which are:
    $[1 : 0 : 0 : 0]$, $[0 : 1 : 0 : 0]$, $[0 : 0 : 1 : 0]$ real independent of any
    constant and $\left[1/d_1 \ : \ 1/d_2 \ : \ 1/d_3 \ : \ 0\right]$ real depending only on the
    parameters (j_1, j_2, j_3). There are two other singular points at infinity
    ($X_4 = 0$) obtained through the method in [15, §55].
(3) Three points $\left[m/d_1 \ : \ -l/d_2 \ : 0 : 1\right]$, $\left[0 : \ n/d_2 \ : \ -m/d_3 \ : \ 1\right]$,
    $\left[-n/d_1 \ : 0 : \ l/d_3 \ : \ 1\right]$ in the real affine space.
(4) Six points of type $X_3 = U_3 = 0$ (and cyclic perturbations) which are
    affine either real or complex conjugated depending on the sign of
    $[(m + n)d_1 + (n + l)d_2]^2 - 4lmd_1d_2$ (and cyclic perturbations).

3.3 Supplementary Coordinates (x_1, x_2)

Inspired by [31], we introduce two supplementary coordinates (x_1, x_2) as the
solutions of the equation

$$\frac{p_1^2}{x - j_1} + \frac{p_2^2}{x - j_2} + \frac{p_3^2}{x - j_3} = 0, \tag{13}$$

which can also be written as

$$x^2 - Ex + F = 0, \tag{14}$$

where $E = \sum_{\alpha=1}^{3}(j_1 + j_2 + j_3 - j_\alpha)p_\alpha^2$, $F = \sum_{\alpha=1}^{3} j_1 j_2 j_3 / j_\alpha \, p_\alpha^2$. Note that
(x_1, x_2) are functions only in p_α, $\alpha = 1, 2, 3$, hence $\{x_1, x_2\} = 0$ and that the
expression of the squared coordinates p_α^2, $\alpha = 1, 2, 3$, are obtained in terms of
(x_1, x_2):

$$\begin{cases} p_1^2 = \dfrac{(j_1 - x_1)(j_1 - x_2)}{(j_1 - j_2)(j_1 - j_3)}, \\[2mm] p_2^2 = \dfrac{(j_2 - x_1)(j_2 - x_2)}{(j_2 - j_3)(j_2 - j_1)}, \\[2mm] p_3^2 = \dfrac{(j_3 - x_1)(j_3 - x_2)}{(j_3 - j_1)(j_3 - j_2)}, \end{cases} \tag{15}$$

where we used $C_2 = p_1^2 + p_2^2 + p_3^2 = 1$.

We also introduce the two solutions j_4, j_5 of the equation

$$\frac{l}{x - j_1} + \frac{m}{x - j_2} + \frac{n}{x - j_3} = 0$$

$$\Longleftrightarrow x^2 - \{(j_2 + j_3)l + (j_1 + j_3)m + (j_2 + j_1)n\} x$$
$$+ (j_2 j_3 l + j_1 j_3 m + j_1 j_2 n) = 0.$$

By $l + m + n = 1$, we have the expression of the parameters l, m, n in terms of j_1, j_2, j_3 as

$$\begin{cases} l & = & \dfrac{(j_1 - j_4)(j_1 - j_5)}{(j_1 - j_2)(j_1 - j_3)}, \\ m & = & \dfrac{(j_2 - j_4)(j_2 - j_5)}{(j_2 - j_3)(j_2 - j_1)}, \\ n & = & \dfrac{(j_3 - j_4)(j_3 - j_5)}{(j_3 - j_1)(j_3 - j_2)}. \end{cases} \tag{16}$$

Putting $K_1 = K_2 = K_3 = 0$ in (5), we have

$$\begin{pmatrix} l \\ m \\ n \end{pmatrix} = A^{-1} \begin{pmatrix} 1 \\ C_3 \\ C_4 \end{pmatrix}, \tag{17}$$

which yields

$$\begin{cases} l & = & -\dfrac{1}{\Delta} \left\{ j_1^2(j_2 - j_3) + C_3 j_1(j_3 - j_2) + C_4(j_2 - j_3) \right\}, \\ m & = & -\dfrac{1}{\Delta} \left\{ j_2^2(j_3 - j_1) + C_3 j_2(j_1 - j_3) + C_4(j_3 - j_1) \right\}, \\ n & = & -\dfrac{1}{\Delta} \left\{ j_3^2(j_1 - j_2) + C_3 j_3(j_2 - j_1) + C_4(j_1 - j_2) \right\}. \end{cases}$$

Therefore, we have

$$(j_2 + j_3)l + (j_1 + j_3)m + (j_2 + j_1)n$$
$$= -\frac{C_3}{\Delta} \left\{ j_1(j_3^2 - j_2^2) + j_2(j_1^2 - j_3^2) + j_3(j_2^2 - j_1^2) \right\} = C_3,$$

$$j_2 j_3 l + j_1 j_3 m + j_1 j_2 n$$
$$= -\frac{C_4}{\Delta} [j_2 j_3(j_2 - j_3) + j_1 j_3(j_3 - j_1) + j_1 j_2(j_1 - j_2)] = C_4$$

and hence j_4, j_5 are the two roots of the equation

$$x^2 - C_3 x + C_4 = 0. \tag{18}$$

Thus, they are independent of the parameters j_1, j_2, j_3 and only depends on C_3, C_4.

Remark 3.3 The parameters (ℓ, m, n) and (c_3, c_4) are in $1 : 1$ correspondence to each other via the linear relation (17). There is also a $1 : 1$

correspondence between (j_4, j_5) and (c_3, c_4) through the quadratic relation (18). ∎

We observe that the two vectors

$$
e_1 = \begin{pmatrix} \frac{p_1}{x_1 - j_1} \\ \frac{p_2}{x_1 - j_2} \\ \frac{p_3}{x_1 - j_3} \end{pmatrix}, \qquad
e_2 = \begin{pmatrix} \frac{p_1}{x_2 - j_1} \\ \frac{p_2}{x_2 - j_2} \\ \frac{p_3}{x_2 - j_3} \end{pmatrix}
$$

are orthogonal to the vector $(p_1, p_2, p_3)^{\mathrm{T}}$ with respect to the standard metric. In general, these two vectors are independent. The vector $(K_1, K_2, K_3)^{\mathrm{T}}$, which is also orthogonal to $(p_1, p_2, p_3)^{\mathrm{T}}$, is thus a linear combination of these two vectors $(K_1, K_2, K_3)^{\mathrm{T}} = Ae_1 + Be_2$, $A, B \in \mathbb{C}$. This yields

$$
\begin{cases}
K_1 = \dfrac{1}{\sqrt{(j_1 - j_2)(j_1 - j_3)}} \left(A\sqrt{\dfrac{x_1 - j_1}{x_2 - j_1}} + B\sqrt{\dfrac{x_2 - j_1}{x_1 - j_1}} \right), \\[2ex]
K_2 = \dfrac{1}{\sqrt{(j_2 - j_3)(j_2 - j_1)}} \left(A\sqrt{\dfrac{x_1 - j_2}{x_2 - j_2}} + B\sqrt{\dfrac{x_2 - j_2}{x_1 - j_2}} \right), \\[2ex]
K_3 = \dfrac{1}{\sqrt{(j_3 - j_1)(j_3 - j_2)}} \left(A\sqrt{\dfrac{x_1 - j_3}{x_2 - j_3}} + B\sqrt{\dfrac{x_2 - j_3}{x_1 - j_3}} \right).
\end{cases} \qquad (19)
$$

Next, we insert these formulae inside (for instance) the first equations of (7) and get

$$
\begin{aligned}
&\frac{(j_1 - x_1)(j_1 - x_2)}{(j_1 - j_2)(j_1 - j_3)} - \frac{(j_1 - j_4)(j_1 - j_5)}{(j_1 - j_2)(j_1 - j_3)} \\[1ex]
&= -\frac{1}{\Delta}\left[-\left[\left\{ A^2\left(\frac{x_1 - j_2}{x_2 - j_2}\right) + B^2\left(\frac{x_2 - j_2}{x_1 - j_2}\right) - 2AB \right\} \right.\right. \\[1ex]
&\qquad\qquad \left.\left. + \left\{ A^2\left(\frac{x_1 - j_3}{x_2 - j_3}\right) + B^2\left(\frac{x_2 - j_3}{x_1 - j_3}\right) - 2AB \right\} \right]\right] \\[1ex]
&= -\frac{1}{\Delta}\left[-A^2\left(\frac{x_1 - j_2}{x_2 - j_2} - \frac{x_1 - j_3}{x_2 - j_3}\right) - B^2\left(\frac{x_2 - j_2}{x_1 - j_2} - \frac{x_2 - j_3}{x_1 - j_3}\right) \right] \\[1ex]
&= -\frac{1}{\Delta}\left[-A^2\left(\frac{(x_2 - x_1)(j_3 - j_2)}{(x_2 - j_2)(x_2 - j_3)}\right) - B^2\left(\frac{(x_1 - x_2)(j_3 - j_2)}{(x_1 - j_2)(x_1 - j_3)}\right) \right],
\end{aligned} \qquad (20)
$$

which yields

$$
\begin{aligned}
&(j_1 - x_1)(j_1 - x_2) - (j_1 - j_4)(j_1 - j_5) \\[1ex]
&= -(x_1 - x_2)\left\{ \frac{A^2}{(x_2 - j_2)(x_2 - j_3)} - \frac{B^2}{(x_1 - j_2)(x_1 - j_3)} \right\}. \qquad (21)
\end{aligned}
$$

Next, from the formulae (19) and the second equation of (7), we obtain

$$(j_2 - x_1)(j_2 - x_2) - (j_2 - j_4)(j_2 - j_5)$$

$$= -(x_1 - x_2) \left\{ \frac{A^2}{(x_2 - j_3)(x_2 - j_1)} - \frac{B^2}{(x_1 - j_3)(x_1 - j_1)} \right\}. \tag{22}$$

We further assume the genericity assumption

$$\det \begin{pmatrix} \frac{1}{(x_2 - j_2)(x_2 - j_3)} & -\frac{1}{(x_1 - j_2)(x_1 - j_3)} \\ \frac{1}{(x_2 - j_2)(x_2 - j_3)} & -\frac{1}{(x_1 - j_2)(x_1 - j_3)} \end{pmatrix}$$

$$= -\frac{(j_1 - j_2)(x_1 - x_2)}{(x_1 - j_1)(x_1 - j_2)(x_1 - j_3)(x_2 - j_1)(x_2 - j_2)(x_2 - j_3)} \neq 0$$

on $(x_1, x_2, j_1, j_2, j_3)$ with which the two last equations (21) and (22) determine uniquely A^2 and B^2 in terms of $(x_1, x_2, j_1, j_2, j_3)$. Then, we have

$$A^2 = -\frac{\Phi(x_2)\Psi(x_1)}{(x_2 - x_1)^2}, \qquad B^2 = -\frac{\Phi(x_1)\Psi(x_2)}{(x_2 - x_1)^2},$$

$$\Phi(x) = (j_1 - x)(j_2 - x)(j_3 - x), \qquad \Psi(x) = (x - j_4)(x - j_5). \tag{23}$$

This also follows from the following identity satisfied by an arbitrary λ:

$$(\lambda - x_1)(\lambda - x_2) - (\lambda - j_4)(\lambda - j_5)$$

$$= \frac{1}{(x_2 - x_1)} \{\Psi(x_1)(x_2 - \lambda) - \Psi(x_2)(x_1 - \lambda)\}. \tag{24}$$

Substituting λ by j_1 and j_2, we have (23).

Next, from (19) and (23), we obtain the formula (7) of [31]:

$$K_1 = \sqrt{\frac{-1}{(j_1 - j_2)(j_1 - j_3)} \frac{1}{x_2 - x_1}}$$

$$\times \left\{ \sqrt{\frac{\Phi(x_2)\Psi(x_1)(x_1 - j_1)}{x_2 - j_1}} - \sqrt{\frac{\Phi(x_1)\Psi(x_2)(x_2 - j_1)}{x_1 - j_1}} \right\},$$

$$K_2 = \sqrt{\frac{-1}{(j_2 - j_3)(j_2 - j_1)} \frac{1}{x_2 - x_1}}$$

$$\times \left\{ \sqrt{\frac{\Phi(x_2)\Psi(x_1)(x_1 - j_2)}{x_2 - j_2}} - \sqrt{\frac{\Phi(x_1)\Psi(x_2)(x_2 - j_2)}{x_1 - j_2}} \right\},$$

$$K_3 = \sqrt{\frac{-1}{(j_3 - j_1)(j_3 - j_2)} \frac{1}{x_2 - x_1}}$$

$$\times \left\{ \sqrt{\frac{\Phi(x_2)\Psi(x_1)(x_1 - j_3)}{x_2 - j_3}} - \sqrt{\frac{\Phi(x_1)\Psi(x_2)(x_2 - j_3)}{x_1 - j_3}} \right\}. \tag{25}$$

At this point, we can observe that these equations induce an algebraic parametrization of the Kummer surface with the coordinates (x_1, x_2). See also [15, §99].

3.4 Linearization of the Hamiltonian Flows

Theorem 3.4 *The Hamiltonian flows associated with the Hamiltonians $\lambda C_3 + \lambda' C_4$ induce linear flows on the Jacobian of the hyperelliptic curves $\mathcal{C}_c$ defined through the equation $y^2 = (j_1 - x)(j_2 - x)(j_3 - x)(j_4 - x)(j_5 - x)$.*

Proof. The equations (2) can be written as:

$$
\begin{cases}
\dot{p}_1 = (\lambda + \lambda' j_2) K_2 p_3 - (\lambda + \lambda' j_3) K_3 p_2, \\
\dot{p}_2 = (\lambda + \lambda' j_3) K_3 p_1 - (\lambda + \lambda' j_1) K_1 p_3, \\
\dot{p}_3 = (\lambda + \lambda' j_1) K_1 p_2 - (\lambda + \lambda' j_2) K_2 p_1.
\end{cases}
\tag{26}
$$

Recall that $n_\alpha = 1/2 I_\alpha$ and $n'_\alpha = 1/2 m_\alpha$.

From (15), we obtain

$$
\frac{dp_1}{dt} = -\frac{1}{2\sqrt{(j_1 - j_2)(j_1 - j_3)}} \left(\sqrt{\frac{j_1 - x_2}{j_1 - x_1}} \frac{dx_1}{dt} + \sqrt{\frac{j_1 - x_1}{j_1 - x_2}} \frac{dx_2}{dt} \right).
$$

On the other hand, from (26), (25), (15), we have

$$
\dot{p}_1 = \sqrt{\frac{-1}{(j_3 - j_1)(j_1 - j_2)} \frac{1}{x_2 - x_1}}
$$
$$
\times \left\{ (\lambda + \lambda' x_2) \sqrt{\frac{x_2 - j_1}{x_1 - j_1}} R(x_1) - (\lambda + \lambda' x_1) \sqrt{\frac{x_1 - j_1}{x_2 - j_1}} R(x_2) \right\},
$$

where we have set $R(x) = \sqrt{(j_1 - x)(j_2 - x)(j_3 - x)(j_4 - x)(j_5 - x)}$. Thus, we have

$$
\sqrt{\frac{j_1 - x_2}{j_1 - x_1}} \frac{dx_1}{dt} + \sqrt{\frac{j_1 - x_1}{j_1 - x_2}} \frac{dx_2}{dt}
$$
$$
= \frac{2}{x_1 - x_2} \left\{ (\lambda + \lambda' x_2) R(x_1) \sqrt{\frac{x_2 - j_1}{x_1 - j_1}} - (\lambda + \lambda' x_1) R(x_2) \sqrt{\frac{x_1 - j_1}{x_2 - j_1}} \right\}.
$$

Similar equations can be obtained via cyclic permutations of $(1, 2, 3)$ for the subindices in j_1, j_2, j_3:

$$\sqrt{\frac{j_2 - x_2}{j_2 - x_1}} \frac{dx_1}{dt} + \sqrt{\frac{j_2 - x_1}{j_2 - x_2}} \frac{dx_2}{dt}$$

$$= \frac{2}{x_1 - x_2} \left\{ (\lambda + \lambda' x_2) R(x_1) \sqrt{\frac{x_2 - j_2}{x_1 - j_2}} - (\lambda + \lambda' x_1) R(x_2) \sqrt{\frac{x_1 - j_2}{x_2 - j_2}} \right\},$$

$$\sqrt{\frac{j_3 - x_2}{j_3 - x_1}} \frac{dx_1}{dt} + \sqrt{\frac{j_3 - x_1}{j_3 - x_2}} \frac{dx_2}{dt}$$

$$= \frac{2}{x_1 - x_2} \left\{ (\lambda + \lambda' x_2) R(x_1) \sqrt{\frac{x_2 - j_3}{x_1 - j_3}} - (\lambda + \lambda' x_1) R(x_2) \sqrt{\frac{x_1 - j_3}{x_2 - j_3}} \right\}.$$

Since j_1, j_2, j_3 (and hence $\sqrt{j_1 - x_2/j_1 - x_1}$, $\sqrt{j_2 - x_2/j_2 - x_1}$, $\sqrt{j_3 - x_2/j_3 - x_1}$) can be assumed to be distinct, this means that the following two linear forms in (ξ, η) are the same:

$$\frac{dx_1}{dt} \xi + \frac{dx_2}{dt} \eta,$$

$$\frac{2}{x_1 - x_2} \left\{ (\lambda + \lambda' x_2) R(x_1) \xi - (\lambda + \lambda' x_1) R(x_2) \eta \right\}.$$

Therefore, the flow induced by the family of Hamiltonian systems $\lambda C_3 + \lambda' C_4$ yields

$$\frac{dx_1}{dt} = 2 \frac{1}{x_1 - x_2} R(x_1)(\lambda' x_2 + \lambda),$$

$$\frac{dx_2}{dt} = 2 \frac{1}{x_2 - x_1} R(x_2)(\lambda' x_1 + \lambda),$$

and hence

$$\begin{cases} \dfrac{1}{R(x_1)} \dfrac{dx_1}{dt} + \dfrac{1}{R(x_2)} \dfrac{dx_2}{dt} = -2\lambda', \\[2mm] \dfrac{x_1}{R(x_1)} \dfrac{dx_1}{dt} + \dfrac{x_2}{R(x_2)} \dfrac{dx_2}{dt} = 2\lambda. \end{cases} \tag{27}$$

This proves the theorem as the two differential forms $dx/R(x)$, $x\,dx/R(x)$ provide a basis of the space of holomorphic 1-forms on the Jacobian of the hyperelliptic curve $\mathcal{C}_c$. □

The set of five parameters $(j_1, j_2, j_3, j_4, j_5)$ is related to the parameters $(I_1, I_2, I_3, m_1, m_2, m_3)$ of the Kirchhoff equations with the Clebsch condition (3) and the values of the first integrals (C_3, C_4). Recall that the values of the Casimir functions (C_1, C_2) are fixed to be $(0, 1)$. Note that these equations (27) induce a uniformization of the Kummer surface; see [15, §99].

4 Some Special Solutions

In this section, we discuss two types of special solutions to the Kirchhoff equations (1), (2) of Clebsch top. Namely, as is pointed out in [14, §2.2, §2.4], there are three special solutions on the three-dimensional vector spaces $p_1 = K_2 = K_3 = 0$, $p_2 = K_3 = K_1 = 0$, $p_3 = K_1 = K_2 = 0$, as well as the solutions on the three-dimensional vector spaces $p_\alpha = \delta_\alpha K_\alpha$, $\alpha = 1, 2, 3$, with suitable constants δ_α, $\alpha = 1, 2, 3$. In fact, assuming e.g. $p_1 = K_2 = K_3$, we can reduce Kirchhoff equations (1), (2) to

$$
\begin{cases}
\dfrac{dK_1}{dt} = \left(\dfrac{1}{m_3} - \dfrac{1}{m_2} \right) p_2 p_3, \\[2mm]
\dfrac{dp_2}{dt} = \dfrac{1}{I_1} p_3 K_1, \\[2mm]
\dfrac{dp_3}{dt} = -\dfrac{1}{I_1} K_1 p_2,
\end{cases}
\tag{28}
$$

while, under the condition $p_\alpha = \delta_\alpha K_\alpha$, $\alpha = 1, 2, 3$, the equations (1), (2) are transformed to

$$
\begin{cases}
\dfrac{dK_1}{dt} = \dfrac{1}{\delta_1} \left(\dfrac{\delta_2}{I_3} - \dfrac{\delta_3}{I_2} \right) K_2 K_3, \\[2mm]
\dfrac{dK_2}{dt} = \dfrac{1}{\delta_2} \left(\dfrac{\delta_3}{I_1} - \dfrac{\delta_1}{I_3} \right) K_3 K_1, \\[2mm]
\dfrac{dK_3}{dt} = \dfrac{1}{\delta_3} \left(\dfrac{\delta_1}{I_2} - \dfrac{\delta_2}{I_1} \right) K_1 K_2.
\end{cases}
\tag{29}
$$

To describe the Hamiltonian structures of these equations (28) and (29), we first consider the modified Lie bracket $[\cdot, \cdot]_M$ on $\mathbb{C}^3$ associated to a diagonal matrix $M = \mathrm{diag}\,(\mu_1, \mu_2, \mu_3)$:

$$
[u, v]_M := M\,(u \times v)\,, u, u \in \mathbb{C}^3.
$$

In association to the modified Lie bracket $[\cdot, \cdot]_M$, we can define the Lie-Poisson bracket $\{\cdot, \cdot\}_M$ through

$$
\{F, G\}_M\,(x) := \langle x, M\,(\nabla F(x) \times \nabla G(x)) \rangle\,,
$$

where $x \in \mathbb{C}^3$ and F, G are differentiable functions on $\mathbb{C}^3$. The Lie bracket $[\cdot, \cdot]_M$ and the associated Lie-Poisson bracket $\{\cdot, \cdot\}_M$ are considered in [13] and used to relate the free rigid body and the simple pendulum. (It is further used in [16] to analyze the quantization problems.)

Since $\{F, G\}_M = \langle (Mx) \times \nabla F(x), \nabla G(x) \rangle$, the Hamiltonian vector field $\Xi_F^{(M)}$ for the Hamiltonian F is given as $\left(\Xi_F^{(M)} \right)_x = (Mx) \times \nabla F(x)$. In particular, if we choose the Hamiltonian F as $F(x) = \dfrac{1}{2} \langle x, \text{diag}\,(f_1, f_2, f_3)\,x \rangle$, then the Hamiltonian vector field is written as

$$\left(\Xi_F^{(M)} \right)_x$$
$$= ((\mu_2 f_3 - \mu_3 f_2)\,x_2 x_3, . \,(\mu_3 f_1 - \mu_1 f_3)\,x_3 x_1, (\mu_1 f_2 - \mu_2 f_1)\,x_1 x_2)^{\mathrm{T}},$$

where $x = (x_1, x_2, x_3)^{\mathrm{T}}$.

Then, one can easily check that, regarding $x_1 = K_1$, $x_2 = p_2$, $x_3 = p_3$, we can recover (28), by setting $\mu_\alpha = 1/I_\alpha$, $\alpha = 1, 2, 3$, $f_1 = 0$, $f_2 = -1$, $f_3 = -1$. On the other hand, (29) can be obtained for $x_\alpha = K_\alpha$, $\alpha = 1, 2, 3$, by substituting as $\mu_1 = \delta_1/\delta_2 \delta_3$, $\mu_2 = \delta_2/\delta_3 \delta_1$, $\mu_3 = \delta_3/\delta_1 \delta_2$, $f_\alpha = \delta_\alpha/I_\alpha$, $\alpha = 1, 2, 3$.

The solutions can be given in terms of elliptic functions because there are two quadratic first integrals $1/2 \langle x, Mx \rangle$, $F(x) = 1/2 \langle x, \text{diag}\,(f_1, f_2, f_3)\,x \rangle$ and because the intersection of the level surfaces of these functions are elliptic curves. This is a parallel argument to the one for the Euler top.

The above two kinds of special solutions can be found in the intersection of four quadric level surfaces for the functions C_1, C_2, C_3, C_4 in (4), but with specific values of C_3 and C_4. (Recall that we assume $C_1 = 0$ and $C_2 = 1$.) About the solution (28), the condition $p_1 = K_2 = K_3 = 0$ implies that $C_1 \equiv 0$, $C_2 = p_2^2 + p_3^2 = 1$, $C_3 = K_1^2 + (j_3 + j_1)p_2^2 + (j_1 + j_2)p_3^2$, $C_4 = j_1 K_1^2 + j_3 j_1 p_2^2 + j_1 j_2 p_3^2$. Now, it is easy to check that $j_1^2 - C_3 j_1^2 + C_4 = 0$. Taking (18) into account, we see that this is equivalent to the case where either $j_1 = j_4$ or $j_1 = j_5$ is satisfied. Clearly, the same type of special solution with the condition $p_2 = K_3 = K_1 = 0$ can be obtained only if $j_2^2 - C_3 j_2 + C_4 = (j_2 - j_4)(j_2 - j_5) = 0$ and the one with $p_3 = K_1 = K_2$ can be only if $j_3^2 - C_3 j_3 + C_4 = (j_3 - j_4)(j_3 - j_5) = 0$.

About the solution (29), the values of C_3 and C_4 must satisfy the condition $j_4 = j_5$. In fact, if we assume the condition $p_\alpha = \delta_\alpha K_\alpha$, $\alpha = 1, 2, 3$, the functions C_1, C_2, C_3, C_4 in (4) amount to

$$C_1 = \sum_{\alpha=1}^{3} \delta_\alpha K_\alpha^2, \quad C_2 = \sum_{\alpha=1}^{3} \delta_\alpha^2 K_\alpha^2,$$

$$C_3 = \sum_{\alpha=1}^{3} \left(1 + (j_\beta + j_\gamma)\,\delta_\alpha^2\right) K_\alpha^2,$$

$$C_4 = \sum_{\alpha=1}^{3} (j_\alpha + j_\beta j_\gamma \delta_\alpha)\,K_\alpha^2. \tag{30}$$

Here, $\{\alpha, \beta, \gamma\} = \{1, 2, 3\}$. For these functions, we have the dependence relation $C_3^2 - 4C_4 = 0$, which is equivalent to say that the quadric equation (18) has one double root, namely $j_4 = j_5$.

To verify the relation $C_3^2 - 4C_4 = 0$, we first show the following lemma.

Lemma 4.1 *For the condition $p_\alpha = \delta_\alpha K_\alpha$, $\alpha = 1, 2, 3$, to be satisfied along an integral curve of the Hamiltonian flow for the Hamiltonian $\lambda C_3 + \lambda' C_4$, it is necessary that there exist two constants $\sigma, \sigma' \in \mathbb{C}$ such that $(j_1 - \sigma')(j_2 - \sigma')(j_3 - \sigma') = \sigma^2$ and that $j_\alpha = \sigma \delta_\alpha + \sigma'$, $\alpha = 1, 2, 3$.*

Proof. For $p_\alpha = \delta_\alpha K_\alpha$, $\alpha = 1, 2, 3$ to be invariant along an integral curve of $\Xi_{\lambda C_3 + \lambda' C_4}$, it is necessary and sufficient that $\{\lambda C_3 + \lambda' C_4, p_\alpha - \delta_\alpha K_\alpha\} = 0$, $\alpha = 1, 2, 3$. Since

$$\{\lambda C_3 + \lambda' C_4, p_1 - \delta_1 K_1\}$$
$$= \{(\lambda + \lambda' j_3)(\delta_1 - \delta_2) - (\lambda + \lambda' j_2)(\delta_1 - \delta_2)$$
$$+ (j_2 - j_3)(\lambda + \lambda' j_1)\delta_1\delta_2\delta_3\} K_2 K_3,$$

we have $(\lambda + \lambda' j_3)(\delta_1 - \delta_2) - (\lambda + \lambda' j_2)(\delta_1 - \delta_2) + (j_2 - j_3)(\lambda + \lambda' j_1) \delta_1\delta_2\delta_3 = 0$. By cyclic change of parameters, we obtain the system of linear equations in $(\delta_1 - \delta_2, \delta_1 - \delta_3, \delta_1\delta_2\delta_3)^T$:

$$\begin{pmatrix} \lambda + j_3\lambda' & -(\lambda + j_2\lambda') & (j_2 - j_3)(\lambda + j_1\lambda') \\ (j_3 - j_1)\lambda' & \lambda + j_1\lambda' & (j_3 - j_1)(\lambda + j_1\lambda') \\ -(\lambda + j_1\lambda') & (j_1 - j_2)\lambda' & (j_1 - j_2)(\lambda + j_3\lambda') \end{pmatrix} \begin{pmatrix} \delta_1 - \delta_2 \\ \delta_1 - \delta_3 \\ \delta_1\delta_2\delta_3 \end{pmatrix} = \begin{pmatrix} 0 \\ 0 \\ 0 \end{pmatrix}.$$

These equations have already appeared in [14, §A.1] and it can be reduced through the Gauß method to the simple equations

$$\begin{pmatrix} 1 & 0 & -(j_1 - j_2) \\ 0 & 1 & -(j_1 - j_3) \\ 0 & 0 & 0 \end{pmatrix} \begin{pmatrix} \delta_1 - \delta_2 \\ \delta_1 - \delta_3 \\ \delta_1\delta_2\delta_3 \end{pmatrix} = \begin{pmatrix} 0 \\ 0 \\ 0 \end{pmatrix}.$$

Thus, there exist $s, s' \in \mathbb{C}$ such that $\delta_\alpha = s j_\alpha + s'$. In other words, we have $j_\alpha = \sigma \delta_\alpha + \sigma'$, $\alpha = 1, 2, 3$, where $\sigma = 1/s$ and $\sigma' = -s'/s$. By the condition $\delta_1\delta_2\delta_3 = s$, we have the condition $(s j_1 + s')(s j_2 + s')(s j_3 + s') = s \iff (j_1 - \sigma')(j_2 - \sigma')(j_3 - \sigma') = \sigma^2$. $\square$

Inserting the condition $j_\alpha = \sigma \delta_\alpha + \sigma'$, $\alpha = 1, 2, 3$, to (30), we have

$$C_3 = \sum_{\alpha=1}^{3} \left\{ 1 + \sigma (\delta_\beta + \delta_\gamma) \delta_\alpha^2 + 2\sigma'\delta_\alpha^2 \right\} K_\alpha^2$$

$$= \frac{1}{s} \sum_{\alpha=1}^{3} \{s + (\delta_\alpha\delta_\beta + \delta_\gamma\delta_\alpha)\delta_\alpha\} K_\alpha^2 + 2\sigma'C_2$$

$$= (\delta_1\delta_2 + \delta_2\delta_3 + \delta_3\delta_1)\sigma C_1 + 2\sigma'C_2 = 2\sigma'$$

and similarly

$$C_4 = \sum_{\alpha=1}^{3} \left\{ \sigma \delta_\alpha + \sigma' + \sigma^2 \delta_\beta \delta_\gamma \delta_\alpha^2 + \sigma \sigma' \left(\delta_\beta + \delta_\gamma \right) \delta_\alpha^2 + \left(\sigma' \right)^2 \delta_\alpha^2 \right\} K_\alpha^2$$

$$= \sigma C_1 + \left(\sigma' \right)^2 C_2 + \sigma^2 \delta_1 \delta_2 \delta_3 \sum_{\alpha=1}^{3} \delta_\alpha K_\alpha^2 + \sum_{\alpha=1}^{3} \left\{ \sigma + \left(\delta_\beta + \delta_\gamma \right) \delta_\alpha^2 \right\} K_\alpha^2$$

$$= \left\{ 2\sigma - \sigma \sigma' \left(\delta_1 \delta_2 + \delta_2 \delta_3 + \delta_3 \delta_1 \right) \right\} C_1 + \left(\sigma' \right)^2 C_2 = \left(\sigma' \right)^2.$$

Here we used the condition $C_1 = 0$ and $C_2 = 1$. Therefore, we have $C_3^2 - 4C_4 = 0$.

5 Explicit Computation of Action-Angle Coordinates

The Clebsch case displays many similarities with the Kowalevski top as there is a linearization on the Jacobian of a genus two curve. Indeed the explicit computation of the action-angle coordinates can be made quite similarly with that made for the Kowalevski top [10] at least for the Weber case ($C_1 = 0$). In this section, we return back to the real coordinates, since the action-angle coordinates are usually considered for real completely integrable Hamiltonian systems.

We first include a detailed report of the general result proved in [10].

We consider an integrable Hamiltonian system with m degrees of freedom, defined on a symplectic manifold (V^{2m}, ω) of dimension $2m$ and where ω is the symplectic form, as the data of m generically independent functions:

$$f_j : V^{2m} \to \mathbb{R}, j = 1, \ldots, m,$$

such that $\{f_i, f_j\} = 0$, where $\{\cdot, \cdot\}$ denotes the Poisson bracket associated with the symplectic form ω.

For instance, the system that we have considered, in restriction to the 4-dimensional symplectic leaf $V^4 : C_1 = 0, C_2 = 1$ defines with the couple C_3, C_4 an integrable Hamiltonian system with $m = 2$ degrees of freedom.

We assume furthermore that $f = (f_1, \ldots, f_m) : V^{2m} \to \mathbb{R}^m$ is proper. Denote D the critical locus of f, then for $c \in \mathbb{R}^m - D$, the connected components of $f^{-1}(c)$ are m-dimensional real tori.

Inspired by several examples (including the Kowalevski top), the following "algebraic linearization" condition (A) was introduced in [10]:

1. There exists m generically independent functions $(x_1, \ldots, x_m)$ defined on V^{2m} so that, together with the collection $(f_1, \ldots, f_m)$ of first integrals, we obtain a system of coordinates on an open dense set of $V^{2m} - f^{-1}(D)$.

2. There is a family of hyperelliptic curves of genus m, $y^2 = P(x, c)$, where P is a polynomial in x of degree $2m$ or $2m + 1$, parameterized by regular values $c \in \mathbb{R}^m \setminus D$ of $(f_1, \ldots, f_m)$, so that

$$\sum_{k=1}^{m} \frac{x_k^{j-1}\{f_i, x_k\}}{\sqrt{P(x_k, c)}} = W_{ij}, \tag{31}$$

where the $m \times m$ matrix $W = (W_{ij})$ is invertible and constant.

In [10], it was shown that the Kowalevski top satisfies the condition (A) with the constant invertible matrix $W = \mathrm{diag}(1, 2)$. We can easily check, from the above computations, that this condition is also verified in the Clebsch case, with $C_1 = 0$.

The following was proved in [10]:

Theorem 5.1 *Assume that an integrable Hamiltonian system satisfies the condition* (A), *then the symplectic form* ω *can be written as:*

$$\omega = \sum_{l=1}^{m} \eta_l \wedge df_l, \quad \eta_l = \sum_{j=1}^{m} A_{jl} df_j + \sum_{j=1}^{m} B_l(x_j) dx_j,$$

$$B_l(x) = \sum_{k=1}^{m} \frac{(W^{-1})_{kl} x^{k-1}}{\sqrt{P(x, c)}}.$$

Consider a family of cycles $\gamma_j(c)$, $(j = 1, \ldots, m)$ which defines a system of generators of the first homology group of the torus $f^{-1}(c)$, where $c \in \mathbb{R}^m \setminus D$ are regular values. We introduce the so-called "period matrix" $\Psi = (\Psi_{ij})$ given by:

$$\Psi_{ij}(c) = \int_{\gamma_j(c)} \eta_i. \tag{32}$$

As the 1-forms η_i are closed when restricted to the tori $f^{-1}(c)$, it is easy to check that the $\Psi_{ij}(c)$ depend only on the homology class of $\gamma_j(c)$ in $H_1(f^{-1}(c), \mathbb{Z})$. By the previous theorem, we obtain:

$$\Psi_{ij}(c) = \int_{\gamma_j(c)} \sum_{l,k} \frac{x_l^{k-1}(W^{-1})_{ki}}{\sqrt{P(x_l, c)}} dx_l.$$

It is known that the curves $\mathcal{C}_c : y^2 = P(w, c)$ is mapped injectively into its Jacobian $\mathrm{Jac}(\mathcal{C}_c) = H^0(\mathcal{C}_c, \Omega^1_{\mathcal{C}_c})^* / H_1(\mathcal{C}_c, \mathbb{Z})$ via the Abel map $S^m(\mathcal{C}_c) \to \mathrm{Jac}(\mathcal{C}_c)$ from the symmetric product $S^m(\mathcal{C}_c)$ of $\mathcal{C}_c$ and that this injection is a quasi-isomorphism, which induces an isomorphism at the level of homology.

So we can assume that the generators $\gamma_j(c)$ are already given as paths on the curve C_c. As consequence, we can suppress the index l in the above formula and write:

$$\Psi_{ij}(c) = \int_{\gamma_j(c)} \sum_k \frac{x^{k-1}(W^{-1})_{ki}}{\sqrt{P(x,c)}} dx. \tag{33}$$

Note that the Riemann theorem on the usual period matrix of hyperelliptic curves yields the invertibility of the "period matrix" $\Psi(c)$. Assume for simplicity that the symplectic form ω is exact, $\omega = d\eta$, then the actions $a_j, j = 1, \ldots, m$ are defined as functions of c (or equivalently of the functions $f_1, \ldots, f_m$) by

$$a_j = f^*(a_j(c)), \qquad a_j(c) = \int_{\gamma_j(c)} \eta,$$

and thus the derivatives are given as

$$\frac{\partial a_j}{\partial f_i} = \Psi_{ij}(c), \qquad i, j = 1, \ldots, m. \tag{34}$$

Clearly, the formula (33) can be applied to the Clebsch case (with $C_1 = 0$) that we have been studying. In that case, the supplementary coordinates (x_1, x_2) are in involution with respect to the Poisson bracket $\{\cdot, \cdot\}$. In this case, we have $P(x,c) = \sqrt{(j_1 - x)(j_2 - x)(j_3 - x)(x^2 - C_3 x + C_4)}$, whose three zeros j_1, j_2, j_3 are independent of C_3, C_4, which is very different for example from the Kowalevski top. In the case $j_1 < j_2 < j_3$ we can choose the two generators $\gamma_1(c), \gamma_2(c)$ associated with the two paths lifted from the complex x-plane to the curve C_c obtained for instance as a path which turns around j_1 and j_2 and another path which encircles j_3 and j_4. The formula (33) yields now a very simple formula

$$\Psi_{i1} = \int_{j_1}^{j_2} \left\{ \left(W^{-1}\right)_{1i} \frac{dx}{P(x,c)} + \left(W^{-1}\right)_{2i} \frac{x dx}{P(x,c)} \right\},$$

$$\Psi_{i2} = \int_{j_3}^{j_4} \left\{ \left(W^{-1}\right)_{1i} \frac{dx}{P(x,c)} + \left(W^{-1}\right)_{2i} \frac{x dx}{P(x,c)} \right\},$$

where the matrix W is diagonal (cf. (27)): $W = \begin{pmatrix} -2 & 0 \\ 0 & 2 \end{pmatrix}$. Thus, we have

$$\Psi_{11} = \int_{j_1}^{j_2} -2\frac{x dx}{R(x,c)}, \quad \Psi_{21} = \int_{j_1}^{j_2} 2\frac{dx}{R(x,c)}, \quad \Psi_{12} = \int_{j_3}^{j_4} -2\frac{dx}{R(x,c)},$$

$$\Psi_{22} = \int_{j_3}^{j_4} 2\frac{x dx}{R(x,c)}.$$

The integration of this "period matrix" is quite easy and it yields

$$a_1 = -2 \int_{j_1}^{j_2} \sqrt{\frac{x^2 - C_3 x + C_4}{(x - j_1)(x - j_2)(x - j_3)}} \, dx,$$

$$a_2 = -2 \int_{j_3}^{j_4} \sqrt{\frac{x^2 - C_3 x + C_4}{(x - j_1)(x - j_2)(x - j_3)}} \, dx.$$

6 Conclusion and Perspectives

As it has been shown in this article, the Clebsch case under Weber's condition (i.e. $C_1 = 0$) provides a quite remarkable example of the deep connections between integrable systems and algebraic geometry. By simple elimination of variables from the quadratic equations given by the constants of motion, it is possible to deduce naturally an associated Kummer surface. It is known in general that there always exists a double covering of a Kummer surface which is the Jacobian of an hyperelliptic curve of genus two. In the case of the Clebsch top under Weber's condition ($C_1 = 0$), such a curve can be derived from the induced Hamiltonian Dynamics. It is an interesting perspective to further develop the analysis of the actions over the singularities of the system and the connection to global aspects such as monodromy. Another important issue would be the extension to the general case ($C_1 \neq 0$), as treated by Kötter [17]. A nice method of the separation of variables in the general case has been recently presented by Skrypnyk in his paper [28]. Further studies on this general case should be done from the dynamical and algebro-geometric viewpoints.

Acknowledgment: We appreciate the comments by the referee which were useful for the improvement of the present article. The second author thanks Jun-ichi Matsuzawa for the stimulating and valuable discussions about Kummer surfaces.

References

[1] M. ADLER, P. VAN MOERBEKE, Geodesic Flow on $so(4)$ and the Intersection of Quadrics, Proc. Natl. Acad. Sci. USA, **81**(14), 4613–4616, 1984.

[2] M. ADLER, P. VAN MOERBEKE, The intersection of Four Quadrics in $\mathbb{P}^6$. Abelian surfaces and their Moduli, Math. Ann., **279**(1), 25–85, 1987.

[3] K. AOMOTO, Rigid body motion in a perfect fluid and θ-formulae – classical works by H. Weber – (in Japanese), RIMS Kôkyûroku, **414**, 98–114, 1981.

[4] V. I. ARNOL'D, *Mathematical Methods of Classical Mechanics*, 2nd ed., Springer-Verlag, New York-Tokyo, 1989.

[5] M. AUDIN, *Spinning Tops*, Cambridge University Press, Cambridge, 1996.

[6] W. BARTH, K. HULEK, C. PETERS, A. VAN DE VEN, *Compact Complex Surfaces*, 2nd ed., Springer-Verlag, Berlin-Heidelberg, 2004.

[7] A. BEAUVILLE, Systèmes Hamiltoniens complétement intégrables associés aux surfaces K3, Problems in the Theory of surfaces and their classification (Cortona, 1988) Symp. Math. vol. 32 , Academic Press, London, 25–31, 1991.

[8] Z. ENOLSKY, YU. N. FEDOROV, Algebraic description of Jacobians isogenous to Certain Prym Varieties with polarization (1,2). Exp. Math., **28**(2), 147–178, 2016.

[9] G. FISCHER, ed., *Mathematical Models*, 2nd ed., **Springer Spektrum**, Springer Nature, Wiesbaden, 2017.

[10] J.-P. FRANÇOISE, Calcul explicite d'Action-Angles, in *Systèmes dynamiques non linéaires: intégrabilité et comportement qualitatif*, Sém. Math. Sup., **102**, Presses Univ. Montréal, Montreal, QC, 101-120, 1986.

[11] J.-P. FRANÇOISE, D. TARAMA, Analytic extension of the Birkhoff normal forms for the free rigid body dynamics on $SO(3)$, Nonlinearity, **28**(5), 1193-1216, 2015.

[12] L. HAINE, Geodesic Flow on $so(4)$ and Abelian surfaces, Math. Ann., **263**(4), 435-472, 1983.

[13] D. D. HOLM, J. E. MARSDEN, The rotor and the pendulum, in *Symplectic Geometry and Mathematical Physics*, P. Donato et al. eds., Birkhäuser, Boston-Basel, 189–203, 1991.

[14] P. HOLMES, J. JENKINS, N. E. LEONARD, Dynamics of the Kirchhoff equations I: coincident centers of gravity and buoyancy, Physica D, **118**(3–4), 311–342, 1998.

[15] R. W. H. T. HUDSON, *Kummer's quartic Surfaces*, Cambridge University Press, Cambridge-New York-Port Chester-Melbourne-Sydney, 1990.

[16] T. IWAI, D. TARAMA, Classical and quantum dynamics for an extended free rigid body, Diff. Geom. Appl., **28**(5), 501–517, 2010.

[17] F. KÖTTER, Ueber die Bewegung eines festen Köpers in einer Flüssigkeit, I, II, J. Reine Angew. Math., **109**, 51–81, 89–111, 1892.

[18] E. KUMMER, Über die Flächen vierten Grades mit sechzehn singulären Punkten, Monatsberichte der Königlichen Preussische Akademie des Wissenschaften zu Berlin, 246-260, 1864.

[19] F. MAGRI, T. SKRYPNYK, The Clebsch System, preprint, arXiv:1512.04872, 2015.

[20] D. MARKUSHEVICH, Some Algebro-Geometric Integrable Systems Versus classical ones, in: *The Kowalewski property (Leeds, 2000)*, CRM proceedings and lecture notes, **32**(1), 197–218, 2002.

[21] J. E. MARSDEN, T. S. RATIU, *Introduction to Mechanics and Symmetry*, 2nd ed., Springer, New York, 2010.

[22] S. MUKAI, Symplectic structure of the moduli of sheaves on an Abelian or K3 surface, Invent. Math., **77**(1), 101–116, 1984.

[23] I. NARUKI, D. TARAMA, Algebraic geometry of the eigenvector mapping for a free rigid body, Diff. Geom. Appl., **29** Supplement 1, S170–S182, 2011.

[24] I. NARUKI, D. TARAMA, Some elliptic fibrations arising from free rigid body dynamics, Hokkaido Math. J., **41**(3), 365–407, 2012.

[25] V. NIKULIN, On Kummer surfaces, Math. USSR. Izvestija, **9** (2), 261–275, 1975.

[26] T. S. RATIU et al., A Crash Course in Geometric Mechanics, in: *Geometric Mechanics and Symmetry: the Peyresq Lectures*, J. Montaldi, T. Ratiu (eds.), Cambridge University Press, Cambridge, 23–156, 2005.

[27] F. SCHOTTKY, Über das analytische Problem der Rotation eines starren Körpers im Raume von vier Dimensionen, Sitzungber. König. Preuss. Akad. Wiss. zu Berlin, **27**, 227–232, 1891.

[28] T. SKRYPNYK, "Symmetric" separation of variables for the Clebsch system, J. Geom. Phys., **135**, 204–218, 2019.

[29] D. TARAMA, J.-P. FRANÇOISE, Analytic extension of Birkhoff normal forms for Hamiltonian systems of one degree of freedom – Simple pendulum and free rigid body dynamics –, RIMS Kôkyûroku Bessatsu, **B52**, 219–236, 2014.

[30] H. WEBER, Ueber die Kummersche Fläche vierter-Ordnung mit sechzehn Knoten-punkten und ihre Beziehung zu den Thetafunctionen mit zwei Veränderlichen, J. Reine Angew. Math. **84**, 332–354, 1878.

[31] H. WEBER, Anwendung der Thetafunctionen zweier Veränderlichen auf die Theorie der Bewegung eines festen Körpers in einer Flüssingkeit, Math. Ann., **14**(2), 173–206, 1879.

Jean-Pierre Françoise
Laboratoire Jacques-Louis Lions,
Sorbonne-Université,
4 Pl. Jussieu, 75252 Paris, France.
E-mail: jean-pierre.francoise@upmc.fr

Daisuke Tarama
Department of Mathematical Sciences Ritsumeikan University,
1-1-1 Nojihigashi, Kusatsu,
Shiga, 525–8577, Japan.
Partially supported by Grant for Basic Science Research
Projects from The Sumitomo Foundation.
E-mail: dtarama@fc.ritsumei.ac.jp

9

An Extension of Delsarte, Goethals and Mac Williams Theorem on Minimal Weight Codewords to a Class of Reed-Muller Type Codes

Cícero Carvalho and Victor G.L. Neumann[1]

Abstract. In 1970 Delsarte, Goethals and Mac Williams published a seminal paper on generalized Reed-Muller codes where, among many important results, they proved that the minimal weight codewords of these codes are obtained through the evaluation of certain polynomials which are a specific product of linear factors, which they describe. In the present paper we extend this result to a class of Reed-Muller type codes defined on a product of (possibly distinct) finite fields of the same characteristic. The paper also brings an expository section on the study of the structure of low weight codewords, not only for affine Reed-Muller type codes, but also for the projective ones.

1 Introduction with a Historical Survey

Let $\mathbb{F}_q$ be a field with q elements, let $K_1, \ldots, K_n$ be a collection of non-empty subsets of $\mathbb{F}_q$, and let

$$\mathcal{X} := K_1 \times \cdots \times K_n := \{(\alpha_1 : \cdots : \alpha_n) \,|\, \alpha_i \in K_i \text{ for all } i\} \subset \mathbb{F}_q^n.$$

Let $d_i := |K_i|$ for $i = 1, \ldots, n$, so clearly $|\mathcal{X}| = \prod_{i=1}^n d_i =: m$, and let $\mathcal{X} = \{\boldsymbol{\alpha}_1, \ldots, \boldsymbol{\alpha}_m\}$. It is not difficult to check that the ideal of polynomials in $\mathbb{F}_q[X_1, \ldots, X_n]$ which vanish on $\mathcal{X}$ is $I_{\mathcal{X}} = (\prod_{\alpha_1 \in K_1}(X_1 - \alpha_1), \ldots, \prod_{\alpha_n \in K_n}(X_n - \alpha_n))$ (see e.g. [25, Lemma 2.3] or [7, Lemma 3.11]). From this we get that the evaluation morphism $\Psi : \mathbb{F}_q[X_1, \ldots, X_n]/I_{\mathcal{X}} \to \mathbb{F}_q^m$ given by $P + I_{\mathcal{X}} \mapsto (P(\boldsymbol{\alpha}_1), \ldots, P(\boldsymbol{\alpha}_m))$ is well-defined and injective. Actually, this is an isomorphism of $\mathbb{F}_q$-vector spaces because for each $i \in \{1, \ldots, m\}$ there exists a polynomial P_i such that $P_i(\boldsymbol{\alpha}_j)$ is equal to 1, if $j = i$, or 0, if $j \neq i$, so that Ψ is also surjective.

[1] Both authors were partially supported by grants from CNPq and FAPEMIG.

Definition 1.1 *Let d be a nonnegative integer. The* affine cartesian code *(of order d) $C_\mathcal{X}(d)$ defined over the sets $K_1, \ldots, K_n$ is the image, by Ψ, of the set of the classes of all polynomials of degree up to d, together with the class of the zero polynomial.*

These codes appeared independently in [25] and [17] (in [17] in a generalized form). In the special case where $K_1 = \cdots = K_n = \mathbb{F}_q$ we have the well-known generalized Reed-Muller code of order d. In [25] the authors prove that we may ignore, in the cartesian product, sets with just one element and moreover may always assume that $2 \le d_1 \le \cdots \le d_n$. They also determine the dimension and the minimum distance of these codes.

For the generalized Reed-Muller codes, the classes of the polynomials whose image are the codewords of minimum weight were first described explicitly by Delsarte, Goethals and Mac Williams in 1970. This result started a series of investigations of the structure of codewords of all weights, not only in generalized Reed-Muller codes, but also in related Reed-Muller type codes. In the present paper we extend the result of Delsarte, Goethals and Mac Williams to affine cartesian codes, in the case where K_i is a field, for all $i = 1, \ldots, n$ and $K_1 \subset K_2 \subset \cdots \subset K_n \subset \mathbb{F}_q$, but before we describe the contents of the next sections of this work, we would like to present a survey of results that pursued the investigation started by Delsarte, Goethals and Mac Williams.

Reed-Muller codes are binary codes defined by Muller ([28]) and were given a decoding algorithm by Reed ([29]), in 1954. In 1968 Kasami, Lin and Peterson ([18]) introduced what they called the generalized Reed-Muller codes, defined over a finite field $\mathbb{F}_q$ with q elements, which coincided with Reed-Muller codes when $q = 2$. Their idea was to consider the $\mathbb{F}_q$-vector space $\mathbb{F}_q[X_1, \ldots, X_n]_{\le d}$ of all polynomials in $\mathbb{F}_q[X_1, \ldots, X_n]$ of degree less than or equal to d, together with the zero polynomial, for some positive integer d, and define the generalized Reed-Muler code of order d as

$$GRM_q(d, n) = \{(f(\alpha_1), \ldots, f(\alpha_{q^n})) \in \mathbb{F}_q^{q^n} \mid f \in \mathbb{F}_q[X_1, \ldots, X_n]_{\le d}\}$$

where $\alpha_1, \ldots, \alpha_{q^n}$ are the points of the affine space $\mathbb{A}^n(\mathbb{F}_q)$. Equivalently, using the fact that $I = (X_1^q - X_1, \ldots, X_n^q - X_n)$ is the ideal of polynomials whose zero set is $\mathbb{A}^n(\mathbb{F}_q)$, we have that $GRM_q(d, n)$ is the image of the linear transformation $\Psi : \mathbb{F}_q[X_1, \ldots, X_n]/I \to \mathbb{F}_q^{q^n}$ given by $P + I_\mathcal{X} \mapsto (P(\alpha_1), \ldots, P(\alpha_{q^n}))$. Kasami et al. proved that if $d \ge n(q - 1)$ then we have $GRM_q(d, n) = \mathbb{F}_q^{q^n}$ hence the minimum distance $\delta_{GRM_q(d,n)}$ of $GRM_q(d, n)$ is 1. For $1 \le d < n(q - 1)$ write $d = k(q - 1) + \ell$ with $0 < \ell \le q - 1$, then $\delta_{GRM_q(d,n)} = (q - \ell)q^{n-k-1}$ (see [18, Thm. 5]). McEliece, studying quadratic forms defined over $\mathbb{F}_q$ (see [26]) described the so-called weight enumerator polynomial for $GRM_q(2, n)$, i.e. described all possible weights for the codewords in $GRM_q(2, n)$, together with the number of codewords of each weight, and also gave canonical forms for the polynomials whose classes produced codewords of all weights.

In 1970 Delsarte, Goethals and Mac Williams published a 40-page seminal paper which started the systematic study of the generalized Reed-Muller codes and other codes related to them. Among the many important results in the paper, there is a description of the polynomials whose evaluation yields the codewords with minimum distance. To state their result, we recall that the affine group of automorphisms of $\mathbb{F}_q[X_1, \ldots, X_n]$ is the one given by transformations of the type $X^t \mapsto AX^t + \boldsymbol{\beta}$, where $X = (X_1, \ldots, X_n)$, A is a $n \times n$ invertible matrix with entries in $\mathbb{F}_q$ and $\boldsymbol{\beta} \in \mathbb{F}_q^n$.

Theorem 1.2 *[13, Theorem 2.6.3] The minimal weight codewords of $GRM_q(d, n)$ come from the evaluation of Ψ in classes $f + I$ of polynomials f which, after a suitable action of an affine automorphism of $\mathbb{F}_q[X_1, \ldots, X_n]$, may be written as*

$$f = \alpha \prod_{i=1}^{k} (X_i^{q-1} - 1) \prod_{i=1}^{\ell} (X_{k+1} - \beta_j)$$

where $d = k(q-1) + \ell$ with $0 < \ell \leq q - 1$, $\alpha \in \mathbb{F}_q^*$ and $\beta_1, \ldots, \beta_\ell$ are distinct elements of $\mathbb{F}_q$ (in the case $k = 0$ we take the first product to be 1).

Since GRM codes arise from the evaluation of polynomials in points of an affine space, there is also an algebraic geometry interpretation for the codewords. In fact, the above theorem shows that the zeros of a minimal weight codeword lie on a special type of hyperplane arrangement. More explicitly, we have the following alternative statement (taken from [1]) for the above result.

Theorem 1.3 *Let V be an algebraic hypersurface in $\mathbb{A}^n(\mathbb{F}_q)$, of degree at most d, with $1 \leq d < n(q-1)$, which is not the whole $\mathbb{A}^n(\mathbb{F}_q)$. Then V has the maximal possible number of zeros if and only if*

$$V = \left(\bigcup_{i=1}^{k} \left(\bigcup_{s=1}^{q-1} V_{i,s} \right) \right) \cup \left(\bigcup_{j=1}^{\ell} W_j \right)$$

where $d = k(q-1) + \ell$ with $0 \leq \ell < q - 1$, the $V_{i,s}$ and W_j are d distinct hyperplanes defined on $\mathbb{F}_q$ such that for each fixed i the $V_{i,s}$ are $q - 1$ parallel hyperplanes, the W_j are ℓ parallel hyperplanes and the $k + 1$ distinct linear forms directing these hyperplanes are linearly independent.

This result was the start of the search for the higher Hamming weights together with the description (algebraic and geometric) of the codewords having these weights, not only for GRMs but in general for all Reed-Muller type codes, like the ones studied in this paper, for the GRMs alone the search is still ongoing.

In 1974 Daniel Erickson, a student of McEliece and Dilworth, devoted his Ph.D. thesis to the determination of the second lowest Hamming weight, also called next-to-minimal weight, of $GRM_q(d, n)$ (see [14]). He succeeded in determining the values of the second weight for many values of d in the relevant range $1 \leq d < n(q - 1)$. For the values that he was not able to determine, following a suggestion by M. Hall, he generalized some of the results of Bruen on blocking sets, which had appeared in [2], and made a conjecture relating the expected value for the missing weights to the cardinality of certain blocking sets in the affine plane $\mathbb{A}^2(\mathbb{F}_q)$. Also, instead of working with the classes of polynomials in $\mathbb{F}_q[X_1, \ldots, X_n]/I$ he worked with a fixed set of representatives called "reduced polynomials" which he noted that were in a one-to-one correspondence with the functions from $\mathbb{F}_q^n$ to $\mathbb{F}_q$. This had an influence on the paper [22] and also the present text, as we will comment later. Unfortunately Erikson's results were not published, and the quest for the next-to-minimal weights of GRM codes went on for many years without his contributions.

In 1976 Kasami, Tokura and Azumi (see [19]) determined all the weights of $GRM_2(d, n)$ (i.e. Reed-Muller codes) which are less than $\frac{5}{2}\delta_{GRM(d,2)}$. They also determined canonical forms for the representatives of the classes whose evaluation produces codewords of these weights, together with the number of such words. In particular, the second weight of Reed-Muller codes was determined. After this paper, there was not much work done on the problem of determining the higher Hamming weights of $GRM_q(d, n)$ during two decades. Then, in 1996 Cherdieu and Rolland (see [12]) determined the second weight of $GRM_q(d, n)$ for d in the range $1 \leq d < q - 1$, provided that q is large enough. They also proved that in this case the zeros of codewords having next-to-minimal weight form an specific type of hyperplane arrangement which they describe. In the following year a work by Sboui (see [35]) proved that the result by Cherdieu and Rolland holds when $d \leq q/2$.

In 2008 Geil (see [15] and [16]) determined the second weight of $GRM_q(d, n)$ for $2 \leq d \leq q - 1$ and $2 \leq n$. Also, for d in the range $(n - 1)(q - 1) < d < n(q - 1)$, he determined the first $d + 1 - (n - 1)(q - 1)$ weights of $GRM_q(d, n)$. His results completely determine the next-to-minimal weight of $GRM_q(d, 2)$, since in this case the relevant range for d is $1 \leq d < 2q$. Geil's theorems were obtained using results from Gröbner basis theory. In 2010 Rolland made a more detailed analysis of the weights also using Gröbner basis theory results, and determined almost all next-to-minimal weights of $GRM_q(d, n)$ (see [34]). In fact, he succeeded in finding the next-to-minimal weights for all values of d, in the range $q \leq d < n(q - 1)$, that can not be written in the form $d = k(q - 1) + 1$. Finally, also in 2010, A. Bruen had his attention directed to Erickson's thesis, and in a note (see [3]) observed that Erickson's conjecture was an easy consequence of results that he, Bruen, had proved in 1992 and 2006 (see [4] and [5]). This finally completed the determination of the next-to-minimal weights $\delta^{(2)}_{GRM_q(d,n)}$ of $GRM_q(d,n)$, and now we know that for $1 \leq d < n(q-1)$, writing $d = k(q-1) + \ell$ with $0 \leq \ell < q-1$, then $\delta_{GRM_q(d,n)} = (q-\ell)q^{n-k-1}$ and $\delta^{(2)}_{GRM_q(d,n)} = \delta_{GRM_q(d,n)} + cq^{n-k-2}$, where

$$
c = \begin{cases}
q & \text{if} \quad k = n-1; \\
\ell - 1 & \text{if} \quad k < n-1 \text{ and } 1 < \ell \leq (q+1)/2; \\
\quad \text{or} & k < n-1 \text{ and } \ell = q-1 \neq 1; \\
q & \text{if} \quad k = 0 \text{ and } \ell = 1; \\
q - 1 & \text{if} \quad q < 4, 0 < k < n-2, \text{ and } \ell = 1; \\
q - 1 & \text{if} \quad q = 3, 0 < k = n-2 \text{ and } \ell = 1; \\
q & \text{if} \quad q = 2, k = n-2 \text{ and } \ell = 1; \\
q & \text{if} \quad q \geq 4, 0 < k \leq n-2 \text{ and } \ell = 1; \\
\ell - 1 & \text{if} \quad q \geq 4, k \leq n-2 \text{ and } (q+1)/2 < \ell.
\end{cases}
$$

In 2012 the 1970's theorem of Delsarte, Goethals and Mac Williams was the subject of a paper by Leducq (see [22]). In their paper, Delsarte et al. prove the theorem on the minimum distance in an Appendix entitled "Proof of Theorem 2.6.3.", which opens with the sentence: "The authors hasten to point out that it would be very desirable to find a more sophisticated and shorter proof." Leducq indeed provides a shorter and less technical proof, treating the codewords as functions from $\mathbb{F}_q^n$ to $\mathbb{F}_q$ and using results from affine geometry. Some of these results appear in the appendix of Delsarte et al. paper, and were also used by Erickson in his work. In the following year, Leducq (see [23]) completed the work of previous researchers, with Sboui, Cherdieu, Rolland and Ballet among them, and proved that the next-to-minimal weights are only attained by codewords whose set of zeros form certain hyperplane arrangements. In the same year Carvalho (see [6]) extended Geils's results of 2008 to affine cartesian codes, also determining a series of higher Hamming weights for these codes.

In 2014 a paper by Ballet and Rolland (see [1]) presented bounds on the third and fourth Hamming weights of $GRM_q(d,n)$ for certain ranges of d. In the following year Leducq (see [24]), pursuing and developing ideas from Erickson's thesis, determined the third weight and characterized the third weight words of $GRM_q(d,n)$ for some values of d. In 2017 Carvalho and Neumann (see [9]) extended many of the results of Rolland, in [34], to affine cartesian codes. They found the second weight of these codes for all values of d which can not be written as $d = \sum_{i=1}^{k}(d_i - 1) + 1$, and they also prove that the weights corresponding to such values of d are attained by codewords whose set of zeros are hyperplane arrangements (yet they don't prove that every word attaining those next-to-minimal weights comes from hyperplane arrangements).

There is a "projective version" of the generalized Reed-Muller codes whose parameters have been studied like those of $GRM_q(d, n)$ and to which they are related. This version was introduced by Lachaud in 1986 (see [20]), but one can find some examples of it already in [39].

Let $\gamma_1, \ldots, \gamma_N$ be the points of $\mathbb{P}^n(\mathbb{F}_q)$, where $N = q^n + \cdots + q + 1$. From e.g. [30] or [27] we get that the homogeneous ideal $J_q \subset \mathbb{F}_q[X_0, \ldots, X_n]$ of the polynomials which vanish in all points of $\mathbb{P}^n(\mathbb{F}_q)$ is generated by $\{X_j^q X_i - X_i^q X_j \mid 0 \le i < j \le n\}$. We denote by $\mathbb{F}_q[X_0, \ldots, X_n]_d$ (respectively, $(J_q)_d$) the $\mathbb{F}_q$-vector subspace formed by the homogeneous polynomials of degree d (together with the zero polynomial) in $\mathbb{F}_q[X_0, \ldots, X_n]$ (respectively, J_q).

Definition 1.4 *Let d be a positive integer and let Θ : $\mathbb{F}_q[X_0, \ldots, X_n]_d/ (J_q)_d \to \mathbb{F}_q^N$ be the $\mathbb{F}_q$-linear transformation given by $\Theta(f + (J_q)_d) = (f(\gamma_1) \ldots, f(\gamma_N))$, where we write the points of $\mathbb{P}^n(\mathbb{F}_q)$ in the standard notation, i.e. the first nonzero entry from the left is equal to 1. The projective generalized Reed-Muller code of order d, denoted by $PGRM_q(n, d)$, is the image of Θ.*

It is easy to check that if one chooses another representation for the projective points the code thus obtained is equivalent to the code defined above. It is also easy to prove that if $d \ge n(q-1)+1$ then Θ is an isomorphism, so the relevant range to investigate the parameters of $PGRM$ codes is $1 \le d \le n(q-1)$.

Lachaud, in [20] presents some bounds for $\delta_{PGRM_q(n,d)}$, the minimum distance for $PGRM_q(n, d)$, and determines the true value in a special case. Serre, in 1989 (see [37]), determined the minimum distance of $PGRM_q(n, d)$ when $d < q$. In 1990 Lachaud (see[21]) presents some properties that some higher weights of $PGRM_q(n, d)$ must have, when $d \le q$ and $d \le n$.

Let $g \in \mathbb{F}_q[X_1, \ldots, X_n]$ be a polynomial of degree $d - 1 \ge 1$ and let ω be the Hamming weight of $\Phi(g + I)$. Let $g^{(h)}$ be the homogenization of g with respect to X_0, then the degree of $X_0 g^{(h)}$ is d and the weight of $\Theta(X_0 g^{(h)} + (J_q)_d)$ is ω. In particular $\delta_{PGRM_q(n,d)} \le \delta_{GRM_q(n,d-1)}$. When $d = 1$ all the codewords of $PGRM_q(n, d)$ have the same number of zeros entries (hence the same weight), which is equal to the number of points of a hyperplane in $\mathbb{P}^n(\mathbb{F}_q)$, this also implies that for $d = 1$ there are no higher Hamming weights. In 1991 Sørensen (see [38]) proved that $\delta_{PGRM_q(n,d)} = \delta_{GRM_q(n,d-1)}$ holds for all d in the relevant range. After this paper, similarly to what had happened with GRM codes, the subject lay dormant for almost two decades. Then, in 2007 Rodier and Sboui (see [31]), under the condition $d(d-1)/2 < q$ determined a Hamming weight of $PGRM_q(n, d)$, which is not the minimal and is only achieved by codewords whose zeros are hyperplane arrangements. In 2008 the same authors (see [32]) proved that for $q/2 + 5/2 \le d < q$ the third weight of PGRM is not only achieved by evaluating Θ in the classes of totally decomposable polynomials but can also be obtained in this case from classes of some polynomials having an irreducible quadric as a factor. Also in 2008, Rolland (see [33]) proved the equivalent of Delsarte, Goethals and Mac Williams theorem for PGRM codes, completely characterizing the codewords of $PGRM_q(n, d)$ which have minimal weights, and proving that they only arise as images by Θ of classes of totally decomposable polynomials, which in a sense may be thought of as the homogenization of the polynomials described by Delsarte et al. In 2009 Sboui ([36]) determined the second and third weights of $PGRM_q(n, d)$ in the range $5 \leq d \leq q/3 + 2$. He proved that codewords which have these weights come only from evaluation of classes of totally decomposable polynomials and calculated the number of codewords having weights equal to the minimal distance, or the second weight, or the third weight. In the already mentioned paper of 2014 (see [1]), Ballet and Rolland find another proof of Rolland's result on minimal weight codewords of PGRM. They also present lower and upper bounds for the second weight of $PGRM_q(n, d)$.

Putting together the reasoning presented in the beginning of the preceding paragraph and Sørensen's result $\delta_{PGRM_q(n,d)} = \delta_{GRM_q(n,d-1)}$, and writing $\delta^{(2)}_{PGRM_q(n,d)}$ for the second Hamming weight of $PGRM_q(n, d)$, we get $\delta^{(2)}_{PGRM_q(n,d)} \leq \delta^{(2)}_{GRM_q(n,d-1)}$ for all $2 \leq d \leq n(q-1) + 1$. In 2016 Carvalho and Neumann (see [8]) determined the second weight of $PGRM_2(n, d)$ for all d in the relevant range, and in 2018 (see [10]) they also determined the second weight of $PGRM_q(n, d)$, for $q \geq 3$ and almost all values of d. For some values of d, in both papers, it happened that $\delta^{(2)}_{PGRM_q(n,d)} < \delta^{(2)}_{GRM_q(n,d-1)}$, and they proved that in all these cases the zeros of the codewords with weight $\delta^{(2)}_{PGRM_q(n,d)}$ are not hyperplane arrangements. They also observed that, writing $d - 1 = k(q - 1) + \ell$, with $0 \leq k \leq n - 1$ and $0 < \ell \leq q - 1$, in the case where $q = 3$, $k > 0$ and $\ell = 1$ we have $\delta^{(2)}_{PGRM_q(n,d)} = \delta^{(2)}_{GRM_q(n,d-1)}$ and there are codewords of weight $\delta^{(2)}_{PGRM_q(n,d)}$ whose set of zeros are hyperplane arrangements and others which do not have this property. The tables below show the current results for $\delta^{(2)}_{PGRM_q(n,d)}$, where we write $d - 1 = k(q - 1) + \ell$ as above. The tables also present the values of $\delta^{(2)}_{GRM_q(n,d-1)}$ so the reader can see the cases where one has $\delta^{(2)}_{PGRM_q(n,d)} < \delta^{(2)}_{GRM_q(n,d-1)}$.

A generalization of PGRM codes was introduced in 2017 by Carvaho, Neumann and López (see [11]), as the class of codes called "projective nested cartesian codes". They determined the dimension of these codes, bounds for the minimum distance and the exact value of this distance in some cases.

In the present paper we extend Delsarte, Goethals and Mac Williams theorem to the class of affine cartesian codes $\mathcal{C}_X(d)$ defined above, in the

Table 9.1. *Second (or next-to-minimal) weights for*
$GRM_q(n, d - 1)$ *and* $PGRM_q(n, d)$ *when* $n \geq 2$ *and* $q = 2$

n	k	ℓ	$\delta^{(2)}_{GRM_2(n,d-1)}$	$\delta^{(2)}_{PGRM_2(n,d)}$
$n \geq 3$	$k = 0$	$\ell = 1$	2^n	$3 \cdot 2^{n-2}$
$n \geq 4$	$1 \leq k < n - 2$	$\ell = 1$	$3 \cdot 2^{n-k-2}$	$3 \cdot 2^{n-k-2}$
$n \geq 2$	$k = n - 2$	$\ell = 1$	4	4
$n \geq 2$	$k = n - 1$	$\ell = 1$	2	2

Table 9.2. *Second (or next-to-minimal) weights for*
$GRM_q(n, d - 1)$ *and* $PGRM_q(n, d)$ *when* $n \geq 1$ *and* $q = 3$

n	k	ℓ	$\delta^{(2)}_{GRM_q(n,d-1)}$	$\delta^{(2)}_{PGRM_q(n,d)}$
$n = 2$	$k = 0$	$\ell = 1$	3^2	3^2
$n \geq 3$	$k = 0$	$\ell = 1$	3^n	$8 \cdot 3^{n-2}$
$n \geq 3$	$1 \leq k \leq n - 2$	$\ell = 1$	$8 \cdot 3^{n-k-2}$	$8 \cdot 3^{n-k-2}$
$n \geq 2$	$0 \leq k \leq n - 2$	$\ell = 2$	$4 \cdot 3^{n-k-2}$	$4 \cdot 3^{n-k-2}$
$n \geq 1$	$k = n - 1$	$\ell = 1, 2$	$4 - \ell$	$4 - \ell$

Table 9.3. *Second (or next-to-minimal) weights for* $GRM_q(n, d - 1)$ *and*
$PGRM_q(n, d)$ *when* $n \geq 1$ *and* $q \geq 4$

n	k	ℓ	$\delta^{(2)}_{GRM_q(n,d-1)}$	$\delta^{(2)}_{PGRM_q(n,d)}$
$n = 2$	$k = 0$	$\ell = 1$	q^2	q^2
$n \geq 3$	$k < n - 2$	$\ell = 1$	q^{n-k}	$q^{n-k} - q^{n-k-2}$
$n \geq 3$	$k = n - 2$	$\ell = 1$	q^2	Unknown
$n \geq 2$	$k \leq n - 2$	$1 < \ell \leq \frac{q+1}{2}$	$(q - 1)(q - \ell + 1) q^{n-k-2}$	$(q - 1)(q - \ell + 1)q^{n-k-2}$
$n \geq 2$	$k \leq n - 2$	$\frac{q+1}{2} < \ell \leq q - 1$	$(q - 1)(q - \ell + 1)q^{n-k-2}$	Unknown
$n \geq 1$	$k = n - 1$	$1 \leq \ell \leq q - 1$	$q - \ell + 1$	$q - \ell + 1$

case where the sets $K_1 \subset \cdots \subset K_n$ are subfields of $\mathbb{F}_q^n$. Our main results are Proposition 3.1 , Proposition 3.2 and Theorem 3.5 which show that, as in the GRM codes, the minimal weight codewords of $\mathcal{C}_{\mathcal{X}}(d)$ come from the evaluation of Ψ in classes $f + I$ of polynomials f which, after a suitable action of an automorphism group, may be written as the product of certain degree one polynomials. In the next section we introduce the concept of code as an $\mathbb{F}_q$-vector space of functions (following [14] and [22]) and define the relevant automorphism group for the main result. We then study the intersection of certain affine subspaces of $\mathbb{F}_q^n$ with $\mathcal{X}$ to find information on the structure of functions that have "few" points in the support (see Corollary 2.11). Then, in the beginning of Section 3, we use these results to determine the structure of the functions (or codewords) of minimal weight, for d within a certain range – in a sense, for the lower values of d (see Proposition 3.1). Finally, after exploring a little further the properties of the intersection of certain hyperplanes with $\mathcal{X}$, we prove our main result (see Theorem 3.5) which generalizes the result by Delsarte, Goethals and Mac Williams.

2 Preliminary Results

Let $C_{\mathcal{X}}(d)$ be the affine cartesian code as in Definition 1.1. We assume from now on that $K_1, \ldots, K_n$ are fields and that $K_1 \subset K_2 \subset \cdots \subset K_n \subset \mathbb{F}_q$. Recall that $|K_i| = d_i$ for $i = 1, \ldots, n$, so $I_{\mathcal{X}} = (X_1^{d_1} - X_1, \ldots, X_n^{d_n} - X_n)$, and observe that, since Ψ is an isomorphism, the code $C_{\mathcal{X}}(d)$ is isomorphic to the $\mathbb{F}_q$-vector space of the classes of polynomials in $\mathbb{F}_q[X_1, \ldots, X_n]/I_{\mathcal{X}}$ of degree up to d (together with the zero class). It is well known that, given a subset $Y \subset \mathbb{F}_q^n$, any function $f : Y \to \mathbb{F}_q$ is given by a polynomial $P \in \mathbb{F}_q[X_1, \ldots, X_n]$ (again, this is a consequence of the fact that given $\boldsymbol{\alpha} \in \mathbb{F}_q^n$ there exists a polynomial $P_{\boldsymbol{\alpha}} \in \mathbb{F}_q[X_1, \ldots, X_n]$ such that $P_{\boldsymbol{\alpha}}(\boldsymbol{\alpha}) = 1$ and $P_{\boldsymbol{\alpha}}(\boldsymbol{\beta}) = 0$ for any $\boldsymbol{\beta} \in \mathbb{F}_q^n \setminus \{\boldsymbol{\alpha}\}$). Denoting by $C_{\mathcal{X}}$ the $\mathbb{F}_q$-algebra of functions defined on $\mathcal{X}$ we clearly have an isomorphism $\Phi : \mathbb{F}_q[X_1, \ldots, X_n]/I_{\mathcal{X}} \to C_{\mathcal{X}}$ hence for each function $f \in C_{\mathcal{X}}$ there exists a unique polynomial $P \in \mathbb{F}_q[X_1, \ldots, X_n]$ such that the degree of P in the variable X_i is less than d_i for all $i = 1, \ldots, n$, and $\Phi(P + I_{\mathcal{X}}) = f$.

Definition 2.1 *We say that P is the reduced polynomial associated to f and we define the* degree of f *as being the degree of P.*

We denote by $C_{\mathcal{X}}(d)$ the $\mathbb{F}_q$-vector space formed by functions of degree up to d, together with the zero function. We saw above that $C_{\mathcal{X}}$ is isomorphic to $\mathbb{F}_q[X_1, \ldots, X_n]/I_{\mathcal{X}}$, and hence to $\mathbb{F}_q^m$, and clearly $C_{\mathcal{X}}(d) \subset C_{\mathcal{X}}$ is isomorphic to the code $\mathcal{C}_{\mathcal{X}}(d) \subset \mathbb{F}_q^m$, so from now on we also call $C_{\mathcal{X}}(d)$ the affine cartesian code of order d. To study the codewords of minimum weight we define the support of a function $f \in C_{\mathcal{X}}$ as the set $\{\boldsymbol{\alpha} \in \mathcal{X} \mid f(\boldsymbol{\alpha}) \neq 0\}$ and we write $|f|$ for its cardinality, which, in this approach, is the Hamming weight of f. Thus the minimum distance of $C_{\mathcal{X}}(d)$ is $\delta_{\mathcal{X}}(d) := \min\{|f| \mid f \in C_{\mathcal{X}}(d) \text{ and } f \neq 0\}$. We denote by

$$Z_{\mathcal{X}}(f) := \{\boldsymbol{\alpha} \in \mathcal{X} \mid f(\boldsymbol{\alpha}) = 0\}$$

the set of zeros of $f \in C_{\mathcal{X}}$, and given functions $g_1, \ldots, g_s$ defined on $\mathbb{F}_q^n$ we denote by $Z(g_1, \ldots, g_s)$ be the set of common zeros, in $\mathbb{F}_q^n$, of these functions.

We write $\mathrm{Aff}(n, \mathbb{F}_q)$ for the affine group of $\mathbb{F}_q^n$, i.e. the transformations of $\mathbb{F}_q^n$ of the type $\boldsymbol{\alpha} \longmapsto A\boldsymbol{\alpha} + \boldsymbol{\beta}$, where $A \in GL(n, \mathbb{F}_q)$ and $\boldsymbol{\beta} \in \mathbb{F}_q^n$.

Definition 2.2 *The affine group associated to $\mathcal{X}$ is*

$$Aff(\mathcal{X}) = \{\varphi : \mathcal{X} \to \mathcal{X} \mid \varphi = \psi_{|\mathcal{X}} \text{ with } \psi \in Aff(n, \mathbb{F}_q) \text{ and } \psi(\mathcal{X}) = \mathcal{X}\}.$$

We say that $f, g \in C_{\mathcal{X}}$ are $\mathcal{X}$-equivalent if there exists $\varphi \in Aff(\mathcal{X})$ such that $f = g \circ \varphi$.

An affine subspace $G \subset \mathbb{F}_q^n$ of dimension r is said to be $\mathcal{X}$-affine if there exists $\psi \in Aff(n, \mathbb{F}_q)$ and $1 \le i_1 < \cdots < i_r \le n$ such that $\psi(\mathcal{X}) = \mathcal{X}$ and $\psi(\langle e_{i_1}, \ldots, e_{i_r}\rangle) = G$, where we write $\{e_1, \ldots, e_n\}$ for the canonical basis of $\mathbb{F}_q^n$. We denote by x_i the coordinate function $x_i(\sum_j a_j e_j) = a_i$ where $\sum_j a_j e_j \in \mathbb{F}_q^n$ (and by abuse of notation we also denote by x_i its restriction to $\mathcal{X}$) for all $i = 1, \ldots, n$. Let $f \in C_{\mathcal{X}}$ be a reduced polynomial of degree one, if there exists $\varphi \in Aff(\mathcal{X})$ and $i \in \{1, \ldots, n\}$ such that $x_i \circ \varphi = f$ on the points of $\mathcal{X}$ then we say that f is $\mathcal{X}$-linear.

Let $\{i_1, \ldots, i_s\} \subset \{1, \ldots, n\}$ and $j \in \{1, \ldots, n\}$, we define $\mathcal{X}_{i_1, \ldots, i_s} := K_{i_1} \times \cdots \times K_{i_s}$, and $\mathcal{X}_{\hat{j}} := K_1 \times \cdots \times K_{j-1} \times K_{j+1} \times \cdots \times K_n$.

Definition 2.3 *Let $j \in \{1, \ldots, n\}$, for every $\alpha \in K_j$ we have an evaluation homomorphism of $\mathbb{F}_q$-algebras given by*

$$
\begin{aligned}
C_{\mathcal{X}} &\longrightarrow C_{\mathcal{X}_{\hat{j}}} \\
f &\longmapsto f(x_1, \ldots, x_{j-1}, \alpha, x_{j+1}, \ldots, , x_n) =: f_\alpha^{(j)}.
\end{aligned}
$$

We now present two results which we will freely use in what follows. The first one states the value of the minimum distance of $C_{\mathcal{X}}(d)$.

Theorem 2.4 *[25, Thm. 3.8] The minimum distance $\delta_{\mathcal{X}}(d)$ of $C_{\mathcal{X}}(d)$ is 1, if $d \ge \sum_{i=1}^n (d_i - 1)$, and for $1 \le d < \sum_{i=1}^n (d_i - 1)$ we have*

$$\delta_{\mathcal{X}}(d) = (d_{k+1} - \ell) \prod_{i=k+2}^n d_i$$

where k and ℓ are uniquely defined by $d = \sum_{i=1}^k (d_i - 1) + \ell$ with $0 < \ell \le d_{k+1} - 1$ (if $k + 1 = n$ we understand that $\prod_{i=k+2}^n d_i = 1$, and if $d < d_1 - 1$ then we set $k = 0$ and $\ell = d$).

The second one is a very useful numerical result, closely related to the above theorem (the link between these two results is explained in [6]).

Lemma 2.5 *[6, Lemma 2.1] Let $0 < d_1 \le \cdots \le d_n$ and $1 \le d \le \sum_{i=1}^n (d_i - 1)$ be integers. Let $m(a_1, \ldots, a_n) = \prod_{i=1}^n (d_i - a_i)$, where $0 \le a_i < d_i$ is an integer for all $i = 1, \ldots, n$. Then*

$$\min\{m(a_1, \ldots, a_n) \mid a_1 + \cdots + a_n \le d\} = (d_{k+1} - \ell) \prod_{i=k+2}^n d_i$$

where k and ℓ are uniquely defined by $d = \sum_{i=1}^{k}(d_i - 1) + \ell$, with $0 < \ell \leq d_{k+1} - 1$ (if $s < d_1 - 1$ then take $k = 0$ and $\ell = d$, if $k + 1 = n$ then we understand that $\prod_{i=k+2}^{n} d_i = 1$).

From Theorem 2.4 we get that the relevant range for d is $1 \leq d < \sum_{i=1}^{n}(d_i - 1)$ (the case $d = 0$ is trivial and if $d \geq \sum_{i=1}^{n}(d_i - 1)$ we have $C_{\mathcal{X}}(d) \cong \mathbb{F}_q^m$). In what follows we will always assume that $1 \leq d < \sum_{i=1}^{n}(d_i - 1)$ and will also freely use the decomposition $d = \sum_{i=1}^{k}(d_i - 1) + \ell$, with $0 < \ell \leq d_{k+1} - 1$ (and $0 \leq k < n$). In many places we consider a nonzero function g defined in $\mathcal{X}_{i_1, \ldots, i_s} \subset \mathbb{F}_q^s$ which belongs to $C_{\mathcal{X}_{i_1, \ldots, i_s}}(d)$, and we want to estimate $|g|$. Applying Theorem 2.4 we get that $|g| \geq 1$ if $d \geq \sum_{t=1}^{s}(d_{i_t} - 1)$ while if $d < \sum_{t=1}^{s}(d_{i_t} - 1)$ then $|g| \geq \delta_{\mathcal{X}_{i_1, \ldots, i_s}}(d)$, and we find $\delta_{\mathcal{X}_{i_1, \ldots, i_s}}(d)$ by a proper application of the formula in Theorem 2.4. Since $\delta_{\mathcal{X}_{i_1, \ldots, i_s}}(d) = 1$ in the case where $d \geq \sum_{t=1}^{s}(d_{i_t} - 1)$, we can always write $|g| \geq \delta_{\mathcal{X}_{i_1, \ldots, i_s}}(d)$.

The following result shows that functions which are related by an affine transformation have the same degree.

Lemma 2.6 *Let $\varphi \in \text{Aff}(\mathcal{X})$ and $f \in C_{\mathcal{X}}$ with $f \neq 0$, then $\deg f = \deg(f \circ \varphi)$.*

Proof. Since $\varphi \in \text{Aff}(\mathcal{X})$ we have that $\varphi(\boldsymbol{\alpha}) = A\boldsymbol{\alpha} + \boldsymbol{\beta}$ where $A \in GL(n, \mathbb{F}_q)$ and $\boldsymbol{\beta} \in \mathbb{F}_q^n$. Let $P \in \mathbb{F}_q[X]$ be the reduced polynomial associated to f, and let's endow $\mathbb{F}_q[X]$ with a degree-lexicographic order. Then the reduced polynomial associated to $f \circ \varphi$ is the remainder, say Q, in the division of $P(AX + \boldsymbol{\beta})$ by $\{X_1^{d_1} - X_1, \ldots, X_n^{d_n} - X_n\}$, where X is a column vector with entries equal to $X_1, \ldots, X_n$. Thus $\deg Q \leq \deg P(AX + \boldsymbol{\beta}) \leq \deg P$, so that $\deg(f \circ \varphi) \leq \deg f$. Applying the argument to φ^{-1} we conclude that $\deg(f \circ \varphi) = \deg f$. $\qquad\square$

The next result, although simple, is the basis for many important results that follow.

Lemma 2.7 *Let $f, h \in C_{\mathcal{X}}$ be nonzero functions. There exists a function $g \in C_{\mathcal{X}}$ such that $f = gh$ if and only if $Z_{\mathcal{X}}(h) \subset Z_{\mathcal{X}}(f)$, i.e. h is a factor of f if and only if f vanishes in $Z_{\mathcal{X}}(h)$. Moreover, if h is $\mathcal{X}$-linear then $\deg g = \deg f - 1$.*

Proof. If $f = gh$ and $h(\boldsymbol{\alpha}) = 0$ then $f(\boldsymbol{\alpha}) = 0$, for all $\boldsymbol{\alpha} \in \mathcal{X}$. Assume now that $Z_{\mathcal{X}}(h) \subset Z_{\mathcal{X}}(f)$, and let $g : \mathcal{X} \to \mathbb{F}_q$ be defined by $g(\boldsymbol{\alpha}) = 0$ if $\boldsymbol{\alpha} \in Z_{\mathcal{X}}(h)$, and $g(\boldsymbol{\alpha}) = f(\boldsymbol{\alpha})/h(\boldsymbol{\alpha})$ if $\boldsymbol{\alpha} \in \mathcal{X} \setminus Z_{\mathcal{X}}(h)$, then clearly $f = gh$ as functions of $C_{\mathcal{X}}$.

Let's assume now that $h \mid f$ and that h is $\mathcal{X}$-linear, so that $h \circ \varphi = x_i$ for some $i \in \{1, \ldots, n\}$ and $\varphi \in \text{Aff}(\mathcal{X})$. Then $f \circ \varphi = (g \circ \varphi)(h \circ \varphi)$ and since from Lemma 2.6 $\deg f = \deg(f \circ \varphi)$ we may simply assume that $h = x_i$.

Let P be the reduced polynomial associated to f and write $P = X_i \cdot Q + R$, where $Q, R \in \mathbb{F}_q[X_1, \ldots, X_n]$ and X_i does not appear in any monomial of R. Observe that for any $j \in \{1, \ldots, n\}$, the degree of X_j in any monomial of Q is at most $d_j - 1$. Let g and t be the functions associated to Q and R, respectively, so $f = x_i g + t$. We must have $t = 0$, otherwise $t(\alpha) \neq 0$ for some $\alpha = (\alpha_1, \ldots, \alpha_n) \in \mathcal{X}$, hence taking $\tilde{\alpha} = (\tilde{\alpha}_1, \ldots, \tilde{\alpha}_n)$, with $\tilde{\alpha}_j = \alpha_j$ for $j \in \{1, \ldots, n\} \setminus \{i\}$ and $\tilde{\alpha}_i = 0$ we get $x_i(\tilde{\alpha}) = 0$ hence $f(\tilde{\alpha}) = 0$ but $t(\tilde{\alpha}) \neq 0$, a contradiction. Since R is the reduced polynomial associated to t we get $R = 0$, and since Q is the reduced polynomial of g we get $\deg g = \deg Q = \deg f - 1$. $\qquad\square$

Lemma 2.8 *Let h be a nonzero function in $C_{\mathcal{X}}(d)$ such that for some $i \in \{1, \ldots, n\}$ and some $\varphi \in \mathrm{Aff}(\mathcal{X})$ we have $h = x_i \circ \varphi$. Then, for $\alpha \in \mathbb{F}_q$, we get that $h - \alpha$ is $\mathcal{X}$-linear if and only if $\alpha \in K_i$. Moreover, let $f \in C_{\mathcal{X}}(d)$, $f \neq 0$ and let $\alpha_1, \ldots, \alpha_s$ be distinct elements of K_i such that $Z_{\mathcal{X}}(h - \alpha_j) \subset Z_{\mathcal{X}}(f)$ for all $j = 1, \ldots, s$, then there exists $g \in C_{\mathcal{X}}(d - s)$ such that $f = g \cdot \prod_{j=1}^{s}(h - \alpha_j)$.*

Proof. Assume that $\alpha \in K_i$ and consider the affine transformation $\tilde{\varphi} : \mathbb{F}_q^n \to \mathbb{F}_q^n$ given by $\tilde{\varphi}(\alpha) = \varphi(\alpha) - \alpha e_i$ for all $\alpha \in \mathbb{F}_q^n$, then one can easily check that $\tilde{\varphi} \in \mathrm{Aff}(\mathcal{X})$ and $x_i \circ \tilde{\varphi} = h - \alpha$. On the other hand, suppose that $h - \alpha$ is $\mathcal{X}$-linear, then $h - \alpha$ must vanish on some point of $\mathcal{X}$. From $h = x_i \circ \varphi$ we get that $h(\mathcal{X}) \subset K_i$ so we must have $\alpha \in K_i$.

Since $h - \alpha_1$ is $\mathcal{X}$-linear and $Z_{\mathcal{X}}(h - \alpha_1) \subset Z_{\mathcal{X}}(f)$ then from Lemma 2.7 we get that $f = g_1(h - \alpha_1)$ with $g_1 \in C_{\mathcal{X}}(d - 1)$. If $s = 1$ we're done, if $s \geq 2$ then from $Z_{\mathcal{X}}(h - \alpha_2) \subset Z_{\mathcal{X}}(f)$ and the fact that $Z_{\mathcal{X}}(h - \alpha_1) \cap Z_{\mathcal{X}}(h - \alpha_2) = \emptyset$ we get that $Z_{\mathcal{X}}(h - \alpha_2) \subset Z_{\mathcal{X}}(g_1)$. From the hypothesis and Lemma 2.7 we get that $g_1 = g_2(h - \alpha_2)$ with $g_2 \in C_{\mathcal{X}}(d - 2)$, this proves the statement in the case where $s = 2$ and if $s > 2$ the assertion is proved after a finite number of similar steps. $\qquad\square$

If G is $\mathcal{X}$-affine and there exists $\psi \in \mathrm{Aff}(n, \mathbb{F}_q)$ and $1 \leq i_1 < \cdots < i_r \leq n$ such that $\psi(\mathcal{X}) = \mathcal{X}$ and $\psi(\langle e_{i_1}, \ldots, e_{i_r} \rangle) = G$ then $\mathcal{X}_G := \mathcal{X}_{i_1, \ldots, i_r}$. The following results states an important property of the support of functions.

Lemma 2.9 *Let $f \in C_{\mathcal{X}}(d)$ be a nonzero function and let S be its support. Then for every $\mathcal{X}$-affine subspace $G \subset \mathbb{F}_q^n$ of dimension r, with $r \in \{1, \ldots, n-1\}$, either $S \cap G = \emptyset$ or $|S \cap G| \geq \delta_{\mathcal{X}_G}(d)$.*

Proof. Since G is an $\mathcal{X}$-affine subspace of dimension r there exists an affine transformation $\psi : \mathbb{F}_q \to \mathbb{F}_q$ such that $\psi(\mathcal{X}) = \mathcal{X}$ and $G = \psi(V)$ where $V = \langle e_{i_1}, \ldots, e_{i_r} \rangle$. Observe that ψ establishes a bijection between the points of $V \cap \psi^{-1}(S)$ and $G \cap S$, we also have that $\psi^{-1}(S)$ is the support of the function $f \circ \psi_{|\chi}$ which belongs to $C_\mathcal{X}(d)$ because $\deg f = \deg(f \circ \psi_{|\chi})$. This shows that, for simplicity, we may assume that $G = \langle e_{i_1}, \ldots, e_{i_r} \rangle$. Suppose that $S \cap G \neq \emptyset$ and let P be the reduced polynomial associated to f, then f induces a nonzero function $\tilde{f}$ defined over $\mathcal{X}_G = \mathcal{X}_{i_1, \ldots, i_r} \subset \mathbb{F}_q^r$ whose reduced polynomial is $\tilde{P}(X_{i_1}, \ldots, X_{i_s})$ obtained from P by making $X_i = 0$ for all $i \in \{1, \ldots, n\} \setminus \{i_1, \ldots, i_s\}$. Clearly $\deg \tilde{f} \leq d$ so that $\tilde{f} \in C_{\mathcal{X}_G}(d)$, also $|S \cap G| = |\tilde{f}|$ and as a consequence of Theorem 2.4 we get $|\tilde{f}| \geq \delta_{\mathcal{X}_G}(d)$. $\qquad\square$

Observe, in the next result, that if S is the support of a function then, from the above result, it already has property (2).

Proposition 2.10 *Let* $1 \leq d < \sum_{i=1}^n (d_i - 1)$ *and write* $d = \sum_{i=1}^k (d_i - 1) + \ell$ *as in Theorem 2.4. Let* $S \subset \mathcal{X}$ *be a nonempty set and assume that* S *has the following properties:*

1. $|S| < (1 + 1/d_{k+1})\, \delta_\mathcal{X}(d) = (1 + 1/d_{k+1})\,(d_{k+1} - \ell)d_{k+2} \cdots d_n.$
2. *For every* $\mathcal{X}$-*affine subspace* $G \subset \mathbb{F}_q^n$ *of dimension* r, *with* $r \in \{0, \ldots, n-1\}$, *either* $S \cap G = \emptyset$ *or* $|S \cap G| \geq \delta_{\mathcal{X}_G}(d)$.

Then there exists an affine subspace $H \subset \mathbb{F}_q^n$, *of dimension* $n - 1$ *and a transformation* $\psi \in \mathrm{Aff}(n, \mathbb{F}_q)$ *such that* $\psi(\mathcal{X}) = \mathcal{X}$, $\psi(V_{k+1}) = H$ *where* V_{k+1} *is the* $\mathbb{F}_q$-*vector space generated by* $\{e_1, \ldots, e_n\} \setminus \{e_{k+1}\}$ *(so, in particular,* H *is* $\mathcal{X}$-*affine) and* $S \cap H = \emptyset$.

Proof. We proceed by induction on n. When $n = 1$ we have $k = 0$, and from the hypothesis we get that $|S| < (1 + 1/d_1)\,(d_1 - \ell) \leq d_1 - 1/d_1$, hence $|S| \leq d_1 - 1$ and $S \subsetneq K_1 \subset \mathbb{F}_q$. A 0-dimensional $\mathcal{X}$-affine subspace is just an element of K_1, so it is enough to take H as a point of $K_1 \setminus S$.

Assume now that the statement is true for all $n < N$, and let $S \subset \mathcal{X} \subset \mathbb{F}_q^N$ as in the hypothesis. For $\alpha \in K_{k+1}$ let

$$G_\alpha = \alpha e_{k+1} + V_{k+1} = \left\{ \beta \in \mathbb{F}_q^N \mid \beta = (\beta_1, \ldots, \beta_N) \text{ and } \beta_{k+1} = \alpha \right\}$$

If for some $\alpha \in K_{k+1}$ we get $S \cap G_\alpha = \emptyset$ then we're done, so assume from now on that $S \cap G_\alpha \neq \emptyset$ for all $\alpha \in K_{k+1}$. If $k = N - 1$ we have $\delta_\mathcal{X}(d) = d_N - \ell$ and

$$d_N \leq \sum_{\alpha \in K_N} |S \cap G_\alpha| = |S| < \left(1 + \frac{1}{d_N}\right)(d_N - \ell)$$

$$\leq \left(1 + \frac{1}{d_N}\right)(d_N - 1) = d_N - \frac{1}{d_N},$$

a contradiction which settles this case. Now we consider the case where $k \leq N - 2$. Since G_α is $\mathcal{X}$-affine we have $|S \cap G_\alpha| \geq \delta_{\mathcal{X}_{\widehat{k+1}}}(d) = (d_{k+2} - \ell)d_{k+3} \cdots d_N$ for every $\alpha \in K_{k+1}$. Thus $d_{k+1}\delta_{\mathcal{X}_{\widehat{k+1}}}(d) \leq |S| < (1 + 1/d_{k+1})\delta_{\mathcal{X}}(d)$ and from the formulas for $\delta_{\mathcal{X}_{\widehat{k+1}}}(d)$ and $\delta_{\mathcal{X}}(d)$ we get $d_{k+1}(d_{k+2} - \ell) < (1 + 1/d_{k+1})(d_{k+1} - \ell)d_{k+2}$. Hence

$$1 - \frac{\ell}{d_{k+2}} < 1 - \frac{\ell}{d_{k+1}^2} - \frac{\ell - 1}{d_{k+1}} \leq 1 - \frac{\ell}{d_{k+1}^2}$$

so that $d_{k+2} < d_{k+1}^2$. Assume that $K_{k+1} \subsetneq K_{k+2}$, since this is a field extension we must have $d_{k+1}^2 \leq d_{k+2}$, a contradiction which settles the case $k \leq N - 2$ and $d_{k+1} < d_{k+2}$.

The last case is when $k \leq N - 2$ and $d_{k+1} = d_{k+2}$, and now we will apply the induction hypothesis. To do that, for $\alpha \in K_{k+1}$, we consider the bijection $\xi_\alpha : G_\alpha \to \mathbb{F}_q^{N-1}$ which acts on an N-tuple $\boldsymbol{\alpha} \in G_\alpha$ by deleting the $(k + 1)$-th entry (which is equal to α). Observe that ξ_α establishes a bijection between affine subspaces of $\mathbb{F}_q^N$ contained in G_α and affine subspaces of $\mathbb{F}_q^{N-1}$. Clearly $\mathcal{X}_{\widehat{k+1}} \subset \mathbb{F}_q^{N-1}$ and we want to show that $\xi_\alpha(S \cap G_\alpha)$ has property (2) of the statement (with $\mathcal{X}_{\widehat{k+1}}$ in place of $\mathcal{X}$). For this, let $L \subset \mathbb{F}_q^{N-1}$ be an r-dimensional $\mathcal{X}_{\widehat{k+1}}$-affine subspace. Then for some $\tilde{\psi} \in \mathrm{Aff}(N - 1, \mathbb{F}_q)$, given by $\tilde{\alpha} \mapsto \tilde{A}\tilde{\alpha} + \tilde{\beta}$, with $\tilde{A} \in GL(N-1, \mathbb{F}_q)$ and $\tilde{\beta} \in \mathbb{F}_q^{N-1}$, we have $\tilde{\psi}(\mathcal{X}_{\widehat{k+1}}) = \mathcal{X}_{\widehat{k+1}}$ and $\tilde{\psi}(L) = \langle \tilde{e}_{i_1}, \ldots, \tilde{e}_{i_r} \rangle$, where $\{\tilde{e}_1, \ldots, \tilde{e}_{N-1}\}$ is the canonical basis for $\mathbb{F}_q^{N-1}$. We claim that $\xi_\alpha^{-1}(L)$ is an $\mathcal{X}$-affine subspace contained in G_α and to see that let A be the matrix obtained from $\tilde{A}$ by adding an $N \times 1$ column of zeros as the $(k + 1)$-th column, an $1 \times N$ line of zeros as the $(k + 1)$-th line and changing the 0 at position $(k+1, k+1)$ to 1. Let β be the $N \times 1$ vector obtained from $\tilde{\beta}$ by adding the entry $-\alpha$ at position $k + 1$. Then, defining $\psi : \mathbb{F}_q^N \to \mathbb{F}_q^N$ by $\boldsymbol{\alpha} \mapsto A\boldsymbol{\alpha} + \beta$ we get that $\psi \in \mathrm{Aff}(N, \mathbb{F}_q)$, and it is easy to check that $\psi(\mathcal{X}) = \mathcal{X}$ and that $\psi(\xi_\alpha^{-1}(L)) = \langle e_{j_1}, \ldots, e_{j_r} \rangle$, with $\{j_1, \ldots, j_r\} \subset \{1, \ldots, n\} \setminus \{k + 1\}$, $j_s = i_s$ whenever $i_s < k + 1$, and $j_s = i_s + 1$ whenever $i_s \geq k+1$, for all $s = 1, \ldots, r$, so that $\{d_{j_1}, \ldots, d_{j_r}\} = \{d_{i_1}, \ldots, d_{i_r}\}$. To show that $\xi_\alpha(S \cap G_\alpha)$ has property (2) of the statement, with $\mathcal{X}_{\widehat{k+1}}$ in place of $\mathcal{X}$, we observe that

$$|(\xi_\alpha(S \cap G_\alpha) \cap L| = |(S \cap G_\alpha) \cap \xi_\alpha^{-1}(L)| = |S \cap \xi_\alpha^{-1}(L)| \geq \delta_{\mathcal{X}_{j_1, \ldots, j_r}}(d)$$

$$= \delta_{(\mathcal{X}_{\widehat{k+1}})_{i_1, \ldots, i_r}}(d).$$

Now we prove that there exists $\alpha \in K_{k+1}$ such that $\xi_\alpha(S \cap G_\alpha)$ also has property (1), with $\mathcal{X}_{\widehat{k+1}}$ in place of $\mathcal{X}$. Indeed, if for all $\alpha \in K_{k+1}$ we have

$$|\xi_\alpha(S \cap G_\alpha)| \geq (1 + 1/d_{k+2})\,\delta_{\mathcal{X}_{\widehat{k+1}}}(d) = (1 + 1/d_{k+2})\,(d_{k+2} - \ell)d_{k+3} \cdots d_N$$

then from $|\xi_\alpha(S \cap G_\alpha)| = |S \cap G_\alpha|$ we get $|S| \geq d_{k+1}\,(1 + 1/d_{k+2})\,(d_{k+2} - \ell)d_{k+3} \cdots d_N = (1 + 1/d_{k+1})\delta_\mathcal{X}(d)$ (because $d_{k+1} = d_{k+2}$) which contradicts property (1). Thus, for some $\alpha \in K_{k+1}$ we get that $\xi_\alpha(S \cap G_\alpha) \subset \mathcal{X}_{\widehat{k+1}} \subset \mathbb{F}_q^{N-1}$ satisfies properties (1) and (2), and from the induction hypothesis there exists an $\mathcal{X}_{\widehat{k+1}}$-affine subspace $L \subset \mathbb{F}_q^{N-1}$ of dimension $N - 2$ and $\tilde{\psi} \in \mathrm{Aff}(N - 1, \mathbb{F}_q)$ such that $\tilde{\psi}(\mathcal{X}_{\widehat{k+1}}) = \mathcal{X}_{\widehat{k+1}}, \; \psi(L)$ is the subspace generated by $\{\tilde{e}_1, \dots, \tilde{e}_{N-1}\} \setminus \{\tilde{e}_{k+1}\}$ and $\xi_\alpha(S \cap G_\alpha) \cap L = \emptyset$. From what we did above we get that $\xi_\alpha^{-1}(L)$ is an $(N - 2)$-dimensional $\mathcal{X}$-affine subspace of $\mathbb{F}_q^N$ and there exists $\psi \in \mathrm{Aff}(N, \mathbb{F}_q)$ such that $\psi(\mathcal{X}) = \mathcal{X}$, $\psi(\xi_\alpha^{-1}(L))$ is the subspace generated by $\{e_1, \dots, e_N\} \setminus \{e_{k+1}, e_{k+2}\}$, and $(S \cap G_\alpha) \cap \xi_\alpha^{-1}(L) = S \cap \xi_\alpha^{-1}(L) = \emptyset$. Thus $\psi(\xi_\alpha^{-1}(L))$ is the subvector space defined by $X_{k+1} = 0$ and $X_{k+2} = 0$, and let $G_{(\gamma_1,\gamma_2)}$ be the hyperplane defined by the equation $\gamma_1 X_{k+1} + \gamma_2 X_{k+2} = 0$, where $(\gamma_1 : \gamma_2) \in \mathbb{P}^1(K_{k+1})$, observe that $G_{(\gamma_1,\gamma_2)} \cap G_{(\gamma_1',\gamma_2')} = \psi(\xi_\alpha^{-1}(L))$ whenever $(\gamma_1 : \gamma_2) \neq (\gamma_1' : \gamma_2')$. One may easily check that for every $(\gamma_1 : \gamma_2) \in \mathbb{P}^1(K_{k+1})$ there exists a linear transformation that takes $G_{(\gamma_1,\gamma_2)}$ onto the subspace defined by $X_{k+1} = 0$, so that $H_{(\gamma_1,\gamma_2)} := \psi^{-1}(G_{(\gamma_1,\gamma_2)})$ is an $\mathcal{X}$-affine subspace of dimension $N - 1$. We claim that for some $(\gamma_1 : \gamma_2) \in \mathbb{P}^1(K_{k+1})$ we must have $S \cap H_{(\gamma_1,\gamma_2)} = \emptyset$. Indeed, if this is not true, then, since $H_{(\gamma_1,\gamma_2)} \cap H_{(\gamma_1',\gamma_2')} = \xi_\alpha^{-1}(L)$ (for any distinct pair $(\gamma_1 : \gamma_2), (\gamma_1' : \gamma_2'), \in \mathbb{P}^1(K_{k+1})$) and $S \cap \xi_\alpha^{-1}(L) = \emptyset$ we get

$$|S| \geq \sum_{(\gamma_1:\gamma_2) \in \mathbb{P}^1(K_{k+1})} |S \cap H_{(\gamma_1,\gamma_2)}| \geq (d_{k+1} + 1)\delta_{\mathcal{X}_{\widehat{k+1}}}(d)$$
$$= (d_{k+1} + 1)(d_{k+2} - \ell)d_{k+3} \cdots d_N,$$

a contradiction with property (1) which, using $d_{k+1} = d_{k+2}$, states that

$$|S| < \left(1 + \frac{1}{d_{k+1}}\right)(d_{k+1} - \ell)d_{k+2} \cdots d_n = (d_{k+1} + 1)(d_{k+2} - \ell)d_{k+3} \cdots d_N,$$

$\square$

The next result combines previous results and gives a first step in the direction of the main result.

Corollary 2.11 *Let f be a nonzero function in $C_\mathcal{X}(d)$ such that $|f| < (1 + 1/d_{k+1})\,\delta_\mathcal{X}(d)$, then f is a multiple of a function h of degree 1 which is $\mathcal{X}$-equivalent to x_{k+1}.*

Proof. Let S be the support of f, from the hypothesis we have that S has property (1) in the statement of Proposition 2.10 and from Lemma 2.9 we get that S also has property (2). Thus, there exists an affine subspace $H \subset \mathbb{F}_q^n$, of dimension $n - 1$ and a transformation $\psi \in \text{Aff}(n, \mathbb{F}_q)$ such that $\psi(\mathcal{X}) = \mathcal{X}$, $\psi(V_{k+1}) = H$ with $V_{k+1} = \{\alpha \in \mathbb{F}_q^n \mid \alpha_{k+1} = 0\}$ and $S \cap H = \emptyset$. Hence $\psi^{-1}(S) \cap V_{k+1} = \emptyset$, and noting that $\psi^{-1}(S)$ is the support of the function $f \circ \psi_{|\mathcal{X}} \in C_{\mathcal{X}}(d)$ we get that $Z_{\mathcal{X}}(x_{k+1}) \subset Z_{\mathcal{X}}(f \circ \psi_{|\mathcal{X}})$. From Lemma 2.7 there exists $g \in C_{\mathcal{X}}(d-1)$ such that $f \circ \psi_{|\mathcal{X}} = g\, x_{k+1}$, hence $f = (g \circ \psi_{|\mathcal{X}}^{-1}) \cdot (x_{k+1} \circ \psi_{|\mathcal{X}}^{-1})$ and we can take $h = x_{k+1} \circ \psi_{|\mathcal{X}}^{-1}$. $\qquad\square$

Recall that we write $d = \sum_{i=1}^k (d_i - 1) + \ell$, with $0 < \ell \leq d_{k+1} - 1$ (and $0 \leq k < n$).

Lemma 2.12 *Let f be a nonzero function in $C_{\mathcal{X}}(d)$, and let $h \in C_{\mathcal{X}}(d)$ be such that $h = x_j \circ \varphi$, where $j \in \{1, \ldots, n\}$ and $\varphi \in \text{Aff}(\mathcal{X})$. If m is the number of $\alpha \in K_j$ such that $Z_{\mathcal{X}}(h - \alpha) \subset Z_{\mathcal{X}}(f)$ then $m \leq d$ and $|f| \geq (d_j - m)\delta_{\mathcal{X}_{\hat{j}}}(d - m)$.*

Proof. Let $\tilde{f} = f \circ \varphi^{-1}$, then $\tilde{f} \in C_{\mathcal{X}}(d)$, $f = \tilde{f} \circ \varphi$ and φ establishes a bijection between the sets $Z_{\mathcal{X}}(h - \alpha)$ and $Z_{\mathcal{X}}(x_j - \alpha)$ for all $\alpha \in K_j$, moreover we get that $Z_{\mathcal{X}}(h - \alpha) \subset Z_{\mathcal{X}}(f)$ if and only if $Z_{\mathcal{X}}(x_j - \alpha) \subset Z_{\mathcal{X}}(\tilde{f})$. This shows that, in the statement, we can take φ to be the identity transformation, without loss of generality. Let $\alpha_1, \ldots, \alpha_m$ be the set of elements $\alpha \in K_j$ such that $Z_{\mathcal{X}}(x_j - \alpha) \subset Z_{\mathcal{X}}(f)$, from Lemma 2.8 we get that $f = g \cdot \prod_{i=1}^m (x_j - \alpha_i)$, with $g \in C_{\mathcal{X}}(d - m)$, and in particular $m \leq d$. Observe that for all $\alpha \in K_j \setminus \{\alpha_1, \ldots, \alpha_m\}$ we get $g_\alpha^{(j)} \neq 0$, so that

$$|f| = \sum_{\alpha \neq \alpha_i} |g_\alpha^{(j)}| \geq (d_j - m)\delta_{\mathcal{X}_{\hat{j}}}(d - m).$$

$\qquad\square$

For our purposes it is important to know when a function $f \in C_{\mathcal{X}}(d)$ has minimal weight, i.e. when $|f| = \delta_{\mathcal{X}}(d)$. Taking into account the previous result, and using its notation, we investigate when $(d_j - m)\delta_{\mathcal{X}_{\hat{j}}}(d - m) \geq \delta_{\mathcal{X}}(d)$ holds, and under which conditions equality holds.

Lemma 2.13 *Let $1 \leq j \leq k+1$. If $d_j > d_{k+1} - \ell$, for $0 < m < \ell + (d_j - d_{k+1})$ we have*

$$(d_j - m)\delta_{\mathcal{X}_{\hat{j}}}(d - m) > \delta_{\mathcal{X}}(d).$$

Proof. Observe that we may write

$$d - m = \sum_{i=1, i \neq j}^{k+1} (d_i - 1) + \ell - m + (d_j - d_{k+1}),$$

and note that $\ell - m + d_j - d_{k+1} \leq \ell - m < \ell < d_{k+1} \leq d_{k+2}$ so that $\delta_{\chi_{\hat{j}}}(d - m) = (d_{k+2} - (\ell - m + d_j - d_{k+1})) \prod_{i=k+3}^{n} d_i$. From $\delta_{\chi}(d) = (d_{k+1} - \ell) \prod_{i=k+2}^{n} d_i$ and

$$(d_j - m)(d_{k+2} - (\ell - m + d_j - d_{k+1})) - (d_{k+1} - \ell)d_{k+2}$$
$$= (\ell - m + d_j - d_{k+1})(d_{k+2} - d_j + m) > 0$$

we get

$$(d_j - m)\delta_{\chi_{\hat{j}}}(d - m) > \delta_{\chi}(d).$$

$\square$

Lemma 2.14 *Let* $1 \leq j \leq k$. *For* $0 < m < d_j$ *we have* $(d_j - m)\delta_{\chi_{\hat{j}}}(d - m) \geq \delta_{\chi}(d)$, *with equality if and only if* $m = d_j - 1$ *or both* $d_j > d_{k+1} - \ell$ *and* $m = \ell + d_j - d_{k+1}$.

Proof. By Lemma 2.13, we may consider $\max\{1, \ell + (d_j - d_{k+1})\} \leq m \leq d_j - 1$. In this case we write

$$d - m = \sum_{i=1, i \neq j}^{k} (d_i - 1) + \ell + (d_j - 1 - m),$$

and we observe that $0 < \ell + d_j - 1 - m \leq d_{k+1} - 1$, so that $\delta_{\chi_{\hat{j}}}(d - m) = (d_{k+1} - (\ell + d_j - 1 - m)) \prod_{i=k+2}^{n} d_i$. From

$$(d_j - m)(d_{k+1} - (\ell + d_j - 1 - m)) - (d_{k+1} - \ell)$$
$$= (m - (\ell + d_j - d_{k+1}))(d_j - 1 - m) \geq 0$$

we get

$$(d_j - m)\delta_{\chi_{\hat{j}}}(d - m) \geq \delta_{\chi}(d),$$

with equality if and only if $m = d_j - 1$ or both $\ell + d_j - d_{k+1} > 0$ and $m = \ell + d_j - d_{k+1}$. $\square$

Lemma 2.15 *For* $0 < m < d_{k+1}$ *we have* $(d_{k+1} - m)\delta_{\chi_{\widehat{k+1}}}(d - m) \geq \delta_{\chi}(d)$, *with equality if and only if* $m = \ell$ *or both* $m = d_{k+1} - 1$ *and* $d_k \geq d_{k+1} - \ell$.

Proof. By Lemma 2.13, we may consider $\ell \leq m \leq d_{k+1} - 1$. In this case we write

$$d - m = \sum_{i=1}^{\tilde{k}} (d_i - 1) + \tilde{\ell},$$

where

$$0 \leq \tilde{k} < k, \quad \tilde{\ell} = \ell - m + \sum_{i=\tilde{k}+1}^{k} (d_i - 1) > 0 \quad \text{and}$$

$$\ell - m + \sum_{i=\tilde{k}+2}^{k} (d_i - 1) \leq 0,$$

hence $\tilde{\ell} \leq d_{\tilde{k}+1} - 1$. We want to prove that

$$(d_{k+1} - m)\delta_{\chi_{\widehat{k+1}}}(d - m) \geq \delta_{\chi}(d) = (d_{k+1} - \ell) \prod_{i=k+2}^{n} d_i,$$

and from $k \geq \tilde{k} + 1$ we get $k + 1 \in \{\tilde{k} + 2, \ldots, n\}$, so that

$$\delta_{\chi_{\widehat{k+1}}}(d - m) = (d_{\tilde{k}+1} - \tilde{\ell}) \prod_{i=\tilde{k}+2, i \neq k+1}^{n} d_i.$$

Thus we must verify that

$$(d_{\tilde{k}+1} - \tilde{\ell}) \left(\prod_{i=\tilde{k}+2}^{k} d_i \right) (d_{k+1} - m) \geq (d_{k+1} - \ell). \tag{1}$$

Let M be the function defined by

$$M(a_{\tilde{k}+1}, \ldots, a_{k+1}) = (d_{\tilde{k}+1} - a_{\tilde{k}+1}) \cdot \cdots \cdot (d_{k+1} - a_{k+1}),$$

where a_i is a nonnegative integer less than d_i, for $i = \tilde{k} + 1, \ldots, k + 1$, and $a_{\tilde{k}+1} + \cdots + a_{k+1} \leq \tilde{\ell} + m$. We have studied this function in [6] and [9]. From $\tilde{\ell} + m = \sum_{i=\tilde{k}+1}^{k} (d_i - 1) + \ell$ and [6, Lemma 2.1] we get $d_{k+1} - \ell$ is the minimum of M so that inequality (1) holds. To find out when (1) is an equality we will use results from [9], and for that we define a tuple $(a_{\tilde{k}+1}, \ldots, a_{k+1})$ to be normalized if whenever $d_{i-1} < d_i = \cdots = d_{i+s} < d_{i+s+1}$ we have $a_i \geq a_{i+1} \geq \cdots \geq a_{i+s}$. From [9, Lemma 2.2] we get that the normalized tuples which reach the minimum of M are exactly of the type:

1. $(a_{\tilde{k}+1}, \ldots, a_{k+1}) = (d_{\tilde{k}+1} - 1, \ldots d_k - 1, \ell)$, or
2. $(a_{\tilde{k}+1}, \ldots, a_{k+1}) = (d_{\tilde{k}+1} - 1, \ldots, d_j - (d_{k+1} - \ell), \ldots, d_{k+1} - 1)$.

Type 2 is only possible if $d_{k+1} - \ell \leq d_j < d_{k+1}$, we also note that if $\ell = d_{k+1} - 1$ then types 1 and 2 are the same so we also assume in type 2 that $\ell < d_{k+1} - 1$. Thus we have equality in (1) if and only if the tuple $(\tilde{\ell}, 0, \ldots, 0, m)$, when normalized, is equal to $(d_{\tilde{k}+1} - 1, \ldots d_k - 1, \ell)$ or $(d_{\tilde{k}+1} - 1, \ldots, d_j - (d_{k+1} - \ell), \ldots, d_{k+1} - 1)$.

In the first case, since we don't have any zero entries in $(d_{\tilde{k}+1} - 1, \ldots d_k - 1, \ell)$ we must have $\tilde{k} + 1 = k$ and the tuple $(\tilde{\ell}, m)$ when normalized is equal to

(d_k-1, ℓ), thus we must have either $(\widetilde{\ell}, m) = (d_k-1, \ell)$ or $(m, \widetilde{\ell}) = (d_k-1, \ell)$. If $(\widetilde{\ell}, m) = (d_k-1, \ell)$ then $m = \ell$, and if $(m, \widetilde{\ell}) = (d_k-1, \ell)$, then $m = d_k-1$ and from the definition of normalized tuple we also must have $d_k = d_{k+1}$. On the other hand if $m = \ell$, from $d = \sum_{i=1}^k (d_i - 1) + \ell$ we get

$$d - m = \sum_{i=1}^{k-1} (d_i - 1) + (d_k - 1)$$

so we must have $\widetilde{k} = k - 1$ and $\widetilde{\ell} = d_k - 1$, hence $(\widetilde{\ell}, m) = (d_k - 1, \ell)$. And if $m = d_k - 1 = d_{k+1} - 1$, from $d = \sum_{i=1}^k (d_i - 1) + \ell$ we get

$$d - m = \sum_{i=1}^{k-1} (d_i - 1) + \ell$$

so we must have $\widetilde{k} = k - 1$ and $\widetilde{\ell} = \ell$, hence $(m, \widetilde{\ell}) = (d_k - 1, \ell)$.

The upshot of this is that $(\widetilde{\ell}, m)$ when normalized is equal to $(d_k - 1, \ell)$ if and only if $m = \ell$ or both $m = d_{k+1} - 1$ and $d_k = d_{k+1}$.

In the second case, since we may have at most only one zero entry in

$$(d_{\widetilde{k}+1} - 1, \ldots, d_j - (d_{k+1} - \ell), \ldots, d_{k+1} - 1),$$

we must have $\widetilde{k} + 1 = k$ or $\widetilde{k} + 2 = k$. If $\widetilde{k} + 1 = k$ then the above tuple is an ordered pair, and since it is a type 2 tuple we must have that $d_k < d_{k+1}$ and that this pair is $(d_k - (d_{k+1} - \ell), d_{k+1} - 1)$. Since $d_k < d_{k+1}$ the tuple $(\widetilde{\ell}, m)$ is already normalized, and if $(\widetilde{\ell}, m) = (d_k - (d_{k+1} - \ell), d_{k+1} - 1)$ then $m = d_{k+1} - 1$ and $\widetilde{\ell} = d_k - (d_{k+1} - \ell)$ so that $d_k - (d_{k+1} - \ell) > 0$. On the other hand if $m = d_{k+1} - 1$ and $d_k - (d_{k+1} - \ell) > 0$, from $d = \sum_{i=1}^k (d_i - 1) + \ell$ we get

$$d - m = \sum_{i=1}^{k} (d_i - 1) + \ell - (d_{k+1} - 1) = \sum_{i=1}^{k-1} (d_i - 1) + d_k - (d_{k+1} - \ell)$$

so we must have $\widetilde{k} = k - 1$ and $\widetilde{\ell} = d_k - (d_{k+1} - \ell)$, hence $(\widetilde{\ell}, m) = (d_k - (d_{k+1} - \ell), d_{k+1} - 1)$.

If $\widetilde{k} + 2 = k$ then we must have $d_k < d_{k+1}$ so the tuple $(\widetilde{\ell}, 0, m)$ is already normalized, and if $(\widetilde{\ell}, 0, m) = (d_{k-1} - 1, d_k - (d_{k+1} - \ell), d_{k+1} - 1)$ then $d_k = d_{k+1} - \ell$ and $m = d_{k+1} - 1$. On the other hand if $m = d_{k+1} - 1$ and $d_k - (d_{k+1} - \ell) = 0$ from $d = \sum_{i=1}^k (d_i - 1) + \ell$ we get

$$d - m = \sum_{i=1}^{k} (d_i - 1) + \ell - (d_{k+1} - 1) = \sum_{i=1}^{k-2} (d_i - 1) + d_{k-1} - 1$$

so we must have $\widetilde{k} = k - 2$ and $\widetilde{\ell} = d_{k-1} - 1$, hence $(\widetilde{\ell}, 0, m) = (d_{k-1} - 1, d_k - (d_{k+1} - \ell), d_{k+1} - 1)$.

Thus we have equality in (1) if and only if $m = \ell$ or both $m = d_{k+1} - 1$ and $d_k \geq d_{k+1} - \ell$.   □

Proposition 2.16 *Let f be a nonzero function in $C_{\mathcal{X}}(d)$, and let $h \in C_{\mathcal{X}}(d)$ be such that $h = x_j \circ \varphi$, where $\varphi \in Aff(\mathcal{X})$ and $1 \leq j \leq k+1$. Let $m > 0$ be the number of $\alpha \in K_j$ such that $Z_{\mathcal{X}}(h - \alpha) \subset Z_{\mathcal{X}}(f)$. Let $g = f \circ \varphi^{-1}$, then $|f| = \delta_{\mathcal{X}}(d)$ if and only if $|g_\alpha^{(j)}| = \delta_{\mathcal{X}_{\hat{j}}}(d - m)$ whenever $g_\alpha^{(j)} \neq 0$, with $\alpha \in K_j$ and m satisfies one of the following:*

1) If $1 \leq j \leq k$ then $m = d_j - 1$ or both $m = \ell + d_j - d_{k+1}$ and $d_j > d_{k+1} - \ell$.
2) If $j = k+1$ then $m = \ell$ or both $m = d_{k+1} - 1$ and $d_k \geq d_{k+1} - \ell$.

Proof. Let $j \in \{1, \ldots, k+1\}$. As in the beginning of the proof of Lemma 2.12 we may assume that φ is the identity, so that $h = x_j$. From the proof of Lemma 2.12 we get

$$|f| = \sum_{\alpha \in K_j} |f_\alpha^{(j)}| \geq (d_j - m)\delta_{\mathcal{X}_{\hat{j}}}(d - m)$$

and equality holds if and only if $|f_\alpha^{(j)}| = \delta_{\mathcal{X}_{\hat{j}}}(d - m)$ whenever $f_\alpha^{(j)} \neq 0$, with $\alpha \in K_j$. From the two previous Lemmas we know that $\delta_{\mathcal{X}_{\hat{j}}}(d - m) \geq \delta_{\mathcal{X}}(d)$ and we also know when equality holds.   □

As mentioned in the paragraph preceding Lemma 2.13 we are investigating when $(d_j - m)\delta_{\mathcal{X}_{\hat{j}}}(d - m) \geq \delta_{\mathcal{X}}(d)$ holds, and under which conditions equality holds. Now we treat the case where $m = 0$.

Lemma 2.17 *Let $1 \leq j \leq k+1$. We have*

$$d_j \delta_{\mathcal{X}_{\hat{j}}}(d) \geq \delta_{\mathcal{X}}(d)$$

with equality if and only if $d_j = d_{k+1} - \ell$ or $d_j = d_{k+2}$.

Proof. If $d_j \leq d_{k+1} - \ell$ we may write

$$d = \sum_{i=1, i \neq j}^{k} (d_i - 1) + \ell + (d_j - 1),$$

so that $\delta_{\mathcal{X}_{\hat{j}}}(d) = (d_{k+1} - (\ell + d_j - 1)) \prod_{i=k+2}^{n} d_i$. From

$$d_j(d_{k+1} - (\ell + d_j - 1)) - (d_{k+1} - \ell)) = (d_j - 1)(d_{k+1} - \ell - d_j) \geq 0$$

we get

$$d_j \delta_{\mathcal{X}_{\hat{j}}}(d) \geq (d_{k+1} - \ell) \prod_{i=k+2}^{n} d_i = \delta_{\mathcal{X}}(d),$$

with equality if and only if $d_j = d_{k+1} - \ell$.

If $d_j > d_{k+1} - \ell$ we may write

$$d = \sum_{i=1, i \neq j}^{k+1} (d_i - 1) + \ell + d_j - d_{k+1},$$

so that $\delta_{\chi_{\hat{j}}}(d) = (d_{k+2} - (\ell + d_j - d_{k+1})) \prod_{i=k+3}^{n} d_i$. From

$$d_j(d_{k+2} - (\ell + d_j - d_{k+1})) - (d_{k+1} - \ell)d_{k+2}$$
$$= (d_j - (d_{k+1} - \ell))(d_{k+2} - d_j) \geq 0$$

we get

$$d_j \delta_{\chi_{\hat{j}}}(d) \geq \delta_\chi(d),$$

with equality if and only if $d_j = d_{k+2}$. $\qquad\square$

Proposition 2.18 *Let $f \in C_\chi(d)$ and suppose that $d_j < d_{k+1} - \ell$ for some $1 \leq j \leq k$. If $|f| = \delta_\chi(d)$ then the number of $\alpha \in K_j$ such that $Z_\chi(x_j - \alpha) \subset Z_\chi(f)$ is $d_j - 1$ and for $\alpha \in K_j$ such that $f_\alpha^{(j)} \neq 0$ we have $|f_\alpha^{(j)}| = |f| = \delta_\chi(d) = \delta_{\chi_{\hat{j}}}(d - (d_j - 1))$.*

Proof. Let m be the number of $\alpha \in K_j$ such that $Z_\chi(x_j - \alpha) \subset Z_\chi(f)$. By Lemma 2.12 we have $|f| \geq (d_j - m)\delta_{\chi_{\hat{j}}}(d - m)$. As $d_j < d_{k+1} - \ell$ and $|f| = \delta_\chi(d)$, from Lemma 2.17 we get $m > 0$ and from Lemma 2.14 we have $m = d_j - 1$ and $\delta_{\chi_{\hat{j}}}(d - (d_j - 1)) = \delta_\chi(d)$. We conclude by observing that for the only element $\alpha \in K_j$ such that $f_\alpha^{(j)} \neq 0$ we have $|f| = |f_\alpha^{(j)}|$. $\qquad\square$

3 Main Results

As in the preceding section we continue to write d as in the statement of Theorem 2.4, namely $d = \sum_{i=1}^{k}(d_i - 1) + \ell$, with $0 < \ell \leq d_{k+1} - 1$ (and $0 \leq k < n$). The next result describes the minimal weight codewords of affine cartesian codes for the lowest range of values of d, meaning the case when $k = 0$.

Proposition 3.1 *Let $1 \leq d < d_1$, the minimal weight codewords of $C_\chi(d)$ are χ-equivalent to the functions*

$$\sigma \prod_{i=1}^{\ell} (x_1 - \alpha_i),$$

with $\sigma \in \mathbb{F}_q^*$, $\alpha_i \in K_1$ and $\alpha_i \neq \alpha_j$ for $1 \leq i \neq j \leq \ell$.

Proof. Let $f \in C_{\mathcal{X}}(d)$ be such that $|f| = \delta_{\mathcal{X}}(d)$. From Corollary 2.11 we get that f has a degree one factor h which is $\mathcal{X}$-equivalent to x_1. Let $m \le d = \ell$ be the number of distinct elements $\alpha \in K_1$ such that $Z_{\mathcal{X}}(x_1 - \alpha) \subset Z_{\mathcal{X}}(f)$.

As $m \le d$, from Proposition 2.16 (2) we have $|f| = \delta_{\mathcal{X}}(d)$ if and only if $m = \ell$. Now the result follows from Lemma 2.8. $\qquad\square$

Now we describe the minimal weight codewords for the case where $\ell = d_{k+1} - 1$ and $0 \le k < n$.

Proposition 3.2 *The minimal weight codewords of $C_{\mathcal{X}}(d)$, for $d = \sum_{i=1}^{k+1}(d_i - 1)$, $0 \le k < n$, are $\mathcal{X}$-equivalent to the functions of the form*

$$\sigma \prod_{i=1}^{k+1}(1 - x_i^{d_i-1}),$$

with $\sigma \in \mathbb{F}_q^*$.

Proof. We will prove the result by induction on k, and we note that the case $k = 0$ is already covered by Proposition 3.1, so we assume $k > 0$ and that the result holds for $k - 1$.

Let $f \in C_{\mathcal{X}}(d)$ be such that $|f| = \delta_{\mathcal{X}}(d)$. From Corollary 2.11 we get that f has a degree one factor h such that $h = x_{k+1} \circ \varphi$, for some $\varphi \in \text{Aff}(\mathcal{X})$. Let $m > 0$ be the number of $\alpha \in K_{k+1}$ such that $Z_{\mathcal{X}}(h - \alpha) \subset Z_{\mathcal{X}}(f)$. From Proposition 2.16 (2) we get $m = d_{k+1} - 1$ (since $\ell = d_{k+1}$). In particular $f_\alpha^{(k+1)} \ne 0$ for only one value of $\alpha \in K_{k+1}$, and without loss of generality, we may assume that φ is the identity transformation and $\alpha = 0$. Hence, from Lemma 2.8 we get

$$f = (1 - x_{k+1}^{d_{k+1}-1})g,$$

for some $g \in C_{\mathcal{X}}(d - (d_{k+1} - 1))$. Let P and Q be the reduced polynomials associated to f and g, respectively. Then

$$P - (1 - X_{k+1}^{d_{k+1}-1})Q$$

is in the ideal $I_{\mathcal{X}} = (X_1^{d_1} - X_1, \dots, X_n^{d_n} - X_n)$. Write $Q = Q_1 + X_{k+1}Q_2$, where Q_1 and Q_2 are reduced polynomials and X_{k+1} does not appear in any monomial of Q_1. Then $P - (1 - X_{k+1}^{d_{k+1}-1})Q_1$ is in $I_{\mathcal{X}}$, and writing g_1 for the function associated to Q_1, we get $f = (1 - x_{k+1}^{d_{k+1}-1})g_1$. Since $\deg(Q_1) = d - (d_{k+1} - 1)$ we have $g_1 \in C_{\mathcal{X}_{\widehat{k+1}}}(d - (d_{k+1} - 1))$, and from $d - (d_{k+1} - 1) = \sum_{i=1}^{k}(d_i - 1)$, $\delta_{\mathcal{X}}(d) = \delta_{\mathcal{X}_{\widehat{k+1}}}(d - (d_{k+1} - 1))$ and $|f| = |g_1|$ we see that g_1 is a minimal weight codeword of $C_{\mathcal{X}_{\widehat{k+1}}}(d - (d_{k+1} - 1))$ so we may apply the induction hypothesis to g_1, which concludes the proof of the Proposition. $\qquad\square$

Lemma 3.3 *Let $d = \sum_{i=1}^{k+1}(d_i - 1)$, $0 \le k < n$ and let $g \in C_{\mathcal{X}}(d)$ be such that $|g| = \delta_{\mathcal{X}}(d)$. Let $h \in C_{\mathcal{X}}(d - s)$, where $0 < s \le d_1 - 1$. If $f = g + h$*

then $|f| \geq (s+1)\delta\chi(d)$ or $|f| = s\delta\chi(d)$. *From the above Proposition there exists* $\varphi \in \text{Aff}(\mathcal{X})$ *such that* $g \circ \varphi^{-1} = \sigma \prod_{i=1}^{k+1}(1 - x_i^{d_i-1})$, *with* $\sigma \in \mathbb{F}_q^*$. *Let* $\widehat{f} = f \circ \varphi^{-1}$, *if* $|f| = s\delta\chi(d)$ *then, for each* $1 \leq j \leq k+1$, *the number of elements* $\alpha \in K_j$ *such that* $Z_{\mathcal{X}}(x_j - \alpha) \subset Z_{\mathcal{X}}(\widehat{f})$ *is either* $d_j - 1$ *or* $d_j - s$.

Proof. As in the proof of Lemma 2.12 we may assume that φ is the identity transformation, so we identify $\widehat{f}$ with f and $g \circ \varphi^{-1}$ with g.

We will make an induction on n. If $n = 1$ then $k = 0$, $d = d_1 - 1$, $j = 1$ and $|g| = 1$. Since $h \in C_{\mathcal{X}}(d_1 - (s+1))$ and $|K_1| = d_1$ we have $|h| \geq s+1$, and a fortiori $|f| \geq s$. If $|f| = s$ then there are $d_1 - s$ elements $\alpha \in K_1$ such that $Z_{\mathcal{X}}(x_1 - \alpha) \subset Z_{\mathcal{X}}(f)$.

We will do an induction on n, so we assume that the result is true for $n - 1$ and let $j \in \{1, \ldots, k+1\}$. From the hypothesis on g and using the notation established in Definition 2.3 we may write $g = (1 - x_j^{d_j-1})g_0^{(j)}$, where $g_0^{(j)} \in C_{\mathcal{X}_{\widehat{j}}}(d - (d_j - 1))$ is a function of minimal weight. We also write

$$|f| = \sum_{\alpha \in K_j} |f_\alpha^{(j)}| = |g_0^{(j)} + h_0^{(j)}| + \sum_{\alpha \in K_j^*} |h_\alpha^{(j)}|.$$

Let's assume that $h_0^{(j)} = 0$, since $\delta\chi_{\widehat{j}}(d - (d_j - 1)) = \prod_{i=k+2}^n d_i = \delta\chi(d)$ and $\delta\chi(d-s) = (d_{k+1} - (d_{k+1} - 1 - s))\prod_{i=k+2}^n d_i$ we get

$$|f| = |g_0^{(j)}| + |h| \geq \delta\chi_{\widehat{j}}(d - (d_j - 1)) + \delta\chi(d-s) = (s+2)\delta\chi(d)$$

which proves the Lemma in this case. Assume now that $h_0^{(j)} \neq 0$, and let m be the number of elements $\alpha \in K_j$ such that $Z_{\mathcal{X}}(x_j - \alpha) \subset Z_{\mathcal{X}}(h)$.

Let's assume that $f_0^{(j)} \neq 0$, in this case m is also the number of elements $\alpha \in K_j$ such that $Z_{\mathcal{X}}(x_j - \alpha) \subset Z_{\mathcal{X}}(f)$ since $g = (1 - x_j^{d_j-1})g_0^{(j)}$. If $m = d_j - 1$ then from Lemma 2.8 we have $h = (1 - x_j^{d_j-1})\widetilde{h}$, with $\widetilde{h} \in C_{\mathcal{X}}(d - (d_j - 1) - s)$. As in the end of the proof of Proposition 3.2 we may assume that $\widetilde{h} \in C_{\mathcal{X}_{\widehat{j}}}(d - (d_j - 1) - s)$ so that $\widetilde{h} = h_0^{(j)}$. We now apply the induction hypothesis to $f_0^{(j)} = g_0^{(j)} + h_0^{(j)}$ and we get

$$|f_0^{(j)}| \geq (s+1)\delta\chi_{\widehat{j}}(d - (d_j - 1)) \text{ or } |f_0^{(j)}| = s\delta\chi_{\widehat{j}}(d - (d_j - 1)).$$

If $|f_0^{(j)}| = s\delta\chi_{\widehat{j}}(d - (d_j - 1))$ then, from the induction hypothesis, we get that for $i \neq j$ there are $d_i - 1$ or $d_i - s$ values of $\alpha \in K_i$ such that $Z_{\mathcal{X}_{\widehat{j}}}(x_j - \alpha) \subset Z_{\mathcal{X}_{\widehat{j}}}(f_0^{(j)})$ and from $f = g + h = (1 - x_j^{d_j-1})g_0^{(j)} + (1 - x_j^{d_j-1})h_0^{(j)} = (1 - x_j^{d_j-1})f_0^{(j)}$ we get the statement of the Lemma for the case where $h_0^{(j)} \neq 0$, $f_0^{(j)} \neq 0$ and $m = d_j - 1$. Still assuming that $h_0^{(j)} \neq 0$ and $f_0^{(j)} \neq 0$, we now treat the case where $0 \leq m < d_j - 1$. From Lemma 2.8 we know that

$h = \prod_{i=1}^{m}(x_j - \alpha_i)\tilde{h}$, where $\alpha_1, \ldots, \alpha_m \in K_j^*$ and $\tilde{h} \in C_{\mathcal{X}}(d - s - m)$ so $h_0^{(j)} = \beta\tilde{h}_0^{(j)}$, with $\beta \in K_j^*$ and we get $h_0^{(j)} \in C_{\mathcal{X}_{\hat{j}}}(d - s - m)$ (note that we also get $h_\alpha^{(j)} \in C_{\mathcal{X}_{\hat{j}}}(d - s - m)$ for all $\alpha \in K_j^* \setminus \{\alpha_1, \ldots, \alpha_m\}$). Thus, from $f_0^{(j)} = g_0^{(j)} + h_0^{(j)}$ we get that the degree of $f_0^{(j)}$ is at most $\max\{d - (d_j - 1), d - (s + m)\}$. We now consider the following cases.

1. Assume that $d_j - 1 < s + m$, so we have that the degree of $f_0^{(j)} = g_0^{(j)} + h_0^{(j)}$ is at most $d - (d_j - 1)$. From $h_0^{(j)} \in C_{\mathcal{X}_{\hat{j}}}(d - (s + m))$ and writing $d - (s + m) = d - (d_j - 1) - (s + m - (d_j - 1))$, we observe that $0 < s + m - (d_j - 1) = s - (d_j - 1 - m) < d_1 - 1$, so we may apply the induction hypothesis on $f_0^{(j)}$ and we get, in particular, that $|f_0^{(j)}| \geq (s + m - (d_j - 1))\delta_{\mathcal{X}_{\hat{j}}}(d - (d_j - 1)) = (s + m + 1 - d_j)\delta_{\mathcal{X}}(d)$. From $|f| = |f_0^{(j)}| + \sum_{\alpha \in K_j^*} |h_\alpha^{(j)}|$ and the fact that $|h_\alpha^{(j)}| \geq \delta_{\mathcal{X}_{\hat{j}}}(d - (s + m))$ for all $\alpha \in K_j^* \setminus \{\alpha_1, \ldots, \alpha_m\}$ we get $|f| \geq (s + m + 1 - d_j)\delta_{\mathcal{X}}(d) + (d_j - m - 1)\delta_{\mathcal{X}_{\hat{j}}}(d - (s + m))$. We claim that

$$\delta_{\mathcal{X}_{\hat{j}}}(d - (s + m)) = (s + m - d_j + 2)\delta_{\mathcal{X}}(d),$$

and to prove this fact we have to consider the cases where $j \leq k$ and $j = k + 1$. We will do the case $j \leq k$ since the proof of the other case is similar to this one. So let $j \leq k$, then

$$d - (s + m) = \sum_{i=1, i \neq j}^{k}(d_i - 1) + (d_{k+1} - 1 + d_j - 1 - (s + m))$$

so

$$\delta_{\mathcal{X}_{\hat{j}}}(d - (s + m)) = (d_{k+1} - (d_{k+1} - 1 + d_j - 1 - (s + m))\prod_{i=k+2}^{n} d_i$$

$$= (s + m - d_j + 2)\delta_{\mathcal{X}}(d).$$

Thus

$$|f| \geq (s + m + 1 - d_j)\delta_{\mathcal{X}}(d) + (d_j - m - 1)(s + m - d_j + 2)\delta_{\mathcal{X}}(d)$$
$$= \left(s + (d_j - m - 1)(s + m + 1 - d_j)\right)\delta_{\mathcal{X}}(d) \geq (s + 1)\delta_{\mathcal{X}}(d)$$

which proves the Lemma in this case.

2. Assume now that $d_j - 1 \geq s + m$, in this case $\deg(g_0^{(j)} + h_0^{(j)}) \leq d - (s + m)$, and we have

$$|f| \geq (d_j - m)\delta_{\chi_{\hat{j}}}(d - (s + m))$$

$$= (d_j - m)(d_{k+1} + s + m + 1 - d_j) \prod_{i=k+2}^{n} d_i$$

$$= \big((s + 1)d_{k+1} + (d_{k+1} + m - d_j)(d_j - 1 - s - m)\big) \prod_{i=k+2}^{n} d_i$$

$$\geq (s + 1)\delta\chi(d).$$

We now consider the case $f_0^{(j)} = g_0^{(j)} + h_0^{(j)} = 0$, so in particular $\deg h_0^{(j)} = \deg g_0^{(j)} = d - (d_j - 1)$. On the other hand $\deg h_0^{(j)} \leq d - (s + m)$, so we get $s + m \leq d_j - 1$. Let $\lambda = ((-1)^m \prod_{i=1}^{m} \alpha_i)^{-1}$, then, using Lemma 2.8 we get that there exists a function $\hat{h}$ such that

$$h - \lambda \left(\prod_{i=1}^{m}(x_j - \alpha_i)\right) h_0^{(j)} = \lambda x_j \left(\prod_{i=1}^{m}(x_j - \alpha_i)\right) \hat{h}$$

Observe that $\deg(h - \lambda(\prod_{i=1}^{m}(x_j - \alpha_i))h_0^{(j)}) \leq d - s$ hence $\deg \hat{h} \leq d - (s + m + 1)$.

Assume that $s + m < d_j - 1$ hence $s + m + 1 \leq d_j - 1$. From

$$h = \lambda \left(\prod_{i=1}^{m}(x_j - \alpha_i)\right) (h_0^{(j)} + x_j \hat{h})$$

Recall that

$$|f| = \sum_{\alpha \in K_j^*} |h_\alpha^{(j)}|$$

and $h_\alpha^{(j)} = \lambda \left(\prod_{i=1}^{m}(\alpha - \alpha_i)\right) (h_0^{(j)} + \alpha \hat{h}_\alpha^{(j)}) \neq 0$ for $d_j - (m + 1)$ values of $\alpha \in K_j^*$. Observe that $\deg h_0^{(j)} = d - (d_j - 1) \leq d - (s + m + 1)$ and since $\deg \hat{h} \leq d - (s + m + 1)$ we get $\deg h_\alpha^{(j)} \leq d - (s + m + 1)$, when $h_\alpha^{(j)} \neq 0$. Then

$$|f| \geq (d_j - (m + 1))\delta_{\chi_{\hat{j}}}(d - (s + m + 1))$$

$$= (d_j - (m + 1))(d_{k+2} - (d_j - (s + m + 2))) \prod_{i=k+3}^{n} d_i$$

$$= \big((s + 1)d_{k+2} + (d_j - (s + m + 2))(d_{k+2} - d_j + m + 1)\big) \prod_{i=k+3}^{n} d_i$$

$$\geq (s + 1)\delta\chi(d).$$

If $s + m = d_j - 1$ then $\deg h_\alpha^{(j)} = d - (d_j - 1)$ whenever $\alpha \in K_j^*$ and $h_\alpha^{(j)} \neq 0$. In this case

$$|f| \geq (d_j - m - 1)\delta_{\chi_j^-}(d - (d_j - 1)) = s\delta_\chi(d),$$

and equality holds if and only if $|f_\alpha^{(j)}| = |h_\alpha^{(j)}| = \delta_{\chi_j^-}(d - (d_j - 1))$, for all $f_\alpha^{(j)} \neq 0$. Observe that in this case the number of elements $\alpha \in K_j$ such that $Z_\chi(x_j - \alpha) \subset Z_\chi(f)$ is $m + 1 = d_j - s$.

Still under the assumption that $s + m = d_j - 1$ we must prove that if $|f| > s\delta_\chi(d)$ then $|f| \geq (s + 1)\delta_\chi(d)$. From the above reasoning we know that if $|f| > s\delta_\chi(d)$ then there exists $\alpha \in K_j^*$ such that $h_\alpha^{(j)} \neq 0$ and $|h_\alpha^{(j)}| > \delta_{\chi_j^-}(d - (d_j - 1)) = \delta_\chi(d)$. We recall that

$$h_\alpha^{(j)} = \lambda \left(\prod_{i=1}^m (\alpha - \alpha_i) \right) (h_0^{(j)} + \alpha \widehat{h_\alpha^{(j)}}),$$

that $h_0^{(j)} = -g_0^{(j)}$ is a function, or codeword, of minimal weight in $C_{\chi_j^-}(d - (d_j - 1))$ and that $\deg(\widehat{h_\alpha^{(j)}}) \leq d - (s + m + 1) = d - (d_j - 1) - 1$. From the induction hypothesis, with $s = 1$, we get from $|h_\alpha^{(j)}| > \delta_{\chi_j^-}(d - (d_j - 1))$ that $|h_\alpha^{(j)}| \geq 2\delta_{\chi_j^-}(d - (d_j - 1))$. Hence, from $|f| = \sum_{\alpha \in K_j^*} |h_\alpha^{(j)}|$ we get

$$|f| \geq (d_j - m - 2)\delta_{\chi_j^-}(d - (d_j - 1)) + 2\delta_{\chi_j^-}(d - (d_j - 1)) = (s + 1)\delta_\chi(d),$$

which completes the proof of the Lemma. □

Lemma 3.4 *Let* $f \in C_\chi(d)$, *where* $d = \sum_{i=2}^{k+1}(d_i - 1)$, $1 \leq k < n$. *If there exist* $\alpha_1, \alpha_2 \in K_1$, $\alpha_1 \neq \alpha_2$, $|f_{\alpha_1}^{(1)}| = |f_{\alpha_2}^{(1)}| = \delta_{\chi_1^-}(d)$ *then there exists* $\varphi \in \mathrm{Aff}(\mathcal{X})$ *such that* $x_1 = x_1 \circ \varphi$ *and* $g_{\alpha_1}^{(1)} = g_{\alpha_2}^{(1)}$, *where* $g = f \circ \varphi$.

Proof. From Proposition 3.2 we may assume without loss of generality that

$$f_{\alpha_1}^{(1)} = \sigma \prod_{i=2}^{k+1} \left(1 - x_i^{d_i - 1}\right),$$

with $\sigma \in \mathbb{F}_q^*$. Since $f \in C_\chi(d)$ there exists $\hat{f} \in C_\chi(d - 1)$ such that $f = f_{\alpha_1}^{(1)} + (x_1 - \alpha_1)\hat{f}$ so that $f_{\alpha_2}^{(1)} = f_{\alpha_1}^{(1)} + (\alpha_2 - \alpha_1)\hat{f}_{\alpha_2}^{(1)}$. Since $|f_{\alpha_2}^{(1)}| = \delta_{\chi_1^-}(d)$, we get from Lemma 3.3 (with $s = 1$) that for each $2 \leq j \leq k + 1$ the number of elements $\alpha \in K_j$ such that $Z_{\chi_1^-}(x_j - \alpha) \subset Z_{\chi_1^-}(f_{\alpha_2}^{(1)})$ is $d_j - 1$. Thus for each $2 \leq j \leq k + 1$ there exists $\beta_j \in K_j$ such that $f_{\alpha_2}^{(1)}$ is a multiple of $\prod_{\alpha \in K_j \setminus \{\beta_j\}}(x_j - \alpha)$. From the equality of the reduced polynomials

$$\prod_{\alpha \in K_j \setminus \{\beta_j\}} (X_j - \alpha) = (X_j - \beta_j)^{d_j - 1} - 1$$

we get, by successive applications of Lemma 2.8, that

$$f_{\alpha_2}^{(1)} = \tau \prod_{i=2}^{k+1} \left(1 - (x_i - \beta_i)^{d_i-1}\right)$$

for some $\tau \in \mathbb{F}_q^*$. Observe that from $(\alpha_2 - \alpha_1) \hat{f}_{\alpha_2}^{(1)} = f_{\alpha_2}^{(1)} - f_{\alpha_1}^{(1)}$ and $\hat{f} \in C_{\mathcal{X}}(d-1)$ we must have $\tau = \sigma$. If $\beta_j = 0$ for all $2 \leq j \leq k+1$ then $f_{\alpha_1}^{(1)} = f_{\alpha_2}^{(1)}$. Otherwise consider a function $\varphi \in \mathrm{Aff}(\mathcal{X})$ such that $x_1 \circ \varphi = x_1$ and

$$x_j \circ \varphi = x_j + \beta_j \frac{x_1 - \alpha_1}{\alpha_2 - \alpha_1} .$$

for all $2 \leq j \leq k+1$. Let $g = f \circ \varphi$, if $x_1 = \alpha_1$ then $x_j \circ \varphi = x_j$, and if $x_1 = \alpha_2$ then $x_j \circ \varphi = x_j + \beta_j$ for all $2 \leq j \leq k+1$. Thus

$$g_{\alpha_1}^{(1)} = (f \circ \varphi)_{\alpha_1}^{(1)} = \sigma \prod_{i=2}^{k+1} \left(1 - ((x_i \circ \varphi)_{\alpha_1}^{(1)})^{d_i-1}\right) = f_{\alpha_1}^{(1)},$$

and

$$g_{\alpha_2}^{(1)} = (f \circ \varphi)_{\alpha_2}^{(1)} = \sigma \prod_{i=2}^{k+1} \left(1 - ((x_i \circ \varphi)_{\alpha_2}^{(1)} - \beta_i)^{d_i-1}\right) = f_{\alpha_1}^{(1)},$$

hence $g_{\alpha_1}^{(1)} = g_{\alpha_2}^{(1)}$. $\qquad\square$

Now we prove the main result of this paper, which generalizes the theorem by Delsarte, Goethals and Mac Williams on minimal weight codewords of $GRM_q(d, n)$ to the minimal weight codewords of $C_{\mathcal{X}}(d)$.

Theorem 3.5 *Let* $d = \sum_{i=1}^{k}(d_i - 1) + \ell$, $0 \leq k < n$ *and* $0 < \ell \leq d_{k+1} - 1$, *the minimal weight codewords of* $C_{\mathcal{X}}(d)$ *are* $\mathcal{X}$*-equivalent to the functions of the form*

$$g = \sigma \prod_{i=1, i \neq j}^{k+1} (1 - x_i^{d_i-1}) \prod_{t=1}^{d_j-(d_{k+1}-\ell)} (x_j - \alpha_t) ,$$

for some $1 \leq j \leq k+1$ *such that* $d_{k+1} - \ell \leq d_j$, *where* $0 \neq \sigma \in \mathbb{F}_q$ *and* $\alpha_1, \ldots, \alpha_{d_j-(d_{k+1}-\ell)}$ *are distinct elements of* K_j *(if* $d_j = d_{k+1} - \ell$ *we take the second product as being equal to 1).*

Proof. If $k = 0$ the $d < d_1$ and the result follows from Proposition 3.1.

We will do an induction on k, so let's assume that the result holds for $k - 1$. If $\ell = d_{k+1} - 1$, then the result follows from Proposition 3.2.

Let $\ell < d_{k+1} - 1$ and let $f \in C_{\mathcal{X}}(d)$ be a minimal weight codeword, i.e. $|f| = \delta_{\mathcal{X}}(d)$. From Corollary 2.11 f has a factor which is $\mathcal{X}$-equivalent to

x_{k+1}. Let $1 \leq j \leq k+1$ be least integer such that f has a factor which is $\mathcal{X}$-equivalent to x_j and and let's assume without loss of generality that $x_j - \alpha$ is a factor of f for some $\alpha \in K_j$. Let $m > 0$ be the number of elements of $\alpha \in K_j$ such that $Z_\mathcal{X}(x_j - \alpha) \subset Z_\mathcal{X}(f)$. From Proposition 2.16 we get $m = d_j - 1$ or $m = d_j - (d_{k+1} - \ell)$.

If $m = d_j - 1$ then, after applying an $\mathcal{X}$-affine transformation if necessary, we write

$$f = (1 - x_j^{d_j-1})g,$$

for some $g \in C_\mathcal{X}(d - (d_j - 1))$, and as in the proof of Proposition 3.2 we show that actually we may write f as

$$f = (1 - x_j^{d_j-1})g_1,$$

with $g_1 \in C_{\widehat{\mathcal{X}_j}}(d-(d_j-1))$. In the case where $1 \leq j \leq k$, since $m = d_j-1$ we get from Lemma 2.14 that $\delta_\mathcal{X}(d) = \delta_{\widehat{\mathcal{X}_j}}(d - (d_j - 1))$ and from $|f| = |g_1|$ we see that g_1 is a minimal weight codeword of $C_{\widehat{\mathcal{X}_j}}(d - (d_j - 1))$, then we may apply the induction hypothesis to get the result. In the case where $j = k + 1$, from Proposition 2.16 we also get $d_k - (d_{k+1} - \ell) \geq 0$ (besides $m = d_{k+1} - 1$) so from Lemma 2.15 we get $\delta_\mathcal{X}(d) = \delta_{\widehat{\mathcal{X}_{k+1}}}(d - (d_{k+1} - 1))$ and from $|f| = |g_1|$ we see that g_1 is a minimal weight codeword of $C_{\widehat{\mathcal{X}_{k+1}}}(d - (d_{k+1} - 1))$. Writing $d-(d_{k+1}-1) = \sum_{j=1}^{k-1}(d_j-1)+(d_k-(d_{k+1}-\ell))$ we see that, as above, we can apply the induction hypothesis to g_1, either because $d_k - (d_{k+1} - \ell) > 0$ or because we get the result from Proposition 3.2 if $d_k = d_{k+1} - \ell$.

Now we assume that $m = d_j - (d_{k+1} - \ell) < d_j - 1$. From Proposition 2.16 we see that there are $d_{k+1} - \ell$ elements in K_j (say, $\beta_1, \ldots, \beta_{d_{k+1}-\ell}$) such that for all $i \in \{1, \ldots, d_{k+1} - \ell\}$ we get $|f_{\beta_i}^{(j)}| = \delta_{\widehat{\mathcal{X}_j}}(\widehat{d})$, with

$$\widehat{d} = d - (d_j - (d_{k+1} - \ell)) = \sum_{i=1,i\neq j}^{k+1} (d_i - 1),$$

while $|f_{\beta_i}^{(j)}| = 0$ for the other elements of K_j (say, $i \in \{d_{k+1}-\ell+1, \ldots, d_j\}$).

From Lemma 2.8 we may write f as

$$f = \widehat{f} \cdot \prod_{i=d_{k+1}-\ell+1}^{d_j} (x_j - \beta_i) \tag{2}$$

with $\widehat{f} \in C_\mathcal{X}(\widehat{d})$.

We treat first the case $j = 1$. From Lemma 3.4, there exists $\psi \in \text{Aff}(\mathcal{X})$ such that $x_1 = x_1 \circ \psi$, and $g_{\beta_1}^{(1)} = g_{\beta_2}^{(1)}$, where $g = \widehat{f} \circ \psi$, and without loss of generality we assume that $\widehat{f} = g$. Observe that $Z_\mathcal{X}(x_i - \beta_i) \subset Z_\mathcal{X}(\widehat{f} - \widehat{f}_{\beta_1}^{(1)}) = Z_\mathcal{X}(\widehat{f} - \widehat{f}_{\beta_2}^{(1)})$ for $i = 1, 2$, so from Lemma 2.8 we may write

$$\widehat{f} = \widehat{f}_{\beta_1}^{(1)} + (x_1 - \beta_1)(x_1 - \beta_2)h \,,$$

with $h \in C_{\mathcal{X}}(\widehat{d} - 2)$. If $d_{k+1} - \ell = 2$, then from $\widehat{f}_{\beta_1}^{(1)} = \widehat{f}_{\beta_2}^{(1)}$ and equation (2) we may write

$$f = \widehat{f}_{\beta_1}^{(1)} \cdot \prod_{i=3}^{d_1} (x_1 - \beta_i) \,,$$

and the result follows from applying Proposition 3.2 to $\widehat{f}_{\beta_1}^{(1)} \in C_{\mathcal{X}_{\widehat{1}}}(\widehat{d})$. If $d_{k+1} - \ell > 2$ then for all $2 < t \leq d_{k+1} - \ell$ we get

$$\widehat{f}_{\beta_t}^{(1)} = \widehat{f}_{\beta_1}^{(1)} + (\beta_t - \beta_1)(\beta_t - \beta_2)h_{\beta_t}^{(1)} \,.$$

If $h_{\beta_t}^{(1)} \neq 0$ then from Lemma 3.3 (taking $s = 2$), we get $|\widehat{f}_{\beta_t}^{(1)}| \geq 2\delta_{\mathcal{X}_{\widehat{1}}}(\widehat{d})$, a contradiction. Hence $\widehat{f}_{\beta_1}^{(1)} = \widehat{f}_{\beta_t}^{(1)}$ for all $1 \leq t \leq d_{k+1} - \ell$, and from equation (2) we may write

$$f = \widehat{f}_{\beta_1}^{(1)} \cdot \prod_{i=d_{k+1}-\ell+1}^{d_1} (x_1 - \beta_i) \,,$$

with $\widehat{f}_{\beta_1}^{(1)} \in C_{\mathcal{X}_{\widehat{1}}}(\widehat{d})$. Again, the result follows from applying Proposition 3.2 to $\widehat{f}_{\beta_1}^{(1)}$, which concludes the case $j = 1$.

Assume now that $j > 1$ and let $\mathcal{X}_{\widehat{1,j}} := K_2 \times \cdots \times K_{j-1} \times K_{j+1} \times \cdots \times K_n$. Then for all $\alpha \in K_1$ we get $Z_{\mathcal{X}}(x_1 - \alpha) \not\subset Z_{\mathcal{X}}(f)$ and from Proposition 2.18 we get $d_1 \geq d_{k+1} - \ell$. From equation (2) we get $|f_{\beta_t}^{(j)}| = |\widehat{f}_{\beta_t}^{(j)}|$, so Proposition 2.16 implies $|\widehat{f}_{\beta_t}^{(j)}| = \delta_{\mathcal{X}_{\widehat{j}}}(\widehat{d})$ for all $t = 1, \ldots, d_{k+1} - \ell$. Thus, in particular, $\widehat{f}_{\beta_1}^{(j)}$ is $\mathcal{X}_{\widehat{j}}$-equivalent to a function of the form $(1 - x_1^{d_1-1})g_1$, where $g_1 \in C_{\mathcal{X}_{\widehat{1,j}}}\left(\sum_{i=2,i \neq j}^{k+1}(d_i - 1)\right)$, and $|g_1| = \delta_{\mathcal{X}_{\widehat{1,j}}}\left(\sum_{i=2,i \neq j}^{k+1}(d_i - 1)\right)$, so we may assume

$$\widehat{f}_{\beta_1}^{(j)} = (1 - x_1^{d_1-1})g_1 \,. \tag{3}$$

Using Lemma 2.7 there exists $h \in C_{\mathcal{X}}(\widehat{d} - 1)$ such that $\widehat{f} = \widehat{f}_{\beta_1}^{(j)} + (x_j - \beta_1)h$, and evaluating both sides at β_t, with $t \in \{2, \ldots, d_{k+1} - \ell\}$, we get $\widehat{f}_{\beta_t}^{(j)} = \widehat{f}_{\beta_1}^{(j)} + (\beta_j - \beta_1)h_{\beta_t}^{(j)}$. We now may apply Lemma 3.3 (replacing f by $\widehat{f}_{\beta_t}^{(j)}$, g by $\widehat{f}_{\beta_1}^{(j)}$, h by $(\beta_j - \beta_1)h_{\beta_t}^{(j)}$), and using that $|\widehat{f}_{\beta_t}^{(j)}| = \delta_{\mathcal{X}_{\widehat{j}}}(\widehat{d})$ we may conclude that there are $d_1 - 1$ elements α in K_1 such that $Z_{\mathcal{X}_{\widehat{j}}}(x_1 - \alpha) \subset Z_{\mathcal{X}_{\widehat{j}}}(\widehat{f}_{\beta_t}^{(j)})$.

From Lemma 2.8, for every $1 \leq t \leq d_{k+1} - \ell$, there exists $\alpha_t \in K_1$ such that

$$\widehat{f}_{\beta_t}^{(j)} = (1 - (x_1 - \alpha_t)^{d_1-1})g_t \,,$$

(here we are using that $((x_1 - \alpha_t)^{d_1-1} - 1)(x_1 - \alpha_t) = x_1^{d_1} - x_1)$, where, as in Proposition 3.2, $g_t \in C_{\mathcal{X}_{\widehat{1,j}}}$ is a minimal weight function of degree $\sum_{i=2, i \neq j}^{k+1} (d_i - 1)$. Note that from (3) we get $\alpha_1 = 0$. We also note that if there exists $\alpha \in K_1$, distinct from α_t for all $t \in \{1, \ldots, d_{k+1} - \ell\}$ then all functions $\widehat{f}_{\beta_t}^{(j)}$ vanish in $x_1 = \alpha$, hence $Z_{\mathcal{X}}(x_1 - \alpha) \subset Z_{\mathcal{X}}(\widehat{f}) \subset Z_{\mathcal{X}}(f)$, a contradiction with the assumption $j > 1$. Thus for all $\alpha \in K_1$ there exists $1 \le t \le d_{k+1} - \ell$ such that $\alpha = \alpha_t$, hence $d_1 \le d_{k+1} - \ell$ and a fortiori $d_1 = d_{k+1} - \ell$.

For each $t \in \{1, \ldots, d_1\}$ let

$$h_t(x_j) = \prod_{i=1, i \neq t}^{d_1} (x_j - \beta_i)$$

and let

$$u = \sum_{i=1}^{d_1} \left(\left(1 - (x_1 - \alpha_i)^{d_1-1} \right) \cdot g_i \cdot \frac{h_i(x_j)}{h_i(\beta_i)} \right) \cdot \prod_{s=d_1+1}^{d_j} (x_j - \beta_s).$$

Clearly, for $d_1 < t \le d_j$, from the definition of u and (2) we get $u_{\beta_t}^{(j)} = 0 = f_{\beta_t}^{(j)}$. For $t \in \{1, \ldots, d_1\}$ we get

$$u_{\beta_t}^{(j)} = \left(1 - (x_1 - \alpha_t)^{d_1-1} \right) g_t \prod_{s=d_1+1}^{d_j} (\beta_t - \beta_s)$$

$$= \widehat{f}_{\beta_t}^{(j)} \prod_{s=d_1+1}^{d_j} (\beta_t - \beta_s) = f_{\beta_t}^{(j)}.$$

Thus we conclude that $u = f$. Letting $x_1 = \alpha_t$, for all $1 \le t \le d_1$ we get

$$f_{\alpha_t}^{(1)} = g_t \cdot \frac{h_t(x_j)}{h_t(\beta_t)} \cdot \prod_{s=d_1+1}^{d_j} (x_j - \beta_s).$$

Observe that $h_t(x_j) \prod_{s=d_1+1}^{d_j}(x_j - \beta_s)$ does not vanish only when $x_j = \beta_t$, so $|f_{\alpha_t}^{(1)}| = |g_t|$. From

$$|g_t| = \delta_{\mathcal{X}_{\widehat{1,j}}} \left(\sum_{i=2, i \neq j}^{k+1} (d_i - 1) \right), \quad \delta_{\mathcal{X}_{\widehat{1}}} \left(\sum_{i=2}^{k+1} (d_i - 1) \right) = \delta_{\mathcal{X}_{\widehat{1,j}}} \left(\sum_{i=2, i \neq j}^{k+1} (d_i - 1) \right)$$

and

$$d = \sum_{i=1}^{k} (d_i - 1) + \ell = \sum_{i=2}^{k+1} (d_i - 1),$$

we get

$$|f_{\alpha_t}^{(1)}| = \delta_{\mathcal{X}_{\hat{1}}} \left(\sum_{i=2}^{k+1} (d_i - 1) \right) = \delta_{\mathcal{X}_{\hat{1}}}(d).$$

Thus we get $f \in C_{\mathcal{X}}(d)$, where $d = \sum_{i=2}^{k+1} (d_i - 1)$ and $|f_{\alpha_1}^{(1)}| = |f_{\alpha_2}^{(1)}| = \delta_{\mathcal{X}_{\hat{1}}}(d)$. From Lemma 3.4, there exists $\theta \in \mathrm{Aff}(\mathcal{X})$ such that $x_1 = x_1 \circ \theta$ and $\tilde{f}_{\alpha_1}^{(1)} = \tilde{f}_{\alpha_2}^{(1)}$, where $\tilde{f} = f \circ \theta$, and without loss of generality we assume that $\tilde{f} = f$. Observe that $Z_{\mathcal{X}}(x_1 - \alpha_i) \subset Z_{\mathcal{X}}(f - f_{\alpha_1}^{(1)}) = Z_{\mathcal{X}}(f - f_{\alpha_2}^{(1)})$ for $i = 1, 2$, so from Lemma 2.8 we may write

$$f = f_{\alpha_1}^{(1)} + (x_1 - \alpha_1)(x_1 - \alpha_2)\overline{f},$$

with $\overline{f} \in C_{\mathcal{X}}(d - 2)$. If $d_1 = 2$, then $f = f_{\alpha_1}^{(1)}$. If $d_1 > 2$ then for all $t \in \{3, \ldots, d_1\}$ we get $f_{\alpha_t}^{(1)} = f_{\alpha_1}^{(1)} + (\alpha_t - \alpha_1)(\alpha_t - \alpha_2)\overline{f}_{\alpha_t}^{(1)}$. If $\overline{f}_{\alpha_t}^{(1)} \neq 0$ then from Lemma 3.3 (taking $s = 2$), we get $|f_{\alpha_t}^{(1)}| \geq 2\delta_{\mathcal{X}_{\hat{1}}}(d)$, a contradiction. Hence we must have $f_{\alpha_t}^{(1)} = f_{\alpha_1}^{(1)}$ for all $1 \leq t \leq d_1$ and the result follows from applying Proposition 3.2 to $f = f_{\alpha_1}^{(1)} \in C_{\mathcal{X}_{\hat{1}}}(d)$. $\qquad \square$

References

[1] S. Ballet, R. Rolland, On low weight codewords of generalized affine and projective Reed-Muller codes. Des. Codes Cryptogr. 73 (2014) 271–297.

[2] A. Bruen, Blocking sets in finite projective planes. SIAM J. Appl. Math. 21 (1971) 380–392.

[3] A. Bruen, Blocking sets and low-weight codewords in the generalized Reed-Muller codes. Error-correcting codes, finite geometries and cryptography, 161–164, Contemp. Math., 523, Amer. Math. Soc., Providence, RI, 2010.

[4] A. Bruen, Polynomial multiplicities over finite fields and intersection sets. J. Combin. Theory Ser. A 60 (1992) 19–33.

[5] A. Bruen, Applications of finite fields to combinatorics and finite geometries. Acta Appl. Math. 93 (2006) 179–196.

[6] C. Carvalho, On the second Hamming weight of some Reed-Muller type codes, Finite Fields Appl. 24 (2013) 88–94.

[7] C. Carvalho, Gröbner bases methods in coding theory. Contemp. Math. 642 (2015) 73–86.

[8] C. Carvalho and V.G.L. Neumann, The next-to-minimal weights of binary projective Reed-Muller codes. IEEE Transactions on Information Theory 62 (2016). 6300–6303.

[9] C. Carvalho and V.G.L. Neumann, On the next-to-minimal weight of affine cartesian codes. Finite Fields Appl. 44 (2017) 113–134.

[10] C. Carvalho and V.G.L. Neumann, On the next-to-minimal weight of projective Reed-Muller codes. Finite Fields Appl. 50 (2018) 382–390.

[11] C. Carvalho, V.G.L. Neumann and H.H. López, Projective nested cartesian codes. Bull. Braz. Math. Soc. (N.S.) 48 (2017) 283–302.

[12] J.P. Cherdieu, R. Rolland R., On the number of points of some hypersurfaces in $\mathbb{F}_q^n$. Finite Field Appl. 2 (1996) 214–224.

[13] P. Delsarte, J.M. Goethals, F.J. Mac Williams, On generalized Reed-Muller codes and their relatives, Inform. Control 16 (1970) 403–442.

[14] D. Erickson, Counting zeros of polynomials over finite fields. PhD Thesis, California Institute of Technology, Pasadena (1974).

[15] O. Geil, On the second weight of generalized Reed-Muller codes. Des. Codes Cryptogr. 48 (2008) 323–330.

[16] O. Geil, Erratum to: On the second weight of generalized Reed-Muller codes. Des. Codes Cryptogr. 73 (2014) 267–267.

[17] O. Geil, C. Thomsen, Weighted Reed-Muller codes revisited. Des. Codes Cryptogr. 66 (2013) 195–220.

[18] T. Kasami, S. Lin, W.W. Peterson, New generalisations of the Reed-Muller codes. Part I: Primitive codes, IEEE Trans. Inform. Theory IT-14 (2) (1968) 189–199.

[19] T. Kasami, N. Tokura, S. Azumi, On the weight enumeration of weights less than 2.5d of Reed-Muller codes. Inform. Control 30(4), 380–395 (1976).

[20] G. Lachaud, Projective Reed-Muller codes. Coding theory and applications (Cachan, 1986), 125–129, Lecture Notes in Comput. Sci., 311, Springer, Berlin, 1988.

[21] G. Lachaud, The parameters of projective Reed-Muller codes, Discrete Math. 81 (1990) 217–221.

[22] E. Leducq. A new proof of Delsarte, Goethals and Mac Williams theorem on minimal weight codewords of generalized Reed-Muller codes. Finite Fields Appl. 18 (2012) 581–586.

[23] E. Leducq. Second weight codewords of generalized Reed-Muller codes. Cryptogr. Commun. 5 (2013) 241–276.

[24] E. Leducq. On the third weight of generalized Reed-Muller codes. Discrete Math. 338 (2015) 1515–1535.

[25] H. H. López, C. Rentería-Márquez, R. H. Villarreal, Affine cartesian codes, Des. Codes Cryptogr. 71 (2014) 5–19.

[26] R.J. McEliece, Quadratic Forms Over Finite Fields and Second-Order Reed-Muller Codes, JPL Space Programs Summary 37-58, vol. III (1969) 28–33.

[27] D.-J. Mercier, R. Rolland, Polynômes homogènes qui s'annulent sur l'espace projectif $\mathbb{P}^m(\mathbb{F}_q)$, J. Pure Appl. Algebra 124 (1998) 227–240.

[28] D. Muller, Application of boolean algebra to switching circuit design and to error detection. IRE Tran. on Electronic Computers EC-3 n.3 (1954) 6–12

[29] I. S. Reed, A class of multiple-error-correcting codes and the decoding scheme. IRE Trans. Information Theory PGIT-4 (1954), 38–49.

[30] C. Rentería, H. Tapia-Recillas, Reed-Muller codes: an ideal theory approach, Commu. Algebra 25 (1997) 401–413.

[31] F. Rodier, A. Sboui, Les arrangements minimaux et maximaux d'hyperplans dans $\mathbb{P}^n(\mathbb{F}_q)$, C. R. Math. Acad. Sci. Paris 344 (2007) 287–290.

[32] F. Rodier, A. Sboui, Highest numbers of points of hypersurfaces over finite fields and generalized Reed-Muller codes. Finite Fields Appl. 14 (2008) 816–822.

[33] R. Rolland, Number of points of non-absolutely irreducible hypersurfaces. Algebraic geometry and its applications, 481–487, Ser. Number Theory Appl., 5, World Sci. Publ., Hackensack, NJ, 2008

[34] R. Rolland, The second weight of generalized Reed-Muller codes in most cases. Cryptogr. Commun. 2 (2010) 19–40.

[35] A. Sboui A., Second highest number of points of hypersurfaces in F_q^n. Finite Fields Appl. 13 (2007) 444–449.

[36] A. Sboui A., Special numbers of rational points on hypersurfaces in the n-dimensional projective space over a finite field, Discret. Math. 309 (2009) 5048-5059.

[37] J.P. Serre, Lettre à M. Tsfasman du 24 Juillet 1989. In: Journées arithmétiques de Luminy 17–21 Juillet 1989, Astérisque, 198–200. Société Mathématique de France (1991).

[38] A. Sørensen, Projective Reed-Muller codes, IEEE Trans. Inform. Theory 37 (1991) 1567–1576.

[39] S.G. Vlǎduţ, Yu.I. Manin, Linear codes and modular curves, J. Soviet Math. 30 (1985) 2611–2643.

Cícero Carvalho and Victor G.L. Neumann
Faculdade de Matemática
Universidade Federal de Uberlândia
Av. J. N. Ávila 2121, 38.408-902 - Uberlândia - MG, Brazil
cicero@ufu.br
victor.neumann@ufu.br

10

A Primer on Lax Pairs

L.M. Bates and R.C. Churchill

*This note is dedicated to Emma Previato on the occasion of
her 65th birthday.*

Abstract. Lax pairs are formulated in differential algebraic terms, their
usefulness in connection with integrals is reviewed, and a general approach
to their construction is suggested. No prior familiarity with these entities is
assumed. Concrete applications are included, the final one involving a curious
geodesic equation which is "almost" in Hamiltonian form w.r.t. an "almost"
symplectic structure.

When X is a smooth vector field on a manifold M a smooth scalar-valued
function g on M is an *integral* of X if $X(g) = 0$. Integrals need not exist,
but when they do Lax pairs can at times offer very elegant packaging. An
elegant example of this for quadratic Hamiltonians may be found in the volume
[A-N]. Unfortunately, useful Lax pairs can be difficult to detect. In this note
we suggest an elementary systematic construction which has proven successful
in several contexts of practical interest. Specific examples are presented. No
previous knowledge of Lax pairs is assumed.

Not covered here is the very important connection between Lax pairs and
algebraic integrability (see, e.g. [Gr] and [Mum]). Roughly speaking, when
one has a Lax pair for an algebraically integrable complex meromorphic
Hamiltonian vector field X_H one can realize the flows on certain level sets of
the Hamiltonian as being contained within complete linear flows on a Jacobi
variety associated with that pair. For a very general treatment of "completions"
of flows see [B].

Our approach is predominantly algebraic. The role of a smooth vector field
X acting on an algebra of smooth functions is subsumed by a derivation

$r \mapsto r'$ acting on an integral domain R of characteristic 0. When convenient this derivation is extended (by the quotient rule) to the quotient field K, and then on occasion to the polynomial algebra $K[y]$ and quotient field $K(y)$ (y being a single indeterminate) by defining $y' = 0$.

1 Basics

In this section n denotes a positive integer. Unless specifically stated to the contrary, "matrix" means an $n \times n$ matrix with entries in R, and "vector" means an $n \times 1$ matrix (i.e. a "column vector") with entries in R.

The derivation on R gives rise to a derivation on the ring of matrices by

$$M' := [m'_{ij}] \quad \text{if} \quad M = [m_{ij}]. \tag{1}$$

If I is the identity matrix, and if M is non-singular, then from $I' = 0$ we see by differentiating $I = MM^{-1}$ that

$$(M^{-1})' = -M^{-1}M'M^{-1}. \tag{2}$$

This identity is used repeatedly in all that follows.

Definition 1.1 *A pair (P, Q) of matrices is a Lax pair (for the given derivation) if*

$$Q' = [P, Q], \tag{3}$$

where $[P, Q]$ denotes the commutator $PQ - QP$. (In [C-F] the definition includes an additional condition, i.e. that $Q' \neq 0$.)

Examples 1.2 *In these examples (adapted from [Cu]) we identify $\mathbb{R}^6 \approx \mathbb{R}^3 \times \mathbb{R}^3$ with the tangent bundle $T(\mathbb{R}^3)$ and write a typical element $x = (x_1, x_2, \ldots, x_6) \in \mathbb{R}^6$ as $(x, v) = (x_1, x_2, x_3, v_1, v_2, v_3)$ accordingly. We denote the usual inner product of vectors $y, z \in \mathbb{R}^3$ by $\langle y, z \rangle$ and set $e_3 := (0, 0, 1) \in \mathbb{R}^3$. (The results also hold on $\mathbb{C}^6$ by assuming the symmetric [as opposed to Hermitian] inner product.)*

The differential equation

$$x' = v$$
$$v' = -e_3 + (\langle q, e_3 \rangle - \langle p, p \rangle)q$$

gives rise to the vector field

$$\widehat{X}(x, v) = \sum_{j=1}^{3} v_j \partial_{x_j} + (\langle q, e_3 \rangle - \langle v, v \rangle) \sum_{j=1}^{3} x_j \partial_{v_j} - \partial_{v_3}$$

on $T(\mathbb{R}^3)$ which in turn restricts to a vector field X on the tangent bundle $T(S^2) := \{(q, v) : |q|^2 = 1 \text{ and } \langle q, v \rangle = 0\}$ of the unit sphere.

The system is related to the Lagrange top (see [C-B, pp. 191–200]).

(a) *A Lax pair for X (and for $\widehat{X}$) is given by*

$$Q := \begin{bmatrix} 0 & 2(x_2v_1 - x_1v_2) \\ 2(x_2v_1 - x_1v_2) & 0 \end{bmatrix},$$

$$P := \begin{bmatrix} 0 & 1 \\ -\langle v, v \rangle + x_3 & 0 \end{bmatrix},$$

(b) *A second Lax pair (for both derivations) is given by*

$$Q := \begin{bmatrix} -2(x_1 + x_2)(v_1 + v_2) & 2(x_1 + x_2)^2 \\ -2(v_1 + v_2)^2 & 2(x_1 + x_2)(v_1 + v_2) \end{bmatrix},$$

$$P := \begin{bmatrix} 0 & 1 \\ -\langle v, v \rangle + x_3 & 0 \end{bmatrix}.$$

(c) *The tangent bundle $T(S^2)$ as defined above admits an obvious identification with the cotangent bundle $T^*(S^2)$, and the latter will be distinguished notationally from the former by replacing each occurrence of (x, v) with (q, p).*

   Assume the usual symplectic structure on $T^*(\mathbb{R}^3) \approx \mathbb{R}^6$ and consider the Hamiltonian system

$$q' = H_p$$
$$p' = -H_q$$

with Hamiltonian

$$H(q, p) = \frac{1}{2}|p|^2 + q_3.$$

The associated vector field on $T^*(\mathbb{R}^3)$ is

$$\widehat{X}_H = \sum_{j=1}^{3} p_j \partial_{q_j} - \partial_{p_3},$$

and the restriction $X_H := \widehat{X}_H|_{T^*(S^2)}$ is not identical with X because $T^*(S^2)$ is not an invariant manifold for $\widehat{X}_H$. The mechanical principle that is used to modify the vector field so that it becomes tangent to $T^*(S^2)$ is called D'Alembert's principle, and it is most easily incorporated by modifying the derivation using the Dirac bracket construction. The standard Poisson bracket is given by

$$\{f, g\} := \sum_{a=1}^{3} \frac{\partial f}{\partial q_a} \frac{\partial g}{\partial p_a} - \frac{\partial g}{\partial q_a} \frac{\partial f}{\partial p_a}$$

and the Hamiltonian derivation is $f' = X_H(f) = \{f, H\}$. *In order to view $T^*(S^2)$ as a constraint, we consider the two functions* $c_1(p, q) := |q|^2$ *and* $c_2(p, q) := \langle p, q \rangle$ *and observe that $T^*(S^2)$ is the joint level set $c_1 = 1$, $c_2 = 0$. Define c as the matrix with entries $c_{ij} = \{c_i, c_j\}$, and let the matrix $C := c^{-1}$. The standard Poisson bracket is modified to give the constrained bracket (see [C-B, p.295–298])*

$$\{f, g\}^* := \{f, g\} - \sum_{k,l=1}^{2} \{f, c_k\}C_{kl}\{c_l, g\}.$$

The constrained bracket has the property that the joint level sets $c_1 = \text{cst} \cap c_2 = \text{cst}$ are invariant manifolds for any Hamiltonian derivation. That is,

$$\{c_1, H\}^* = \{c_2, H\}^* \equiv 0.$$

It follows that

$$\{q_a, q_b\}^* = 0,$$

$$\{q_a, p_b\}^* = \delta_{ab} - \frac{1}{c_1}q_a q_b,$$

$$\{p_a, p_b\}* = \frac{1}{c_1}(p_a q_b - p_b q_a),$$

and so a short computation yields that for the Hamiltonian $H(q, p) = \frac{1}{2}|p|^2 + q_3$

$$q' = \{q, H\}^* = p_a - \frac{c_2}{c_1}q_a,$$

$$p' = \{p, H\}^* = -\delta_{3a} + \frac{1}{c_1}(c_2 p_a + (q_3 - |p|^2)q_a).$$

These equations simplify dramatically on the level set $c_1 = 1$, $c_2 = 0$. It is also important to note that this derivation does not coincide with the expression given by Richard Cushman, except on the constraint set. This is due to the fact that he cleverly multiplies by terms in the ideal generated by c_1 and c_2 so as to obtain a polynomial vector field on the ambient space. For this example it is easy to see that one can multiply by c_1, and then substitute the values $c_1 = 1$ and $c_2 = 0$. This leaves the vector field on the constraint set unchanged, but gives a more convenient expression to work with in the ambient space.

Major applications of Lax pairs are based on the following observation and the subsequent corollary. The results are particular consequences of work by Peter Lax on nonlinear evolution equations [Lax].

Theorem 1.3 *When $Q' = [P, Q]$ and m is an any positive integer the trace of Q^m is a constant, i.e.*

$$(\mathrm{tr}(Q^m))' = 0. \tag{4}$$

The result is obvious when $Q' = 0$, since every entry of Q is then a constant. The interesting cases occur when $Q' \neq 0$.

Proof. One has

$$(Q^m)' = \sum_{j=0}^{m} Q^{m-j} Q' Q^{j-1}$$

$$= \sum_{j=0}^{m} Q^{m-j}(PQ - QP)Q^{j-1}$$

$$= \sum_{j=0}^{m} \left(\left(Q^{m-1} P Q^{j-1} \right) Q - Q \left(Q^{m-1} P Q^{j-1} \right) \right),$$

and from the standard identities $\mathrm{tr}(A + B) = \mathrm{tr}(A) + \mathrm{tr}(B)$ and $\mathrm{tr}(AB) = \mathrm{tr}(BA)$ it follows that (i) holds. □

Note that $\mathrm{tr}(Q^m) = \sum_{j=1}^{n} \lambda_j^m$, where $\lambda_1, \lambda_2, \ldots, \lambda_n$ are the (not necessarily distinct) eigenvalues of Q. It is a classical result ("Newton's Formulas") that every elementary symmetric function of these eigenvalues can be expressed as polynomial (with integer coefficients) in these power sums, and vice-versa (see, e.g. [Usp, Chapter XI, §2, pp. 260–262]). It is in turn well-known that every symmetric function in R can be expressed as a polynomial (with integer coefficients) in these elementary symmetric functions. (These facts require our standing characteristic 0 hypothesis.) The two collections therefore generate the same subfield of K. Since the coefficients described in (b) of the following corollary comprise the elementary symmetric functions, that result is a direct consequence of (a).

Corollary 1.4 *For every Lax pair (P, Q) the following statements hold.*

(a) *Every element of the subfield of K generated by the symmetric functions of the eigenvalues $\lambda_1, \lambda_2, \ldots, \lambda_n$ of Q is a constant.*
(b) *The coefficients appearing in the characteristic polynomial of Q are constants.*

That the eigenvalues should be constants can be seen in the special case that we are working over the complex numbers. Suppose $Q(t)$ is a curve in a conjugacy class, so $Q(t) = M(t)CM^{-1}(t)$ for some $M(t)$, with $Q(0) = C$, and it follows from the Jordan normal form that the eigenvalues are invariant. Such a curve gives rise to a Lax pair as follows.

Proposition 1.5 *Let C and M be matrices such that*

$$C' = 0, \tag{5}$$

$$M' \neq 0, \tag{6}$$

and M is non-singular. Then for

$$Q := MCM^{-1} \tag{7}$$

and

$$P := M'M^{-1} \tag{8}$$

one has

$$Q' = [P, Q]. \tag{9}$$

In particular, (P, Q) is a Lax pair for the derivation.

Proof. One has

$$
\begin{aligned}
Q' &= M'CM^{-1} + MC'M^{-1} + MC(M^{-1})' \\
&= M'CM^{-1} + M \cdot 0 \cdot M^{-1} + MC(-M^{-1}M'M^{-1}) \quad \text{(by(i))} \\
&= M'M^{-1}MCM^{-1} + 0 - MCM^{-1}M'M^{-1} \\
&= PQ - QP \\
&= [P, Q].
\end{aligned}
$$

$\square$

Example 1.6 *If $H = \frac{1}{2}|p|^2 + q_3$ is the Hamiltonian of Example 1.2(c), and if*

$$
Q =
\begin{bmatrix}
0 & 0 & 0 & 0 & 0 \\
0 & 0 & 0 & 0 & 0 \\
0 & 0 & 0 & 0 & 0 \\
0 & 0 & 0 & 0 & 0 \\
0 & -H & q_1 \cdot H & 0 & H
\end{bmatrix}
+
\begin{bmatrix}
0 & 0 & 0 & 0 & 0 \\
0 & 0 & 0 & 0 & 0 \\
0 & 0 & 0 & 0 & 0 \\
12 & -12q_3 p_2 & 12q_1 q_3 p_2 & 12 & 0 \\
0 & 0 & 0 & 0 & 0
\end{bmatrix}
\cdot y
$$

$$
\begin{bmatrix}
0 & 0 & 0 & 0 & 0 \\
0 & 0 & q_1(q_1 p_2 - q_2 p_1) & 0 & 0 \\
0 & 0 & q_1 p_2 - q_2 p_1 & 0 & 0 \\
0 & 0 & 0 & 0 & 0 \\
0 & 0 & 0 & 0 & 0
\end{bmatrix}
\cdot y^2 +
\begin{bmatrix}
0 & q_3 p_2 & -q_1 q_3 p_2 & 0 & 0 \\
0 & 1 & -q_1 & 0 & 0 \\
0 & 0 & 0 & 0 & 0 \\
0 & 0 & 0 & 0 & 0 \\
0 & 1 & -1 & 0 & 0
\end{bmatrix}
\cdot y^3 +
$$

$$+ \begin{bmatrix} 1 & -q_3 p_2 & q_1 q_2 p_2 & 0 & 0 \\ 0 & 0 & 0 & 0 & 0 \\ 0 & 0 & 0 & 0 & 0 \\ -1 & q_3 p_2 & -q_1 q_3 p_2 & 0 & 0 \\ 0 & 0 & 0 & 0 & 0 \end{bmatrix} \cdot y^4 \ and \ P = \begin{bmatrix} 0 & p_2 p_3 & q_1 q_3 p_2 & 0 & 0 \\ 0 & 0 & p_1 & 0 & 0 \\ 0 & 0 & 0 & 0 & 0 \\ 0 & 0 & 0 & 0 & 0 \\ 0 & 0 & 0 & 0 & 0 \end{bmatrix},$$

then (P, Q) is a Lax pair for the Hamiltonian vector field X_H. This particular pair was constructed as in Proposition 1.5 using

$$C := \begin{bmatrix} y^4 & 0 & 0 & 0 & 0 \\ 0 & y^3 & 0 & 0 & 0 \\ 0 & 0 & (q_1 p_2 - q_2 p_1)y^2 & 0 & 0 \\ 0 & 0 & 0 & 12y & 0 \\ 0 & 0 & 0 & 0 & H \end{bmatrix} \ and \ M = \begin{bmatrix} 1 & p_2 p_3 & 0 & 0 & 0 \\ 0 & 1 & q_1 & 0 & 0 \\ 0 & 0 & 1 & 0 & 0 \\ -1 & 0 & 0 & 1 & 0 \\ 0 & 1 & 0 & 0 & 1 \end{bmatrix},$$

with the choice for M quite arbitrary. (Our particular choice was easy to invert, gave P a very simple form, and gave Q a non-trivial appearance. The characteristic polynomial of Q (which is obviously that of C),is

$$(\lambda - y^4)(\lambda - y^3)(\lambda - (q_1 p_2 - q_2 p_1))(\lambda - 12y)(\lambda - H),$$

which clearly displays the two fundamental integrals of X_H.

In practice constructing Lax pairs can appear to be somewhat of an art. We see from the preceding example how they provide information about integrals, but in that example the integrals were already known. When this is not the case we will offer some suggestions, with no claim regarding originality, indicating how to proceed.

2 Generalities

We continue with the notation and terminology used in the previous section.

For any given $n \times n$ matrix P one can produce matrix solutions to $Q' = [P, Q]$ by tensoring (column vector) solutions of the linear system

$$x' = Px \tag{10}$$

with solutions of the adjoint equation

$$y' = -P^\tau y \tag{11}$$

(where the τ denotes transposition). We recall the details.

The *tensor product* $x \otimes y$ of (column) vectors[1] x and y is defined to be the $n \times n$ matrix with ij-entry $x_i y_j$, i.e.

$$x \otimes y := [x_i y_j]. \tag{12}$$

If x and y satisfy (10) and

$$y' = Sy \tag{13}$$

respectively we see from (12), (10) and (13) that

$$\begin{aligned}(x \otimes y)' &= x \otimes y' + x' \otimes y \\ &= x \otimes (Sy) + (Px) \otimes y.\end{aligned} \tag{14}$$

Write $P = [p_{ij}]$ and $S = [s_{ij}]$. Focusing on the ij-entries of these last two terms we see that

$$(x \otimes (Sy))_{ij} = x_i (Sy)_j = x_i \sum_k s_{jk} y_k = \sum_k x_i y_k s_{jk}$$
$$= \sum_k [x \otimes y]_{ik} s_{jk} = ((x \otimes y)S^\tau)_{ij},$$

and that

$$((Px) \otimes y)_{ij} = (Px)_i y_j = \left(\sum_k p_{ik} x_k\right) y_j = \sum_k p_{ik} x_k y_j$$
$$= \sum_k p_{ik} [x \otimes y]_{kj} = (P(x \otimes y))_{ij},$$

and as a result of (14) that

$$(x \otimes y)' = P(x \otimes y) + (x \otimes y)S^\tau. \tag{15}$$

By choosing $S = -P^\tau$ we now see from (11) that

$$(x \otimes y)' = P(x \otimes y) - (x \otimes y)P = [P, x \otimes y], \tag{16}$$

and we have thereby produced solutions of $Q' = [P, Q]$.

To apply the procedure just described one would like to begin with P in a simple form. To achieve such a form we need to understand how a Lax pair transforms under a change of basis. See also proposition 1.5.

Proposition 2.1 *Suppose (P, Q) is a Lax pair and M is a nonsingular matrix. Define*

$$\widehat{Q} := MQM^{-1} \quad and \quad \widehat{P} := MPM^{-1} + M'M^{-1}. \tag{17}$$

Then $(\widehat{Q}, \widehat{P})$ is also a Lax pair.

---

[1] The same definition applies when y is an $m \times 1$-column vector, and all the results of the paragraph containing (13) remain true, but we have no need for the greater generality.

The mappings $(M, P) \mapsto MPM^{-1}$ and $(M, P) \mapsto M \cdot P := MPM^{-1} + M'M^{-1}$ are left actions of $GL(n, K)$ on the space of $n \times n$ matrices with entries in K. The second is exactly the gauge action, and can be used to construct cyclic vectors for a given matrix P [C-K]. As a consequence we see that if one can produce a Lax pair for a given derivation, then one can construct such a pair $(\widehat{Q}, \widehat{P})$ in which $\widehat{P}$ has the companion matrix form

$$
\begin{bmatrix}
0 & 1 & 0 & \cdots & 0 \\
0 & 0 & 1 & \ddots & \vdots \\
\vdots & \ddots & \ddots & \ddots & 0 \\
0 & 0 & \ddots & 0 & 1 \\
c_1 & c_2 & \cdots & c_{n-1} & c_n
\end{bmatrix}.
\tag{18}
$$

Proof. First note as a result of the differentiation formula $(M^{-1})' = -M^{-1}M'M^{-1}$ that

$$
M(M^{-1})' = -M'M^{-1}.
\tag{19}
$$

Now observe from the initial definition in (i) that

$$
M'QM^{-1} = M'M^{-1}MQM^{-1} = M'M^{-1}\widehat{Q}
\tag{20}
$$

and that

$$
MQ(M^{-1})' = MQM^{-1}M(M^{-1})' = \widehat{Q}M(M^{-1})'.
\tag{21}
$$

To finish the proof simply check that

$$
\begin{aligned}
\widehat{Q}' &= (MQM^{-1})' \\
&= M'QM^{-1} + MQ'M^{-1} + MQ(M^{-1})' \\
&= M'QM^{-1} + M[P, Q]M^{-1} + MQ(M^{-1})' \quad \text{(because } Q' = [P, Q]) \\
&= [MPM^{-1}, \widehat{Q}] + [M'M^{-1}, \widehat{Q}] \\
&= [MPM^{-1} + M'M^{-1}, \widehat{Q}] \\
&= [\widehat{P}, \widehat{Q}] \text{ (by (i))}.
\end{aligned}
$$

$\square$

The advantage of the companion matrix form for $\widehat{P}$ seen in (18) of Proposition 2.1 is that the first-order linear system

$$
x' = \widehat{P}x
\tag{22}
$$

is "equivalent" to the nth-order linear differential equation

$$y^{(n)} = \sum_{j=1}^{n} c_j y^{(j-1)} \tag{23}$$

in the sense that a column vector

$$x = \begin{bmatrix} x_1 \\ x_2 \\ \vdots \\ x_n \end{bmatrix} \tag{24}$$

of elements of K is a solution of (22) if and only if $y := x_1$ is a solution of (23) and

$$x_{j+1} = x_n^{(j)} \qquad \text{for} \qquad j = 1, 2, \ldots, n-1. \tag{25}$$

The nth-order form is often more amenable to standard methods for producing explicit solutions, e.g. that of 'undetermined coefficients'.

The analogue of the results of the previous paragraph for the adjoint equation

$$x' = -\widehat{P}^\tau x \tag{26}$$

of (22) is that (26) is "equivalent" to the n-order linear differential equation

$$(-1)^n y^{(n)} + (-1)^{(n-1)}(c_1 y)^{(n-1)} + \cdots + (-1)(c_{n-1} y)' + c_n y = 0 \tag{27}$$

in the sense that a column vector x as in (24) is a solution of (26) if and only if $y = x_n$ is a solution of (27) and

$$x_{n-j} = (-1)^j x_n^{(j)} + \sum_{i=0}^{j-1}(-1)^i(c_{j-i}x_n)^{(i)} \qquad \text{for} \qquad j = 1, 2, \ldots, n-1. \tag{28}$$

For obvious reasons one refers to equation (27) as the *adjoint equation* of (23). For example, the adjoint equation of the second-order linear differential equation

$$y'' - c_2 y' - c_1 y = 0 \tag{29}$$

is

$$y'' - c_1 y' + (c_2 - c_1')y = 0. \tag{30}$$

3 Particulars in the 2 × 2 Case

We begin with a generalization of Proposition 1.5 of [C-F].

Theorem 3.1 *For any matrices*

$$Q := \begin{bmatrix} * & 2q \\ * & * \end{bmatrix}, \qquad P := \begin{bmatrix} 0 & 1 \\ p & s \end{bmatrix}$$

over K (or $K(T)$) the following two assertions are equivalent:

(a) $Q' = [P, Q]$; and
(b) Q has the form

$$Q = \begin{bmatrix} -q' - qs + c & 2q \\ -q'' - sq' + (2p - s')q & q' + qs + c \end{bmatrix}, \tag{31}$$

wherein $c \in K$ is an arbitrary constant, and p, q and s satisfy

$$q''' - \big(4p + (s^2 - 2s')\big)q' - \big(2p' + \tfrac{1}{2}(s^2 - 2s')'\big)q = 0. \tag{32}$$

Moreover, when either assertion holds one has

(c) $Q' = 0$ *if and only if*

$$q' = 0 \quad \text{and either} \quad \begin{cases} q = 0 \ \text{ or} \\ p' = 0 = s'; \end{cases} \tag{33}$$

(d) *the characteristic polynomial of Q is*

$$\lambda^2 - 2c\lambda + 2qq'' - (q')^2 + (2s' - s^2 - 4p)q^2 + c^2; \tag{34}$$

and

(e) *the entity*

$$2qq'' - (q')^2 + (2s' - s^2 - 4p)q^2 \tag{35}$$

is a constant.

If a Hamiltonian derivation X_H admits a 2×2 Lax pair, then by Proposition 2.1 it admits one in which P has the indicated form. In that sense Theorem 3.1 covers all possibilities for 2×2 Lax pairs.

Of course there is no guarantee that equation (33) of Theorem 3.1(c) admits non-trivial solutions. However, if K posesses a non-constant element t and an element s satisfying $s' = \tfrac{1}{2}s^2$ then one can easily check that $p := t''/t$, $q := t^2$ and s will provide such a solution. Unfortunately, for this choice the constant (35) of Theorem 3.1(e) is 0, so this particular solution gives no information regarding previously undetected constants of the derivation.

The 2 appearing in the matrix Q is a normalization which simplifies the calculations in the following proof.

Proof. By writing Q in the form

$$Q = \begin{bmatrix} x & 2q \\ y & z \end{bmatrix} \tag{36}$$

one sees directly that

$$Q' = [P, Q] \quad \Leftrightarrow \quad \begin{bmatrix} x' & 2q' \\ y' & z' \end{bmatrix} = \begin{bmatrix} y - 2pq & z - x - 2sq \\ p(x-z) + sy & 2pq - y \end{bmatrix}. \tag{37}$$

(a) $\Rightarrow$ (b) : By comparing the diagonal entries in (37) one sees that $z' = -x'$, hence that $(x+z)' = 0$, and as a result that

$$2c := x + z \tag{38}$$

is a constant. (The 2 is another convenient normalization.) Comparing the 1,2-entries in (37) gives $2q' = z - x - 2sq$, hence $-2q' - 2sq + 2c = -(z-x) + x + z = 2x$, and therefore

$$x = -q' - sq + c. \tag{39}$$

Equality (38) now gives $2c = (pq' - sq + c) + z$; hence that

$$z = q' + sq + c. \tag{40}$$

From (40) and the 2,2-entries in (37) we have $2pq - y = z' = (q'+sq+c)' = q'' + sq' + s'q$, leading immediately to

$$y = -q'' - sq' + (2p - s')q. \tag{41}$$

This establishes (31) of (b) for the constant c arising in (38). However, since for any constant k one has $Q' = [P, Q]$ if and only if $(Q + kI)' = [P, Q + kI]$, it follows that (31) of (b) holds for any constant c.

From (39)–(41) we can write the equality of the lower left entries in (37) as

$$
\begin{aligned}
y' &= p(x - z) + sy \\
&= p\big((-q' - sq + c) - (q' + sq + c)\big) + s\big(-q'' + 2pq - sq' - s'q\big) \\
&= -2pq' - 2spq - sq'' + 2spq - s^2q' - ss'q \\
&= -sq'' - (2p + s^2)q' - ss'q.
\end{aligned}
$$

On the other hand, if we compute y' from the lower left entry of Q (as expressed in (31) of (b)) we obtain

$$\left\{
\begin{aligned}
y' &= (-q'' + 2pq - (sq)')' \\
&= -q''' + (2pq)' - (sq)'' \\
&= -q''' + 2p'q + 2pq' - s''q - 2s'q' - sq'' \\
&= -q''' - sq'' + 2(p - s')q' + (2p' - s'')q.
\end{aligned}
\right. \tag{42}$$

Combining the two calculations for y' gives

$$-q''' - sq'' + 2(p - s')q' + (2p' - s'')q = -sq'' - (2p + s^2)q' - ss'q,$$
(43)

which quickly reduces to the linear differential equation exhibited in (33) of (c).

(b) $\Rightarrow$ (a): From (b) and (i) we see that

$$x = -q' - sq + c$$

$$y = -q'' - sq' + (2p - s')q$$

$$z = q' + sq + c,$$

whence from (37) that

$$
\begin{cases}
Q' = \begin{bmatrix} x' & (2q)' \\ y' & z' \end{bmatrix} \\[2mm]
= \begin{bmatrix} (-q' - sq + c)' & 2q' \\ (-q'' - sq' + (2p - s')q)' & (q' + sq + c)' \end{bmatrix} \\[2mm]
= \begin{bmatrix} -q'' - sq' - s'q & 2q' \\ -q''' - sq'' + 2(p - s')q' + (2p' - s'')q & q'' + sq' + s'q \end{bmatrix}
\end{cases}
$$
(44)

and that $[P, Q]$ is equal to

$$
\begin{bmatrix} y - 2pq & z - x - 2sq \\ p(x - z) + sy & 2pq - y \end{bmatrix}
$$

$$
= \begin{bmatrix} (-q'' - sq' + (2p - s')q) - 2pq & 2q' + 2sq - 2sq \\ p(-2q' - 2sq) + s(-q'' - sq' + (2p - s')q) & 2pq - (-q'' - sq' + (2p - s')q) \end{bmatrix}
$$

$$
= \begin{bmatrix} -q'' - sq' - s'q & 2q' \\ -sq'' - (2p + s^2)q' - ss'q & q'' + sq' + s'q \end{bmatrix}.
$$

With a second appeal to (37) we conclude that $Q' = [P, Q]$ if and only if (43) holds, which we have previously noted is equivalent to (37).

(c) By (44) we have

$$
Q' = \begin{bmatrix} -q'' - sq' - s'q & 2q' \\ -q''' - sq'' + 2(p - s')q' + (2p' - s'')q & q'' + sq' + s'q \end{bmatrix},
$$

from which one sees immediately that a necessary condition for $Q' = 0$ is that

$$q' = 0.$$
(45)

Since $q' = 0$ implies $q'' = q''' = 0$, assuming (45) reduces Q' to

$$Q' = \begin{bmatrix} -s'q & 0 \\ (2p' - s'')q & s'q \end{bmatrix},$$

in which case one sees that $Q' = 0$ if and only if $q = 0$ or $s' = p' = 0$.

(d) The verification is by straightforward calculation.

(e) This is immediate from (b) of Corollary 1.4. ☐

Corollary 3.2 *If there is a 2 × 2 Lax pair* (P, Q) *with P and Q having the forms appearing in* Theorem 3.1, *then* $(\widehat{P}, Q)$ *is also a Lax pair, where*

$$\widehat{P} = \begin{bmatrix} 0 & 1 \\ p + \frac{1}{4}(s^2 - 2s') & 0 \end{bmatrix}. \tag{46}$$

The practical consequence is: when attempting to produce a 2 × 2 Lax pair (P, Q) for a derivation, one should begin by assuming P and Q have the forms

$$P = \begin{bmatrix} 0 & 1 \\ p & 0 \end{bmatrix} \quad \text{and} \quad Q = \begin{bmatrix} -q' + c & 2q \\ -q'' + 2pq & q' + c \end{bmatrix}, \tag{47}$$

preferably with the constant c equal to 0.

Proof. By Theorem 3.1 one has $Q' = [P, Q]$ if and only if

$$q''' - \left(4p + (s^2 - 2s')\right)q' - \left(2p' + \frac{1}{2}(s^2 - 2s')'\right)q = 0,$$

which by defining $\widehat{p} := p + \frac{1}{4}(s^2 - 2s')$ may be expressed as

$$q''' - 4\widehat{p}q' - 2\widehat{p}'q = 0. \tag{48}$$

Since $\widehat{P} = \begin{bmatrix} 0 & 1 \\ \widehat{p} & 0 \end{bmatrix}$, we conclude from (48) and a second appeal to Theorem 3.1 that $Q' = [\widehat{P}, Q]$. To establish the final assertion simply note that the Lax pair condition for a pair (P, Q) satisfying $Q' = [P, Q]$ is a requirement on Q, and the same Q occurs in both pairs under consideration. ☐

Corollary 3.3 *Suppose P and Q are matrices of the forms*

$$P = \begin{bmatrix} 0 & 1 \\ p & 0 \end{bmatrix} \quad \text{and} \quad Q = \begin{bmatrix} * & 2q \\ * & * \end{bmatrix}, \tag{49}$$

and $\text{trace}(Q) = 0$. *Then* (P, Q) *is a Lax pair for the derivation if and only if*

$$q''' = 4pq' + 2p'q, \tag{50}$$

and when that is the case

$$Q = \begin{bmatrix} -q' & 2q \\ -q'' + 2pq & q' \end{bmatrix}. \tag{51}$$

Moreover, the entity

$$2qq'' - (q')^2 - 4pq^2 \tag{52}$$

will be a constant.

If $t \in K$ is a non-constant then $p := t''/t$ and $q := t^2$ will satisfy (50) of Corollary 3.3, but the constant appearing in (52) will be 0. To obtain interesting constants one may need to pass to a differential extension field of K or $K(y)$.

Corollary 3.3 reduces the problem of constructing 2×2 Lax pairs to that of producing solutions $(q, p) = (a, b)$ of

$$\begin{aligned} q''' &= 4pq' + 2p'q \\ &= 2(pq' + (pq)'). \end{aligned} \tag{53}$$

In practice one soon realizes that the rewards are greater if one can uncover solutions in $K[y] \setminus K$ (or $K(y) \setminus K$) rather than in K, where $K[y]$ is a differential ring extension of K generated by an indeterminate y satisfying $y' = 0$. One therefore looks for solutions of (53) of the form

$$a = a_0 + a_1 y + a_2 y^2 + \cdots \qquad \text{and} \qquad b = b_0 + b_1 y + b_2 y^2 + \cdots. \tag{54}$$

In the examples we restrict attention to the case

$$a = a_0 + a_1 y + y^2 \qquad \text{and} \qquad b = b_0 + b_1 y + y^2. \tag{55}$$

Of course this barely scratches the surface of possibilities, but it is hopefully sufficient to both convey the basic ideas and to generate a reasonable collection of examples. If solutions a and b of the stated form are uncovered, the constant (52) of Corollary 3.3 will be the product of $-\frac{1}{4}$ and the polynomial

$$\left\{ \begin{aligned} & y^6 + (2a_1 + b_1)y^5 + (a_1^2 + 2a_1b_1 + 2a_0 + b_0)y^4 \\ & \quad + (2a_0(a_1 + b_1) + 2a_1b_0 + a_1^2b_1 - \tfrac{1}{2}a_1'')y^3 \\ & \quad + (a_1^2b_0 + 2a_0b_0 - a_1''a_1 + \tfrac{1}{4}(a_1')^2 + a_0^2 - \tfrac{1}{2}a_0'' + 2a_0a_1b_1)y^2 \\ & \quad + (a_0^2b_1 - \tfrac{1}{2}a_0a_1' - \tfrac{1}{2}a_0''a_1 + \tfrac{1}{2}a_0'a_1' + 2a_0a_1b_0)y \\ & \quad + a_0^2b_0 + \tfrac{1}{4}(a_0')^2 - \tfrac{1}{2}a_0a_0''. \end{aligned} \right. \tag{56}$$

In this instance (53) reduces to

$$\left\{ \begin{aligned} q_0''' + q_1''' y = {} & 2\big(p_0 q_0' + (p_0 q_0)'\big) \\ & + \big(4 p_1 q_0' + 2 p_1' q_0 + 4 p_0 q_1' + 2 p_0' q_1\big) y \\ & + \big(4 q_0' + 2 p_0' + 4 p_1 q_1' + 2 p_1' q_1\big) y^2 \\ & + \big(4 q_1' + 2 p_1'\big) y^3, \end{aligned} \right. \tag{57}$$

and the task is therefore to solve the four equations

$$\left\{ \begin{aligned} \text{(a)} \quad & q_0''' = 2\big(p_0 q_0' + (p_0 q_0)'\big) \\ \text{(b)} \quad & q_1''' = 2\big(p_1 q_0' + (p_1 q_0)'\big) + 2\big(p_0 q_1' + (p_0 q_1)'\big) \\ \text{(c)} \quad & 0 = 2 q_0' + p_0' + 2 p_1 q_1' + p_1' q_1 \\ \text{(d)} \quad & 0 = 2 q_1' + p_1'. \end{aligned} \right. \tag{58}$$

Note that when p_0, p_1, q_0, q_1 satisfy (58) one has

$$\begin{aligned} 2 p_1 q_1' + p_1' q_1 &= 2 p_1 (-\tfrac{1}{2} p_1') + (-2 q_1') q_1 \quad \text{(by (d))} \\ &= -p_1 p_1' - 2 q_1 q_1' \\ &= -\big(\tfrac{1}{2}(p_1^2) + (q_1^2)\big)', \end{aligned}$$

and equations (58) are therefore equivalent to

$$\left\{ \begin{aligned} \text{(a)} \quad & q_0''' = 2\big(p_0 q_0' + (p_0 q_0)'\big) \\ \text{(b)} \quad & q_1''' = 2\big(p_1 q_0' + p_0 q_1' + (p_1 q_0 + p_0 q_1)'\big) \\ \text{(c)} \quad & 2(q_0 + p_0)' = \big(\tfrac{1}{2}(p_1^2) + (q_1^2)\big)', \\ \text{(d)} \quad & 2 q_1' + p_1' = 0. \end{aligned} \right. \tag{59}$$

Note that $p_0, p_1, q_0, q_1 = $ cst is a solution of these equations. We can summarize the previous paragraph as follows.

Proposition 3.4 *Suppose (P, Q) is a Lax pair as in (49) of Corollary 3.3 and a a and b as in (55) are solutions $((a, b) \sim (q, p))$ of (58). Then the y-coefficients appearing in (56) are constants of the derivation.*

In practice one might want a known constant c to be the coefficient of, say, y^2. Then from (56) one sees that one should seek a solution (a, b) of (58) which also satisfies

$$a_1^2 b_0 + 2 a_0 b_0 - a_1'' a_1 + \frac{1}{4}(a_1')^2 + a_0^2 - \frac{1}{2} a_0'' + 2 a_0 a_1 b_1 = c.$$

4 Examples

4.1 The Hénon-Heiles Family

The title refers to the two-degree of freedom Hamiltonian system on $\mathbb{R}^4$ defined by

$$H(q, p) = \frac{1}{2}|p|^2 + \frac{1}{2}(Aq_1^2 + Bq_2^2) + \frac{1}{3}q_1^3 + Lq_1q_2^2, \qquad L \geq 0, \qquad (60)$$

with the symplectic structure being the usual one. The corresponding vector field X_H is integrable for $L = 0, \frac{1}{6}$ and 1. The system is non-integrable for all other values of L, although establishing this for the case $L = \frac{1}{2}$ required a significant generalization of the non-integrability method used with the remaining parameter values (see [M-R-S].) In this section we construct Lax pairs as in Corollary 3.3 and the associated integrals (see (52) of that corollary) for the integrable cases by specifying solutions $(q, p) = (a_0, b_0)$ of the equation

$$q''' = 4pq' + 2p'q. \qquad (61)$$

All these solutions were obtained by first expressing a_0 and b_0 as polynomials of total degree at most 2 in q_1, q_2, p_1, p_2 with undetermined coefficients, substituting these into (61), and choosing one solution (generally of many) of the resulting equations determined by MAPLE. Another approach to Lax pairs in this family was described by Ravoson, Gavrilov and Caboz in [R-G-C].

4.1.1 The Case $L = 0$

The choices

$$a_0 := \tfrac{\sqrt{2}}{2}(p^2 + 2 + Bq_2^2)$$
$$b_0 := \tfrac{1}{2}(p_1^2 + Bq_2^2)$$

for (q, p) result in the Lax pair

$$Q := \begin{bmatrix} 0 & \frac{\sqrt{2}}{2}(p_2^2 + Bq_2^2) \\ \frac{\sqrt{2}}{2}((p_2^2 + Bq_2^2) & 0 \end{bmatrix},$$

$$P := \begin{bmatrix} 0 & 1 \\ \frac{1}{2}(p_2^2 + Bq_2^2) & 0 \end{bmatrix}$$

and the functionally independent (from H) integral

$$(p_2^2 + Bq_2^2)^3$$

for X_H.

The special case $B = 0$ of this example is Example 3 in §2 of [C-F].

4.1.2 The Case $L = 1/6$

This case is Example 1 in §2 of [C-F]. The choices

$$a_0 := Y^2 + \tfrac{1}{6}q_1 Y - \tfrac{1}{64}(4B - A)^2 + \tfrac{1}{48}(4B - A)q_1 - \tfrac{1}{144}q_2^2$$
$$b_0 := Y - \tfrac{1}{3}q_1 - \tfrac{1}{8}(4B + A)$$

for (q, p) result in a rather cumbersome Lax pair which we do not record. The corresponding integral ((52) of Corollary 3.3) is 0, whereas one can extract the independent integral

$$\frac{3}{2}(4B - A)(Bq_2^2 + p_2^2) + Bq_1 q_2^2 + p_2(q_2 p_1 - q_1 p_2) + \frac{1}{6}q_2^2\left(q_1^1 + \frac{1}{4}q_2^2\right)$$

from the constant term of the characteristic polynomial. (See Example 1 in §2 of [C-F].)

4.1.3 The Case $L = 1$

This is also covered in [C-F] (it is Example 2 of §2), although in a slightly different way. The choice

$$a_0 = q_1 + q_2$$
$$b_0 = -\tfrac{1}{3}q_1 + \tfrac{1}{3}q_2 - \tfrac{1}{4}B$$

only satisfies (61) if $B = A$, and we therefore impose that restriction. One then obtains

$$Q = \begin{bmatrix} p_1 - p_2 & -2q_1 + 2q_2 \\ -\tfrac{1}{3} + \tfrac{1}{6}(-6A + 3B + 4q_2)q_1 + \tfrac{B}{2}q_2 - \tfrac{1}{3}q_2^2 & p_2 - p_1 \end{bmatrix},$$

$$P = \begin{bmatrix} 0 & 1 \\ \tfrac{1}{3}(q_2 - q_1) - \tfrac{B}{4} & 0 \end{bmatrix},$$

and the corresponding integral (52) of Corollary 3.3 is

$$\frac{1}{4}(p_1 - p_2)^2 + \frac{1}{12}(q_1 - q_2)^2(3A + 2(q_1 - q_2)),$$

which is easily checked to be functionally independent of H.

4.1.4 The Case $L = 1/2$

This is by far the most interesting choice for the parameter. When $A = B = 1$ the corresponding complex analytic flow passes the Painlevé test and admits a completion [B, §5], both of which suggest integrability, and yet the complex system is not integrable [M-R-S].

The value $L = 1/2$ arises in our context when one attempts a Lax pair construction for (60), at first with no restriction on L, using polynomials a_0 and

b_0 of total degree at most 2 in q_1, q_2, p_1, p_2 with undetermined coefficients. Substitution into

$$q''' = 4pq' + 2p'q \tag{62}$$

leads to a collection of relations between these coefficients which must be satisfied if (a_0, b_0) is a solution. One machine-produced solution imposes no restriction on L, whereas in all others one has either $L = 0, L = 1/6$, or $L = 1$. The solution with no restrictions leads to the choice

$$a_0 = (1/2)q_2^2$$
$$b_0 = -B - 2q_1,$$

and when this is substituted into (61) the result is $(2 - 4L)q_1q_2^2 + (8 - 16L)p_1q_1q_2 = 0$, thereby forcing $L = \frac{1}{2}$. Unfortunately, for the corresponding Lax pair

$$Q := \begin{bmatrix} -p_2q_2 & q_2^2 \\ -p_2^2 & p_2q_2 \end{bmatrix}, \qquad P := \begin{bmatrix} 0 & 1 \\ -B - q_1 & 0 \end{bmatrix}$$

the characteristic polynomial $\det(Q - TI)$ is $T^2 + 4(1 - 2L)q_1q_2^4 = T^2$, and as a result this particular pair offers no information about integrals other than those generated by H. The integral given by (52) of Corollary 3.3 is also 0. If one wants the characteristic polynomial to return H one can replace Q above by $\widehat{Q} = \begin{bmatrix} -2p_2q_2 + H & 2q_2^2 \\ -2p_2^2 & 2p_2q_2 + H \end{bmatrix}$, in which case that polynomial becomes $T^2 - 2HT + H^2$.

4.2 A Curious Connection

Consider the connection in the plane that is defined by declaring that the two vector fields

$$X = \cos x \frac{\partial}{\partial x} + \sin x \frac{\partial}{\partial y}, \quad Y = -\sin x \frac{\partial}{\partial x} + \cos x \frac{\partial}{\partial y}$$

form an absolute parallelism. The corresponding second order equation for the geodesics of this connection is

$$\ddot{x} + \dot{x}\dot{y} = 0, \qquad \ddot{y} - \dot{x}^2 = 0.$$

Since the equation is invariant by translation in x and y, one can reduce the order and immediatey integrate using hyperbolic functions. One can also note that since the connection is a metric connection for the standard Euclidean metric in the plane (but not the Levi-Civita connection!) the quantity

$$L = \frac{1}{2}(\dot{x}^2 + \dot{y}^2)$$

is a first integral. It is of interest to note that this system may be also written in the almost-Hamiltonian form

$$X \lrcorner \, \omega = dH$$

with Hamiltonian $H = \frac{1}{2}(p_1^2 + p_2^2)$ and almost-symplectic form

$$\omega = p_1 dq_1 \wedge dq_2 + dq_1 \wedge dp_1 + dq_2 \wedge dp_2$$

where $p_1 = \dot{x}$, $q_1 = x$, $p_2 = \dot{y}$ and $q_2 = y$. The interest here in the integrability is that neither of the symmetry vector fields ∂/∂_x or ∂/∂_y is the Hamiltonian vector field of a momentum function, which is what one would expect in the Hamiltonian theory.

A Lax pair for this equation can be constructed by using

$$q = (5M + 2)p_1 + p_2, \qquad p = Mp_1^2 + Mp_1 p_2 - \left(M + \frac{1}{4}\right)p_2^2$$

where M is either (complex) root of the quadratic $5z^2 + 4z + 1$. By (52) of corollary 3.3 we obtain the integral

$$(p_1^2 + p_2^2)^2.$$

This work was completed while one of us (RC) was a visiting professor at the University of Calgary.

References

[A-N] V. Arnol'd and S. Novikov, *Dynamical systems IV*, Encyclopedia of mathematical sciences, volume 4, Springer, 1989.

[B] A. Baider, *Completion of Vector Fields and the Painlevé Property*, J. Diff. Equations, **97** (1992), pp. 27–53.

[C-F] R.C. Churchill and G. Falk, Lax pairs in the Hénon-Heiles and related families, in *Hamiltonian Dynamical Systems: History, Theory, and Applications* (H.S. Dumas, K.R. Meyer and D.S. Schmidt, eds.), The IMA Volumes in Mathematics and Its Applications **63**, Springer-Verlag, New York, 1995.

[C-K] R.C. Churchill and J.J. Kovacic, Cyclic Vectors, in *Differential Algebra and Related Topics*, (Li Guo, P. Cassidy, W. Keigher and W. Sit, eds.), World Scientific, Singapore, 2002.

[C-B] R.H. Cushman and L.M. Bates, *Global Aspects of Classical Integrable Systems*, Birkhäuser, Basel, 1997.

[Cu] R. Cushman, *A new Lax pair for the spherical pendulum*, Letter to L. Gavrilov, 1990.

[Gr] P. Griffiths, *Linearizing Flows and a Cohomological Interpretation of Lax Equations*, Am. J. Math. **107** (1985), pp, 1445–85.

[Lax] P. Lax, *Integrals of nonlinear equations of evolution and solitary waves*, Comm. Pure Applied Math., **21** (5) (1968), 467–490.

[M-R-S] J.J. Morales-Ruiz, J.-P. Ramis & C. Simó, Integrability of Hamiltonian systems and differential Galois groups of higher variational equations, *Ann. Sci. Éc. Norm. Supér.* (4) 40 No. 6 (2007), 845–884.

[Mum] D. Mumford, *Tata Lectures on Theta II*, Progress in Math. 43, Birkhäuser.

[R-G-C] V. Ravoson, L. Gavrilov and R. Caboz, *Separability and Lax pairs for Hénon-Heiles system*, Journal of Mathematical Physics 34, (1993) 2385–2393.

[Usp] J.V. Uspensky, *Theory of Equations*, McGraw-Hill, New York, 1948.

Larry Bates
Department of mathematics
University of Calgary
Calgary, AB
Canada
bates@ucalgary.ca

Richard Churchill
Department of mathematics
Hunter College and The Graduate Center, CUNY
New York, NY
USA
rchurchi@hunter.cuny.edu

11

Lattice-Theoretic Characterizations of Classes of Groups

Roland Schmidt

Dedicated to Emma Previato on her 65th birthday

Abstract. One of the most interesting problems in the field Subgroup Lattices of Groups is to find lattice-theoretic characterizations of interesting classes of groups. After explaining this problem by looking at the classical characterizations of cyclic and of finite soluble groups, we first present lattice-theoretic characterizations of some classes of infinite soluble groups. Then for a finite group G we try to determine in its subgroup lattice $L(G)$ the Fitting length of G and properties defined by arithmetical conditions. For this we use some new ideas to determine minimal normal subgroups and the orders of minimal subgroups in $L(G)$.

1 Introduction

For every group G, we shall denote by $L(G)$ the lattice of all subgroups of G. A lattice-theoretic characterization of a class $\mathfrak{X}$ of groups is some class $\mathfrak{Y}$ of lattices such that for every group G,

$$G \in \mathfrak{X} \text{ if and only if } L(G) \in \mathfrak{Y}. \tag{1}$$

The oldest result of this type is the following beautiful

Theorem (Ore 1938) *The group G is locally cyclic if and only if $L(G)$ is distributive.*

Actually, Ore did not so much want to characterize the class of locally cyclic groups but rather to determine the groups with distributive subgroup

2010 *Mathematics Subject Classification.* Primary 20D30; Secondary 20E15
Key words and phrases. Group, subgroup lattice, classes of groups, lattice-theoretic characterizations

lattice; but as stated, this and the following corollary contain lattice-theoretic characterizations of interesting classes of groups.

Corollary *The group G is cyclic if and only if L(G) is distributive and satisfies the maximal condition. In particular, a finite group G is cyclic if and only if L(G) is distributive.*

If G and $\bar{G}$ are groups, an isomorphism from $L(G)$ onto $L(\bar{G})$ is also called a *projectivity* from G to $\bar{G}$. And we say that a class $\mathfrak{X}$ of groups is *invariant under projectivities* if for all groups G and $\bar{G}$ with $L(G) \simeq L(\bar{G})$,

$$G \in \mathfrak{X} \text{ implies } \bar{G} \in \mathfrak{X}. \tag{2}$$

It is trivial that the existence of a class $\mathfrak{Y}$ of lattices satisfying (1) implies (2) for $\mathfrak{X}$. And if $\mathfrak{X}$ satisfies (2), then the class $\mathfrak{Y}$ of all lattices which are isomorphic to $L(H)$ for some $H \in \mathfrak{X}$ clearly satisfies (1). So, in fact, both properties of $\mathfrak{X}$ are equivalent. However, we are looking for a lattice-theoretic description of such a class $\mathfrak{Y}$ of lattices satisfying (1) as in Ore's theorem or rather its corollary: distributive and maximal condition. In general, it is much easier to prove that a class $\mathfrak{X}$ is invariant under projectivities. So the usual situation is that this is known for $\mathfrak{X}$ and we are looking for a lattice-theoretic description of some class $\mathfrak{Y}$ satisfying (1).

But there are also important classes of groups which are not invariant under projectivities, for example, the classes of abelian groups, nilpotent groups, p-groups for some prime p, and many more. For the classes mentioned this is already shown by the elementary abelian group of order 9 and the nonabelian group of order 6 which have isomorphic subgroup lattices. For such group classes $\mathfrak{X}$, it is interesting to determine the *projective closure* $\mathfrak{X}^*$ of $\mathfrak{X}$, that is, the class of all groups which are images of groups in $\mathfrak{X}$ under projectivities. Clearly, $\mathfrak{X}^*$ is the smallest class of groups containing $\mathfrak{X}$ which is invariant under projectivities. We shall also try to find lattice-theoretic characterizations of such classes $\mathfrak{X}^*$.

Most of our notation is standard and can be found in [9]; for definitions and properties concerning lattices and subgroup lattices we refer to [11]. In particular, if L is a complete lattice, the smallest and the largest element of L are denoted by 0 and I, respectively, a minimal nontrivial element of L is an atom of L, elements with the dual property are called antiatoms, and for $a, b \in L$ such that $a \leq b$, we let $[b/a] = \{x \in L \mid a \leq x \leq b\}$. If G is a group, we write $H \cup K$ for the group generated by the subgroups H and K of G; we call G L-indecomposable if $L(G)$ is directly indecomposable. If G is finite, p a prime and π a set of primes, we write $O_p(G)$ (and $O_\pi(G)$) for the largest normal p-subgroup (π-subgroup) of G and $O^p(G)$ (and $O^\pi(G)$) for the smallest normal subgroup N of G with G/N a p-group (π-group). Finally, for the definition and structure of P-groups, the groups lattice-isomorphic to noncyclic elementary abelian groups, we refer the reader to Section 2.2 of [11].

2 The Translation Method

There is a rather natural way for trying to obtain a lattice-theoretic characterization of a class $\mathfrak{X}$ of groups.

Remark 2.1 We take a suitable definition of the groups in $\mathfrak{X}$ and replace concepts appearing in it by lattice-theoretic concepts that are equivalent to them or nearly so. That is, we somehow "translate" the group-theoretic definition of $\mathfrak{X}$ into lattice theory and hope to get a lattice-theoretic characterization of $\mathfrak{X}$ this way.

Since a group G is finite if and only if so is $L(G)$, we translate "finite group" by "finite lattice"; and in view of Ore's theorem we should translate "cyclic factor group H/K" by "distributive interval $[H/K]$ in $L(G)$ satisfying the maximal condition". But in many definitions there appear properties of normal subgroups and abelian factor groups. As mentioned in the introduction, these concepts cannot be translated into lattice theory. But we can approximate them by lattice-theoretic concepts which we recall from [11, 2.1].

Definition 2.2 *Let L be a lattice.*

(a) *L is modular if for all $x, y, z \in L$ the modular law holds:*

$$\text{If } x \leq z, \text{ then } x \cup (y \cap z) = (x \cup y) \cap z.$$

(b) *An element $m \in L$ is modular in L, if it satisfies the modular law in the place of x and also in place of y.*

(c) *A subgroup M of a group G is called modular in G if M is modular in $L(G)$.*

It is easy to see that every normal subgroup of a group G is modular in $L(G)$ and that every abelian group has modular subgroup lattice. But it is difficult to describe the deviation of modular subgroups from normality, and the structure of finite groups with modular subgroup lattice is rather complicated. However, the first nontrivial result on modular subgroups of finite groups states that if M is a maximal proper modular subgroup of G, then either $M \trianglelefteq G$ or G/M_G is nonabelian of order pq for primes p and q [11, 5.1.2]. This immediately yields the following classical result in which the translation method works in its purest form and for three different definitions of the class $\mathfrak{X}$.

Theorem 2.3 (Schmidt 1968) *The following properties of the finite group G are equivalent.*

(a) G is soluble.
(b) There are subgroups G_i of G such that $1 = G_0 \leq \cdots \leq G_r = G$, G_i is
    modular in G and $[G_{i+1}/G_i]$ is modular $(i = 0, \ldots, r - 1)$.
(c) There are subgroups G_i of G such that $1 = G_0 \leq \cdots \leq G_s = G$, G_i is
    modular in G_{i+1} and $[G_{i+1}/G_i]$ is modular $(i = 0, \ldots, s - 1)$.
(d) There are subgroups G_i of G such that $1 = G_0 < \cdots < G_t = G$, G_i is
    maximal and modular in G_{i+1} $(i = 0, \ldots, t - 1)$.

The classes of finite simple, perfect, and supersoluble groups can be characterized in similar ways. However, the situation is much more complicated
for infinite groups. The main reason for this is the existence of the so called
Tarski groups, first examples of which were constructed in 1979 by A. Yu.
Olshanskii. These are infinite groups G for which every nontrivial, proper
subgroup has prime order. Then $L(G)$ consists of G, 1, and an infinite antichain
of subgroups of prime order; hence $L(G)$ is modular but G is far from
being soluble. Therefore in order to extend the translation method to infinite
soluble groups one has to add some finiteness conditions to the translations of
normal subgroups and abelian groups. It is easily seen that normal and even
permutable subgroups, that is, subgroups $H \leq G$ such that $HX = XH$ for
every $X \leq G$, have the following property: if $X \leq G$ is cyclic and $[U/V]$
is a finite interval between HX and H, then V has finite index in U. This
motivated G. Zacher to introduce in [19] the following concept which I called
"permodular" in [11] since it actually translated "permutable subgroup" into
lattice theory.

Definition 2.4 *A subgroup M of the group G is permodular in G if*

(1) *M is modular in G and*
(2) *for every cyclic subgroup X of G and every $Y \leq G$ such that
    $M \leq Y \leq M \cup X$ and $[M \cup X/Y]$ is a finite lattice, the index
    $|M \cup X : Y|$ is finite.*

By Ore's theorem, we can detect cyclic subgroups of G in $L(G)$ and since
there exist lattice-theoretic characterizations of the finiteness of the index of
a subgroup [11, p. 287], condition (2) can be regarded as a lattice-theoretic
property. The translation of "abelian group" in the infinite case is simpler.

Definition 2.5 *A lattice L is permodular if it is modular and for all $a, b \in L$
such that $b \leq a$, $[a/b]$ is a finite lattice whenever it has finite length.*

It is easy to show (see [11, 6.2.2 and 6.4.3]) that every permutable, in
particular every normal, subgroup is permodular in G, and $L(G)$ is permodular
if and only if every subgroup of G is permodular in G. Furthermore, if G is

abelian, then $L(G)$ is permodular; conversely, if $L(G)$ is permodular, then G is metabelian. Therefore it was possible to use the translation method with these concepts to generalize the characterizations of the classes of finite groups mentioned above to the classes of infinite simple, perfect, supersoluble, and hyperabelian groups and to obtain characterizations of other related classes of groups as well. Unfortunately, however, the method did not yield the desired result for the class of soluble groups, the most interesting case. On the one hand, a result of Emma Previato from 1978 in this new terminology took the following form.

Theorem 2.6 (Previato 1978) *Let G be a finitely generated group. Then G is soluble if and only if there exist subgroups M_i of G such that $1 = M_0 \leq \cdots \leq M_n = G$, M_i is permodular in G and $[M_{i+1}/M_i]$ is permodular ($i = 0, \ldots, n-1$).*

On the other hand, already in 1973, S. Stonehewer [15] had constructed a nonsoluble group G with a permutable subgroup M such that $[M/1]$ and $[G/M]$ are permodular. So we see that the translation method using permodular subgroups and lattices worked perfectly for finitely generated but not for arbitrary soluble groups. And we had the "usual" situation mentioned in the introduction: B. V. Yakovlev [17] had shown already in 1970 that the class of soluble groups is invariant under projectivities, but a lattice-theoretic characterization was not known. This problem finally was solved by G. Zacher.

Theorem 2.7 (Zacher 1999) *The group G is soluble if and only if there exist a group lattice L and elements $a, b \in L$ with the following properties.*

(a) *a and b are permodular in L, $a \cap b = 0$, $a \cup b = I$, $[a/0] \simeq L(G)$ and $[b/0] \simeq L(\mathbb{Q})$ where $\mathbb{Q}$ is the additive group of rational numbers.*

(b) *There exists a finite chain $0 = s_0 \leq s_1 \leq \cdots \leq s_n = a$ such that s_i is permodular in $[b \cup s_{i+1}/0]$ and $[s_{i+1}/s_i]$ is permodular for all $i = 0, \ldots, n-1$.*

To realize that this indeed is a lattice-theoretic characterization of the class of soluble groups, one first of all has to know that there are lattice-theoretic characterizations of free groups and their normal subgroups which together yield a lattice-theoretic characterization of subgroup lattices of groups; this is due to Yakovlev [18] and can be found in [11, pp. 345–358]. So "group lattice" is a lattice-theoretic concept. Furthermore, $L(\mathbb{Q})$ can be characterized by the fact that it is distributive and every nontrivial interval $[H/0]$ and $[\mathbb{Q}/K]$ in it is infinite. Finally, permodular elements in lattices are defined as in 2.4 for subgroup lattices where one has to replace the group-theoretic terms "cyclic subgroups" and "finite index" by their lattice-theoretic characterizations. But clearly, it would be nice if we could solve the following

Problem 2.8 *Find lattice-theoretic approximations of "normal subgroup" and "abelian group" for which the translation method yields a lattice-theoretic characterization of the class of infinite soluble groups.*

3 The Classes $\mathfrak{N}^k$ and $\mathfrak{N}^{k-1}\mathfrak{A}$ for $k \geq 2$

It is well-known that many classes of finite soluble groups are invariant under projectivities among them, for $2 \leq k \in \mathbb{N}$, the classes $\mathfrak{N}^k$ of groups with nilpotent length at most k and $\mathfrak{N}^{k-1}\mathfrak{A}$ of groups with commutator subgroup in $\mathfrak{N}^{k-1}$. We refer the reader to [11, 7.5] where one can also find lattice-theoretic characterizations for some of these classes. However, for the classes $\mathfrak{N}^k$ and $\mathfrak{N}^{k-1}\mathfrak{A}$ it was an open problem to find such characterizations, even in the simplest cases $\mathfrak{N}^2$ of nilpotent-by-nilpotent and $\mathfrak{N}\mathfrak{A}$ of nilpotent-by-abelian groups. In [2] M. Costantini and G. Zacher used their work on the subgroup lattice index problem to characterize the classes $\mathfrak{N}^k$. After seeing their paper, I found a simpler method to characterize the classes $\mathfrak{N}^{k-1}\mathfrak{A}$ which, in fact, also worked for the other problem. To describe this solution let us start to look at the class $\mathfrak{N}$ of finite nilpotent groups. Since every such group is the direct product of its Sylow subgroups and since the subgroup lattice of a direct product of groups with pairwise coprime orders is the direct product of the subgroup lattices of these direct factors, we only have to look at groups of prime power order. We define some classes of lattices which nearly describe the subgroup lattices of these groups.

Definition 3.1 *Let L be a finite lattice and write $\Phi(L)$ for the intersection of all the antiatoms in L. We say that*

(a) $L \in \mathcal{K}$ if L is a chain,
(b) $L \in \mathcal{P}$ if $L \notin \mathcal{K}$ and L is modular, complemented, and directly indecomposable,
(c) $L \in \mathcal{Q}$ if $L \notin \mathcal{K} \cup \mathcal{P}$ and L has the following properties.

   (i) L is lower semimodular.
   (ii) Every interval in L is directly indecomposable.
   (iii) $[I/\Phi(L)] \in \mathcal{P}$.
   (iv) Let $q + 1$ be the number of atoms in an interval of length 2 in $[I/\Phi(L)]$. If $q > 2$, then for every $a \in L$, either $[a/0] \in \mathcal{K}$ or the number m of atoms in $[a/0]$ satisfies $m \equiv 1 + q \pmod{q^2}$.

(d) $L \in \mathcal{P}_p$ (respectively, $L \in \mathcal{Q}_p$) for a prime p if $L \in \mathcal{P}$ (respectively, $L \in \mathcal{Q}$) and L has an interval isomorphic to M_{p+1}, the lattice of length 2 with $p + 1$ atoms.

The following results are well-known.

Theorem 3.2 *Let G be a finite group and p be a prime.*

(a) $L(G) \in \mathcal{K}$ if and only if G is cyclic of prime power order.
(b) $L(G) \in \mathcal{P}$ if and only if G is a P-group.
(c) $L(G) \in \mathcal{Q}$ if and only if G is a group of prime power order which is neither cyclic nor elementary abelian.
(d) $L(G) \in \mathcal{P}_p$ if and only if $G \in P(n, p)$ for some $n \in \mathbb{N}$ with $n \geq 2$.
(e) $L(G) \in \mathcal{Q}_p$ if and only if G is a p-group which is neither cyclic nor elementary abelian.

Proof. See [11, 1.2.7] for (a), [11, 2.4.3 and 1.6.4] for (b), and [16, p. 352] for (c); (d) and (e) are trivial consequences. □

This theorem solves the problem mentioned at the end of the introduction for the classes of groups of prime power order and for $\mathfrak{N}$. We only note

Corollary 3.3 *The class of finite direct products of lattices in $\mathcal{K} \cup \mathcal{P} \cup \mathcal{Q}$ is a lattice-theoretic characterization of the projective closure $\mathfrak{N}^*$ of $\mathfrak{N}$.*

So it is clear that $\mathcal{K}$, $\mathcal{P}$, and $\mathcal{Q}$ have to be used for the desired lattice-theoretic characterizations of $\mathfrak{N}^k$ and $\mathfrak{N}^{k-1}\mathfrak{A}$. But first of all we should note that it is not difficult to see how to get such characterizations inductively for every $k \geq 2$ if we have those for $\mathfrak{N}^2$ and $\mathfrak{N}\mathfrak{A}$, respectively. Therefore this is the crucial case and here we have two well-known theorems of W. Gaschütz that help. The first says that all classes $\mathfrak{X}$ considered are saturated formations [5]. Therefore $G \in \mathfrak{X}$ if and only if $G/\Phi(G) \in \mathfrak{X}$ where $\Phi(G)$ is the Frattini subgroup of G, that is, the intersection of all the maximal subgroups of G. Clearly, $\Phi(G) = \Phi(L(G))$ can be determined in $L(G)$ and therefore we may assume that $\Phi(G) = 1$. The other theorem of Gaschütz [4] says that in such a group the Fitting subgroup $F(G)$, the maximal nilpotent normal subgroup of G, is the product of the abelian minimal normal subgroups of G. And here we again can use our translation method.

Definition 3.4 *For a finite lattice L, we define $\mathcal{M}(L)$ to be the set of all $m \in L$ such that $m \neq 0$, m is modular in L and $[m/0]$ is a modular lattice. Let $\mathcal{M}(L)_*$ be the set of minimal elements in $\mathcal{M}(L)$ and define*

$$\mathcal{F}(L) = \bigcup \{m \mid m \in \mathcal{M}(L)_*\}.$$

Using well-known properties of modular subgroups of finite groups [11, 5.1 and 5.2] it is not difficult to obtain the following results.

Lemma 3.5 *Let G be a finite group.*

(a) *If N is an abelian minimal normal subgroup of G, then $N \in \mathcal{M}(L(G))_*$.*
(b) *Let $\Phi(G) = 1$, $L(G/F(G)) \in \mathcal{P} \cup \mathcal{Q}$ and assume that there exists $S \leq G$ such that $L(S) \in \mathcal{P} \cup \mathcal{Q}$ and $G = F(G)S$. Then M is a minimal normal subgroup of G if and only if $M \in \mathcal{M}(L(G))_*$.*

Since every P-group lies in $\mathfrak{N}\mathfrak{A}$, it suffices to characterize P-indecomposable groups, that is, groups which have no P-group as a coprime direct factor. In such a group, $F(G)$ is invariant under projectivities [11, 4.3.2]; and if $\Phi(G) = 1$, Gaschütz's theorem can be translated into a lattice-theoretic characterization of $F(G)$.

Proposition 3.6 *If G is P-indecomposable and $\Phi(G) = 1$, then $F(G) = \mathcal{F}(L(G))$.*

For the final characterization of $\mathfrak{N}\mathfrak{A}$ and $\mathfrak{N}^2$ we clearly only have to look at directly indecomposable lattices and therefore only at L-indecomposable groups. By Gaschütz's theorem, we may assume that $\Phi(G) = 1$ and hence by Proposition 3.6, $F(G)$ is visible in $L(G)$. Clearly, G can only be contained in $\mathfrak{N}^2$ if $G/F(G) \in \mathfrak{N}^*$, and hence, by Corollary 3.3, $[G/F(G)]$ is a direct product of lattices out of $\mathcal{K} \cup \mathcal{P} \cup \mathcal{Q}$. So by [11, 1.6.5], we finally have to decide for every one of the corresponding direct factors $G_i/F(G)$ whether it is abelian (respectively, nilpotent) or not. For this we need the concept of a centralizer in a lattice L. Recall that for $a, b \in L$, (a, b) is a distributive pair in L if the distributive law $c \cap (a \cup b) = (c \cap a) \cup (c \cap b)$ is satisfied for every $c \in L$. Then for $x, y \in L$ we let

$$C_{L;x}(y) = \bigcup \{z \leq x \mid z \cap y = 0 \text{ and } (z, y) \text{is a distributive pair in} L\}.$$

It is easy to show that if $H, K \leq G$, then $C_{L(G);H}(K) = O^{\pi}(C_H(K))$ where $\pi = \pi(K)$ is the set of prime divisors of $|K|$. We also write $C_{L(G)}(K)$ for $C_{L(G);G}(K)$. Now we can state the final result for $\mathfrak{N}\mathfrak{A}$ and $\mathfrak{N}^2$; for a proof see [12].

Theorem 3.7 *Suppose that G is a finite L-indecomposable group with $\Phi(G) = 1$ and that $L(G) \notin \mathcal{P}$. Write $F = \mathcal{F}(L(G))$ and suppose that*

$$[G/F] \simeq L_1 \times \cdots \times L_n$$

with directly indecomposable lattices L_i; let $\sigma : L_1 \times \cdots \times L_n \to [G/F]$ be an isomorphism and let $L_i{}^{\sigma} = G_i/F$.

(a) $G \in \mathfrak{M}\mathfrak{A}$ *if and only if for every* $i \in \{1, \ldots, n\}$, *either* $L_i \in \mathcal{K}$ *or there exists* $S_i \leq G_i$ *such that*

(i) $G_i = F \cup S_i$ *and* $L(S_i) \in \mathcal{P} \cup \mathcal{Q}$,

(ii) $[S_i/C_{L(G_i);S_i}(M)] \in \mathcal{K}$ *for all* $M \in \mathcal{M}(L(G_i))_*$ *with* $M \cap S_i = 1$, *and*

(iii) $\bigcap \{C_{L(G_i);S_i}(M) \mid M \in \mathcal{M}(L(G_i))_* \text{ and } M \cap S_i = 1\} = F \cap S_i$.

(b) $G \in \mathfrak{N}^2$ *if and only if for every* $i \in \{1, \ldots, n\}$, $L_i \in \mathcal{K} \cup \mathcal{P} \cup \mathcal{Q}$ *and if* $L_i \in \mathcal{P}$, *then there exists* $S_i \leq G_i$ *such that* (i)–(iii) *hold.*

4 *p*-Atoms

In this section we want to determine in $L(G)$, as far as possible, the orders of the subgroups of a finite group G. For this we use a method which was introduced in [3] and further developed in [1] and [14] where the notion of a p-atom (p a prime) was introduced. This is an atom $a \in L$ satisfying one of certain lattice-theoretic conditions which guarantee that if $\sigma : L \to L(G)$ is an isomorphism, then a^σ is a subgroup of order p in G. Clearly, if the orders of all minimal subgroups of $H \leq G$ are known, then a subgroup of H maximal with respect to having only minimal subgroups of order p is a Sylow p-subgroup P of H and its order is p^d where d is the length of $L(P)$; so we can determine the order $|H|$ of H in $L(G)$.

Unfortunately, the definition of a p-atom is rather complicated. By Theorem 3.2, the order of a minimal subgroup X of G is visible in $L(G)$ if X is contained in a (Sylow-) subgroup P with $L(P) \in \mathcal{Q}$. But if the Sylow subgroups of G containing X are cyclic or elementary abelian, we need additional information to be able to determine $|X|$ in $L(G)$. This is the idea in the next three definitions and in the final definition of a p-atom.

Definition 4.1 *Let x be an atom in L. We say that*

(a) $x \in \mathcal{P}_L(p)$ *if there exists $h \in L$ such that $x \leq h$ and $[h/0] \in \mathcal{P}_p$,*

(b) $x \in \mathcal{Q}_L(p)$ *if there exists $h \in L$ such that $x \leq h$ and $[h/0] \in \mathcal{Q}_p$,*

(c) $x \in \mathcal{S}_L(p)$ *if there exists $y \in L$ such that $[y/0]$ is a chain, $\phi(y) \neq 0$, $[x \cup \phi(y)/0]$ is distributive and $[x \cup y/\phi(y)] \simeq M_{p+1}$.*

Lemma 4.2 *If $X \in \mathcal{Q}_{L(G)}(p)$ or $X \in \mathcal{S}_{L(G)}(p)$, then $|X| = p$.*

Proof. As mentioned above, it is clear that $|X| = p$ if $X \in \mathcal{Q}_{L(G)}(p)$. The other assertion was proved in [3] and I want to present this proof in order to show how the properties in the definition of $\mathcal{S}_{L(G)}(p)$ and, similarly, in the other definitions work. So suppose that $X \in \mathcal{S}_{L(G)}(p)$. Then by definition there exists $Y \leq G$ such that $L(Y)$ is a chain of length at least 2 and $L(X \cup \Phi(Y))$

is distributive. Thus Y is cyclic of prime power order and by Ore's theorem, $X \cup \Phi(Y)$ is cyclic. Since $[X \cup Y/\Phi(Y)] \simeq M_{p+1}$, $X \not\leq Y$ and hence $X \cup \Phi(Y) = X \times \Phi(Y)$; in particular, X and Y have coprime orders. By [11, 2.2.4], $X \cup Y/\Phi(Y)$ is nonabelian of order pq where $p > q \in \mathbb{P}$. If X would normalize Y, it would centralize it since $X \cup \Phi(Y)$ is abelian; but $X \cup Y/\Phi(Y)$ is nonabelian. It follows that $|Y/\Phi(Y)| = q$ and $|X| = p$.   □

Condition (c) of Definition 4.1 is all we need if the Sylow p-subgroups of G are cyclic. It remains to consider the case that the Sylow p-subgroups are elementary abelian and not cyclic. Then there exists $S \leq G$ such that $X \leq S \in P(2, p)$; and in the next definition we present lattice-theoretic conditions which imply that in such a situation $|X| = p$.

Definition 4.3 *Let x be an atom in L. We say that $x \in \mathcal{U}_L(p)$ if there exist $s, h \in L$ such that $x < s \leq h$, $[s/0] \simeq M_{p+1}$ and one of the following holds.*

(a) *For every atom $y \leq s$ with $x \neq y$ there exists a prime $q < p$ such that $y \in \mathcal{P}_L(q) \cup \mathcal{Q}_L(q) \cup \mathcal{S}_L(q)$.*
(b) *$[x/0]$ is a maximal chain in $[h/0]$ and for every atom $y \leq s$ with $x \neq y$ there exists $u \leq h$ such that $y < u$ and $[u/0]$ is a chain.*
(c) *For every atom $y \leq s$ with $x \neq y$, we have $C_{[h/0]}(x) < C_{[h/0]}(y)$.*
(d) *x is modular in $[h/0]$ and no other atom y in $[s/0]$ is modular in $[h/0]$.*

Lemma 4.4 *If $X \in \mathcal{U}_{L(G)}(p)$, then $|X| = p$.*

Proof. This is trivial since every one of these four conditions (a)–(d) implies that X cannot be conjugate in S or H to any other minimal subgroup of S. But S is elementary abelian of order p^2 or nonabelian of order pr with $p > r \in \mathbb{P}$; hence X would have p conjugates in S if $|X| \neq p$.   □

We come to the third and last property that enters into the definition of a p-atom. Here we have to use some deeper results on modular subgroups of finite groups.

Definition 4.5 *Let x be an atom in L. We say that $x \in \mathcal{W}_L(p)$ if there exists $s \in L$ such that $x < s$, $[s/0] \in \mathcal{P}_p$ and one of the following holds.*

(a) *$p = 2$.*
(b) *There exist elements $m_i, y_i, h_i \in L$ such that $s = \bigcup\{m_i \mid i = 1, \dots, n\}$ and for all $i \in \{1, \dots, n\}$,*

  (i) *$y_i \leq s \leq h_i$,*
  (ii) *m_i is a minimal nontrivial modular element in $[h_i/0]$,*
  (iii) *y_i is not modular in $[h_i/0]$.*

Lemma 4.6 *If $X \in \mathcal{W}_{L(G)}(p)$, then $|X| = p$.*

Proof. This is clear if (a) holds since every group S with $L(S) \in \mathcal{P}_2$ is an elementary abelian 2-group. Suppose that S, M_i, X_i, H_i satisfy (b). Then the structure theorem on modular subgroups of finite groups [11, 5.1.14] implies that if M_i is not a p-group, then $H_i = M_i{}^{H_i} \times K_i$ where $M_i{}^{H_i}$ is a P-group and $(|M_i{}^{H_i}|, |K_i|) = 1$. Since $Y_i \leq S = M_i{}^S \leq M_i{}^{H_i}$, Y_i is modular in H_i, a contradiction. Hence all M_i are p-groups and so is the P-group S they generate. Since $X \leq S$, $|X| = p$. $\qquad\square$

Definition 4.7 *Let p be a prime, L a finite lattice, and G a finite group.*

(a) *An element $x \in \mathcal{Q}_L(p) \cup \mathcal{S}_L(p) \cup \mathcal{U}_L(p) \cup \mathcal{W}_L(p)$ is called a p-atom of L.*

(b) *A subgroup X of G is a p-atom of G if X is a p-atom of $L(G)$.*

Proposition 4.8 *If X is a p-atom of G, then $|X| = p$.*

Proof. This follows from Lemmas 4.2, 4.4, and 4.6. $\qquad\square$

In general, of course, if $X \leq H \leq G$ and X is a p-atom of G, then X need not be a p-atom of H - for example, if $X = H$. But since $\mathcal{X}_{L(H)}(p) \subseteq \mathcal{X}_{L(G)}(p)$ for all $\mathcal{X} \in \{\mathcal{Q}, \mathcal{S}, \mathcal{U}, \mathcal{W}\}$, we have the following useful property.

Lemma 4.9 *If $H \leq G$ and X is a p-atom of H, then X is a p-atom of G.*

Now the question is: Are these p-atoms really a good approximation to subgroups of order p? The following answer was given in [14].

Theorem 4.10 *Let p be a prime, G a finite group, and suppose that $X \leq G$ with $|X| = p$. Then X is not a p-atom of G if and only if one of the following holds.*

(1) *$p > 2$ and $G = S \times T$ where $S \in \mathcal{P}_p$ and $(|S|, |T|) = 1$.*

(2) *G is p-nilpotent with cyclic Sylow p-subgroups.*

(3) *$p > 2$, G is p-nilpotent, and if $X \leq P \in Syl_p(G)$, then P is elementary abelian, $P = (P \cap Z(G)) \times X$ and $P \cap Z(G) = O_p(G) \neq 1$.*

Remark 4.11 The p-atoms of the groups in (3) are the minimal subgroups of $O_p(G)$. The groups in (1) and (2) do not possess p-atoms; the only other such groups are the p'-groups.

The proof of Theorem 4.10 is too long to be presented here. The theorem describes exactly the bad situations in which there exist subgroups of order p which are not p-atoms. It shows that in these cases the structure of G is very much restricted. Therefore we can usually handle these exceptional situations in which the concepts of "p-atom" and "subgroup of order p" do not coincide.

Remark 4.11 in particular describes all finite groups which do not contain p-atoms and states that the groups in (1) and (2) are the only groups of this type with order divisible by p.

The obvious application of Theorem 4.10 is in the study of the problem mentioned at the beginning of this section, the so called subgroup lattice index problem. For this, we introduce the following concept.

Definition 4.12 *Let L be a lattice, G a group.*

(a) An element h of L is called regular in L if every atom of $[h/0]$ is a p-atom of L for some prime p.

(b) $Reg(L) = \bigcup \{h \in L \mid h$ is regular in $L\}$.

(c) A subgroup H of G is called regular in G if H is regular in $L(G)$; furthermore, $R(G) := Reg(L(G))$.

As mentioned above, it is clear that if H is regular in G, then we can determine the order of every subgroup of H in $L(G)$. An easy consequence of Theorem 4.10 is

Theorem 4.13 *Let G be a finite group.*

(a) $R(G)$ is regular in G.

(b) If G is P-indecomposable, then $R(G)$ has a cyclic complement in G.

Part (a) of the theorem shows that $R(G)$ is the unique maximal subgroup of G in which our p-atom method works, that is, in which it is possible to compute orders and indices of subgroups using p-atoms. And part (b) shows that $R(G)$ in general is rather large.

5 $O_p(G)$, $O_{p'}(G)$, p-Soluble Groups, and $\mathfrak{N}^*$

In this final section we want to use our translation method with "p-atom" as approximation to "subgroup of order p". This works very well in a number of different situations. We just report some of the results which are contained in [14]. By Sylow's theorem, a subgroup of a finite group is a p-group if every minimal subgroup of it has order p, and it is a p'-group if no minimal subgroup of it has order p. The obvious translation is given in (a) of

Definition 5.1 *Let p be a prime, L a finite lattice, and G a finite group.*

(a) An element x of L is called a p-element of L if every atom $a \in L$ which is contained in x is a p-atom of L; x is a p'-element of L if no atom $a \in L$ with $a \leq x$ is a p-atom of L.

(b) $O_p(L)$ is the intersection of all maximal p-elements and $O_{p'}(L)$ is the intersection of all maximal p'-elements of L.

(c) *For short, a p-element of $L(G)$ is called a p^*-subgroup of G and a*
   p'-element of $L(G)$ is a $\bar{p}$-subgroup of G; furthermore,
   $O_{p^*}(G) := O_p(L(G))$ and $O_{\bar{p}}(G) := O_{p'}(L(G))$.

Clearly, (b) translates $O_p(G)$ and $O_{p'}(G)$. Using Theorem 4.10 and Remark 4.11, it is easy to determine $O_{p^*}(G)$ and $O_{\bar{p}}(G)$. If G has no p-atom, then the trivial subgroup is the unique p^*-subgroup and G is a $\bar{p}$-subgroup of G; hence $O_{p^*}(G) = 1$ and $O_{\bar{p}}(G) = G$. If G is p-*regular* (that is, every subgroup of order p of G is a p-atom), then a subgroup H of G is a p-group (respectively, a p'-group) if and only if H is a p^*-subgroup (respectively, a $\bar{p}$-subgroup) of G. Hence, trivially, $O_{p^*}(G) = O_p(G)$ and $O_{\bar{p}}(G) = O_{p'}(G)$. Finally, if G is p-singular (that is, not p-regular) and has a p-atom, then by Remark 4.11, (3) of Theorem 4.10 holds. Therefore the p^*-subgroups of G are the subgroups of $O_p(G)$ and hence $O_{p^*}(G) = O_p(G)$. And every maximal $\bar{p}$-subgroup of G has the form NX where N is the normal p-complement of G and X has order p but is not a p-atom. Therefore $O_{\bar{p}}(G) = N = O_{p'}(G)$ and we have proved the following theorem of which (a) is a lattice-theoretic characterization of $O_p(G)$ and $O_{p'}(G)$ in groups having a p-atom.

Theorem 5.2 *Let G be a finite group.*

(a) *If G has a p-atom, then $O_{p^*}(G) = O_p(G)$ and $O_{\bar{p}}(G) = O_{p'}(G)$.*
(b) *If G has no p-atom, then $O_{p^*}(G) = 1$ and $O_{\bar{p}}(G) = G$.*

Recall that a finite group G is p-soluble if its upper p-series

$$1 = P_0 \trianglelefteq N_0 \trianglelefteq P_1 \trianglelefteq \cdots \trianglelefteq P_m \trianglelefteq N_m \trianglelefteq \ldots \tag{3}$$

defined by $N_i/P_i = O_{p'}(G/P_i)$ and $P_{i+1}/N_i = O_p(G/N_i)$ reaches G. In this case, the length of this series, the number of p-quotients P_{i+1}/N_i with $N_i < G$ in it, is called the p-length of G and denoted by $l_p(G)$. It is well-known [11, p. 183] that the class of p-soluble groups is invariant under projectivities and that if the p-length $l_p(G) \geq 2$, then it is preserved by every projectivity. If we use $O_{p^*}(G)$ and $O_{\bar{p}}(G)$ to translate the above definition, we obtain

Definition 5.3 *The group G is p^*-soluble if its upper p^*-series*

$$1 = P_0^* \trianglelefteq N_0^* \trianglelefteq P_1^* \trianglelefteq \cdots \trianglelefteq P_m^* \trianglelefteq N_m^* \trianglelefteq \ldots \tag{4}$$

defined by $N_i^*/P_i^* = O_{\bar{p}}(G/P_i^*)$ and $P_{i+1}^*/N_i^* = O_{p^*}(G/N_i^*)$ reaches G. In this case, the length of this series, the number of quotients P_{i+1}^*/N_i^* with $N_i^* < G$ in it, is called the p^*-length of G and denoted by $l_{p^*}(G)$.

Here we have to count every factor group P_{i+1}^*/N_i^* even if it is trivial. Unlike in the upper p-series this can happen in the upper p^*-series without this series becoming stationary. For example, if G is the semidirect product of a group P of order p with a cyclic group Q of order q^2 inducing an automorphism of order q in P (p and q primes), then $P \in S_{L(G)}(p)$ is a p-atom and hence $N_0^* = \Omega(Q)$, but $P_1^* = N_0^*$ and $N_1^* = G$ since $G/\Omega(Q)$ has no p-atom. Thus G is p^*-soluble and $l_{p^*}(G) = 1 = l_p(G)$.

Theorem 5.4 *The finite group G is p-soluble if and only if it is p^*-soluble.*

This is a lattice-theoretic characterization of the class of finite p-soluble groups. If G has a p-atom, the proof of Theorem 5.4 also yields a lattice-theoretic characterization of $l_p(G)$. It turns out that in this situation either $l_p(G) = l_{p^*}(G)$ or $l_p(G) = l_{p^*}(G) + 1$ and it can be determined in $L(G)$ which of the two equations holds for G. This even is a slight improvement of the result on projectivities of p-soluble groups mentioned above since if $l_p(G) \geq 2$, then by Remark 4.11, G clearly has a p-atom.

We finally look at lattice-theoretic characterizations of the projective closure $\mathfrak{N}^*$ of $\mathfrak{N}$. A first one, translating the fact that the finite nilpotent groups are precisely the groups which are direct products of their Sylow subgroups, was given in Corollary 3.3. It is a remarkable fact that if we translate other standard definitions of $\mathfrak{N}$, like the normalizer condition or the existence of a central series, we do not get lattice-theoretic characterizations of $\mathfrak{N}^*$; see [11, 5.3.10 and 5.3.11] and [13]. But there are further possible definitions which we can try to translate. For example, a finite group G also is nilpotent if and only if for every prime p, the set of p-subgroups (p'-subgroups) of G has a unique maximal element. Here we can translate p-subgroup and p'-subgroup by p^*-subgroup and $\bar{p}$-subgroup, respectively, and obtain quite different results. If r and q are primes, every semidirect product $G = R \rtimes Q$ of a nonabelian r-group R by a cyclic q-group Q has the considered property for p^*-subgroups (the maximal q^*-subgroup is the trivial group) but G only belongs to $\mathfrak{N}^*$ if the operation of Q on R is trivial. For $\bar{p}$-subgroups, the situation is better.

Theorem 5.5 *The following properties of a finite group G are equivalent.*

(a) $G \in \mathfrak{N}^*$.
(b) *For every $p \in \mathbb{P}$, $O_{p^*}(G)O_{\bar{p}}(G) = G$.*
(c) *For every $p \in \mathbb{P}$, the set of $\bar{p}$-subgroups of G has a unique maximal element.*

Finally, G is nilpotent if and only if any two elements of coprime (prime power) order in G commute. To translate this property, we call a p-element (p'-element) x of a lattice L a p-chain (p'-chain) of L if $[x/0]$ is a chain; and a p-chain (p'-chain) X of $L(G)$ is called a p^*-chain ($\bar{p}$-chain) of G. So a nontrivial p^*-chain of G is a cyclic p-subgroup X for which $\Omega(X)$ is a p-atom and a $\bar{p}$-chain of G is a cyclic subgroup Y of prime power order for which $\Omega(Y)$ is not a p-atom of G.

Theorem 5.6 *The following properties of a finite group G are equivalent.*

(a) $G \in \mathfrak{N}^*$.

(b) *For every prime p, p^*-subgroup A and $\bar{p}$-subgroup B of G,*
$A \leq C_{L(G)}(B)$.

(c) *For every prime p, p^*-chain X and $\bar{p}$-chain Y of G, $X \leq C_{L(G)}(Y)$.*

References

[1] S. Andreeva, *Verbandstheoretische Analoga gruppentheoretischer Sätze von Baer, Ito und Redei*, Dissertation, Kiel, 2011.

[2] M. Costantini and G. Zacher, *On the subgroup index problem in finite groups*, Israel J. Math. 185 (2011), 293–316.

[3] M. De Falco, F. de Giovanni, C. Musella and R. Schmidt, *Detecting the index of a subgroup in the subgroup lattice*, Proc. Amer. Math. Soc. 133 (2004), 979–985.

[4] W. Gaschütz, *Über die Φ-Untergruppe endlicher Gruppen*, Math. Z. 58 (1953), 160–170.

[5] W. Gaschütz, *Zur Theorie der endlichen auflösbaren Gruppen*, Math. Z. 80 (1963), 300–305.

[6] A. Yu. Olshanskii, *Infinite groups with cyclic subgroups*, Soviet Math. Dokl. 20 (1979), 343–346.

[7] O. Ore, *Structures and group theory II*, Duke Math. J. 4 (1938), 247–269.

[8] E. Previato, *Una caratterizzazione reticolare dei gruppi risolubili finitamente generati*, Istit. Veneto Sci. Lett. Arti Atti Cl. Sci. Fis. Mat. Natur. 136 (1978), 7–11.

[9] D.J.S. Robinson, *A course in the theory of groups*, GTM 80, Springer, New York Heidelberg Berlin, 1982.

[10] R. Schmidt, *Eine verbandstheoretische Charakterisierung der auflösbaren und der überauflösbaren endlichen Gruppen*, Archiv Math. 19 (1968), 449–452.

[11] R. Schmidt, *Subgroup lattices of groups*, Expositions in Mathematics 14, de Gruyter, Berlin New York, 1994.

[12] R. Schmidt, *Lattice-theoretic characterizations of the group classes $\mathfrak{N}^{k-1}\mathfrak{A}$ and $\mathfrak{N}^k$ for $k \geq 2$*, Archiv Math. 96 (2011), 31–37.

[13] R. Schmidt, *Finite groups with modular chains*, Colloquium Mathematicum 131 (2013), 195–208.

[14] R. Schmidt, *On the subgroup lattice index problem in finite groups*, Israel J. Math. 200 (2014), 141–169.

[15] S. Stonehewer, *Permutable subgroups of some finite p-groups*, J. Austral. Math. Soc. 16 (1973), 90–97.

[16] M. Suzuki, *On the lattice of subgroups of finite groups*, Trans. Amer. Math. Soc. 70 (1951), 345–371.

[17] B.V. Yakovlev, *Lattice isomorphisms of solvable groups*, Algebra and Logic 9 (1970), 210–222.

[18] B.V. Yakovlev, *Conditions under which a lattice is isomorphic to the lattice of subgroups of a group*, Algebra and Logic 13 (1974), 400–412.

[19] G. Zacher, *Una relazione di normalità sul reticolo dei sottogruppi di un gruppo*, Ann. Mat. Pura Appl. (4) 131 (1982), 57–73.
[20] G. Zacher, *Reticoli di sottogruppi e struttura normale di gruppi*, Istit. Veneto Sci. Lett. Arti Atti Cl. Sci. Fis. Mat. Natur. 157 (1999), 37–41.

Roland Schmidt
Mathematisches Seminar
Universität Kiel
Ludewig-Meyn-Str. 4
D-24098 Kiel, Germany
E-mail address: schmidt@math.uni-kiel.de

12

Jacobi Inversion Formulae for a Curve in Weierstrass Normal Form

Jiyro Komeda and Shigeki Matsutani

Dedicated for Emma Previato's 65th birthday.

Abstract. We consider a pointed curve (X, P) which is given by the Weierstrass normal form, $y^r + A_1(x)y^{r-1} + A_2(x)y^{r-2} + \cdots + A_{r-1}(x)y + A_r(x)$ where x is an affine coordinate on $\mathbb{P}^1$, the point ∞ on X is mapped to $x = \infty$, and each A_j is a polynomial in x of degree $\leq js/r$ for certain coprime positive integers r and s ($r < s$) so that its Weierstrass non-gap sequence at ∞ is a numerical semigroup. It is a natural generalization of Weierstrass' equation in the Weierstrass elliptic function theory. We investigate such a curve and show the Jacobi inversion formulae of the strata of its Jacobian using the result of Jorgenson [Jo].

1 Introduction

The Weierstrass σ function is defined for an elliptic curve of Weierstrass' equation $y^2 = 4x^3 - g_2 x - g_3$ in Weierstrass' elliptic function theory [WW]; $\left(\wp(u) = -\frac{d^2}{du^2}\ln\sigma(u), \frac{d\wp(u)}{du}\right)$ is identical to a point (x, y) of the curve. The σ function is related to Jacobi's θ function. As the θ function was generalized by Riemann for any Abelian variety, its equivalent function Al was defined for any hyperelliptic Jacobian by Weierstrass [W1] and was refined by Klein [Kl1, Kl2] as a generalization of the elliptic σ function such that it satisfied a modular invariance under the action of $\mathrm{Sp}(2g, \mathbb{Z})$ (up to a root of unity).

Recently the studies on the σ functions in the nineteenth century have been reevaluated and reconsidered. Grant and Ônishi gave their modern perspective and showed precise structures of hyperelliptic Jacobians from a viewpoint of

2010 *Mathematics Subject Classification*. Primary, 14K25, 14H40. Secondary, 14H55, 14H50.
Key words and phrases. Weierstrass normal form, Weierstrass semigroup, Jacobi inversion formulae, non-symmetric numerical semigroup

383

number theory using the σ functions [G, O], whereas Buchstaber, Enolskiĭ and Leĭkin [BEL1] and Eilbeck, Enolskiĭ and Leĭkin [EEL] investigated the hyperelliptic σ functions for their applications to the integrable system. One of these authors also applied Baker's results to the dynamics of loops in a plane [Pr3, MP6] associated with the modified Korteweg-de Vries hierarchy [M1, M3, MP6]. Further Buchstaber, Leĭkin and Enolskiĭ [BLE] and Eilbeck, Enolskiĭ and Leĭkin [EEL] generalized the σ function to more general plane curves. Nakayashiki connected these σ functions with the θ functions in Fay's study [F] and the τ functions in the Sato universal Grassmannian theory [N1, N2].

Weierstrass' elliptic function theory provides the concrete and explicit descriptions of the geometrical, algebraic and analytic properties of elliptic curves and their related functions [WW], and thus it has strong effects on various fields in mathematics, physics and technology. We have reconstructed the theory of Abelian functions of curves with higher genus to give the vantage point like Weierstrass's elliptic function theory [MK, MP1-5, KMP2-4].

In the approach of the σ function, the pointed curve (X, P) is crucial, since the relevant objects are written in terms of $H^0(X, \mathcal{O}(*P))$. The representation of the affine curve $X \setminus P$ is therefore also relevant, and so is the Weierstrass semigroup (W-semigroup) at P. It is critical to find the proper basis of $H^0(X, \mathcal{O}(*P))$ to connect the (transcendental) θ (σ) function with the (algebraic) functions and differentials of the curve; in Mumford's investigation, the basis corresponds to U function of the Mumford triplet, UVW [Mu], which is identical to a square of Weierstrass al function [W1]. The connection means the Abel-Jacobi map and its inverse correspondence, or the Jacobi inversion formula.

Using the connection, we represent the affine curve $X \setminus P$ in terms of the σ (θ) functions and find these explicit descriptions of the geometrical, algebraic and analytic properties of the curve and their related functions. These properties are given as differential equations and related to the integrable system as in [B2, Mu, Pr1, Pr2, EEMOP2]. Since the additional structure of the hyperelliptic curves [EEMOP1] is closely related to Toda lattice equations and classical Poncelet's problem, the additional structures were also revealed as dynamical systems [KMP1].

Using the σ functions, we have the Jacobi inversion formulae of the Jacobians, e.g., for hyperelliptic curves [W1, B1], for cyclic (n, s) curves (super-elliptic curves) including the strata of the Jacobians [MP1, MP4], for curves with "telescopic" Weierstrass semigroups [A] and cyclic trigonal curves [MK, KMP2, KMP3, KMP4]. In this paper, we consider the Jacobi inversion problem for a general compact Riemann surface. The main effort in these studies is directed toward explicitness.

Since it is known that every compact Riemann surface is birationally-equivalent to a curve given by the Weierstrass normal form [W2, B1, Ka], in this paper, we investigate the Jacobi inversion formula of such a curve, which is called Weierstrass curve; its definition is given in Proposition 2.1. We reevaluate Jorgenson's results for the θ divisor of a Jacobian by restricting ourselves to the Jacobian of a Weierstrass curve. Then we show explicit Jacobi inversion formulae even for the strata in the θ divisor in Theorem 5.2 as our main theorem of this paper.

This approach enables us to investigate the properties of image of the Abelian maps of a compact Riemann surface more precisely. In the case of the hyperelliptic Jacobian, the sine-Gordon equation gives the relation among the meromorphic functions on the Jacobian, which is proved by the Jacobi inversion formula. Further the Jacobi inversion formula of the stratum in the hyperelliptic Jacobian generalizes the sine-Gordon equation to differential relations in the stratum [M2], Very recently, Ayano and Buchstaber found novel differential equations that characterizes the stratum in the hyperelliptic Jacobian of genus three [AB]. Accordingly, it is expected that the results in this paper might also lead a generalization of the integrable system and bring out the geometric and algebraic data of the strata in the Jacobians.

We use the word "curve" for a compact Riemann surface: on occasion, we use a singular representation of the curve; since there is a unique smooth curve with the same field of meromorphic function, this should not cause confusion. We let the non-negative integer denoted by $\mathbb{N}_0$ and the positive integer by $\mathbb{N}$.

The contents of this paper are organized as follows: in Section 2, we collect definitions, properties, and examples of Weierstrass curves and Weierstrass semigroups. Section 3 provides the Abelian map (integral) and the θ functions as the transcendental properties of the Abelian map (integral) image. We also mentioned Jorgenson's result there. In Section 4 we show the properties of holomorphic one forms of Weierstrass curve (X, P) in terms of the proper basis of $H^0(X, \mathcal{O}(*P))$ as the algebraic property of the curve X. In Section 5, we applied these data to the result of Jorgenson and show our main theorem, the explicit representations of Jacobi inversion formulae of a general Weierstrass curve in Theorem 5.2.

Acknowledgments: The second named author thanks Professor Victor Enol-skii for the notice on the paper of Jorgenson [Jo], though he said that it was learned from Professor Emma Previato. The authors were supported by the Grant-in-Aid for Scientific Research (C) of Japan Society for the Promotion of Science, Grant No. 18K04830 and Grant No. 16K05187 respectively. Finally, the authors would like to thank Professor Emma Previato for her extraordinary generosity and long collaboration, and dedicate this paper to her.

2 Weierstrass Normal Form and Weierstrass Curve

We set up the notations for Weierstrass semigroups and Weierstrass normal form, and recall the results we use.

2.1 Numerical Semigroups and Weierstrass Semigroups

A pointed curve is a pair (X, P), with P a point of a curve X; the Weierstrass semigroup for X at P, which we denote by $H(X, P)$, is the complement of the Weierstrass gap sequence L, namely the set of natural numbers $\{\ell_0 < \ell_1 < \cdots < \ell_{g-1}\}$ such that $H^0(\mathcal{K}_X - \ell_i P) \neq H^0(\mathcal{K}_X - (\ell_i - 1)P)$, for $\mathcal{K}_X$ a representative of the canonical divisor (we identify divisors with the corresponding sheaves). By the Riemann-Roch theorem, $H(X, P)$ is a numerical semigroup. In general, a numerical semigroup H has a unique (finite) minimal set of generators, $M = M(H)$ and the finite cardinality g of $L(M) = \mathbb{N}_0 \backslash H$; g is the genus of H or L. We let $a_{\min}(H)$ be the smallest positive integer of $M(H)$. For example,

$$L(M) = \{1, 2, 3, 4, 6, 8, 9, 11, 13, 16, 18, 23\}, \ a_{\min} = 5, \text{ for } M = \langle 5, 7 \rangle,$$

$$L(M) = \{1, 2, 3, 4, 6, 8, 9, 13\}, \ a_{\min} = 5, \text{ for } M = \langle 5, 7, 11 \rangle, \text{ and}$$

$$L(M) = \{1, 2, 4, 5\}, \ a_{\min} = 3, \text{ for } M = \langle 3, 7, 8 \rangle.$$

The Schubert index of the set $L(M(H))$ is

$$\alpha(H) := \{\alpha_0(H), \alpha_1(H), \ldots, \alpha_{g-1}(H)\}, \tag{1}$$

where $\alpha_i(H) := \ell_i - i - 1$ [EH]. By letting the row lengths be $\Lambda_i = \alpha_{g-i} + 1$, $i \leq g$, we have the Young diagram of the semigroup, $\Lambda := (\Lambda_1, \ldots, \Lambda_g)$. If for a numerical semigroup H, there exists a curve whose Weierstrass non-gap sequence is identical to H, we call the semigroup H *Weierstrass*. It is known that every numerical semigroup is not Weierstrass. A Weierstrass semigroup is called symmetric when $2g - 1$ occurs in the gap sequence. It implies that $H(X, P)$ is symmetric if and only if its Young diagram is symmetric, in the sense of being invariant under reflection across the main diagonal.

2.2 Weierstrass Normal Form and Weierstrass Curve

We now review the "Weierstrass normal form", which is a generalization of Weierstrass' equation for elliptic curves [WW]. Baker [B1, Ch. V, §§60–79] gives a complete review, proof and examples of the theory, though he calls it "Weierstrass canonical form". Here we refer to Kato [Ka], who also produces this representation, with proof.

Let $m = a_{\min}(H(X, P))$ and let n be the least integer in $H(X, P)$ which is prime to m. By denoting $P = \infty$, X can be viewed as an $m : 1$ cover of $\mathbb{P}^1$ via an equation of the following form: $f(x, y) = y^m + A_1(x)y^{m-1} + \cdots + A_{m-1}(x)y + A_m(x) = 0$, where x is an affine coordinate on $\mathbb{P}^1$, the point ∞ on X is mapped to $x = \infty$, and each A_j is a polynomial in x of degree $\leq jn/m$, with equality being attained only for $j = m$. The algebraic curve Spec $\mathbb{C}[x, y]/f(x, y)$, is, in general, singular and we denote by X its unique normalization. Since [Ka] is only available in Japanese, and since the meromorphic functions constructed in his proof will be used in our examples in 2.3 below, we reproduce his proof in sketch: it shows that the affine ring of the curve $X\backslash\infty$ can be generated by functions that have poles at ∞ corresponding to the g non-gaps.

Proposition 2.1 ([Ka]) *For a pointed curve* (X, ∞) *with Weierstrass semi-group* $H(X, \infty)$ *for which* $a_{\min}(H(X, \infty)) = m$, *we let* $m_i := \min\{h \in H(X, \infty) \setminus \{0\} \mid i \equiv h \mod m\}$, $i = 0, 1, 2, \ldots, m - 1$, $m_0 = m$ *and* $n = \min\{m_j \mid (m, j) = 1\}$. (X, ∞) *is defined by an irreducible equation,*

$$f(x, y) = 0, \qquad (2)$$

for a polynomial $f \in \mathbb{C}[x, y]$ *of type,*

$$f(x, y) := y^m + A_1(x)y^{m-1} + A_2(x)y^{m-2} + \cdots + A_{m-1}(x)y + A_m(x), \qquad (3)$$

where the $A_i(x)$'s *are polynomials in* x,

$$A_i = \sum_{j=0}^{\lfloor in/m \rfloor} \lambda_{i,j} x^j,$$

and $\lambda_{i,j} \in \mathbb{C}$, $\lambda_{m,m} = 1$.

We call the pointed curve given in Proposition 2.1 *Weierstrass curve* in this paper.

Proof. We let $I_m := \{m_1, m_2, \ldots, m_{m-1}\} \setminus \{m_{i_0}\}, =: \{m_{i_1}, m_{i_2}, \ldots, m_{i_{m-2}}\}$, where i_0 is such that $n = m_{i_0}$. Let y_{m_i} be a meromorphic function on X whose only pole is ∞ with order m_i, taking $x = y_m$ and $y = y_n$. From the definition of X, we have, as $\mathbb{C}$-vector spaces,

$$H^0(X, \mathcal{O}_X(*\infty)) = \mathbb{C} \oplus \sum_{i=0}^{m-1} \sum_{j=0} \mathbb{C}x^j y_{m_i}.$$

Thus for every $i_j \in I_m$ ($j = 1, 2, \ldots, m - 2$), we obtain the following equations

$$\left\{ \begin{array}{l} y_n y_{m_{i_1}} = A_{1,0} + A_{1,1} y_{m_{i_1}} + \cdots + A_{1,m-2} y_{m_{i_{m-2}}} + A_{1,m-1} y_n, \\ y_n y_{m_{i_2}} = A_{2,0} + A_{2,1} y_{m_{i_1}} + \cdots + A_{2,m-2} y_{m_{i_{m-2}}} + A_{2,m-1} y_n, \\ \qquad \cdots \\ y_n y_{m_{i_{m-2}}} = A_{m-2,0} + A_{m-2,1} y_{m_{i_1}} + \cdots + A_{m-2,m-2} y_{m_{i_{m-2}}} + A_{m-2,m-1} y_n, \end{array} \right. \qquad (4)$$

$$y_n^2 = A_{m-1,0} + A_{m-1,1} y_{m_{i_1}} + \cdots + A_{m-1,m-2} y_{i_{m-2}} + A_{m-1,m-1} y_n, \qquad (5)$$

where $A_{i,j} \in \mathbb{C}[x]$.

When $m = 2$, (2) equals (5). We assume that $m > 2$ and then (4) is reduced to

$$
\begin{pmatrix}
A_{1,1} - y_n & A_{1,2} & \cdots & A_{1,m-2} \\
A_{2,1} & A_{2,2} - y_n & \cdots & A_{2,m-2} \\
\vdots & \vdots & \ddots & \vdots \\
A_{m-2,1} & A_{m-2,2} & \cdots & A_{m-2,m-2} - y_n
\end{pmatrix}
\begin{pmatrix}
y_{m_{i_1}} \\
y_{m_{i_2}} \\
\vdots \\
y_{m_{i_{m-2}}}
\end{pmatrix}
$$

$$
= -
\begin{pmatrix}
A_{1,0} + A_{1,m-1} y_n \\
A_{2,0} + A_{2,m-1} y_n \\
\vdots \\
A_{m-2,0} + A_{m-2,m-1} y_n
\end{pmatrix}.
\tag{6}
$$

One can check that the determinant of the matrix on the left-hand side of (6) is not equal to zero by computing the order of pole at ∞ of the monomials $B_i y_n^{m-2-i}$ in the expression,

$$
P(x, y_n) :=
\begin{vmatrix}
A_{1,1} - y_n & A_{1,2} & \cdots & A_{1,m-2} \\
A_{2,1} & A_{2,2} - y_n & \cdots & A_{2,m-2} \\
\vdots & \vdots & \ddots & \vdots \\
A_{m-2,1} & A_{m-2,2} & \cdots & A_{m-2,m-2} - y_n
\end{vmatrix}
$$

$$
= y_n^{m-2} + B_1 y_n^{m-3} + \cdots + B_{m-3} y_n + B_{m-4},
$$

which is $n(m - 2 - i) + m \cdot \deg B_i$. The fact that $(m, n) = 1$ shows that $n(m - 2 - i) + m \cdot \deg B_i \neq n(m - 2 - j) + m \cdot \deg B_j$ for $i \neq j$.

Hence by solving equation (6) we have

$$
y_{j_i} = \frac{Q_i(x, y_n)}{P(x, y_n)},
\tag{7}
$$

where $j_i \in I_m$, $Q_i(x, y_n) \in \mathbb{C}[x, y_n]$ and a polynomial of order at most $m - 2$ in y_n. Note that the equations (7) are not independent in general but in any cases the function field of the curve can be generated by these y_{j_i}'s, and its affine ring can be given by $\mathbb{C}[x, y_n, y_{a_3}, \ldots, y_{a_\ell}]$ for $j_i \in M_g$, where $M_g := \{a_1, a_2, \ldots, a_\ell\} \subset \mathbb{N}^\ell$ with $(a_{k'}, a_k) = 1$ for $k' \neq k$, $a_1 = a_{\min} = m$, $a_2 = n$, is a minimal set of generators for $H(X, \infty)$.

By putting (7) into (5), we have (3) since (3) is irreducible. □

Remark 2.2 Since every compact Riemann surface of genus g has a Weierstrass point whose Weierstrass gap sequence with genus g [ACGH], it is characterized by the behavior of the meromorphic functions around the point and thus there is a Weierstrass curve which is birationally equivalent to the compact Riemann surface. The Weierstrass curve admits a local $\mathbb{Z}/m\mathbb{Z}$-action at ∞, in the following sense. We assume that a curve in this article is a Weierstrass curve which is given by (3). We consider the generator M_g of the Weierstrass semigroup $H = H(X, \infty)$ in the proof. For a polynomial ring $\mathbb{C}[Z] := \mathbb{C}[Z_1, Z_2, \cdots, Z_\ell]$ and a ring homomorphism,

$$\varphi : \mathbb{C}[Z] \to \mathbb{C}[t^{a_1}, t^{a_2}, \cdots, t^{a_\ell}], \quad (\varphi(Z_i) = t^{a_i}, \quad a_i \in M_g),$$

we consider a monomial ring $B_H := \mathbb{C}[Z]/\ker\varphi$. The action is defined by sending Z_i to $\zeta_m^{a_i} Z_i$, where ζ_m is a primitive m-th root of unity. Sending Z_1 to $1/x$ and Z_i to $1/y_{a_i}$, the monomial ring B_H determines the structure of gap sequence [He, Pi].

There are natural projections,

$$\varpi_x : X \to \mathbb{P}, \quad \varpi_y : X \to \mathbb{P},$$

to obtain x-coordinate and y-coordinate such that $\varpi_x(\infty) = \infty$ and $\varpi_y(\infty) = \infty$.

2.3 Examples: Pentagonal, Trigonal Non-Cyclic

We clarify that the words "trigonal" and "pentagonal" are used here only as an indication of the fact that the pointed curve (X, ∞) has Weierstrass semigroup of type three, five, respectively; a different convention requires that k-gonal curves not be j-gonal for $j < k$ (as in trigonal curves, which by definition are not hyperelliptic); of course this cannot be guaranteed in our examples. Baker [B1, Ch. V, §70] also points this out.

I. Nonsingular affine equation (3):
$y^5 + A_4 y^4 + A_3 y^3 + A_2 y^2 + A_1 y + A_0 = 0$, $n = 7$ case: (6) corresponds to

$$\begin{pmatrix} -y & 0 & 1 \\ 1 & -y & 0 \\ -A_3 & -A_2 & -A_4 - y \end{pmatrix} \begin{pmatrix} y^3 \\ y^2 \\ y^4 \end{pmatrix} = \begin{pmatrix} 0 \\ 0 \\ A_1 y + A_0 \end{pmatrix},$$

(7) becomes

$$y^{2+i} = -\frac{(A_1 y + A_0) y^i}{y^3 + A_4 y^2 + A_3 y + A_2}.$$

The curve has 5-semigroup at ∞ but is not necessarily cyclic.

II. Singular affine equation (3): $y^5 = k_2(x)^2 k_3(x)$, where $k_2(x) = (x - b_1)(x - b_2)$, $k_3(x) = (x - b_3)(x - b_4)(x - b_5)$ for pairwise distinct $b_i \in \mathbb{C}$: (6) corresponds to

$$\begin{pmatrix} -y & 1 & 0 \\ 0 & -y & 0 \\ 0 & 0 & -y \end{pmatrix} \begin{pmatrix} w \\ yw \\ y^2 \end{pmatrix} = \begin{pmatrix} 0 \\ -k_2 k_3 \\ -k_2 w \end{pmatrix}.$$

The affine ring is $H^0(\mathcal{O}_X(*\infty)) = \mathbb{C}[x, y, w]/(y^3 - k_2w, w^2 - k_3y, y^2w - k_2k_3)$. Here (7) is reduced to

$$w = \frac{k_2k_3}{y^2}, \quad yw = \frac{k_2k_3}{y}, \quad y^2 = \frac{k_2w}{y}.$$

This is a pentagonal cyclic curve (X, ∞) with $H(X, \infty) = \langle 5, 7, 11 \rangle$.

III. Singular affine equation (3):
$y^3 + a_1k_2(x)y^2 + a_2\tilde{k}_2(x)k_2(x)y + k_2(x)^2k_3(x) = 0$, where $k_2(x) = (x - b_1)(x - b_2)$, $k_3(x) = (x - b_3)(x - b_4)(x - b_5)$, $\tilde{k}_2(x) = (x - b_6)(x - b_7)$ for pairwise distinct $b_i \in \mathbb{C}$ and a_j generic constants. Here (5) and (6) correspond to

$$y^2 + a_1k_2y + k_2a_2\tilde{k}_2 + k_2w = 0, \quad yw = k_2k_3.$$

Multiplying the first equation by y and using the second equation gives the curve's equation. Multiplying that by w^2 gives

$$w^3 + a_2\tilde{k}_2w^2 + a_1k_2k_3w + k_2k_3^2 = 0,$$

which is reduced to the less order relation with respect to w,

$$w^2 + a_2\tilde{k}_2w + a_1k_2k_3 + k_3y = 0.$$

This curve is trigonal with $H(X, \infty) = \langle 3, 7, 8 \rangle$ but not necessarily cyclic.

2.4 Weierstrass Non-Gap Sequence

Let the commutative ring R be that of the affine part of (X, ∞), $R := H^0(X, \mathcal{O}(*\infty))$, which is also obtained by several normalizations of $\mathbb{C}[x, y]/(f(x, y))$ of (3) as a normal ring.

Proposition 2.3 *The commutative ring R has the basis $S_R := \{\phi_i\}_i \subset R$ as a set of monomials of R such that*

$$R = \bigoplus_{i=0} \mathbb{C}\phi_i$$

and $\mathrm{wt}_\infty\phi_i < \mathrm{wt}_\infty\phi_j$ *for $i < j$, where* $\mathrm{wt}_\infty : R \to \mathbb{Z}$ *is the order of the pole at ∞.*

The examples are following.

Table 12.1. *The φ's of Examples § 2.3*

	0	1	2	3	4	5	6	7	8	9	10	11	12	13	14
I	1	-	-	-	-	x	-	y	-	-	x^2	-	xy	-	y^2
II	1	-	-	-	-	x	-	y	-	-	x^2	w	xy	-	y^2
III	1	-	-	x	-	-	x^2	y	w	x^3	xy	xw	x^4	x^2y	y^2

	15	16	17	18	19	20	21	22	23	24
	x^3	-	x^2y	-	xy^2	x^4	y^3	x^3y	-	x^2y^2
	x^3	xw	x^2y	wy	xy^2	x^4	y^3	x^3y	xyw	x^2y^2
	yw	w^2	xy^2	xyw	xw^2	x^2y^2	x^2yw	x^2w^2	x^3y	x^3w^2

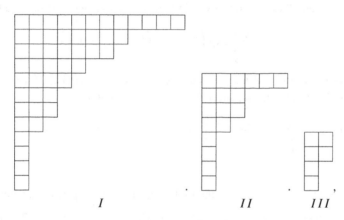

I $\qquad\qquad$ II $\qquad$ III

3 Abelian Integrals and θ Functions

3.1 Abelian Integrals and θ Functions

Let us consider a compact Riemann surface X of genus g and its Jacobian $\mathcal{J}^\circ(X) := \mathbb{C}^g / \Gamma^\circ$ where $\Gamma^\circ := \mathbb{Z}^g + \tau\mathbb{Z}^g$; $\kappa : \mathbb{C}^g \to \mathcal{J}^\circ(X)$. Let $\widetilde{X}$ be an Abelian covering of X ($\varpi : \widetilde{X} \to X$). Since the covering space $\widetilde{X}$ is constructed by a quotient space of path space (contour of integral) fixing a point $P \in X$, for $\gamma_{P',P} \in \widetilde{X}$ whose ending points are $P' \in X$ and P, $\varpi\gamma_{P',P}$ is equal to P', and we consider an embedding X into $\widetilde{X}$ by a map $\iota : X \to \widetilde{X}$ such that $\varpi \circ \iota = id$. We fix ι. For $\gamma_{P',P} \in \widetilde{X}$, we define the Abelian integral $\widetilde{w}^\circ$ and the Abel-Jacobi map w°,

$$\widetilde{w}^\circ : \mathcal{S}^k\widetilde{X} \to \mathbb{C}^g, \quad \widetilde{w}^\circ(\gamma_{P_1,P}, \cdots, \gamma_{P_k,P}) = \sum_{i=1}^{k} \widetilde{w}^\circ(\gamma_{P_i,P}) = \sum_{i=1}^{k} \int_{\gamma_{P_i,P}} v^\circ,$$

$$w^\circ := \kappa \circ \widetilde{w}^\circ \circ \iota : \mathcal{S}^k X \to \mathcal{J}^\circ(X),$$

where $\mathcal{S}^k\widetilde{X}$ and $\mathcal{S}^k X$ are k-symmetric products of $\widetilde{X}$ and X respectively, and v° is the column vector of the normalized holomorphic one forms $v^\circ{}_i$ ($i = 1, 2, \ldots, g$) of X; for the standard basis $(\alpha_i, \beta_i)_{i=1,\ldots,g}$ of the homology of X such that $\langle \alpha_i, \alpha_j \rangle = \langle \beta_i, \beta_j \rangle = 0$, $\langle \alpha_i, \beta_j \rangle = \delta_{ij}$, we have

$$\int_{\alpha_i} v^\circ{}_j = \delta_{ij}, \quad \int_{\beta_i} v^\circ{}_j = \tau_{ij}.$$

The Abel theorem shows $\kappa \circ \widetilde{w}^\circ = w^\circ \circ \varpi$. We fix the point $P \in X$ as a marked point ∞ of (X, ∞).

The map w° embeds X into $\mathcal{J}^\circ(X)$ and generalizes to a map from the space of divisors of X into $\mathcal{J}^\circ(X)$ as $w^\circ\left(\sum_i n_i P_i\right) := \sum_i n_i w^\circ(P_i)$, $P_i \in X$, $n_i \in \mathbb{Z}$. Similarly we define $\widetilde{w}^\circ(\iota D)$ for a divisor D of X.

The Riemann θ function, analytic in both variables z and τ, is defined by

$$\theta(z, \tau) = \sum_{n \in \mathbb{Z}^g} \exp\left(2\pi\sqrt{-1}\left({}^t n z + \frac{1}{2}{}^t n \tau n\right)\right).$$

The zero-divisor of θ modulo Γ° is denoted by $\Theta := \kappa \operatorname{div}(\theta) \subset \mathcal{J}^\circ(X)$.

The θ function with characteristic $\delta', \delta'' \in \mathbb{R}^g$ is defined as:

$$\theta\begin{bmatrix} \delta' \\ \delta'' \end{bmatrix}(z, \tau) = \sum_{n \in \mathbb{Z}^g} \exp\left[\pi\sqrt{-1}\{{}^t(n + \delta')\tau(n + \delta') + 2\,{}^t(n + \delta')(z + \delta'')\}\right].$$

$$(8)$$

If $\delta = (\delta', \delta'') \in \{0, \frac{1}{2}\}^{2g}$, then $\theta[\delta](z, \tau) := \theta\begin{bmatrix} \delta' \\ \delta'' \end{bmatrix}(z, \tau)$ has definite parity in z, $\theta[\delta](-z, \tau) = e(\delta)\theta[\delta](z, \tau)$, where $e(\delta) := e^{4\pi i \delta''\delta''}$. There are 2^{2g} different characteristics of definite parity.

In the paper [KMP3], we showed that the characteristics are defined for a Weierstrass curve by shifting the Abelian integrals and the Riemann constant as in Proposition 4.14. We recall the basic fact by Lewittes [Le]:

Proposition 3.1 (Lewittes [Le])
(1) Using the Riemann constant $\xi \in \mathbb{C}^g$, we have the relation between the θ divisor $\Theta := \operatorname{div}(\theta)$ and the standard θ divisor $w^\circ(\mathcal{S}^{g-1}X)$,

$$\Theta = w^\circ(\mathcal{S}^{g-1}X) + \xi \quad modulo \quad \Gamma^\circ,$$

i.e., for $P_i \in X$, $\theta(\widetilde{w}^\circ \circ \iota(P_1, \ldots, P_{g-1}) + \xi) = 0$.
(2) The canonical divisor $\mathcal{K}_X$ of X and the Riemann constant ξ have the relation,

$$w^\circ(\mathcal{K}_X) + 2\xi = 0 \quad modulo \quad \Gamma^\circ.$$

Further in [Jo], Jorgenson found a crucial relation by considering the analytic torsion on the Jacobian:

Proposition 3.2 (Jorgenson) [23, Theorem 1] *For $P_1, P_2, \ldots, P_\ell \in X$, $\ell = g - 1$, and general complex numbers a_i and b_i ($i = 1, 2, \ldots, g$), the following holds:*

$$\frac{\begin{vmatrix} v^\circ_1(P_1) & v^\circ_2(P_1) & \cdots & v^\circ_g(P_1) \\ v^\circ_1(P_2) & v^\circ_2(P_2) & \cdots & v^\circ_g(P_2) \\ \vdots & \vdots & \ddots & \vdots \\ v^\circ_1(P_\ell) & v^\circ_2(P_\ell) & \cdots & v^\circ_g(P_\ell) \\ a_1 & a_2 & \cdots & a_g \end{vmatrix}}{\begin{vmatrix} v^\circ_1(P_1) & v^\circ_2(P_1) & \cdots & v^\circ_g(P_1) \\ v^\circ_1(P_2) & v^\circ_2(P_2) & \cdots & v^\circ_g(P_2) \\ \vdots & \vdots & \ddots & \vdots \\ v^\circ_1(P_\ell) & v^\circ_2(P_\ell) & \cdots & v^\circ_g(P_\ell) \\ b_1 & b_2 & \cdots & b_g \end{vmatrix}} = \frac{\displaystyle\sum_{i=1}^{g} a_i \frac{\partial}{\partial z_i} \theta(\widetilde{w}^\circ \circ \iota(P_1, P_2, \cdots, P_\ell) + \xi)}{\displaystyle\sum_{i=1}^{g} b_i \frac{\partial}{\partial z_i} \theta(\widetilde{w}^\circ \circ \iota(P_1, P_2, \cdots, P_\ell) + \xi)}.$$

In this paper, we investigate this formula for a Weierstrass curve.

4 Holomorphic One Forms of a Weierstrass Curve

In this section we show several basic properties of a Weierstrass curve (X, ∞) with Weierstrass semigroup $H(X, \infty)$, including non-symmetric ones. We consider the sheaf of the holomorphic one-from $\mathcal{A}_X$ and its sections $H^0(X, \mathcal{A}_X(*\infty))$.

The normalized holomorphic one-forms $\{v^\circ_i \mid i = 1, \ldots, g\}$ on the Weierstrass curve (X, P) hold the following relation:

Lemma 4.1 *There are $h \in R$, a finite positive integer N and a subset $\widehat{S}_h := \{\phi_{\ell_i} \in S_R \mid i = 1, \ldots, N\}$ satisfying*

$$v^\circ_i = \frac{\sum_{j=1}^{N} \alpha_{ij} \phi_{\ell_j} dx}{h}, \quad i = 1, 2, \ldots, g$$

where α_{ij} is a certain complex number for $i = 1, \ldots, g$ and $j = 1, \ldots, N$.

Proof. The Weierstrass curve has the natural projection $\varpi_x : X \to \mathbb{P}$ and Nagata's Jacobian criterion shows that dx is the natural one-form. Thus each v°_i has an expression $v^\circ_i = g_i dx / h_i$ for certain $g_i, h_i \in R$. The h is the least common multiple of h_i's. Due to Riemann-Roch theorem, the numerator must have the finite order of the singularity at ∞ and thus there exists $N < \infty$. $\square$

For the set $\{(h, \widehat{S}_h)\}$ whose element satisfies the condition in Lemma 4.1, we employ the pair $(h, \widehat{S}_h)$ whose h has the least weight and fix it from here. The Riemann-Roch theorem enables us to find the basis of $H^0(X, \mathcal{A}_X(*\infty))$.

Definition 4.2 *Let us define the subset $\widehat{S}_R$ of R whose element is given by a finite sum of $\phi_j \in S_R$ with coefficients $a_{ij} \in \mathbb{C}$, $\widehat{\phi}_i := \sum_j a_{ij}\phi_j$, satisfying the following relations,*

(1) for the maximum j in non-vanishing a_{ij}, $a_{ij} = 1$,
(2) $\mathrm{wt}_\infty \widehat{\phi}_i < \mathrm{wt}_\infty \widehat{\phi}_j$ for $i < j$, and
(3) $H^0(X, \mathcal{A}_X(*\infty)) = \bigoplus_{i=0} \mathbb{C}\widehat{\phi}_i \frac{dx}{h}$.

The Riemann-Roch theorem mean that $\left\{ v_i := \frac{\widehat{\phi}_{i-1} dx}{h} \right\}_{i=1,\ldots,g}$ is another basis of $H^0(X, \mathcal{A}_X)$. Let $\widehat{S}_R^{(g)} := \{\widehat{\phi}_i\}_{i=0,1,\ldots,g-1}$ which gives the canonical embedding of the curve and we call v_i ($i = 1,\ldots,g$) unnormalized holomorphic one forms.

Definition 4.3 *(1) The matrix ω' is defined by*

$$v = 2\omega'v^\circ, \quad for \quad v := \begin{pmatrix} v_1 \\ v_2 \\ \vdots \\ v_g \end{pmatrix}, \quad v^\circ := \begin{pmatrix} v^\circ_1 \\ v^\circ_2 \\ \vdots \\ v^\circ_g \end{pmatrix}.$$

(2) The unnormalized Abelian integral and the unnormalized Abel-Jacobi map are defined by

$$\widetilde{w}(P_1,\ldots,P_k) = 2\omega'\widetilde{w}^\circ(P_1,\ldots,P_k),$$
$$w(P_1,\ldots,P_k) = 2\omega'w^\circ(P_1,\ldots,P_k).$$

From the definition, the following lemma is obvious:

Lemma 4.4 *By letting $\omega'' := \omega'\tau$, we have*

$$\left(\int_{\alpha_i} v_j \right) = 2\omega', \quad \left(\int_{\beta_i} v_j \right) = 2\omega''.$$

We introduce the unnormalized lattice $\Gamma := \langle \omega', \omega'' \rangle_{\mathbb{Z}}$ and Jacobian $\mathcal{J}(X) := \mathbb{C}^g / \Gamma$.

Then we have the following lemma:

Lemma 4.5 *(1) The matrix ω' is regular.*
(2) The divisor of $\frac{dx}{h}$ is expressed by $(2g - 2 + d_1)\infty - \mathfrak{B}$ for a certain effective divisor $\mathfrak{B}$ whose degree is $d_1 \geq 0$, i.e.,
$\left(\frac{dx}{h} \right) = (2g - 2 + d_1)\infty - \mathfrak{B}$.
(3) $\mathrm{wt}_\infty \widehat{\phi}_{g-1}$ is $(2g - 2) + d_1$.

Table 12.2. *The $\widehat{\phi}$'s in $\widehat{S}_R^{(g)}$ of Examples § 2.3*

	0	1	2	3	4	5	6	7	8	9	10	11
I	1	-	-	-	-	x	-	y	-	-	x^2	-
II	-	-	-	-	-	-	-	y	-	-	-	w
III	-	-	-	-	-	-	-	y	w	-	xy	xw

	12	13	14	15	16	17	18	19	20	21	22
	xy	-	y^2	x^3	-	x^2y	-	xy^2	x^4	y^3	x^3y
	xy	-	y^2	-	xw	x^2y	wy	xy^2	-	-	-
	-	-	-	-	-	-	-	-	-	-	-

(4) $(\widehat{\phi_i}) \geq (\mathfrak{B} - (2g - 2 + d_1)\infty)$ *for every* $\widehat{\phi_i} \in \widehat{S}_R^{(g)}$, $(i = 0, 1, 2, \ldots,$
$\quad g - 1)$.
(5) $(\widehat{\phi_i}) \geq (\mathfrak{B} - (g - 1 + d_1 + i)\infty)$ *for every* $\widehat{\phi_i} \in \widehat{S}_R$, $(i \geq g)$.

Proof. (1) is trivial. (2) is obvious because degree of meromorphic one-form is $2g - 2$ and h is the element of R. From the Riemann-Roch theorem, $\mathrm{wt}_\infty \nu_g = 0$ and thus, we obtain (3), (4) and (5). □

Remark 4.6 The degree of $\mathfrak{B}$ vanishes, if and only if the Weierstrass semigroup is symmetric and $2g - 1$ is a gap.

Now we introduce meromorphic functions over $S^n X$, which can be regarded as a natural generalization of U-function of Mumford triplet UVW [Mu].

Definition 4.7 *We define the Frobenius-Stickelberger (FS) matrix with entries in $\widehat{S}_R$: let n be a positive integer and $P_1, \ldots, P_n$ $(1 \leq n)$ be in $X \backslash \infty$,*

$$\widehat{\Psi}_n(P_1, P_2, \ldots, P_n) := \begin{pmatrix} \widehat{\phi}_0(P_1) & \widehat{\phi}_1(P_1) & \widehat{\phi}_2(P_1) & \cdots & \widehat{\phi}_{n-1}(P_1) \\ \widehat{\phi}_0(P_2) & \widehat{\phi}_1(P_2) & \widehat{\phi}_2(P_2) & \cdots & \widehat{\phi}_{n-1}(P_2) \\ \vdots & \vdots & \vdots & \ddots & \vdots \\ \widehat{\phi}_0(P_n) & \widehat{\phi}_1(P_n) & \widehat{\phi}_2(P_n) & \cdots & \widehat{\phi}_{n-1}(P_n) \end{pmatrix}.$$

The Frobenius-Stickelberger (FS) determinant is

$$\widehat{\psi}_n(P_1, \ldots, P_n) := \det(\widehat{\Psi}_n(P_1, \ldots, P_n)).$$

We define the meromorphic function,

$$\widehat{\mu}_n(P) := \widehat{\mu}_n(P; P_1, \ldots, P_n) := \lim_{P_i' \to P_i} \frac{1}{\widehat{\psi}_n(P_1', \ldots, P_n')} \widehat{\psi}_{n+1}(P_1', \ldots, P_n', P),$$

where the P_i' are generic, the limit is taken (irrespective of the order) for each i, and the meromorphic functions $\widehat{\mu}_{n,k}$'s by

$$\widehat{\mu}_n(P) = \widehat{\phi}_n(P) + \sum_{k=0}^{n-1} (-1)^{n-k} \widehat{\mu}_{n,k}(P_1, \ldots, P_n) \widehat{\phi}_k(P),$$

with the convention $\mu_{n,n}(P_1, \ldots, P_n) \equiv \widehat{\mu}_{n,n}(P_1, \ldots, P_n) \equiv 1$.

Remark 4.8   When X is a hyperelliptic curve, by letting $P \in X$ and $P_i \in X$ expressed by (x, y) and (x_i, y_i), $\widehat{\mu}_k$ is identical to U in [Mu]:

$$\widehat{\mu}_k(P; P_1, \ldots, P_k) = (x - x_1)(x - x_2) \cdots (x - x_k)$$

and each $\widehat{\mu}_{k,i}$ is an elementary symmetric polynomial of x_i's.

We mention the behavior of the meromorphic function $\mu_{n,k}(P_1, \ldots, P_n)$, which is obvious:

Lemma 4.9   *For $k < n$, the order of pole of $\mu_{n,k}(P_1, \ldots, P_n)$ as a function of P_n at ∞ is $\mathrm{wt}_\infty \widehat{\phi}_n - \mathrm{wt}_\infty \widehat{\phi}_{n-1}$.*

Using $\mu_n(P : P_1, \ldots, P_n)$, we have an addition structure in Jacobin $\mathcal{J}(X)$ as the linear system; "$\sim$" means the linear equivalence.

Proposition 4.10   *The divisor of $\widehat{\mu}_{g-1}(P; P_1, \ldots, P_{g-1})$ with respect to P is given by*

$$(\widehat{\mu}_{g-1}) = \sum_{i=1}^{g-1} P_i + \sum_{i=1}^{g-1} Q_i + \mathfrak{B} - (2g - 2 + d_1)\infty,$$

where Q_i's are certain points of X.

Proof. It is obvious that there is a non-negative number N satisfying the relation,

$$(\widehat{\mu}_{g-1}) = \sum_{i=1}^{g-1} P_i + \sum_{i=1}^{N} Q_i + \mathfrak{B} - (g - 1 + N + d_1)\infty,$$

and Lemma 4.5 (3) means $N = g - 1$.   □

Lemma 4.11 *There is a certain divisor* $\mathfrak{B}_0$ *such that* $\mathfrak{B}_1 - d_1\infty = 2(\mathfrak{B}_0 - d_0\infty)$ *and thus we have the following linear equivalence,*

$$\sum_{i=1}^{g-1} P_i + \mathfrak{B}_0 - 2(g-1+d_0)\infty \sim -\left(\sum_{i=1}^{g-1} Q_i + \mathfrak{B}_0 - 2(g-1+d_0)\infty\right).$$

Proof. The Abel-Jacobi theorem implies that since the Abel-Jacobi map is surjective, the existence $\mathfrak{B}_0$ is obvious. As the linear equivalence, we have

$$\sum_{i=1}^{g-1} P_i + \sum_{i=1}^{g-1} Q_i + 2\mathfrak{B}_0 - 2(g-1+d_0)\infty \sim 0,$$

where Q_i's are certain points of X. $\qquad\square$

Lemma 4.11 implies that the image of the Abel-Jacobi map has the symmetry of the Jacobian as in the following lemma:

Lemma 4.12 *By defining the shifted Abel-Jacobi map,*

$$w_s(P_1, \ldots, P_k) := w(P_1, \ldots, P_k) + w(\mathfrak{B}_0),$$

the following relation holds:

$$-w_s(S^{g-1}X) = w_s(S^{g-1}X).$$

Proof. Lemma 4.11 means $w_s(P_1, \ldots, P_{g-1}) = -w_s(Q_1, \ldots, Q_{g-1})$ and thus $-w_s(S^{g-1}X) \subset w_s(S^{g-1}X)$. It leads the relation. $\qquad\square$

Proposition 4.13 *The canonical divisor* $\mathcal{K}_X \sim (2g-2+2d_0)\infty - 2\mathfrak{B}_0$.

For $(\gamma_1, \gamma_2, \ldots, \gamma_k) \in S^k \widetilde{X}$, we also define the shifted Abelian integral $\widetilde{w}_s$ by

$$\widetilde{w}_s(\gamma_1, \gamma_2, \ldots, \gamma_k) := \widetilde{w}(\gamma_1, \gamma_2, \ldots, \gamma_k) + \widetilde{w} \circ \iota(\mathfrak{B}_0).$$

We recall our previous results on the Riemann constant [KMP3]:

Proposition 4.14 *(1) If* $d_1 > 0$, *the Riemann constant* ξ *is not a half period of* Γ°.

(2) The shifted Riemann constant $\xi_s := \xi - \frac{1}{2}\omega'^{-1}\widetilde{w} \circ \iota(\mathfrak{B}_0)$ *is the half period of* Γ°.

(3) By using the shifted Abel-Jacobi map, we have

$$\Theta = \frac{1}{2}\omega'^{-1}w_s(S^{g-1}X) + \xi_s \quad modulo \quad \Gamma^\circ,$$

i.e., for $P_i \in X$, $\theta\left(\frac{1}{2}\omega'^{-1}\widetilde{w}_s \circ \iota(P_1, \ldots, P_{g-1}) + \xi_s\right) = 0$.

(4) There is a θ-*characteristic* δ *of a half period which represents the shifted Riemann constant* ξ_s, *i.e.,* $\theta[\delta]\left(\frac{1}{2}\omega'^{-1}\widetilde{w}_s \circ \iota(P_1, \ldots, P_{g-1})\right) = 0$.

In order to describe the Jacobi inversion formulae for the strata in the Jacobian, we define the Wirtinger variety $\mathcal{W}_k := w_s(\mathcal{S}^k X)$ and

$$\Theta_k := \mathcal{W}_k \bigcup [-1]\mathcal{W}_k,$$

where $[-1]$ is the minus operation on the Jacobian $\mathcal{J}(X) := \Theta_g$. We also define the strata: $\mathcal{W}^k_{s,1} := w_s(\mathcal{S}^k_1 X) \ (\mathcal{W}^k_1 := (\mathcal{S}^k_1 X))$, where $\mathcal{S}^n_m(X) := \{D \in \mathcal{S}^n(X) \mid \dim |D| \ge m\}$.

The Abel-Jacobi theorem and Lemma 4.12 mean that $\mathcal{W}_g = \mathcal{J}(X)$ and $[-1]\mathcal{W}_i = \mathcal{W}_i$ for $i = g-1, g$ but in general, $[-1]\mathcal{W}_i$ does not equal to $\mathcal{W}_i$ for $i < g-1$.

5 Jacobi Inversion Formulae for any Weierstrass Curve

The μ function enables us to rewrite Jorgenson's relation in Proposition 3.2. Since any element $u = {}^t (u_1, u_2, \ldots, u_g)$ of $\mathbb{C}^g$ is given as $u = \widetilde{w}(\gamma_1, \ldots, \gamma_g)$ for a certain element $(\gamma_1, \ldots, \gamma_g) \in S^g \widetilde{X}$, we consider the differential $\partial_i := \frac{\partial}{\partial u}$ of a function on $\mathbb{C}^g$ and have the following proposition:

Proposition 5.1 *For $P_1, P_2, \ldots, P_\ell \in X$ and $\ell = g - 1$, we have*

$$\widehat{\mu}_g(P; P_1, \ldots, P_\ell) = \widehat{\phi}_\ell(P)$$

$$+ \sum_{i=1}^{g} \frac{\partial_i \theta\left(\frac{1}{2}\omega'^{-1}\widetilde{w} \circ \iota(P_1, P_2, \cdots, P_\ell) + \xi\right)}{\partial_\ell \theta\left(\frac{1}{2}\omega'^{-1}\widetilde{w} \circ \iota(P_1, P_2, \cdots, P_\ell) + \xi\right)} \times \widehat{\phi}_{i-1}(P).$$

The right hand side does not depend on the choice of the embedding ι.

Proof. First we rewrite Proposition 3.2 in terms of the unnormalized holomorphic one forms. It is obvious that for general complex numbers a_i and b_i $(i = 1, 2, \ldots, g)$, we have

$$\frac{\begin{vmatrix} v_1(P_1) & v_2(P_1) & \cdots & v_g(P_1) \\ v_1(P_2) & v_2(P_2) & \cdots & v_g(P_2) \\ \vdots & \vdots & \ddots & \vdots \\ v_1(P_\ell) & v_2(P_\ell) & \cdots & v_g(P_\ell) \\ a_1 & a_2 & \cdots & a_g \end{vmatrix} \sum_{i=1}^{g} a_i \partial_i \theta\left(\frac{1}{2}\omega'^{-1}\widetilde{w} \circ \iota(P_1, P_2, \cdots, P_\ell) + \xi\right)}{\begin{vmatrix} v_1(P_1) & v_2(P_1) & \cdots & v_g(P_1) \\ v_1(P_2) & v_2(P_2) & \cdots & v_g(P_2) \\ \vdots & \vdots & \ddots & \vdots \\ v_1(P_\ell) & v_2(P_\ell) & \cdots & v_g(P_\ell) \\ b_1 & b_2 & \cdots & b_g \end{vmatrix} \sum_{i=1}^{g} b_i \partial_i \theta\left(\frac{1}{2}\omega'^{-1}\widetilde{w} \circ \iota(P_1, P_2, \cdots, P_\ell) + \xi\right)}.$$

By letting $(a_1, \ldots, a_g) = (v_1(P), \ldots, v_g(P))$ and $(b_1, \ldots, b_g) = (0, \ldots, 0, v_g(P))$ for a generic point $P \in X$, the relation is reduced to the proposition. $\qquad\square$

Using Jorgenson's result with the ordering of $\widehat{\phi}_i$, we show our main theorem:

Theorem 5.2 *For* $(P_1, \ldots, P_k) \in \mathcal{S}^k X \setminus \mathcal{S}_1^k X$ $(k < g)$ *and a positive integer* $i \leq k$, *we have the relation,*

$$
\widehat{\mu}_{k,i-1}(P_1, \ldots, P_k) = \frac{\partial_i \theta \left(\frac{1}{2}\omega'^{-1}\widetilde{w} \circ \iota(P_1, P_2, \cdots, P_k) + \xi \right)}{\partial_{k+1} \theta \left(\frac{1}{2}\omega'^{-1}\widetilde{w} \circ \iota(P_1, P_2, \cdots, P_k) + \xi \right)},
$$

and for a certain θ-characteristic δ for ξ_s,

$$
\widehat{\mu}_{k,i-1}(P_1, \ldots, P_k) = \frac{\partial_i \theta[\delta] \left(\frac{1}{2}\omega'^{-1}\widetilde{w}_s \circ \iota(P_1, P_2, \cdots, P_k) \right)}{\partial_{k+1} \theta[\delta] \left(\frac{1}{2}\omega'^{-1}\widetilde{w}_s \circ \iota(P_1, P_2, \cdots, P_k) \right)},
$$

especially $k = 1$ case,

$$
\frac{\widehat{\phi}_1(P_1)}{\widehat{\phi}_0(P_1)} = \frac{\partial_1 \theta[\delta] \left(\frac{1}{2}\omega'^{-1}\widetilde{w}_s \circ \iota(P_1) \right)}{\partial_2 \theta[\delta] \left(\frac{1}{2}\omega'^{-1}\widetilde{w}_s \circ \iota(P_1) \right)}.
$$

Proof. Since these $\widehat{\phi}_i$'s have the natural ordering for $P_k \to \infty$, we could apply the proof of Theorem 5.1 (3) in [MP1] to this case. It means that we prove it inductively. For $k < g$ and $i < k$, using the property of Lemma 4.9, we consider $\widehat{\mu}_{k,i-1}(P_1, P_2, \ldots, P_k)/\widehat{\mu}_{k,k-1}(P_1, P_2, \ldots, P_k)$ and its limit $P_k \to \infty$. Then we have the relation for $\widehat{\mu}_{k-1,i-1}$. $\qquad\square$

Theorem 5.2 means the Jacobi inversion formula of Θ_k for a general Weierstrass curve. Using the natural basis of the Weierstrass non-gap sequence of the Weierstrass curve, we can consider the precise structure of strata Θ_k in the Jacobian. The relation contains the cyclic (r, s) curve cases and trigonal curve cases reported in [MP1, MP4, KMP4].

Since for every compact Riemann surface Y, there exists a Weierstrass curve (X, ∞) which is birationally-equivalent to Y, Theorem 5.2 gives the structure of the strata in the Jacobian of Y.

Remark 5.3 Nakayashiki investigated precise structure of the θ function for a pointed compact Riemann surface whose Weierstrass gap is associated with a Young diagram [N2]. He refined the Riemann-Kempf theory [ACGH] in terms of the unnormalized holomorphic one-forms. Using the results, it is easy to rewrite the main theorem using non-vanishing quantities as in [MP4, KMP4].

Remark 5.4 Though it is well-known that the sine-Gordon equation is an algebraic relations of the differentials of μ_g-function in the hyperelliptic Jacobian $\mathcal{J}(X)$ [Mu, Pr2], the equation can be generalized to one in strata in the Jacobian $\mathcal{J}(X)$ using the μ_k function $k < g$ [M2]. Ayano and Buchstaber investigated similar structure of stratum of the hyperelliptic Jacobian of genus three to find a novel differential equation which characterizes the stratum [AB]. In other words, it is expected that these relations in our theorem might show some algebraic relations in the strata in the Jacobian of a general compact Riemann surface.

For example, for a Weierstrass curve X whose $\widehat{\phi}_1/\widehat{\phi}_0$ equals to x, we have the Burgers equation,

$$\frac{\partial}{\partial u_i} \frac{\partial_1\theta[\delta]\left(\frac{1}{2}\omega'^{-1}w_s(P)\right)}{\partial_2\theta[\delta]\left(\frac{1}{2}\omega'^{-1}w_s(P)\right)} - \frac{\widehat{\phi}_j(P)}{\widehat{\phi}_i(P)} \frac{\partial}{\partial u_j} \frac{\partial_1\theta[\delta]\left(\frac{1}{2}\omega'^{-1}w_s(P)\right)}{\partial_2\theta[\delta]\left(\frac{1}{2}\omega'^{-1}w_s(P)\right)} = 0.$$

Remark 5.5 Korotkin and Shramchenko defined the σ function for a general compact Riemann surface based on the Klein's investigation [KS]; they assume that the θ characteristic δ as a half period of Jacobian Γ^o is given by the Riemann constant ξ with $w(D_s)$ of the divisor D_s of the spin structure of the curve X. The spin structure corresponds to $\mathfrak{B}_0$ in Propositions 4.14 and 4.13 [At].

We have constructed the σ functions for trigonal curves [MP1, MP4, KMP4] using the EEL-construction proposed by Eilbeck, Enolskiĭ and Leĭkin [EEL] whose origin is Baker [B1]; the EEL-construction differs from Klein's approach. Due to the arguments on this paper, Emma Previato posed a problem whether we could reproduce the σ function of Korotkin and Shramchenko by means of the EEL-construction for the Weierstrass curve. It means an extension of the EEL-construction of the σ function.

Further using the extension and Nakayashiki's investigation in [N2], we could connect our results with Sato Grassmannian in the Sato universal Grassmannian theory and investigate the σ function of a Weierstrass curve more precisely.

Since it is known that there are infinitely many numerical semigroups which are not Weierstrass, but the distribution of Weierstrass semigroups in the set of the numerical semigroups is not clarified. Thus it is a natural question how the Weierstrass curves are characterized in Sato Grassmannian manifolds.

6 An Example

6.1 Curve X of Example II in § 2.3

Let us consider the curve X of Example II in § 2.3. For a point $P_i = (x_i, y_i, w_i)$ $(i = 1, 2, \ldots, 7)$ of X, the $\widehat{\mu}_7$ is given by

$$\widehat{\mu}_7(P, P_1, P_2, \ldots, P_7) = \frac{\begin{vmatrix} y_1 & w_1 & x_1 y_1 & y_1^2 & x_1 w_1 & x_1^2 y_1 & w_1 y_1 & x_1 y_1^2 \\ y_2 & w_2 & x_2 y_2 & y_2^2 & x_2 w_2 & x_2^2 y_2 & w_2 y_2 & x_2 y_2^2 \\ y_3 & w_3 & x_3 y_3 & y_3^2 & x_3 w_3 & x_3^2 y_3 & w_3 y_3 & x_3 y_3^2 \\ y_4 & w_4 & x_4 y_4 & y_4^2 & x_4 w_4 & x_4^2 y_4 & w_4 y_4 & x_4 y_4^2 \\ y_5 & w_5 & x_5 y_5 & y_5^2 & x_5 w_5 & x_5^2 y_5 & w_5 y_5 & x_5 y_5^2 \\ y_6 & w_6 & x_6 y_6 & y_6^2 & x_6 w_6 & x_6^2 y_6 & w_6 y_6 & x_6 y_6^2 \\ y_7 & w_7 & x_7 y_7 & y_7^2 & x_7 w_7 & x_7^2 y_7 & w_7 y_7 & x_7 y_7^2 \\ y & w & xy & y^2 & xw & x^2 y & wy & xy^2 \end{vmatrix}}{\begin{vmatrix} y_1 & w_1 & x_1 y_1 & y_1^2 & x_1 w_1 & x_1^2 y_1 & w_1 y_1 \\ y_2 & w_2 & x_2 y_2 & y_2^2 & x_2 w_2 & x_2^2 y_2 & w_2 y_2 \\ y_3 & w_3 & x_3 y_3 & y_3^2 & x_3 w_3 & x_3^2 y_3 & w_3 y_3 \\ y_4 & w_4 & x_4 y_4 & y_4^2 & x_4 w_4 & x_4^2 y_4 & w_4 y_4 \\ y_5 & w_5 & x_5 y_5 & y_5^2 & x_5 w_5 & x_5^2 y_5 & w_5 y_5 \\ y_6 & w_6 & x_6 y_6 & y_6^2 & x_6 w_6 & x_6^2 y_6 & w_6 y_6 \\ y_7 & w_7 & x_7 y_7 & y_7^2 & x_7 w_7 & x_7^2 y_7 & w_7 y_7 \end{vmatrix}}.$$

Thus we have the following Proposition:

Proposition 6.1 *1) For* $(P, P_1, P_2, \ldots, P_k) \in X \times \left(\mathcal{S}^k(X) \setminus \mathcal{S}_1^k(X) \right)$ *and* $k < 8$, *we have*

$$\widehat{\mu}_{k,i-1}(P_1, \ldots, P_k) = \frac{\partial_i \theta[\delta] \left(\frac{1}{2} \omega'^{-1} w_s(P_1, P_2, \cdots, P_k) \right)}{\partial_8 \theta[\delta] \left(\frac{1}{2} \omega'^{-1} w_s(P_1, P_2, \cdots, P_k) \right)}$$

especially

$$\frac{\partial_1 \theta[\delta] \left(\frac{1}{2} \omega'^{-1} w_s(P_1) \right)}{\partial_2 \theta[\delta] \left(\frac{1}{2} \omega'^{-1} w_s(P_1) \right)} = \frac{w_1}{y_1}.$$

References

[ACGH] E. Arbarello, M. Cornalba, P.A. Griffiths and J. Harris, Geometry of Algebraic Curves. Vol. I, Springer-Verlag, New York, 1985.

[At] M. F. Atiyah, *Riemann surfaces and spin structure*, Ann. sci. l'É.N.S., **4** (1971) 47–62.

[A] T. Ayano, *On Jacobi Inversion Formulae for Telescopic Curves*, SIGMA **12** (2016), Paper No. 086, 21 pp.

[AB] T. Ayano and V. Buchstaber *Construction of two parametric deformation of KdV-hierarchy and solution in terms of meromorphic functions on the sigma divisor of a hyperelliptic curve of genus 3*, arXiv:1811.07138.

[B1] H.F. Baker, Abelian functions. Abel's theorem and the allied theory of theta functions Cambridge University Press, Cambridge, 1995, Reprint of the 1897 original.

[B2] H.F. Baker, *On a system of differential equations leading to periodic functions*, Acta Math., **27** (1903) 135–156

[BEL1] V.M. Buchstaber, V.Z. Enolskiĭ and D.V. Leĭkin, *Kleinian functions, hyperelliptic Jacobians and applications*, Reviews in Mathematics and Mathematical Physics, **10** (1997) 1–103.

[BLE] V.M. Buchstaber, D.V. Leĭkin and V.Z. Enolskiĭ, *σ-functions of (n, s)-curves*, Uspekhi Mat. Nauk, **54** (1999), no. 3 (327), 155–156; translation in Russian Math. Surveys 54 (1999), no. 3, 628–629.

[EEL] J.C. Eilbeck, V.Z. Enolskii and D.V. Leykin, *On the Kleinian construction of abelian functions of canonical algebraic curves*, SIDE III—symmetries and integrability of difference equations (Sabaudia, 1998), 121–138, CRM Proc. Lecture Notes, 25, Amer. Math. Soc., Providence, RI, 2000.

[EEMOP1] J.C. Eilbeck, V.Z. Enol'skii, S. Matsutani, Y. Ônishi, and E. Previato, *Abelian functions for trigonal curves of genus three*, Int. Math. Res. Notices, **2007** (2007) 1–38,

[EEMOP2] J.C. Eilbeck, V.Z. Enol'skii, S. Matsutani, Y. Ônishi, and E. Previato, *Addition formulae over the Jacobian pre-image of hyperelliptic Wirtinger varieties*, J. Reine Angew Math., **619** (2008) 37–48.

[EH] D. Eisenbud and J. Harris, *Existence, decomposition, and limits of certain Weierstrass points*, Invent. Math., **87** (1987) 495–515.

[F] J. D. Fay, Theta functions on Riemann surfaces, Lectures Notes in Mathematics, vol. **352**, Springer, Berlin, 1973.

[He] J. Herzog, *Generators and relations of Abelian semigroup and semigroup ring*, Manuscripta Math., **3** (1970) 175–193.

[G] D. Grant, *Formal groups in genus two*, J. reine angew. Math., **411** (1990) 96–121.

[Jo] J. Jorgenson, *On directional derivatives of the theta function along its divisor*, Israel J. Math., **77** (1992) 273–284.

[Ka] T. Kato, *Weierstrass normal form of a Riemann surface and its applications*, (in Japanese) Sû,gaku 32 (1980), no. 1, 73–75.

[Kl1] F. Klein, *Ueber hyperelliptische sigmafunctionen*, Math. Ann., **27** (1886) 431–464.

[Kl2] F. Klein, *Ueber hyperelliptische sigmafunctionen (Zweite Abhandlung)*, Math. Ann., **32** 1888) 351–380

[KMP1] Y. Kodama, S. Matsutani, and E. Previato, *Quasi-periodic and periodic solutions of the Toda lattice via the hyperelliptic sigma function,,* Annales de l'institut Fourier, (2012) **63** (2013) 655–688

[KMP2] J. Komeda, S. Matsutani and E. Previato, *The sigma function for Weierstrass semigroups* $\langle 3, 7, 8 \rangle$ *and* $\langle 6, 13, 14, 15, 16 \rangle$, Internat. J. Math., **24** (2013) no. 11, 1350085, 58 pp.

[KMP3] J. Komeda, S. Matsutani and E. Previato, *The Riemann constant for a nonsymmetric Weierstrass semigroup*, Arch. Math. (Basel) **107** (2016), no. 5, 499–509.

[KMP4] J. Komeda, S. Matsutani, and E. Previato, *The sigma function for trigonal cyclic curves,,* Lett. Math. Phys., **109** (2019) 423–447.

[KS] D. Korotkin and V. Shramchenko, *On higher genus Weierstrass sigmafunction.* Physica D **241** (2012), 2086–2094.

[Le] J. Lewittes, *Riemann surfaces and the theta functions*, Acta Math., **111** (1964) 37–61.

[M1] S. Matsutani, *Hyperelliptic solutions of modified Kortweg-de Vries equation of genus g: essentials of Miura transformation,,* J. Phys. A: Math. & Gen.,, **35** (2002) 4321–4333,

[M2] S. Matsutani, *Relations of al Functions over Subvarieties in a Hyperelliptic Jacobian*, CUBO A Math. J, **7** (2005) 75–85,

[M3] S. Matsutani, *Euler's Elastica and Beyond,,* J. Geom. Symm. Phys, **17** (2010) 45–86,

[MK] S. Matsutani and J. Komeda, *Sigma functions for a space curve of type (3,4,5)*, J. Geom. Symmetry Phys. **30** (2013), 75–91.

[MP1] S. Matsutani and E. Previato, *Jacobi inversion on strata of the Jacobian of the C_{rs} curve $y^r = f(x)$ I* J. Math. Soc. Japan, **60** (2008) 1009–1044.

[MP2] S. Matsutani and E. Previato, *A generalized Kiepert formula for C_{ab}, curves* Israel J. Math., **171** (2009) 305–323,

[MP3] S. Matsutani and E. Previato, *A class of solutions of the dispersionless KP equation*, Phys. Lett. A, **373** (2009) 3001–3004.

[MP4] S. Matsutani and E. Previato, *Jacobi inversion on strata of the Jacobian of the C_{rs} curve $y^r = f(x)$ II* J. Math. Soc. Japan, **66** (2014) 647–692.

[MP5] S. Matsutani and E. Previato, *The al function of a cyclic trigonal curve of genus three*, Collectanea Mathematica,, **66** 3, (2015) 311–349,

[MP6] S. Matsutani and E. Previato, *From Euler's elastica to the mKdV hierarchy, through the Faber polynomials*, J. Math. Phys., **57** (2016) 081519; arXiv:1511.08658

[Mu] D. Mumford, Tata Lectures on Theta, Vol.s I, II Birkhäuser 1981, 1984.

[N1] A. Nakayashiki, *On algebraic expansions of sigma functions for (n, s) curves*, Asian J. Math., **14** (2010) 175–212.

[N2] A. Nakayashiki, *Tau function approach to theta functions*, Int. Math. Res. Not. IMRN (2016) 2016 (17): 5202–5248.

[O] Y. Ônishi, *Determinant expressions for hyperelliptic functions*, Proc. Edinburgh Math. Soc., **48** (2005) 705–742.

[Pi] H. C. Pinkham, *Deformation of algebraic varieties with G_m action*, Astérisque, **20** (1974) 1–131.

[Pr1] E. Previato, *Hyperelliptic quasi-periodic and soliton solutions of the nonlinear Schrdinger equation*, Duke Math. J., **52** (1985) 329–377.

[Pr2]　E. Previato, *A particle-system model of the sine-Gordon hierarchy*, Physica D: Nonlinear Phenomena, **18** (1986) 312–314.

[Pr3]　E. Previato, *Geometry of the modified KdV equation, in* Geometric and Quantum Aspects of Integrable Systems: Proceedings of the Eighth Scheveningen Conference Scheveningen, The Netherlands, August 16–21, 1992, edited. by G. F. Helminck, LMP 424, Speinger, 1993, 43–65.

[WW]　E.T. Whittaker and G.N. Watson, A Course of Modern Analysis, Cambridge University Press 1927.

[W1]　K. Weierstrass, *Zur Theorie der Abelschen Functionen*, J. Reine Angew. Math., **47** (1854) 289–306.

[W2]　K. Weierstrass, *Uber Normalformen algebraischer Gebilde*, Mathematische Werke III, 297–307, Georg Ohms and Johnson reprint (1967).

Jiryo Komeda
Department of Mathematics,
Center for Basic Education and Integrated Learning,
Kanagawa Institute of Technology,
1030 Shimo-Ogino, Atsugi, Kanagawa 243-0292, JAPAN
E-mail address: komeda@gen.kanagawa-it.ac.jp

Shigeki Matsutani
Industrial Mathematics,
National Institute of Technology, Sasebo College,
1-1 Okishin, Sasebo, Nagasaki, 857-1193, JAPAN
moved to
Faculty of Electrical, Information and Communication Engineering,
Graduate School of Natural Science & Technology,
Kanazawa University,
Kakuma Kanazawa, 920-1192, JAPAN
E-mail address: s-matsutani@se.kanazawa-u.ac.jp

13

Spectral Construction of Non-Holomorphic Eisenstein-Type Series and their Kronecker Limit Formula

James Cogdell, Jay Jorgenson[1] and Lejla Smajlović

Dedicated to Emma Previato, on the occasion of her 65th birthday

Abstract. Let X be a smooth, compact, projective Kähler variety and D be a divisor of a holomorphic form F, and assume that D is smooth up to codimension two. Let ω be a Kähler form on X and K_X the corresponding heat kernel which is associated to the Laplacian that acts on the space of smooth functions on X. Using various integral transforms of K_X, we will construct a meromorphic function in a complex variable s whose special value at $s = 0$ is the log-norm of F with respect to μ. In the case when X is the quotient of a symmetric space, then the function we construct is a generalization of the so-called elliptic Eisenstein series which has been defined and studied for finite volume Riemann surfaces.

1 Introduction

1.1 Kronecker's Limit Formula

The discrete group $\mathrm{PSL}_2(\mathbb{Z})$ acts on the upper half plane $\mathbb{H}$, and the quotient space $\mathrm{PSL}_2(\mathbb{Z})\backslash\mathbb{H}$ has one cusp which can be taken to be at $i\infty$ by identifying $\mathrm{PSL}_2(\mathbb{Z})\backslash\mathbb{H}$ with its fundamental domain. Associated to the cusp is a non-holomorphic Eisenstein series $\mathcal{E}_\infty^{\mathrm{par}}(z, s)$ which initially is defined as a Poincaré series for $\mathrm{Re}(s) > 1$ but can be shown to admit a meromorphic continuation to all $s \in \mathbb{C}$. One realization of the classical Kronecker limit formula is the asymptotic expansion that

$$\mathcal{E}_\infty^{\mathrm{par}}(z, s) = \frac{3}{\pi(s-1)} - \frac{1}{2\pi}\log\big(|\Delta(z)|\mathrm{Im}(z)^6\big) + C + O_z(s-1) \text{ as } s \to 1$$

where $C = 6(1 - 12\,\zeta'(-1) - \log(4\pi))/\pi$. An elegant proof of Kronecker's limit formula can be found in [Si80], though the normalization used in [Si80] is

[1] The second named author acknowledges grant support PSC-CUNY.

slightly different than in [JST16] from which we quote the above formulation. The series $\mathcal{E}_\infty^{\mathrm{par}}(z, s)$ has a well-known functional equation which allows one to restate Kronecker's limit formula as

$$\mathcal{E}_\infty^{\mathrm{par}}(z, s) = 1 + \log\big(|\Delta(z)|^{1/6}\mathrm{Im}(z)\big)s + O_z(s^2) \text{ as } s \to 0.$$

There are many results in the mathematical literature which develop and explore analogues of Kronecker's limit formula. One particularly motivating study is given in [KM79] in which the authors define a non-holomorphic hyperbolic Eisenstein series $\mathcal{E}_\gamma^{\mathrm{hyp}}(z, s)$ associated to any hyperbolic subgroup, generated by a hyperbolic element γ, of an arbitrary co-finite discrete subgroup Γ of $\mathrm{PSL}_2(\mathbb{R})$. The Kronecker limit formula obtained in [KM79] states that the Poincaré series which defines $\mathcal{E}_\gamma^{\mathrm{hyp}}(z, s)$ admits a meromorphic continuation to $s \in \mathbb{C}$, and the value of $\mathcal{E}_\gamma^{\mathrm{hyp}}(z, s)$ at $s = 0$ is, in effect, the harmonic one-form which is dual to the geodesic on $\Gamma\backslash\mathbb{H}$ associated to γ.

Abelian subgroups of discrete groups Γ which act on $\mathbb{H}$ are classified as parabolic, hyperbolic and elliptic, so it remained to define and study non-holomorphic Eisenstein series associated to any elliptic subgroup of an arbitrary discrete group Γ. Any elliptic subgroup can be viewed as the stabilizer group of a point w on the quotient $\Gamma\backslash\mathbb{H}$, where in all but a finite number of cases the elliptic subgroup consists solely of the identity element of Γ. One can envision the notion of a non-holomorphic elliptic Eisenstein series $\mathcal{E}_w^{\mathrm{ell}}(z, s)$ which, if the above examples serve as forming a pattern, will admit a mermorphic continuation and whose special value at $s = 0$ will be associated to a harmonic form of some type specified by w. Indeed, such series were studied in [vP10] and, in fact, the Kronecker limit function is the log-norm of a holomorphic form which vanishes only at w.

1.2 A Unified Approach

The article [JvPS16] developed a unified construction of the hyperbolic, elliptic and parabolic Eisenstein series mentioned above for any finite volume quotient of $\mathbb{H}$; of course, if the quotient is compact, then parabolic Eisenstein series do not exist. The goal of [JvPS16] was to devise a means, motivated by a type of pre-trace formula, so that the various Eisenstein series could be obtained by employing different test functions. As one would expect, there were numerous technical considerations which arose, especially in the case when Γ was not co-compact. In the end, one can view the approach developed in [JvPS16] as starting with a heat kernel, and then undertaking a sequence of integral transforms until one ends up with each of the above mentioned Eisenstein series. Whereas the article [JvPS16] did provide a unified approach to the construction of parbolic, hyperbolic and elliptic Eisenstein series for hyperbolic Riemann surfaces, the analysis did employ the geometry of $\mathrm{SL}_2(\mathbb{R})$ quite extensively.

1.3 Our Results

The goal of the present paper is to understand the heat kernel construction of non-holomorphic elliptic Eisenstein series in a more general setting. We consider a smooth, complex, projective variety X of complex dimension N. We fix a smooth Kähler metric on X, which we denote by the $(1, 1)$ form ω. In general terms, let us now describe the approach we undertake to define and study what we call elliptic Eisenstein series.

Let t be a positive real variable, and let z and w be points on X. Let $K_X(z, w; t)$ be the heat kernel acting on smooth functions on X associated to the Laplacian Δ_X corresponding to ω; see, for example, [Ch84] or [BGV91]. One of the key properties of $K_X(z, w; t)$ is that it satisfies the heat equation, meaning that

$$(\Delta_z + \partial_t) K_X(z, w; t) = 0.$$

We compute the integral transform in t of $K_X(z, w; t)$ after multiplying by a function $G(t, u)$ which satisfies the differential equation $(\partial_t - \partial_u^2)G(t, u) = 0$. By what amounts to integration by parts, we get a function $(K_X * G)(z, w; u)$ which satisfies the equation

$$\left(\Delta_z - \partial_u^2\right)(K_X * G)(z, w; u) = 0.$$

If one formally replaces u by iu, one gets the kernel function associated to the wave equation. However, this substitution is only formal because of convergence considerations; nonetheless, one is able to use the language of distributions in order to achieve the desired result which is to obtain a wave kernel $W_X(z, w; u)$. At this point, one would like to integrate the wave kernel against the test function $(\sinh u)^{-s}$ for a complex variable s to yield, as in [JvPS16], the elliptic Eisenstein series. Again, however, technical problems occur because of the vanishing of $\sinh(u)$ when $u = 0$. Instead, we integrate the wave kernel against $(\cosh u)^{-s}$, for which there is no such technical issue. We then replace s by $s + 2k$ and sum over k, in a manner dictated by, of all things, the binomial theorem, thus allowing us to mimic the use of $(\sinh u)^{-s}$. In doing so, we arrive at the analogue of the elliptic Eisenstein series $E_X(z, w; s)$, where s is a complex variable, initially required to have real part $\mathrm{Re}(s)$ sufficiently large, z is a variable on X, and w is a fixed point on X. Though w may be referred to as the elliptic point, it is, in the case X is smooth, simply a chosen point on X.

As a final step, we let D be the divisor of a holomorphic form F on X, and assume that D is smooth up to codimension two. We show that the integral of $E_X(z, w; s)$ with respect of the metric $\mu_D(w)$ on D induced from the Kähler form ω has an expansion in s at $s = 0$, and the second order term in s is the log-norm of F. This result is the analogue of the classical Kronecker limit formula.

Thus far, all results are obtained by using the spectral expansion of the heat kernel associated to the Laplacian Δ_X. We can equally well reconsider all of the above steps for the operator $\Delta_X - Z$ for any complex number Z, in which case we do not begin with the heat kernel $K_X(z, w; t)$ but rather we begin with $K_X(z, w; t)e^{-Zt}$. If, for whatever reason, there is a means by which we have another expression for the heat kernel, and also have a compelling reason to choose a specific Z, then we may end up with another expression for $E_X(z, w; s)$. Such a situation occurs when, for instance, X is the quotient of a symmetric space G/K by a discrete group Γ. In this case, the heat kernel can be obtained as the inverse spherical transform of an exponential function. In that setting, it is natural to take $Z = -\rho_0^2$ where ρ_0 is essentially the norm, with respect to the Killing form, of half the sum of the positive roots. (In the notation of Gangoli [Ga68], our ρ_0 would be his $|\rho_*|$.) Finally, we note that one can, without loss of generality, re-scale the time variable t by a positive constant c, so then all begins with the function $K_X(z, w; t/c)e^{-Zt/c}$. In the development of our results, it will be evident that it is necessary to both translate the Laplacian Δ_X and re-scale time t, where it will become evident that as long as $\rho_0^2 \neq 0$, then it is natural to take $c = 1/(4\rho_0^2)$, which would have the effect of scaling ρ_0 to be $1/2$, or ρ_0^2 to be $1/4$. (See section 2.7 below.)

The full development of these considerations in all instances would take a great deal of time and space, so for the purpose of the present article we will focus on the results obtainable by studying the spectral decomposition of the heat kernel in the case of a compact Kähler variety X. However, it is possible to give an indication of what will follow when an additional expression for the heat kernel is available. For instance, if X is an abelian variety, we obtain an expression for the heat kernel on X by viewing X as a complex torus. As an example of our analysis, we can take D to be the divisor of the Riemann theta function θ, so then our construction expresses the log-norm of the Riemann theta function θ as a type of Kronecker limit function.

1.4 Outline of the Paper

The article is organized as follows. In section 2 we establish notation and recall some known results. In section 3, we define the wave distribution associated to a certain space of smooth functions on X. In section 4 we apply the wave distribution to the test function $\cosh^{-(s-\rho_0)}(u)$, for a suitably chosen constant ρ_0, yielding a function $K_{X;\rho_0^2}(z, w; s)$. In section 5 we define two series formed from $K_{X;\rho_0^2}(z, w; s)$, one producing a formula for the resolvent kernel $G_{X;\rho_0^2}(z, w; s)$, which is the integral kernel that inverts the operator $\Delta_X + s(s - \rho_0)$. The second series $E_{X;\rho_0^2}(z, w; s)$ is the analogue of the elliptic Eisenstein series. The analogue of Kronecker's limit formula is given in section 6. Finally, in section 7, we conclude with some examples. In our opinion, each example is of independent interest. Admittedly, the discussion in section 7 is somewhat speculative; however, we elected to include the material in an attempt to illustrate some of the directions we believe our results can apply.

In an unavoidable mishap of notation, the heat kernel on X is denoted by $K_X(z, w; t)$, and the function obtained by applying the wave distribution to $\cosh^{-(s-\rho_0)}(u)$ is $K_{X;\rho_0^2}(z, w; s)$. Similarly, Γ will sometimes signify the Gamma function and sometimes signify a discrete group acting on a symmetric space. In each case, the meaning will be clear from the context of the discussion.

2 Background Material

In this section we establish notation and state certain elementary results which will be used throughout the article. The contents in this section are given in no particular order of importance.

2.1 Stirling's Approximation

Stirling's approximation for the logarithm $\log \Gamma(s)$ of the classical gamma function is well-known, and we will use the form which states that

$$\log \Gamma(s) = s \log(s) - s + \frac{1}{2} \log(2\pi/s) + \sum_{n=1}^{M} \frac{B_{2n}}{2n(2n-1)s^{2n-1}} + h_M(s),$$

(1)

where B_n is the n-th Bernoulli number and $h_M(s)$ is a holomorphic function in the half-plane $\mathrm{Re}(s) \gg 0$ and $h_M(s) = O_M(s^{-2M-1})$ as $s \to \infty$. The proof of (1) is found in various places in the literature; see, for example, [JLa93]. Going further, the proof from [JLa93] extends to show that one can, in effect, differentiate the above asymptotic formula. More precisely, for any integer $\ell \geq 0$, one has that

$$\partial_s^\ell \log \Gamma(s) = \partial_s^\ell \left(s \log(s) - s + \frac{1}{2} \log(2\pi/s) + \sum_{n=1}^{M} \frac{B_{2n}}{2n(2n-1)s^{2n-1}} \right)$$
$$+ \partial_s^\ell h_M(s),$$

(2)

where $\partial_s^\ell h_M(s) = O_{M,\ell}(s^{-2M-\ell-1})$, as $s \to \infty$.

We will use the notational convenience of the Pochhammer symbol $(s)_n$, which is defined as

$$(s)_n := \frac{\Gamma(s+n)}{\Gamma(s)}.$$

2.2 Elementary Integrals

For any real number $r \in \mathbb{R}$ and complex number v with $\mathrm{Re}(v) > 0$, we have, from 3.985.1 of [GR07] the integral formula

$$\int_0^\infty \cos(ur)\cosh^{-v}(u)\,du = \frac{2^{v-2}}{\Gamma(v)}\Gamma\left(\frac{v-ir}{2}\right)\Gamma\left(\frac{v+ir}{2}\right). \qquad (3)$$

If $r = 0$, then we get the important special case that

$$\int_0^\infty \cosh^{-v}(u)\,du = \frac{2^{v-2}\Gamma^2(v/2)}{\Gamma(v)}, \qquad (4)$$

which is stated in 3.512.1 of [GR07]. Additionally, we will use that for any $r \in \mathbb{C}$ with $\mathrm{Re}(r^2) > 0$ and $u \in \mathbb{C}$ with $\mathrm{Re}(u) > 0$, one has that

$$\frac{u}{\sqrt{4\pi}}\int_0^\infty e^{-r^2 t}e^{-u^2/(4t)}t^{-1/2}\frac{dt}{t} = e^{|r|u}. \qquad (5)$$

For any real valued function g and $r \in \mathbb{C}$, we define $H(r, g)$ as

$$H(r, g) := 2\int_0^\infty \cos(ur)g(u)\,du. \qquad (6)$$

This is a purely formal definition; the conditions on g under which we consider $H(r, g)$ are stated in section 3 below.

2.3 An Asymptotic Formula

For the convenience of the reader, we state here a result from page 37 of [Er56]. Let $(\alpha, \beta) \subset \mathbb{R}$. Let g be a real-valued continuous function, and h be a real-valued continuously differentiable function, on (α, β) such that the integral

$$\int_\alpha^\beta g(t)e^{xh(t)}\,dt$$

exists for sufficiently large x. Assume there is an $\eta > 0$ such that $h'(t) < 0$ for $t \in (\alpha, \alpha + \eta)$. In addition, for some $\epsilon > 0$, assume that $h(t) \le h(\alpha) - \epsilon$ for $t \in (\alpha + \eta, \beta)$. Suppose that

$$h'(t) = -a(t-\alpha)^{v-1} + o((t-\alpha)^{v-1})$$

and

$$g(t) = b(t-\alpha)^{\lambda-1} + o((t-\alpha)^{\lambda-1})$$

as $t \to \alpha^+$ for some positive λ and ν. Then

$$\int_\alpha^\beta g(t)e^{xh(t)}dt = \frac{b}{\nu}\Gamma(\lambda/\nu)(\nu/(ax))^{\lambda/\nu}e^{xh(\alpha)}(1+o(1)) \quad \text{as } x \to \infty. \quad (7)$$

Note that the assumptions on h hold if $\alpha = 0$, $h(0) = 0$ and is monotone decreasing, which is the setting in which we will apply the above result. In this case, we will use that the above integral is $O(x^{-\lambda/\nu})$.

2.4 Geometric Setting

Let X be a compact, complex, smooth projective variety of complex dimension N. Fix a smooth Kähler metric μ on X, which is associated to the Kähler $(1, 1)$ form ω. Let ρ denote a (local) potential for the metric μ. If we choose local holomorphic coordinates $z_1, \ldots, z_N$ in the neighborhood of a point on X, then one can write ω as

$$\omega = \frac{i}{2}\sum_{j,k=1}^N g_{j,\bar k}dz_j \wedge d\bar z_k = \frac{i}{2}\partial_z\partial_{\bar z}\rho$$

If M is a subvariety of X, the induced metric on M will be denoted by μ_M. In particular, the induced metric on X itself is μ_X, which we will simply write as μ. In a slight abuse of notation, we will also write μ_M, or μ in the case $M = X$, for the associated volume form against which one integrates functions.

The corresponding Laplacian Δ_X which acts on smooth functions on X is

$$\Delta_X = -\sum_{j,k=1}^N g^{j,\bar k}\frac{\partial^2}{\partial z_j \partial \bar z_k},$$

where, in standard notation, $(g^{j,\bar k}) = (g_{j,\bar k})^{-1}$; see page 4 of [Ch84]. An eigenfunction of the Laplacian Δ_X is an *a priori* C^2 function ψ_j which satisfies the equation

$$\Delta_X\psi_j - \lambda_j\psi_j = 0$$

for some constant λ_j, which is the eigenvalue associated to ψ_j. It is well-known that any eigenfunction is subsequently smooth, and every eigenvalue is greater than zero except when ψ_j is a constant whose corresponding eigenvalue is zero. As is standard, we assume that each eigenfunction is normalized to have L^2 norm equal to one.

Weyl's law asserts that

$$\#\{\lambda_j | \lambda_j \leq T\} = (2\pi)^{-2N}\text{vol}_N(\mathbb{B})\text{vol}_\omega(X)T^N + O(T^{N-1/2}) \quad \text{as } T \to \infty$$

and $\text{vol}_\omega(X)$ is the volume of X under the metric μ induced by ω, and $\text{vol}_N(\mathbb{B})$ is the volume of the unit ball in $\mathbb{R}^{2N}$. As a consequence of Weyl's law, one has that for any $\varepsilon > 0$,

$$\sum_{k=1}^{\infty} \lambda_k^{-N-\varepsilon} < \infty; \tag{8}$$

see, for example, page 9 of [Ch84]. The eigenfunction ψ_j corresponding to the eigenvalue λ_j satisfies a sup-norm bound on X, namely that

$$\|\psi_j\|_\infty = O_X\left(\lambda_j^{N/2-1/4}\right); \tag{9}$$

see [SZ02] and references therein.

2.5 Holomorphic Forms

By a holomorphic form F we mean a holomorphic section of a power of the canonical bundle Ω on X; see page 146 of [GH78]. (Note: On page 146 of [GH78], the authors denote the canonical bundle by K_X, which we will not since this notation is being used both for the heat kernel and the function obtained by applying the wave distribution to hyperbolic cosine.) The weight n of the form equals the power of the bundle of which F is a section. Let D denote the divisor of F, and assume that D is smooth up to codimension two. In the case X is a quotient of a symmetric space G/K by a discrete group Γ, then F is a holomorphic automorphic form on G/K with respect to Γ. With a slight gain in generality, and with no increase in complication of the analysis, we can consider sections of the canonical bundle obtained by considering the tensor product of the canonical bundle with a flat line bundle. The Kähler form ω will induce a norm on F, which we denote by $\|F\|_\omega$; see [GH78] for a general discussion as well as section 2 of [JK01]. We can describe the norm as follows.

As in the notation of section 2 of [JK01], let U be an element of an open cover of X. Once we trivialize Ω on U, we can express the form F in local coordinates $z_1, \ldots, z_N$. Also, we have the existence of a Kähler potential ρ of the Kähler form ω. Up until now, there has been no natural scaling of ω. We do so now, by scaling ω by a multiplicative constant c so that $c\omega$ is a Chern form of Ω; see page 144 of [GH78] as well as chapter 2 of [Fi18]. In a slight abuse of notation, we will denote the re-scaled Kähler form by ω.

With this scaling of ω, one can show that $|F(z)|e^{-n\rho(z)}$ is invariant under change of coordinates; see section 2.3 of [JK01]. With this, one defines

$$\|F\|_\omega^2(z) := |F(z)|^2 e^{-2n\rho(z)}, \tag{10}$$

where n is the weight of the form. The formula is local for each U in the open cover, but its invariance implies that the definition extends independently of the various choices made. Following the discussion of Chapter 1 of [La88], the above equation can be written in differential form as

$$dd^c \log \|F\|_\omega^2 = n(\delta_D - \omega) \tag{11}$$

where δ_D denotes the Dirac delta distribution supported on D.

Kähler metrics have the property that the associated Laplacian of a function does not involve derivatives of the metric, as stated on page 75 of [Ba06]. Exercise 1.27.3(a) of [Fi18] states the formula

$$\frac{1}{2}\Delta_{X,d} f \omega^n = ni\partial\bar\partial f \wedge \omega^{n-1} \tag{12}$$

where $\Delta_{X,d}$ is the Laplacian stemming from the differential d and f is a smooth function, subject to the normalizations of various operators as stated in [Fi18]. As a corollary of (12), one can interpret (11) as asserting that $\Delta_X \log \|F\|_\omega^2$ is a non-zero constant away from D.

2.6 The Heat and Poisson Kernel

The heat kernel acting on smooth functions on X can be defined formally as

$$K_X(z, w; t) = \sum_{k=0}^\infty e^{-\lambda_k t} \psi_k(z)\overline{\psi_k}(w),$$

where $\{\psi_j\}$ are eigenfunctions associated to the eigenvalue λ_j. As a consequence of Weyl's law and the sup-norm bound for eigenfunctions, the series which defines the heat kernel converges for all $t > 0$ and $z, w \in X$. Furthermore, if $z \neq w$, then the heat kernel has exponential decay when t approaches zero; see page 198 of [Ch84].

For any $Z \in \mathbb{C}$ with $\mathrm{Re}(Z) \geq 0$, the *translated by* $-Z$ *Poisson Kernel* $\mathcal{P}_{X,-Z}(z, w; u)$, for $z, w \in X$ and $u \in \mathbb{C}$ with $\mathrm{Re}(u) \geq 0$ is defined by

$$\mathcal{P}_{X,-Z}(z, w; u) = \frac{u}{\sqrt{4\pi}} \int_0^\infty K_X(z, w; t)e^{-Zt}e^{-u^2/(4t)}t^{-1/2}\frac{dt}{t}. \tag{13}$$

The translated Poisson kernel $\mathcal{P}_{X,-Z}(z, w; u)$ is a fundamental solution associated to the differential operator $\Delta_X + Z - \partial_u^2$. For certain considerations to come, we will choose a constant $\rho_0 \geq 0$, which will depend on the geometry of X, and write each eigenvalue of Δ_X as $\lambda_j = \rho_0^2 + t_j^2$. Thus, we divide the spectral expansion of the heat kernel K_X into two subsets: The finite sum for

$\lambda_j < \rho_0^2$, so then $t_j \in (0, i\rho_0]$, and the sum over $\lambda_j \geq \rho_0^2$, so then $t_j \geq 0$. Using (5), we get the spectral expansion

$$\mathcal{P}_{X,-Z}(z, w; u) = \sum_{\lambda_k < \rho_0^2} e^{-u\sqrt{\lambda_k + Z}} \psi_k(z)\overline{\psi}_k(w) + \sum_{\lambda_k \geq \rho_0^2} e^{-u\sqrt{\lambda_k + Z}} \psi_k(z)\overline{\psi}_k(w).$$

(14)

By Theorem 5.2 and Remark 5.3 of [JLa03], $\mathcal{P}_{X,-Z}(z, w; u)$ admits an analytic continuation to $Z = -\rho_0^2$. In analogy with [JvPS16], we can deduce that that the continuation of $\mathcal{P}_{X,-Z}(z, w; u)$ for $Z = -\rho_0^2$, with $\text{Re}(u) > 0$ and $\text{Re}(u^2) > 0$ is given by

$$\mathcal{P}_{X,\rho_0^2}(z, w; u) = \sum_{\lambda_k < \rho_0^2} e^{-u\sqrt{\lambda_k - \rho_0^2}} \psi_k(z)\overline{\psi}_k(w) + \sum_{\lambda_k \geq \rho_0^2} e^{-ut_k} \psi_k(z)\overline{\psi}_k(w),$$

(15)

where $\sqrt{\lambda_k - \rho_0^2} = t_k \in (0, i\rho_0]$ is taken to be the branch of the square root obtained by analytic continuation through the upper half-plane.

As stated, if $z \neq w$, then the heat kernel has exponential decay as t approaches zero. From this, one can show that the Poisson kernel $\mathcal{P}_{X,-Z}(z, w; u)$ is bounded as u approaches zero for any Z.

At this point, we would like to define the (translated by ρ_0^2) wave kernel by defining

$$W_{X,\rho_0^2}(z, w; u) = \mathcal{P}_{X,\rho_0^2}(z, w; iu) + \mathcal{P}_{X,\rho_0^2}(z, w; -iu),$$

for some branch of the meromorphic continuation of $\mathcal{P}_{X,\rho_0^2}(z, w; u)$ to all $u \in \mathbb{C}$. However, because of convergence issues, we cannot simply replace u by iu in the expression for the Poisson kernel. As a result, we define the wave distribution via the spectral expansion for the analytic continuation of $\mathcal{P}_{X,\rho_0^2}$ in (15).

2.7 An Elementary, yet Important, Rescaling Observation

By writing the Laplacian as in the beginning of section 2.3, we have established specific conventions regarding various scales, or multiplicative constants, in our analysis. However, there is one additional scaling which could be employed. Specifically, one could consider the heat equation $\Delta_z + c\partial_t$ for any positive constant c. The associated heat kernel would be $K_X(z, w; t/c)$, if the heat kernel associated to $\Delta_z + \partial_t$ is $K_X(z, w; t)$. In doing so, we would replace (13) by

$$\mathcal{P}_{X,-Z}(z, w; u) = \frac{u}{\sqrt{4\pi}} \int_0^\infty K_X(z, w; t/c) e^{-Zt/c} e^{-u^2/(4t)} t^{-1/2} \frac{dt}{t} \quad (16)$$

for some positive constant c. In effect, we are changing the coordinate for the positive real axis $\mathbb{R}^+$ from the parameter t to t/c for any positive constant c. In this manner, we rescale the data from the beginning of our consideration so that when we study the translation of the heat kernel, we can, provided $\rho_0^2 > 0$, choose c appropriately so that translation ρ_0^2/c is always equal to $1/4$.

In the examples we develop, the choice of the ρ_0^2 will be determined by a "non-spectral" representation of the heat kernel, after which we choose $c = 1/(4\rho_0^2)$, provided $\rho_0 \neq 0$. As it turns out, the translation by $1/4$ matters. This point will become relevant in section 6 below.

3 The Wave Distribution

For $z, w \in X$ and function $g \in C_c^\infty(\mathbb{R}^+)$, we formally define the wave distribution $\mathcal{W}_{X,\rho_0^2}(z, w)(g)$ applied to g by the series

$$\mathcal{W}_{X,\rho_0^2}(z, w)(g) = \sum_{\lambda_j \geq 0} H(t_j, g) \psi_j(z) \overline{\psi}_j(w), \quad (17)$$

where $H(t_j, g)$ is given by (6) and $t_j = \sqrt{\lambda_j - \rho_0^2}$ if $\lambda_j \geq \rho_0^2$, otherwise $t_j \in (0, i\rho_0]$.

Definition 3.1 *For $a \in \mathbb{R}^+$ and $m \in \mathbb{N}$, let $S'_m(\mathbb{R}^+, a)$ be the set of Schwartz functions on $\mathbb{R}^+$ with $g^{(k)}(0) = 0$ for all odd integers k with $0 \leq k \leq m+1$ and where $e^{ua}|g(u)|$ is dominated by an integrable function on $\mathbb{R}^+$.*

The following proposition addresses the question of convergence of (17).

Theorem 3.2 *Fix $z, w \in X$, with $z \neq w$, there exists a continuous, real-valued function $F_{z,w}(u)$ on $\mathbb{R}^+$ and an integer m sufficiently large such that the following assertions hold.*

(i) *One has that $F_{z,w}(u) = (-1)^{m+1} \sum_{\lambda_j < \rho_0^2} e^{u\sqrt{\rho_0^2 - \lambda_j}} \cdot t_j^{-(m+1)} \psi_j(z)$*
   $\overline{\psi}_j(w) + O(u^{m+1})$ as $u \to \infty$.

(ii) *For any non-negative integer $j \leq m$, we have the bound $\partial_u^j F_{z,w}(u) = O(u^{m+1-j})$ as $u \to 0^+$.*

(iii) For any $g \in S'_m(\mathbb{R}^+, \rho_0)$ such that $\partial_u^j g(u) \exp(\rho_0 u)$ has a limit as $u \to \infty$ and is bounded by some integrable function on $\mathbb{R}^+$ for all non-negative integers $j \leq m + 1$, we have

$$\mathcal{W}_{X,\rho_0^2}(z, w)(g) = \int_0^\infty F_{z,w}(u)\partial_u^{m+1} g(u)du. \tag{18}$$

The implied constants in the error terms in statements (i) and (ii) depend on m and the distance between z and w.

Proof. Choose an integer $m \geq 4N + 1$, where N is the complex dimension of X. To begin, we claim the following statement: For every integer k with $0 \leq k \leq m$, there is a polynomial $h_{k,m}(x)$ of degree at most m such that

$$h_{k,m}(\sin(x)) = \frac{x^k}{k!} + O_m(x^{m+1}) \quad \text{as } x \to 0.$$

Indeed, one begins by initially setting $h_{k,m}^{(0)}(x) = x^k/k!$. The function $h_{k,m}^{(0)}(\sin(x))$ has a Taylor series expansion near zero of the form

$$h_{k,m}^{(0)}(\sin(x)) = \frac{x^k}{k!} + c_\ell x^\ell + O_\ell(x^{\ell+1}) \quad \text{as } x \to 0$$

for some real number c_ℓ and integer $\ell \geq k + 1$. Now set $h_{k,m}^{(1)}(x)$ to be $h_{k,m}^{(1)}(x) = h_{k,m}^{(0)}(x) - c_\ell x^\ell$ so then

$$h_{k,m}^{(1)}(\sin(x)) = \frac{x^k}{k!} + c_p x^p + O_p(x^{p+1}) \quad \text{as } x \to 0$$

for some real number c_p and integer $p \geq \ell + 1 \geq k + 2$. One can continue to subtract multiplies of monomials of higher degree, thus further reducing the order of the error term until the claimed result is obtained; it is elementary to complete the proof of the assertion with the appropriate proof by induction argument.

Having proved the above stated assertion, we then have for any $\zeta \in \mathbb{C} \setminus \{0\}$ that

$$e^{-t\zeta} - \sum_{k=0}^m h_{k,m}(\sin(t))(-\zeta)^k = O_\zeta(t^{m+1}) \quad \text{as } t \to 0. \tag{19}$$

For $t > 0$ we define

$$P_m(t, \zeta) := \frac{e^{-t\zeta} - \sum_{k=0}^m h_{k,m}(\sin(t))(-\zeta)^k}{(-t)^{m+1}} \tag{20}$$

and set $P_m(0, \zeta) = \lim_{t \to 0} P_m(t, \zeta)$; the existence of this limit is ensured by (19). For $\text{Re}(\zeta) \geq 0$, we have $P_m(t, \zeta) = O(t^{-m-1})$ as $t \to \infty$. Hence, the bounds for the eigenvalue growth (8) and sup-norm for eigenfunctions (9), together with the choice of m, imply that the series

$$\tilde{F}_{z,w}(\zeta) = \sum_{\lambda_j \geq 0} P_m(t_j, \zeta) \psi_j(z) \overline{\psi}_j(w) \tag{21}$$

converges uniformly and absolutely on X for ζ in the closed half-plane $\mathrm{Re}(\zeta) \geq 0$.

Furthermore, any of the first $m + 1$ derivatives of $\tilde{F}_{z,w}(\zeta)$ in ζ converges uniformly and absolutely when $\mathrm{Re}(\zeta) > 0$. This allows us to differentiate the series above term by term, and by doing so $m + 1$ times and using that

$$\frac{d^{m+1}}{d\zeta^{m+1}} P_m(t_j, \zeta) = e^{-t_j \zeta}$$

we conclude that for $\mathrm{Re}(\zeta) > 0$, one has the identity

$$\frac{d^{m+1}}{d\zeta^{m+1}} \tilde{F}_{z,w}(\zeta) = \mathcal{P}_{M,\rho_0^2}(z, w; \zeta), \tag{22}$$

where $\mathcal{P}_{M,\rho_0^2}$ is defined in (15). Set $\mathcal{P}^{(0)}(z, w; \zeta) = \mathcal{P}_{X,\rho_0^2}(z, w; \zeta)$, and define inductively for $\mathrm{Re}(\zeta) > 0$ the function

$$\mathcal{P}^{(k)}(z, w; \zeta) = \int_0^\zeta \mathcal{P}^{(k-1)}(z, w; \xi) \, d\xi,$$

where the integral is taken over a ray contained in the upper half plane. Note that

$$\mathcal{P}^{(k)}(z, w; \zeta) = O_{z,w,k}(\zeta^k) \quad \text{as } \zeta \to 0. \tag{23}$$

From (22), we have that

$$\mathcal{P}^{(m+1)}(z, w; \zeta) - \tilde{F}_{z,w}(\zeta) = q_m(z, w; \zeta), \tag{24}$$

where $q_m(z, w; \zeta)$ is a degree m polynomial in ζ with coefficients which depend on z and w. Using this, for $u \in \mathbb{R}^+$ we define

$$F_{z,w}(u) = \frac{1}{2i} \left[\left(\tilde{F}_{z,w}(iu) + q_m(z, w; iu) \right) - \left(\tilde{F}_{z,w}(-iu) + q_m(z, w; -iu) \right) \right] \tag{25}$$

Assertions (i) and (ii) follow immediately from the above construction of F and its relation to the Poisson kernel. Since the expansion (21) converges uniformly for $\mathrm{Re}(\zeta) = 0$, property (iii) will follow directly from (i), (ii) and $(m + 1)$ term-by-term integration by parts. □

4 A Basic Test Function

The building block for our Kronecker limit formula is obtained by applying the wave distribution to the function $\cosh^{-(s-\rho_0)}(u)$. As stated above, we will choose ρ_0 depending on the geometry of X and then re-scale the time variable in the heat kernel so that ultimately we have either $\rho_0 = 0$ or $\rho_0 = 1/2$. For the time being, let us work out the results for a general ρ_0.

Proposition 4.1 *For* $s \in \mathbb{C}$ *with* $\mathrm{Re}(s) > 2\rho_0$, *the wave distribution of* $g(u) = \cosh^{-(s-\rho_0)}(u)$ *exists and admits the spectral expansion*

$$\mathcal{W}_{X,\rho_0^2}(z,w)(\cosh^{-(s-\rho_0)}) = \sum_{\lambda_j \geq 0} c_{j,(s-\rho_0)} \psi_j(z)\overline{\psi}_j(w), \qquad (26)$$

where

$$c_{j,(s-\rho_0)} = \frac{2^{s-\rho_0-1}}{\Gamma(s-\rho_0)}\Gamma\left(\frac{s-\rho_0-it_j}{2}\right)\Gamma\left(\frac{s-\rho_0+it_j}{2}\right). \qquad (27)$$

Furthermore, for any $z, w \in X$, *the series* (26) *converges absolutely and uniformly in* s *on any compact subset of the half-plane* $\mathrm{Re}(s) > 2\rho_0$.

Proof. If $\mathrm{Re}(s) > 2\rho_0$, then the conditions of Theorem 3.2 apply. The spectral coefficients are computed using (3) and (4). Finally, Stirling's formula (2) implies that the factor (27) decays exponentially as $t_j \to \infty$. When combined with the sup-norm bound (9) on the eigenfunctions ψ_j, the assertion regarding uniform convergence follows. $\square$

Corollary 4.2 *For* $z, w \in X$ *with* $z \neq w$ *and* $s \in \mathbb{C}$ *with* $\mathrm{Re}(s) > 2\rho_0$, *let*

$$K_{X;\rho_0^2}(z,w;s) := \frac{\Gamma(s-\rho_0)}{\Gamma(s)}\mathcal{W}_{X,\rho_0^2}(z,w)(\cosh^{-(s-\rho_0)}).$$

Then the function $\Gamma(s)\Gamma^{-1}(s-\rho_0)K_{X;\rho_0^2}(z,w;s)$ *admits a meromorphic continuation to all* $s \in \mathbb{C}$ *with poles at points of the form* $s = \rho_0 \pm it_j - 2m$ *for any integer* $m \geq 0$. *Furthermore, the function* $K_{X;\rho_0^2}(z,w;s)$ *satisfies the differential-difference equation*

$$(\Delta_X + s(s-2\rho_0))K_{X;\rho_0^2}(z,w;s) = s(s+1)K_{X;\rho_0^2}(z,w;s+2). \qquad (28)$$

Proof. By Proposition 4.1, $K_{X;\rho_0^2}(z,w;s)$ is well defined for $\mathrm{Re}(s) > 2\rho_0$. Keeping $\mathrm{Re}(s) > 2\rho_0$, we have

$$K_{X;\rho_0^2}(z,w;s) = \frac{\Gamma(s-\rho_0)}{\Gamma(s)}\sum_{\lambda_j \geq 0} H(t_j, \cosh^{-(s-\rho_0)})\psi_j(z)\overline{\psi}_j(w). \qquad (29)$$

For $v \in \mathbb{C}$, $\text{Re}(v) > \rho_0$, $r \in \mathbb{R}^+$ or $r \in [0, i\rho_0]$ and non-negative integer n, one has that

$$H(r, \cosh^{-(v+2n)}) = H(r, \cosh^{-v}) \frac{2^{2n} \left(\frac{v+ir}{2}\right)_n \left(\frac{v-ir}{2}\right)_n}{(v)_{2n}}. \tag{30}$$

Indeed, the evaluation of $H(t_j, \cosh^{-(s-\rho_0)})$ in terms of the Gamma function is stated in (27). One then can use that $\Gamma(s+1) = s\Gamma(s)$ and the definition of the Pochammer symbol $(s)_n$ to arrive at (30). With this, we can write, for any positive integer n,

$$\frac{2^{2n}\Gamma(s)}{\Gamma(s-\rho_0)} K_{X;\rho_0^2}(z, w; s)$$
$$= \frac{\Gamma(s-\rho_0+2n)}{\Gamma(s-\rho_0)} \sum_{\lambda_j \geq 0} \frac{H(t_j, \cosh^{-(s-\rho_0+2n)})}{Q_n(t_j, s-\rho_0)} \psi_j(z)\overline{\psi_j}(w), \tag{31}$$

where

$$Q_n(r, v) = \left(\frac{v+ir}{2}\right)_n \left(\frac{v-ir}{2}\right)_n.$$

For $n \geq \lfloor \rho_0 \rfloor + 1$, the right-hand-side of (31) defines a meromorphic function in the half-plane $\text{Re}(s) > 2\rho_0 - 2n$ with possible poles at the points $s = \rho_0 \pm it_j - 2l$, for $l \in 0, \ldots, n-1$. Therefore, the function $\Gamma(s)\Gamma^{-1}(s-\rho_0)K_{X;\rho_0^2}(z, w; s)$ admits a meromorphic continuation to all $v \in \mathbb{C}$ with poles at points $s = \rho_0 \pm it_j - 2m$ for integers $m \geq 0$.

It remains to prove the difference-differential equation. As stated, the right-hand-side of the equation (29) converges absolutely and uniformly on compact subsets of the right half plane $\text{Re}(s) > 2\rho_0$. When viewed as a function of $z \in X$, the convergence is uniform on X. Therefore, when restricting s to $\text{Re}(s) > 2\rho_0$, we can interchange the action of Δ_X and the sum in (29) to get

$$\Delta_X K_{X;\rho_0^2}(z, w; s) = \frac{\Gamma(s-\rho_0)}{\Gamma(s)} \sum_{\lambda_j \geq 0} (t_j^2 + \rho_0^2) H(t_j, \cosh^{-(s-\rho_0)}) \psi_j(z)\overline{\psi_j}(w).$$

Applying (30) with $n = 1$ we can write the above equation, for sufficiently large $\text{Re}(s)$ as

$$\Delta_X K_{X;\rho_0^2}(z, w; s)$$
$$= \frac{\Gamma(s+2-\rho_0)}{\Gamma(s)} \sum_{\lambda_j \geq 0} \frac{(t_j^2 + \rho_0^2)}{(s-\rho_0)^2 + t_j^2} H(t_j, \cosh^{-(s+2-\rho_0)}) \psi_j(z)\overline{\psi_j}(w).$$

Let $n = 1$ in (31) and multiply by $2^{-2}s(s - 2\rho_0)\Gamma(s - \rho_0)\Gamma^{-1}(s)$ to get

$$s(s - 2\rho_0)K_{X;\rho_0^2}(z, w; s) = s(s + 1)\frac{\Gamma(s + 2 - \rho_0)}{\Gamma(s + 2)}$$

$$\times \sum_{\lambda_j \geq 0} H(t_j, \cosh^{-(s+2-\rho_0)})\frac{s^2 - 2s\rho_0}{(s - \rho_0)^2 + t_j^2}\psi_j(z)\overline{\psi}_j(w).$$

Adding up the last two equations, we obtained the desired result for sufficiently large $\mathrm{Re}(s)$, and then for all s by meromorphic continuation. $\qquad\square$

Remark 4.3   It is necessary to assume that $z \neq w$ when considering the wave distribution of the test function $g(u) = \cosh^{-(s-\rho_0)}(u)$. Only after one computes the spectral expansion of $K_{X;\rho_0^2}(z, w; s)$ is one able to extend the function to $z = w$.

5 Two Series Expansions

We will define two series using the function $K_{X;\rho_0^2}(z, w; s)$. The first, in the next Theorem, is shown to equal the resolvent kernel, which is integral kernel that inverts the operator $(\Delta_X + s(s - 2\rho_0))$ for almost all values of s. As a reminder, the resolvent kernel can be realized as an integral transform of the heat kernel $K_X(z, w; t)$, namely

$$\int_0^\infty K_X(z, w; t)e^{-s(s-2\rho_0)t}\,dt,$$

provided $z \neq w$ and $\mathrm{Re}(s(s - 2\rho_0)) > 0$. In that instance, the heat kernel decays exponentially as t approaches zero, so then

$$\int_0^\infty K_X(z, w; t)e^{-s(s-2\rho_0)t}\,dt$$

$$= \lim_{\epsilon \to 0}\int_\epsilon^\infty K_X(z, w; t)e^{-s(s-2\rho_0)t}\,dt$$

$$= \lim_{\epsilon \to 0}\sum_{\lambda_j \geq 0}\frac{1}{(s - \rho_0)^2 + t_j^2}\psi_j(z)\overline{\psi}_j(w) \cdot e^{-\epsilon(s(s-\rho_0)+\lambda_j)}.$$

$$(32)$$

Theorem 5.1 *For $z, w \in X$, $z \neq w$ and $s \in \mathbb{C}$ with $\mathrm{Re}(s) > 2\rho_0$, consider the function*

$$G_{X;\rho_0^2}(z, w; s) = \frac{2^{-s-1+\rho_0}\Gamma(s)}{\Gamma(s + 1 - \rho_0)}\sum_{k=0}^\infty \frac{\left(\frac{s}{2}\right)_k\left(\frac{s}{2} + \frac{1}{2}\right)_k}{k!\,(s + 1 - \rho_0)_k}K_{X;\rho_0^2}(z, w; s + 2k).$$

$$(33)$$

Then we have the following results.

i) *The series defining $G_{X;\rho_0^2}(z, w; s)$ is holomorphic in the half-plane* $\text{Re}(s) > 2\rho_0$ *and continues meromorphically to the whole s-plane.*

ii) *The function $G_{X;\rho_0^2}(z, w; s)$ admits the spectral expansion*

$$G_{X;\rho_0^2}(z, w; s) = \sum_{\lambda_j \geq 0} \frac{1}{(s - \rho_0)^2 + t_j^2} \psi_j(z)\overline{\psi_j}(w)$$

which is conditionally convergent, in the sense of (32), for $z \neq w$ and for all $s \in \mathbb{C}$ provided $s(s - \rho_0) + \lambda_j \neq 0$ for some λ_j.

iii) *The function $G_{X;\rho_0^2}(z, w; s)$ satisfies the equation*

$$(\Delta_X + s(s - 2\rho_0))G_{X;\rho_0^2}(z, w; s) = 0, \tag{34}$$

for all $s \in \mathbb{C}$ provided $s(s - \rho_0) + \lambda_j \neq 0$ for some λ_j.

Proof. Let us study each term in the series (33), which is

$$\frac{2^{-s-1+\rho_0}\Gamma(s)}{\Gamma(s + 1 - \rho_0)} \frac{\left(\frac{s}{2}\right)_k \left(\frac{s}{2} + \frac{1}{2}\right)_k}{k! (s + 1 - \rho_0)_k} K_{X;\rho_0^2}(z, w; s + 2k)$$

$$= 2^{-1+\rho_0} \frac{2^{-(s+2k)}\Gamma(s + 2k - \rho_0)}{\Gamma(k + 1)\Gamma(s + 1 - \rho_0 + k)}$$

$$\times \mathcal{W}_{X,\rho_0^2}(z, w)(\cosh^{-(s+2k-\rho_0)}). \tag{35}$$

For now, let us assume that $\text{Re}(s) > 2\rho_0$. A direct computation using Stirling's formula (1) yields that

$$2^{-1+\rho_0} \frac{2^{-(s+2k)}\Gamma(s + 2k - \rho_0)}{\Gamma(k + 1)\Gamma(s + 1 - \rho_0 + k)} = O_s(k^{-3/2}) \quad \text{as } k \to \infty. \tag{36}$$

It remains to determine the asymptotic behavior of the factor in (35) involving the wave distribution. For this, Theorem 3.2 implies that for any $\delta > 0$, there is a $C > 1$ depending upon the distance between z and w such that we have the bound

$$\int_\delta^\infty F_{z,w}(u)\partial_u^{m+1}\left(\cosh^{-(s+2k-\rho_0)}(u)\right) du = O_{s,z,w}(C^{-(s+2k-\rho_0)}) \quad \text{as } k \to \infty,$$

where the implied constant depends on the distance between z and w and $m \geq 4N + 1$ is a sufficiently large, fixed integer. Since $z \neq w$, we can combine equations (22), (23), (24) and (25) together with integration by parts to write, for some $C_1 > 1$

$$\int_0^\delta F_{z,w}(u) \partial_u^{m+1} \left(\cosh^{-(s+2k-\rho_0)}(u) \right) du$$

$$= (-1)^{m+1} \int_0^\delta \left(\partial_u^{m+1} F_{z,w}(u) \right) \cosh^{-(s+2k-\rho_0)}(u) du$$

$$+ O_{s,z,w}(C_1^{-(s+2k-\rho_0)}) \quad \text{as } k \to \infty. \tag{37}$$

In essence, the use of Theorem 3.2 ensures that the boundary term at $u = 0$ vanishes, so then the constant C_1 comes from the evaluation of the boundary terms at $u = \delta$ and depends on the distance between z and w. To finish, we may use (7) where $h(t) = -\log(\cosh(t))$, so then $\lambda = 1$ and $\nu = 2$, to conclude that

$$\int_0^\delta \left(\partial_u^{m+1} F_{z,w}(u) \right) \cosh^{-(s+2k-\rho_0)}(u) du = O_{s,z,w}(k^{-1/2}) \quad \text{as } k \to \infty,$$

$$\tag{38}$$

where, again, the implied constant depends on the distance between z and w, and we assume that $z \neq w$. If we combine (36), (37), and (38),we obtain that (35) is of order $O_{s,z,w}(k^{-2})$. Therefore, the series (33) converges uniformly and absolutely for s in compact subsets in a right half plane $\mathrm{Re}(s) > 2\rho_0$, and $z, w \in X$ provided z and w are uniformly bounded apart.

At this point, we have the convergence of the series defining $G_{X;\rho_0^2}(z, w; s)$ for $\mathrm{Re}(s) > 2\rho_0$. In order to obtain the meromorphic continuation of (33), re-write the series as a finite sum of terms for $k \leq n$ and an infinite sum for $k > n$, for any integer n. For the finite sum, the meromorphic continuation is established in Corollary 4.2. For the infinite sum, the above argument applies to prove the convergence in the half-plane $\mathrm{Re}(s) > 2\rho_0 - n$. With this, we have completed the proof of assertion (i).

Going further, one can follow the argument given above using (2) for any positive integer ℓ and conclude that for $z \neq w$, s in some compact subset of the half-plane $\mathrm{Re}(s) > 2\rho_0$, we have

$$\partial_s^\ell \left(\frac{2^{-s-1+\rho_0} \Gamma(s)}{\Gamma(s+1-\rho_0)} \frac{(\frac{s}{2})_k (\frac{s}{2}+\frac{1}{2})_k}{k! (s+1-\rho_0)_k} K_{X;\rho_0^2}(z, w; s+2k) \right)$$

$$= O_{s,z,w}(k^{-2-\ell/2}) \quad \text{as } k \to \infty,$$

where the implied constant depends on the distance between z and w and the compact set which contains s. Namely, repeated differentiation of Gamma factors ℓ times reduces the exponent by ℓ, while differentiation of (37), after application of formula (7), reduces the exponent by $\ell/2$.

The convergence of the series (33) for $\operatorname{Re}(s) > 2\rho_0$ as well as the series of derivatives allows us to interchange differentiation and summation. Therefore, for $z \neq w$, $\operatorname{Re}(s) > 2\rho_0$ and any $\ell \geq 0$ we get

$$
\partial_s^\ell \left(\sum_{k=0}^{\infty} \frac{\left(\frac{s}{2}\right)_k \left(\frac{s}{2}+\frac{1}{2}\right)_k}{k!\,(s+1-\rho_0)_k} K_{X;\rho_0^2}(z,w;s+2k) \right)
$$

$$
= \sum_{k=0}^{\infty} \partial_s^\ell \left(\frac{\left(\frac{s}{2}\right)_k \left(\frac{s}{2}+\frac{1}{2}\right)_k}{k!\,(s+1-\rho_0)_k} K_{X;\rho_0^2}(z,w;s+2k) \right). \tag{39}
$$

Now, we would like to include the case $z = w$. Recall that

$$
K_{X;\rho_0^2}(z,w;s) = \frac{\Gamma(s-\rho_0)}{\Gamma(s)} \sum_{\lambda_j \geq 0} c_{j,(s-\rho_0)} \psi_j(z) \overline{\psi}_j(w),
$$

where the spectral coefficients $c_{j,(s-\rho_0)}$ are given by (27). The coefficients $c_{j,(s-\rho_0)}$ are exponentially decreasing in $t_j = \sqrt{\lambda_j - \rho_0^2}$, as $j \to \infty$ and differentiable with respect to s, with the derivatives also exponentially decreasing in t_j. Moreover, the application of the sup-norm bound for the eigenfunctions ψ_j and the Stirling formula for coefficients $c_{j,(s+2k-\rho_0)}$ shows that, uniformly in $z, w \in X$ for s in a compact subset of the half-plane $\operatorname{Re}(s) > 2\rho_0$, one has

$$
\left| \mathcal{W}_{X,\rho_0^2}(z,w)(\cosh^{-(s+2k-\rho_0)}) \right| = O_{s,z,w}(k^{N+1}),
$$

where N is the complex dimension of X. In addition, repeated differentiation of the coefficients with respect to s reduces the exponent of k by one each time. Therefore, for sufficiently large ℓ

$$
\sum_{k=0}^{\infty} \left| \partial_s^\ell \left(\frac{\left(\frac{s}{2}\right)_k \left(\frac{s}{2}+\frac{1}{2}\right)_k}{k!\,(s+1-\rho_0)_k} K_{X;\rho_0^2}(z,w;s+2k) \right) \right| = O_{s,z,w}(1),
$$

where $\mathrm{Re}(s) > 2\rho_0$, and the bound is uniform in $z, w \in X$. Hence, we may interchange the sum and the integral to get, for sufficiently large ℓ

$$
\int_X \partial_s^\ell \left(\sum_{k=0}^\infty \frac{\left(\frac{s}{2}\right)_k \left(\frac{s}{2} + \frac{1}{2}\right)_k}{k!\,(s+1-\rho_0)_k} K_{X;\rho_0^2}(z, w; s+2k) \right) \psi_j(w)\mu(w)
$$

$$
= \sum_{k=0}^\infty \partial_s^\ell \left(\frac{\left(\frac{s}{2}\right)_k \left(\frac{s}{2} + \frac{1}{2}\right)_k}{k!\,(s+1-\rho_0)_k} \int_X K_{X;\rho_0^2}(z, w; s+2k)\psi_j(w)\mu(w) \right)
$$

$$
= \partial_s^\ell \left(\sum_{k=0}^\infty \frac{\left(\frac{s}{2}\right)_k \left(\frac{s}{2} + \frac{1}{2}\right)_k}{k!\,(s+1-\rho_0)_k} \int_X K_{X;\rho_0^2}(z, w; s+2k)\psi_j(w)\mu(w) \right),
$$

where the last equation above follows from the absolute and uniform convergence of the series over k, derived in the previous lines.

From the spectral expansion of $K_{X;\rho_0^2}(z, w; s)$ we immediately get

$$
\int_X K_{X;\rho_0^2}(z, w; s+2k)\psi_j(w)\mu(w)
$$

$$
= \frac{2^{s+2k-\rho_0-1}}{\Gamma(s+2k)} \Gamma\left(\frac{s+2k-\rho_0-it_j}{2} \right) \Gamma\left(\frac{s+2k-\rho_0+it_j}{2} \right) \psi_j(z)
$$

$$
= \frac{2^{s+2k-\rho_0-1}}{\Gamma(s+2k)} \left(\frac{s-\rho_0-it_j}{2} \right)_k \left(\frac{s-\rho_0+it_j}{2} \right)_k \Gamma\left(\frac{s-\rho_0-it_j}{2} \right)
$$

$$
\times \Gamma\left(\frac{s-\rho_0+it_j}{2} \right) \psi_j(z).
$$

An application of the doubling formula for the Gamma function yields that

$$
\frac{2^{s+2k-\rho_0-1} \left(\frac{s}{2}\right)_k \left(\frac{s}{2}+\frac{1}{2}\right)_k}{\Gamma(s+2k)} = \frac{2^{s-1-\rho_0}}{\Gamma(s)}.
$$

Therefore,

$$
\frac{\left(\frac{s}{2}\right)_k \left(\frac{s}{2}+\frac{1}{2}\right)_k}{k!\,(s+1-\rho_0)_k} \int_X K_{X;\rho_0^2}(z, w; s+2k)\psi_j(w)\mu(w)
$$

$$
= \Gamma\left(\frac{s+2k-\rho_0-it_j}{2} \right) \Gamma\left(\frac{s+2k-\rho_0+it_j}{2} \right) \psi_j(z)
$$

$$
= \frac{2^{s-1-\rho_0}}{\Gamma(s)} \frac{\left(\frac{s-\rho_0-it_j}{2}\right)_k \left(\frac{s-\rho_0+it_j}{2}\right)_k}{k!\,(s+1-\rho_0)_k} \Gamma\left(\frac{s-\rho_0-it_j}{2} \right) \Gamma\left(\frac{s-\rho_0+it_j}{2} \right) \psi_j(z).
$$

Observe that $\text{Re}\left(\frac{s-\rho_0-it_j}{2} + \frac{s-\rho_0+it_j}{2} - (s+1-\rho_0)\right) = -1 < 0$, so then the hypergeometric function

$$\sum_{k=0}^{\infty} \frac{\left(\frac{s-\rho_0-it_j}{2}\right)_k \left(\frac{s-\rho_0+it_j}{2}\right)_k}{k!\,(s+1-\rho_0)_k} = F\left(\frac{s-\rho_0-it_j}{2}, \frac{s-\rho_0+it_j}{2}, s+1-\rho_0; 1\right)$$

is uniformly and absolutely convergent. From [GR07], formula 9.122.1 we get

$$F\left(\frac{s-\rho_0-it_j}{2}, \frac{s-\rho_0+it_j}{2}, s+1-\rho_0; 1\right)$$

$$= \frac{\Gamma(s+1-\rho_0)}{\Gamma\left(\frac{s-\rho_0+it_j}{2}\right)\Gamma\left(\frac{s-\rho_0-it_j}{2}\right)} \cdot \frac{4}{(s-\rho_0)^2 + t_j^2}.$$

Therefore,

$$\sum_{k=0}^{\infty} \frac{\left(\frac{s}{2}\right)_k \left(\frac{s}{2}+\frac{1}{2}\right)_k}{k!\,(s+1-\rho_0)_k} \int_X K_{X;\rho_0^2}(z, w; s+2k)\psi_j(w)\mu(w)$$

$$= \frac{2^{s+1-\rho_0}\Gamma(s+1-\rho_0)}{\Gamma(s)((s-\rho_0)^2 + t_j^2)} \psi_j(z).$$

This, together with the definition of the function $G_{X;\rho_0^2}(z, w; s)$ yields

$$\int_X \partial_s^\ell \left(G_{X;\rho_0^2}(z, w; s)\right) \psi_j(w)\mu(w) = \partial_s^\ell \left(\frac{1}{(s-\rho_0)^2 + t_j^2}\right) \psi_j(z), \quad (40)$$

for sufficiently large positive integer ℓ.

The above computations are valid provided $\text{Re}(s) > 2\rho_0$. The arguments could be repeated with the portion of the series in (33) with $k > n$, for an arbitrary positive integer n, from which one would arrive at a version of (40) where the right-hand-side would have a finite sum of terms subtracted with the restriction that $\text{Re}(s) > 2\rho_0 - n$. However, there is no problem interchanging sum and differentiation for the finite sum of terms in (40) obtained by considering those with $k \leq n$, from which we conclude that (40) holds for all s with $\text{Re}(s) > 2\rho_0 - n$ provided s is not a pole of (33).

There is a unique meromorphic function $\tilde{G}(z, w; s)$ which is symmetric in z and w and satisfies $(\Delta_X + s(s - 2\rho_0))\tilde{G}(z, w; s) = 0$. Indeed, for $\text{Re}(s(s - 2\rho_0)) > 0$ one can express $\tilde{G}(z, w; s)$ as an integral transform of the heat kernel, namely

$$\tilde{G}(z, w; s) = \int_0^\infty K_X(z, w; t)e^{-s(s-2\rho_0)t}\, dt.$$

At this point, we have that $\tilde{G}(z, w; s) = G_{X;\rho_0^2}(z, w; s) + p_\ell(s)$, where $p_\ell(s)$ is a polynomial of degree ℓ. The asymptotic behavior as s tends to infinity can be computed for $G_{X;\rho_0^2}(z, w; s)$ using Stirling's formula, and that of $\tilde{G}(z, w, s)$ using the above integral expression. By combining, we get that $p_\ell(s) = o(1)$ as s tends to infinity, thus $p_\ell(s) = 0$.

This proves that $G_{X;\rho_0^2}(z, w; s)$ coincides with the conditionally convergent series given as a limit (32) for $\text{Re}(s) > 2\rho_0$. Moreover, since both $G_{X;\rho_0^2}(z, w; s)$ and the resolvent kernel $\tilde{G}(z, w; s)$ possess meromorphic continuation to the whole complex $\mathbb{C}-$plane, they must coincide.

With all this, assertions (ii) and (iii) are established.   □

Theorem 5.2 *Let*

$$E_{X;\rho_0^2}(z, w; s) = \frac{\Gamma((s + 1 - 2\rho_0)/2)}{\Gamma(s/2)} \sum_{k=0}^{\infty} \frac{\left(\frac{s}{2}\right)_k}{k!} K_{X;\rho_0^2}(z, w; s + 2k)$$

for $z, w \in X$, $z \neq w$. Then $E_{X;\rho_0^2}(z, w; s)$ converges to a meromorphic function for $\text{Re}(s) < 0$ away from the poles of any $K_{X;\rho_0^2}(z, w; s + 2k)$ and negative integers. Furthermore, $E_{X;\rho_0^2}(z, w; s)$ extends to a meromorphic function for all s and satisfies the differential-difference equation

$$(\Delta_X + s(s - 2\rho_0))E_{X;\rho_0^2}(z, w; s) = -s^2 E_{X;\rho_0^2}(z, w; s + 2). \tag{41}$$

Proof. The estimates in the proof of Theorem 5.1, namely (36), (37), and (38) combine to show that

$$\frac{\left(\frac{s}{2}\right)_k}{k!} K_{X;\rho_0^2}(z, w; s + 2k) = O_{s,z,w}(k^{s/2-\rho_0-3/2}) \quad \text{as } k \to \infty,$$

where the implied constant depends upon s and upon the distance between points z and w. Therefore, the series converges for s with $\text{Re}(s) < 0$ provided no term has a pole. Set

$$\tilde{E}_{X;\rho_0^2}(z, w; s) = \sum_{k=0}^{\infty} \frac{\left(\frac{s}{2}\right)_k}{k!} K_{X;\rho_0^2}(z, w; s + 2k).$$

Using the difference-differential equation for $K_{X;\rho_0^2}(z, w; s)$, as established in Corollary 4.2, we can prove such an equation for $\tilde{E}_{X;\rho_0^2}(z, w; s)$. Indeed, for $\text{Re}(s) \ll 0$ begin by writing

$$(\Delta_X + s(s - 2\rho_0))\tilde{E}_{X;\rho_0^2}(z, w; s)$$

$$= \sum_{k=0}^{\infty} \left(\frac{\left(\frac{s}{2}\right)_k}{k!} \Delta_X K_{X;\rho_0^2}(z, w; s + 2k) + \frac{\left(\frac{s}{2}\right)_k}{k!} s(s - 2\rho_0) K_{X;\rho_0^2}(z, w; s + 2k) \right)$$

$$= \sum_{k=0}^{\infty} \frac{\left(\frac{s}{2}\right)_k}{k!} \left(-(s + 2k)(s + 2k - 2\rho_0) K_{X;\rho_0^2}(z, w; s + 2k) \right)$$

$$+ (s + 2k)(s + 2k + 1)K_{X;\rho_0^2}(z, w; s + 2k + 2)\Big)$$

$$+ \sum_{k=0}^{\infty} \frac{\left(\frac{s}{2}\right)_k}{k!} s(s - 2\rho_0) K_{X;\rho_0^2}(z, w; s + 2k)$$

$$= \sum_{k=1}^{\infty} \frac{\left(\frac{s}{2}\right)_k}{k!} \left(-(s + 2k)(s + 2k - 2\rho_0) + s(s - 2\rho_0)\right) K_{X;\rho_0^2}(z, w; s + 2k)$$

$$+ \sum_{k=0}^{\infty} \frac{\left(\frac{s}{2}\right)_k}{k!} (s + 2k)(s + 2k + 1)K_{X;\rho_0^2}(z, w; s + 2k + 2)$$

$$= \sum_{n=0}^{\infty} \frac{\left(\frac{s}{2}\right)_{n+1}}{(n + 1)!} \left(-(s + 2n + 2)(s + 2n + 2 - 2\rho_0)\right.$$

$$+ s(s - 2\rho_0)\Big) K_{X;\rho_0^2}(z, w; s + 2n + 2)$$

$$+ \sum_{n=0}^{\infty} \frac{\left(\frac{s}{2}\right)_n}{n!} (s + 2n)(s + 2n + 1)K_{X;\rho_0^2}(z, w; s + 2n + 2).$$

Since

$$- (s + 2n + 2)(s + 2n + 2 - 2\rho_0) + s(s - 2\rho_0)$$
$$= -(2n + 2)(2s + 2n + 2 - 2\rho_0),$$

the coefficient of $K_{X;\rho_0^2}(z, w; s + 2n + 2)$ in the last expression is

$$-\frac{\left(\frac{s}{2}\right)_{n+1}}{(n + 1)!}(2n + 2)(2s + 2n + 2 - 2\rho_0) + \frac{\left(\frac{s}{2}\right)_n}{n!}(s + 2n)(s + 2n + 1).$$

Using the definition of the Pochhammer symbol, it is elementary to show that

$$-\frac{\left(\frac{s}{2}\right)_{n+1}}{(n + 1)!}(2n + 2)(2s + 2n + 2 - 2\rho_0) + \frac{\left(\frac{s}{2}\right)_n}{n!}(s + 2n)(s + 2n + 1)$$

$$= \frac{\left(\frac{s+2}{2}\right)_n}{n!}(-s(s + 1 - 2\rho_0)),$$

hence we arrive at the equation

$$(\Delta_X + s(s - 2\rho_0))\tilde{E}_{X;\rho_0^2}(z, w; s) = -s(s + 1 - 2\rho_0)\tilde{E}_{X;\rho_0^2}(z, w; s + 2).$$

Notice that

$$E_{X;\rho_0^2}(z, w; s) = \frac{\Gamma((s + 1 - 2\rho_0)/2)}{\Gamma(s/2)}\tilde{E}_{X;\rho_0^2}(z, w; s),$$

so then

$$(\Delta_X + s(s - 2\rho_0))E_{X;\rho_0^2}(z, w; s)$$

$$= \frac{\Gamma((s+1-2\rho_0)/2)}{\Gamma(s/2)}(-s(s+1-2\rho_0))\,\tilde{E}_{X;\rho_0^2}(z, w; s+2)$$

$$= -s^2 \frac{\Gamma((s+1-2\rho_0)/2)}{\Gamma(s/2)} \frac{(s+1-2\rho_0)/2}{s/2}\tilde{E}_{X;\rho_0^2}(z, w; s+2)$$

$$= -s^2 \frac{\Gamma(((s+2)+1-2\rho_0)/2)}{\Gamma((s+2)/2)}\tilde{E}_{X;\rho_0^2}(z, w; s+2)$$

$$- s^2 E_{X;\rho_0^2}(z, w; s+2),$$

as asserted.   □

Remark 5.3 The motivation of the series in Theorem 5.2 is the following elementary formula, which was first employed in the context of elliptic Eisenstein series in [vP10]. For any x with $|x| < 1$ and complex s, one has the convergent Taylor series

$$(1 - x)^{-s/2} = \sum_{k=0}^{\infty} \frac{\left(\frac{s}{2}\right)_k}{k!} x^k.$$

By setting $x = (\cosh u)^{-2}$, one then gets that

$$(1 - (\cosh u)^{-2})^{-s/2} = \sum_{k=0}^{\infty} \frac{\left(\frac{s}{2}\right)_k}{k!} (\cosh u)^{-2k}.$$

Now write

$$(1 - (\cosh u)^{-2})^{-s/2} = (\cosh u)^s ((\cosh u)^2 - 1)^{-s/2} = (\cosh u)^s (\sinh u)^{-s}$$

from which we obtain the identity

$$\sinh^{-s}(u) = \sum_{k=0}^{\infty} \frac{\left(\frac{s}{2}\right)_k}{k!} \cosh^{-(s+2k)}(u).$$

In this way, we can study the function obtained by applying the wave distribution to $g(u) = \sinh^{-s}(u)$, even though this function does not satisfy the conditions of Theorem 3.2. Indeed, this observation is the motivation behind the definition of $E_{X;\rho_0^2}(z, w; s)$.

6 Kronecker Limit Formulas

We now prove the Kronecker limit formulas for $G_{X;\rho_0^2}(z, w; s)$ and $E_{X;\rho_0^2}(z, w; s)$, meaning we analyze the first two terms in the Laurent series at $s = 0$. We will continue assuming that $\rho_0 \geq 0$ is arbitrary. The choice of ρ_0

plays no role in the analysis of $G_{X;\rho_0^2}(z, w; s)$. However, in the approach taken in this section, the case when $\rho_0 = 1/2$ will be particularly interesting when studying $E_{X;\rho_0^2}(z, w; s)$.

Corollary 6.1 *If $\rho_0 \neq 0$, then the function $G_{X;\rho_0^2}(z, w; s)$ has the asymptotic behavior*

$$G_{X;\rho_0^2}(z, w; s) = \frac{-1/(2\rho_0)}{\mathrm{vol}_\omega(X)} s^{-1} + G_{X;\rho_0^2}(z, w) + \frac{1/(2\rho_0)^2}{\mathrm{vol}_\omega(X)} + O(s) \ \text{as } s \to 0$$

where $G_{X;\rho_0^2}(z, w)$ is the Green's function associated to the Laplacian Δ_X acting on the space of smooth functions on X which are orthogonal to the constant functions.

If $\rho_0 = 0$, then we have the expansion

$$G_{X;\rho_0^2}(z, w; s) = \frac{1}{\mathrm{vol}_\omega(X)} s^{-2} + G_{X;\rho_0^2}(z, w) + O(s) \ \text{as } s \to 0.$$

Proof. The result follows directly from part (ii) of Theorem 5.1 having noted that the eigenfunction associated to the zero eigenvalue is $1/(\mathrm{vol}_\omega(X))^{1/2}$ and that

$$\frac{1}{s(s - 2\rho_0)} = \frac{-1/(2\rho_0)}{s} + \frac{-1/2\rho_0}{s - 2\rho_0} = \frac{-1/(2\rho_0)}{s} + \frac{1}{(2\rho_0)^2} + O(s) \ \text{as } s \to 0$$

in the case $\rho_0 \neq 0$. If $\rho_0 = 0$, the assertion follows immediately from part (ii) of Theorem 5.1. $\square$

Remark 6.2 Corollary 6.1 is, in some sense, elementary and well-known. Indeed, using (32), we can write

$$G_{X;\rho_0^2}(z, w; s) = \int_0^\infty \left(K_X(z, w; t) e^{-s(s-2\rho_0)t} - \frac{1}{\mathrm{vol}_\omega(X)} \right) dt$$

$$+ \frac{1}{s(s - 2\rho_0)} \frac{1}{\mathrm{vol}_\omega(X)}.$$

As s approaches zero while $\mathrm{Re}(s(s - 2\rho_0)) > 0$, the above integral converges to the Green's function $G_{X;\rho_0^2}(z, w)$. Nonetheless, the novel aspect of Theorem 5.1 is the expression of the resolvent kernel as a series.

Remark 6.3 The statement of Corollary 6.1 highlights the difference between the cases when $\rho_0 = 0$ and $\rho_0 \neq 0$. The difference determines the order of the singularity of the resolvent kernel $G_{X;\rho_0^2}(z, w; s)$ at $s = 0$. Of course, one could re-write Corollary 6.1 as

$$G_{X;\rho_0^2}(z, w; s) = \frac{1}{\mathrm{vol}_\omega(X)} \frac{1}{s(s - 2\rho_0)} + G_{X;\rho_0^2}(z, w) + O(s) \ \text{as } s \to 0,$$

which includes both $\rho_0 = 0$ and $\rho_0 \neq 0$.

Theorem 6.4 *Let D be the divisor of a holomorphic form F_D on X, and assume that D is smooth up to codimension two in X. Then, for $z \notin D$, there exist constants c_0 and c_1 such that*

$$\int_D G_{X;\rho_0^2}(z, w; s)\mu_D(w) = \frac{\text{vol}_\omega(D)}{\text{vol}_\omega(X)} \frac{1}{s(s - 2\rho_0)}$$
$$+ c_0 \log \|F_D(z)\|_\omega^2 + c_1 + O(s) \quad \text{as } s \to 0.$$
$$(42)$$

Proof. For now, assume that $z \notin D$ and $w \in D$. By part (i) of Theorem 5.1, the function $G_{X;\rho_0^2}(z, w; s)$ is holomorphic in s for $\text{Re}(s) > 2\rho_0$, so then the integral in (42) exists for $\text{Re}(s) > 2\rho_0$. The integral has a meromorphic continuation in s, again by part (i) of Theorem 5.1, and the Laurent expansion of the integral near $s = 0$ can be evaluated integrating over D the expansion given in Corollary 6.1. The singularity of the Green's function $G_{X;\rho_0^2}(z, w)$ as z approaches w is known; see, for example, page 94 of [Fo76] as well as [JK98] and [JK01]. In the latter references, the authors carefully evaluate the integrals of functions with Green's function type singularities; see section 3 of [JK98]. From those arguments, we conclude that

$$\int_D G_{X;\rho_0^2}(z, w; s)\mu_D(w)$$

has a logarithmic singularity as z approaches D.

Throughout this discussion the Laplacian Δ_X acts on the variable z. From the equation

$$(\Delta_X + s(s - 2\rho_0))G_{X;\rho_0^2}(z, w; s) = 0,$$

as proved in Theorem 5.1, and the expansion in Corollary 6.1, we conclude that for $z \neq w$, we have

$$\Delta_X G_{X;\rho_0^2}(z, w) = \frac{2}{\text{vol}_\omega(X)}. \qquad (43)$$

Let us consider the difference

$$\int_D G_{X;\rho_0^2}(z, w; s)\mu_D(w) - \frac{\text{vol}_\omega(D)}{\text{vol}_\omega(X)} \frac{1}{s(s - 2\rho_0)} - c_0 \log \|F_D(z)\|_\omega^2 \quad (44)$$

near $s = 0$. For any c_0, the difference is holomorphic in s near $s = 0$. From section 2.5, we have that $\Delta_X \log \|F_D(z)\|_\omega^2$ is a non-zero constant. Choose c_0 so that

$$c_0 \Delta_X \log \|F_D(z)\|_\omega^2 = \frac{2}{\text{vol}_\omega(X)}.$$

By combining (43) and (12), we conclude that

$$\mathrm{dd}^c \int_D G_{X;\rho_0^2}(z, w)\mu_D(w) = \delta_{D'} - \omega$$

where D' is a divisor whose support is equal to the support of D. It remains to show that $D' = D$.

Consider the difference

$$R_{X;\rho_0^2}(z; D) := \int_D G_{X;\rho_0^2}(z, w)\mu_D(w) - \frac{c_0}{\mathrm{vol}_\omega(X)} \log \|F_D(z)\|_\omega^2. \quad (45)$$

which satisfies

$$\mathrm{dd}^c R_{X;\rho_0^2}(z; D) = \delta_{D'} - n\delta_D,$$

which means that $R_{X;\rho_0^2}(z; D)$ is harmonic away from the support of D and has logarithmic growth as z approaches D. If X is an algebraic curve, then D is a finite sum of points, say $D = \sum m_j D_j$, with multiplicities m_j. In this case,

$$\int_D G_{X;\rho_0^2}(z, w)\mu_D(w) = \sum m_j G_{X;\rho_0^2}(z, D_j).$$

It follows that $D' = D$. By the Riemann removable singularity theorem, the difference (45) is harmonic on all of X, hence bounded, which implies that $R_{X;\rho_0^2}(z; D)$ is a constant. The argument for general X is only slightly different. Again, write $D = \sum m_j D_j$ where each D_j is irreducible. Choose a smooth point P on D, hence on some D_j. One can express the integral in (45) near P using suitably chosen local coordinates in X, as in section 3 of [JK98]. By doing so, one again concludes that the value of the integral of $G_{X;\rho_0^2}(z, w)$ as P approaches D is equal to the coefficient of D_j. Therefore, the difference (44) is bounded as z approaches P. Since D is smooth in codimension two, we can again apply the Riemann removable singularity theorem (see Corollary 7.3.2, page 262 of [Kr82]) to conclude that $R_{X;\rho_0^2}(z; D)$ is bounded and harmonic on X, hence constant. $\qquad\square$

Remark 6.5   The constant c_0 can be expressed as a function of the weight of the form F_D. The constant c_1 of (42) can be determined by integrating both sides of (42) with respect to z, using that the integral of the Green's function $G_{X;\rho_0^2}(z, w)$ is zero. This will express c_1 as an integral of $\log \|F_D(z)\|_\omega^2$.

Remark 6.6   In effect, the proof of Theorem 6.4 requires that the norm $\|F_D\|_\omega$ is such that the Laplacian Δ_X of its logarithm is constant, so then a certain linear combination of the Green's function and $\log \|F_D\|_\omega$ has Laplacian equal to zero away from D. This statement can hold in settings not covered by the conditions stated of Theorem 6.4. In this setting, the forms

F_D one studies are determined by the condition that $\log |F_D|$ are harmonic even in the setting when a complex structure does not exist. In fact, this requirement is true for quotients of hyperbolic n-spaces of any dimension $n \geq 2$.

Remark 6.7 Suppose we are given a codimension one subvariety D of X, and assume that D is smooth in codimension one. Then one can realize the log norm of the form F_D which vanishes along D by (42). In this manner, we can construct F_D when D has been given. The form F_D need not be a holomorphic form, but rather a section of the canonical bundle twisted by a flat line bundle. The parameters of the flat line bundle can be viewed as a generalization of Dedekind sums since classical Dedekind sums stem from attempting to drop the absolute values from the Kronecker limit function associated to the parabolic Eisenstein series for $\mathrm{PSL}(2, \mathbb{Z})$.

Let us now extend the development of Kronecker limit functions to $E_{X;\rho_0^2}(z, w; s)$. To do so, we first proof that for certain ρ_0^2, the functions $G_{X;\rho_0^2}(z, w; s)$ and $E_{X;\rho_0^2}(z, w; s)$ have the same expansion at $s = 0$ out to $O(s^2)$.

Proposition 6.8 For $z, w \in X$, $z \neq w$ consider the difference

$$D_{X;\rho_0^2}(z, w; s) := E_{X;\rho_0^2}(z, w; s) - 2^{s+1-\rho_0} \frac{\Gamma(s + 1 - \rho_0)\Gamma((s + 1 - 2\rho_0)/2)}{\Gamma(s)\Gamma(s/2)}$$
$$\times G_{X;\rho_0^2}(z, w; s).$$

Then for all $\rho_0 \geq 0$, such that $\rho_0 \neq m$ or $\rho_0 \neq m + 1/2$, for integers $m \geq 1$ we have that $D_{X;\rho_0^2}(z, w; s) = O(s^2)$ as $s \to 0$.

Proof. The factor

$$2^{s+1-\rho_0} \frac{\Gamma(s + 1 - \rho_0)\Gamma((s + 1 - 2\rho_0)/2)}{\Gamma(s)\Gamma(s/2)} \qquad (46)$$

of $G_{X;\rho_0^2}(z, w; s)$ was chosen so that the $k = 0$ term in the series expansions for $E_{X;\rho_0^2}(z, w; s)$ and $G_{X;\rho_0^2}(z, w; s)$ agree. For any $k \geq 1$, the k-th term in the series expansion for the difference $D_{X;\rho_0^2}(z, w; s)$ is

$$\frac{\Gamma((s + 1 - 2\rho_0)/2)}{\Gamma(s/2)} \frac{\left(\frac{s}{2}\right)_k}{k!} \left(1 - \frac{\left(\frac{s}{2} + \frac{1}{2}\right)_k}{(s + 1 - \rho_0)_k}\right) K_{X;\rho_0^2}(z, w; s + 2k). \qquad (47)$$

From the spectral expansion in Proposition 4.1, the function $K_{X;\rho_0}(z, w; s + 2k)$ is holomorphic at $s = 0$ for any integer $k \geq 1$. When ρ_0 is distinct from any positive half-integer, the function $\Gamma((s + 1 - 2\rho_0)/2)$ is also holomorphic at $s = 0$, while $\frac{\left(\frac{s}{2}+\frac{1}{2}\right)_k}{(s+1-\rho_0)_k}$ is holomorphic at $s = 0$ for ρ_0 distinct from any positive integer. Since there is a factor of $\Gamma^{-2}(s/2)$ in the above coefficient, it follows that the function $D_{X;0}(z, w; s)$ is $O(s^2)$ as s approaches zero, for all ρ_0 different from positive integers or half-integers.

It remains to prove the statement for $\rho_0 = 1/2$, in which case (47) becomes

$$\frac{\left(\frac{s}{2}\right)_k}{k!}\left(1 - \frac{\left(\frac{s}{2}+\frac{1}{2}\right)_k}{(s+1/2)_k}\right) K_{X;1/4}(z, w; s + 2k)$$

$$= \frac{\left(\frac{s}{2}\right)_k}{k!}\left(\frac{(s+1/2)_k - \left(\frac{s}{2}+\frac{1}{2}\right)_k}{(s+1/2)_k}\right) K_{X;1/4}(z, w; s + 2k).$$

For $k \geq 1$, the factor $(s/2)_k$ vanishes at $s = 0$, as does the difference $(s + 1/2)_k - (s/2 + 1/2)_k$, so it follows that the function $D_{X;1/4}(z, w; s)$ is $O(s^2)$ as s approaches zero. □

Corollary 6.9 *For $z, w \in X$, $z \neq w$ and $\rho_0 > 0$ which is not equal to a positive integer or half-integer, $E_{X;\rho_0^2}(z, w; s) = O(s)$, as $s \to 0$. When $\rho_0 = 1/2$, there are constants b_0, b_1 and b_2 such that the function $E_{X;1/4}(z, w; s)$ has the asymptotic behavior*

$$E_{X;1/4}(z, w; s) = b_0 + (b_1 + b_2 G_{X;1/4}(z, w))s + O(s^2) \quad \text{as } s \to 0$$

where $G_{X;\rho_0^2}(z, w)$ is the Green's function associated to the Laplacian Δ_X acting on the space of smooth functions on X which are orthogonal to the constant functions.

Proof. When $\rho_0 > 0$ is not equal to a positive integer or half-integer, the functions $\Gamma(s+1-\rho_0)$ and $\Gamma((s+1-2\rho_0)/2)$ are holomorphic at $s = 0$, hence the factor (46) is $O(s^2)$ as s approaches zero. Combining this with Proposition 6.8 and Corollary 6.1 yields the statement.

When $\rho_0 = 1/2$

$$2^{s+1-\rho_0}\frac{\Gamma(s+1-\rho_0)\Gamma((s+1-2\rho_0)/2)}{\Gamma(s)\Gamma(s/2)} = 2^{s+1/2}\frac{\Gamma(s+1/2)}{\Gamma(s)}$$

$$= a_1 s + a_2 s^2 + O(s^3)$$

so the statement of Proposition 6.8 becomes

$$E_{X;1/4}(z, w; s)$$

$$= (a_1 s + a_2 s^2 + O(s^3))\left(\frac{1}{\text{vol}_\omega(X)}s^{-1} + G_{X;1/4}(z, w) + O(s)\right) + O(s^2),$$

as $s \to 0$. Multiplying the above expression we deduce the statement. □

Remark 6.10 When $\rho_0 = 0$ the term in (46) is $\sqrt{\pi}s^2 + O(s^3)$ as $s \to 0$. When combined with Proposition 6.8 and Corollary 6.1, one gets that

$$E_{X;0}(z, w; s) = \sqrt{\pi}/\text{Vol}_\omega(X) + O(s^2) \quad \text{as } s \to 0.$$

Remark 6.11 In the case when X is a hyperbolic Riemann surface, the result of Proposition 6.8 is stated in Corollary 7.4 of [vP16], with a slightly different renormalization constant $\sqrt{2\pi}$ in front of the Green's function, which stems from a different constant term in the definition of the corresponding series. However, in their proof, the author used special function identities which are specific to that setting.

Remark 6.12 The constants b_0, b_1 and b_2 in case when $\rho_0 = 1/2$ are easily evaluated using asymptotic behavior of the factor (46) near $s = 0$, which are not so significant to us at this point. What does matter is that for $\rho_0 = 1/2$, we have that $E_{X;1/4}(z, w; s)$ admits a Kronecker limit formula. In the notation of Theorem 6.4, there are constants c_0, c_1 and c_2 such that

$$\int_D E_{X;1/4}(z, w; s) d\mu_D(w) = c_0 \text{vol}_\omega(D) + \left(c_1 \log \|F_D(z)\|_\mu^2 + c_2 \right) s$$
$$+ O(s^2) \quad \text{as } s \to 0.$$

Remark 6.13 It is important to note that we have not excluded the possibility of a "nice" Kronecker limit function for $E_{X;\rho_0^2}(z, w; s)$ when $\rho_0 \neq 1/2$. The approach we took in this article was to compare the Kronecker limit function of $E_{X;\rho_0^2}(z, w; s)$ to that of the resolvent kernel $G_{X;\rho_0^2}(z, w; s)$. We find it quite interesting that the comparison yields a determination of the Kronecker limit function of $E_{X;\rho_0^2}(z, w; s)$ only in the case when $\rho_0 = 1/2$.

Remark 6.14 Ultimately, we are interested in the cases when X is the quotient of a symmetric space G/K. In this setting, ρ_0 is zero only when G/K is Euclidean. In all other cases, ρ_0^2 is positive. (See section 1.3.)

7 Examples

As stated above, we began our analysis with the heat kernel and obtained our results using its spectral expansion. As one could imagine, any other representation of the heat kernel has the potential of combining with our results to yield formulas of possible interest. We will proceed along these lines and introduce three examples. It is our opinion that each example is of independent interest. Rather than expanding upon any one example, we will present, in rather broad strokes, the type of formulas which will result, and we will leave a detailed analysis for future work.

7.1 Abelian Varieties

Let Ω be an $N \times N$ complex matrix which is symmetric and whose imaginary part is postive definite. Let Λ_Ω denote the $\mathbb{Z}$-lattice formed by vectors in $\mathbb{Z}^N$ and $\Omega\mathbb{Z}^N$. Let X be an abelian variety whose complex points form the N-dimension complex torus $\mathbb{C}^N/(\mathbb{Z}^N \otimes \Omega\mathbb{Z}^N)$. Assume that X is equipped with its natural flat metric induced from the Euclidean metric on $\mathbb{C}^N$. It can be shown that all eigenfunctions on the associated Laplacian are exponential functions. In addition, the heat kernel on X can be obtained by periodizing over Λ_Ω the heat kernel on $\mathbb{C}^N$. By the uniqueness of the heat kernel on X, one obtains a formula of the type

$$K_X(z, w; t) = \sum_{k=0}^{\infty} e^{-\lambda_k t} \psi_k(z) \psi_k(w) = \sum_{v \in \Lambda_\Omega} \frac{1}{(4\pi t)^N} e^{-\|z-w-v\|^2/(4t)}$$

where $\| \cdot \|$ denotes the absolute value in $\mathbb{C}^N$. In effect, the identity obtained by equating the above two expressions for the heat kernel is the Poisson summation formula. In the setting of section 3, we take $\rho_0^2 = 0$, so then the Poisson kernel (13) becomes

$$\begin{aligned}
\mathcal{P}_{X,0}(z, w; u) &= \frac{u}{\sqrt{4\pi}} \int_0^\infty K_X(z, w; t) e^{-u^2/(4t)} t^{-1/2} \frac{dt}{t} \\
&= \sum_{v \in \Lambda_\Omega} \frac{u\Gamma(N + 1/2)}{\pi(u^2 + \|z - w - v\|^2)^{N+1/2}}.
\end{aligned} \tag{48}$$

As is evident, one cannot simply replace u by iu in (48) since then the sum would have singularities whenever $u^2 = \|z - w - v\|^2$. However, this is where the distribution theory approach is necessary and, indeed, one will obtain the function $K_{X;0}(z, w; s)$. For now, one can formally express $K_{X;0}(z, w; s)$ as the integral of $\cosh^{-s}(u)$. In the notation of Theorem 6.4, one can take D to be the theta divisor of the Riemann theta function Θ on X. The Kronecker limit formula for $\log \|\Theta\|$ then could be viewed as coming from the series over Λ_Ω. Upon exponentiation, one would have a product formula, or regularized product, formula for $\|\Theta\|^2$. Certainly, the exploration of this example is worthy of study.

7.2 Complex Projective Space

Let ω_{FS} denote the Fubini-Study metric on complex projective space $\mathbb{CP}^n$. The authors in [HI02] derived an explicit expression for the heat kernel $K_{\mathbb{CP}^n}$

associated to the Laplacian of the Fubini-Study metric on $\mathbb{CP}^n$. Specifically, it is proved that

$$K_{\mathbb{CP}^n}(z, w; t) = \frac{e^{n^2 t}}{2^{n-2} \pi^{n+1}} \int_r^{\pi/2} \frac{-d(\cos u)}{\sqrt{\cos^2 r - \cos^2 u}} \left(-\frac{1}{\sin u} \frac{d}{du} \right)^n [\Theta_{n+1}(t, u)],$$

$$(49)$$

where $z, w \in \mathbb{CP}^n$, $t > 0$, and $r = \text{dist}_{g_{FS}}(z, w) = \tan^{-1}(|z - w|)$, and the function $\Theta_{n+1}(t, u)$ is given by

$$\Theta_{n+1}(t, u) = \sum_{\ell=0}^{\infty} e^{-4t(\ell+n/2)^2} \cos((2\ell + n)u).$$

Equivalently, one can write

$$K_{\mathbb{CP}^n}(z, w, t) = \sum_{\ell=0}^{\infty} e^{-\lambda_\ell t} \theta_\ell(r), \tag{50}$$

where $\lambda_\ell = 4\ell(\ell + n)$, and

$$\theta_\ell(r) = \frac{1}{2^{n-2} \pi^{n+1}} \int_r^{\pi/2} \frac{\sin \tau}{\sqrt{\cos^2 r - \cos^2 \tau}} \left(-\frac{1}{\sin \tau} \frac{d}{d\tau} \right)^n \cos((2\ell + n)\tau) \, d\tau.$$

As in the previous example, the formula for the heat kernel is explicit, and all integral transforms leading up to the resolvent kernel $G_{X;\rho^2}(z, w; s)$ and $E_{X;\rho^2}(z, w; s)$ can be evaluated, at least formally. It seems as if one would also take $\rho_0^2 = 0$ in this case, though it would be worthwhile to consider $\rho_0^2 = 1/2$ as well. Of course, the divisors to consider would be the zeros of homogenous polynomials in N-variables, and the norm of homogenous polynomials would be with respect to the Fubini-Study metric.

7.3 Compact Quotients of Symmetric Spaces

Let G be a connected, non-compact semisimple Lie group with finite centrer, and let K be its maximal compact subgroup. Let Γ be a discrete subgroup of G such that the quotient $\Gamma \backslash G$ is compact. Then the quotient space $X = \Gamma \backslash G / K$ is also compact.

On page 160 of [Ga68], the author presents a formula for the heat kernel on G. In general terms, the heat kernel $K_G(g; t)$ with singularity when g is the identity, is equal to the inverse spherical transform of a Gaussian; see, Proposition 3.1 as well as [JLa01] in the case $G = \text{SL}_n(\mathbb{R})$.

In the case that G is complex, the inverse transform can be computed and the resulting formula is particularly elementary; see Proposition 3.2 of [Ga68]. In this case, one has that ρ_0^2 is equal to the norm of $1/2$ of the sum of the positive roots of the Lie algebra of G.

The heat kernel on X can be written, as in the notation of (4.2) of [Ga68], as the series

$$K_X(z, w; t) = \sum_{\gamma \in \Gamma} K_G(z^{-1} \gamma w; t).$$

The expressions from Proposition 3.1 and Proposition 3.2 of [Ga68] are such that the integral in (13) can be computed term-by-term. As discussed in section 2.5, one should replace t by $t/(4\rho_0^2)$ so then one has the Kronecker limit theorem as in Remark 6.12. One can be optimistic that the case of general G will not be significantly different from $G = \mathrm{SL}_2(\mathbb{R})$.

7.4 Concluding Remarks

Though we began with the assumption that X is a Kähler variety, one could review the proofs we developed and relax this condition. For example, if X is a hyperbolic n-manifold, meaning the compact quotient of $\mathrm{SO}(n, 1)$, then the structure of the Laplacian associated to the natural hyperbolic metric is such that all aspects of our proofs apply. In this case, the Kronecker limit function associated to $G_{X;\rho_0^2}(z, w; s)$ would be a harmonic form with a singularity when z approaches w. Furthermore, the heat kernel on the hyperbolic n-space has a particularly elementary expression; see, for example, [DGM76] who attribute the result to Millson. In this case, $\rho_0^2 \neq 0$, so then would expect, as in the case when $n = 2$, a generalization of the elliptic Eisenstein series as a sum over the uniformizing group. The study of Poincaré series associated to $\mathrm{SO}(n, 1)$ is developed in [CLPS91], and it will be interested to connect those results with the non-L^2 series $E_{X;\rho_0^2}(z, w; s)$.

Finally, we began with the heat kernel acting on smooth functions. Certainly, one could follow the same construction when using a form-valued heat kernel. By doing so, one would perhaps not consider the resolvent kernel, but rather focus on $K_{X;\rho_0^2}(z, w; s)$. In this case, one would integrate one of the variables over a cycle γ on X, as in section 5 of [JvPS16], and study the resulting Kronecker limit function. It seems plausible to expect that in this manner one would obtain a direct generalization of [KM79], whose series admitted a Kronecker limit function which was the Poincaré dual to the γ.

References

[Ba06] Ballmann, W.: *Lectures on Kähler Manifolds*. ESI Lectures in Mathematics and Physics. European Mathematical Society (EMS), Zürich, 2006.

[BGV91] Berline, N., Getzler, E., and Vergne, M.: *Heat Kernels and Dirac Operators*, Sprniger, New York, 1991.

[Ch84]   Chavel, I.: *Eigenvalues in Riemannian Geometry*, Academic Press, New York, 1984.

[CLPS91]   Cogdell, J., Li, J.-S., Piatetski-Shapiro, I., and Sarnak, P.: *Poincaré series for $SO(n, 1)$*, Acta Math. **167** (1991), 229-285.

[DGM76]   Debiard, A. Gaveau, B., and Mazet, E.: *Théorèmes de comparaison en géomtrie riemannienne*, Publ. Res. Inst. Math. Sci. **12** (1976/77), 391-425.

[Er56]   Erdélyi, A.: *Asymptotic Expansions*, Dover Publications, New York, 1956.

[Fi18]   Fine, J.: *A rapid introduction to Kähler geometry*. Available at http://homepages.ulb.ac.be/~joelfine/papers.html.

[Fo76]   Folland, G.: *Introduction to Partial Differential Equations*, Princeton University Press, Princeton, NJ, 1976.

[Fr82]   Friedlander, F.: *Introduction to the Theory of Distributions*, Cambridge University Press, Cambridge, England, 1982.

[Ga68]   Gangolli, R.: *Asymptotic behavior of spectra of compact quotients of certain symmetric spaces*. Acta Math. **121** (1968), 151-192.

[Ge81]   Gérardin, P.: *Formes automorphes associées aux cycles géodésiques des surfaces de Riemann hyperboliques (d'après S. Kudla et J. Millson)*. Bourbaki Seminar, Vol. 1980/81, pp. 23-35, Lecture Notes in Math., **901**, Springer, Berlin-New York, 1981.

[GR07]   Gradshteyn, I. S. and Ryzhik, I. M.: *Table of Integrals, Series and Products*. Elsevier Academic Press, Amsterdam, 2007.

[GH78]   Griffiths, P. and Harris, J.: *Principles of Algebraic Geometry*. John Wiley & Sons, New York, 1978.

[HI02]   Hafoud, A. and Intissar, A.: *Représentation intégrale de noyau de la chaleur sur l'espace projectif complexe $P^n(C)$, $n \geq 1$*. C. R. Math. Acad. Sci. Paris **335** (2002), no. 11, 871–876.

[JK98]   Jorgenson, J. and Kramer, J.: *Towards the arithmetic degree of line bundles on abelian varieties*, Manuscripta Math. **96** (1998), 335–370.

[JK01]   Jorgenson, J. and Kramer, J.: *Star products of Green's currents and automorphic forms*, Duke Math. J. **106** (2001), 553-580.

[JLa93]   Jorgenson, J. and Lang, S.: *Basic Analysis of Regularized Products and Series*, Springer Lecture Notes in Mathematics **1564** (1993).

[JLa01]   Jorgenson, J. and Lang, S.: *Spherical Inversion on $SL_n(\mathbb{R})$*. Springer-Verlag Monographs in Mathematics, Springer Verlag, New York, 2001.

[JLa03]   Jorgenson, J. and Lang, S.: *Analytic continuation and identities involving heat, Poisson, wave and Bessel kernels*. Math. Nachr. **258** (2003), 44–70.

[JvPS16]   Jorgenson, J., von Pippich, A.-M., and Smajlović, L.: *On the wave representation of elliptic and hyperbolic Eisenstein series*, Advances in Math. **288** (2016), 887–921.

[JvPS18]   Jorgenson, J., von Pippich, A.-M., and Smajlović, L.: *Applications of Kronecker's limit formula for elliptic Eisenstein series*, Annales Mathématiques du Québec **43**(2019), 99–124.

[JST16]   Jorgenson, J, Smajlović, L., and Then, H.: *Kronecker's limit formula, holomorphic modular functions and q-expansions on certain arithmetic groups*. Experimental Mathematics **54** (2016), 295–320.

[Kr82]   Krantz, S.: *Function Theory of Several Complex Variables*, John Wiley & Sons Inc., New York, 1982.

[KM79]   Kudla, S. S. and Millson, J. J.: *Harmonic differentials and closed geodesics on a Riemann surface*. Invent. Math. **54** (1979), 193–211.

[La88] Lang, S.: *Introduction to Arakelov Theory.* Springer, New York, 1988.

[vP10] von Pippich, A.-M.: *The arithmetic of elliptic Eisenstein series.* PhD thesis, Humboldt-Universität zu Berlin, 2010.

[vP16] von Pippich, A.-M.: *A Kronecker limit type formula for elliptic Eisenstein series.* https://arxiv.org/abs/1604.00811

[Si80] Siegel, C. L.: *Advanced Analytic Number Theory.* Tata Institute of Fundamental Research Studies in Mathematics, **9**, Tata Institute of Fundamental Research, Bombay, 1980.

[SZ02] Sogge, C. D. and Zelditch, S.: *Riemannian manifolds with maximal eigenfunction growth,* Duke Math. J. **114** (2002), 387–437

James W. Cogdell
Department of Mathematics
Ohio State University
231 W. 18th Ave.
Columbus, OH 43210
U.S.A.
e-mail: cogdell@math.ohio-state.edu

Jay Jorgenson
Department of Mathematics
The City College of New York
Convent Avenue at 138th Street
New York, NY 10031
U.S.A.
e-mail: jjorgenson@mindspring.com

Lejla Smajlović
Department of Mathematics
University of Sarajevo
Zmaja od Bosne 35, 71 000 Sarajevo
Bosnia and Herzegovina
e-mail: lejlas@pmf.unsa.ba

14

Some Topological Applications of Theta Functions

Mauro Spera

Abstract. In this paper some recent topological applications of Riemann surface theory and especially of their associated theta functions (in different geometric incarnations) are surveyed, taking the circle of ideas around geometric quantization as a vantage point. They include classical and quantum monodromy of 2d-integrable systems and the construction of unitary Riemann surface braid group representations (aimed, in particular, at devising a mathematical interpretation of the Laughlin wave functions emerging in condensed matter physics). The noncommutative version of theta functions due to A. Schwarz is briefly discussed, showing in particular its efficacy in Fourier-Mukai-Nahm computations.

1 Introduction

Theta functions (Jacobi, Riemann) constitute a major creation of 19th century mathematics, and their crucial role in most mathematical developments, both pure and applied, has been steadily growing since then. As a result, it would be a hopeless task for the present author to even partially track the immense richness and ramifications of the present day theory. Therefore, in this survey we shall concentrate on some basic geometric features exhibited by theta functions (and some of their generalisations), stressing their connection with the Weyl-Heisenberg group, together with a review of a few of their applications to topology and mathematical physics related to the author's activity (in particular, [95, 105]). In contrast, the so-called non-abelian theta functions will not be addressed here (but see e.g. [113] for a wide-ranging overview). The general background being well known and expounded in

Keywords: Theta functions, Riemann surface braid groups, stable holomorphic vector bundles, prime form, Laughlin wave functions, noncommutative geometry, classical and quantum hamiltonian monodromy.
MSC 2010: 81S10, 20F36, 14H42, 14H40, 14J60 58B34, 37J35, 53D50, 81R60, 81P68.

several excellent treatises (see e.g. [79, 110, 64, 81, 76, 46, 120], and [42] as well), brief digressions elaborating on it for the benefit of a wider audience are included, with some amplifications and technical details deferred to an Appendix.

The outline of the paper is the following. In Section 2, meant as a prelude, we give a concise discussion of the physical motivation of the ensuing purely mathematical developments, within the framework of geometric quantization [123, 29, 65, 98, 69, 48, 113]: specifically, we discuss aspects of classical and quantum monodromy ([38, 36]), together with condensed matter issues ([77, 30]).

In Section 3 we gather together the basics of theta function theory, tailoring the exposition to our needs. In particular we review the geometric theory of theta functions à la Matsushima ([75]), where they arise as holomorphic sections of Hermitian-Einstein vector bundles over Abelian varieties, emphasizing their connection with representations of the Weyl-Heisenberg group (building on and generalising [99]), mostly focussing on the genus one case. Then we turn to a noncommutative geometric approach the above bundles ([105]), originating in the seminal Connes paper [34] and discuss theta functions à la Schwarz [96]. The above apparatus will lead to a kind of *duality*, naturally interpreted within an appropriate Fourier-Mukai-Nahm transform ([14]).

In Section 4, the above notions are used, in conjunction with the prime form on Riemann surfaces (RS - cf. [79]), to geometrically construct unitary representations of the RS-braid group of [105], building on Bellingeri's presentation of the latter ([16, 17]), together with a formulation of Laughlin wave functions ([70, 51, 30]).

Section 5 deals with the theta function approach to classical and quantum monodromy ([38, 36, 85, 86, 94, 39, 124, 125, 126]) developed in [95], leading in particular to a quick derivation of the monodromy of the spherical pendulum.

The last Section (6) summarises our conclusions and points out possible future research directions.

2 Overview: Some Topological Issues in Classical and Quantum Mechanics

This section briefly analyses some typical effects of topology in classical and quantum mechanics from a geometrical standpoint and it is meant to lend the underlying physical motivation for the mathematical problems discussed in this paper.

The prototype of topological phenomenon in quantum mechanics is the celebrated Aharonov-Bohm effect, [1, 2] showing that the vector potential (connection) of the electromagnetic field (curvature) has a physical meaning in that it modifies the diffraction pattern of an electron wave passing past a solenoid, giving rise to a (purely topological) Berry's phase ([22, 97]): outside the solenoid the field (i.e. curvature) is zero, and we have a flat connection manifesting its presence via holonomy (see e.g. [33]).

2.1 Geometric Quantization

In this subsection we concisely review the basics of *Geometric Quantization* (GQ), referring to [123, 29, 65, 98, 69, 48, 113] for a complete account. Here we closely follow the expositions given in [95, 102, 104, 106]. GQ is a quite elegant and powerful method, essentially arising from generalizing Dirac's approach to the magnetic monopole and casting it into the appropriate mathematical framework of differential geometry and topology of complex line bundles, which allows for a neat geometrical understanding of important topics such as group representation theory (e.g. the Borel-Weil theorem and its extensions, see e.g. [48, 91]) and it is crucial in many modern physical theories such as, among others, integral and fractional quantum Hall effects, conformal field theories and Chern-Simons-Witten theory [66, 91, 68, 13, 122].

2.1.1 Prequantization

Recall that if (M, ω) is a symplectic manifold of (real) dimension $2n$ such that the ensuing cohomology class $\left[\frac{1}{2\pi}\omega\right] \in H^2(M, \mathbb{Z})$ (*integrality*, i.e. integral of ω over any 2-cycle is an integer) then the *Weil-Kostant Theorem* states that there exists a complex line bundle (L, ∇, h) over M equipped with a hermitian metric h and a compatible connection ∇ with curvature $F_\nabla = \omega$. Hence $[\omega] = c_1(L)$, the first Chern class of $L \to M$. The result holds in the pre-symplectic case as well, i.e. one may drop non-degeneracy. The integrality condition arises as a consistency condition coming from computing, via the Stokes theorem, the parallel transport of the connection on a loop in M bounded by two different 2-chains building up a 2-cycle. The connection ∇ is called a *prequantum connection* and $L \to M$ the *prequantum line bundle*. The different choices of $L \to M$ and ∇ are parametrized by the first cohomology group $H^1(M, S^1)$ (see e.g. [123], Ch.8 and the following subsection for an amplification). In particular, if M is simply connected, this cohomology group it trivial and the connection is unique.

2.1.2 Kähler Quantization

Given a Kähler polarization J on (M, ω), so that M is indeed a Kähler manifold, we can endow the complex line bundle $L \to M$ with the structure of a *holomorphic* line bundle. This leads to the so-called *holomorphic quantization*, whereby one takes the space of holomorphic sections $H^0(L, J)$ of a holomorphic prequantum line bundle, provided it is not trivial, (this being achieved by Kodaira vanishing conditions) as the Hilbert space of the theory, see e.g. [56] for details. As a holomorphic line bundle, $L \to M$ varies with J, whilst its topological type is fixed. In this case there is a canonically defined connection, called the *Chern,* or *Chern-Bott connection,* compatible with both the hermitian and the holomorphic structure (cf. [46]). Independence of polarization (i.e. of the complex structure, in this case) is achieved once one finds a (projectively) flat connection on the vector bundle $V \to \mathcal{T}$ with fibre $H^0(L, J)$ (of constant dimension, under suitable assumptions provided by the Kodaira vanishing theorem) over the (Teichmüller) space of complex structures $\mathcal{T}$ (see [56]). An important example, crucial for what follows, is provided by the *k-level theta functions,* which can be viewed as (a basis of) the space of holomorphic sections of a holomorphic line bundle (the kth tensor product of the theta line bundle) defined on a principally polarized Abelian variety ([46, 64, 122, 13, 56]). By the Riemann-Roch theorem, this space has (complex) dimension k^g. In real dimension two, the role of $\mathcal{T}$ is played by the Poincaré upper half plane $\mathbb{H}$, a complex structure on a torus being labelled by $\tau \in \mathbb{H}$; the nonequivalent complex structures are parametrized by the orbits of the modular group $PSL(2, \mathbb{Z}) = SL(2, \mathbb{Z})/\pm I$, see [76] and also below. The covariant constancy of the thetas is ascribed to their fulfilment of the (holomorphic) *heat equation.* More on this in the sequel.

2.1.3 Geometric Quantization of Integrable Systems

In the case of a completely integrable classical system, one can restrict the prequantum data to a Lagrangian submanifold Λ (an n-torus) provided the *Bohr-Sommerfeld* (BS) conditions on the action variables $I_k, k = 1, 2 \ldots n$ are satisfied thereon (θ being a local potential for ω: $\omega = d\theta$):

$$I_k = \frac{1}{2\pi} \int_{\gamma_k} \theta = \frac{1}{2\pi i} \log(hol(\nabla|_\Lambda, \gamma_k)) \in \mathbb{Z},$$

with the $\gamma_k, k = 1, 2 \ldots n$ providing a basis of integral cycles for Λ. Notice the explicit dependence of the action variables on ∇.

Then one abuts at quantum one-dimensional Hilbert spaces generated by a covariantly constant section

$$s(p) = hol(\nabla, \gamma)$$

for $p \in \Lambda$, γ being any path connecting a chosen point $p_0 \in \Lambda$ to p. We refer to [95, 123, 113] for full details. Notice that here and in [95] we overlooked Maslov-type phenomena. However, we do not delve on this point any further in this paper (see e.g. [123] for further information).

2.2 Classical and Quantum Monodromy

Topological issues also arise in classical mechanics: in particular, we have the phenomenon of (Hamiltonian) monodromy for integrable systems possessing (isolated) singular Liouville tori: in quite crude terms, in a Liouville-Arnol'd context ([4, 5]), execution of a circuit in the parameter space B (whose points are labelled by local action variables: one has a Lagrangian fibration via a moment map $f : M \to B$) leads to a non trivial action on the homology basis given by fundamental cycles on a model n-torus, and the result does not depend on the homotopy class of the circuit (in algebraic geometry this is a manifestation of the *Gauß-Manin connection* (see e.g. [72, 62, 63])). The action variables I_k, $k = 1, 2 \ldots n$, can be properly interpreted as *covariantly constant sections* of a canonical connection ∇^{can} ([95]). The archetypical example of this phenomenon is provided by the *spherical pendulum*, see [38, 15, 36].

Upon (geometric) quantization, monodromy essentially arises via the non-uniqueness of the prequantum connection. This point of view has been pursued in [95], to which we refer for a precise formulation of the problem and full details (see also [52]). In that work, acting within the framework of Bohr-Sommerfeld (BS) and Kählerian geometric quantization, we discussed classical and quantum monodromy from several viewpoints, all related to parallel transport via suitable flat connections. As anticipated above, Hamiltonian monodromy (actually, its linear part) is just the holonomy of the $\mathbb{Z}^n$-bundle over the set of regular values of a natural moment map, whose typical fibre is by definition the integral homology of the n-torus. We restricted from the outset to *linear* monodromy, instead of dealing with the full, i.e. affine, monodromy. Hamiltonian monodromy, together with the so-called Chern-Duistermaat class, provides an obstruction to the global definition of action-angle variables for completely integrable Hamiltonian systems ([38, 84]). The latter is absent in the framework of [95]. Our specific contributions consisted, first of all, in reinterpreting the canonical Ehresmann-Weinstein connection ([119]) arising from the traditional treatment (see e.g. [38, 15, 36, 124, 125, 126]) in vector bundle terms, abutting at the canonical connection ∇^{can}. Subsequently, we related monodromy to the freedom of choice of a prequantum connection, and in particular we found that it may be viewed as the obstruction to patching together geometric prequantization bundles equipped with local

"BS-adapted" connections showing, in addition, that it can be detected via a shift of the quantum action operators (constructed via the recipe of geometric quantization). This is consistent with the fact that, experimentally, monodromy manifests itself via a shift of the energy levels ([94, 39, 31, 32]).

Nevertheless, and this is what we shall be concerned with in the present article, in the case of completely integrable Hamiltonian systems with two degrees of freedom (with focus-focus type singularities), we related monodromy to theta function theory, via the differential geometric interpretation of the *heat equation* fulfilled by the k-level theta functions going back to [122, 13, 56]. Referring to the sequel for a more detailed treatment, we proved existence of a representation of the fundamental group $\pi_1(B)$ of the base space B of the Lagrangian fibration in tori pertaining to a Hamiltonian completely integrable system with two degrees of freedom, via the holonomy of a flat connection living on a natural complex vector bundle (of rank 2) made up of the (2-level) theta functions (pulled back) over B. The appearance of theta functions in this context is quite natural from a mechanical point of view: anticipating the forthcoming discussion a little bit, we can explain this as follows. Given a basis of cycles on a Liouville torus, constructed as in [85, 86] - see also below - a natural *complex structure* thereon is determined upon setting

$$\tau = -\Theta + iT,$$

where Θ is the rotation number and $T > 0$ is the (Poincaré) first return time of a point on one of the basis cycles, denoted by γ_1 (the rotation number is essentially the discrepancy, measured on the cycle γ_1 - corresponding to one of the actions - between the final and initial position of the aforementioned point, and one can easily manufacture a cycle γ_2 from such an arrangement (the roles of the γ's in [95] and [85, 86] are interchanged). Therefore, each Liouville torus comes equipped with a polarization making it an Abelian variety, and hence with a Kähler structure (clearly, the original symplectic form vanishes when restricted to a Liouville torus), and we have a family of (unobstructed) geometric quantizations of such tori, yielding precisely the theta functions of level k as their quantum Hilbert space (also, they can be adjusted so as to yield orthonormal bases thereof). Upon varying τ on the Poincaré upper half-plane $\mathbb{H}$, one gets a vector bundle whose generic fibre is given by the 2-level theta functions, which has a natural flat connection (that we termed *heat connection*) for which the latter are the covariantly constant sections. The ensuing parallel transport translates into the heat equation fulfilled by the thetas. This "universal" construction, when pulled back to B, yields the above mentioned flat connection, which incorporates monodromy. We shall also discuss the direct relationship between the variation of the rotation number (producing monodromy) and the braiding of the Weierstraß roots of the elliptic curve associated to τ, again via theta functions, together with a quick derivation of the monodromy of the spherical pendulum (see e.g. [38, 36]). This quite naive approach naturally fits within the Gauß-Manin connection context.

2.3 Quantum Hall Effects and Braiding Statistics

It is well known that the non trivial topology of the configuration space of a physical system is a source of ambiguities in the quantization prescriptions, leading to consideration of irreducible unitary representations of its fundamental group, that is, holonomies of flat connections (governing the adiabatic motion of the particles entering the system) on the vector bundles whose sections yield the wave functions of the system, and this impacts on the statistical properties of the latter. For 2d-configuration spaces braid groups emerge, leading to *exotic (anyon) statistics* in addition to the standard Fermi-Bose ones [44, 45, 121, 43, 59, 58]. An impressive instance of this is provided by Quantum Hall Effects (QHE), integral and fractional (see e.g. the monographs [77, 30]). In the QHE framework, different Aharonov-Bohm type boundary conditions will give rise to families of Landau vector potentials and ultimately to what is called a *spectral (vector) bundle over the Jacobian of a Riemann surface*: physically relevant quantities receive a topological interpretation. This is elaborated on right below.

In the paper [105] we studied the simplest unitary representations of the braid group associated to a general Riemann surface from a geometrical standpoint, with a view to possible applications in topological quantum computing ([83, 118]), where unitary braid group representations are employed for constructing quantum gates (topology would then automatically enforce robustness and fault tolerance), with the Fractional Quantum Hall Effect (FQHE) possibly yielding the physical clue to its practical implementation ([83]).

Recall that the FQHE arises for a (Coulomb) interacting spin-polarised 2d-electron gas, at low temperature and in the presence of a strong magnetic field. It is usually observed in semiconductor structures, such as electrons trapped in a thin layer of GaAs surrounded by AlGaAs, Si-MOSFETs (see e.g. [30]) and it has been recently detected in graphene ([28]) as well. The ground state of such a system can be approximately (but most effectively) described by a *Laughlin wave function* of the form (in a plane geometry, [70, 30]):

$$\prod_{i<j}(z_i - z_j)^m e^{-\sum_{i=1}^{N}|z_i|^2}$$

Here N is the number of electrons in the sample, m is an odd positive integer (this ensuring Fermi statistics). One notes the appearance of the ground state of a quantum harmonic oscillator. The quantity $v = \frac{1}{m}$ is the *filling factor* intervening in the fractional quantization of the *Hall conductance*:

$$\sigma_H = v \frac{e^2}{h}$$

and, in the limit $N \to \infty$, equals the electron density per state: $v = \frac{N}{N_S}$ with N_S the number of magnetic flux quanta: $N_S = B \cdot \mathcal{A}/\Phi_0$ (B is the modulus of the constant magnetic field (acting perpendicularly to the layer), $\mathcal{A}$ is the area of the given sample, whereas $\Phi_0 := hc/e$ is the flux quantum). The number N_S also gives the degeneracy of the lowest Landau level (for the free system) - i.e. the ground state - which appears as a degenerate ground state of a quantum harmonic oscillator.

On the mathematical side, Landau levels admit elegant algebro-geometric descriptions along the lines of geometric quantization: for instance, if the layer is a (closed) Riemann surface of genus g, the lowest Landau level is the space of holomorphic sections of a suitable holomorphic line bundle ([57, 47, 87, 107, 108]); on a torus ($g = 1$) it can be realised as a space of theta functions, see e.g. [47, 87], and also [99] and Appendix). A recent reformulation of the GQ-approach to Landau levels, in different guises and for the standard plane geometry, has been developed in [41].

Now, *on the one hand*, it turns out that the elementary excitations around the Laughlin ground state are *quasiparticles/holes* having *fractional charge* $\pm ve$ ([70, 30]) and *anyon statistics* $(-1)^v$ ([51, 30]), and this leads to considering the *braid group* associated to the N-point configuration space of the given layer (N now being the number of quasiparticles/holes). Wave functions for quasiparticles/holes can be cast in the form given above, with v replacing m (see [51, 30]).

On the other hand, the filling factor $v = 1/m$ (together with others) for a *torus* sample has been interpreted as the *slope* (that is, degree over rank) of a *stable* holomorphic vector bundle over the corresponding "spectral", or "Brillouin manifold" (which is again a torus, parametrising all admissible (Ahronov-Bohm) boundary conditions, see [50, 115, 47] and below); therefore, the filling factor has a *topological* meaning. (For v integral one recovers the interpretation of the integral Quantum Hall Effect via the first Chern class of a line bundle over the Brillouin manifold, see e.g. [109, 77, 30].)

We showed in [105] that the above coincidence has an abstract braid group theoretical origin: we considered a general closed oriented Riemann surface – so that the role of the Brillouin manifold was now played by the *Jacobian* of the surface (cf. [107, 108]) – and its associated braid group, with the Bellingeri presentation ([16, 17]); then the equalities, in the genus one case,

$v :=$ filling factor = statistical parameter = slope of a stable vector bundle have been derived from a group theoretical perspective (see e.g. [61] for a recent comprehensive coverage of braid groups.)

We outline the basic thread of the argument. Our first observation is that, in our context, braiding can approached via representations of the Weyl-Heisenberg group corresponding to the (rational) statistical parameter ν, both infinite dimensional and finite dimensional. Then, generalising [99], we observe that the infinite dimensional representations can be constructed geometrically on L^2-sections of holomorphic Hermitian stable bundles over the Jacobian of the Riemann surface under consideration. Stable bundles are irreducible holomorphic vector bundles over Kähler manifolds admitting a *Hermitian-Einstein structure (HE)* – namely a (unique) Hermitian connection with central constant curvature – in view of the Donaldson-Uhlenbeck-Yau theorem ([80, 37, 114, 67]). Specifically, the representation of the Weyl-Heisenberg group we look for stems from suitable parallel transport operators associated to the HE-connection (which will have constant curvature, essentially given by the statistical parameter ν). The solution is actually reduced to finding suitable *projectively flat HE-bundles over Jacobians*, which can be obtained via the classical Matsushima construction ([75, 53, 67]). In particular, we get a "slope-statistics" formula $\mu = \nu g!$ (with μ denoting the slope of a holomorphic vector bundle). We also show that, at least for genera $g = 2,3$, which allow for totally split Jacobians (cf. [40]), one can take box products of bundles on elliptic curves.

The other important geometrical ingredient needed to describe the statistical behaviour governing "particle" exchange is the Klein prime form on a Riemann surface ([79]), manufactured via theta function theoretic tools. The problem of extracting general roots of a line bundle then arises and it is circumvented by exploiting a universal property of the prime form. Then we define, following Halperin [70, 51, 30], (Laughlin type) vector valued wave functions obeying, in general, *fractional* statistics and having their "centre of mass" part represented by holomorphic sections of the above bundles (see also [50, 57, 115, 23]). Theorem 3.2 in [105] summarizes the whole construction of the RS-braid group representations we aimed at.

The successive developments in [105] portrayed a possibly interesting "braid duality" and run as follows. Focussing in particular on the $g = 1$ case, we showed that everything can be made even more explicit by resorting to A. Connes' noncommutative geometric setting ([34, 35]) for noncommutative tori and to the notion of noncommutative theta vector introduced by A. Schwarz ([96]), encompassing the classical notions. The upshot is that the "centre of mass" parts of Laughlin wave functions are precisely the Schwarz theta vectors. A notable feature is now the following: the space of theta vectors naturally determines a finite dimensional braid group representation corresponding to the reciprocal parameter $\nu' = 1/\nu$, which, via Matsushima, gives rise to a projectively flat HE-bundle with the corresponding slope. Therefore a (Matsushima-Connes (MC)) "duality" emerges, essentially coinciding with the one given by the Fourier-Mukai-Nahm (FMN) transform ([14]). In particular, the noncommutative theta vector approach can be used to calculate the Nahm-transformed connection explicitly.

For the reader's benefit we provide an Appendix sketching the explicit example of a toral configuration space together with its geometric quantization, giving rise to the lowest Landau level and to its accompanying spectral manifold (again a torus) and spectral bundle (whose holomorphic sections are given by the k-level theta functions). We also describe its corresponding Laughlin wave functions, originally devised by Haldane and Rezayi ([50]), following [47].

3 Theta Functions and Hermitian-Einstein Bundles

3.1 The Canonical (Weyl-Heisenberg) Commutation Relations

The celebrated *Weyl-Heisenberg Canonical Commutation Relations (CCR)* ([117], see also e.g. [89, 92]) read, on an infinite dimensional complex separable Hilbert space $\mathcal{H}$, up to obvious inessential notational changes:

$$V(\vec{\beta}) \, U(\vec{\alpha}) = e^{2\pi \sqrt{-1} \cdot \nu \, \vec{\alpha} \cdot \vec{\beta}} U(\vec{\alpha}) \, V(\vec{\beta})$$

with $\vec{\alpha}, \vec{\beta} \in \mathbb{R}^g$, $\nu \in \mathbb{R}$, $U(\cdot)$, $V(\cdot)$ unitary operators on $\mathcal{H}$.

Notice that we are actually blurring the difference between the standard CCR and their integrated version (Weyl form of the CCR). No confusion should arise.

Irreducible finite dimensional unitary representations also exist, corresponding to the finite version of Weyl-Heisenberg commutation relations and giving rise to particular rational *noncommutative tori* (see also below). Their construction is quite standard: set $\nu := q/r$, with q and r relatively prime positive integers, and consider, for $g = 1$, unitaries U_i, $i = 1, 2$ on $\mathbb{C}^r$ satisfying

$$U_1 U_2 = e^{-2\pi \sqrt{-1} \nu} U_2 U_1$$

Irreducibility easily entails $U_j^r = 1$, $j = 1, 2$ (so the cyclic group $\mathbb{Z}/r\mathbb{Z}$ is involved). Concretely, one may take $U_1 = \text{diag}(1, e(\nu), e(2 \cdot \nu), \ldots e((r-1) \cdot \nu))$ and $U_2 = $ matrix of the shift map $e_i \to e_{i-1}$, $i = 1, 2, \ldots r$, $e_0 = e_r$, with $(e_1, \ldots, e_r)$ being the canonical basis of $\mathbb{C}^r$ and where we defined, for real x, $e(x) := e^{2\pi \sqrt{-1} x}$.

Both kinds of representations arise in theta function theory and in its applications to Riemann surface braid groups: the finite version being well known in various guises (see e.g. [79, 75]), the infinite dimensional one quite implicitly, see [99, 96, 105] and below for treatments featuring this approach.

3.2 Elliptic Integrals, Theta Functions and Geometric Quantization

In this subsection we collect some facts about elliptic integrals and theta functions in one and several variables, in view of future use. The theory is thoroughly expounded in many classical texts, see e.g. among others [79, 76, 110, 120, 46, 64].

The Weierstraß canonical forms of the elliptic integrals of the first, second and third kind read, respectively:

$$I_1 = \int \frac{dz}{\sqrt{P(z)}}, \quad I_2 = \int \frac{z\,dz}{\sqrt{P(z)}}, \quad I_3 = \int \frac{dz}{(z-c)\sqrt{P(z)}},$$

where

$$P(x) := 4x^3 - g_2 x - g_3 = 4(x - e_1)(x - e_2)(x - e_3),$$

with $e_1 + e_2 + e_3 = 0$ (the e_i's are all distinct); in I_3, c is required not to be a root of P. One abuts at the elliptic curve $\mathcal{C}$ with equation

$$y^2 = P(x).$$

The elliptic integral I_1 above is explicitly inverted by the celebrated Weierstraß function $\wp = \wp(z, g_2, g_3) \equiv \wp(z, \tau)$, fulfilling the preceding equation with $x = \wp$, $y = \wp'$. Then $\mathcal{C} \cong \mathbb{C}/\mathbb{Z} + \mathbb{Z}\tau$, the torus defined by quotienting $\mathbb{C}$ by a normalized lattice $\mathbb{Z} + \mathbb{Z}\tau$, where $\tau = \frac{\omega'}{\omega} \in \mathbb{C}, \Im\tau > 0$) (ratio of periods). One has $e_i = \wp(\omega_i)$, where $\omega_1 = \omega$, $\omega_2 = \omega + \omega'$, $\omega_3 = \omega'$. The (Jacobi) modulus (squared) k^2 (with $k \in \mathbb{C} \setminus \{0, 1\}$) together with its complementary modulus k' fulfilling $k'^2 = 1 - k^2$, can be interpreted as the simple ratio of the three roots of P (see below). The standard theta function reads

$$\vartheta(z, \tau) = \sum_{n \in \mathbb{Z}} e^{i\pi n^2 \tau + 2\pi i\, nz}.$$

Let us also record the expressions for theta function with 2-characteristics (using Mumford's notation ([79]):

$$\vartheta_{ab}(z, \tau) = e^{\pi i a^2 \tau + 2\pi i a(z+b)} \vartheta(z + a\tau + b, \tau),$$

where $a, b \in \frac{1}{2}\mathbb{Z}$. Comparison with traditional notations yields $\vartheta_{00} = \vartheta_3$, $\vartheta_{0\frac{1}{2}} \equiv \vartheta_{01} = \vartheta_4$, $\vartheta_{\frac{1}{2}0} \equiv \vartheta_{10} = \vartheta_2$, $\vartheta_{\frac{1}{2}\frac{1}{2}} \equiv \vartheta_{11} = \vartheta_1$.

The Jacobi modulus k of the attached elliptic curve can be recovered from τ via the formula

$$k^2 = \frac{\vartheta_2{}^4(0, \tau)}{\vartheta_3{}^4(0, \tau)} = \frac{e_2 - e_3}{e_1 - e_3}$$

(this is the very motivation which led Jacobi to devising theta functions).

Indeed, let us recall the following expressions relating the Weierstraß roots to theta functions:

$$e_2 - e_3 = \left(\frac{\pi}{2\omega}\right)^2 \vartheta_2{}^4(0, \tau) \qquad e_1 - e_2 = \left(\frac{\pi}{2\omega}\right)^2 \vartheta_4{}^4(0, \tau),$$

following from

$$
\begin{aligned}
e_1 &= \frac{\pi^2}{12\omega^2}[\vartheta_3{}^4(0, \tau) + \vartheta_4{}^4(0, \tau)] \\
e_2 &= \frac{\pi^2}{12\omega^2}[\vartheta_2{}^4(0, \tau) - \vartheta_4{}^4(0, \tau)] \\
e_3 &= -\frac{\pi^2}{12\omega^2}[\vartheta_2{}^4(0, \tau) + \vartheta_3{}^4(0, \tau)],
\end{aligned}
$$

the beautiful Jacobi formula:

$$\vartheta_2{}^4(0, \tau) + \vartheta_4{}^4(0, \tau) = \vartheta_3{}^4(0, \tau)$$

and, most important for us, the following transformation law:

$$
\begin{aligned}
\vartheta_1(z, \tau + 1) &= e^{i\frac{\pi}{4}}\,\vartheta_1(z, \tau) \\
\vartheta_2(z, \tau + 1) &= e^{i\frac{\pi}{4}}\,\vartheta_2(z, \tau) \\
\vartheta_3(z, \tau + 1) &= \vartheta_4(z, \tau) \\
\vartheta_4(z, \tau + 1) &= \vartheta_3(z, \tau)
\end{aligned}
$$

Let us now consider the following modified theta function:

$$\widetilde{\vartheta}(z, \tau) = e^{\frac{\pi}{2}(\Im\tau)^{-1}z^2}\vartheta(z, \tau).$$

Notice that the prefactor $e^{\frac{\pi}{2}(\Im\tau)^{-1}z^2}$ is invariant with respect to the trasformation $\tau \mapsto \tau + 1$. It is this modified theta function that, in the algebro-geometric literature (see e.g. [64]) gives rise to the unique (up to a constant) holomorphic section of the theta line bundle associated to a complex torus (and, in general, to a principally polarized Abelian variety), which is actually a prequantum bundle, as anticipated ([122, 13, 56], see also [99]). This is readily generalized to the k-level theta functions, which (up to constants) yield an orthonormal basis for the (k-dimensional, by Riemann-Roch) quantum Hilbert space (see [64, 71]).

We record the relevant formulae, for definiteness:

$$\widetilde{\theta}_{k,j}(z, \tau) = e^{k\frac{\pi}{2}(\Im\tau)^{-1}z^2}\sum_{n\in\mathbb{Z}} e^{\frac{i}{k}\pi(kn+j)^2\tau + 2\pi i(kn+j)z} \equiv e^{k\frac{\pi}{2}(\Im\tau)^{-1}z^2}\theta_{k,j}(z, \tau)$$

for $j = 0, \ldots, k - 1$. A crucial fact for what follows is that the k-level theta functions $\theta_{k,j}$ fulfil the (holomorphic) *heat equation*

$$\left[\frac{\partial}{\partial \tau} + \frac{1}{4 \pi k} \frac{\partial^2}{\partial z^2} \right] \theta_{k,j} = 0.$$

Now, a straightforward computation shows that, under the trasformation $\tau \mapsto \tau + 1$, the 2-level theta functions $\vartheta_{2,0}$ and $\vartheta_{2,1}$, *together with their "tilded" analogues*, behave as follows

$$\theta_{2,0}(z, \tau + 1) = \theta_{2,0}(z, \tau), \qquad \theta_{2,1}(z, \tau + 1) = e^{i \frac{\pi}{2}} \theta_{2,1}(z, \tau).$$

Consider the vector bundle $V \to \mathbb{H}$, with V_τ (fibre at τ) given by the 2-dimensional complex vector space of 2-level theta functions with fixed parameter τ. It comes equipped with the *heat connection* ∇, and the 2-level theta functions provide a basis of covariantly constant sections thereof, this being expressed by fulfilment of the heat equation. An important consequence is that, in particular, the natural $SL(2, \mathbb{Z})$-action on $\mathbb{H}$ given by

$$\tau \mapsto \frac{a\tau + b}{c\tau + d} \equiv Z(\tau)$$

$(ad - bc = 1)$, yields, in turn, a parallel displacement map $Q(Z) : V_\tau \to V_{Z \cdot \tau}$, for $Z \in SL(2, \mathbb{Z})$ (along any path connecting the two points). Specifically, for the matrix Z_0 associated to the map $\tau \mapsto \tau + 1$, i.e.

$$Z_0 = \begin{pmatrix} 1 & 1 \\ 0 & 1 \end{pmatrix}$$

one has the ("phase gate" [33]) matrix, whereby we rephrase the transformation formula for the $\theta_{2,j}$'s and the $\widetilde{\theta}_{2,j}$'s :

$$Q(Z_0) = \begin{pmatrix} 1 & 0 \\ 0 & e^{i \frac{\pi}{2}} \end{pmatrix}$$

acting on the theta vector $\widetilde{\theta}_2(z) = (\widetilde{\theta}_{2,0}(z, \tau), \widetilde{\theta}_{2,1}(z, \tau))^T$. Notice that $Q(Z_0)^4 = Id_2$. The matrix Z_0 can in turn be identified, with an abuse of language, with the holonomy $hol(\nabla^{can})$ of the canonical connection ∇^{can} hinted at in Subsection 2.2. See Section 5 for further elaboration of this issue.

We also remark that, by virtue of the preceding formulae, the map $\tau \mapsto \tau + 1$ determines a switch of the roots e_2 and e_3. This will be important in the sequel (Section 5).

Finally we notice that, for ϑ_3, one has

$$\vartheta_3(0, \tau + 1) = \vartheta_4(0, \tau) = \frac{\vartheta_4(0, \tau)}{\vartheta_3(0, \tau)} \vartheta_3(0, \tau) = \sqrt{k'} \, \vartheta_3(0, \tau)$$

(by the Jacobi formula), yielding a sort of differential geometric interpretation of the Jacobi modulus.

Let us also briefly recall, in addition to the above, some basic concepts in Riemann surface theory needed in the sequel, together with the several variable theta function, just to establish notation, at the cost of some repetitions. For a thorough treatment we refer e.g. to [46, 64, 79, 81]. Consider a closed Riemann surface Σ_g – a complex structure and a (Kähler) metric thereon are thus understood – and a canonical dissection thereof in terms of a $4g$-gon, leading to a basis of 1-cycles $\alpha_j, \beta_j, j = 1, 2, \ldots g$. One then finds a basis of Abelian differentials (holomorphic 1-forms) $\omega_j, j = 1, 2, \ldots g$ such that

$$\int_{\alpha_i} \omega_j = \delta_{ij}, \qquad \int_{\beta_i} \omega_j = Z_{ij},$$

with the $g \times g$ *symmetric matrix* $Z = (Z_{ij})$ having *positive-definite* imaginary part $\Im Z$.

Let Λ be the lattice in $\mathbb{C}^g$ generated by the columns of the *period matrix*

$$\Pi = (I, Z)$$

(I being the $g \times g$ identity matrix). The Jacobian $J(\Sigma_g)$ attached to the Riemann surface Σ_g is the g-dimensional complex torus

$$J(\Sigma_g) := H^1(\Sigma_g, \mathbb{R})/H^1(\Sigma_g, \mathbb{Z}) \cong \mathbb{C}^g/\Lambda.$$

Actually, $J(\Sigma_g)$ is a Kähler manifold, and its Kähler form ω can be cast in the form

$$\omega = \sum_j dq_j \wedge dp_j = \frac{\sqrt{-1}}{2} \sum_{\alpha,\beta} W_{\alpha\beta} \, dz_\alpha \wedge d\bar{z}_\beta$$

with respect to suitable (Darboux) symplectic coordinates $(q_1, p_1, q_2, p_2, \ldots q_g, p_g)$ (and their complex counterparts) of $J(\Sigma_g)$, and where $W := (\Im Z)^{-1}$.

The above conditions for a complex torus (Riemann conditions), without any reference to a Riemann surface, define a general *principally polarized Abelian variety*. It is well known that Abelian varieties, that is, complex tori admitting a polarization, namely, a positive Hermitian form on a complex vector space V such that its imaginary part, restricted to a full lattice $\Lambda \subset V$, is *integral*, are exactly the complex tori embeddable in projective space. See also the section on Fourier-Mukai for further developments.

Subsequently, let us briefly recall the *Abel map* $\mathcal{A}$ ([46, 79, 81]):

$$\mathcal{A} : \Sigma_g \longrightarrow J(\Sigma_g)$$

$$x \mapsto \mathcal{A}(x) := (\textstyle\int_{x_0}^x \omega_i)_{i=1,2\ldots,g} \text{ mod periods}$$

(with a choice of a base point x_0). More generally, one has a map (denoted by the same symbol)

$$\mathcal{A} : C_n(\Sigma_g) \quad \longrightarrow \quad J(\Sigma_g)$$

$$(x_1, x_2, \ldots, x_n) \mapsto \mathcal{A}\left(\sum_{j=1}^{n} x_j\right) := \left(\sum_{j=1}^{n} \int_{x_0}^{x_j} \omega_i\right)_{i=1,2\ldots,g} \qquad \text{mod periods.}$$

The notation $\sum_{j=1}^{n} x_j$ stands for the *(positive) divisor* on Σ_g consisting of the points x_j, $j = 1, 2, \ldots n$. The configuration space $C_n(\Sigma_g)$ is a complex n-dimensional manifold and $\mathcal{A}$ is a holomorphic map (cf. [81]).

We also record the *theta functions* (with rational characteristics, for $\vec{a}, \vec{b} \in \mathbb{Q}^g$):

$$\vartheta\left(\begin{bmatrix} \vec{a} \\ \vec{b} \end{bmatrix}\right)(\vec{z}, Z) = \sum_{\vec{n} \in \mathbb{Z}^g} \exp\left[\pi\sqrt{-1}(\vec{n} + \vec{a})^T Z(\vec{n} + \vec{a})\right.$$
$$\left. + 2\pi\sqrt{-1}(\vec{n} + \vec{a})^T (\vec{z} + \vec{b})\right].$$

For $\vec{a} = \vec{b} = \vec{0}$ one gets the standard theta function (which is indeed defined, together with the other ones, for a general Abelian variety, for Z fulfilling the same conditions as above).

A theta function corresponds to the unique holomorphic section (up to a scalar) of a holomorphic line bundle $\Theta \to J(\Sigma_g)$ actually coming from (holomorphic) geometric quantization of $(J(\Sigma_g), \omega)$, as a consequence of the Riemann-Roch theorem (see e.g. [46, 79, 81]). The first Chern class $c_1(\Theta)$ equals the de Rham class of the Kähler form ω.

We recall at this point, for further repeated use, that, given a smooth Hermitian vector bundle (E, h) over a complex manifold M, equipped with a connection $\nabla : \Lambda^0 \to \Lambda^1$ whose curvature Ω_∇ is of type $(1, 1)$, then E possesses a natural holomorphic structure – i.e. the $(0, 1)$-component of the connection (it fulfils the integrability condition $\overline{\nabla}^2 = 0$) – such that ∇ coincides with the canonical (Chern-Bott) connection, the latter being characterized by compatibility with both the metric h and the holomorphic structure (i.e. in a local holomorphic frame one has $\overline{\nabla} = \bar{\partial}$), see e.g. [10], Theorem 5.1. We record for clarity and future use the explicit local formula

$$\nabla = d + h^{-1}\partial h, \qquad \Omega_\nabla = h^{-1}\bar{\partial}\partial h - (h^{-1}\bar{\partial}h) \wedge (h^{-1}\partial h)$$

(stemming from the general local formula for a connection $\nabla = d + \omega$, $\Omega_\nabla = d\omega + \omega \wedge \omega$).

Now, the important point for us is that the line bundle Θ, equipped with its canonical connection carries an irreducible representation of the CCR, by Riemann-Roch combined with the von Neumann Uniqueness Theorem

([117]), since theta appears essentially as the ground state of a quantum harmonic oscillator: the lowering (annihilation) operators read, in fact

$$A_j \sim \nabla_{\bar{z}_j} = \frac{\partial}{\partial \bar{z}_j}$$

(see [99] for details and also below). Actually one has a family of such representations, labelled by the Picard group $Pic^0(J(\Sigma_g)) \cong J(\Sigma_g)$, namely, the receptacle of the isomorphism classes of holomorphic line bundles having vanishing first Chern class. The same observation applies to the tensor powers Θ^k of the theta line bundle, whose holomorphic sections are the k-level theta functions, and yield a k^g-dimensional space. Variation on $J(\Sigma_g)$ yields a rank k-holomorphic vector bundle (the *spectral bundle*) having a crucial physical role (see below for further discussion).

3.3 The Matsushima Construction

For further background material related to this Subsection we mostly refer to [67]. Let M be a n-dimensional compact Kähler manifold, with Kähler metric $\tilde{g}$ and associated Kähler form ω, $\mathcal{E} \to M$ a vector bundle over it, with rank $\mathrm{rk}(\mathcal{E})$. Recall that the *degree* $\deg(\mathcal{E})$ of $\mathcal{E}$ is given by

$$\deg(\mathcal{E}) = \int_M c_1(\mathcal{E}) \wedge \omega^{n-1},$$

where the first Chern class $c_1(\mathcal{E}) \in H^2(M, \mathbb{Z})$ can be computed via the curvature form Ω of any connection ∇ on $\mathcal{E}$; with a slight notational abuse:

$$c_1(\mathcal{E}) = \mathrm{Tr}_{\mathrm{End}(\mathcal{E})}\left(-\frac{\Omega}{2\pi\sqrt{-1}}\right),$$

where $\mathrm{End}(\mathcal{E})$ is the endomorphism bundle associated with $\mathcal{E}$, and Tr denotes trace.

The *Chern Character* (with values in $H^{\mathrm{even}}(M, \mathbb{Q})$) reads

$$\mathrm{Ch}(\mathcal{E}) = \mathrm{Tr}_{\mathrm{End}(\mathcal{E})}\left(\exp\left[-\frac{\Omega}{2\pi\sqrt{-1}}\right]\right)$$

and can be organised through the *Chern character vector*

$$(\mathrm{Ch}_0(\mathcal{E}) = \mathrm{rk}(\mathcal{E}), \mathrm{Ch}_1(\mathcal{E}) = c_1(\mathcal{E}), \ldots, \mathrm{Ch}_n\mathcal{E}).$$

The Chern character satisfies the ring homomorphism properties

$$\mathrm{Ch}(\mathcal{E}_1 \oplus \mathcal{E}_2) = \mathrm{Ch}(\mathcal{E}_1) + \mathrm{Ch}(\mathcal{E}_2), \qquad \mathrm{Ch}(\mathcal{E}_1 \otimes \mathcal{E}_2) = \mathrm{Ch}(\mathcal{E}_1) \cdot \mathrm{Ch}(\mathcal{E}_2).$$

The *slope* $\mu(\mathcal{E})$ of $\mathcal{E}$ is by definition

$$\mu(\mathcal{E}) := \frac{\deg(\mathcal{E})}{\mathrm{rk}(\mathcal{E})}.$$

A holomorphic vector bundle $\mathcal{E} \to M$ is said to be *semistable* if for every holomorphic subbundle $\mathcal{F}$ one has $\mu(\mathcal{F}) \le \mu(\mathcal{E})$, *stable* if strict equality holds (for a proper subbundle) ([80, 67]). The above condition can be phrased in differential geometric terms in view of the Donaldson-Uhlenbeck-Yau theorem (see [80, 37, 114, 67]): briefly, this goes as follows. Recall that a Hermitian metric h on $\mathcal{E}$ is called a *Hermitian-Einstein* (HE) metric if, given its Chern-Bott connection ∇, one has

$$\sqrt{-1}\Lambda\Omega = \lambda I_{\mathrm{End}(\mathcal{E})},$$

with Λ denoting here contraction with the Kähler form ω and λ a real constant.

Then the *Kobayashi-Hitchin correspondence* states that an irreducible holomorphic vector bundle (i.e. without proper holomorphic direct summands) admits a HE-metric if and only if it is stable; it has been eventually proved for any compact Kähler manifold in [114]. In particular, a projectively flat holomorphic vector bundle admitting a projectively flat Hermitian structure, i.e. equipped with a Hermitian metric whose corresponding canonical connection has constant curvature, is stable. We are thus naturally led to look for a Hermitian holomorphic vector bundle $\mathcal{E} \to J(\Sigma_g)$ over the Jacobian $J(\Sigma_g)$ of the Riemann surface in question, equipped with a HE-connection ∇ having constant curvature equal (up to a $2\pi\sqrt{-1}$ factor) to a rational number ν - called henceforth *statistical parameter* (in view of its future braid theoretic interpretation) and this will give rise to a holomorphic stable bundle with slope $\mu(\mathcal{E}) \propto \nu$ (we shall soon verify that the precise factor will be g!).

The construction of such bundles over a generic Abelian variety is classical (cf. in particular [75, 53, 67]) and we briefly review it below for the case $g = 1$. Letting $\nu = q/r$, g.c.d.$(q, r) = 1$, start from a r-dimensional representation of the finite Weyl-Heisenberg group corresponding to a 2-lattice $\Gamma \subset \mathbb{C}$, giving rise to the complex -torus $\mathbb{C}/\Gamma$:

$$U(\gamma + \gamma') = U(\gamma) \cdot U(\gamma')\, e^{\frac{\sqrt{-1}}{2r} A(\gamma', \gamma)}$$

with A being the imaginary part of a Hermitian form R on $\mathbb{C}$ such that

$$\frac{1}{2\pi} A(\gamma, \gamma') \in \mathbb{Z}.$$

In our setting we take (abuse of notation, and in standard coordinates)

$$A = -2\pi q \cdot \omega = -2\pi q \cdot dx \wedge dy = -\sqrt{-1}\pi q \cdot dz \wedge d\bar{z}.$$

Equivalently

$$U_1 U_2 = U_2 U_1\, e^{-2\pi\sqrt{-1}\frac{q}{r}}.$$

Therefore

$$R(z, w) = \pi q \cdot z\, \bar{w}.$$

We get a factor of automorphy (theta factor)

$$j(\gamma, z) = U(\gamma) \cdot \exp\left[\frac{1}{2r} R(z, \gamma) + \frac{1}{4r} R(\gamma, \gamma)\right], \qquad (\gamma, z) \in \Gamma \times \mathbb{C},$$

satisfying

$$j(\alpha + \beta, z) = j(\alpha, \beta + z) j(\beta, z) \qquad \forall \alpha, \beta \in \Gamma, \, z \in \mathbb{C}.$$

The rank r (stable) vector bundle is then

$$\mathcal{E}_\nu = (\mathbb{C}/\Gamma \times \mathbb{C}^r)/\Gamma,$$

with the action of Γ specified via

$$\gamma \cdot (z, \zeta) := (\gamma + z, j(\gamma, z)\zeta), \qquad \gamma \in L, \, (z, \zeta) \in \mathbb{C}/\Gamma \times \mathbb{C}^r.$$

The vector bundle $\mathcal{E}_\nu$ comes equipped with the Hermitian metric

$$h(z) = \exp\left[-\frac{1}{2r} R(z, z)\right] I_r,$$

leading to a Chern-Bott connection ∇ having the required constant curvature

$$\Omega_\nabla = \sqrt{-1}\, \frac{A}{r}\, I_r = -2\pi \sqrt{-1}\, \nu \cdot e_1 \wedge e_2\, I_r.$$

The space of holomorphic sections $H^0(\mathcal{E}_\nu)$ is then q-dimensional, giving rise to the *Matsushima theta functions* and, in the case $r = 1$, we retrieve the q-level theta functions. Of course the above treatment is consistent with the early findings of Atiyah ([8]). Notice that the Matsushima theta functions are *vector valued* and this is relevant for their braid group theoretical intepretation. However, they can be made to correspond to the q-level theta functions ([75]).

Therefore, the conclusion is that *a projectively flat HE-bundle on the Jacobian $J(\Sigma_g)$ with HE-connection with curvature $-2\pi\sqrt{-1}\nu\,\omega$ can be manufactured via the Matsushima construction*, and, by a result of Hano ([53]), this is essentially the only way to achieve this. Indeed, the problem of identifying the projectively flat HE-vector bundles one can employ for representing the RS-braid group can be addressed via the Riemann-Roch-Hirzebruch (RRH)-theorem, together with a cohomology vanishing theorem. We refer to [21, 55] for background and full details, and also below for further applications.

Let us consider, on the Clifford module $\Lambda(T^{0,1}M)^* \otimes W$, the Dolbeault-Dirac (or spin^c) operator

$$D_W := \sqrt{2}(\bar{\partial}_W + \bar{\partial}_W^*)$$

associated to the Dolbeault complex attached to a Hermitian holomorphic vector bundle W over a Kähler manifold M. The Hodge theorem implies then that

$$\mathrm{Ker}(D_W) \cong H^{\bullet}(M, \mathcal{O}(W))$$

(the cohomology of W). As usual, we set $h^i(M, W)) := \dim H^i(M, \mathcal{O}(W))$. The index "ind" of the spin^c operator is the dimension of the above kernel, looked upon as a *superspace*:

$$\mathrm{ind}(D_W) = \dim(\mathrm{Ker}\, \overline{\partial}_W) - \dim(\mathrm{Ker}\, \overline{\partial}_W^*).$$

The RRH-theorem (viewed à la Atiyah-Singer) yields then the formula

$$\mathrm{ind}(D_W) = \chi(W) = \sum_{i=0}^{n} (-1)^i h^i(M, W) = \int_M \mathrm{Ch}(W)\mathrm{Td}(M)$$

($\chi(W)$ is the holomorphic Euler characteristic of W, $\mathrm{Ch}(W)$ its Chern Character, and $\mathrm{Td}(M)$ the Todd class of M – we shall not need the explicit expression for the latter). In [105] we prove the following general result.

Theorem 3.1 *([105]) (i) Let $\mathcal{E}$ be a projectively flat holomorphic vector bundle over $J(\Sigma_g)$ (or, more generally, over an Abelian variety) carrying a HE-connection ∇ with constant curvature $\Omega_{\nabla} = -2\pi\sqrt{-1}\, v \cdot \omega := -2\pi\sqrt{-1}\, q/r \cdot \omega$, (with $r > 0$, $q > 0$ and g.c.d$(r, q) = 1$). Then one has,*

$$R := \mathrm{rk}(\mathcal{E}) = k\, r^g, \qquad h^0(\mathcal{E}) = k\, q^g,$$

with k a positive integer.
(ii) The following slope-statistics formula *holds:*

$$\mu(\mathcal{E}) = v\, g!\,.$$

The proof makes use of the Weitzenböck formula, and can be effectively phrased in Fermion algebra terms, see [105] for details. Therefore, we see that Matsushima construction produces examples for $k = 1$. Upon taking the direct sum of $k > 1$ copies of a Matsushima bundle one produces the more general examples (as an immediate consequence of the properties of the Chern character).

Explicit examples for higher genera via products of elliptic curves $\Sigma_1 \cong J(\Sigma_1)$ are constructed in [105].

Now consider the projectively flat HE-vector bundles $\mathcal{E}$ of the preceding Subsection and take $H_1 := L^2(\mathcal{E})$, namely the L^2- sections of $\mathcal{E}$ obtained by completing its smooth sections with respect to the inner product

$$\langle \cdot, \cdot \rangle := \int_{J(\Sigma_g)} h(\cdot, \cdot) \frac{\omega^g}{g!}.$$

Specifically, again with respect to the standard (Darboux) symplectic coordinates $(q_1, p_1, q_2, p_2, \ldots q_g, p_g)$ of $J(\Sigma_g)$, we have

$$\left[\nabla_{\frac{\partial}{\partial q_j}}, \nabla_{\frac{\partial}{\partial q_k}} \right] = \left[\nabla_{\frac{\partial}{\partial p_j}}, \nabla_{\frac{\partial}{\partial p_k}} \right] = 0; \qquad \left[\nabla_{\frac{\partial}{\partial q_j}}, \nabla_{\frac{\partial}{\partial p_k}} \right] = -2\pi\sqrt{-1}\, \nu\, \delta_{jk} \cdot I$$

for $j, k = 1, 2 \ldots g$.

Notice in fact that, by periodicity and the compatibility of ∇ with h, one has

$$0 = \int_{J(\Sigma_g)} X\, h(\cdot, \cdot) \frac{\omega^g}{g!} = \int_{J(\Sigma_g)} \left[h(\nabla_X \cdot, \cdot) + h(\cdot, \nabla_X \cdot) \right] \frac{\omega^g}{g!}$$

with $X = \partial/\partial q_j, \partial/\partial p_j$, thus the operators $\nabla_{\frac{\partial}{\partial q_j}}$, $\nabla_{\frac{\partial}{\partial p_j}}$ are formally skew-hermitian. By classical functional analytic arguments they are skew-adjoint (cf. [92]).

Thus, under the above assumptions, we get an infinite dimensional representation of the Weyl-Heisenberg Commutation relations (generalising [99]) with multiplicity q^g.

3.4 Noncommutative Geometry and Schwarz Theta Functions

In this section we resort to noncommutative geometry (NCG) and we shall devise, giving details in the genus one case, a construction of stable bundles, together with their holomorphic sections, via suitable noncommutative theta vectors in the sense of A. Schwarz ([96]). The noncommutative setup will allow the emergence of a natural "statistical" duality which will be related to the Fourier-Mukai-Nahm transform (FMN).

Applications of NCG to the QHE, both integral and fractional, in which NCG is used in its full strength at the very foundational level, are well known (see e.g. [18, 73, 74]). Our approach pursues a different, more classical route in the sense that NCG plays an auxiliary, though relevant, role. We concentrate on the $g = 1$ case and we use noncommutative geometry to sketch a construction of "classical" stable bundles thereon. Everything is ultimately based on A. Connes' seminal paper ([34], see also [93, 35]), to which we refer for complete details. We are tacitly identifying, via Swan's theorem, sections of vector bundles with finitely generated projective modules over the algebra of smooth functions on the base manifold: it is precisely this interpretation that renders the transition to a noncommutative environment (i.e. general algebras and finitely generated projective modules thereon) possible, and the classical differential geometric apparatus (connections, curvature and so on) carries through to this new situation.

Recall that the noncommutative torus A_ϑ, $\vartheta \in \mathbb{R}/\mathbb{Z}$, is the universal unital C^*-algebra generated by unitary operators U_j, $j = 1, 2$ satisfying the commutation relation

$$U_1 U_2 = e^{2\pi\sqrt{-1}\vartheta} U_2 U_1$$

and it is to be thought of as a deformation of the standard commutative algebra of continuous functions on a torus - via Fourier theory.

Actually we shall use its smooth subalgebra $\mathbb{T}^2_\vartheta$ consisting of all series with rapidly decreasing coefficients:

$$\sum a_{mn} U_1^m U_2^n, \qquad \{a_{mn}\} \in \mathcal{S}(\mathbb{Z}^2).$$

Natural (Hermitian) $\mathbb{T}^2_\theta$-right modules (the analogues of vector bundles) are given by

$$\mathcal{E}_{p,q} = \mathcal{S}(\mathbb{R}, \mathbb{C}^q)$$

(vector valued Schwartz functions), p and q positive integers, g.c.d.$(p, q) = 1$, having "rank" (à la Murray-von Neumann)

$$\mathrm{rk}(\mathcal{E}_{p,q}) := \tau_{\mathrm{End}(\mathcal{E}_{p,q})}(I) = p - \vartheta q$$

(the trace on the endomorphism algebra being involved); also assume $p - \vartheta q > 0$. In detail, the module structure, is given by first setting

$$(V_1 \xi)(s) = e(s)\xi(s), \qquad (V_2 \xi)(s) = \xi(s - (p/q - \vartheta)), \qquad \xi \in \mathcal{E}_{p,q}.$$

Then, one considers the following finite Weyl-Heisenberg commutation relations for $\mathbb{Z}/q\mathbb{Z}$:

$$w_1 w_2 = \bar{e}(p/q) w_2 w_1, \qquad w_1^q = w_2^q = 1.$$

Concretely, one may take, as we already observed: $w_1 = \mathrm{diag}(1, e(p/q), e(2 \cdot p/q), \ldots e((q-1) \cdot /q))$ and $w_2 =$ matrix of the shift map $e_i \to e_{i-1}$, $i = 1, 2, \ldots q$, $e_0 = e_q$, with $(e_1, \ldots, e_q)$ being the canonical basis of $\mathbb{C}^q$.

Then one sets

$$\xi U_i = (V_i \otimes w_i)\xi, \qquad i = 1, 2.$$

The *Connes connection* ([34, 35]) reads:

$$(\nabla_1 \xi)(s) = 2\pi\sqrt{-1}\frac{q}{p - \vartheta q} s\, \xi(s), \qquad (\nabla_2 \xi)(s) = \xi'(s), \qquad s \in \mathbb{R}$$

(Indices 1 and 2 refer to the canonical basis of $\mathbb{R}^2$ viewed as the Lie algebra of an ordinary torus acting on $\mathcal{A}_\vartheta$). It actually fulfils the appropriate version of the Leibniz rule, it is hermitian with respect to a natural metric, it has constant curvature

$$\Omega := [\nabla_1, \nabla_2] e_1 \wedge e_2 = -2\pi\sqrt{-1}\frac{q}{p - \vartheta q} e_1 \wedge e_2$$

and *first Chern class* $c_1(\mathcal{E}_{p,q})$ given by the usual expression

$$c_1(\mathcal{E}_{p,q}) = \frac{1}{2\pi\sqrt{-1}}\ \tau_{\mathrm{End}(\mathcal{E}_{p,q})}(I)\left(2\pi\sqrt{-1}\ \frac{q}{p-\vartheta q}\right) = q.$$

We now introduce, mimicking the classical case, a *holomorphic structure* ([100, 101, 90]) on $\mathcal{E}_{p,q}$ via the $\overline{\nabla}$ operator as

$$\overline{\nabla} := \nabla_1 + \sqrt{-1}\nabla_2.$$

The kernel of the $\overline{\nabla}$ operator is called the space of *noncommutative theta vectors* ([96], see [99] as well) and it easily seen to be q-dimensional:

$$\xi = \xi(s) = e^{-\pi\,\frac{q}{p-\vartheta q}s^2}\cdot v \qquad s\in\mathbb{R}, \qquad v\in\mathbb{C}^q$$

i.e. a (vector valued) one-dimensional harmonic oscillator ground state. The adjoint $\overline{\nabla}^*$, by contrast, has trivial kernel

$$Ker(\overline{\nabla}^*) = \{0\}.$$

This follows easily from the commutation relation

$$[\overline{\nabla}, \overline{\nabla}^*] = 4\pi\ \frac{q}{p-\vartheta q}I$$

(indeed, the operators $\overline{\nabla}$, resp. $\overline{\nabla}^*$ are (up to constants) lowering and raising operators). The above result can by cast in the form of an index formula:

$$\mathrm{ind}(\overline{\nabla}) := \dim Ker(\overline{\nabla}) - \dim Ker(\overline{\nabla}^*) = q$$

Now observe that the modules $\mathcal{E}_{p,q}$, as vector spaces, are actually *independent of* p. Thus, upon taking $p = r$ a positive integer and $\vartheta = 0$, we obtain a "classical" rank r Hermitian holomorphic vector bundle over a complex torus, which, in view of the above discussion, is indeed stable, with slope = statistical parameter = $-1/(2\pi\sqrt{-1})\cdot$ curvature = q/r. Summing up, we have the following:

Theorem 3.2 ([105]) *(i) The above modules $\mathcal{E}_{r,q}$, with g.c.d$(r, q) = 1$, are the modules of sections of a Hermitian holomorphic stable vector bundle (denoted by the same symbol) over a complex torus Σ_1 equipped with a Hermitian connection with constant curvature, having slope*

$$\mu(\mathcal{E}_{r,q}) = \frac{q}{r} = \nu.$$

(ii) The space $H^0(\mathcal{E}_{r,q})$ of holomorphic sections of $\mathcal{E}_{r,q}$ has dimension

$$h^0(\mathcal{E}_{r,q}) = q.$$

*(iii) In particular, the case $q = r = 1$ yields back the theta function line
       bundle.*

*(iv) ("Schwarz = Laughlin"). The "centre of mass" part of generalised
      Laughlin wave functions can be realised via suitable Schwarz theta
      vectors.*

Part (iv) will be clearer once we discuss RS-braid group representations
in the following Section. Here we just anticipate that the above expression
refers to the holomorphic sections of a HE-bundle, entering in the construction
of generalized Laughlin wave functions (extending the one discussed in
Subsection 2.3).

Actually, a whole *family* of constant curvature connections may be found,
and it is labelled by the torus itself (viewed again as $Pic^0(\Sigma_1)$). So, *de facto*,
these stable vector bundles essentially coincide with the Matsushima bundles
$\mathcal{E}_\nu$ previously discussed.

Remarks 1. A similar construction of stable bundles over an elliptic curve can
    be extracted via [90], where a noncommutative Fourier-Mukai transform is
    discussed (see below for applications of FM).

2. The previous noncommutative geometric construction of "classical"
    holomorphic stable bundles can be easily generalised upon resorting to
    the higher dimensional noncommutative tori $\mathbb{T}_\vartheta^{2m}$ following e.g. [93, 101,
    7, 96]. One deals with a "standard module" $\mathcal{E}$, i.e. a projective $\mathbb{T}_\vartheta^{2m}$-module
    equipped with a constant curvature connection ∇. The bundle is equipped
    with a holomorphic structure, and the kernel of the antiholomorphic part
    $\overline{\nabla}$ of the connection ∇ is the $h^0(\mathcal{E})$-dimensional space of the theta vectors
    in the sense of [96]. The connection naturally gives rise to a representation
    of the CCR, direct sum of $h^0(\mathcal{E})$ copies of the Schrödinger representation,
    again in view of the von Neumann uniqueness theorem ([96],
    [99] and above). The number $h^0(\mathcal{E})$ also equals, according to Schwarz,
    the K-theory class $\tilde{\mu}(\mathcal{E})$ of the module $\mathcal{E}$ evaluated on an appropriate
    symplectic basis of the Lie algebra of the (commutative) torus acting
    on $\mathbb{T}_\vartheta^{2m}$. In a "commutative" environment, one again gets the Matsushima
    stable holomorphic Hermitian vector bundles previously discussed.

3.5 Duality and Fourier-Mukai-Nahm-Transform

The preceding noncommutative approach sheds light on a notable symmetry.

Let as above $\nu = q/r$, and take one of the above stable bundles $\mathcal{E}_\nu$,
constructed à la Matsushima; then set $\nu' = r/q = 1/\nu$ and consider the
previous "dual" representation of the finite CCR:

$$w_1 w_2 = e^{-2\pi\sqrt{-1}\cdot\frac{r}{q}} w_2 w_1$$

realised on $H^0(\mathcal{E}) \cong \mathbb{C}^q$, corresponding to the *rational* noncommutative torus $\mathbb{T}^2_{\frac{r}{q}}$. We write, functorially:

$$C : \mathcal{E}_\nu \to \mathbb{T}^2_{\nu'}$$

(C stands for Connes). This representation can be promoted to a stable bundle $\mathcal{E}_{\nu'}$ via the Matsushima (M) construction:

$$M : \mathbb{T}^2_\nu \to \mathcal{E}_\nu,$$

so the upshot is the following:

Theorem 3.3 *([105]) (double interpretation of theta vectors). The space of theta vectors pertaining to the above geometrical (infinite dimensional) unitary representation of the CCR with "statistics parameter"* $\sigma = e^{\pi\sqrt{-1}\cdot\frac{q}{r}} = (-1)^\nu$ *also determines a finite dimensional unitary CCR-representation with "dual" statistics parameter* $\sigma' = e^{\pi\sqrt{-1}\cdot\frac{r}{q}} = (-1)^{\nu'}$. *We have an ensuing "Matsushima-Connes (MC) duality", written succinctly*

$$M : \mathbb{T}^2_\nu \to \mathcal{E}_\nu, \quad C : \mathcal{E}_\nu \to \mathbb{T}^2_{\nu'}, \quad MC : \mathcal{E}_\nu \to \mathcal{E}_{\nu'}, \quad CM : \mathbb{T}^2_\nu \to \mathbb{T}^2_{\nu'}$$

Remark The Matsushima construction matches exactly with the treatment of Landau levels on a torus given by [87] in terms of "symmetry breaking" of the genuine Weyl-Heisenberg group (magnetic translations) to a finite one. The noncommutative geometrical approach improves on this picture, rendering it quite natural.

We now reinterpret the above duality in terms of the Fourier-Mukai-Nahm transform. For a comprehensive treatment of the latter see e.g. [14, 54]. Here we confine ourselves to cursory remarks. Let $M = V/\Lambda$ be an Abelian variety as before, and $\widehat{M}$ its dual one. Let us briefly recall its construction: one takes the dual conjugate space $\bar{V}^*$ and the dual lattice $\Lambda^* = \{\ell \in \bar{V}^* | \ell(\Lambda) \subset \mathbb{Z}\}$; then $\widehat{M} = \bar{V}^*/\Lambda^*$. Let $\pi_1 : M \times \widehat{M} \to M$ and $\pi_2 : M \times \widehat{M} \to \widehat{M}$ denote the obvious projections. Let $\mathcal{P} \to M \times \widehat{M}$ be the *Poincaré bundle*, which is uniquely characterised by the fact that its restriction $\mathcal{P}_\xi := \mathcal{P}_{M\times\{\xi\}} \to M$ is the line bundle pertaining to $\xi \in \widehat{M}$, together with the requirement that, restricted to $\widehat{M} \cong \pi_M^{-1}(0)$, it is trivial. It can be also described via the Hermitian form H on $V \times \bar{V}^*$ given by:

$$H(v, w, \alpha, \beta) = \overline{\beta(v)} + \alpha(w) \qquad v, w \in V, \ \alpha, \beta \in \bar{V}^*$$

and the semicharacter

$$\chi(\lambda, \mu) = e^{\sqrt{-1}\pi\mu(\lambda)} \qquad \lambda \in \Lambda, \ \mu \in \Lambda^*.$$

Let R be the *direct image functor*. The Fourier-Mukai (FM) transform of a holomorphic vector bundle on M (or more generally, a sheaf – as before, one describes a vector bundle via its sections) is the following sheaf on $\widehat{M}$:

$$(FM)(\mathcal{E} \to M) = [R\pi_{2*}(\pi_1^*(\mathcal{E}) \otimes \mathcal{P})] \to \widehat{M}.$$

In favourable cases one obtains a bona fide holomorphic vector bundle: we present a concrete instance of this phenomenon in the genus one case, for the (Matsushima) bundles $\mathcal{E} = \mathcal{E}_\nu$ previously treated. The Jacobian $J(\Sigma_1)$ is a self-dual abelian variety and it also parametrizes the flat line bundles over itself, in the sense that its points label holonomies of flat connections thereon, and at the same time the holomorphically inequivalent degree zero (i.e. flat) line bundles $\mathcal{P}_x$, $x \in J(\Sigma_1)$ (by the very definition of the Poincaré bundle). Upon tensoring $\mathcal{E}$ with $\mathcal{P}_\xi$, one obtains a vector bundle $\mathcal{E}_\xi = \mathcal{E} \otimes \mathcal{P}_\xi$ fulfilling the same conditions as $\mathcal{E}$, by the RRH-theorem. Then by the *index theorem for families*, the cohomology spaces $H^0(\mathcal{E}_x)$, $x \in J(\Sigma_1)$ become the fibres of the sought-for FM-transformed holomorphic vector bundle $FM(\mathcal{E}) \to J(\Sigma_1)$, which is known to be *stable*. (The direct image functor R involves in this case just the 0-cohomology: we are in the so-called IT_0 situation, [14, 54]). Upon subsequent dualization (denoted by *), the first Chern class changes sign and stability persists. Explicitly, one has, for their respective Chern character vectors:

$$\big(\mathrm{Ch}_0(\mathcal{E}), \mathrm{Ch}_1(\mathcal{E})\big) = (r, q), \qquad \big(\mathrm{Ch}_0(FM^*(\mathcal{E})), \mathrm{Ch}_1(FM^*(\mathcal{E}))\big) = (q, r).$$

Notice then that *the Chern Character vectors of the MC-vector bundle and of the FM-transformed bundle are the same.*

One has also the *Nahm transform* $\widehat{\mathcal{E}}$ of the HE-vector bundle $\mathcal{E}$. It comes equipped with a metric and compatible connection $\widehat{\nabla}$ with curvature of type $(1, 1)$, which endows it with a holomorphic structure, and one has $\widehat{\mathcal{E}} \cong FM(\mathcal{E})$ (see e.g. [107]). Let us briefly review this construction, for a Hermitian vector bundle $E \to \Sigma_1$ equipped with a unitary connection ∇; since the curvature form Ω_∇ is of type $(1, 1)$, $E \to \Sigma_1$ becomes a holomorphic vector bundle, with Chern-Bott connection ∇. Let us again consider the canonical spinor bundle $\mathbf{S} = \Lambda^{0, \bullet} T^* \Sigma_1$ with the natural splitting

$$\mathbf{S} = \mathbf{S}^+ \oplus \mathbf{S}^-, \qquad \mathbf{S}^+ = \Lambda^{0,0} T^* \Sigma_1, \qquad \mathbf{S}^- = \Lambda^{0,1} T^* \Sigma_1.$$

Upon tensoring with the Poincaré line bundle $\mathcal{P} \to \Sigma_1 \times J(\Sigma_1)$ one gets a family of spin^c Dirac operators

$$D_\xi : \Lambda^0(\Sigma_1, \mathbf{S}^+ \otimes E_\xi) \to \Lambda^1(\Sigma_1, \mathbf{S}^- \otimes E_\xi)$$

coinciding with the Dolbeault operators

$$D_\xi = \sqrt{2}\,(\bar{\partial}_{E_\xi}^* + \bar{\partial}_{E_\xi}).$$

The Atiyah-Singer theorem for families gives rise to an element $\text{ind}(D) = [\text{Ker}D] - [\text{Coker }D]$ in the K-theory of $J(\Sigma_1)$; actually, in view of the previous RRH considerations, one abuts at a bona fide vector bundle manufactured from the holomorphic sections

$$\text{Ker}D_\xi = \text{Ker }\bar{\partial}_{E_\xi} = H^0(E_\xi) \Rightarrow h^0(E_\xi) = q \quad \forall \xi \in J(\Sigma_1).$$

Let us recall the construction of the Nahm connection $\hat{\nabla}$. Upon taking L^2-completions, one has a short exact sequence

$$0 \to \hat{E}_\xi \to L^2(E_\xi \otimes \mathbf{S}^+) \overset{D_\xi}{\to} L^2(E_\xi \otimes \mathbf{S}^-) \to 0.$$

The L^2-spaces in the middle are typical fibres of trivial vector bundles over $J(\Sigma_1)$, equipped with a standard flat connection. Subsequent projection defines the Nahm connection $\hat{\nabla}$ (it is actually a Grassmann connection). We write, concisely, $FMN(\mathcal{E})$ for the FM-bundle, equipped with the Nahm transformed connection. However, by Uhlenbeck-Yau, $\hat{\nabla}$ coincides with its unique (up to a scalar) HE-connection. *Therefore, up to moduli (and dualization), the MC-vector bundle agrees with the FMN-transformed bundle.* The Nahm-transformed connection turns out to have constant curvature. We are now going to check all this explicitly via a noncommutative geometric approach.

First, consider the constant curvature connections (parametrized by a torus)

$$\tilde{\nabla}_1 = \nabla_1 - 2\pi\sqrt{-1}\,\alpha\,I, \qquad \tilde{\nabla}_2 = \nabla_2 - 2\pi\sqrt{-1}\,\beta\,I, \qquad \alpha, \beta \in [0,1].$$

We remark in passing that they represent Yang-Mills gauge equivalence classes. (Here one has a manifestation of the general principle "symplectic quotient = Mumford quotient", see [48, 80, 9, 13] and [93, 100, 101] as well).

Then consider their associated holomorphic structures

$$\overline{\tilde{\nabla}} = \tilde{\nabla}_1 + \sqrt{-1}\tilde{\nabla}_2 = \overline{\nabla} - 2\pi\sqrt{-1}zI$$

(setting $z = \alpha + \sqrt{-1}\,\beta$). If we compute the corresponding theta vectors (all together, they consist in the standard *coherent states*, see e.g. [89, 106]), we easily get

$$\xi_z \equiv \xi_{\alpha,\beta} = e^{-\pi\frac{r}{q}(\frac{q}{r}s-z)^2}v, \qquad v \in \mathbb{C}^q.$$

One finds, successively (working with scalar functions, since this clearly suffices)

$$|\xi_z|^2 = e^{2\pi\frac{r}{q}\beta^2}e^{-2\pi\frac{r}{q}(\frac{q}{r}s-\alpha)^2}$$

which, after using, for $\gamma > 0$, $\int_\mathbb{R} e^{-\gamma x^2}dx = \sqrt{\pi/\gamma}$, leads to L^2-normalized theta vectors

$$\tilde{\xi}_z = e^{-\pi\frac{r}{q}\beta^2}(2q/r)^{\frac{1}{4}}\xi_z.$$

Now, according to the general recipe, regard the theta vectors as fibres of the Nahm-Fourier-Mukai transformed vector bundle over the torus parametrized by z and compute the Nahm-transformed connection, or, in physical terminology, adiabatic *Berry-Simon connection* (form) ([22, 97, 33]), (it is enough to work componentwise):

$$z \mapsto A_z = \langle \tilde{\xi}_z, d\,\tilde{\xi}_z \rangle = \langle \tilde{\xi}_z, \partial_\alpha \tilde{\xi}_z \rangle d\alpha + \langle \tilde{\xi}_z, \partial_\beta \tilde{\xi}_z \rangle d\beta.$$

Upon making use of $\int_{\mathbb{R}} e^{-\gamma x^2} x\,dx = 0$, we find:

$$\langle \tilde{\xi}_z, \partial_\alpha \tilde{\xi}_z \rangle = -2\pi\sqrt{-1}\,\frac{r}{q}\,\beta, \qquad \langle \tilde{\xi}_z, \partial_\beta \tilde{\xi}_z \rangle = 0,$$

whence (reintroducing vectors):

$$A = -2\pi\sqrt{-1}\,\frac{r}{q}\,\beta\,d\alpha \cdot I_q,$$

having (constant) curvature

$$\Omega = 2\pi\sqrt{-1}\,\frac{r}{q}\,d\alpha \wedge d\beta \cdot I_q$$

as in the Matsushima construction and leading to the correct values $(q, -r)$ for the FMN-transformed bundle.

Remark In physical literature, what is called the adiabatic curvature is actually the trace of Ω, $\mathrm{Tr}(\Omega)$, and it is the curvature form of the determinant line bundle $\det(\hat{\mathcal{E}})$:

$$\mathrm{Tr}(\Omega) = 2\pi\sqrt{-1}\,r\,d\alpha \wedge d\beta.$$

The constancy of $\mathrm{Tr}(\Omega)$ would also follow from the general Bismut-Gillet-Soulé theory (BGS) ([24, 25, 26]) used as in [107, 108], taking into account the fact that for tori, no *analytic torsion* phenomena arise and one can compute via the Quillen metric. Our simple-minded use of noncommutative geometry allowed us to avoid usage of the formidable BGS apparatus.

Alternatively, one could have resorted to Varnhagen's approach via standard theta function theory, but this would have been much more complicated (see [115]).

We recapitulate the preceding discussion by means of the following:

Theorem 3.4 *([105]) (i) Set $v = q/r$, $v' = r/q = 1/v$. The MC-correspondence and the FMN^*-correspondence (Fourier-Mukai-Nahm, plus dualization):*

$$FMN^* : \mathcal{E}_v \to \mathcal{E}_{v'}$$

agree up to moduli.

(ii) *The FMN^*-correspondence can be explicitly realised via noncommutative theta vectors by means of the direct calculation performed above.*

4 Riemann Surface Braid Group Representations via Theta Functions

4.1 The Riemann Surface Braid Group

The braid group on n strands $B(X, n)$ pertaining to a topological space X is by definition the fundamental group of the associated configuration space $C_n(X)$ consisting of all n-ples of distinct points, up to order or, equivalently, of all n-point subsets of X. In our case $X = \Sigma_g$, a closed orientable surface (actually, a Riemann surface) of genus $g \geq 1$. Its associated braid group $B(\Sigma_g, n)$ (termed *RS-braid group*) admits, among others, the following presentation due to P. Bellingeri ([16, 17]). The generators are $\sigma_1, \ldots, \sigma_{n-1}; a_1, \ldots, a_g, b_1, \ldots, b_g$ (the former are the standard braid group generators, the latter have a simple geometric interpretation, in terms of the natural dissection of the surface by means of a $4g$-gon, see [16]). The presence of the extra generators is natural: if, say, two points loop around each other, their trajectories can at the same time wind around the handles of the surface.

In the presentation, in addition to the ordinary braid relations

$$(B1): \sigma_j\sigma_{j+1}\sigma_j = \sigma_{j+1}\sigma_j\sigma_{j+1}, \qquad j = 1, 2, \ldots n - 2$$
$$(B2): \sigma_i\sigma_j = \sigma_j\sigma_i, \qquad |i - j| > 1,$$

one has "mixed" relations

$$(R1): \quad a_r\sigma_i = \sigma_i a_r, \qquad 1 \leq r \leq g, \quad i \neq 1$$
$$b_r\sigma_i = \sigma_i b_r, \qquad 1 \leq r \leq g, \quad i \neq 1$$
$$(R2): \quad \sigma_1^{-1}a_r\sigma_1^{-1}a_r = a_r\sigma_1^{-1}a_r\sigma_1^{-1}, \qquad 1 \leq r \leq g,$$
$$\sigma_1^{-1}b_r\sigma_1^{-1}b_r = b_r\sigma_1^{-1}b_r\sigma_1^{-1}, \qquad 1 \leq r \leq g,$$
$$(R3): \quad \sigma_1^{-1}a_s\sigma_1 a_r = a_r\sigma_1^{-1}a_s\sigma_1, \qquad s < r$$
$$\sigma_1^{-1}b_s\sigma_1 b_r = b_r\sigma_1^{-1}b_s\sigma_1, \qquad s < r$$
$$\sigma_1^{-1}a_s\sigma_1 b_r = b_r\sigma_1^{-1}a_s\sigma_1, \qquad s < r$$
$$\sigma_1^{-1}b_s\sigma_1 a_r = a_r\sigma_1^{-1}b_s\sigma_1, \qquad s < r$$
$$(R4): \quad \sigma_1^{-1}a_r\sigma_1^{-1}b_r = b_r\sigma_1^{-1}a_r\sigma_1, \qquad 1 \leq r \leq g$$
$$(TR): \quad [a_1, b_1^{-1}]\cdots[a_g, b_g^{-1}] = \sigma_1\sigma_2\cdots\sigma_{n-1}^2\cdots\sigma_2\sigma_1$$

with the usual group theoretical convention $[a, b] = aba^{-1}b^{-1}$, used in this Subsection only. We shall restrict ourselves to *unitary* representations

$$\rho : B(\Sigma_g, n) \to U(\mathcal{H})$$

$(U(\mathcal{H})$ being the unitary group on a complex separable Hilbert space $\mathcal{H}$) with $\rho(\sigma_j) = \sigma I$, $j = 1, \ldots, n - 1$ (I being the identity operator on $\mathcal{H}$). One writes

$$\sigma = e^{2\pi\sqrt{-1}\theta} \equiv e^{\pi\sqrt{-1}\nu} = (-1)^\nu$$

and calls θ (a priori defined up to integers) the *statistics parameter* (same for $\nu = 2\theta$ or even σ itself, with abuse of language). The relations B1, B2, R1 and R2 are automatically fulfilled, the relations R3 become:

$$[\rho(a_s), \rho(a_r)] = [\rho(b_s), \rho(b_r)] = I, \qquad r, s = 1, \ldots, g \qquad (1)$$

whereas R4 yields:

$$[\rho(a_r), \rho(b_r)] = \sigma^2 I, \qquad r = 1, \ldots, g.$$

Condition TR gives, in turn, after checking that

$$[\rho(a_r), \rho(b_r^{-1})] = \sigma^{-2} I, \qquad (2)$$

the constraint

$$\sigma^{2(n-1+g)} = 1,$$

yielding (for $n - 1 + g \neq 0$)

$$\theta = \frac{q}{2(n-1+g)}, \quad q \in \mathbb{Z} \quad \text{or, equivalently} \quad \nu = \frac{q}{n-1+g},$$

that is, *fractional statistics*, in general. Notice that if $\sigma^2 = 1$, that is $\theta \in 1/2 \cdot \mathbb{Z}$ (slight abuse of notation) we recover ordinary Fermi-Bose statistics (see also below).

The above steps where carried out in the paper [58] as well, but the connection with the CCR was apparently not observed.

Next we introduce the following tensor product Hilbert space:

$$\mathcal{H} := H_1 \otimes H_2,$$

with H_1 carrying a representation of the *Weyl-Heisenberg Canonical Commutation Relations*, see Subsection 3.1, and where H_2 is one-dimensional. Clearly $H_1 \otimes H_2 \cong H_1$, but we keep the distinction in view of our physical applications. Now take, after denoting by $(e_1, e_2, \ldots e_g)$ the canonical basis of $\mathbb{R}^g$:

$$\rho_1(a_r) = U(e_r), \qquad \rho_1(b_r^{-1}) = V(e_r), \qquad r = 1, 2, \ldots g.$$

Upon setting

$$\rho(a_r) = \rho_1(a_r) \otimes I_{H_2}, \quad \rho(b_r^{-1}) = \rho_1(b_r^{-1}) \otimes I_{H_2}, \qquad r = 1, \ldots, g$$
$$\rho(\sigma_j) = I_{H_1} \otimes \sigma I_{H_2} \equiv I_{H_2} \otimes \rho_2(\sigma_j), \qquad j = 1, \ldots, n - 1$$

and in view of (1) and (2), we immediately get:

Theorem 4.1 *([105]) (i) Any representation of the Weyl-Heisenberg Commutation relations yields, via the map*

$$\rho : B(\Sigma_g, n) \to U(\mathcal{H})$$

defined above, an infinite dimensional unitary representation of the Riemann surface braid group $B(\Sigma_g, n)$ on the Hilbert space $\mathcal{H}$.
(ii) Irreducible finite dimensional unitary RS-braid group representations $\hat{\rho}$ also exist, stemming from the finite version of Weyl-Heisenberg commutation relations.

The proof is clear in view of the preceding discussion. The representations in (ii) correspond in fact to particular rational *noncommutative tori*, as we have seen.

4.2 Geometric Construction of Riemann Surface Braid Group Representations

We are now ready to exhibit our promised geometrically constructed RS-braid group representations. Building on the material presented before, and resuming the Hilbert space H_1, we have that the braid group generators a_i and b_i, $i = 1, \ldots, g$ are then represented as

$$\rho_1(a_r) := \exp\left(\nabla_{\frac{\partial}{\partial q_r}}\right), \quad \rho_1(b_r^{-1}) := \exp\left(\nabla_{\frac{\partial}{\partial p_r}}\right), \quad r = 1, 2, \ldots g.$$

Remark   The holomorphic Hermitian stable bundle $(\mathcal{E}, h, \nabla) \to J(\Sigma_g)$ can be pulled-back (via the Abel map) to

$$(\mathcal{A}^*\mathcal{E}, \mathcal{A}^*h, \mathcal{A}^*\nabla) \to C_n(\Sigma_g)$$

equipped with the pulled-back metric $\mathcal{A}^*h$ and connection $\mathcal{A}^*\nabla$. The corresponding pulled-back representation is well defined on pulled-back sections. The Hilbert space H_1 will be the receptacle, because of the Abel map, of "centre of mass" wave functions, cf. [50, 115, 47], and below.

We now recall the basic ingredient needed for a physically relevant realisation of the one-dimensional space H_2, namely the so-called *prime form* on a Riemann surface. We closely follow [79], to which we refer for full details. The prime form $E(x, y)$ (for $x, y \in \Sigma_g$) on $\Sigma_g \times \Sigma_g$ is the unique – up to a scalar – holomorphic section of the holomorphic line bundle $\mathcal{O}(\Delta) \to \Sigma_g \times \Sigma_g$ (Δ being the diagonal in $\Sigma_g \times \Sigma_g$) or, alternatively, a differential form of weight $(-1/2, -1/2)$ on its universal cover $\tilde{\Sigma}_g \times \tilde{\Sigma}_g$. Here $\mathcal{O}(D)$ denotes, as usual, the holomorphic line bundle pertaining to the divisor D (roughly speaking, a hypersurface in the base manifold). The prime form E provides a generalization to its genus zero counterpart

$$E_0(x, y) = \frac{x - y}{\sqrt{dx}\,\sqrt{dy}}$$

to which it reduces, up to a third order term, upon working in local coordinates. A bit more explicitly (in terms of theta functions with characteristics)

$$E(x, y) = \frac{\vartheta[\beta](\mathcal{A}(y - x))}{\sqrt{\zeta(x)}\,\sqrt{\zeta(y)}} = -E(y, x)$$

for a non singular, odd theta characteristic β, which, in turn, is interpreted both as a divisor and as an element

$$\beta = \begin{bmatrix} \beta' \\ \beta'' \end{bmatrix} \in \frac{\frac{1}{2}\mathbb{Z}^{2g}}{\mathbb{Z}^{2g}}.$$

The 1-form

$$\zeta = \sum_{i=1}^{g} \frac{\partial\,\vartheta[\beta]}{\partial z_i}(0)\omega_i$$

is the (essentially) unique 1-form vanishing on β: actually its divisor is 2β, whence its square root $\sqrt{\zeta}$ is well-defined. Notice that this arrangement ensures that $E(x, y)$, as a function of x, vanishes only at y, since the theta function $\vartheta[\beta]$ also possesses extra zeros at specific points p_j, $j = 1, 2, \dots g - 1$ (they are therefore absent in the genus one case).

Set $\Sigma_g^n := \Sigma_g \times \Sigma_g \times \cdots \times \Sigma_g$ (n copies). Define the following holomorphic line bundle :

$$\mathcal{L} := \prod_{i<j}^{\otimes} \pi_{ij}^* \mathcal{O}(\Delta_{ij}) \to \Sigma_g^n,$$

where

$$\pi_{ij} : \Sigma_g^n \to (\Sigma_g)_i \times (\Sigma_g)_j$$

(with $(\Sigma_g)_j$ denoting the j-th copy of Σ_g, and Δ_{ij} the diagonal in $(\Sigma_g)_i \times (\Sigma_g)_j$).

The bundle $\mathcal{L} \to \Sigma_g^n$ naturally descends to a bundle (same notation) $\mathcal{L} \to C_n(\Sigma_g)$ whereupon the braid group $B(\Sigma_g, n)$ naturally acts; however, one ends up with a representation of the symmetric group S_n, yielding ordinary statistics. If one wishes to implement fractional statistics then one faces the problem of extracting *roots* of line bundles and this cannot be achieved in general for non trivial line bundles. Thus, in order to circumvent this difficulty, we adopt a "minimalistic" approach and resort to a local description, which however retains an intrinsic character with respect to braiding: define the Hilbert space

$$H_2 = \mathbb{C}\psi_v, \qquad \psi_v := \prod_{i<j}(\zeta_i - \zeta_j)^v$$

with ζ being a local coordinate (the behaviour of the prime form near the diagonal is however independent of the choice of the local coordinate). Branching is then produced. Actually, ψ_v is the "topological" part of the Laughlin wave function discussed in [23, 103], upon regarding the coordinates ζ_i as global coordinates on the configuration space $C_n(\mathbb{C})$.

A scalar product can be introduced in H_2 in an obvious manner. The function ψ_v then manifestly enjoys the correct transformation law under the exchange of two points $x_i \leftrightarrow x_j$:

$$\psi_v \mapsto (-1)^v \psi_v = \sigma \psi_v.$$

The upshot is that we may devise generalised "ground state" *Laughlin wave functions* [70, 51, 50, 57] in $\mathcal{H} = H_1 \otimes H_2$ as follows:

$$\Psi(x_1, \ldots x_n) := \psi_v \cdot \xi,$$

where $x_1, \ldots x_n$ are distinct points in Σ_g, and ξ is a *holomorphic section* of the Matsushima stable bundle involved (depending on a centre of mass coordinate). These holomorphic sections play the role of the ground states, or lowest Landau levels, see also the Appendix. Notice that they are not invariant under the action of the "full" braid group, since parallel transport does not preserve the holomorphic structure, in general.

In this way we have also generalised the geometric treatment given for the standard braid group by A. Besana and the author [23, 103] as well.

We summarise the developments of this Section in the following theorem/definition.

Theorem 4.2 *([105]) Let $\mathcal{E} \to J(\Sigma_g)$ be a Matsushima HE-holomorphic vector bundle with slope $\mu(\mathcal{E}) = vg! = q/r \cdot g!$. The representation ρ_1 of the CCR on the Hilbert space $H_1 = L^2(\mathcal{E})$ – built up as above via parallel transport operators associated with the canonical HE-connection ∇ – together with the position*

$$\rho_2(\sigma_j)\psi := (-1)^v \psi, \qquad \psi \in H_2,$$

gives rise to a unitary representation

$$\rho : B(\Sigma_g, n) \to U(\mathcal{H}),$$

where $n = r + 1 - g$, $\mathcal{H} = H_1 \otimes H_2$. The representation ρ_1 has multiplicity $h^0(\mathcal{E}) = q^g$. The vectors $\psi = \psi_v \xi$, $\xi \in H^0(\mathcal{E})$ (ξ is then a Matsushima theta vector) are called Laughlin generalised wave functions.

Another proposal for the Hilbert space H_2 is $H_2 = \langle \widetilde{\Psi} \rangle$, with

$$\widetilde{\Psi}(x_1, \ldots, x_n) = \prod_{i<j} \frac{E(x_i, x_j)^{2\theta}}{E(x_i, x_0)^{\theta} E(x_j, x_0)^{\theta}}$$

(still a local expression); the correct transformation law under the exchange of two points $x_i \leftrightarrow x_j$ still holds:

$$\widetilde{\Psi} \mapsto (-1)^{2\theta} \, \widetilde{\Psi} = \sigma \, \widetilde{\Psi}.$$

This is motivated by the fact that, fixing say, x_j, and taking an arbitrary point $x_0 \neq x_j$, the 1-form

$$\omega_{x_j - x_0}(x_i) := d_{x_i} \log \frac{E(x_i, x_j)}{E(x_i, x_0)}$$

is a differential of the third kind with residues ± 1 at x_j and x_0, respectively. This generalises the complex plane situation, with the form $dz_i/(z_i - z_j)$, which has residues ± 1 at x_j and ∞, respectively.

In the genus one case one can add the following observation: the tangent bundle $T\Sigma_1$ is trivial, and, since $\Delta \cong \Sigma_1$, one has that the restriction $\mathcal{O}(\Delta)|_{\Delta} \cong T\Delta$ is trivial. Thus the prime form bundle is trivial when restricted to a tubular neighbourhood of the diagonal Δ, which can be taken as an open dense set in $\Sigma_1 \times \Sigma_1$. Restriction to this open dense set renders root extraction possible.

Remark   The root extraction problem deserves further investigation: an important step would be the determination of the second cohomology group $H^2(C_n(\Sigma_g), \mathbb{Z})$. The rational cohomology groups of configuration spaces of surfaces have been studied in [27].

The case $\nu = 2\theta = 1$ gives back Fermi-Dirac statistics, and one can safely employ the prime form bundle (actually $\mathcal{L} \to C_n(\Sigma_g)$) as it stands. Of course one may take tensor powers thereof as well. As for the centre of mass part, also in view of the above considerations, one retrieves the ordinary theta line bundle, having first Chern class (and slope) equal to one, together with the geometric theory of Landau levels discussed in [47, 87], see also [99] and below.

Clearly, the previous duality is transmitted to the corresponding RS-braid group representations, ultimately giving rise to a *ν-anyon - ν'-anyon correspondence* ([105]).

5 Hamiltonian Monodromy via Theta Functions

5.1 The Heat Connection

Resuming the discussion in Subsections 2.2 and 3.2, the 2-torus bundle f : $M \longrightarrow B$ has non trivial monodromy if and only if the holonomy of the canonical Ehresmann-Weinstein connection ∇^{can} on B is non trivial ([95]). Now we can relate the monodromy of the fibration f to the holonomy of the *heat connection* introduced right below.

Define a map $\tau_U : B \supset U \to \mathbb{H}$ via $\tau(b) := -\Theta(b) + i\, T(b)$ (notice that $\Im(\tau) > 0$) using a basis (γ_1, γ_2) for the cycles as in [85] (with the roles of the γ_i interchanged, also cf. Subsection 2.2). Note that this is the crucial point wherein two-dimensionality intervenes.

Resuming the 2-level theta vector bundle $V \to \mathbb{H}$, one constructs the pulled-back bundle

$$\tau_U^* V \to U$$

equipped with a flat connection $\nabla_U = \tau_U^* \nabla$ (∇ is the "old" heat connection on the theta bundle). Gluing these local bundles together one ends up with a (smooth) vector bundle $\mathcal{V} \to B$ - such that $\mathcal{V} |_U = \tau_U^* V$ - also endowed with a flat connection, called again heat connection and denoted by ∇^{heat}. The following holds.

Theorem 5.1 ([95]) *Let (M, ω, h) be a completely integrable Hamiltonian system with two degrees of freedom, possessing a finite number of singularities of focus-focus type. Then:*

(i) *The holonomy of the heat connection on $\mathcal{V} \to B$, the pulled-back 2-level theta vector bundle, relates to the holonomy of the canonical connection ∇^{can} in the following guise*

$$hol(\nabla^{heat}) = Q(hol(\nabla^{can})),$$

namely, the holonomy of the heat connection comes from applying the theta parallel displacement map Q to the holonomy of the canonical connection, i.e. the (classical) monodromy, see also below, Remark 2.

(ii) *As a corollary, the system has non trivial monodromy if the holonomy of the heat connection $hol(\nabla^{heat})$ is non trivial.*

Remarks 1. The tracing of a non trivial path in $\pi_1(B)$ can be interpreted as an adiabatic motion, causing the variation of the basis of cycles and thence of the parameter τ. The overall action on the theta space given by $Q(Z_0)$ (cf. Subsection 3.2) can then be viewed as a "topological" Berry phase (see e.g. [22, 97, 33]) signalling monodromy. This peculiarity pertains to the 2d-environment only.

2. We also stress the fact that the map $Q(Z_0)$ yields a *unitary operator* (the crucial fact is that $\Im\tau$ does not change): indeed, if we read the map Z_0 classically, then $Q(Z_0)$ is precisely its *quantum* counterpart and takes the form of a "phase gate", familiar from quantum computing in the qubit space $\mathbb{C}^2$ ([33]). We notice in passing that the appearance of a finite group like $\mathbb{Z}_4$ ($Q(Z_0)^4 = Id_2$) is to be expected on general grounds (cf. [71, 46]).

3. We point out an important difference between our approach and Tyurin's one ([113]): in the latter case the BS-torus becomes the *real* part of an Abelian variety; in our case we have a 2d-BS-torus endowed with a complex structure. The latter is then holomorphically quantized via 2-level theta functions (the natural substitute for the BS-covariantly constant section whereupon the map $\tau \mapsto \tau + 1$ acts à la Berry.) Hence there is no need of complexifying the manifold, study the ensuing complex monodromy and then coming back to the (mechanically relevant) real picture (see also [12, 116]).

5.2 The 3-Strand Braid Group, Monodromy and the Spherical Pendulum

In this Subsection we discuss the relationship between monodromy and the braiding of the Weierstraß roots. More details on the braid group and its relationship with the modular group $PSL(2, \mathbb{Z})$ and, in particular, on related representations, can be found e.g. in [78, 111, 112, 20].

A (non trivial) representation of the braid group B_3 on $\mathbb{C}^2$ via $SL(2, \mathbb{Z})$ can be realized by the matrices

$$b_1 = \begin{pmatrix} 1 & 0 \\ -1 & 1 \end{pmatrix} \qquad b_2 = \begin{pmatrix} 1 & 1 \\ 0 & 1 \end{pmatrix}.$$

Indeed one immediately checks the defining relation $b_1 b_2 b_1 = b_2 b_1 b_2$. Notice that

$$b_2 = (b_1)^{-T}.$$

Also, one recognizes that *the transformation $\tau \mapsto \tau + 1$ can be represented by b_2.* Thus the braid group generators are dual to each other, from the point of view of classical-quantum monodromy. This, in turn, can be read on the fundamental cycles, on the thetas, and on the Weierstraß roots of the associated elliptic curve: $e_1 \mapsto e_1, e_2 \mapsto e_3, e_3 \mapsto e_2$ (see Subsection 3.2 and e.g. [110], [78]). The following reference formulae are helpful in making this point:

$$2\omega = \oint_{\gamma_1} \frac{dz}{\sqrt{P(z)}} \qquad 2\omega' = \oint_{\gamma_3} \frac{dz}{\sqrt{P(z)}}$$

with $\omega' = \tau\omega$, and where γ_1 surrounds e_2 and e_3 - passing to the other sheet of the ramified double cover, through the cut joining e_3 to ∞ - and γ_3 encircles e_1 and e_2 (also cf. [110], p.85).

The above considerations lead us to the following

Theorem 5.2 *([95]) In the case of an isolated focus-focus singularity, the variation of the rotation number is tantamount to the (multiple) switching of the roots e_2 and e_3 (with the above conventions). More precisely, if $[\gamma]$ is a generator for $\pi_1(B) \cong \mathbb{Z}$, then* classical *monodromy is represented via*

$$m \cdot [\gamma] \leftrightarrow b_2{}^m$$

whereas quantum *monodromy is given by*

$$m \cdot [\gamma] \leftrightarrow b_1{}^m$$

The rotation number transforms accordingly as

$$\Theta \mapsto \Theta - m$$

(recall that $\tau = -\Theta + iT$). This situation seems again to be specific to the 2d-case.

We finally sketch a derivation of the monodromy of the spherical pendulum (the prototype of monodromic behaviour, see also e.g. [15]) via root braiding. We refer to [36] for background and notation. The central object is the polynomial

$$P(x) = 2(h - x)(1 - x^2) - j^2.$$

The point $(j, h) = (0, 1)$ is the only critical point in a (punctured) open "shield" (i.e. the parameter space B for the spherical pendulum). Consider the circuit

$$j = \varepsilon \cos t, \quad h = 1 + \varepsilon \sin t, \tag{3}$$

for $t \in [0, 2\pi)$ and $\varepsilon > 0$ small enough. The roots of P can be guessed via an ε-power series expansion, which immediately leads to the (exact!) expressions below:

$$x^- = -1 + \varepsilon^2 \frac{\cos^2 t}{8}, \quad x^+ = 1 - \frac{\varepsilon}{2}(1 - \sin t), \quad x^0 = 1 + \frac{\varepsilon}{2}(1 + \sin t).$$

As for $\tau = -\Theta + iT$, one has, for the spherical pendulum

$$\Theta = 2j \int_{x^-}^{x^+} \frac{dx}{(1 - x^2)\sqrt{P(x)}}, \quad T = 2 \int_{x^-}^{x^+} \frac{dx}{\sqrt{P(x)}}.$$

The first integral is a sum of elliptic integrals of the third kind, whereas the second one is of the first kind. A direct inspection of their behaviour upon going along the above circuit (3) leads to

Theorem 5.3 *([95]) T is single-valued, whereas the variation of* Θ *along the circuit* (3) *equals* -1.

This recovers monodromy for the spherical pendulum ([38, 36] in a Gauß-Manin fashion.

The reader is referred to [20] for further relations existing in the framework of geometric quantum mechanics (see e.g. [19]) among elliptic integrals and $SU(2)$- representations of B_3.

6 Conclusions and Outlook

In this paper we have surveyed recent topological and physico-mathematical applications of modern theta function theory carried out by the present author and collaborators, specifically concerning the geometric construction of the simplest unitary RS-braid group representations, together with an application to quantum monodromy.

In conclusion, we wish to signal some possibly interesting questions left open by the preceding analysis. First, the problem of root extraction of line bundles over configuration space requires further effort, in order to render the construction of Laughlin wave functions fully geometrical. Also, it seems that no analogue of the Knizhnik-Zamolodchikov-Kohno (KZK) connection for Riemann surfaces (see e.g. [68]), together with their prospective Arnol'd's relations ([3, 68]) has been found to-date. The duality issue we discussed above could, in turn, be related to other forms of duality found in the literature (condensed matter or string theoretical, see e.g. the comprehensive recent essay [82]). Establishment of contact with the theta function approach to knot theory devised in [42], and with the braid approach of [60] is also necessary. The intriguing geometric approach to anyons outlined in [11] deserves further pursuit as well. Theta functions are heavily used in the quantum vortex theory proposed in [88], and the RS-Green function techniques employed therein could be of help in elucidation of the above issues. Also, in the present survey we did not deal with coherent states, a most useful tool both in mathematics and in physics and also related to theta functions ([89] and e.g. [104, 106]). We hope to be able to report on progress on some of these questions in a near future.

The field of condensed matter physics is presently witnessing a tremendous experimental and theoretical impetus (see e.g. Haldane's superb Nobel Lecture [49] and the recent fascinating article [6]), and appears to be naturally amenable to algebro-geometric approaches, which, in conjunction with an array of techniques coming from a variety of mathematical domains, could prove essential for future progress: this would comply with the all-round daring and openminded research attitude of *Emma* throughout her brilliant academic career.

Appendix

Lowest Landau Level and Laughlin Wave Functions on a Torus

We briefly summarize the exposition of [47] for Landau levels of a free electron, slightly changing their notation. They start from an elliptic curve

$$\Sigma_1 = \mathbb{C}/(\mathbb{Z} \oplus \tau\mathbb{Z}).$$

Set

$$z = x + \sqrt{-1}y, \qquad \tau = (L_2/L_1)e^{\sqrt{-1}\psi}, \Im\tau > 0$$

(L_1 and L_2 being the dimensions of the rectangular sample employed). Holomorphic geometric quantization is then carried out, yielding a holomorphic line bundle $\Theta^k \to \Sigma_1$, k an odd positive integer, with connection and curvature

$$A = -\frac{2\pi k}{\Im\tau}ydx, \qquad F_A = \frac{2\pi k}{\Im\tau}dx \wedge dy$$

and first Chern class

$$c_1(\Theta^k) = \frac{1}{2\pi}\int_{\Sigma_1} F_A = k \in \mathbb{Z}, \qquad 2\pi k = \frac{eB}{hc}L_1^2$$

(magnetic flux quantization). The raising ($a^\dagger$) and lowering operators (a) read

$$a = \sqrt{\frac{\Im\tau}{\pi k}}\left(\frac{\partial}{\partial\bar{z}} + \sqrt{-1}\frac{\pi k}{\Im\tau}\Im z\right), \quad a^\dagger = -\sqrt{\frac{\Im\tau}{\pi k}}\left(\frac{\partial}{\partial z} + \sqrt{-1}\frac{\pi k}{\Im\tau}\Im z\right),$$

and the Hamiltonian is

$$H = \hbar\omega_c\left(a^\dagger a + \frac{1}{2}\right), \qquad \omega_c = \frac{\hbar}{m_e}\frac{2\pi k}{\Im\tau}, \qquad [a, a^\dagger] = 1$$

(m_e denoting the mass of the electron). The ground state space can be identified with the holomorphic sections of the above bundle, i.e. with the k-level theta functions; explicitly

$$f_l(z, \tau) = \vartheta\begin{bmatrix} 0 \\ l/k \end{bmatrix}(z|\tau/k), \quad l = 0, 1, \ldots, k-1.$$

The Laughlin wave functions in this situation are the Haldane-Razayi (HR)-wave functions, written for generic boundary conditions, corresponding to phases associated to a double Bohm-Aharonov device (see e.g. [109, 47]) and with Z being the centre of mass coordinate:

$$F_l^{HR} = \vartheta\begin{bmatrix} (l + \phi_1)/k \\ \phi_2 \end{bmatrix}(kZ|k\tau)\prod_{i<j}\vartheta^k\begin{bmatrix} 1/2 \\ 1/2 \end{bmatrix}(z_i - z_j|\tau),$$

$$l = 0, 1, \ldots, k.$$

Now consider a HR-two-quasihole-two electron wave function (see [47], p.317, (34), with slight notational changes and a possible typo corrected); we separately record the centre of mass part

$$(F_l^{2QH})_{cm} = \vartheta \begin{bmatrix} l/k \\ 0 \end{bmatrix} (k(z_1 + z_2) + \eta_1 + \eta_2 | k\tau)$$

the purely electronic part

$$(F_l^{2QH})_{el} = \vartheta^k \begin{bmatrix} 1/2 \\ 1/2 \end{bmatrix} (z_1 - z_2 | \tau) \leftrightarrow (-1)^k = (-1)^{\nu'}$$

and the purely quasi-hole part

$$(F_l^{2QH})_{qh} = \vartheta^{\frac{1}{k}} \begin{bmatrix} 1/2 \\ 1/2 \end{bmatrix} (\eta_1 - \eta_2 | \tau) \leftrightarrow (-1)^{\nu}.$$

The other ingredients are the "mixed" terms $\vartheta \begin{bmatrix} 1/2 \\ 1/2 \end{bmatrix} (z_i - \eta_j | \tau)$.

Subsequently, set $\xi = (\phi_1, \phi_2)$ and consider the family $\Theta_\xi^k \to J(\Sigma_1)$ of k-level theta line bundles. One has $h^0(\Theta_\xi^k) = k$. The corresponding FMN^*-transformed vector bundle (spectral bundle) has rank k, degree 1, slope $1/k$. The formula for theta functions with characteristics recalled in Section 3 allows one to interpret the Bohm-Aharonov angular parameters as quasihole coordinates (varying adiabatically). Therefore, the spectral bundle attached to electrons, encoding the centre of mass dynamics, can be interpreted as a quasihole centre of mass bundle, this being signalled by the correct statistical parameter $\nu = 1/k$.

Acknowledgements

It is a great pleasure for the author to gratefully dedicate this work to *Emma Previato*, superb mathematician and most dear colleague and friend. He acknowledges enlightening discussions and fruitful collaboration with her along the years, fostered by her indomitable enthusiasm and vision. He is also indebted to all his past and present collaborators for helping him in casting his present understanding of the topics discussed in the present work. He warmly thanks the Editors for the honour bestowed on him with their kind invitation to contribute to the present Volume, and the Referee for his/her careful reading, criticism and suggestions. Support from Unicatt local D1-funds is also gratefully acknowledged. This work has been carried out under the activities of INDAM.

References

[1] Y. Aharonov and D. Bohm, Significance of electromagnetic potentials in quantum theory, *Phys.Rev.* **115** (1959), 485–491.

[2] Y. Aharonov and D. Bohm, Further considerations on electromagnetic potentials in the quantum theory, *Phys.Rev.* **123** (1961), 1511–1524.

[3] V.I. Arnol'd, The cohomology ring of colored braids, *Mat. Zametki* **5** No 2 (1969), 227–231. (Russian) English transl. in *Trans. Moscow Math.Soc.* **21** (1970), 30–52.

[4] V.I. Arnold, *Mathematical Methods of Classical Mechanics*, 2nd edition, Springer-Verlag, 1989.

[5] V.I. Arnold and A.B. Givental, Symplectic Geometry. *Dynamical Systems IV*, 1–138, Encyclopaedia Math. Sci. 4, Springer, Berlin, 2001.

[6] M. Asorey, Space, matter and topology, *Nature Physics* **12** (2016), 616–618.

[7] A. Astashkevich and A. Schwarz, Projective modules over non-commutative tori: classification of modules with constant curvature connection, *J. Operator Theory* **46** (2001), 619–634.

[8] M.F. Atiyah, Vector bundles over an elliptic curve, *Proc. London Math.Soc.* **7** (1957), 414–452.

[9] M.F. Atiyah and R. Bott, The Yang-Mills equations over a Riemann surface, *Phil. Trans. R. Soc. Lond. A* **308** (1982), 523–615.

[10] M.F. Atiyah, N.J. Hitchin and I.M. Singer, Self-duality in four-dimensional Riemannian geometry, *Proc. R. Soc. Lond. A* **362** (1978), 425–461.

[11] M.F. Atiyah and M. Marcolli, Anyons in geometric models of matter, *arXiv:1611.04047v1 [math-ph]*.

[12] M. Audin, Hamiltonian Monodromy via Picard-Lefschetz Theory, *Commun.Math.Phys.* **229** (2002), 459–489.

[13] S. Axelrod, S. Della Pietra. and E. Witten, Geometric quantization of Chern-Simons gauge theory, *J. Diff. Geom.* **33** (1991), 787–902.

[14] C. Bartocci, U. Bruzzo and D. Hernandez Ruiperez, *Fourier-Mukai and Nahm Transforms and Applications in Mathematical Physics*, Progress in Mathematics 276, Birkäuser, Basel, 2009.

[15] L.M. Bates, Monodromy in the champagne bottle, *ZAMP* **42** (1991), 837–847.

[16] P. Bellingeri, On presentation of Surface Braid Groups, *J. Algebra* **274** (2004), 543–563.

[17] P. Bellingeri, E. Godelle, J. Guaschi, Abelian and metabelian quotients of surface braid groups, *arXiv:1404.0629v1 [math.GR]*.

[18] J. Bellissard, H. Schulz-Baldes and A. van Elst, The Non Commutative Geometry of the Quantum Hall Effect *J.Math.Phys.* **35** (1994), 5373–5471.

[19] A. Benvegnù, N. Sansonetto and M. Spera: Remarks on Geometric Quantum Mechanics, *J.Geom.Phys.* **51** (2004), 229–243.

[20] A. Benvegnù and M. Spera, On Uncertainty, Braiding and Entanglement in Geometric Quantum Mechanics, *Rev.Math.Phys.* **18** (2006), 1075–1102.

[21] N. Berline, E. Getzler and M. Vergne: *Heat kernels and Dirac Operators*, Springer, Berlin, 1992.

[22] M. Berry: Quantal phase factors accompanying adiabatic changes, *Proc.Roy.Soc. London A* **392** (1984), 45–57.

[23] A. Besana and M. Spera, On some symplectic aspects of knots framings, *J. Knot Theory Ram.* **15** (2006), 883–912.

[24] J-M. Bismut, H. Gillet and C. Soulé, Analytic Torsion and Holomorphic Determinant Line Bundles I. Bott-Chern Forms and Analytic Torsion, *Commun.Math.Phys.* **115** (1988), 49–78.

[25] J-M. Bismut, H. Gillet and C. Soulé, Analytic Torsion and Holomorphic Determinant Line Bundles II. Direct Images and Bott-Chern Forms, *Commun.Math.Phys.* **115** (1988), 79–126.

[26] J-M. Bismut, H. Gillet and C. Soulé, Analytic Torsion and Holomorphic Determinant Line Bundles III. Quillen Metrics and Holomorphic Determinants, *Commun.Math.Phys.* **115** (1988), 301–351.

[27] M. Bödigheimer and F. Cohen, Rational cohomology of configuration spaces of surfaces, *Algebraic Topology and Transformation Groups* (Göttingen 1987), 7–13, *Lecture Notes in Mathematics* **1361**, Springer-Verlag, Berlin-New York, 1988.

[28] K.I. Bolotin, F. Ghahari, M.D. Shulman, H.L. Stormer and P. Kim, Observation of the fractional quantum Hall effect in graphene, *Nature* **462** (2009), 196–199; corrigendum: *Nature* **475** (2011), 122.

[29] J-L. Brylinski, *Loop Spaces, Characteristic Classes and Geometric Quantization*, Birkhäuser, Basel, 1993.

[30] T. Chakraborty and P. Pietiläinen, *The Quantum Hall Effects - Fractional and Integral*, Springer, Berlin, 1995.

[31] M.S. Child, Quantum states in Champagne Bottle, *J.Phys.A* **31** (1998), 657–670.

[32] M.S. Child, T. Weston and J. Tennyson, Quantum monodromy in the spectrum of H_2O and other species: new insight into the quantum level structure of quasilinear molecules, *J.Mol.Phys.* **96** (1999), 371–379.

[33] D. Chruściński and A. Jamiołkowski, *Geometric Phases in Classical and Quantum Mechanics*, Birkhäuser, Boston, 2004.

[34] A. Connes, C* algèbres et géométrie différentielle, *C.R. Acad. Sc. Paris* **290** (1980), 599–604.

[35] A. Connes, *Noncommutative geometry*, Academic Press, London, 1994.

[36] R.H. Cushman and L. Bates, *Global Aspects of Classical Integrable Systems*, Birkhäuser, Basel, 1997.

[37] S. K. Donaldson, A New Proof of a Theorem of Narasimhan and Seshadri, *J.Diff.Geom.* **18** (1983), 269–277.

[38] J.J. Duistermaat, On global action-angle coordinates, *Comm. Pure Appl.Math.* **33** (1980), 687–706.

[39] K. Efstathiou, M. Joyeux, D.A. Sadovskií, Global bending quantum number and absence of monodromy in the HCN ↔ CNH molecule, *Phys.Rev. A* **69** (2004) 032504.

[40] T. Ekedahl and J-P. Serre, Examples de courbes algébriques à jacobienne complètement décomposable, *C.R.Acad.Sci. Paris* **317** Série I, (1993), 509–513.

[41] A. Galasso and M. Spera, Remarks on the geometric quantization of Landau levels, *Int.J.Geom.Meth.Mod.Phys.* **13** (2016), 1650122.

[42] R. Gelca, *Theta functions and Knots*, World Scientific Publishing Co. Pte. Ltd., Singapore, 2014.

[43] G. Goldin, Parastatistics, θ-statistics, and Topological Quantum Mechanics from Unitary Representations of Diffeomorphism Groups, *Proceedings of the XV International Conference on Differential Geometric Methods in Physics*, H.D. Doebner and J.D. Henning (eds), World Scientific, Singapore, 1987, 197–207.

[44] G.A. Goldin, R. Menikoff, and D.H. Sharp, Particle statistics from induced representa- tions of a local current group. *J.Math.Phys.* **21** (1980), 650–664.

[45] G.A. Goldin, R. Menikoff, and D.H. Sharp, Representations of a local current algebra in non-simply connected space and the Aharonov-Bohm effect, *J.Math.Phys.* **22** (1981), 1664–1668.

[46] P. Griffiths and J. Harris, *Principles of Algebraic Geometry*, Wiley, New York, 1978.

[47] J.M. Guilarte, J.M. Mufioz Porras and M. de la Torre Mayado, Elliptic theta functions and the fractional quantum Hall effects, *J.Geom.Phys.* **27** (1998), 297–332.

[48] V. Guillemin and S. Sternberg, Geometric quantization and multiplicity of group representations, *Inv.Math.* **67** (1982), 515–538.

[49] F.D.M. Haldane, Nobel Lecture: Topological quantum matter, *Rev.Mod.Phys.* **89** (2017), 040502-1–10.

[50] F.D.M. Haldane and E.H. Rezayi, Periodic Laughlin-Jastrow wave functions for the fractional quantized Hall effect, *Phys.Rev.B* **31** (1985), 2529–2531.

[51] B.I. Halperin, Statistics of Quasiparticles and the Hierarchy of Fractional Quantized Hall States, *Phys.Rev.Lett.* **52** (1984), 1583–1586. Erratum: *Phys.Rev.Lett.* **52** (1984), 2390.

[52] M. Hamilton, Classical and quantum monodromy via action-angle variables, *J.Geom.Phys.* **115** (2017), 37–44.

[53] J. Hano, A geometrical characterization of a class of holomorphic vector bundles over a complex torus, *Nagoya Math.J.* **61** (1976), 197–201.

[54] D. Hernández Ruipérez and C. Tejero Prieto, Fourier-Mukai transforms for coherent systems on elliptic curves, *J. London Math.Soc.* **77** (2008),15–32.

[55] F. Hirzebruch, *Topological Methods in Algebraic Geometry*, Springer, New York, 1966, 1978.

[56] N. Hitchin, Flat Connections and Geometric quantization, *Commun.Math.Phys.* **131** (1990), 347–380.

[57] R. Iengo and D. Li, Quantum mechanics and quantum Hall effect on Riemann surfaces *Nucl.Phys.B* **413** (1994), 735–753.

[58] T.D. Imbo and J. March-Russell, Exotic statistics on surfaces, *Phys.Lett.B* **252** (1990), 84–90.

[59] T.D. Imbo, C. Shah Imbo, E.C.G. Sudarshan, Identical particles, exotic statistics and braid groups, *Phys.Lett.B* **234** (1990), 103–107.

[60] J. Jacak, R. Gonczarek, L. Jacak, I. Józwiac, *Composite Fermion Structure - Application of Braid Groups in 2D Hall System Physics*, World Scientific Publishing Co. Pte. Ltd., Singapore, 2012.

[61] C. Kassel and V. Turaev, *Braid Groups*, Springer, New York, 2008.

[62] N.M. Katz, On the differential equations satisfied by period matrices, *Publ.Math. IHES* **35** (1968), 71–106.

[63] N.M. Katz and T. Oda, On the differentiation of de Rham cohomology classes with respect to parameters, *J.Math. Kyoto Univ.* **8** (1968), 199–213.

[64] G.R. Kempf, *Complex Abelian Varieties and Theta Functions*, Springer, Berlin, 1991.

[65] A.A. Kirillov, Geometric Quantization, *Dynamical Systems IV*, 139–176, Encyclopaedia Math. Sci. 4, Springer, Berlin, 2001.

[66] J. Klauder and E. Onofri, Landau levels and geometric quantization, *Int.J.Mod.Phys.* **4** (1989), 3939–3949.

[67] S. Kobayashi, *Differential geometry of complex vector bundles*, Iwanami Shoten Publishers, Tokyo – Princeton University Press, Princeton, 1987.

[68] T. Kohno, *Conformal Field Theory and Topology*, AMS, Providence, RI, 2002.

[69] B. Kostant, Quantization and unitary representations. I. Prequantization. *Lectures in modern analysis and applications, III*, pp. 87–208. Lecture Notes in Math., Vol. 170, Springer, Berlin, 1970.

[70] R.B. Laughlin, Anomalous Quantum Hall Effect: An Incompressible Quantum Fluid with Fractionally Charged Excitations, *Phys.Rev.Lett.* **50** (1983), 1395–1398.

[71] A. Loi, The function epsilon for complex Tori and Riemann surfaces, *Bull. Belgian Math.Soc. Simon Stevin*, **7** (2000), 229–236.

[72] Y. Manin, Algebraic curves over fields with differentiation *Transl. Amer. Math. Soc.* **37** (1964), 59–78; [Y. Manin, *Izv.Akad.Nauk. SSSR Ser.Mat.* **22** (1958), 737–756].

[73] M. Marcolli and V. Mathai, Twisted index theory on good orbifolds I: Noncommutative Bloch theory, *Communications in Contemporary Mathematics*, **1** (1999), 553–587.

[74] M. Marcolli and V. Mathai, Twisted index theory on good orbifolds II: Fractional quantum numbers, *Commun.Math.Phys.* **217** (2001), 55–87.

[75] Y. Matsushima, Heisenberg groups and holomorphic vector bundles over a complex torus, *Nagoya Math.J.* **61** (1976), 161–195.

[76] H. McKean and V. Moll, *Elliptic curves*, Cambridge University Press, Cambridge, 1999.

[77] G. Morandi, *Quantum Hall Effect*, Bibliopolis, Naples, 1988.

[78] G.D. Mostow, Braids, hypergeometric functions, and lattices, *Bull.Am.Math.Soc.* **16** (1987), 225–246.

[79] D. Mumford, *Tata Lectures on Theta I-III*, Birkhäuser, Basel, 1983, 1984, 1991.

[80] D. Mumford, J. Fogarty and F. Kirwan, *Geometric Invariant Theory*, Springer, Berlin, 1994.

[81] R. Narasimhan, *Lectures on Riemann Surfaces*, Birkhäuser, Basel, 1994.

[82] H. Năstase, *String Theory Methods for Condensed Matter Physics*, Cambridge University Press, Cambridge, 2017.

[83] C. Nayak, S.H. Simon, A. Stern, M. Freedman and S.D. Sarma, Non-Abelian anyons and topological quantum computation, *Rev.Mod.Phys.* **80** (2008), 1083–1159.

[84] N.N. Nekhoroshev, Action-angle variables, and their generalizations (Russian) *Trudy Moskov.Mat.Obšč.* **26** (1972), 181–198. English translation: Transactions of the Moscow Mathematical Society for the year 1972 (Vol. 26). American Mathematical Society, Providence, R. I., 1974.

[85] S. Vũ Ngoc, Quantum monodromy in integrable systems, *Commun.Math.Phys.* **203** (1999), 465–479.

[86] S. Vũ Ngoc, Quantum Monodromy and Bohr-Sommerfeld rules, *Lett.Math.Phys.* **55** (2001), 205–217.

[87] E. Onofri, Landau Levels on a Torus, *Int.J.Theor.Phys.* **40** (2001), 537–549.

[88] V. Penna and M. Spera, Remarks on quantum vortex theory on Riemann surfaces, *J.Geom.Phys.* **27** (1998), 99–112.

[89] A. Perelomov, *Generalized Coherent States and Their Applications*, Springer, Berlin, 1986.

[90] A. Polishchuk and A. Schwarz, Categories of Holomorphic Vector Bundles on Noncommutative Two-Tori, *Commun.Math.Phys.* **236** (2003), 135–159.

[91] A. Pressley and G. Segal, *Loop groups.* Oxford University Press, Oxford, 1986.

[92] M. Reed, B. Simon, *Methods of Modern Mathematical Physics I,II,III*, Academic Press, New York/London, 1972–5;1980;1979.

[93] M.A. Rieffel, Non-commutative tori - A case study of non-commutative differentiable manifolds, *Contemporary Mathematics* **105** (1990), 191–211.

[94] D.A. Sadovskií and B.I. Zhilinskií, Quantum monodromy, its generalizations and molecular interpretations, *Mol.Phys.* **104** (2006), 2595–1615.

[95] N. Sansonetto and M. Spera, Hamiltonian monodromy via geometric quantization and theta functions, *J.Geom.Phys.* **60** (2010), 501–512.

[96] A. Schwarz, Theta functions on noncommutative tori, *Lett.Math.Phys.* **58** (2001), 81–90.

[97] B. Simon, Holonomy, the quantum adiabatic theorem, and Berry's phase, *Phys.Rev.Lett.* **51** (1983), 2167–2170.

[98] J-M. Souriau, *Structure des systèmes dynamiques* (French), Dunod, Paris 1970.

[99] M. Spera, Quantization on Abelian varieties, *Rend.Sem.Mat.Univers. Politecn. Torino* **44** (1986), 383–392.

[100] M. Spera, Yang Mills theory in non commutative differential geometry, *Rend.Sem.Fac. Scienze Univ. Cagliari Suppl.* **58** (1988), 409–421.

[101] M. Spera, A symplectic approach to Yang Mills theory for non commutative tori, *Canad.J.Math.* **44** (1992), 368–387.

[102] M. Spera, On Kählerian coherent states, in *Geometry, integrability and quantization (Varna, 1999)*, 241–256, Coral Press Sci. Publ., Sofia, 2000.

[103] M. Spera, A survey on the differential and symplectic geometry of linking numbers, *Milan J.Math.* **74** (2006), 139–197.

[104] M. Spera, Geometric Methods in Quantum Mechanics, *J.Geom.Sym.Phys.* **24** (2011), 1–44.

[105] M. Spera, On the geometry of some unitary Riemann surface braid group representations and Laughlin-type wave functions, *J.Geom.Phys.* **94** (2015), 120–140.

[106] M. Spera, On Some Geometric Aspects of Coherent States, in *Coherent States and Their Applications* (J-P. Antoine, F. Bagarello, J-P. Gazeau eds), Ch.8 (16 pp.) Springer Proceedings in Physics 205 (2018) [CIRM, Luminy, 14th-18th November 2016.], Springer, Cham.

[107] C. Tejero Prieto, Fourier-Mukai Transform and Adiabatic Curvature of Spectral Bundles for Landau Hamiltonians on Riemann Surfaces, *Commun.Math.Phys.* **265** (2006), 373–396.

[108] C. Tejero Prieto, Quantum Hall effect on Riemann surfaces, *J.Phys.: Conf. Ser.* **175** (2009), 012014

[109] D.J. Thouless, Topological interpretations of quantum Hall conductance, *J. Math. Phys.* **35** (1994), 5362–5372.

[110] F. Tricomi, *Funzioni ellittiche*, Zanichelli, Bologna, 1937.

[111] I. Tuba, Low-dimensional representations of B_3, *Proc.Amer.Math.Soc.* **129** (2001), 2597–2606.

[112] I. Tuba and H. Wenzl, Representations of the braid group B_3 and of $SL(2, \mathbb{Z})$, *Pacific J.Math.* **197** (2001), 491–509.

[113] A. Tyurin *Quantization, Classical and Quantum Field Theory and Theta Functions*, CRM Monograph Series 21, AMS, Providence, RI, 2003.

[114] K. Uhlenbeck and S. T. Yau, On the Existence of Hermitian-Yang-Mills Connections in Stable Vector Bundles, *Commun. Pure Appl.Math.* Vol. XXXIX, Suppl. (1986), 257–293 and: A Note on Our Previous Paper: On the Existence of Hermitian-Yang-Mills Connections in Stable Vector Bundles, *Commun. Pure Appl.Math.* Vol. XLII (1989), 703–707.

[115] R. Varnhagen, Topology and Fractional Quantum Hall Effect, *Nucl.Phys.B* **443** (1995), 501–515.

[116] O. Vivolo, The monodromy of the Lagrange top and the Picard-Lefschetz formula, *J.Geom.Phys.* **46** (2003), 99–124.

[117] J. von Neumann, Die Eindeutigkeit der Schrödingerschen Operatoren, *Math. Ann.* **104** (1931), 570–578.

[118] Z. Wang, *Topological Quantum Computation*, American Mathematical Society, Providence, 2010.

[119] A. Weinstein, Symplectic manifolds and their Lagrangian submanifolds, *Adv.Math.* **16** (1971), 329–346.

[120] E.T. Whittaker and G.N. Watson, *A course of Modern Analysis*, Cambridge University Press, Cambridge, 1927, (4th Edition, reprinted 1980).

[121] F. Wilczek, Quantum mechanics of fractional-spin particles, *Phys.Rev.Lett.* **49** (1982), 957–959.

[122] E. Witten Quantum field theory and the Jones polynomial, *Commun.Math.Phys.* **121** (1989), 351–399.

[123] N. Woodhouse, *Geometric Quantization*, Oxford University Press, Oxford, 1992.

[124] M. Zou, Monodromy in two degrees of freedom in integrable systems. Ph.D. Thesis, University of Arizona, 1992.

[125] M. Zou, Monodromy in two degrees of freedom in integrable systems, *J.Geom.Phys.* **10** (1992), 37–45.

[126] N.T. Zung, A note on focus-focus singularities, *Diff.Geom.Appl.* **7** (1997), 123–130.

Note added in proof

For a recent application of theta function techniques to topological insulators, see the present author's paper, Spin structures, theta functions and topological insulators, *Albanian J. Math.* 13 (2019), 201–210.

Mauro Spera
Dipartimento di Matematica e Fisica "Niccolò Tartaglia"
Università Cattolica del Sacro Cuore
Via dei Musei 41, 25121 Brescia, Italia
E-mail: mauro.spera@unicatt.it

15

Multiple Dedekind Zeta Values are Periods of Mixed Tate Motives

Ivan Horozov

Abstract. Recently, the author defined multiple Dedekind zeta values [5] associated to a number K field and a cone C. These objects are number theoretic analogues of multiple zeta values. In this paper we prove that every multiple Dedekind zeta value over any number field K is a period of a mixed Tate motive. Moreover, if K is a totally real number field, then we can choose a cone C so that every multiple Dedekind zeta associated to the pair $(K; C)$ is unramified over the ring of algebraic integers in K. In [7], the author proves similar statements in the special case of a real quadratic fields for a particular type of multiple Dedekind zeta values.

The mixed motives are defined over K in terms of the Deligne-Mumford compactification of the moduli space of curves of genus zero with n marked points.

1 Introduction

The Riemann zeta function

$$\zeta(s) = \sum_{n>0} \frac{1}{n^s}$$

is widely used in number theory, algebraic geometry and quantum field theory. Euler's multiple zeta values

$$\zeta(s_1, \ldots, s_m) = \sum_{0 < n_1 < \cdots < n_m} \frac{1}{n_1^{s_1} \ldots n_m^{s_m}},$$

where $s_1, \ldots, s_m$ are positive integers and $s_m \geq 2$, appear as values of some Feynman amplitudes, and in algebraic geometry, as periods of mixed Tate

MSC 2010: 11M32, 11R42, 14G10, 14G25

Keywords: Multiple zeta values, Dedekind zeta values, mixed Tate motives, unramified motive, periods

motives over $Spec(\mathbb{Z})$ (see [4], [3], [1], [8]). The value $s_1 + \cdots + s_m$ is called
the weight of a multiple zeta value. It is an important invariant useful for
interpretations in algebraic geometry and mixed Hodge structures.

Dedekind zeta values

$$\zeta_K(s) = \sum_{\mathfrak{a} \neq (0)} \frac{1}{N(\mathfrak{a})^s},$$

are a generalization of the Riemann zeta function to a number field K. In some
Feynman amplitudes one of the summands is $\log(1 + \sqrt{2})$ or $\log\left(\frac{1+\sqrt{5}}{2}\right)$.
These values are essentially the residues at $s = 1$ of Dedekind zeta functions
over $\mathbb{Q}(\sqrt{2})$ and over $\mathbb{Q}(\sqrt{5})$, respectively. For $s = 2, 3, 4, \ldots$ the values
$\zeta_K(s)$ are periods of mixed Tate motives over the ring of algebraic integers in
K with ramification only at the discriminant of K (see [2]).

In [5], the author has constructed multiple Dedekind zeta values, which are
a generalization of Euler's multiple zeta values to number fields in the same
way as Dedekind zeta values generalizes Riemann zeta values. For a quadratic
number field K, the key examples of multiple Dedekind zeta values are

$$\zeta_{K;C}(s_1, \ldots, s_m) = \sum_{\beta_1, \ldots, \beta_m \in C} \frac{1}{N(\beta_1)^{s_1} N(\beta_1 + \beta_2)^{s_2} \cdots N(\beta_1 + \cdots + \beta_m)^{s_m}},$$

(1)

where $s_1, \ldots, s_m$ are positive integers and $s_m \geq 2$ and C is a cone generated
by totally positive algebraic integers $e_1, \ldots, e_n$ in K defined by

$$C = \mathbb{N}\{e_1, \ldots, e_n\} = \{\gamma \in K \mid \gamma = a_1 e_1 + \ldots a_i e_i, \text{ for positive integers } a_i\}.$$

The weight of a multiple Dedekind zeta value is $s_1 + \cdots + s_m$. Similar types
of cones were considered by Zagier in [9] and [10].

In [5], the author has proven that multiple Dedekind zeta values can be
interpolated to multiple Dedekind zeta functions, which have meromorphic
continuation to all complex values of the variables $s_1, \ldots, s_m$.

In this paper we prove the following two theorems.

Theorem 1.1 *Every multiple Dedekind zeta over any number field K is a
period of a mixed Tate motive over K.*

Theorem 1.2 *If K is a totally real field, then we can find a cone C such
that every multiple Dedekind zeta $\zeta_{K;C}(s_1, \ldots, s_m)$ is a period of mixed Tate
motive, which is unramified over any prime.*

2 Background

2.1 Multiple Zeta Values and Iterated Integrals

The Riemann zeta function at the value $s = 2$ can be expressed in term of an iterated integral in the following way

$$\int_0^1 \left(\int_0^y \frac{dx}{1-x} \right) \frac{dy}{y} = \int_0^1 \left(\int_0^y (1 + x + x^2 + x^3 \dots) dx \right) \frac{dy}{y}$$

$$= \int_0^1 \left(y + \frac{y^2}{2} + \frac{y^3}{3} + \frac{y^4}{4} + \dots \right) \frac{dy}{y}$$

$$= y + \frac{y^2}{2^2} + \frac{y^3}{3^2} + \frac{y^4}{4^2} \dots \Big|_{y=0}^{y=1}$$

$$= 1 + \frac{1}{2^2} + \frac{1}{3^2} + \frac{1}{4^2} \dots = \zeta(2).$$

Let us examine the domain of integration of the iterated integral. Note that $0 < x < y$ and $0 < y < 1$. We can put both inequalities together. Then we obtain the domain $0 < x < y < 1$, which is a simplex. Thus, we can express the iterated integral as

$$\zeta(2) = \int_0^1 \left(\int_0^y \frac{dx}{1-x} \right) \frac{dy}{y} = \int_{0<x<y<1} \frac{dx}{1-x} \wedge \frac{dy}{y}.$$

Moreover, Goncharov and Manin [4] have expressed all multiple zeta values of weight M as periods of motives related to the moduli space of curves of genus zero with $M + 3$ marked points, $\mathcal{M}_{0,M+3}$. In particular, $\zeta(2)$ can be expressed as a period of the motive $H^2(\overline{\mathcal{M}}_{0,5} - A, B - A \cap B)$ by pairing of $[\Omega_A] \in Gr_4^W H^2(\overline{\mathcal{M}}_{0,5} - A)$ for $\Omega_A = \frac{dx}{1-x} \wedge \frac{dy}{y}$, with $[\Delta_B] \in (Gr_0^W H^2(\overline{\mathcal{M}}_{0,5} - B))^\vee$. The Deligne-Mumford compactification $\overline{\mathcal{M}}_{0,5}$ of the moduli space $\mathcal{M}_{0,5}$ can be obtained by three blow-ups of $\mathbb{P}^1 \times \mathbb{P}^1$ at the points $(0,0)$, $(1,1)$ and (∞, ∞). Let us name the exceptional divisors at the three points by E_0, E_1 and E_∞, respectively. Then $A = (x = 1) \cup (y = 0) \cup (x = \infty) \cup (y = \infty) \cup E_\infty$ and $B = (x = 0) \cup (x = y) \cup (y = 1) \cup E_0 \cup E_1$.

Similarly, one can express $\zeta(3)$ and $\zeta(1, 2)$ as iterated integrals

$$\zeta(3) = \int_0^1 \left(\int_0^z \left(\int_0^y \frac{dx}{1-x} \right) \frac{dy}{y} \right) \frac{dz}{z}$$

$$= \int_{0<x<y<z<1} \frac{dx}{1-x} \wedge \frac{dy}{y} \wedge \frac{dz}{z},$$

$$\zeta(1, 2) = \int_0^1 \left(\int_0^z \left(\int_0^y \frac{dx}{1-x} \right) \frac{dy}{1-y} \right) \frac{dz}{z}$$

$$= \int_{0<x<y<z<1} \frac{dx}{1-x} \wedge \frac{dy}{1-y} \wedge \frac{dz}{z}.$$

Again, $\zeta(3)$ and $\zeta(1, 2)$ can be expressed as periods of motives related to $\mathcal{M}_{0,6}$. In the same paper, Goncharov and Manin prove that the motives associated to multiple zeta values (MZVs) are mixed Tate motives unramified over $Spec(\mathbb{Z})$.

A few years later, Francis Brown [1] proved that periods of mixed Tate motives unramified over $Spec(\mathbb{Z})$ can be expressed as a $\mathbb{Q}$-linear combination of MZVs times an integer power of $2\pi i$.

2.2 Multiple Dedekind Zeta Values (MDZVs) and Iterated Integrals on Membranes

We recall the construction of MDZVs. Let $\mathcal{O}_K$ be the ring of integers in a number field K of degree n over $\mathbb{Q}$.

Denote by $\sigma_1, \ldots, \sigma_n$ all the embeddings of K into the complex numbers $\mathbb{C}$. And let $e_1, \ldots, e_n$ be elements of $\mathcal{O}_K$ such that

(1) $e_i \in \mathcal{O}_K$ for all i
(2) $(e_1, \ldots, e_n)$ forms a basis of K over $\mathbb{Q}$
(3) $Re(\sigma_j(e_i)) \geq 0$ for all i and j

Let C be the cone defined as $\mathbb{N}$-linear combinations of $e_1, \ldots, e_2$, that is,

$$C = \{\alpha \in \mathcal{O}_K \mid \gamma = a_1 e_1 + \cdots + a_n e_n, \text{ for } a_1, \ldots, a_n \in \mathbb{N}\}.$$

Let

$$f_0(C; t_1, \ldots, t_n) = \sum_{\alpha \in C} \exp\left[-\sum_{j=1}^{n} t_j \sigma_j(\alpha)\right].$$

We express $\zeta_{K;C}(2)$, $\zeta_{K;C}(3)$ and $\zeta_{K;C}(1, 2)$ as iterated integrals on a membrane. See [5] and [6], for more examples and properties of these constructions. We have

$$\int_0^\infty \cdots \int_0^\infty \left(\int_{u_1}^\infty \cdots \int_{u_n}^\infty f_0(C; t_1, \ldots, t_n) dt_1 \ldots dt_n\right) du_1 \ldots du_n$$

$$= \int_0^\infty \cdots \int_0^\infty \left(\int_{u_1}^\infty \cdots \int_{u_n}^\infty \sum_{\alpha \in C} \exp\left[-\sum_{j=1}^{n} t_j \sigma_j(\alpha)\right] dt_1 \ldots dt_n\right)$$

$$\times du_1 \ldots du_n$$

$$= \int_0^\infty \cdots \int_0^\infty \left(\sum_{\alpha \in C} \left(\prod_{j=1}^{n} \exp\left[-t_j \sigma_j(\alpha)\right]\right) dt_1 \ldots dt_n\right) du_1 \ldots du_n$$

$$= \int_0^\infty \cdots \int_0^\infty \sum_{\alpha \in C} \prod_{j=1}^n \frac{\exp\left[-u_j\sigma_j(\alpha)\right]}{\prod_j \sigma_j(\alpha)} du_1 \dots du_n$$

$$= \sum_{\alpha \in C} \frac{1}{(\prod_j \sigma_j(\alpha))^2} = \sum_{\gamma \in C} \frac{1}{N(\alpha)^2}$$

$$= \zeta_{K;C}(2). \tag{2}$$

Similarly,

$$\int_0^\infty \cdots \int_0^\infty \left(\int_{v_1}^\infty \cdots \int_{v_n}^\infty \left(\int_{u_1}^\infty \cdots \int_{u_n}^\infty \right.\right.$$

$$f_0(C; t_1, \dots, t_n) dt_1 \cdots dt_n \Big) du_1 \cdots du_n \Big) dv_1 \cdots dv_n$$

$$= \int_0^\infty \cdots \int_0^\infty \left(\int_{v_1}^\infty \cdots \int_{v_n}^\infty \left(\int_{u_1}^\infty \cdots \int_{u_n}^\infty \right.\right.$$

$$\times \sum_{\alpha \in C} \exp\left[-\sum_{j=1}^n t_j\sigma_j(\alpha) \right] dt_1 \dots dt_n \Big) du_1 \dots du_n \Big) dv_1 \cdots dv_n$$

$$= \sum_{\alpha \in C} \int_0^\infty \cdots \int_0^\infty \left(\int_{v_1}^\infty \cdots \int_{v_n}^\infty \prod_{j=1}^n \frac{\exp\left[-u_j\sigma_j(\alpha)\right]}{\sigma_j(\alpha)} du_1 \dots du_n \right)$$

$$\times dv_1 \dots dv_n$$

$$= \sum_{\alpha \in C} \int_0^\infty \cdots \int_0^\infty \prod_{j=1}^n \frac{\exp\left[-v_j\sigma_j(\alpha)\right]}{\sigma_j(\alpha)^2} dv_1 \dots dv_n = \sum_{\alpha \in C} \frac{1}{N(\alpha)^3}$$

$$= \zeta_{K;C}(3).$$

We recall the simplest type of multiple Dedekind zeta value

$$\int_0^\infty \cdots \int_0^\infty \left(\int_{v_1}^\infty \cdots \int_{v_n}^\infty \left(\int_{u_1}^\infty \cdots \int_{u_n}^\infty \right.\right.$$

$$f_0(C; t_1, \dots, t_n) dt_1 \cdots dt_n \Big) f_0(C; u_1, \dots, u_n) du_1 \cdots du_n \Big) dv_1 \cdots dv_n$$

$$= \int_0^\infty \cdots \int_0^\infty \left(\int_{v_1}^\infty \cdots \int_{v_n}^\infty \left(\int_{u_1}^\infty \cdots \int_{u_n}^\infty \right.\right.$$

$$\times \sum_{\alpha \in C} \exp\left[-\sum_{j=1}^n t_j\sigma_j(\alpha) \right] dt_1 \dots dt_n \Big)$$

$$\times \sum_{\beta \in C} \exp\left[-\sum_{j=1}^{n} u_j \sigma_j(\beta)\right] du_1 \ldots du_n\right) dv_1 \cdots dv_n$$

$$= \sum_{\alpha,\beta \in C} \int_0^\infty \cdots \int_0^\infty \times \left(\int_{v_1}^\infty \cdots \int_{v_n}^\infty \prod_{j=1}^{n} \frac{\exp\left[-u_j \sigma_j(\alpha)\right]}{\sigma_j(\alpha)}\right.$$

$$\times \exp\left[-u_j \sigma_j(\beta)\right] \times du_1 \ldots du_n\right) dv_1 \ldots dv_n$$

$$= \sum_{\alpha,\beta \in C} \int_0^\infty \cdots \int_0^\infty \left(\int_{v_1}^\infty \cdots \int_{v_n}^\infty \prod_{j=1}^{n} \frac{\exp\left[-u_j \sigma_j(\alpha+\beta)\right]}{\sigma_j(\alpha)}\right.$$

$$\times du_1 \ldots du_n\right) dv_1 \ldots dv_n$$

$$= \sum_{\alpha,\beta \in C} \int_0^\infty \cdots \int_0^\infty \prod_{j=1}^{n} \frac{\exp\left[-v_j \sigma_j(\alpha+\beta)\right]}{\sigma_j(\alpha)\sigma_j(\alpha+\beta)} dv_1 \ldots dv_n$$

$$= \sum_{\alpha \in C} \frac{1}{N(\alpha)^1 N(\alpha+\beta)^2}$$

$$= \zeta_{K;C}(1,2).$$

3 Transition to Algebraic Geometry

We may write the infinite sum in the definition of f_0 as a product of n geometric series as follows.

Lemma 3.1 Let $x_i = e^{-t_i}$ for $i = 1, 2, \ldots, n$. Then $e^{-t_j \sigma_j(e_i)} = x_j^{\sigma_j(e_i)}$ and

$$f_0(C; t_1, \ldots, t_n) = \prod_{i=1}^{n} \left(\frac{\prod_{j=1}^{n} x_j^{\sigma_j(e_i)}}{1 - \prod_{j=1}^{n} x_j^{\sigma_j(e_i)}}\right) \tag{3}$$

Proof. We simplify the function f_0 by expressing it in terms of products:

$$f_0(C; t_1, \ldots, t_n) = \sum_{\alpha \in C} \exp\left[-\sum_{j=1}^{n} \sigma_j(\alpha) t_j\right]$$

$$= \sum_{a_1=1}^{\infty} \cdots \sum_{a_n=1}^{\infty} \exp\left[-\sum_{j=1}^{n} t_j[a_1 \sigma_j(e_1) + \cdots + a_n \sigma_j(e_n)]\right]$$

$$= \sum_{a_1=1}^{\infty} \cdots \sum_{a_n=1}^{\infty} \exp[-a_1[t_1\sigma_1(e_1) + \cdots + t_n\sigma_n(e_1)]]$$

$$\times \cdots \times \exp[-a_n[t_1\sigma_1(e_n) + \cdots + t_n\sigma_n(e_n)]]$$

$$= \frac{\exp[-[t_1\sigma_1(e_1) + \cdots + t_n\sigma_n(e_1)]]}{1 - \exp[-[t_1\sigma_1(e_1) + \cdots + t_n\sigma_n(e_1)]]}$$

$$\times \cdots \times \frac{\exp[-[t_1\sigma_1(e_n) + \cdots + t_n\sigma_n(e_n)]]}{1 - \exp[-[t_1\sigma_1(e_n) + \cdots + t_n\sigma_n(e_n)]]}$$

$$= \prod_{i=1}^{n} \frac{\exp\left[-\sum_{j=1}^{n} t_j\sigma_j(e_i)\right]}{1 - \exp\left[-\sum_{j=1}^{n} t_j\sigma_j(e_i)\right]}$$

$$= \prod_{i=1}^{n} \frac{\prod_{j=1}^{n} \exp[-t_j\sigma_j(e_i)]}{1 - \prod_{j=1}^{n} \exp[-t_j\sigma_j(e_i)]}.$$

$\square$

3.1 The Algebraic Exponent

We are going to define new variables x_{ij}, so that when we express $f_0(C; t_1, \ldots, t_n)$ in terms of x_{ij}, then we obtain a rational function. Intuitively $x_{ij} = x_j^{\sigma_j(e_i)}$. To achieve that, we need to define algebraically $f^{\sigma(\gamma)}(x) = x^{\sigma(\gamma)}$ where $\gamma \in \mathcal{O}_K$ and σ is an embedding of K into $\mathbb{C}$. We follow similar ideas as in the announcement [7]. Let $\mathcal{O}_K = \mathbb{Z}\{\mu_1, \mu_2, \ldots, \mu_n\}$ as a $\mathbb{Z}$-module, where $\mu_1 = 1$. Let (c_{ij}) by the $n \times n$-matrix associated to γ in the basis $(\mu_1, \ldots, \mu_n)$, that is,

$$\gamma\mu_i = \sum_{k=1}^{n} c_{ik}\mu_k.$$

We define a function $f^{\sigma(\gamma)}$ corresponding to raising to a power $\sigma(\gamma)$ by sending monomials in the variables $y_1, \ldots, y_n$ to monomials in the same variables. Let

$$f^{\sigma(\gamma)}(y_i) = \prod_{k=1}^{n} y_k^{c_{ik}}.$$

Lemma 3.2 *Iterated application of the above definition of exponentiation has the following property:*

$$f^{\sigma(\beta)} f^{\sigma(\alpha)}(y_i) = f^{\sigma(\alpha\beta)}(y_i).$$

Proof. It follows from the fact that $\gamma \mapsto (c_{ik})$ is a representation of the ring $\mathcal{O}_K$ as an endomorphism of $\mathbb{Z}^n$. More precisely, let $\alpha \mapsto (a_{ij})$, $\beta \mapsto (b_{jk})$ and $\gamma \mapsto (c_{ik})$. If $\alpha\beta = \gamma$ then $\sum_j a_{ij}b_{jk} = c_{ik}$. Thus,

$$f^{\sigma(\beta)} f^{\sigma(\alpha)}(y_i) = f^{\sigma(\beta)} \left(\prod_j y_j^{a_{ij}} \right) = \prod_{j,k} y_k^{a_{ij} b_{jk}} = \prod_k y_k^{\sum_j a_{ij} b_{jk}}$$

$$= \prod_{j,k} y_k^{c_{ik}} = f^{\sigma(\alpha\beta)}(y_i). \qquad \Box$$

Now that we have defined an algebraic power of a variable, we return to expressing f_0 as a rational function. Let $\{e_1, \ldots, e_n\}$ be a basis of $\mathcal{O}_K$ considered as a $\mathbb{Z}$-module with $e_1 = 1$. Let $\gamma = \sum_i c_i e_i$ be an element of $\mathcal{O}_K$. Note that $c_1, \ldots, c_n$ are integers. Let σ_j a fixed embedding of K into C and let x_j be a variable. Similarly to the above construction, using variables $x_{1j}, \ldots x_{nj}$, we define algebraic powers of x_j by

$$f^{\sigma(\gamma)}(x_j) = \prod_{i=1}^n x_{ij}^{c_i}.$$

Note that $x_{1j} = x_j$. Then $x_{ij} = f^{\sigma_j(e_i)}(x_{1j}) = f^{\sigma_j(e_i)}(x_j)$. Intuitively, $x_{ij} = x_j^{\sigma_j(e_i)}$ Thus, we have n^2 variables x_{ij} where both i and j vary from 1 to n.

The reason for defining the above algebraic exponentiation is to express the function f_0 as a rational function

$$f_0(C; t_1, \ldots, t_n) = \prod_{i=1}^n \left(\frac{\prod_{j=1}^n x_j^{\sigma_j(e_i)}}{1 - \prod_{j=1}^n x_j^{\sigma_j(e_i)}} \right).$$

It can be written formally as

$$f_0(C; t_1, \ldots, t_n) = \prod_{i=1}^n \left(\frac{\prod_{j=1}^n f^{\sigma_j(e_i)}(x_j)}{1 - \prod_{j=1}^n f^{\sigma_j(e_i)}(x_j)} \right) = \prod_{i=1}^n \left(\frac{\prod_{j=1}^n x_{ij}}{1 - \prod_{j=1}^n x_{ij}} \right). \tag{4}$$

4 Multiple Dedekind Zeta Values, Differential Forms and Rational Functions

Let us recall the definition of a multiple Dedekind zeta value (see [5]). Let

$$\alpha_0(t_1, \ldots, t_n) = dt_1 \wedge \cdots \wedge dt_n$$
$$\alpha_1(t_1, \ldots, t_n) = f_0(t_1, \ldots, t_n) dt_1 \wedge \cdots \wedge dt_n.$$

The definition of a multiple Dedekind zeta value (1) is as follows.

$$\zeta_{K;C}(s_1, \ldots, s_m) = \int_\Delta \bigwedge_{k=1}^M \alpha_{\epsilon_k}(t_{1,k}, \cdots, t_{n,k}), \tag{5}$$

where

(1) $M = s_1 + \cdots + s_d$;
(2) $\Delta = \Delta_1 \times \cdots \times \Delta_n$, is an n-fold product of M-simplices $\Delta_1, \ldots, \Delta_n$
(3) Δ_j is a simplex consisting of points $(t_{j,1}, t_{j,2}, \cdots, t_{j,M})$ such that
$t_{j,1} > t_{j,2} > \cdots > t_{j,M} > 0$;
(4) the indecies ϵ_k have values 0 or 1 and

$$\epsilon_1 = 1, \qquad \epsilon_1 = \cdots = \epsilon_{s_1} = 0$$
$$\epsilon_{s_1+1} = 1, \qquad \epsilon_{s_1+2} = \cdots = \epsilon_{s_1+s_2} = 0$$
$$\cdots$$
$$\epsilon_{s_1+\cdots+s_{d-1}+1} = 1, \; \epsilon_{s_1+\cdots+s_{d-1}+2} = \cdots = \epsilon_{s_1+\cdots+s_d} = 0.$$

Definition 4.1 Let $z_i = \prod_{j=1}^{n} x_{ij}$. We define the differential forms ω_0 and ω_1 by

$$\omega_0(z_1, \ldots, z_n) = \bigwedge_{i=1}^{n} \frac{dz_i}{z_i} \tag{6}$$

$$\omega_1(z_1, \ldots, z_n) = \bigwedge_{i=1}^{n} \frac{dz_i}{1 - z_i}. \tag{7}$$

Proposition 4.2 Evaluate x_{ij} at $e^{-t_j^{\sigma e_i}}$. Then

$$\omega_0(z_1, \ldots, z_n) = \sqrt{\Delta} \cdot \alpha_0(t_1, \ldots, t_n)$$
$$\omega_1(z_1, \ldots, z_n) = \sqrt{D} \cdot \alpha_1(t_1, \ldots, t_n),$$

where $\sqrt{D} = \det(\sigma_j(e_i))$ and D is an integer multiple of the discriminant.

Proof. If we evaluate x_{ij} at $e^{-t_j \sigma_j(e_i)}$, then $z_i = \prod_j x_{ij} = \prod_j e^{-t_j \sigma_j(e_i)} = e^{-\sum_j t_j \sigma_j(e_i)}$. Then

$$\omega_0(z_1, \ldots, z_n) = \bigwedge_i \frac{dz_i}{z_i} = det(\sigma_j(e_i)) \bigwedge dt_i$$
$$= \sqrt{D} \cdot \alpha_0(t_1, \ldots, t_n). \tag{8}$$

In that case, we also have

$$f_0(C; t_1, \ldots, t_n) = \prod_i \frac{z_i}{1 - z_i}.$$

Therefore,

$$\omega_1(z_1, \ldots, z_n) = \bigwedge_i \frac{dz_i}{1 - z_i} = \left(\prod_i \frac{z_i}{1 - z_i}\right) \bigwedge_i \frac{dz_i}{z_i}$$

$$= \det(\sigma_j(e_i)) f_0(t_1, \ldots, t_n) \bigwedge dt_i$$

$$= \sqrt{D} \cdot \alpha_1(t_1, \ldots, t_n). \tag{9}$$

Proposition 4.3 *We have the following relation between multiple Dedekind zeta values of the differential forms ω_0 and ω_1*

$$\int_\Delta \bigwedge_{k=1}^M \omega_{\epsilon_k}(t_{1,k}, \cdots, t_{n,k}) = \left(\sqrt{D}\right)^M \zeta_{K;C}(s_1, \ldots, s_m).$$

Proof. It follows directly from Equations (5), (8) and (9).   □

5 Tangential Base Points

Let $y = e^{-bt}$ and $z = e^{-ct}$, where b and c are complex numbers such that $Re(b) > 0$ and $Re(c) > 0$, and $|b| \neq |c|$. Then

$$\lim_{t \to +\infty} \frac{dy}{dz} = \lim_{t \to +\infty} \frac{de^{-bt}}{de^{-ct}} = \lim_{t \to +\infty} \frac{de^{ct}}{de^{bt}} = q,$$

where

$$p = \begin{cases} +\infty & \text{or } [0:1] & \text{if } b < c \\ 0 & \text{or } [0:1] & \text{if } b > c \end{cases}.$$

Also

$$\lim_{t \to 0} \frac{dy}{dz} = \lim_{t \to 0} \frac{de^{-bt}}{de^{-ct}} = \lim_{t \to 0} \frac{-be^{-bt}}{-ce^{-ct}} = \frac{b}{c}.$$

Let $\gamma : [0, 1] \to \mathcal{M}_5$, by sending t to (y, z), where $y = e^{-bt}$ and $z = e^{-ct}$. For a vector $v = (a, b)$, consider $[v] = [a : b]$ as an element of $\mathbb{P}^1$.

We have proven the following lemma.

Lemma 5.1 *(a)*

$$\lim_{t \to \infty} \left[\frac{dy}{dt}\right] = \begin{cases} [0:1] & \text{if } b < c \\ [1:0] & \text{if } b > c \end{cases}$$

Moreover, the limit is well defined for for any distinct positive real numbers b and c.

(b)

$$\lim_{t \to 0} \left[\frac{dy}{dt}\right] = [b : c].$$

Let $t = t_j$. Let also $b = \sigma_j(e_i)$, $c = \sigma_j(e_k)$. Then

$$\lim_{t_j \to 0} \left[\frac{dx_{i_1,j}}{dx_{i_2,j}} \right] = [\sigma_j(e_i) : \sigma_j(e_k)].$$

We define

$$[q(i,k)] = [\sigma_j(e_i) : \sigma_j(e_k)].$$

And

$$\lim_{t_j \to \infty} \left[\frac{dx_{i,j}}{dx_{k,j}} \right] = \begin{cases} [0:1] & \text{if } \sigma_j(e_i) < \sigma_j(e_k) \\ [1:0] & \text{if } \sigma_j(e_i) > \sigma_j(e_k). \end{cases}$$

Let

$$[p(i,k)] = \begin{cases} [0:1] & \text{if } \sigma_j(e_i) < \sigma_j(e_k) \\ [1:0] & \text{if } \sigma_j(e_i) > \sigma_j(e_k). \end{cases}$$

More generally, let $[p(i_0, \ldots, i_r)] = [a_0 : \cdots : a_r]$ be a point on $\mathbb{P}^r(\mathbb{Q})$ with a_i being 0 or 1, where all a_i's are zero except the one whose index c, for a_c, is such that $|\sigma_j(e_c)|$ is a maximal value among the elements $|\sigma_j(e_0)|, \ldots, |\sigma_j(e_r)|$.

6 Multiple Dedekind Zeta Values and the Moduli Space $\mathcal{M}_{0,n^2M+3}$

Let $z_{ik} = \prod_{j=1}^k x_{ij}$. there are n^2 such variables. If we consider multiple Dedekind zeta value of weight $M = s_1 + \cdots + s_m$ then we need $n^2 M$ variables $(z_{ikd})_{i,k,d=1,1,1}^{n,n,M} \in \mathcal{M}_{0,n^2M+3}$. The dimension of the differential form $\Omega(A)$ is nM. Let ϵ_d be 0 or 1 for $d = 1, 2, \ldots, M$. Let also $\epsilon_1 = 1$ and $\epsilon_M = 0$.

$$\Omega(A) = \bigwedge_{d=1}^M \omega_{\epsilon_d}$$

where

$$A = (z_{ind} = \epsilon_d)_{i,d=1,1}^{n,M} \cup (z_{ind} = \infty)_{i,d=1,1}^{n,M}$$

and $B = B_1 \cup B_2$, where $codim B_r = r$. De define the divisors

$$B_1 = (z_{i,1,1} = 0)_{i=1}^n \cup (z_{i,j,d} = z_{i,j,d+1})_{i,j,d=1,1,1}^{n,n,M-1} \cup (z_{i,1,M} = 1)_{i=1}^n$$

together with the intersection of boundary components of $\overline{\mathcal{M}}_{0,n^2M+3} - \mathcal{M}_{0,n^2M+3}$ containing the same variable or the same constant 0 or 1. Besides codimension 1 components, B also contains a codimension 2 components.

Let B'_{i_1,i_2} be a quiasi-subvariety in the boundary of the Deligne-Mumford compactification that has coordinates with

$$[z_{i_1,1,1} : z_{i_2,1,1}] = [p(i_1, i_2)].$$

in the blow-up of the intersection $(z_{i_1,j,1} = 0) \cap (z_{i_2,j,1} = 0)$. Let B''_{i_1,i_2} be a cycle in the boundary of the Deligne-Mumford compactification above the intersection $(z_{i,1,M} = 1) \cap (z_{k,j,M} = 1)$ such that the coordinated of B'_{i_1,i_2} in the blowup are

$$[1 - z_{i_1,1,1} : 1 - z_{1_2,1,1}] = [\sigma_j(e_{i_1}) : \sigma_j(e_{i_2})].$$

The codimension 2 components of B are the union of all B'_{i_1,i_2} and B''_{i_1,i_2}. That is

$$B_2 = \bigcup_{i_1 < i_2} \left(B'_{i_1,i_2} \cup B''_{i_1,i_2} \right).$$

Now let us write $\omega_0(x_1, x_2)$ and $\omega_1(x_1, x_2)$, when we want to specify the dependence on the variables. In fact, both forms depend also on y_1 and y_2; however, we will take care of that by choosing a region of integration together with tangential base points.

Theorem 6.1 (a) *Every multiple Dedekind zeta value over a field K times a suitable multiple of a power of the discriminant of K is a period of a mixed Tate motive over K. More precisely,*

$$\left(\sqrt{D}\right)^M \zeta_{K;C}(s_1, \ldots, s_m) = \int_\Delta \bigwedge_{k=1}^M \omega_{\epsilon_k}(t_{1,k}, \cdots, t_{n,k})$$

is a period of

$$H^{nM}(\overline{\mathcal{M}}_{0,n^2M+3} - A, B - (A \cap B)),$$

$\overline{\mathcal{M}}_{0,n^2M+3}$ *is the Delingne-Mumford compactification of the moduli space of curves of genus zero with $n^2M + 3$ marked points, and A and B that consist of a union of lower dimensional moduli spaces of curves of genus zero with marked points.*

(b) *For any field K of degree n over $\mathbb{Q}$, with the property that K has n units $e_1, \ldots, e_n$ that are linearly independent over $\mathbb{Q}$ and $|\sigma_j(e_i)| \neq |\sigma_j(e_k)|$, we have the following stronger statement. In particular, when K is a totally real number field, we choose a cone $C = \mathbb{N}\{e_1, \ldots, e_n\}$. Then for any positive integers $s_1, \ldots, s_m$ with $s_m \geq 2$, we have that*

$$\left(\sqrt{D}\right)^M \zeta_{K;C}(s_1, \ldots, s_m)$$

is a period of an unramified mixed Tate motive over the ring of algebraic integers $\mathcal{O}_K$ in the field K.

Proof. In this proof we are going to follow closely the paper by Goncharov and Manin [4]. The period will be a pairing between $[\Omega_A] \in Gr_{2nM}^W H^{nM}(\overline{\mathcal{M}}_{0,n^2M+3} - A)$ and $[\Delta_B] \in \left(Gr_0^W H^{nM}(\overline{\mathcal{M}}_{0,n^2M+3} - B)\right)^\vee$ associated to a mixed Tate motive $H^{nM}(\overline{\mathcal{M}}_{0,n^2M+3} - A; B - A \cap B)$.

We have that A and B_1 are defined over $\mathbb{Z}$. Moreover, any component and any intersection of components of A and B are isomorphic to a moduli space $\mathcal{M}_{0,N}$ for some N. The component $B_2, B_3, \ldots$ are defined over the field K. Moreover, any intersection is isomorphic to $\mathcal{M}_{0,N/K}$ for some N. Thus all multiple Dedekind zeta values are mixed Tate motives over the field of definition K.

If $e_1, \ldots, e_n$ are unit in $\mathcal{O}_K$, which are linearly independent over $\mathbb{Q}$, then all $[q(i_1, i_2, j)]$, $[q(i_1, i_2, i_3, j)]$ etc., have coordinates 0 or units. Then, the component $B_2, B_3, \ldots$ are defined over the ring $\mathcal{O}_K$. Moreover, any intersection is isomorphic to $\mathcal{M}_{0,N/\mathcal{O}_K}$ for some N.

We have that $H^i(\overline{\mathcal{M}}_{0,N})$ is a mixed Tate motive over $Spec(\mathbb{Z})$. This implies that $H^i(\overline{\mathcal{M}}_{0,N})$ is a mixed Tate motive over $Spec(\mathcal{O}_K)$. we obtain that the motivic cohomology of the components of B are mixed Tate motives. Using Proposition 1.7 from Deligne and Goncharov, [3], we conclude that for $l \neq char(v)$ the l-adic cohomology of the reduction of B_j modulo v of the motive $H^i(B_j)$ is unramified for any component B_j of B, since B_j is isomorphic to $\overline{\mathcal{M}}_{0,N}$ over $Spec(\mathcal{O}_K)$ for some N. We conclude that for $l \neq char(v)$ the l-adic cohomology of the reduction modulo any $v \in Spec(\mathcal{O}_K)$ of the motive $H^{nM}(\overline{\mathcal{M}}_{0,n^2M+3} - A; B - A \cap B)$ is unramified. Thus, $H^{nM}(\overline{\mathcal{M}}_{0,n^2M+3} - A; B - A \cap B)$ is a mixed Tate motive unramified over $Spec(\mathcal{O}_K)$.

Acknowledgements

I would like to express great respect to Emma Previato for her work and style in research and in communication. Without her interest in my work, this paper would not have been completed. I am very thankful to Yuri Manin and Don Zagier for conversations during the preparation of this paper and for their interest. I would also to thank the referee whose questions and suggestions improved the article.

Part of this work was finished at the Max Planck Institute for Mathematics and I am grateful for the excellent working conditions and the financial support provided. Support for this project was also provided by a PSC-CUNY Award, jointly funded by The Professional Staff Congress and The City University of New York.

References

[1] Brown, F.: *Mixed Tate motives over* $\mathbb{Z}$. Ann. of Math. (2) 175 (2012), no. 2, 949–976.

[2] Brown, F.: *Dedekind zeta motives for totally real number fields*. Invent. Math. 194 (2013), no. 2, 257–311.

[3] Deligne, P. and Goncharov, A: *Groupes fondamentaux motiviques de Tate mixte.* (French) [Mixed Tate motivic fundamental groups], Ann. Sci. École Norm. Sup. (4) 38 (2005), no. 1, 1–56.

[4] Goncharov, A. and Manin, Yu.: *Multiple ζ-motives and moduli spaces $\overline{\mathcal{M}}_{0,n}$.* Compos. Math. 140 (2004), no. 1, 1–14.

[5] Horozov, I.: *Multiple Dedekind Zeta Functions,* J. Reine Angew. Math. 722 (2017), 65–104. (Crelle's Journal), DOI: 10.1515/crelle-2014-0055

[6] Horozov, I.: *Non-commutative Hilbert modular symbols,* Algebra and Number Theory Vol. 9 (2015), No. 2, 317–370 DOI: 10.2140/ant.2015.9.317

[7] Horozov, I.: *Periods of mixed Tate motives over real quadratic number rings.* Geometric methods in physics XXXV, 181–190, Trends Math., Birkhauser/Springer, Cham, 2018.

[8] Kontsevich, M. and Zagier, D.: *Periods,* Mathematics unlimited - 2001 and beyond, 771–808, Springer, Berlin, 2001.

[9] Zagier, D.: *A Kronecker limit formula for real quadratic fields.* Math. Ann. 213 (1975), 153–184.

[10] Zagier, D.: *On the values at negative integers of the zeta-function of a real quadratic field,* L'Enseignement Mathmatique, 22 (1976) 55–95.

Ivan Horozov
City University of New York,
Bronx Community College,
2155 University Avenue, Bronx,
NY 10453, U.S.A.
and
Graduate Center at City University in New York,
365 Fifth Ave, New York City,
NY 10016, U.S.A.
ivan.horozov@bcc.cuny.edu

16

Noncommutative Cross-Ratio and Schwarz Derivative

Vladimir Retakh, Vladimir Rubtsov and Georgy Sharygin

Dedicated to Emma Previato.

Abstract. We present here a theory of noncommutative cross-ratio, Schwarz derivative and their connections and relations to the operator cross-ratio. We apply the theory to "noncommutative elementary geometry", relate it to noncommutative integrable systems and provide noncommutative versions of triple cross-ratios, the leapfrog map, and the celebrated "pentagramma mirificum".

1 Introduction

Cross-ratio and Schwarz derivative are some of the most famous invariants in mathematics (see [15, 18, 19]). Different versions of their noncommutative analogs and various applications of these constructions to integrable systems, control theory and other subjects were discussed in several publications including [4]. In this paper, which is the first of a series of works, we recall some of these definitions, revisit the previous results and discuss their connections with each other and with noncommutative elementary geometry. In the forthcoming papers we shall further discuss the role of noncommutative cross ratio in the theory of noncommutative integrable models and in topology.

The present paper is organized as follows. In Sections 2, 3 we recall a definition of noncommutative cross-ratios based on the theory of noncommutative quasi-Plücker invariants (see [8, 9]), in Section 4 the theory of quasideterminants (see [7]) is used to obtain noncommutative versions of Menelaus's and Ceva's theorems. In Section 5 we compare our definition of cross-ratio with the operator version used in control theory [24] and show how Schwarz derivatives appear as the infinitesimal analogs of noncommutative cross-ratios. In section 6 we revisit an approach to noncommutative Schwarz derivative from [21] and section 7 deals with possible applications of the developed theory. It should be also mentioned that the present paper continues the constructions and ideas, first outlined in [21].

It is our pleasure to dedicate this paper to Emma Previato, whose intelligence, erudition, interest in various areas of our field are spectacular and her friendship is constant and loyal. Her results ([4]) were an important motive which inspired us to think once more about the role of non-commutative cross-ratios.

Acknowledgements. The authors are grateful to B. Khesin, V. Ovsienko and S. Tabachnikov for helpful discussions. This research was started during V. Retakh's visit to LAREMA and Department of Mathematics, University of Angers. He is thankful to the project DEFIMATH for its support and LAREMA for hospitality. V. Roubtsov thanks the project IPaDEGAN (H2020-MSCA-RISE-2017), Grant Number 778010 for support of his visits to CRM, University of Montreal where the paper was finished and the CRM group of Mathematical Physics for hospitality. He is partly supported by the Russian Foundation for Basic Research under the Grants RFBR 18-01-00461. G. Sharygin is thankful to IHES and LAREMA for hospitality during his visits. His research is partly supported by the Russian Science Foundation, Grant No. 16-11-10069 and by the Simons foundation.

2 Quasi-Plücker Coordinates

We begin with a list of basic properties of noncommutative cross-ratios introduced in [20]. To this end we first recall the definition and properties of quasi-Plücker coordinates; observe that we shall only deal with the quasi-Plücker coordinates for $2 \times n$-matrices over a noncommutative division ring $\mathcal{R}$. The corresponding theory for general $k \times n$-matrices is presented in [8, 9].

Recall (see [6, 7] and subsequent papers) that for a matrix $\begin{pmatrix} a_{1k} & a_{1i} \\ a_{2k} & a_{2i} \end{pmatrix}$ one can define four quasideterminants provided the corresponding elements are invertible:

$$\begin{vmatrix} \boxed{a_{1k}} & a_{1i} \\ a_{2k} & a_{2i} \end{vmatrix} = a_{1k} - a_{1i}a_{2i}^{-1}a_{2k}, \quad \begin{vmatrix} a_{1k} & \boxed{a_{1i}} \\ a_{2k} & a_{2i} \end{vmatrix} = a_{1i} - a_{1k}a_{2k}^{-1}a_{2i},$$

$$\begin{vmatrix} a_{1k} & a_{1i} \\ \boxed{a_{2k}} & a_{2i} \end{vmatrix} = a_{2k} - a_{2i}a_{1i}^{-1}a_{1k}, \quad \begin{vmatrix} a_{1k} & a_{1i} \\ a_{2k} & \boxed{a_{2i}} \end{vmatrix} = a_{2i} - a_{2k}a_{1k}^{-1}a_{1i}.$$

Let $A = \begin{pmatrix} a_{11} & a_{12} & \cdots & a_{1n} \\ a_{21} & a_{22} & \cdots & a_{2n} \end{pmatrix}$ be a matrix over $\mathcal{R}$.

Lemma 2.1 *Let $i \neq k$. Then*

$$\begin{vmatrix} a_{1k} & \boxed{a_{1i}} \\ a_{2k} & a_{2i} \end{vmatrix}^{-1} \begin{vmatrix} a_{1k} & \boxed{a_{1j}} \\ a_{2k} & a_{2j} \end{vmatrix} = \begin{vmatrix} a_{1k} & a_{1i} \\ a_{2k} & \boxed{a_{2i}} \end{vmatrix}^{-1} \begin{vmatrix} a_{1k} & a_{1j} \\ a_{2k} & \boxed{a_{2j}} \end{vmatrix}^{-1}$$

if the corresponding expressions are defined.

Note that in the formula the boxed elements on the left and on the right must be in the same row.

Definition 2.2 *We call the expression*

$$q_{ij}^k(A) = \begin{vmatrix} a_{1k} & \boxed{a_{1i}} \\ a_{2k} & a_{2i} \end{vmatrix}^{-1} \begin{vmatrix} a_{1k} & \boxed{a_{1j}} \\ a_{2k} & a_{2j} \end{vmatrix} = \begin{vmatrix} a_{1k} & a_{1i} \\ a_{2k} & \boxed{a_{2i}} \end{vmatrix}^{-1} \begin{vmatrix} a_{1k} & a_{1j} \\ a_{2k} & \boxed{a_{2j}} \end{vmatrix}^{-1}$$

the quasi-Plücker coordinates of matrix A.

Our terminology is justified by the following observation. Recall that in the commutative case the expressions

$$p_{ik}(A) = \begin{vmatrix} a_{1i} & a_{1k} \\ a_{2i} & a_{2k} \end{vmatrix} = a_{1i}a_{2k} - a_{1k}a_{2i}$$

are the Plücker coordinates of A. One can see that in the commutative case

$$q_{ij}^k(A) = \frac{p_{jk}(A)}{p_{ik}(A)},$$

i.e. quasi-Plücker coordinates are ratios of Plücker coordinates.

Let us list here the properties of quasi-Plücker coordinates over (noncommutative) division ring $\mathcal{R}$. For the sake of brevity we shall sometimes write q_{ij}^k instead of $q_{ij}^k(A)$ where it cannot lead to a confusion.

(1) Let g be an invertible matrix over $\mathcal{R}$. Then

$$q_{ij}^k(g \cdot A) = q_{ij}^k(A).$$

(2) Let $\Lambda = \text{diag}(\lambda_1, \lambda_2, \dots, \lambda_n)$ be an invertible diagonal matrix over $\mathcal{R}$. Then

$$q_{ij}^k(A \cdot \Lambda) = \lambda_i^{-1} \cdot q_{ij}^k(A) \cdot \lambda_j.$$

(3) If $j = k$ then $q_{ij}^k = 0$; if $j = i$ then $q_{ij}^k = 1$ (we always assume $i \neq k$).
(4) $q_{ij}^k \cdot q_{j\ell}^k = q_{i\ell}^k$. In particular, $q_{ij}^k q_{ji}^k = 1$.
(5) "Noncommutative skew-symmetry": For distinct i, j, k

$$q_{ij}^k \cdot q_{jk}^i \cdot q_{ki}^j = -1.$$

One can also rewrite this formula as $q_{ij}^k q_{jk}^i = -q_{ik}^j$.

(6) "Noncommutative Plücker identity": For distinct i, j, k, ℓ

$$q_{ij}^k q_{ji}^\ell + q_{i\ell}^k q_{\ell i}^j = 1.$$

One can easily check two last formulas in the commutative case. In fact,

$$q_{ij}^k \cdot q_{jk}^i \cdot q_{ki}^j = \frac{p_{jk} p_{ki} p_{ij}}{p_{ik} p_{ji} p_{kj}} = -1$$

because Plücker coordinates are skew-symmetric: $p_{ij} = -p_{ji}$ for any i, j.

Also, assuming that $i < j < k < \ell$

$$q_{ij}^k q_{ji}^\ell + q_{i\ell}^k q_{\ell i}^j = \frac{p_{jk} p_{i\ell}}{p_{ik} p_{j\ell}} + \frac{p_{\ell k} p_{ij}}{p_{ik} p_{\ell j}}.$$

Since $\frac{p_{\ell k}}{p_{\ell j}} = \frac{p_{k\ell}}{p_{j\ell}}$, the last expression is equal to

$$\frac{p_{jk} p_{i\ell}}{p_{ik} p_{j\ell}} + \frac{p_{k\ell} p_{ij}}{p_{ik} p_{j\ell}} = \frac{p_{ij} p_{k\ell} + p_{i\ell} p_{jk}}{p_{ik} p_{j\ell}} = 1$$

due to the celebrated Plücker identity

$$p_{ij} p_{k\ell} - p_{ik} p_{j\ell} + p_{i\ell} p_{jk} = 0.$$

Remark 2.3 We present here the theory of the *left* quasi-Plücker coordinates for 2 by n matrices where $n > 2$. The theory of the *right* quasi-Plücker coordinates for n by 2 or, more generally, for n by k matrices where $n > k$ can be found in [8, 9].

3 Definition and Basic Properties of Cross-Ratios

3.1 Non-Commutative Cross-Ratio: Basic Definition

We define cross-ratios over (noncommutative) division ring $\mathcal{R}$ by imitating the definition of classical cross-ratios in homogeneous coordinates. Namely, if four points in (real or complex) projective plane can be represented in homogeneous coordinates by vectors a, b, c, d such that $c = a + b$ and $d = ka + b$, then their cross-ratio is k.

Let

$$x = \begin{pmatrix} x_1 \\ x_2 \end{pmatrix}, \quad y = \begin{pmatrix} y_1 \\ y_2 \end{pmatrix}, \quad z = \begin{pmatrix} z_1 \\ z_2 \end{pmatrix}, \quad t = \begin{pmatrix} t_1 \\ t_2 \end{pmatrix}$$

be four vectors in $\mathcal{R}^2$. We define their cross-ratio $\kappa = \kappa(x, y, z, t)$ by equations

$$\begin{cases} t = x\alpha + y\beta \\ z = x\alpha\gamma + y\beta\gamma \cdot \kappa \end{cases}$$

where $\alpha, \beta, \gamma, \kappa \in \mathcal{R}$.

In order to obtain explicit formulas consider the matrix

$$\begin{pmatrix} x_1 & y_1 & z_1 & t_1 \\ x_2 & y_2 & z_2 & t_2 \end{pmatrix}.$$

We shall identify its columns with x, y, z, t. The following theorem was proved in [20].

Theorem 3.1

$$\kappa(x, y, z, t) = q_{zt}^y \cdot q_{tz}^x .$$

Note that in the generic case

$$\kappa(x, y, z, t) = \begin{vmatrix} y_1 & \boxed{z_1} \\ y_2 & z_2 \end{vmatrix}^{-1} \begin{vmatrix} y_1 & \boxed{t_1} \\ y_2 & t_2 \end{vmatrix} \cdot \begin{vmatrix} x_1 & \boxed{t_1} \\ x_2 & t_2 \end{vmatrix}^{-1} \begin{vmatrix} x_1 & \boxed{z_1} \\ x_2 & z_2 \end{vmatrix}$$

$$= z_2^{-1}(z_1 z_2^{-1} - y_1 y_2^{-1})^{-1}(t_1 t_2^{-1} - y_1 y_2^{-1})(t_1 t_2^{-1} - x_1 x_2^{-1})^{-1}$$
$$\times (z_1 z_2^{-1} - x_1 x_2^{-1})z_2$$

which shows that $\kappa(x, y, z, t)$ coincides with the standard cross-ratio in commutative case and also demonstrates the importance of conjugation in the noncommutative world.

Corollary 3.2 *Let* x, y, z, t *be vectors in* $\mathcal{R}$, g *be a 2 by 2 matrix over* $\mathcal{R}$ *and* $\lambda_i \in \mathcal{R}$, $i = 1, 2, 3, 4$. *If the matrix g and elements* λ_i *are invertible then*

$$\kappa(gx\lambda_1, gy\lambda_2, gz\lambda_3, gt\lambda_4) = \lambda_3^{-1}\kappa(x, y, z, t)\lambda_3 . \tag{1}$$

Again, as expected, in the commutative case the right hand side of (3.1) equals $\kappa(x, y, z, t)$.

Remark 3.3 Note that the group $GL_2(\mathcal{R})$ acts on vectors in $\mathcal{R}^2$ by multiplication from the left: $(g, x) \mapsto gx$, and the group $\mathcal{R}^\times$ of invertible elements in $\mathcal{R}$ acts by multiplication from the right: $(\lambda, x) \mapsto x\lambda^{-1}$. These actions determine the action of $GL_2(\mathcal{R}) \times T_4(\mathcal{R})$ on $P_4 = \mathcal{R}^2 \times \mathcal{R}^2 \times \mathcal{R}^2 \times \mathcal{R}^2$ where $T_4(\mathcal{R}) = (\mathcal{R}^\times)^4$. The cross-ratios are *relative invariants* of the action.

The following theorem generalizes the main property of cross-ratios to the noncommutive case (see [20]).

Theorem 3.4 *Let* $\kappa(x, y, z, t)$ *be defined and* $\kappa(x, y, z, t) \neq 0, 1$. *Then* *4-tuples* (x, y, z, t) *and* (x', y', z', t') *from* P_4 *belong to the same orbit of* $GL_2(\mathcal{R}) \times T_4(\mathcal{R})$ *if and only if there exists* $\mu \in \mathcal{R}^\times$ *such that*

$$\kappa(x, y, z, t) = \mu \cdot \kappa(x', y', z', t') \cdot \mu^{-1} . \tag{2}$$

The following theorem shows that the cross-ratios we defined satisfy *cocycle conditions* (see [15]).

Theorem 3.5 *For all vectors x, y, z, t, w the following equations hold*

$$\kappa(x, y, z, t) = \kappa(w, y, z, t)\kappa(x, w, z, t)$$
$$\kappa(x, y, z, t) = 1 - \kappa(t, y, z, x),$$

if all the cross-ratios in these formulas exist.

The last statement can also be generalized as follows:

Corollary 3.6 *For all vectors $x, x_1, x_2, \ldots x_n, z, t \in \mathcal{R}^2$ one has*

$$\kappa(x, x, z, t) = 1$$

and

$$\kappa(x_{n-1}, x_n, z, t)\kappa(x_{n-2}, x_{n-1}, z, t)\ldots\kappa(x_1, x_2, z, t) = \kappa(x_1, x_n, z, t)$$

where we assume that all the cross-ratios exist.

3.2 Noncommutative Cross-Ratios and Permutations

There are 24 cross-ratios defined for vectors $x, y, z, t \in \mathcal{R}^2$, if we permute them. They are related by the following formulas:

Proposition 3.7 *Let $x, y, z, t \in \mathcal{R}$. Then*

$$q_{tz}^x \kappa(x, y, z, t) q_{zt}^x = q_{tz}^y \kappa(x, y, z, t) q_{zt}^y = \kappa(y, x, t, z); \tag{3}$$
$$q_{xz}^y \kappa(x, y, z, t) q_{zx}^y = q_{xz}^t \kappa(x, y, z, t) q_{zx}^t = \kappa(z, t, x, y); \tag{4}$$
$$q_{yz}^x \kappa(x, y, z, t) q_{zy}^x = q_{yz}^t \kappa(x, y, z, t) q_{zy}^t = \kappa(t, z, x, y); \tag{5}$$
$$\kappa(x, y, z, t)^{-1} = \kappa(y, x, z, t). \tag{6}$$

Note again the appearance of conjugation in the noncommutative case; this happens since q_{ij}^k and q_{ji}^k are inverse to each other. Also observe that using Proposition 3.7 and the cocycle condition (corollary 3.5) one can get all 24 formulas for cross-ratios of x, y, z, t knowing just one of them.

3.3 Noncommutative Triple Ratio

Let $\mathcal{R}$ be a division ring as above; we shall work with the "right $\mathcal{R}$-plane" $\mathcal{R}^2$, i.e. we use right multiplication of vectors by the elements from $\mathcal{R}$. Consider the triangle with vertices $O(0, 0)$, $X(x, 0)$, and $Y(0, y)$ in $\mathcal{R}^2$. Let $A(a_1, a_2)$ be a point on side XY, $B(b, 0)$ be a point on side OX and $C(0, c)$ be a point on side OY. Recall that the geometric condition $A \in XY$ means

$$x^{-1}a_1 + y^{-1}a_2 = 1.$$

Let $P(p_1, p_2)$ be the point of intersection of XC and YB. Then one has

$$p_1 = (y - c)(yb^{-1} - cx^{-1})^{-1}, \quad p_2 = (x - b)(by^{-1} - xc^{-1})^{-1}.$$

Let Q be the point of intersection of OP and XY. The non-commutative cross ratio for Y, A, Q, X is equal to

$$x^{-1}(1 - p_1 p_2^{-1} a_2 a_1^{-1})^{-1} x.$$

By changing the order of Y, A, Q, X we get up to a conjugation

$$\begin{aligned} p_1 p_2^{-1} a_2 a_1^{-1} &= -(y - c)(yb^{-1} - cx^{-1})^{-1}(x - b)^{-1} xc^{-1} \\ &\times (yb^{-1} - cx^{-1})bx^{-1}(x - a_1)a_1^{-1}. \end{aligned} \quad (7)$$

In the commutative case (up to a sign) we have

$$p_1 p_2^{-1} a_2 a_1^{-1} = (y - c)c^{-1}b(x - b)^{-1}(x - a_1)a_1^{-1}$$

(compare it with the Ceva theorem in elementary geometry).

Note that $(x - a_1)^{-1}a_1^{-1} = YA/AX$ and (3.7) is a (non-commutative analogue of) triple cross-ration (see section 6.5 in the book by Ovsienko and Tabachnikov [18])

3.4 Noncommutative Angles and Cross-Ratios

Let $\mathcal{R}$ be a noncommutative division ring. Recall that noncommutative angles (or noncommutative λ-lengths) $T_i^{jk} = T_i^{kj}$ for vectors $A_1, A_2, A_3, A_4 \in \mathcal{R}^2$, $A_i = (a_{1i}, a_{2i})$ are defined by the formulas

$$T_i^{jk} = x_{ji}^{-1} x_{jk} x_{ik}^{-1}.$$

Here $x_{ij} = a_{1j} - a_{1i}a_{2i}^{-1}a_{2j}$, or $x_{ij} = a_{2j} - a_{2i}a_{1i}^{-1}a_{1j}$ (see [3]). On the other hand the cross-ratio $\kappa(A_1, A_2, A_3, A_4) = \kappa(1, 2, 3, 4)$ (see definition 2.2) is

$$\kappa(1, 2, 3, 4) = q_{34}^2 q_{43}^1.$$

It implies that

$$\kappa(1, 2, 3, 4) = x_{43}^{-1}(T_4^{23})^{-1}T_4^{31}x_{43}.$$

In other words, cross-ratio is a ratio of two angles up to a conjugation.

Under the transformation $x_{ij} \mapsto \lambda_i x_{ij}$ we have

$$T_i^{jk} \mapsto T_i^{jk} \cdot \lambda_i^{-1}.$$

Also note that

$$T_i^{jk}(T_i^{mk})^{-1} = q_{ik}^j q_{ki}^m$$

i.e. $T_i^{jk}(T_i^{mk})^{-1}$ is a cross-ratio. Further details on the properties of T_i^{jk} can be found in [3].

4 Noncommutative Menelaus' and Ceva's Theorems

4.1 Higher Rank Quasi-Determinants: Reminder

Let $A = (a_{ij})$, $i, j = 1, 2, \ldots, n$ be a matrix over a ring. Denote by A^{pq} the submatrix of matrix A obtained from A by removing the p-th row and the q-th column. Let $r_p = (a_{p1}, a_{p2}, \ldots, \hat{a}_{pq}, \ldots a_{pn})$ be the row submatrix and $c_q = (a_{1q}, a_{2q}, \ldots, \hat{a}_{pq}, \ldots a_{nq})^T$ be the column submatrix of A. Following [6] we say that the quasideterminant $|A|_{pq}$ is defined if and only if the submatrix A^{pq} is invertible. In this case

$$|A|_{pq} = a_{pq} - r_p (A^{pq})^{-1} c_q \, .$$

In the commutative case $|A|_{pq} = (-1)^{p+q} \det A / \det A^{pq}$. It is sometimes convenient to use the notation

$$|A|_{pq} = \begin{vmatrix} \cdots & \cdots & a_{1q} & \cdots \\ \cdots & \cdots & \cdots & \cdots \\ a_{p1} & \cdots & \boxed{a_{pq}} & \cdots \\ \cdots & \cdots & \cdots & \cdots \end{vmatrix} \, .$$

4.2 Commutative Menelaus and Ceva Theorems

We follow the affine geometry proof. Let the points D, E, F lie on the straight lines AB, BC and AC respectively (see Figure 16.1(a)). Denote by λ_D the coefficient for homothety with center D sending B to C, by λ_E the coefficient for homothety with center E sending C to A, and by λ_F the coefficient for homothety with center F sending A to B. Note that (in a generic case)

$$\lambda_D = (b_i - d_i)^{-1}(c_i - d_i), \quad i = 1, 2,$$
$$\lambda_E = (c_i - e_i)^{-1}(a_i - e_i), \quad i = 1, 2,$$
$$\lambda_F = (a_i - f_i)^{-1}(b_i - f_i), \quad i = 1, 2.$$

Here (a_1, a_2) are the coordinates of A etc. We shall omit the indices and write $\lambda_D = (b - d)^{-1}(c - d)$, etc.

Theorem 4.1 *Points E, D, F belong to a straight line if an only if*

$$(a - f)^{-1}(b - f) \cdot (c - e)^{-1}(a - e) \cdot (b - d)^{-1}(c - d) = 1 \, .$$

This is the Menelaus theorem in the commutative case.

Proof. The composition of transformations $\lambda_D, \lambda_E, \lambda_F$ leaves the point B unchanged, thus it is equal to a homothety with center B. On the other hand,

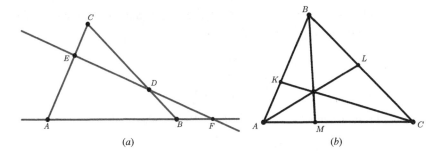

Figure 16.1 The classical Menelaus (part (*a*)) and Ceva (part (*b*)) theorems.

if points belong to the same straight line then the center should belong to this line too, so the composition is equal to identity. So

$$\lambda_F \lambda_E \lambda_D = 1$$

i.e.

$$(a - f)^{-1}(b - f) \cdot (c - e)^{-1}(a - e) \cdot (b - d)^{-1}(c - d) = 1.$$

The opposite statement can be proved by contradiction.   □

Somewhat dually, one obtains Ceva's theorem (see Figure 16.1 (*b*)):

Theorem 4.2 *Lines AD, BE and CF intersect each other in a point O if and only if*

$$(e - a)^{-1}(e - c) \cdot (f - b)^{-1}(f - a) \cdot (d - c)^{-1}(d - b) = -1 .$$

This is the Ceva theorem in the commutative case.

4.3 Non-Commutative Menelaus and Ceva Theorems

Let $\mathcal{R}$ be a noncommutative division ring. Consider $\mathcal{R}^2$ as the right vector space over $\mathcal{R}$. For a point $X \in \mathcal{R}^2$ denote by x_i its i-th coordinate, $i = 1, 2$. Here and below we shall use the properties of quasideterminants, see [6, 7]:

Proposition 4.3 *Let that points X and Y are in generic position, i.e. that matrix*

$$\begin{pmatrix} x_1 & y_1 \\ x_2 & y_2 \end{pmatrix}$$

is invertible. Then the points $X, Y, Z \in \mathcal{R}^2$ belong to the same straight line (in the sense of linear algebra) if and only if

$$\begin{vmatrix} x_1 & y_1 & z_1 \\ x_2 & y_2 & z_2 \\ 1 & 1 & \boxed{1} \end{vmatrix} = 0 .$$

Proof. From the general theory of quasideterminants it follows that that

$$\begin{vmatrix} x_1 & y_1 & z_1 \\ x_2 & y_2 & z_2 \\ 1 & 1 & \boxed{1} \end{vmatrix} = 1 - \lambda - \mu$$

where $\lambda, \mu \in R$ satisfy the equation $X\lambda + Y\mu = Z$. Note that X, Y and Z belong to the same straight line if and only if there exists $\lambda + \mu = 1$. □

Corollary 4.4 *Assume that $x_i - y_i \in \mathcal{R}$, $i = 1, 2$ are invertible. Then X, Y, Z belong to one straight line if and only if*

$$(y_1 - x_1)^{-1}(z_1 - x_1) = (y_2 - x_2)^{-1}(z_2 - x_2) .$$

Proof. Note that

$$\begin{vmatrix} x_1 & y_1 & z_1 \\ x_2 & y_2 & z_2 \\ 1 & 1 & \boxed{1} \end{vmatrix} = - \begin{vmatrix} x_1 & y_1 - x_1 \\ \boxed{x_2} & y_2 - x_2 \end{vmatrix}^{-1} \cdot \begin{vmatrix} y_1 - x_1 & z_1 - x_1 \\ y_2 - x_2 & \boxed{z_2 - x_2} \end{vmatrix}$$

and that $(y_1 - x_1)^{-1}(z_1 - x_1) = (y_2 - x_2)^{-1}(z_2 - x_2)$ if and only if

$$\begin{vmatrix} y_1 - x_1 & z_1 - x_1 \\ y_2 - x_2 & \boxed{z_2 - x_2} \end{vmatrix} = 0 .$$

□

4.4 NC Analogue of Konopelchenko Equations

Let again $\mathcal{R}$ be a division ring. Consider $\mathcal{R}^2$ as the right module over $\mathcal{R}$.

Proposition 4.5 *Let $F_1 = (x_1, y_1)$, $F_2 = (x_2, y_2)$ be two points in $\mathcal{R}^2$ in a generic position. Then the equation of the straight line L_{12} passing through F_1 and F_2 is*

$$(y_2 - y_1)^{-1}(y - y_1) = (x_2 - x_1)^{-1}(x - x_1).$$

Corollary 4.6 *An equation of the line L'_{12} parallel to L and passing through $(0, 0)$ is*

$$(y_2 - y_1)^{-1}y = (x_2 - x_1)^{-1}x,$$

i.e. any point F_{12} on L'_{12} has coordinates $((x_2 - x_1)f_{12}, (y_2 - y_1)f_{12})$.

The proposition and the corollary are both straightforward consequences of Proposition 4.3 and Corollary 4.4.

Denote by L_{ij} the straight line passing through $F_i = (x_i, y_i)$ and $F_j = (x_j, y_j)$ and by L'_{ij} the parallel line though $(0, 0)$. Consider now (additionaly to the line L'_{12} and to a point $(x_2 - x_1)f_{12}, (y_2 - y_1)f_{12}$ on it) points $F_{23} = ((x_3 - x_2)f_{23}, (y_3 - y_2)f_{23})$ on line L'_{23} and $F_{31} = (x_1 - x_3)f_{31}, (y_1 - y_3)f_{31})$ on line L'_{31}.

Proposition 4.7 *For generic points F_1, F_2, F_3 the points F_{12}, F_{23}, F_{31} belong to a straight line iff*

$$f_{12}^{-1} + f_{23}^{-1} + f_{31}^{-1} = 0.$$

Warning: Note that before we considered points with coordinates (x_1, x_2), (y_1, y_2), (z_1, z_2) and now with coordinates (x_i, y_i).

Proof. According to Proposition 4.3 in order to show that F_{12}, F_{23}, F_{31} lie on the same straight line it is necessary and sufficient to check that

$$\theta := \begin{vmatrix} (x_2 - x_1)f_{12} & (x_3 - x_2)f_{23} & (x_1 - x_3)f_{31} \\ (y_2 - y_1)f_{12} & (y_3 - y_2)f_{23} & (y_1 - x_3)f_{31} \\ 1 & 1 & \boxed{1} \end{vmatrix} = 0.$$

According to the standard properties of quasideterminants

$$\theta = \begin{vmatrix} (x_2 - x_1) & (x_3 - x_2) & (x_1 - x_3) \\ (y_2 - y_1) & (y_3 - y_2) & (y_1 - y_3) \\ f_{12}^{-1} & f_{23}^{-1} & \boxed{f_{31}^{-1}} \end{vmatrix} f_{31}.$$

Adding the first two columns to the third one does not change θ, so

$$\theta = \begin{vmatrix} (x_2 - x_1) & (x_3 - x_2) & 0 \\ (y_2 - y_1) & (y_3 - y_2) & 0 \\ f_{12}^{-1}1 & f_{23}^{-1} & \boxed{f_{12}^{-1} + f_{21}^{-1} + f_{31}^{-1}} \end{vmatrix} f_{31}$$

$$= (f_{12}^{-1} + f_{21}^{-1} + f_{31}^{-1})f_{31}.$$

$\square$

This is a noncommutative generalization of formula (32) from Konopelchenko ([14])

4.5 Noncommutative Menelaus Theorem and Quasi-Plücker Coordinates

Let $\mathcal{R}$ be a noncommutative division ring as above. Consider $\mathcal{R}^2$ as the right vector space over $\mathcal{R}$. For a point $X \in \mathcal{R}^2$ denote by x_i its i-th coordinate, $i = 1, 2$. Recall that points X, Y, Z are collinear if and only if

$$\begin{vmatrix} x_1 & y_1 & z_1 \\ x_2 & y_2 & z_2 \\ 1 & 1 & \boxed{1} \end{vmatrix} = 0 \, .$$

Proposition 4.8 *Points* X, Y, Z *are collinear if and only if*

$$(y_1 - x_1)^{-1}(z_1 - x_1) = (y_2 - x_2)^{-1}(z_2 - y_2)$$

or, equivalently,

$$(x_2 - y_2)^{-1}(z_2 - y_2) = q_{XZ}^Y \, .$$

Remark 4.9   The second identity is equivalent to the equality

$$\begin{vmatrix} x_1 & y_1 & \boxed{z_1} \\ x_2 & y_2 & z_2 \\ 1 & 1 & 1 \end{vmatrix} = 0 \, .$$

Proposition 4.10 *Let* A, B, C *be non-collinear points in* $\mathcal{R}^2$. *Then any point* $P \in \mathcal{R}^2$ *can be uniquely written as*

$$P = At + Bu + Cv, \quad t, u, v \in \mathcal{R}, \quad t + u + v = 1 \, .$$

We will write $P = [t, u, v]$.

Proposition 4.11 *Let* $P_i = [t_i, u_i, v_i]$, $i = 1, 2, 3$. *Then* P_1, P_2, P_3 *are collinear if and only if*

$$\begin{vmatrix} t_1 & t_2 & \boxed{t_3} \\ u_1 & u_2 & u_3 \\ v_1 & v_2 & v_3 \end{vmatrix} = 0 \, .$$

We follow now the book by Kaplansky, [12], see pages 88–89. Consider a triangle ABC (vertices go anti-clock wise). Take point R at line AB, point P at line BC, and point Q at line AC. Then

$$P = B(1 - t) + Ct, \quad Q = C(1 - u) + Au, \quad R = A(1 - v) + Bv \, .$$

Proposition 4.11 implies

Theorem 4.12 *Points* A, B, C *are collinear if and only if*

$$u(1 - u)^{-1}t(1 - t)^{-1}v(1 - v)^{-1} = -1 \, .$$

Note that $t(1 - t)^{-1} = (c_1 - p_1)^{-1}(p_1 - b_1)$. Corollary 4.8 implies that

$$t(1 - t)^{-1} = -q_{CB}^P$$

where q_{CB}^P is a quasi-Plücker coordinate. Similarly,

$$u(1 - u)^{-1} = -q_{AC}^Q, \quad v(1 - v)^{-1} = -q_{BA}^R$$

and Theorem 4.12 implies

Theorem 4.13

$$q_{AC}^Q q_{CB}^P q_{BA}^R = 1 \, .$$

5 Relation with Matrix Cross-Ratio

In this section we discuss the cross ratio in noncommutative algebras, introduced above in terms of *quasideterminants* and its relation with the *operator cross-ratio* of Zelikin (see [24] and also Chapter 5 of [23]). We also consider the infinitesimal part of the cross-ratio in this case, which just like in the classical case will lead us to the construction of the noncommutative Schwarz derivative. Further details about the constructions of Schwarzian will be given in the next section.

Recall that we defined the cross ratio of four elements $a, b, c, d \in \mathcal{R}^{\oplus 2}$ by explicit formulas as follows:

$$\kappa(a, b, c, d) = \begin{vmatrix} b_1 & \boxed{d_1} \\ b_2 & d_2 \end{vmatrix}^{-1} \begin{vmatrix} b_1 & \boxed{c_1} \\ b_2 & c_2 \end{vmatrix} \begin{vmatrix} a_1 & \boxed{c_1} \\ a_2 & c_2 \end{vmatrix}^{-1} \begin{vmatrix} a_1 & \boxed{d_1} \\ a_2 & d_2 \end{vmatrix},$$

under the assumption that all these expressions exist (in fact, except for the existence of the inverse elements of a_2 and b_2, it is enough to assume further that the matrices $\begin{pmatrix} a_1 & c_1 \\ a_2 & c_2 \end{pmatrix}$ and $\begin{pmatrix} b_1 & d_1 \\ b_2 & d_2 \end{pmatrix}$ are invertible).

The expression $\kappa(a, b, c, d)$ has various algebraic properties (see sections 1–3). We are going now to compare it with the *operator cross ratio* of Zelikin (see [24]). To this end we begin with the description of his construction.

Let $\mathcal{H}$ be an even-dimensional (possibly infinite-dimensional) vector space; let us fix its polarization $\mathcal{H} = V_0 \oplus V_1$, where the subspaces V_0, V_1 have the same dimension (in infinite dimensional case one can assume that there is a fixed isomorphism $\psi : V_0 \to V_1$ between them); let $(\mathscr{P}_1, \mathscr{P}_2)$ and $(\mathscr{Q}_1, \mathscr{Q}_2)$ be two other pairs of subspaces, polarizing $\mathcal{H}$, i.e. $\mathscr{P}_i$, $\mathscr{Q}_i$ are isomorphic to V_j and $\mathscr{P}_1$ (resp. $\mathscr{Q}_1$) is transversal to $\mathscr{P}_2$ (resp. to $\mathscr{Q}_2$). Then the cross ratio of these two pairs (or of the spaces $\mathscr{P}_1 \, \mathscr{P}_2$, $\mathscr{Q}_1$, $\mathscr{Q}_2$ is the operator

$$\mathrm{DV}(\mathscr{P}_1, \mathscr{P}_2, \mathscr{Q}_1, \mathscr{Q}_2) = (\mathscr{P}_1 \xrightarrow{\mathscr{P}_2} \mathscr{Q}_1 \xrightarrow{\mathscr{Q}_2} \mathscr{P}_1)$$

Here we use the notation from [9], where $\mathscr{P}_1 \xrightarrow{\mathscr{P}_2} \mathscr{Q}_1$ denotes the projection of $\mathscr{P}_1$ to $\mathscr{Q}_1$ along $\mathscr{P}_2$ and similarly for the second arrow.

In the cited paper the following explicit formula for DV was proved: let $\mathscr{P}_i$ be given by the graph of an operator $P_i : V_0 \to V_1$, $i = 1, 2$ and similarly for $\mathscr{Q}_j$, then the following formula holds:

$$\mathrm{DV}(\mathscr{P}_1, \mathscr{P}_2, \mathscr{Q}_1, \mathscr{Q}_2) = (P_1 - P_2)^{-1}(P_2 - Q_1)(Q_1 - Q_2)^{-1}$$
$$\times (Q_2 - P_1) : V_0 \to V_0.$$

The invertibility of the operators $P_1 - P_2$ and $Q_1 - Q_2$ is provided by the transversality of $\mathscr{P}_1$ and $\mathscr{P}_2$ (resp. $\mathscr{Q}_1$ and $\mathscr{Q}_2$).

Proposition 5.1 *The operator cross ratio* $\mathrm{DV}(\mathscr{Q}_2, \mathscr{P}_2, \mathscr{Q}_1, \mathscr{P}_1)$ *(if it exists) is equal to* $\kappa(p_1, p_2, q_1, q_2)$, *for* $p_1 = \begin{pmatrix} 1 \\ P_1' \end{pmatrix}$, $p_2 = \begin{pmatrix} 1 \\ P_2' \end{pmatrix}$, $q_1 = \begin{pmatrix} 1 \\ Q_1' \end{pmatrix}$, $q_2 = \begin{pmatrix} 1 \\ Q_2' \end{pmatrix}$, *where* 1 *is the identity operator on* V_0 *and we identify* V_0 *and* V_1 *using the fixed map* ψ *so that* $P_i' = \psi^{-1} \circ P_i : V_0 \to V_0$.

Proof. This is a direct computation based on the explicit formula:

$$\kappa(p_1, p_2, q_1, q_2) = \begin{vmatrix} 1 & \boxed{1} \\ P_2' & Q_2' \end{vmatrix}^{-1} \begin{vmatrix} 1 & \boxed{1} \\ P_2' & Q_1' \end{vmatrix} \begin{vmatrix} 1 & \boxed{1} \\ P_1' & Q_1' \end{vmatrix}^{-1} \begin{vmatrix} 1 & \boxed{1} \\ P_1' & Q_2' \end{vmatrix}$$

$$= (1 - (P_2')^{-1}Q_2')^{-1}(1 - (P_2')^{-1}Q_1')(1 - (P_1')^{-1}Q_1')^{-1}(1 - (P_1')^{-1}Q_2')$$
$$= (1 - P_2^{-1}Q_2)^{-1}(1 - P_2^{-1}Q_1)(1 - P_1^{-1}Q_1)^{-1}(1 - P_1^{-1}Q_2)$$
$$= (P_2 - Q_2)^{-1}P_2 P_2^{-1}(P_2 - Q_1)(P_1 - Q_1)^{-1}P_1 P_1^{-1}(P_1 - Q_2)$$
$$= (Q_2 - P_2)^{-1}(P_2 - Q_1)(Q_1 - P_1)^{-1}(P_1 - Q_2)$$
$$= \mathrm{DV}(\mathscr{Q}_2, \mathscr{P}_2, \mathscr{Q}_1, \mathscr{P}_1).$$

□

Note that the role of ψ is insignificant here: in effect, one can define the quasideterminants in the context of categories, i.e. for A being a matrix of morphisms in certain category with its entries a_{ij} being maps from the i-th object to the j-th object (see [9]). This makes the use of ψ redundant.

5.1 Cocycle Identity: Cross-Ratio and "Classifying Map"

It is shown in [24] that the following equality holds for the DV: let $(\mathscr{P}_1, \mathscr{P}_2)$ be a polarizing pair, and $\mathscr{X}$, $\mathscr{Y}$, $\mathscr{Z}$ three hyperplanes, then

$$\mathrm{DV}(\mathscr{P}_1, \mathscr{X}, \mathscr{P}_2, \mathscr{Y})\,\mathrm{DV}(\mathscr{P}_1, \mathscr{Y}, \mathscr{P}_2, \mathscr{Z})\,\mathrm{DV}(\mathscr{P}_1, \mathscr{Z}, \mathscr{P}_2, \mathscr{X}) = 1, \quad (8)$$

or, using the algebraic properties of DV,

$$\mathrm{DV}(\mathscr{P}_1, \mathscr{X}, \mathscr{P}_2, \mathscr{Y})\,\mathrm{DV}(\mathscr{P}_1, \mathscr{Y}, \mathscr{P}_2, \mathscr{Z}) = \mathrm{DV}(\mathscr{P}_1, \mathscr{X}, \mathscr{P}_2, \mathscr{Z}) \quad (9)$$

if all three terms are well-defined. This relation can be reinterpreted topologically, by saying that the operator cross ratio corresponds to the change of coordinates function in the tautological fibre bundle over the Grassmanian space of polarizations of $\mathcal{H}$ (see [24]).

Recall now that the noncommuative cross-ratio $\kappa(a, b, c, d)$ satisfies a similar relation

$$\kappa(y, x, p_2, p_1)\kappa(z, y, p_2, p_1) = \kappa(z, x, p_2, p_1), \tag{10}$$

see Section 2 and 3 above for a purely algebraic proof. One can ask, if there exists an analogous topological interpretation of κ, i.e. if one can construct an analog of tautological bundle in the purely algebraic case. Here we shall sketch a construction, intended to answer this question, postponing the details to a forthcoming paper, dealing with the topological applications of the noncommutative cross-ratio.

In order to give an interpretation of relations (8)–(10), let us fix a vector $\omega = \begin{pmatrix} \omega_1 \\ \omega_2 \end{pmatrix} \in \mathcal{R}^{\oplus 2}$; let $\mathcal{R}^2_\omega$ denote the set of the elements $\begin{pmatrix} a_1 \\ a_2 \end{pmatrix} \in \mathcal{R}^{\oplus 2}$ such that the matrix

$$\begin{pmatrix} \omega_1 & a_1 \\ \omega_2 & a_2 \end{pmatrix}$$

is invertible. Let $a \in \mathcal{R}^2_\omega$ and let $x \in \mathcal{R}^{\oplus 2}$ be such that both matrices

$$\begin{pmatrix} a_1 & x_1 \\ a_2 & x_2 \end{pmatrix} \text{ and } \begin{pmatrix} \omega_1 & x_1 \\ \omega_2 & x_2 \end{pmatrix}$$

are invertible. It is clear that the set of such x is equal to the intersection $\mathcal{R}^2_\omega \cap \mathcal{R}^2_a$; we shall denote it by $\mathcal{R}^2_{a,\omega} = \tilde{\mathcal{R}}^2_a$ since ω is fixed.

Consider now a Čech type simplicial complex $\check{C}_\cdot(\mathcal{R}^2)$: its set of n-simplices is spanned by the disjoint union of the intersections

$$\check{C}_n(\mathcal{R}^2; \omega) = \coprod_{a_0, \ldots, a_n} \tilde{\mathcal{R}}^2_{a_0} \cap \tilde{\mathcal{R}}^2_{a_1} \cap \cdots \cap \tilde{\mathcal{R}}^2_{a_n},$$

and the faces/degeneracies are given by omitting/repeating the terms in the intersections respectively.

Then the formula

$$\phi = \{\phi_{a_0, a_1}\} : \check{C}_1(\mathcal{R}^2; \omega) \to \mathcal{R}^*, \quad \phi_{a_0, a_1}(x) = \kappa(a_1, a_0, x, \omega),$$

determines a map on the second term of this complex. Note that the cocycle condition now can be interpreted as the statement that ϕ can be extended to a simplicial map from $\check{C}_\cdot(\mathcal{R}^2; \omega)$ to the bar-resolution of the group $\mathcal{R}^*$ of invertible elements in $\mathcal{R}$. Namely: put

$$\phi_0 = 1 : \check{C}_0(\mathcal{R}^2; \omega) \to [1] = B_0(\mathcal{R}^*);$$
$$\phi_1 = \phi : \check{C}_1(\mathcal{R}^2; \omega) \to \mathcal{R}^* = B_1(\mathcal{R}^*);$$

and for all other $n \geq 2$

$$\phi_n : \check{C}_n(\mathcal{R}^2; \omega) \to (\mathcal{R}^*)^{\times n} = B_n(\mathcal{R}^*)$$

given by the formula

$$\phi_n(x) = [\phi_{a_0, a_1}(x) | \phi_{a_1, a_2}(x) | \dots | \phi_{a_{n-1}, a_n}(x)],$$

for all $x \in \tilde{\mathcal{R}}^2_{a_0, \dots, a_n}$. Then

Proposition 5.2 *The collection of maps* $\{\phi_n\}_{n \geq 0}$ *determine a simplicial map from* $\check{C}.(\mathcal{R}^2, \omega)$ *to* $B.(\mathcal{R}^*)$.

Remark 5.3   The construction we just described bears striking similarity with the well-known Goncharov complex (see [11]), so one can wonder if there is any relation with the actual Goncharov Grassmannian complex and higher cross ratios/polylogarithms in this case?

5.2 Schwarzian Operator

In classical theory the Schwarzian operator is a quadratic differential operator, measuring the "non-projectivity" of a diffeomorphism of (real or complex) projective line. Applying the same ideas to the noncommutative plane, we shall obtain an analog of operator as an infinitesimal part of the deformation of the cross-ratio. It follows from the construction that this operator is invariant with respect to the action of $GL_2(\mathcal{R})$ from the left and multiplication by invertible elements from $\mathcal{R}$ from the right.

First, following the ideas by Zelikin [24] we consider a smooth one-parameter family $Z(t) = \begin{pmatrix} Z(t)_1 \\ Z(t)_2 \end{pmatrix}$ of elements in $\mathcal{R}^{\oplus 2}$, such that for all different t_1, t_2, t_3, t_4 the cross ratio $\kappa(Z(t_1), Z(t_2), Z(t_3), Z(t_4))$ is well defined. Then, let us consider the function

$$f(t, t_1, t_2, t_3) = \kappa(Z(t_3), Z(t_1), Z(t), Z(t_2))$$
$$= (z(t_2) - z(t_1))^{-1}(z(t_1) - z(t))(z(t) - z(t_3))^{-1}(z(t_3) - z(t_2)),$$

where $z(t) = Z(t)_1^{-1} Z(t)_2$. Fix $t = 0$, and let $t_2 \to 0$. Then $f(0, t_1, t_2, t_3) \to 1$ and

$$\frac{\partial f}{\partial t_2}(0, t_1, 0, t_3) = -(z(0) - z(t_1))^{-1} z'(0) + (z(0) - z(t_3))^{-1} z'(0).$$

Thus,

$$f(t, t_1, t_2, t_3) = 1 - (t_2 - t)\left((z(t) - z(t_1))^{-1} z'(t) - (z(t) - z(t_3))^{-1} z'(t) \right)$$
$$+ o(t_2 - t)$$

If $t_1 = t_3$, the derivative on the right vanishes. Consider now the second partial derivative:

$$\frac{\partial^2 f}{\partial t_3 \partial t_2}(0, t_1, 0, t_1) = -(z(0) - z(t_1))^{-1} z'(t_1)(z(0) - z(t_1))^{-1} z'(0),$$

so that

$$f(t, t_1, t_2, t_3) = 1 - (t_2 - t)(t_3 - t_1)(z(t) - z(t_1))^{-1} z'(t_1)(z(t)$$
$$- z(t_1))^{-1} z'(t) + o((t_2 - t)(t_1 - t_3)).$$

This expression has a singularity at $t_1 = 0$. Now, using the Taylor series for $z(t)$ we compute for $t_1 \to 0$:

$$\frac{\partial^2 f}{\partial t_3 \partial t_2}(0, t_1, 0, t_1)$$
$$= t_1^{-2} \left(1 + \frac{t_1^2 (z'(0))^{-1} z'''(0)}{6} - \frac{t_1^2 ((z'(0))^{-1} z''(0))^2}{4} + \dots \right)$$

where $\dots$ denote the terms of degrees 3 and higher in t_1. Therefore,

$$\frac{\partial^2 f}{\partial t_3 \partial t_2}(0, t_1, 0, t_1) = t_1^{-2}(1 + 6t_1^2 S(Z) + \dots) + \dots \tag{11}$$

where we put

$$S(Z) = (z'(0))^{-1} z'''(0) - \frac{3}{2}((z'(0))^{-1} z''(0))^2.$$

Here Z and z are related as explained above. This differential operator is well-defined on functions with values in $\mathcal{R}^{\oplus 2}$, it is invariant with respect to the action of $GL_2(\mathcal{R})$ and is conjugated by $\lambda \in \mathcal{R}^\times$, when Z is multiplied by it on the right.

Thus we come up with the following statement:

Proposition 5.4 *If we have a 1-parameter family of elements in the projective noncommutative plane $\mathcal{R}^{\oplus 2}$, then the infinitesimal part of the cross-ratio of four generic points in this family is equal to the noncommutative Schwarzian $S(Z)$.*

Proof. Above we have given a short explanation of the formula for $S(Z)$; however, formula (11) was obtained in a slightly artificial way. Consider now the formal Taylor expansion of $z(t_i)$ near $t_i = 0$: $z(t_i) = z(0) + z'(0)t_i + \frac{1}{2}z''(0)t_i^2 + \frac{1}{6}z'''(0)t_i^3 + \dots$, $t_i = t, t_1, t_2, t_3$. Then (omitting the argument (0) from our notation)

$$z(t_i) - z(t_j) = z'(t_i - t_j) + \frac{1}{2}z''(t_i^2 - t_j^2) + \frac{1}{6}z'''(t_i^3 - t_j^3) + \dots$$
$$= (t_i - t_j)\left(z' + \frac{1}{2}z''(t_i + t_j) + \frac{1}{6}z'''(t_i^2 + t_i t_j + t_j^2) \right) + \dots$$

and similarly

$$(z(t_i) - z(t_j))^{-1}(z(t_k) - z(t_l)) =$$

$$= \frac{t_k - t_l}{t_i - t_j}\left(1 + \frac{1}{2}(z')^{-1}z''(t_i + t_j) + \frac{1}{6}(z')^{-1}z'''(t_i^2 + t_i t_j + t_j^2)\right)^{-1}$$

$$\times \left(1 + \frac{1}{2}(z')^{-1}z''(t_k + t_l) + \frac{1}{6}(z')^{-1}z'''(t_k^2 + t_k t_l + t_l^2)\right) + \ldots$$

$$= \frac{t_k - t_l}{t_i - t_j}\left(1 - \frac{1}{2}(z')^{-1}z''(t_i + t_j) - \frac{1}{6}(z')^{-1}z'''(t_i^2 + t_i t_j + t_j^2)\right.$$

$$+ \frac{1}{4}((z')^{-1}z'')^2(t_i + t_j)^2\right)\left(1 + \frac{1}{2}(z')^{-1}z''(t_k + t_l)\right.$$

$$\left. + \frac{1}{6}(z')^{-1}z'''(t_k^2 + t_k t_l + t_l^2)\right) + \ldots$$

$$= \frac{t_k - t_l}{t_i - t_j}\left(1 + \frac{1}{2}(z')^{-1}z''(t_k + t_l - t_i - t_j) + \frac{1}{6}(z')^{-1}z'''\right.$$

$$\times (t_k^2 + t_k t_l + t_l^2 - t_i^2 - t_i t_j - t_j^2)$$

$$\left. + \frac{1}{4}((z')^{-1}z'')^2((t_i + t_j)^2 - (t_k + t_l)(t_i + t_j))\right) + \ldots$$

where we use … to denote the elements of degree 3 and higher in t_i. In particular, taking $t_i = t_2$, $t_j = t_1$, $t_k = t_1$, $t_l = t$, we obtain

$$(z(t_2) - z(t_1))^{-1}(z(t_1) - z(t)) =$$

$$= \frac{t_1 - t}{t_2 - t_1}\left(1 + (t - t_2)\left(\frac{1}{2}(z')^{-1}z'' + \frac{1}{6}(z')^{-1}z'''(t + t_1 + t_2)\right.\right.$$

$$\left.\left. - \frac{1}{4}((z')^{-1}z'')^2(t_2 + t_1)\right)\right) + \ldots$$

Similarly, with $t_i = t$, $t_j = t_3$, $t_k = t_3$, $t_l = t_2$, we have:

$$(z(t) - z(t_3))^{-1}(z(t_3) - z(t_2)) =$$

$$= \frac{t_3 - t_2}{t - t_3}\left(1 + (t_2 - t)\left(\frac{1}{2}(z')^{-1}z'' + \frac{1}{6}(z')^{-1}z'''(t + t_2 + t_3)\right.\right.$$

$$\left.\left. - \frac{1}{4}((z')^{-1}z'')^2(t + t_3)\right)\right) + \ldots$$

Finally, taking the product of these two expressions we obtain

$$f(t, t_1, t_2, t_3) = \frac{(t_1 - t)(t_3 - t_2)}{(t_2 - t_1)(t - t_3)}\left(1 + (t_2 - t)(t_3 - t_1)\right.$$

$$\left. \times \left(\frac{1}{6}(z'(0))^{-1}z'''(0) - \frac{1}{4}((z(0)')^{-1}z(0)'')^2\right)\right)$$

$\square$

Compare this formula with the formula (4.7) from the paper [1].

We call the expression $Sch(z) = (z')^{-1}z''' - \frac{3}{2}((z')^{-1}z'')^2$ *the noncommutative Schwarzian of* $z(t)$. Just like the classical Schwarz derivative, this operator is invariant (up to conjugations) with respect to the Möbius transformations in $\mathcal{R}^2$: this is the direct consequence of the method we derived this formula from the (operator) cross-ratio.

5.3 Infinitesimal Ceva Ratio

The following expression may be viewed as a 2-dimensional analog of the Schwarzian operator. More accurately, the Schwarz derivative can be regarded as the infinitesimal transformation of the cross-ratio under a diffeomorphism of the projective line. It is natural to assume that the role of cross-ratio in projective plane should in some sense be played by the Ceva theorem (see Figure 16.1, part (b)). Thus, here we find the infinitesimal part of the transformation of the Ceva ratio under a diffeomorphism; in a general case this is quite a difficult question, so we do it under certain additional conditions.

Let ξ, η be two commuting vector fields on a manifold M, and let f : $M \to M$ be a self-map of M such that $df(\xi) = \kappa \cdot \xi$, $df(\eta) = \kappa \cdot \eta$ for some smooth function $\kappa \in C^\infty(M)$. It follows from this condition, that f maps integral trajectories of both fields and of the fields, equal to their linear combinations with constant coefficients. One can imagine this map as a "change of coordinates along the 2-dimensional net", or a generalized conformal map. However, we do not assume that these fields are linearly independent, they can even be proportional to each other.

Let us consider the following expression: take any point x; let $\phi(t)$ and $\psi(s)$ be the one-parameter diffeomorphism families, generated by ξ and η respectively. Since these fields commute, the composition $\phi(-r) \circ \psi(r) = \psi(r) \circ \phi(-r) =: \theta(r)$ is the one-parameter family, corresponding to their difference $\zeta = \eta - \xi$. Consider now the infinitesimal "triangle" at x: first we move from x to $\phi(\epsilon)(x)$, then from this point to $\psi(2\epsilon)(x)$; then we apply to this point $\theta(\epsilon)$ and $\theta(2\epsilon)$; and finally we apply twice the diffeomorphism $\psi(-\epsilon)$. By definition, we come to the point x again, having spun a "curvilinear triangle" ABC ($A = x$, $B = \phi(2\epsilon)(x)$, $C = \psi(2\epsilon)(x)$) with points K, L, M on its sides ($K = \phi(\epsilon)(x)$, $L = (\phi(\epsilon) \circ \psi(\epsilon))(x)$, $M = \psi(\epsilon)(x)$). If we use the inherent "time" along the trajectories of the vector fields to measure length along these trajectories, then the points K, L and M will be midpoints of the sides of ABC and the standard Ceva relation will be trivially 1:

$$c(A, B, C; K, L, M) = \frac{AK}{KB} \cdot \frac{BL}{LC} \cdot \frac{CM}{MA} = 1.$$

Consider now the image of triangle ABC under f: the points K, L and M will again fall on the "sides" of this image, however the lengths will be somehow distorted (in fact even the fields $df(\xi) = \kappa \cdot \xi$ and $df(\eta) = \kappa \cdot \eta$ need not be commuting). Let us now explore this "distortion" up to the degree 2 in ϵ:

Proposition 5.5 *Up to degree 2 the difference between the distorted Ceva relation and 1 is trivial; we put*

$$c(f(A), f(B), f(C); f(K), f(L), f(M)) - 1 =: \epsilon^2 S_3(f, \xi, \eta; x) + o(\epsilon^2),$$

then

$$S_3(f, \xi, \eta; x) = \frac{5}{6} \frac{\kappa''_{\eta\eta}(x) - \kappa''_{\xi\xi}(x)}{\kappa(x)},$$

where we use the standard notation $\kappa'_\xi = \xi(\kappa), \ \kappa'_\eta = \eta(\kappa)$.

Proof. Note that

$$f(A)f(K) = \epsilon\kappa(x) + \frac{1}{2}\epsilon^2\kappa'_\xi(x) + \frac{1}{6}\epsilon^3\kappa''_{\xi\xi}(x) + o(\epsilon^3),$$

$$f(K)f(B) = \epsilon\kappa(x + \epsilon\xi) + \frac{1}{2}\epsilon^2\kappa'_\xi(x + \epsilon\xi) + \frac{1}{6}\epsilon^3\kappa''_{\xi\xi}(x + \epsilon\xi) + o(\epsilon^3)$$

$$= \epsilon\kappa(x) + \frac{3}{2}\epsilon^2\kappa'_\xi(x) + \frac{5}{3}\epsilon^3\kappa''_{\xi\xi}(x) + o(\epsilon^3),$$

$$f(M)f(A) = -\epsilon\kappa(x) - \frac{1}{2}\epsilon^2\kappa'_\eta(x) - \frac{1}{6}\epsilon^3\kappa''_{\eta\eta}(x) + o(\epsilon^3),$$

$$f(C)f(M) = -\epsilon\kappa(x) - \frac{3}{2}\epsilon^2\kappa'_\eta(x) - \frac{5}{3}\epsilon^3\kappa''_{\eta\eta}(x) + o(\epsilon^3),$$

$$f(B)f(L) = \epsilon\kappa(x + 2\epsilon\xi) + \frac{1}{2}\epsilon^2\kappa'_\zeta(x + 2\epsilon\xi) + \frac{1}{6}\epsilon^3\kappa''_{\zeta\zeta}(x + 2\epsilon\xi) + o(\epsilon^3)$$

$$= \epsilon\kappa(x) + 2\epsilon^2\kappa'_\xi(x) + 2\epsilon^3\kappa''_{\xi\xi}(x)$$

$$+ \frac{1}{2}\epsilon^2\kappa'_\zeta(x) + \epsilon^3\kappa''_{\xi\zeta}(x) + \frac{1}{6}\epsilon^3\kappa''_{\zeta\zeta}(x) + o(\epsilon^2)$$

$$= \epsilon\kappa(x) + \frac{3}{2}\epsilon^2\kappa'_\xi(x) + \frac{1}{2}\epsilon^2\kappa'_\eta(x)$$

$$+ \frac{5}{6}\epsilon^3\kappa''_{\xi\xi}(x) + \frac{1}{6}\epsilon^3\kappa''_{\eta\eta}(x) + \frac{2}{3}\epsilon^3\kappa''_{\xi\eta}(x) + o(\epsilon^3)$$

$$f(L)f(C) = \epsilon\kappa(x + \epsilon(\xi + \eta)) + \frac{1}{2}\epsilon^2\kappa'_\zeta(x + \epsilon(\xi + \eta))$$

$$+ \frac{1}{6}\epsilon^3\kappa''_{\zeta\zeta}(x + \epsilon(\xi + \eta)) + o(\epsilon^3)$$

$$= \epsilon\kappa(x) + \frac{3}{2}\epsilon^2\kappa'_\eta(x) + \frac{1}{2}\epsilon^2\kappa'_\xi(x)$$

$$+ \frac{1}{2}\epsilon^3\kappa''_{\xi\xi}(x) + \frac{1}{2}\epsilon^3\kappa''_{\eta\eta}(x) + \epsilon^3\kappa''_{\xi\eta}(x)$$

$$+ \frac{1}{2}\epsilon^3\kappa''_{\eta\eta}(x) - \frac{1}{2}\epsilon^3\kappa''_{\xi\xi}(x)$$

$$+ \frac{1}{6}\epsilon^3 \kappa''_{\xi\xi}(x) + \frac{1}{6}\epsilon^3 \kappa''_{\eta\eta}(x) - \frac{1}{3}\epsilon^3 \kappa''_{\xi\eta}(x) + o(\epsilon^3)$$
$$= \epsilon\kappa(x) + \frac{3}{2}\epsilon^2 \kappa'_\eta(x) + \frac{1}{2}\epsilon^2 \kappa'_\xi(x)$$
$$+ \frac{1}{6}\epsilon^3 \kappa''_{\xi\xi}(x) + \frac{5}{6}\epsilon^3 \kappa''_{\eta\eta}(x) + \frac{2}{3}\epsilon^3 \kappa''_{\xi\eta}(x) + o(\epsilon^3).$$

Plugging these expressions into the formula for $c(A, B, C; K, L, M)$, we obtain the expression we need. □

The analogy between this expression and the Schwarz derivative is quite evident. One can ask if it is possible to extend it in any reasonable way to a more general situation with less restrictions on the diffeomorphism, and also if there exist a non-commutative version of this operator. We are going to address these questions in forthcoming papers.

6 Non-commutative Schwarzian and Differential Relations

In this section we present an alternative construction of the Schwarz derivative: it is obtained as an invariant of a system of "differential equations" on an algebra. In the commutative case the relation of Schwarzian and differential equations is well-known, see for example [19]. It is remarkable, that this construction also can be phrased in purely algebraic terms. We discuss below some properties of this construction.

Consider the following system of linear "differential equations":

$$\begin{cases} f_1'' + af_1' + bf_1 = 0 \\ f_2'' + af_2' + bf_2 = 0. \end{cases} \tag{12}$$

Here a, b, f_1, f_2 are elements of a division ring $\mathcal{R}$, and $'$ denotes a linear differentiation in this ring, i.e. a linear endomorphism of $\mathcal{R}$ verifying the noncommutative Leibniz identity (a model example is the algebra of smooth operator-valued functions of one (real) variable, however one can plug in arbitrary algebra with a differentiation of any sort on it).

We assume below that all the elements we deal with are invertible if necessary. Using this assumption it is not difficult to solve the equations (12) as a linear system on a and b: multiplying the equations by f_1^{-1} and f_2^{-1} respectively and subtracting the second one from the first one we obtain (see [3])

$$a = -(f_1'' f_1^{-1} - f_2'' f_2^{-1})(f_1' f_1^{-1} - f_2' f_2^{-1})^{-1}$$

and similarly

$$b = -(f_1''(f_1')^{-1} - f_2''(f_2')^{-1})(f_1(f_1')^{-1} - f_2(f_2')^{-1})^{-1}.$$

We can rewrite these formulas as

$$a = -(f_1'' - f_2''f_2'^{-1}f_1')(f_1' - f_2'f_2'^{-1}f_1')^{-1}$$
$$b = -(f_1'' - f_2''(f_2')^{-1}f_1')(f_1 - f_2(f_2')^{-1}f_1')^{-1}$$

so that now it is evident that a and b can be expressed as $a = -q_{32}^1$, $b = -q_{31}^2$, where q_{jk}^i are right quasi-Plücker coordinates of the 3×2-matrix $\begin{pmatrix} f_1 & f_1' & f_1'' \\ f_2 & f_2' & f_2'' \end{pmatrix}^T$. See section 1 for details.

Observe that in the process of solving (12) we obtained the expression:

$$-b = af_1'f_1^{-1} + f_1''f_1^{-1} = af_2'f_2^{-1} + f_2''f_2^{-1}. \qquad (13)$$

(The expression is a special case for the formula from Proposition 4.8.1 from [2] rewritten for right quasi-Plücker coordinates. The proposition connects quasi-Plücker coordinates for matrices of different sizes.)

Thus

$$af_1'f_1^{-1}f_2 + f_1''f_1^{-1}f_2 = af_2' + f_2''.$$

Hence

$$af_1(f_1^{-1}f_1'f_1^{-1}f_2 - f_1^{-1}f_2') = f_1(f_1^{-1}f_2'' - f_1^{-1}f_1''f_1^{-1}f_2). \qquad (14)$$

Now one has

$$(f^{-1})'' = -(f^{-1}f'f^{-1})' = 2f^{-1}f'f^{-1}f'f^{-1} - f^{-1}f''f^{-1},$$

and

$$(fg)'' = f''g + 2f'g' + fg'',$$

for all $f, g \in A$; so

$$(f^{-1}g)'' = 2f^{-1}f'f^{-1}f'f^{-1}g - 2f^{-1}f'f^{-1}g' - f^{-1}f''f^{-1}g + f^{-1}g''.$$

Thus on the right hand side of (14) we have

$$
\begin{aligned}
f_1(f_1^{-1}f_2'' - f_1^{-1}f_1''f_1^{-1}f_2) &= f_1(2f_1^{-1}f_1'f_1^{-1}f_1'f_1^{-1}f_2 - 2f_1^{-1}f_1'f_1^{-1}f_2' \\
&\quad - f_1^{-1}f_1''f_1^{-1}f_2 + f_1^{-1}f_2'') \\
&\quad - 2f_1(f_1^{-1}f_1'f_1^{-1}f_1'f_1^{-1}f_2 - 2f_1^{-1}f_1'f_1^{-1}f_2') \\
&= f_1(f_1^{-1}f_2)'' - 2f_1'(f_1^{-1}f_1'f_1^{-1}f_2 - f_1^{-1}f_2') \\
&= f_1(f_1^{-1}f_2)'' + 2f_1'(f_1^{-1}f_2)'
\end{aligned}
$$

On the other hand, on the left hand side of (14) we have $-af_1(f_1^{-1}f_2)'$, so denoting $\varphi = f_1^{-1}f_2$ we get:

$$af_1 = -2f_1' - f_1\varphi''(\varphi')^{-1}, \tag{15}$$

or equivalently

$$a = -2f_1'f_1^{-1} - f_1\varphi''(\varphi')^{-1}f_1^{-1}. \tag{16}$$

Here's a simple corollary of the formula (16):

Proposition 6.1 *When the elements* $f_i \in A$ *are replaced by* $\tilde{f}_i = hf_i$, $i = 1, 2$ *for some* $h \in \mathcal{R}$, *then* a *in the system (12) should be replaced by* $\tilde{a} = -2h'h^{-1} + hah^{-1}$.

Proof. Observe that φ is not affected by the coordinate change $f_i \leftrightarrow \tilde{f}_i$, $i = 1, 2$. Now direct calculation with formula (16) shows

$$\begin{aligned}
\tilde{a} &= -2\tilde{f}_1'\tilde{f}_1^{-1} - \tilde{f}_1\varphi''(\varphi')^{-1}\tilde{f}_1^{-1} \\
&= -2(h'f_1 + hf_1')f_1^{-1}h^{-1} - h(f_1\varphi''(\varphi')^{-1}f_1^{-1})h^{-1} \\
&= -2h'h^{-1} + h(-2f_1'f_1^{-1} - f_1\varphi''(\varphi')^{-1}f_1^{-1})h^{-1}.
\end{aligned}$$

$\square$

Remark 6.2 It is worth observing a striking similarity of the expression in proposition 6.1 and the gauge transformation of a linear connection (the unnecessary 2 in front of $h'h^{-1}$ can be eliminated by considering $\alpha = \frac{1}{2}a$).

It is now our purpose to find the way b changes, when f_1, f_2 are multiplied by h, at least under some additional assumptions on h. We begin with the simple observation:

Corollary 6.3 *If* h *verifies the "differential equation"* $h' = \frac{1}{2}ha$ *then* $\tilde{a} = 0$.

Further, there's another simple consequence of the formula (15):

Proposition 6.4 *Assume that* h *verifies the equation* $h' = \frac{1}{2}ha$; *denote* $\tilde{f}_1 = hf_1$, $\theta = \varphi''(\varphi')^{-1}$. *Then*

$$\tilde{f}_1' = -\frac{\tilde{f}_1}{2}\theta.$$

Proof.

$$\tilde{f}_1' = (hf_1)' = \frac{1}{2}haf_1 + hf_1' = \text{using equation (15)}$$

$$= -hf_1' - \frac{1}{2}hf_1\theta + hf_1' = -\frac{\tilde{f}_1}{2}\theta.$$

$\square$

Repeating the differentiation we see:

$$\tilde{f}_1'' = -\left(\frac{\tilde{f}_1}{2}\theta\right)' = \frac{\tilde{f}_1}{4}\theta - \frac{\tilde{f}_1}{2}\theta'. \qquad (17)$$

Finally, substituting these formulas in the first expression of (13) we obtain the following result:

Theorem 6.5 *If h satisify the equation $h' = \frac{1}{2}ha$ then the coordinate change $f_i \mapsto \tilde{f}_i = hf_i$, $i = 1, 2$ transforms the system (12) in such a way that*

$$a \mapsto 0$$

$$b \mapsto \frac{1}{2}\tilde{f}_1\left(\theta' - \frac{1}{2}\theta\right)\tilde{f}_1^{-1}$$

where $\theta = \varphi''(\varphi')^{-1}$, $\varphi = f_1^{-1}f_2$ and the equation $2\tilde{f}_1' + \tilde{f}_1\theta = 0$ holds.

Proof. Since $a \mapsto 0$, we obtain from (13):

$$b = -\tilde{f}_1''\tilde{f}_1^{-1} = \text{using (17)} = -\left(\frac{\tilde{f}_1}{4}\theta - \frac{\tilde{f}_1}{2}\theta'\right)\tilde{f}_1^{-1}.$$

$\square$

Remark 6.6 Observe that in the commutative case the expression $\theta' - \frac{1}{2}\theta$ coincides with the classical Schwarz differential of φ.

6.1 Generalized NC Schwarzian

Let f and g be two (invertible) elements of a division ring $\mathcal{R}$, equipped with a derivation $'$ (see previous section). We suppose that they satisfy so-called *left coefficients equations* $f'' = F_1 f$, $g'' = F_2 g$ for some $F_1, F_2 \in \mathcal{R}$. We set $h := fg^{-1}$ and $G := F_1 h - h F_2$.

Theorem 6.7 *If $G = 0$ then we have the following relation:*

$$h''' = (3/2)h''(h')^{-1}h'' - 2h'F_2 \qquad (18)$$

(a non-commutative analogue of the Schwarzian equation).

Proof.

$$h' = f'g^{-1} - hg'g^{-1},$$

$$h'' = G - 2h'g'g^{-1},$$

$$h''' = G' - 2h''g'g^{-1} - 2h'F_2 + 2h'(g'g^{-1})^2.$$

One can express $g'g^{-1} = (1/2)(h')^{-1}(G - h'')$ and get

$$h''' = (3/2)h''(h')^{-1}h'' - 2h'F_2 - (3/2)h''(h')^{-1}G + (1/2)G(h')^{-1}(G - h'').$$

Let $f^{-1}f'' = g^{-1}g''$, i.e. f, g are solutions of the same differential equation with *right coefficients*. Let $g'' = Fg$, i.e. g is also a solution of a differential equation with a left coefficient. Let $h = fg^{-1}$. Then

$$h''' - (3/2)h''(h')^{-1}h'' = -2h'F.$$

Note that the left-hand side is stable under Möbius transform

$$h \to (ah + b)(ch + d)^{-1}$$

where $a' = b' = c' = d' = 0.$ □

Remark 6.8 Consider the commutative analogue of (18)

$$(h')^{-1}h''' = (3/2)(h')^{-2}h''^2 - 2F_2. \tag{19}$$

This equality can be regarded as yet another definition of the Schwarzian $\mathrm{Sch}(h)$

$$\mathrm{Sch}(h) := (h')^{-1}h''' - (3/2)(h')^{-2}h''^2 = -2F_2.$$

Hence, we obtain one more justification for calling *a NC Schwarzian of h* the following expression

$$\mathrm{NCSch}(h) := (h')^{-1}h''' - (3/2)(h')^{-1}h''(h')^{-1}h''. \tag{20}$$

Remark 6.9 In commutative case there exist the following famous version of KdV equation

$$h_t = (h')\mathrm{Sch}(h). \tag{21}$$

It is invariant under the projective action of SL_2 and, when written as an evolution on the invariant Sch(h) it becomes the "usual" KdV

$$\mathrm{Sch}(h)_t = \mathrm{Sch}(h)''' + 3\mathrm{Sch}(h)'\mathrm{Sch}(h). \tag{22}$$

Introducing two commuting derivatives $\partial_x ='$ and ∂_t of our skew-field $\mathcal{R}$ with respect to two distinguished elements x and t one can write the analogs of (21):

$$h_t = (h')\mathrm{NSch}(h) = h''' - (3/2)h''(h')^{-1}h''. \tag{23}$$

Remark 6.10 The equation (23) has an interesting geometric interpretation (specialisation) as the *Spinor Schwarzian-KdV equation* (see the equation (4.6) in [2]).

7 Some Applications of Noncommutative Cross-Ratios

Let us briefly describe a few possible applications of noncommutative cross-ratios inspired by the classical constructions.

7.1 Noncommutative Leapfrog Map

Let $\mathbb{P}^1$ be the projective line over a noncommutative division ring $\mathcal{R}$. Consider five points $S_{i-1}, S_i, S_{i+1}, S_i^-$ and S_i^+ on $\mathbb{P}^1$. The theory of noncommutative cross-ratios (see theorem 3.4) implies that there exists a projective transformation sending

$$\left(S_{i-1}, S_i, S_{i+1}, S_i^-\right) \to \left(S_{i+1}, S_i, S_{i-1}, S_i^+\right)$$

(in this order!) if and only if the corresponding cross-ratios coincide:

$$\left(S_{i+1} - S_i\right)^{-1} \left(S_i^- - S_i\right) \left(S_i^- - S_{i-1}\right)^{-1} (S_{i+1} - S_{i-1}) =$$
$$= \lambda^{-1} (S_{i-1} - S_i)^{-1} \left(S_i^+ - S_i\right) \left(S_i^+ - S_{i+1}\right)^{-1} (S_{i-1} - S_{i+1}) \lambda$$

where $\lambda \in R$.

Note that the factor $(S_{i+1} - S_{i-1})$ appears in both sides of the equation but with the different signs. It shows that in the commutative case one gets the identity (5.14) from [10]; this map is integrable and constitutes a part of the pentagramm family of maps, see the next paragraph.

Problem 7.1 *It is very intriguing if the same properties exist in noncommutative case.*

7.2 Noncommutative Cross-Ratios and the Pentagramma Mirificum

7.2.1 Classical 5-Recurrence

There is a wonderful observation (known as the *Gauss Pentagramma mirificum*) that when a pentagramma is drawn on a unit sphere in $\mathbb{R}^3$ (see Figure 16.2 where we do not observe the orthogonality of great circles) with successively orthogonal great circles with the lengths of inner side arcs α_i, $i = 1, \ldots, 5$ and one takes $y_i := \tan^2(\alpha_i)$, then the following recurrence relation satisfies:

$$y_i y_{i+1} = 1 + y_{i+3}, \quad \text{mod} \quad \mathbb{Z}_5. \tag{24}$$

Gauss has observed that the first three equations for $i = 1, 2, 3$ in 24 completely define the last two equations for $i = 4, 5$.

It was discussed in [17] (which is our main source of the classical data for the Gauss Pentagramma Mirificum) that the variables y_i can be expressed via the classical cross-ratios:

$$y_i = [p_{i+1}, p_{i+2}, p_{i+3}, p_{i+4}] = \frac{(p_{i+4} - p_{i+1})(p_{i+3} - p_{i+2})}{(p_{i+4} - p_{i+3})(p_{i+2} - p_{i+1})},$$

where $p_i = p_{i+5}$ are five points on real or complex projective line.

Proposition 7.2 *Suppose that two consecutive points y_i and y_{i+1} (cyclically) are differents. Then the five cross-ratios y_i satisfy the relation (24).*

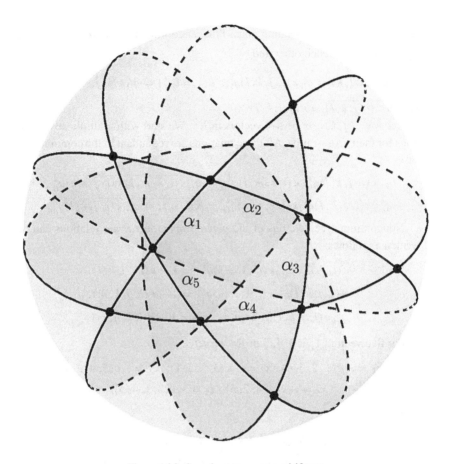

Figure 16.2 Gauss' pentagramma mirificum

It was remarked in [17] that after renaming $x_1 = y_1$, $x_2 = y_4$, $x_3 = y_2$, $x_4 = y_5$, $x_5 = y_3$, the variables x_i, $i = 1, \ldots, 5$ satisfy the famous *pentagon recurrence*:

$$x_{i-1}x_{i+1} = 1 + x_i.$$

It is also known, that this construction is closely related to cluster algebras, see [5] for further details.

7.2.2 Non-Commutative Analogues

Let $\mathcal{R}$ be an associative division ring. In [6] (see sections 2 and 3) we defined the cross-ratio $\kappa(i, j, k, l)$ four vectors $i, j, k, l \in \mathcal{R}^2$. Recall that

$$\kappa(i, j, k, l) = q_{kl}^{j} q_{lk}^{i}$$

where q_{ij}^k is the corresponding quasi-Plücker coordinate. In particular, q_{ij}^k and q_{ji}^k are inverse to each other and

$$\kappa(j, i, l, k) = q_{lk}^i \kappa(i, j, k, l) q_{kl}^i, \quad \kappa(k, l, i, j) = q_{ik}^j \kappa(i, j, k, l) q_{ki}^j.$$

We set $\overline{\kappa(i, j, k, l)} = \kappa(j, i, k, l)$.

Let now i, j, k, l, m be five vectors in $\mathcal{R}^2$. We start with multiplicative relations for their cross-ratios. All these relations are redundant in the commutative case.

$$\kappa(i, j, k, l) q_{km}^i \kappa(i, k, m, l) q_{mk}^i = q_{kl}^j \overline{\kappa(i, k, m, l)} \, \overline{\kappa(i, j, k, l)} q_{lk}^j,$$

$$q_{mk}^l \kappa(i, j, k, l) q_{ki}^l \kappa(l, k, i, m) q_{im}^l = \overline{\kappa(l, k, i, m)} q_{ml}^k \overline{\kappa(i, j, k, l)} q_{lm}^k.$$

Noncommutative versions of the *pentagramma mirificum* relations can be written as follows:

$$\kappa(i, j, k, l) q_{kj}^i \kappa(m, l, j, i) q_{jk}^i = 1 - \kappa(m, j, k, i),$$

$$\kappa(i, j, k, l) q_{ki}^l \kappa(l, k, i, m) q_{jk}^i = 1 - \kappa(l, j, k, m),$$

$$q_{jk}^l \kappa(i, j, k, l) q_{kj}^l \kappa(m, l, j, i) = 1 - \kappa(i, k, j, m).$$

For five vectors $1, 2, 3, 4, 5$ in $\mathcal{R}^2$ we set

$$x_1 = -\kappa(1, 2, 3, 4), \quad x_2 = -\kappa(5, 2, 3, 1), \quad x_3 = -\kappa(5, 4, 2, 1),$$
$$x_4 = -\kappa(3, 4, 2, 5), \quad x_5 = -\kappa(3, 1, 4, 5).$$

Then

$$x_1 q_{32}^1 x_3 q_{23}^1 = 1 + x_2, \quad x_4 q_{23}^5 x_2 q_{32}^5 = 1 + x_3,$$
$$x_3 q_{24}^5 x_5 q_{42}^5 = 1 + x_4, \quad x_6 q_{42}^3 x_4 q_{24}^3 = 1 + x_5,$$
$$x_5 q_{41}^3 x_7 q_{14}^3 = 1 + x_6,$$

where $x_6 := \bar{x}_1$ and $x_7 := \bar{x}_2$. Note the different order for even and odd left hand sides. So, we have *5-antiperiodicity*, i.e. the periodicity up to the anti-involution $x_{k+5} = \bar{x}_k$

Also, the relations with odd left hand parts imply the relations for even left hand parts as in the commutative case.

Remark 7.3 There is an important "continuous limit" of "higher pentagramma" maps on polygons in $\mathbb{P}^n$ which is the Boussinesq (or generalized $(2, n + 1)$−KdV hierarchy) equation ([13]).

Problem 7.4 *What is a non-commutative "higher analogue" of pentagramma recurrences? Is there a related non-commutative integrable analogue of the Boussinesq equation?*

We hope to return to these questions in our future paper devoted to new examples of NC integrable systems ([22]).

References

[1] A. Agrachev, I. Zelenko. *Geometry of Jacobi curves, I,* Journal of Dynamical and Control Systems, vol **8**, Issue 1, pp 93–140 (January 2002)

[2] G. Marí Beffa, Moving frames, geometric Poisson brackets and KdV-Schwarzian evolution of pure spinors, *Ann. Inst. Fourier, Grenoble,* **61**,6, (2011), pp. 2405–2434;

[3] A. Berenstein, V. Retakh, *Noncommutative marked surfaces,* Advances in Math., vol **328**, pp 1010–1087 (2018),

[4] M. J. Dupré, J. F. Glazebrook, E. Previato. Differential algebras with Banach-algebra coefficients II: The operator cross-ratio tau-function and the Schwarzian derivative, *Complex Analysis and Operator Theory,* vol **7**, Issue 6, pp 1713 –1734 (December 2013)

[5] S. Fomin, N. Reading, Root systems and generalized associahedra, *Geometric combinatorics,* 63–131, IAS/Park City Math. Ser., 13, Amer. Math. Soc., 2007.

[6] I. Gelfand, V. Retakh, Determinants of matrices over noncommutative rings, *Funct. Anal. Appl.* **25** (1991), no. 2, pp. 91–102.

[7] I. Gelfand, V. Retakh, A theory of noncommutative determinants and characteristic functions of graphs, *Funct. Anal. Appl.* **26** (1992), no. 4, pp. 1–20.

[8] I. Gelfand, V. Retakh, Quasideterminants, I, *Selecta Math.* **3** (1997), no. 4, pp. 517–546.

[9] I. Gelfand, S. Gelfand, V. Retakh, R. Wilson, Quasideterminants, *Advances in Math.,* **193** (2005), no 1, pp. 56–141.

[10] M. Gekhtman, M. Shapiro, S. Tabachnikov, A. Vainshtein, Higher pentagram maps, weighted directed networks, and cluster dynamics. *Electron. Res. Announc. Math. Sci.* **19** (2012), 1–17

[11] A. Goncharov, Geometry of configuration, polylogarithms and motivic cohomology. *Adv. Math.* **114** (1995), 197–318

[12] I. Kaplansky, Linear algebra and geometry. A second course. Allyn and Bacon, Inc., Boston, Mass. 1969 xii+139 pp.

[13] B. Khesin, F.Soloviev Integrability of higher pentagram maps, *Math. Ann.* (2013) **357**:1005?1047.

[14] Konopelchenko, B. G. Menelaus relation, Hirota-Miwa equation and Fay's trisecant formula are associativity equations. *J. Nonlinear Math. Phys.* 17 (2010), no. 4, 591–603.

[15] F. Labourie, What is a Cross Ratio, *Notices of the AMS,* **55** (2008), no. 10, pp.1234–1235.

[16] F. Labourie, G. McShane, Cross ratios and identities for higher Teichmüller-Thurston theory, *Duke Math. J.,* **149** (2009), no. 2, pp. 209–410.

[17] S. Morier-Genoud, Coxeter's frieze patterns at the crossroads of algebra, geometry and combinatorics, *Bull. Lond. Math. Soc.* **47**, no. 6, pp 895–938 (2015)

[18] V. Ovsienko, S. Tabachnikov. Projective differential geometry old and new. From the Schwarzian derivative to the cohomology of diffeomorphism groups. Cambridge Univ. Press, Cambridge, 2005.

[19] V. Ovsienko, S. Tabachnikov. What is the Schwarzian derivative? *Notices Amer. Math. Soc.* **56** no. 1 p. 34–36 (2009)

[20] V. Retakh, Noncommutative cross-ratios, *J. of Geometry and Physics,* **82**, (August 2014)

[21] V. Retakh, V. Shander, The Schwarz derivative for noncommutative differential algebras, in: *Advances in Soviet Math.*, vol **17** (1993), pp. 139–154

[22] V. Retakh, V. Rubtsov, G. Sharygin, Non- commutative integrable systems revisited, *in progress.*

[23] Zelikin, M. I. Control theory and optimization. I. Homogeneous spaces and the Riccati equation in the calculus of variations. Encyclopaedia of Mathematical Sciences, 86. Springer-Verlag, Berlin, 2000. xii+284 pp.

[24] M. I. Zelikin, Geometry of operator cross ratio, *Sb. Math.*, 197:1 (2006), 37–51

Vladimir Retakh
Department of Mathematics
Rutgers University
Piscataway, New Jersey 08854, USA
e-mail: vretakh@math.rutgers.edu

Vladimir Rubtsov
Maths Department, University of Angers
Building I
Lavoisier Boulevard
Angers, 49045, CEDEX 01, France
e-mail: volodya@univ-angers.fr

Georgy Sharygin
Department of Mathematics and Mechanics
Moscow State (Lomonosov) University
Leninskie Gory, d. 1
Moscow 119991, Russia
e-mail: sharygin@itep.ru